Fundamental Planetary Sciences
Physics, Chemistry and Habitability

Updated Edition

A quantitative introduction to the Solar System and planetary systems science for advanced undergraduate students, this engaging new textbook explains the wide variety of physical, chemical and geological processes that govern the motions and properties of planets. The authors provide an overview of our current knowledge and discuss some of the unanswered questions at the forefront of research in planetary science and astrobiology today. They combine knowledge of the Solar System and the properties of extrasolar planets with astrophysical observations of ongoing star and planet formation, offering a comprehensive model for understanding the origin of planetary systems. The book concludes with an introduction to the fundamental properties of living organisms and the relationship that life has to its host planet. With more than 200 exercises to help students learn how to apply the concepts covered, this textbook is ideal for a one-semester or two-quarter course for undergraduate students.

JACK J. LISSAUER is a Space Scientist at NASA's Ames Research Center in Moffett Field, California, and an adjunct professor at Stanford University. His primary research interests are the formation of planetary systems, detection of extrasolar planets, planetary dynamics and chaos, planetary ring systems and circumstellar/protoplanetary disks. He is lead discoverer of the six-planet Kepler-11 system, co-discoverer of the first four planets found to orbit about faint M dwarf stars and co-discoverer of two broad tenuous dust rings and two small inner moons orbiting the planet Uranus. He is serving as President of the International Astronomical Union's Commission F2 Exoplanets and the Solar System, 2018–2021.

IMKE DE PATER is a Professor in the Astronomy Department and the Department of Earth and Planetary Science at the University of California, Berkeley, and is affiliated with the Delft Institute of Earth Observation and Space Systems at the Delft University of Technology, Netherlands. She began her career observing and modeling Jupiter's synchrotron radiation, followed by detailed investigations of the planet's thermal radio emission. In 1994 she led a worldwide campaign to observe the impact of Comet D/Shoemaker–Levy 9 with Jupiter. Currently, she is exploiting adaptive optics techniques in the infrared range to obtain high angular resolution data of bodies in our Solar System.

Fundamental Planetary Sciences

PHYSICS, CHEMISTRY AND HABITABILITY

UPDATED EDITION

Jack J. Lissauer

NASA Ames Research Center

Imke de Pater

University of California, Berkeley

CAMBRIDGE
UNIVERSITY PRESS

CAMBRIDGE
UNIVERSITY PRESS

University Printing House, Cambridge CB2 8BS, United Kingdom

One Liberty Plaza, 20th Floor, New York, NY 10006, USA

477 Williamstown Road, Port Melbourne, VIC 3207, Australia

314–321, 3rd Floor, Plot 3, Splendor Forum, Jasola District Centre, New Delhi – 110025, India

79 Anson Road, #06–04/06, Singapore 079906

Cambridge University Press is part of the University of Cambridge.

It furthers the University's mission by disseminating knowledge in the pursuit of education, learning and research at the highest international levels of excellence.

www.cambridge.org
Information on this title: www.cambridge.org/9781108411981
DOI: 10.1017/9781108304061

First published 2013
Updated edition published 2019

A catalog record for this publication is available from the British Library.

Library of Congress Cataloging-in-Publication Data
Names: Lissauer, Jack Jonathan, author. | De Pater, Imke, 1952– author.
Title: Fundamental planetary science : physics, chemistry, and habitability / Jack Lissauer (NASA Ames Research Center), Imke de Pater (University of California, Berkeley).
Description: Updated edition. | Cambridge ; New York, NY : Cambridge University Press, 2019. | Includes bibliographical references and index.
Identifiers: LCCN 2019009474 | ISBN 9781108411981 (alk. paper)
Subjects: LCSH: Planetary theory. | Planets – Geology. | Planets – Atmospheres. | Planetary rings.
Classification: LCC QB361 .L57 2019 | DDC 523.201–dc23
LC record available at https://lccn.loc.gov/2019009474

ISBN 978-1-108-41198-1 Paperback

Additional resources for this publication at www.cambridge.org/lissauer.

CONTENTS

Tables

Preface

Astronomy compels the soul to look upwards
and leads us from this world to another.
Plato (427–347 BCE), *The Republic*

The wonders of the night sky, the Moon and the Sun have fascinated mankind for many millennia. We now know that objects akin to the Earth that we walk on are to be found in the heavens. What are these bodies like? What shaped them? How are they similar to our Earth, and how do they differ? And are any of them inhabited by living beings?

This text is written to provide college students majoring in the sciences with an overview of current knowledge in these areas, and the context and background to seek out and understand more detailed treatments of particular issues. We discuss what has been learned and some of the unanswered questions that remain at the forefront of planetary sciences and astrobiology research today. Topics covered include:

- the orbital, rotational and bulk properties of planets, moons and smaller bodies
- gravitational interactions, tides and resonances between bodies
- thermodynamics and other basic physics for planetary sciences
- properties of stars and formation of elements

- energy transport
- vertical structure, chemistry, dynamics and escape of planetary atmospheres
- planetary surfaces and interiors
- magnetospheres
- giant planets
- terrestrial planets
- moons
- meteorites, asteroids and comets
- planetary rings
- the new and rapidly blossoming field of extrasolar planet studies

We then combine this knowledge of current Solar System and extrasolar planet properties and processes with astrophysical data and models of ongoing star and planet formation to develop models for the origin of planetary systems. Planetary science is a key component in the new discipline of astrobiology, and a basic understanding of life is useful to planetary scientists. We therefore conclude with:

- fundamental properties of living organisms
- the relationship that life has to the planet(s) on which it forms and evolves

Parts of this book are based on the recently published second edition of our graduate textbook

Planetary Sciences. However, we have substantially modified the presentation to be more suitable for undergraduate students.

One year of calculus is required to understand all of the equations herein. Basic high school classes in physics, chemistry and, for the final chapter of this book, biology are assumed. A college-level class designed for majors in at least one of these sciences (or in geology/geophysics or meteorology) is also expected. A small number of sections and subsections require additional background or are especially difficult; these sections are denoted with an asterisk following the section number.

The learning of concepts in the physical sciences is greatly enhanced when students 'get their hands dirty' by solving problems. Working through such exercises enables students to obtain a deeper understanding of Solar System properties. Thus, we have included an extensive collection of exercises at the end of each chapter in this text. We denote problems with a higher degree of conceptual difficulty with an asterisk.

We have used black-and-white illustrations throughout the book, augmented with a section of color plates that repeats figures for which color is most essential to show the appearance of an object or to convey other important information. Color versions of many of the illustrations within the book are also available on the book's webpage at www.cambridge.org/lissauer. This website also includes updates, movies, answers to selected problems (for instructors only) and links to various Solar System information sites.

Various symbols are commonly used to represent variables and constants in both equations and the text. Some variables are represented by a single standard symbol throughout the literature, and other variables are represented by differing symbols by different authors; many symbols have multiple uses. The interdisciplinary nature of the planetary sciences and astrobiology exacerbates the problem because standard notation differs between fields. We have endeavored to minimize confusion within the text and to provide the student with the greatest access to the literature by using standard symbols, sometimes augmented by nonstandard subscripts or printed using calligraphic fonts in order to avoid duplication of meanings when practical.

A list of the symbols used in this book is presented as Appendix A. Acronyms are common in our field, so we list the ones used in this book in Appendix B. Tables of physical and astronomical constants are provided in Appendix C. Appendix D is the Periodic Table of Elements. Tabulations of various properties of Solar System objects are presented in Appendix E. Because the resurgence in planetary studies during the past half century is due primarily to spacecraft sent to make close-up observations of distant bodies, we present an introduction to rocketry and list the most significant lunar and planetary missions in Appendix F.

The breadth of the material covered in the text extends well beyond the areas of expertise of the authors. As such, we benefited greatly from comments by many of our colleagues. Those who provided input for the first 15 chapters are acknowledged in our graduate text *Planetary Sciences*, but the following group either provided significant new comments for this book or were so helpful on that text that they merit recognition here as well: Larry Esposito, Ron Greeley, Andy Ingersol, Mark Marley and Bert Vermeersen. Especially helpful suggestions for Chapter 16 were provided by Roger Linfield, Rocco Mancinelli, Frances Westall and, last but not least, Kevin Zahnle.

Preface to the Updated Edition

Planetary sciences is an active research field, and our knowledge of the planets and smaller bodies in the Solar System is increasing very rapidly. The newer discipline of exoplanet research is expanding at an even faster pace. Thus, no compendium on this subject can be completely up to date. In this Updated Edition, Chapter 14, Extrasolar Planets, has been substantially revised and updated. Elsewhere within the main text, we have made corrections and various small updates, but no major modifications. Substantial new Solar System material, which would require repagination if placed within Chapters 1–13 and 15–16, leading to higher textbook cost, is presented in Appendix G.

Jack J. Lissauer and Imke de Pater

(a)

(b)

Plate 1.3

Plate 1.6

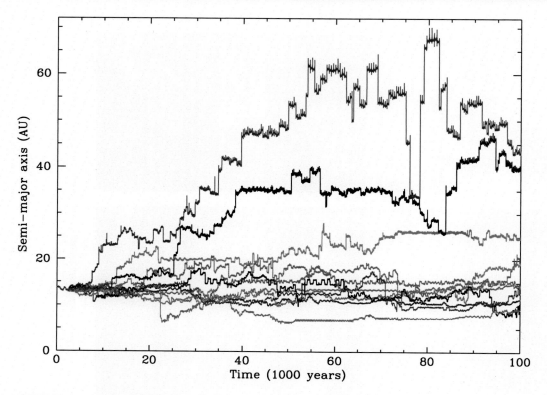

Plate 2.11

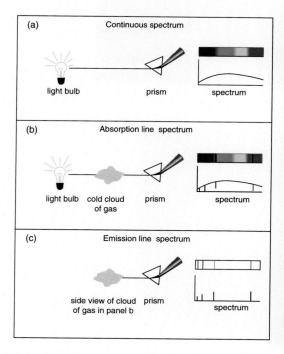

Plate 4.7

Plate 5.17

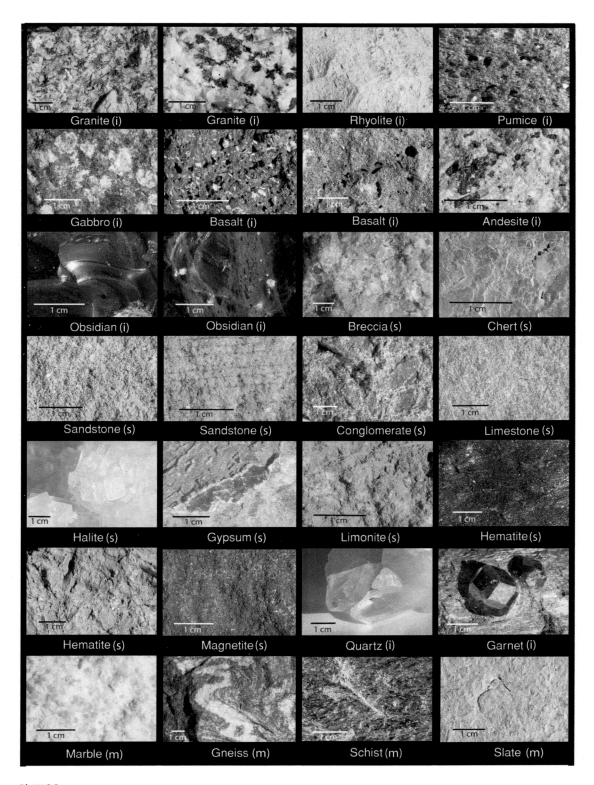

Granite (i)   Granite (i)   Rhyolite (i)   Pumice (i)

Gabbro (i)   Basalt (i)   Basalt (i)   Andesite (i)

Obsidian (i)   Obsidian (i)   Breccia (s)   Chert (s)

Sandstone (s)   Sandstone (s)   Conglomerate (s)   Limestone (s)

Halite (s)   Gypsum (s)   Limonite (s)   Hematite (s)

Hematite (s)   Magnetite (s)   Quartz (i)   Garnet (i)

Marble (m)   Gneiss (m)   Schist (m)   Slate (m)

Plate 6.2

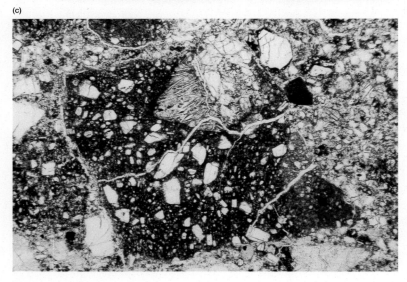

Plate 6.3

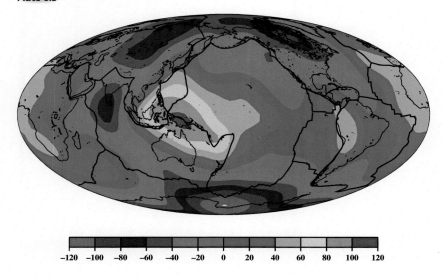

-120 -100 -80 -60 -40 -20 0 20 40 60 80 100 120

Plate 6.6

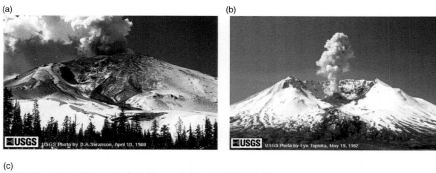

Plate 6.13

Plate 6.14

Plate 6.15

(a)   (b)

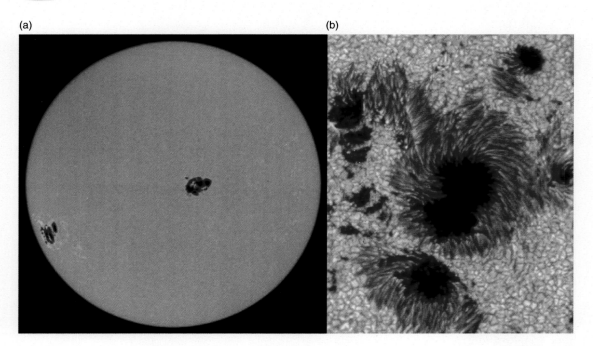

Plate 7.1

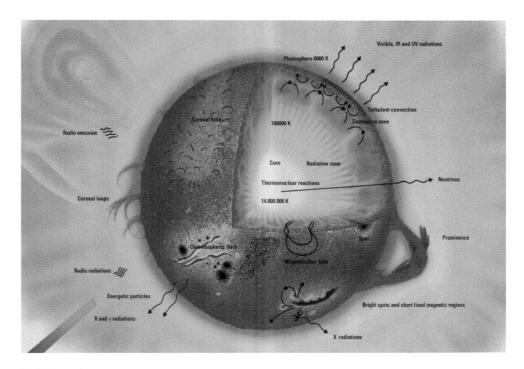

Plate 7.2

Plate 7.7

(a)

Plate 7.16

Plate 8.1

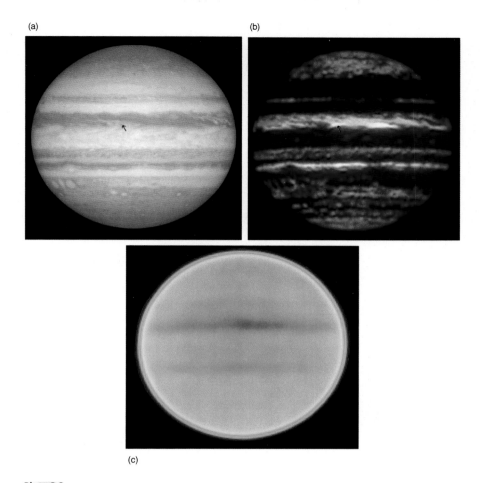

(a)

(b)

(c)

Plate 8.2

Plate 8.5

Plate 8.19

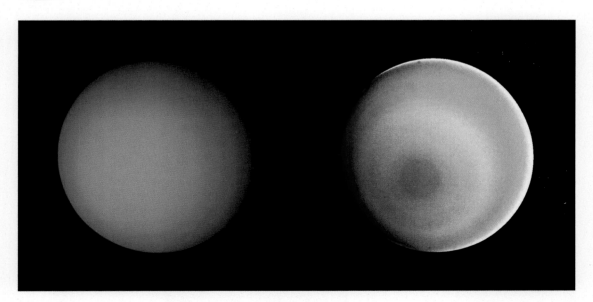

Plate 8.22

(a)

(b)

Plate 9.1

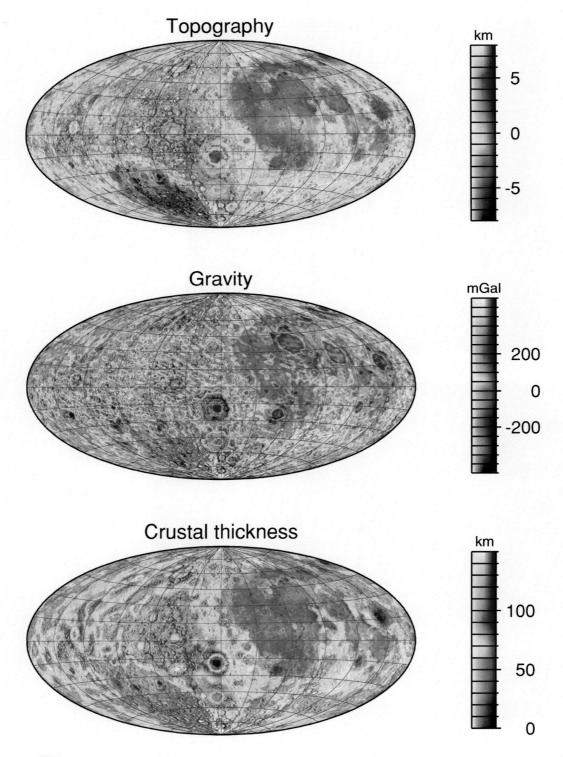

Plate 9.4

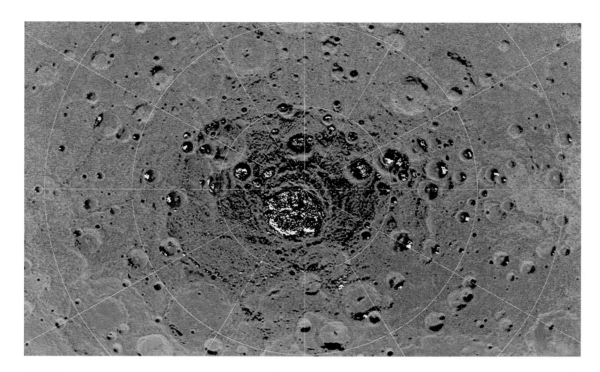

Plate 9.10

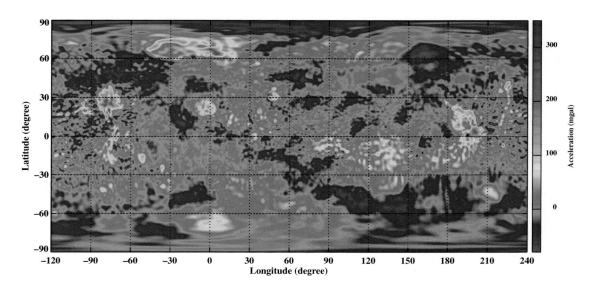

Plate 9.14

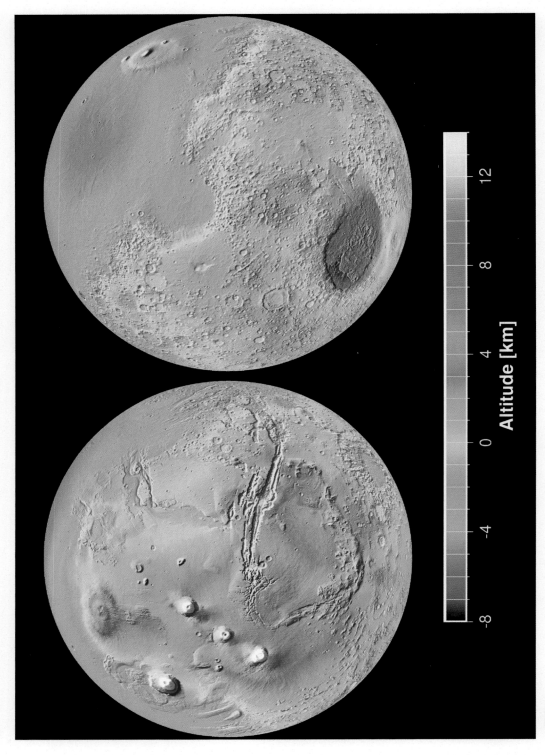

Altitude [km]

Plate 9.20

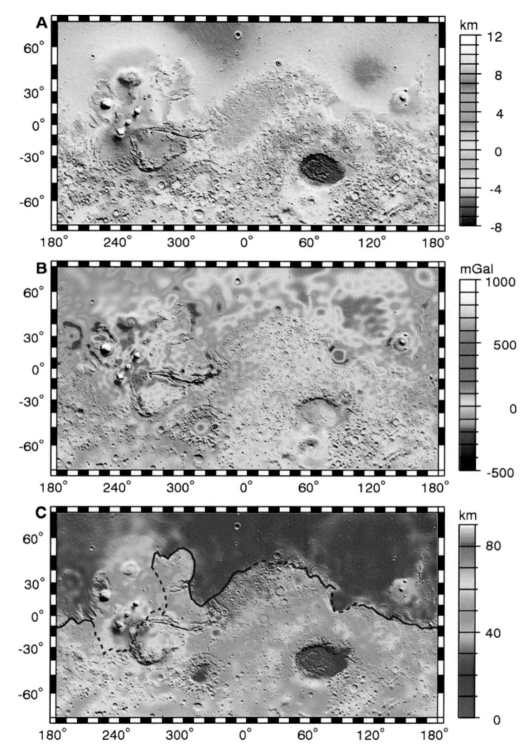

Plate 9.21

Plate 9.23

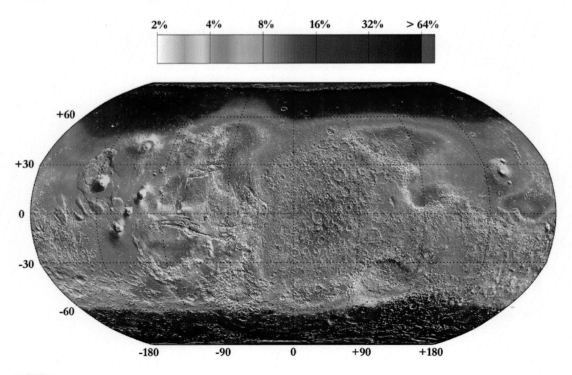

Plate 9.26

Plate 9.31

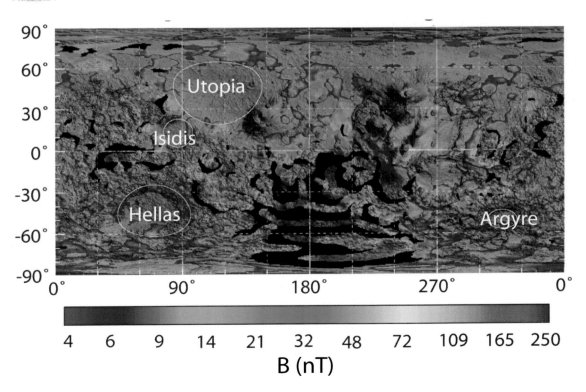

Plate 9.36

Plate 10.3

(a)

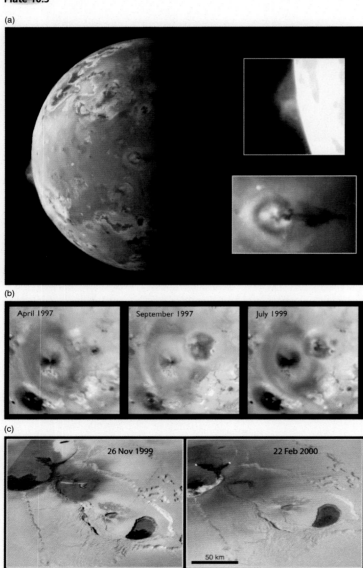

(b)

April 1997 | September 1997 | July 1999

(c)

26 Nov 1999 | 22 Feb 2000

50 km

Plate 10.4

(a)

(b)

(c)

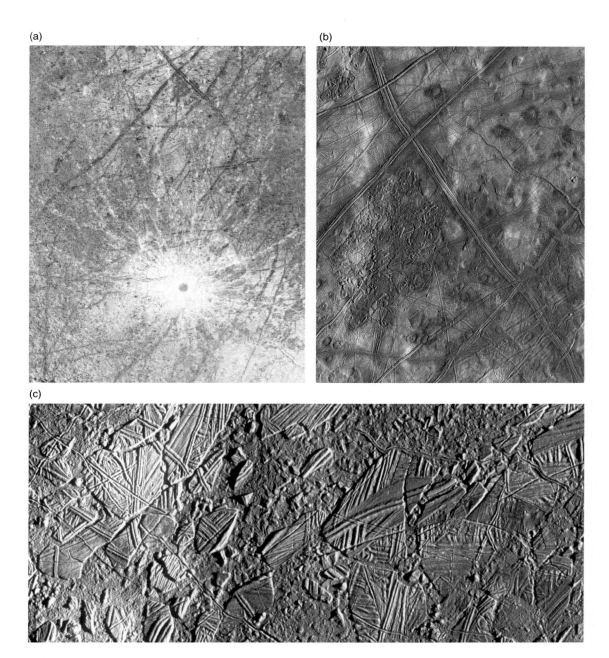

Plate 10.9

Plate 10.25

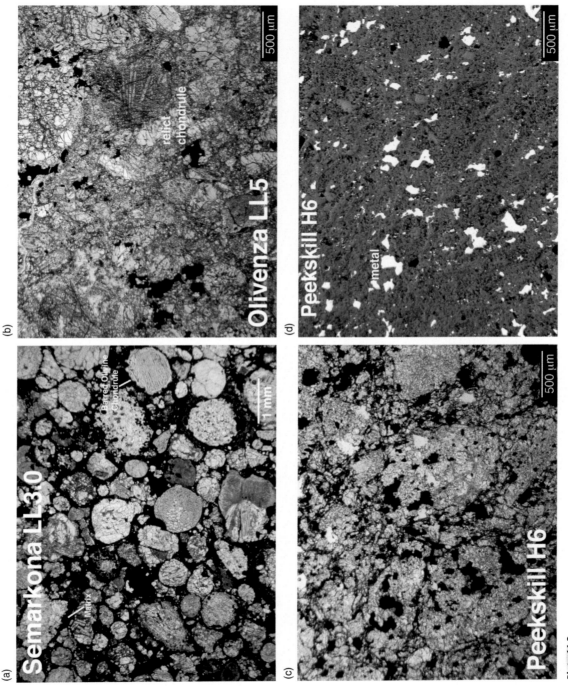

(a) Semarkona LL3.0 — Matrix, Barred Olivine Chondrule, 1 mm

(b) Olivenza LL5 — relict chondrule, 500 μm

(c) Peekskill H6 — 500 μm

(d) Peekskill H6 — metal, 500 μm

Plate 11.3

Plate 12.1

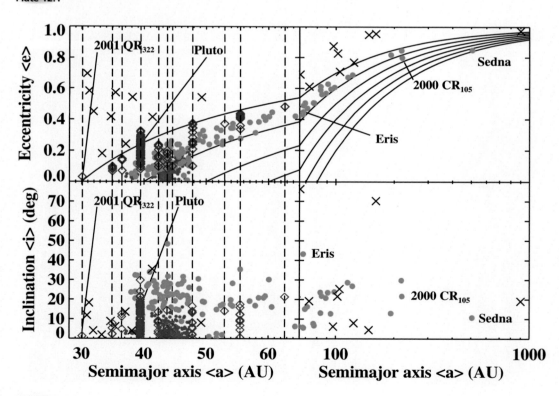

Plate 12.4

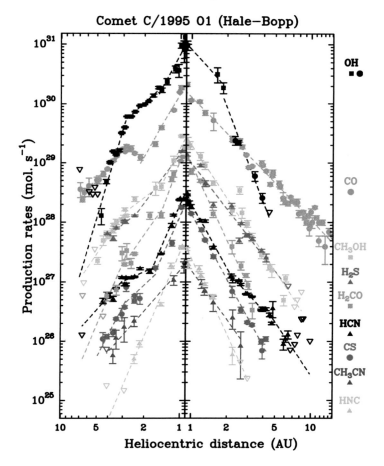

Comet C/1995 O1 (Hale–Bopp)

Production rates (mol. s^{-1}) vs. Heliocentric distance (AU)

OH
CO
CH$_3$OH
H$_2$S
H$_2$CO
HCN
CS
CH$_3$CN
HNC

Plate 12.33

Plate 13.5

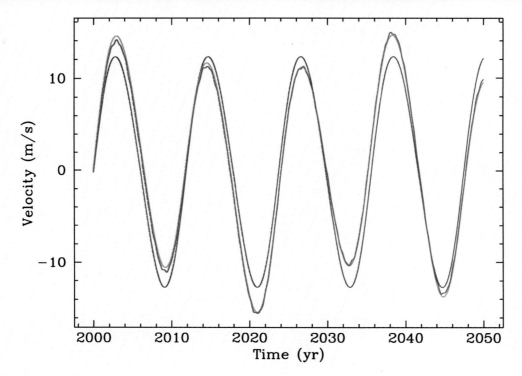

Plate 14.1

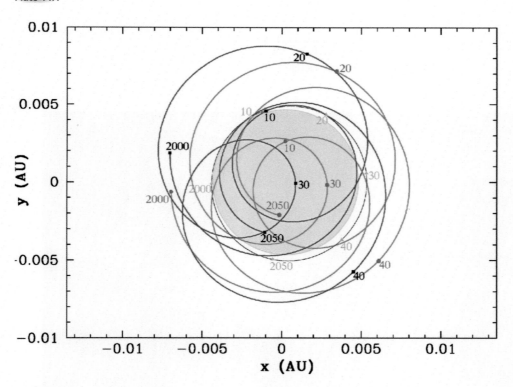

Plate 14.2

(a)

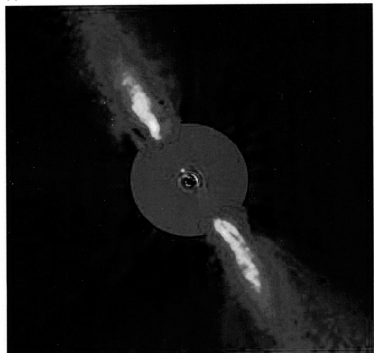

(b)

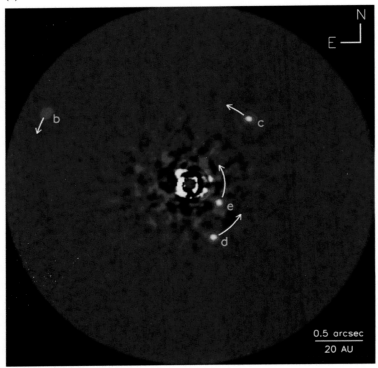

Plate 14.19

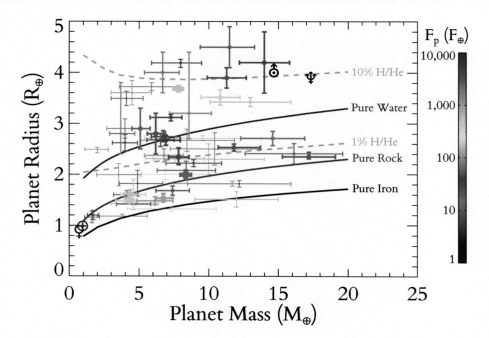

Plate 14.24

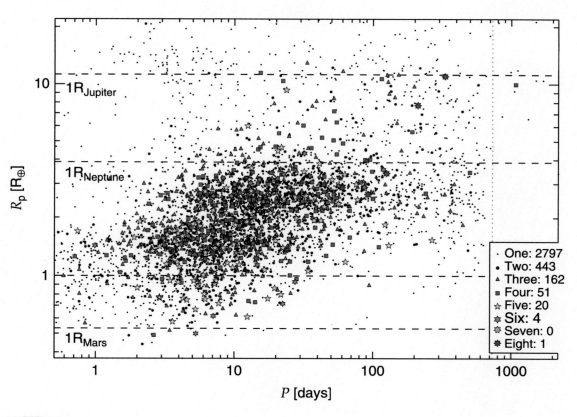

Plate 14.29

$\rho > \rho_\oplus$ $\rho = \rho_\oplus$ $\rho < \rho_\oplus$

Plate 14.34

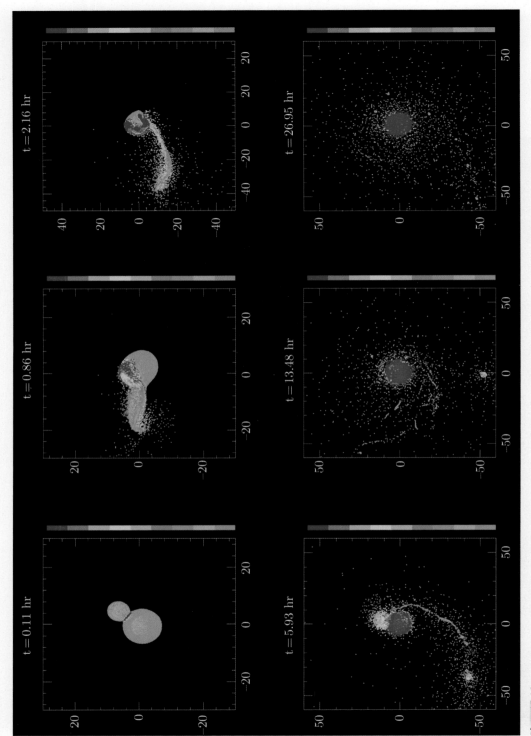

Plate 15.14

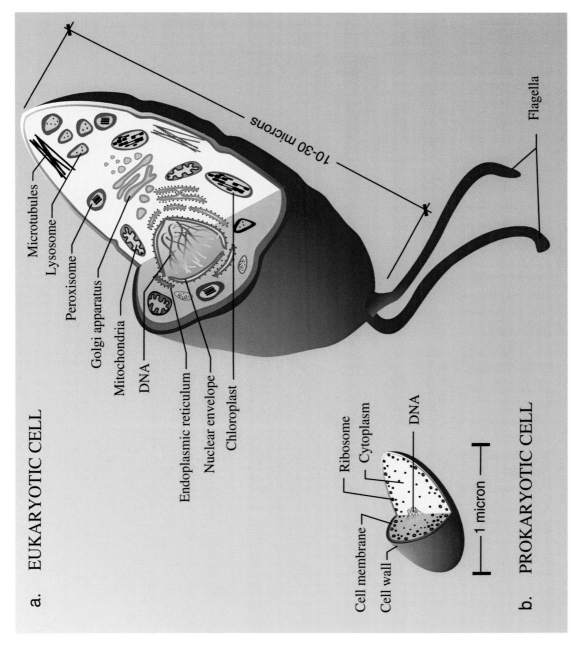

a. EUKARYOTIC CELL

Microtubules
Lysosome
Peroxisome
Golgi apparatus
Mitochondria
DNA
Endoplasmic reticulum
Nuclear envelope
Chloroplast

10-30 microns

Flagella

Ribosome
Cytoplasm
DNA

Cell membrane
Cell wall

1 micron

b. PROKARYOTIC CELL

Plate 16.1

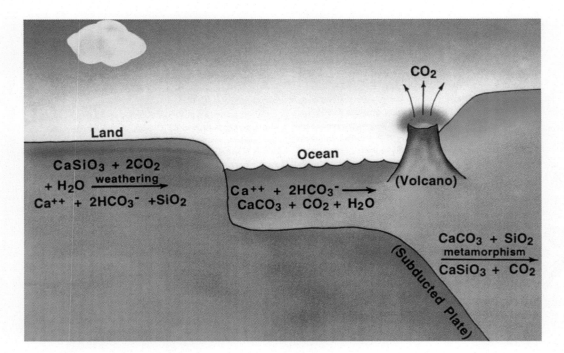

$$CaSiO_3 + 2CO_2 + H_2O \xrightarrow{\text{weathering}} Ca^{++} + 2HCO_3^- + SiO_2$$

Land

Ocean

$$Ca^{++} + 2HCO_3^- \longrightarrow CaCO_3 + CO_2 + H_2O$$

CO_2

(Volcano)

$$CaCO_3 + SiO_2 \xrightarrow{\text{metamorphism}} CaSiO_3 + CO_2$$

(Subducted Plate)

Plate 16.5

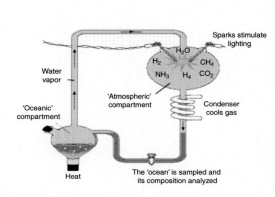

Sparks stimulate lighting

Water vapor

H_2 H_2O CH_4

NH_3 H_4 CO_2

'Atmospheric' compartment

'Oceanic' compartment

Condenser cools gas

Heat

The 'ocean' is sampled and its composition analyzed

Plate 16.10

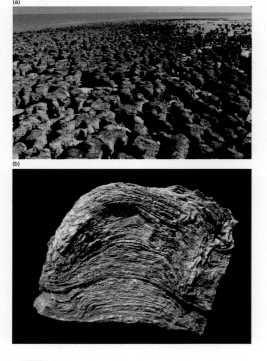

(a)

(b)

Plate 16.14

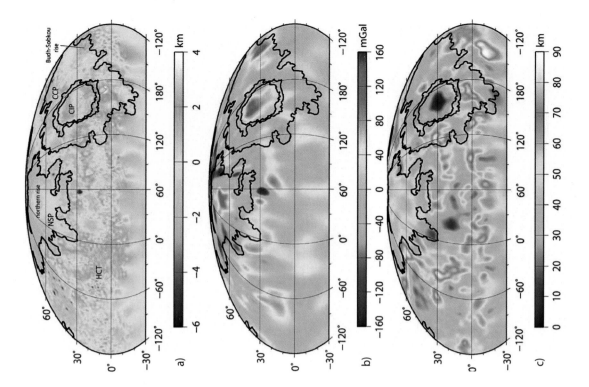

Plate G.1

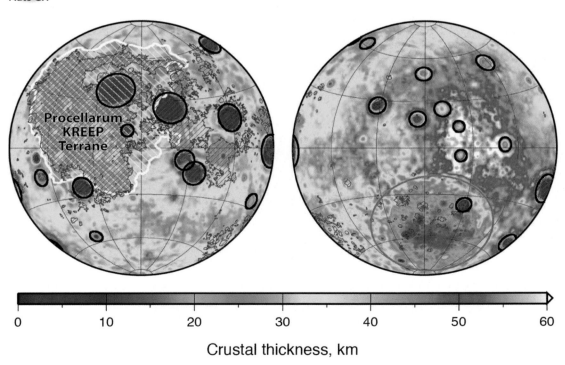

Crustal thickness, km

Plate G.5

K abundance   Mg/Si ratio   Al/Si ratio

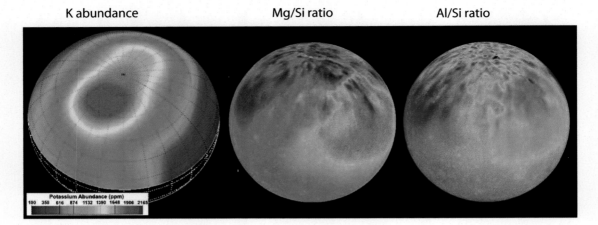

Plate G.2

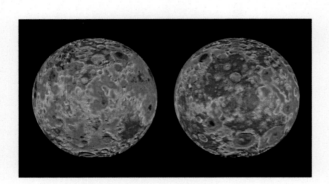

Plate G.20

Plate G.44

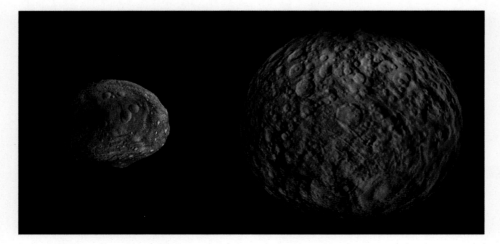

Plate G.23

CHAPTER 1

Introduction

There are in fact two things, science and opinion; the former begets knowledge, the latter ignorance.

Hippocrates, *Law*, 460–377 BCE

Why are we so fascinated by planets? After all, planets make up a tiny fraction (probably substantially less than 1%) of ordinary matter in the Universe[1]. And why do terrestrial planets, which contain less than 1% of the planetary mass within our Solar System, hold a particular place in our hearts? The simple answer is that we live on a terrestrial planet. But there is a broader, more inclusive, version of that answer: To the best of our knowledge, planets or moons with solid surfaces are the only places where life can begin and evolve into advanced forms.

In this chapter, we introduce the subject of planetary sciences and provide some background needed for the remainder of the book. The history of planetary observations dates back thousands of years, and the prehistory likely extends much, much further back; we present a brief overview in the next section. We then give an inventory of objects in our Solar System in §1.2. This is followed in §1.3 by a discussion of definitions of the word 'planet' and of words describing various smaller and larger objects.

Despite the far larger number of planets known around other stars, most of our knowledge of planetary sciences was developed from observations of bodies within our own planetary system. This information is far from complete, and understanding observables is key to assessing the reliability of data; §1.4 discusses what aspects of planetary bodies we can observe.

Many lower-level planetary textbooks begin by covering the formation of our Solar System because that makes the most sense from a chronological perspective. However, although we can observe distant circumstellar disks that appear to be planetary nurseries, our observations of these disks are far less precise than those of objects orbiting the Sun. Furthermore, the accretion of planets takes a long time compared with the few decades since such observations began. Therefore, most of our understanding of planetary formation comes from a synthesis of theoretical modeling with data from our own Solar System and extrasolar planets. We thus defer our main discussion of this subject, which is among the most intellectually challenging in planetary science, until near the end of this book. Nonetheless, scientists have modeled the origin of planets for hundreds of years, and our understandings of this process have provided the best estimates of certain planetary properties that are not directly observable, such as interior composition. Because interpretation of data and planetary formation models often go hand in hand, we present a brief summary of current models of planetary formation in the final section of this chapter.

1.1 A Brief History of the Planetary Sciences

The sky appears quite spectacular on a clear night away from the light of modern cities. Ancient civilizations were particularly intrigued by several brilliant 'stars' that move among the far more numerous 'fixed' (stationary) stars. The Greeks called these objects **planets**, or wandering stars. Old drawings and manuscripts by people from all over the world, including the Chinese, Greeks and Anasazi, attest to their interest in comets, solar eclipses and other celestial phenomena. And observations of planets surely date to well before the dawn of writing and historical records, perhaps predating humanity itself. Some migratory birds use the patterns of stars in the night sky to guide their journeys and might be aware that a few of these objects move relative to the others. Indeed, some sharp-eyed and keen-witted dinosaurs may have realized that a few points of light in the night

---

[1] **Dark matter**, most of which is **nonbaryonic** (i.e., not composed of protons or neutrons), is more than five times as abundant as ordinary matter, which is also referred to as **luminous matter**. **Dark energy** has more than twice the mass-energy density of all types of matter in the Universe combined.

sky moved relative to the fixed pattern produced by most 'stars' more than 100 million years ago, but as dinosaurs never (to our knowledge) developed a written language, it is unlikely that such speculation will ever be confirmed.

The Copernican–Keplerian–Galilean–Newtonian revolution in the sixteenth and seventeenth centuries completely changed humanity's view of the dimensions and dynamics of the Solar System, including the relative sizes and masses of the bodies and the forces that make them orbit about one another. Gradual progress was made over the next few centuries, but the next revolution had to await the space age.

The age of planetary exploration began in October of 1959, with the Soviet Union's spacecraft *Luna 3* returning the first pictures of the far side of Earth's Moon (Fig. F.1). Over the next three decades, spacecraft visited all eight known terrestrial and giant planets in the Solar System, including our own. These spacecraft have returned data concerning the planets, their rings and moons. Spacecraft images of many objects showed details never suspected from earlier Earth-based pictures. Spectra from γ-rays to radio wavelengths revealed previously undetected gases and geological features on planets and moons, and radio detectors and magnetometers transected the giant magnetic fields surrounding many of the planets. The planets and their satellites have become familiar to us as individual bodies. The immense diversity of planetary and satellite surfaces, atmospheres and magnetic fields has surprised even the most imaginative researchers. Unexpected types of structure were observed in Saturn's rings, and whole new classes of rings and ring systems were seen around all four giant planets. Some of the new discoveries have been explained, but others remain mysterious.

Five comets and ten asteroids have thus far been explored close up by spacecraft (Table F.2), and there have been several missions to study the Sun and the solar wind. The Sun's gravitational domain extends thousands of times the distance to the farthest known planet, Neptune. Yet the vast outer regions of the Solar System are so poorly explored that many bodies remain to be detected, possibly including some of planetary size.

Hundreds of planets are now known to orbit stars other than the Sun. Although we know far less about any of these extrasolar planets than we do about the planets in our Solar System, it is clear that many of them have gross properties (orbits, masses, radii) quite different from any object orbiting our Sun, and they are thus causing us to revise some of our models of how planets form and evolve.

Biologists have redrawn the tree of life over the past few decades. We have learned of the interrelationships between all forms of life on Earth and of life's great diversity. This diversity enables some species to live in environments that would be considered quite extreme to humans and suggests that conditions capable of sustaining life exist on other planets and moons in our Solar System and beyond.

The renewed importance of the planetary sciences as a subfield of astronomy implies that some exposure to Solar System studies is an important component to the education of astronomers. Planetary sciences' close relationship to geophysics, atmospheric and space sciences means that the study of the planets offers the unique opportunity for comparison available to Earth scientists. The properties of planets are key to astrobiology, and understanding the basics of life is useful to planetary scientists.

1.2 Inventory of the Solar System

What is the **Solar System**? Our naturally geocentric view gives a highly distorted picture; thus, it is better to phrase the question as: What is seen by an objective observer from afar? The **Sun**, of course; the Sun has a luminosity 4×10^8 times

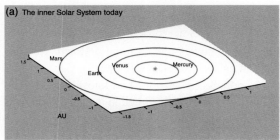 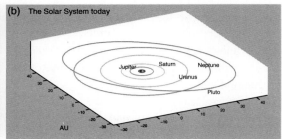

Figure 1.1 The orbits of (a) the four terrestrial planets and (b) all eight major planets in the Solar System and Pluto are shown to scale. The axes are in AU. The movies show variations in the orbits over the past 3 million years; these changes are caused by mutual perturbations among the planets (see Chapter 2). Figure 2.12 presents plots of the variations in planetary eccentricities from the same integrations. (Illustrations courtesy Jonathan Levine)

as large as the total luminosity (reflected plus emitted) of Jupiter, the second brightest object in the Solar System. The Sun also contains >99.8% of the mass of the known Solar System. By these measures, the Solar System can be thought of as the Sun plus some debris. However, by other measures, the planets are not insignificant. More than 98% of the angular momentum in the Solar System lies in orbital motions of the planets. Moreover, the Sun is a fundamentally different type of body from the planets – a ball of plasma powered by nuclear fusion in its core – but the smaller bodies in the Solar System are composed of molecular matter, some of which is in the solid state. This book focuses on the debris in orbit about the Sun, although we do include a summary of the properties of stars, including our Sun, in §3.3, and an overview of the outer layers of the Sun and its effect on the interplanetary medium in §§7.1 and 7.2. The debris encircling the Sun is composed of the giant planets, the terrestrial planets and numerous and varied smaller objects.

Figures 1.1 to 1.3 present three differing views of the Solar System. The orbits of the major planets and Pluto are diagrammed in Figure 1.1. Two different levels of reduction are displayed because of the relative closeness of the four terrestrial planets and the much larger spacings in the outer Solar

System. Note the high inclination of Pluto's orbit relative to the orbits of the major planets. Figure 1.2 plots the sizes of various classes of Solar System objects as a function of location. The jovian (giant) planets dominate the outer Solar System, and the terrestrial planets dominate the inner Solar System. Small objects tend to be concentrated in regions where orbits are stable or at least long lived. Images of the planets and the largest planetary satellites are presented to scale in Figure 1.3. Figure 1.4 shows close-up views of those comets and asteroids that had been imaged by interplanetary spacecraft as of 2010.

1.2.1 Giant Planets

Jupiter dominates our planetary system. Its mass, 318 Earth masses ($M_\oplus$), exceeds twice that of all other known Solar System planets combined. Thus, as a second approximation, the Solar System can be viewed as the Sun, Jupiter and some debris. The largest of this debris is **Saturn**, with a mass of nearly $100 M_\oplus$. Saturn, similar to Jupiter, is made mostly of hydrogen (H) and helium (He). Each of these planets probably possesses a heavy element 'core' of mass $\sim 10 M_\oplus$. The third and fourth largest planets are **Neptune** and **Uranus**, each having a mass roughly one-sixth that of Saturn. These planets belong to a different class, with

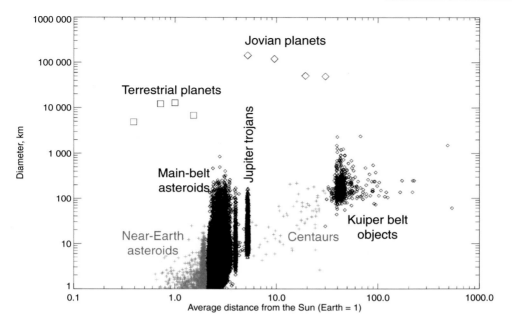

Figure 1.2 Inventory of objects orbiting the Sun. Small bodies are discussed in Chapter 12. The orbits of Jupiter Trojans are described in §2.2.1 and those of Centaurs are discussed in §12.2.2. (Courtesy John Spencer)

most of their masses provided by a combination of three common astrophysical 'ices', water (H_2O), ammonia (NH_3), methane (CH_4), together with 'rock', high temperature condensates consisting primarily of silicates and metals, yet most of their volumes are occupied by relatively low mass (1–4 $M_\oplus$) H–He dominated atmospheres. The four largest planets are known collectively as the **giant planets**; Jupiter and Saturn are called **gas giants**, with radii of $\sim$70 000 km and 60 000 km, respectively, and Uranus and Neptune are referred to as **ice giants** (although the 'ices' are present in fluid rather than solid form), with radii of $\sim$25 000 km. All four giant planets possess strong magnetic fields. These planets orbit the Sun at distances of approximately 5, 10, 20 and 30 AU, respectively. (One **astronomical unit**, 1 AU, is defined to be the semimajor axis of a massless [test] particle whose orbital period about the Sun is one year. As our planet has a finite mass, the semimajor axis of Earth's orbit is slightly larger than 1 AU.)

1.2.2 Terrestrial Planets

The mass of the remaining known 'debris' totals less than one-fifth that of the smallest giant planet, and their orbital angular momenta are also much smaller. This debris consists of all of the solid bodies in the Solar System, and despite its small mass, it contains a wide variety of objects that are interesting chemically, geologically, dynamically and, in at least one case, biologically. The hierarchy continues within this group, with two large **terrestrial**[2] planets, **Earth** and **Venus**, each with a radius of about 6000 km, at approximately 1 and 0.7 AU from the Sun, respectively. Our Solar System also contains two small terrestrial planets, **Mars** with a radius of $\sim$3500 km and orbiting at

[2] In this text, the word 'terrestrial' is used to mean Earth-like or related to the planet Earth, as is the convention in planetary sciences and astronomy. Geoscientists and biologists generally use the same word to signify a relationship with land masses.

(a)

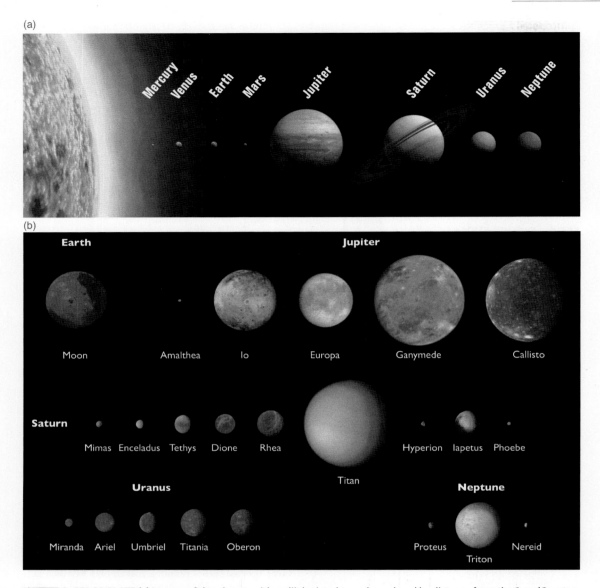

(b)

Figure 1.3 COLOR PLATE (a) Images of the planets with radii depicted to scale, ordered by distance from the Sun. (Courtesy International Astronomical Union/Martin Kornmesser) (b) Images of the largest satellites of the four giant planets and Earth's Moon, which are depicted in order of distance from their planet. Note that these moons span a wide range of size, albedo (reflectivity) and surface characteristics; most are spherical, but some of the smallest objects pictured are quite irregular in shape. (Courtesy Paul Schenk)

$\sim$1.5 AU and **Mercury** with a radius of $\sim$2500 km orbiting at $\sim$0.4 AU.

All four terrestrial planets have atmospheres. Atmospheric composition and density vary widely among the terrestrial planets, with Mercury's atmosphere being exceedingly thin. However, even the most massive terrestrial planet atmosphere, that of Venus, is minuscule by giant planet standards.

Figure 1.4 Views of the first four comets (lower right) and nine asteroid systems that were imaged close-up by interplanetary spacecraft, shown at the same scale. The object name and dimensions, as well as the name of the imaging spacecraft and the year of the encounter, are listed below each image. Note the wide range of sizes. Dactyl is a moon of Ida.

Earth and Mercury each have an internally generated magnetic field, and evidence suggests that Mars possessed one in the distant past.

1.2.3 Minor Planets and Comets

The **Kuiper belt** is a thick disk of ice/rock bodies beyond the orbit of Neptune. The two largest members of the Kuiper belt to have been sighted are **Eris**, whose **heliocentric distance**, the distance from the Sun, oscillates between 38 and 97 AU, and **Pluto**, whose heliocentric distance varies from 29 to 50 AU. The radii of Eris and Pluto exceed 1000 km. Pluto is known to possess an atmosphere. Numerous smaller members of the Kuiper belt

have been cataloged, but the census of these distant objects is incomplete even at large sizes. **Asteroids**, which are minor planets that all have radii <500 km, are found primarily between the orbits of Mars and Jupiter.

Smaller objects are also known to exist elsewhere in the Solar System, for example as moons in orbit around planets, and as comets. Comets are ice-rich objects that shed mass when subjected to sufficient solar heating. Comets are thought to have formed in or near the giant planet region and then been 'stored' in the **Oort cloud**, a nearly spherical region at heliocentric distances of ~ 1–5×10^4 AU, or in the Kuiper belt or the **scattered disk**. Scattered disk objects (SDOs) have moderate to high eccentricity orbits that lie in whole or in part within the Kuiper belt. Estimates of the total number of comets larger than 1 km in radius in the entire Oort cloud range from $\sim 10^{12}$ to $\sim 10^{13}$. The total number of Kuiper belt objects (KBO) larger than 1 km in radius is estimated to be $\sim 10^8$–10^{10}. The total mass and orbital angular momentum of bodies in the scattered disk and Oort cloud are uncertain by more than an order of magnitude. The upper end of current estimates places as much mass in distant unseen icy bodies as is observed in the entire planetary system.

The smallest bodies known to orbit the Sun, such as the dust grains that together produce the faint band in the plane of the planetary orbits known as the **zodiacal cloud**, have been observed collectively but not yet individually detected via remote sensing.

1.2.4 Satellite and Ring Systems

Some of the most interesting objects in the Solar System orbit about the planets. Following the terrestrial planets in mass are the seven major moons of the giant planets and Earth. Two planetary satellites, Jupiter's moon Ganymede and Saturn's moon Titan, are slightly larger than the planet Mercury, but because of their lower densities, they are less than half as massive. Titan's atmosphere is denser than that of Earth. Triton, by far the largest moon of Neptune, has an atmosphere that is much less dense, yet it has winds powerful enough to strongly perturb the paths of particles ejected from geysers on its surface. Very tenuous atmospheres have been detected about several other planetary satellites, including Earth's Moon, Jupiter's Io and Saturn's Enceladus.

Natural satellites have been observed in orbit about most of the planets in the Solar System, as well as many Kuiper belt objects and asteroids. The giant planets all have large satellite systems, consisting of large- and/or medium-sized satellites (Fig. 1.3b) and many smaller moons and rings.

Most of the smaller moons orbiting close to their planet were discovered from spacecraft flybys. All major satellites, except Triton, orbit the respective planet in a **prograde** manner (i.e., in the direction that the planet rotates) close to the planet's equatorial plane. Small, close-in moons are also exclusively in low-inclination, low-eccentricity orbits, but small moons orbiting beyond the main satellite systems can travel around the planet in either direction, and their orbits are often highly inclined and eccentric. Earth and Pluto each have one large moon: our Moon has a little over 1% of Earth's mass, and Charon's mass is just over 10% that of Pluto. These moons probably were produced by giant impacts on the Earth and Pluto when the Solar System was a small fraction of its current age. Two tiny moons travel on low-inclination, low-eccentricity orbits about Mars.

The four giant planets all have ring systems, which are primarily located within about 2.5 planetary radii of the planet's center. However, in other respects, the characters of the four ring systems differ greatly. Saturn's rings are bright and broad, full of structure such as density waves, gaps and 'spokes'. Jupiter's ring is very tenuous and composed mostly of small particles. Uranus has nine narrow opaque rings plus broad regions of tenuous dust orbiting close to the plane defined by the planet's equator. Neptune has four rings, two narrow ones and two faint broader rings; the most remarkable part of Neptune's ring system is the ring arcs, which are bright segments within one of the narrow rings.

1.2.5 Tabulations

The orbital and bulk properties of the eight 'major' planets are listed in Tables E.1 to E.3. Symbols for each of these planets, which we often use as subscripts on masses and radii, are also given in Table E.1. Table E.4 gives orbital elements and brightnesses of all inner moons of the eight planets, as well as those outer moons whose

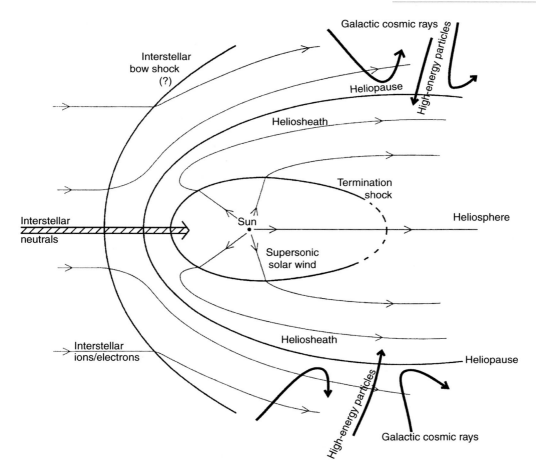

Figure 1.5 Sketch of the teardrop-shaped heliosphere. Within the heliosphere, the solar wind flows radially outwards until it encounters the heliopause, the boundary between the solar wind–dominated region and the interstellar medium. Weak cosmic rays are deflected away by the heliopause, but energetic particles penetrate the region down to the inner Solar System. (Adapted from Gosling 2007)

radii are estimated to be $\gtrsim 10$ km. Many of the orbital parameters listed in the tables are defined in §2.1. Rotation rates and physical characteristics of these satellites, whenever known, are given in Table E.5. Properties of some the largest 'minor planets', asteroids and Kuiper belt objects are given in Tables E.6 and E.7, and densities of some minor planets are listed in Table E.8.

The brightness of a celestial body is generally expressed as the **apparent magnitude** at visual wavelengths, $m_{\rm v}$. A $6^{\rm th}$ magnitude ($m_{\rm v} = 6$) star is just visible to the naked eye in a dark sky. The magnitude scale is logarithmic (mimicking the perception of human vision), and a difference of 5 magnitudes equals a factor of 100 in brightness (i.e., a star with $m_{\rm v} = 0$ is 100 times brighter than one with $m_{\rm v} = 5$). The apparent magnitudes of planetary satellites are listed in Table E.4. Those moons with $m_{\rm v} > 20$ can only be detected with a large telescope or nearby spacecraft.

1.2.6 Heliosphere

All planetary orbits lie within the **heliosphere**, the region of space containing magnetic fields and plasma of solar origin. Figure 1.5 diagrams key components of the heliosphere. The **solar wind** consists of **plasma** (ionized gas) traveling outward from the Sun at supersonic speeds. The solar wind merges with the **interstellar medium** at the **heliopause**, the boundary of the heliosphere.

The composition of the heliosphere is dominated by solar wind protons and electrons, with a typical density of 5×10^6 protons m^{-3} at 1 AU from the Sun, decreasing as the reciprocal distance squared. These particles move outwards at speeds of ~ 400 km s^{-1} near the solar equator but ~ 700–800 km s^{-1} closer to the solar poles. In contrast, the local interstellar medium, at a density of less than 1×10^5 atoms m^{-3}, contains mainly hydrogen and helium atoms. The Sun's motion relative to the mean motion of neighboring stars is roughly 26 km s^{-1}. Hence, the heliosphere moves through the interstellar medium at about this speed. The heliosphere is thought to be shaped like a teardrop, with a tail in the downwind direction (Fig. 1.5). Interstellar ions and electrons generally flow around the heliosphere because they cannot cross the solar magnetic fieldlines. Neutrals, however, can enter the heliosphere, and as a result interstellar H and He atoms move through the Solar System in the downstream direction with a typical speed of ~ 22 (for H) to 26 (for He) km s^{-1}.

Just interior to the heliopause is the **termination shock**, where the solar wind is slowed down. Because of variations in solar wind pressure, the location of this shock moves radially with respect to the Sun in accordance with the 11-year solar activity cycle. The *Voyager 1* spacecraft crossed the termination shock in December 2004 at a heliocentric distance of 94.0 AU; *Voyager 2* crossed the shock (multiple times) in August 2007 at ~ 83.7 AU. Both *Voyagers* spent approximately one decade in the **heliosheath**, between the termination shock and the heliopause. *Voyager 1* entered interstellar space on 25 August 2012, when it was at a heliocentric distance of 121 AU, and *Voyager 2* followed in late 2018.

1.3 What is a Planet?

The ancient Greeks referred to all moving objects in the sky as planets. To them, there were seven such objects, the Sun, the Moon, Mercury, Venus, Mars, Jupiter and Saturn. The Copernican revolution removed the Sun and Moon from the planet club, but added the Earth. Uranus and Neptune were added as soon as they were discovered in the eighteenth and nineteenth centuries, respectively.

Pluto, by far the brightest Kuiper belt object (KBO) and the first that was discovered, was officially classified as a planet from its discovery in 1930 until 2006; 1 Ceres, the first detected (in 1801) and by far the largest member of the asteroid belt, was also once considered to be a planet, as were the next few asteroids that were discovered. With the detection of other KBOs, debates began with regard to the classification of Pluto as a planet, culminating in August 2006 with the resolution by the International Astronomical Union (IAU):

- A **planet** is a celestial body that (1) is in orbit around the Sun, (2) has sufficient mass for its self-gravity to overcome rigid body forces so that it assumes a hydrostatic equilibrium (nearly round) shape and (3) has cleared the neighborhood around its orbit.
- A **dwarf planet** is a celestial body that (1) is in orbit around the Sun, (2) has sufficient mass for its self-gravity to overcome rigid body forces so that it assumes a hydrostatic equilibrium (nearly round) shape, (3) has not cleared the neighborhood around its orbit, and (4) is not a satellite.

Just as the discoveries of small bodies orbiting the Sun have forced astronomers to decide how small an object can be and still be worthy of being classified as a planet, detections of substellar objects orbiting other stars have raised the question of an upper size limit to planethood. We adopt the following definitions, which are consistent with current IAU nomenclature:

- **Star:** self-sustaining fusion is sufficient for thermal pressure to balance gravity ($\gtrsim 0.075$ $M_\odot$ ≈ 80 M_{2_+} for solar composition; the minimum mass for an object to be a star is often referred to as the **hydrogen burning limit**)
- **Stellar remnant:** dead star – no more fusion (or so little that the object is no longer supported primarily by thermal pressure)
- **Brown dwarf:** substellar object with substantial deuterium fusion – more than half of the object's original inventory of deuterium is ultimately destroyed by fusion
- **Planet:** negligible fusion ($\lesssim 0.012$ $M_\odot$ ≈ 13 M_{2_+}, with the precise value again depending on initial composition), plus it orbits one or more stars and/or stellar remnants

1.4 Planetary Properties

All of our knowledge regarding specific characteristics of Solar System objects, including planets, moons, comets, asteroids, rings and interplanetary dust, is ultimately derived from observations, either astronomical measurements from the ground or Earth-orbiting satellites, or from close-up (often *in situ*) measurements obtained by interplanetary spacecraft. One can determine the following quantities more or less directly from observations:

(1) Orbit
(2) Mass, distribution of mass
(3) Size
(4) Rotation rate and direction
(5) Shape
(6) Temperature
(7) Magnetic field
(8) Surface composition
(9) Surface structure
(10) Atmospheric structure and composition

With the help of various theories, these observations can be used to constrain planetary properties such as bulk composition and interior structure, two attributes that are crucial elements in modeling the formation of the Solar System.

1.4.1 Orbit

In the early part of the seventeenth century, Johannes Kepler deduced three 'laws' of planetary motion directly from observations:

(1) All planets move along elliptical paths with the Sun at one focus.
(2) A line segment connecting any given planet and the Sun sweeps out area at a constant rate.
(3) The square of a planet's orbital period about the Sun, $P_{\rm orb}$, is proportional to the cube of its semimajor axis, a, i.e., $P_{\rm orb}^2 \propto a^3$.

A Keplerian orbit is uniquely specified by six orbital elements: a (semimajor axis), e (eccentricity), i (inclination), ω (argument of periapse; or ϖ for the longitude of periapse), Ω (longitude of ascending node) and f (true anomaly). These orbital elements are defined graphically in Figure 2.1 and discussed in more detail in §2.1. The first few of these elements are more fundamental than the last: a and e fully define the size and shape of the orbit, i gives the tilt of the orbital plane to some reference plane, the longitudes ϖ and Ω determine the orientation of the orbit and f (or, indirectly, t_ϖ, the time of periapse passage) tells where the planet is along its orbit at a given time. Alternative sets of orbital elements are also possible; for instance, an orbit is fully specified by the planet's location and velocity relative to the Sun at a given time (again, six independent scalar quantities), provided the masses of the Sun and planet are known.

Kepler's laws (or more accurate versions thereof) can be derived from Newton's laws of motion and of gravity, which were formulated later in the seventeenth century (§2.1). Relativistic effects also affect planetary orbits, but they are small compared with the gravitational perturbations that the planets exert on one other (Problem 2-5).

All planets and asteroids revolve around the Sun in the direction of solar rotation. Their orbital planes generally lie within a few degrees of each other and close to the solar equator. For observational convenience, inclinations are usually measured relative to the Earth's orbital plane, which is known as the **ecliptic plane**. The Sun's equatorial plane is inclined by 7° with respect to the ecliptic plane. Among the eight major planets, Mercury's orbit is the most tilted, with $i = 7°$. (However, because inclination is effectively a vector, the similarity of these two inclinations does not imply that Mercury's orbit lies within the plane of the solar equator. Indeed, Mercury's orbit is inclined by 3.4° relative to the Sun's equatorial plane.) Similarly, most major satellites orbit their planet close to its equatorial plane. Many smaller objects that orbit the Sun and the planets have much larger orbital inclinations. In addition, some comets, minor satellites and Neptune's large moon Triton orbit the Sun or planet in a **retrograde** sense (opposite to the Sun's or planet's rotation). The observed 'flatness' of most of the planetary system is explained by planetary formation models that hypothesize that the planets grew within a disk that was in orbit around the Sun (see Chapter 15).

1.4.2 Mass

The mass of an object can be deduced from the gravitational force that it exerts on other bodies.

- Orbits of moons: The orbital periods of natural satellites, together with Newton's generalization of Kepler's third law (eq. 2.18), can be used to solve for mass. The result is actually the sum of the mass of the planet and moon (plus, to a good approximation, the masses of moons on orbits interior to the one being considered), but except for the Earth/Moon and various minor planets, including Pluto/Charon, the secondaries' masses are very small compared with that of the primary. The major source of uncertainty in this method results from measurement errors in the semimajor axis; timing errors are negligible.

- What about planets without moons? The gravity of each planet perturbs the orbits of all other planets. Because of the large distances involved, the forces are much smaller, so the accuracy of this method is not high. Note, however, that Neptune was discovered as a result of the perturbations that it forced on the orbit of Uranus. This technique is still used to provide the best (albeit in some cases quite crude) estimates of the masses of some large asteroids. The perturbation method can actually be divided into two categories: short-term and long-term perturbations. The extreme example of short-term perturbations includes single close encounters between asteroids. Trajectories can be computed for a variety of assumed masses of the body under consideration and fit to the observed path of the other body. Long-term perturbations are best exemplified by masses derived from periodic variations in the relative positions of moons locked in stable orbital resonances (§2.3.2).

- Spacecraft tracking data provide the best means of determining masses of planets and moons visited because the Doppler shift and periodicity of the transmitted radio signal can be measured very precisely. The long time baselines afforded by **orbiter** missions allow much higher accuracy than **flyby** missions. The best estimates for the masses of some of the outer planet moons are those obtained by combining accurate short-term perturbation measurements from *Voyager* images with *Voyager* tracking data and/or resonance constraints from long timeline ground-based observations.

Figure 1.6 COLOR PLATE Simulated 3-D renderings of the eight planets within our Solar System. (www.lesud.com © 2011)

- The best estimates of the masses of some of Saturn's small inner moons were derived from the amplitude of spiral density waves they resonantly excite in Saturn's rings or of density wakes that they produce in nearby ring material. These processes are discussed in §13.4.
- Crude estimates of the masses of some comets have been made by estimating nongravitational forces, which result from the asymmetric escape of released gases and dust (§12.2.4), and comparing them with observed orbital changes.

The gravity field of a mass distribution that is not spherically symmetric differs from that of a point source of identical mass. Such deviations, combined with the knowledge of the rotation period, can be used to estimate the degree of central concentration of mass in rotating bodies (§6.2.2). The deviation of the gravity field of an asymmetric body from that of a point mass is most pronounced, and thus most easily measured, closest to the body (§2.6). To determine the precise gravity field, one can make use of both spacecraft tracking data and the orbits of moons and/or eccentric rings.

1.4.3 Size

Bodies in the Solar System exhibit a wide range of sizes and shapes. Figure 1.6 illustrates the vast dynamic range of just the bodies considered to be planets. The size of an object can be measured in various ways:

- The diameter of a body is the product of its angular size (measured in radians) and its distance from the observer. Solar System distances are simple to estimate from orbits; however, limited resolution from Earth results in large uncertainties in angular size. Thus, other techniques often give the best results for bodies that have not been imaged at close distances by interplanetary spacecraft.
- The diameter of a Solar System body can be deduced by observing a star as it is occulted by the body. The angular velocity of the star relative to the occulting body can be calculated from orbital data, including the effects of the Earth's orbit and rotation. Multiplying the duration of an occultation as viewed from a particular observing site by both its angular velocity and its distance gives the length of a chord of the body's projected silhouette. Three well-separated chords suffice for a spherical planet. Many chords are needed if the body is irregular in shape, and observations of the same event from many widely spaced telescopes are necessary. This technique is particularly useful for small bodies that have not been visited by spacecraft. Occultations of sufficiently bright stars are infrequent and require appropriate predictions as well as significant observing campaigns in order to obtain enough chords.
- Radar echoes can be used to determine radii and shapes. The radar signal strength drops as $1/r^4$ ($1/r^2$ going to the object and $1/r^2$ returning to the antenna), so only relatively nearby objects may be studied with radar. Radar is especially useful for studying solid planets, asteroids and cometary nuclei.
- An excellent way to measure the radius of an object is to send a lander and triangulate using it together with an orbiter. This method, as well as the radar technique, also works well for

terrestrial planets and satellites with substantial atmospheres.

- The size and the albedo of a body can be estimated by combining photometric observations at visible and infrared (IR) wavelengths. At visible wavelengths, one measures the sunlight reflected off the object, but at infrared wavelengths, one observes the thermal radiation from the body itself (see Chapter 4 for a detailed discussion).

The mean density of an object can be trivially determined after its mass and size are known. The density of an object gives a rough idea of its composition, although compression at the high pressures that occur in planets and large moons must be taken into account, and the possibility of significant void space should be considered for small bodies. The low density ($\sim$1000 kg m^{-3}) of the four giant planets, for example, implies material with low mean molecular weight. Terrestrial planet densities of 3500–5500 kg m^{-3} imply rocky material, including some metal. Most of the medium and large satellites around the giant planets have densities between 1000 and 2000 kg m^{-3}, suggesting a combination of ices and rock. Comets have densities of roughly 1000 kg m^{-3} or less, indicative of rather loosely packed dirty ices.

In addition to the density, one can also calculate the escape velocity using the mass and size of the object (eq. 2.24). The escape velocity, together with temperature, can be used to estimate the ability of the planetary body to retain an atmosphere.

1.4.4 Rotation

Simple rotation is a vector quantity, related to spin angular momentum. The **obliquity** (or **axial tilt**) of a planetary body is the angle between its spin angular momentum and its orbital angular momentum. Bodies with obliquity $<90°$ are said to have **prograde** rotation, and planets with obliquity $>90°$ have **retrograde** rotation. The rotation of an object can be determined using various techniques:

- The most straightforward way to determine a planetary body's rotation axis and period is to observe how markings on the surface move around with the disk. Unfortunately, not all planets have such features; moreover, if atmospheric features are used, winds may cause the deduced period to vary with latitude, altitude and time.
- Planets with sufficient magnetic fields trap charged particles within their magnetospheres. These charged particles are accelerated by electromagnetic forces and emit radio waves. Because magnetic fields are not uniform in longitude and because they rotate with (presumably the bulk of) the planet, these radio signals have a periodicity equal to the planet's rotation period. For planets without detectable solid surfaces, the magnetic field period is viewed as more fundamental than the periods of cloud features (see, however, §7.3.4).
- The rotation period of a body can often be determined by periodicities observed in its **lightcurve**, which gives the total disk brightness as a function of time. Lightcurve variations can be the result of differences in albedo or, for irregularly shaped bodies, in projected area. Whereas irregularly shaped bodies produce lightcurves with two very similar maxima and two very similar minima per revolution, albedo variations have no such preferred symmetry. Thus, ambiguities of a factor of two sometimes exist in spin periods determined by lightcurve analysis. Most asteroids have double-peaked lightcurves, indicating that the major variations are due to shape, but the peaks are distinguishable from each other because of minor variations in hemispheric albedo and local topography.
- The measured Doppler shift across the disk can give a rotation period and a crude estimate of the rotation axis, provided the body's radius is known. This can be done passively in visible light or actively using radar.

The rotation periods of most objects orbiting the Sun are of the order of three hours to a few days. Mercury and Venus, both of whose rotations have almost certainly been slowed by solar tides, form exceptions with periods of 59 and 243 days, respectively. Six of the eight planets rotate in a prograde sense with obliquities of 30° or less. Venus rotates in a retrograde direction with an obliquity of 177°, and the rotation axis of Uranus is so tilted that it lies close to this planet's orbital plane. Most planetary satellites rotate synchronously with their orbital periods as a result of planet-induced tides (§2.7.2).

1.4.5 Shape

Figure 1.7a shows a close-up image of the jagged small martian moon Phobos in silhouette against the smooth limb of Mars. Many different forces together determine the shape of a body. Self-gravity tends to produce bodies of spherical shape, a minimum for gravitational potential energy. Material strength maintains shape irregularities, which may be produced by accretion, impacts or internal geological processes. Because self-gravity increases with the size of an object, larger bodies tend to be rounder. Typically, bodies with mean radii larger than ∼200 km are fairly round. Smaller objects may be quite oddly shaped.

There is a relationship between a planet's rotation and its oblateness because the rotation introduces a centrifugal pseudo-force, which causes a planet to bulge out at the equator and to flatten at the poles. A perfectly fluid planet would be shaped as an oblate spheroid. Polar flattening is greatest for planets that have a low density and rapid rotation. In the case of Saturn, the flattening parameter, $\epsilon \equiv (R_e - R_p)/R_e$, where R_e and R_p are the equatorial and polar radii, respectively, is ∼0.1, and polar flattening is easily discernible on some images of the planet, such as that shown in Figure 1.7b.

The shape of an object can be determined from:

- Direct imaging, from either the ground or spacecraft
- Length of chords observed by stellar occultation experiments at various sites (see §1.4.3)

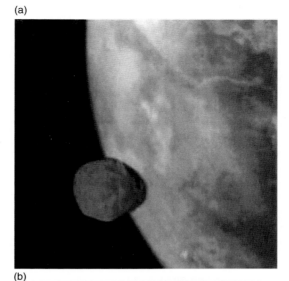

(a)

(b)

Figure 1.7 (a) Image of the small irregularly shaped moon Phobos against the background of the limb of the nearly spherical planet Mars. Phobos appears much larger relative to Mars than it actually is because the Soviet spacecraft *Phobos 2* was much closer to the moon than to the planet when it took this image. (b) *Hubble Space Telescope* image of Saturn taken on 24 February 2009 less than five months before saturnian equinox passage. The rings are seen at a low tilt angle, with the ring shadow appearing across the planet just above the rings. Four moons are seen to be transiting (partially eclipsing the planet); from left to right, they are Enceladus, Dione, Titan and Mimas; the shadows of Enceladus and Dione can also be seen. Note the pronounced oblateness of this low-density, rapidly rotating planet. (NASA/STScI/Hubble Heritage)

- Analysis of radar echoes
- Analysis of lightcurves. Several lightcurves obtained from different viewing angles are required for accurate measurements
- The shape of the **central flash**, which is observed when the center of a body with an atmosphere passes in front of an occulted star. The central flash results from the focusing of light rays refracted by the atmosphere and can be seen only under fortuitous observing circumstances.

1.4.6 Temperature

The equilibrium temperature of a planet can be calculated from the energy balance between solar insolation and reradiation outward (see Chapter 4). However, internal heat sources provide a significant contribution to the energy balance of many planets. Moreover, there may be diurnal, latitudinal and seasonal variations in the temperature. The **greenhouse effect**, a thermal 'blanket' caused by an atmosphere that is more transparent to visible radiation (the Sun's primary output) than to infrared radiation from the planet, raises the surface temperature on some planets far above the equilibrium blackbody value. For example, because of the high albedo of its clouds, Venus actually absorbs less solar energy per unit area than does Earth; thus (as internal heat sources on these two planets are negligible compared with solar heating), the effective radiating temperature of Venus is lower than that of Earth. Nonetheless, as a consequence of the greenhouse effect, Venus's surface temperature is raised up to $\sim 730\,\mathrm{K}$, well above the surface temperature on Earth.

Direct *in situ* measurements with a thermometer can provide an accurate estimate of the temperature of the accessible (outer) parts of a body. The thermal infrared spectrum of a body's emitted radiation is also a good indicator of the temperature of its surface or cloud tops. Most solid and liquid planetary material can be characterized as a nearly perfect blackbody radiator with its emission peak at near- to mid-infrared wavelengths. Analysis of emitted radiation sometimes gives different temperatures at differing wavelengths. This could be attributable to a combination of temperatures from different locations on the surface, such as pole-to-equator differences, albedo variations, or volcanic hot spots such as those seen on Io (§10.2.1). Also, the opacity of an atmosphere varies with wavelength, which allows us to remotely probe different altitudes in a planetary atmosphere.

1.4.7 Magnetic Field

Magnetic fields are created by moving charges. Currents moving through a solid medium decay quickly (unless the medium is a superconductor, which is unreasonable to expect at the high temperatures found in planetary interiors). Thus, internally generated planetary magnetic fields must either be produced by a (poorly understood) **dynamo** process, which can only operate in a fluid region of a planet (§7.4.2) or be caused by **remanent ferromagnetism**, which is a result of charges that are bound to atoms of a solid locked in an aligned configuration. Remanent ferromagnetism is not viewed to be a likely cause of large fields because, in addition to the fact that it is expected to decay away on timescales short compared with the age of the Solar System, it would require the planet to have been subjected to a nearly constant (in direction) magnetic field during the long period in which the bulk of its iron cooled through its **Curie point**. (At temperatures below the Curie point of a ferromagnetic material, the magnetic moments are partially aligned within magnet domains.) Magnetic fields may also be induced through the interaction between the solar wind (which is composed predominantly of charged particles) and conducting regions within the planet or its ionosphere.

A magnetic field may be detected directly using an *in situ* magnetometer or indirectly via radiation (radio emissions) produced by accelerating charges. The presence of localized **aurorae**, luminous disturbances caused by charged particle precipitation in a planet's upper atmosphere, is also indicative of a magnetic field. The magnetic fields of the planets can be approximated by dipoles, with perturbations to account for their irregularities. All four giant planets, as well as Earth, Mercury and Jupiter's moon Ganymede, have magnetic fields generated in their interiors. Venus and comets have magnetic fields induced by the interaction between the solar wind and charged particles in their atmosphere/ionosphere, whereas Mars and the Moon have localized crustal magnetic fields. Perturbations in Jupiter's magnetic field near Europa and Callisto are indicative of salty oceans in the interiors of these moons (§10.2). Geyser activity on Enceladus perturbs Saturn's magnetic field (§10.3.3).

1.4.8 Surface Composition

The composition of a body's surface can be derived from:

- Spectral reflectance data. Such spectra may be observed from Earth; however, spectra at ultraviolet wavelengths can only be obtained above the Earth's atmosphere.
- Thermal infrared spectra and thermal radio data. Although difficult to interpret, these measurements contain information about a body's composition.
- Radar reflectivity. Such observations can be carried out from Earth or from spacecraft that are near the body.
- X-ray and γ-ray fluorescence. These measurements may be conducted from a spacecraft in orbit around the planet (or, in theory, even a flyby spacecraft) if the body lacks a substantial atmosphere. Detailed measurements require landing a probe on the body's surface.
- Chemical analysis of surface samples. This can be performed on samples brought to Earth by

natural processes (meteorites) or spacecraft, or (in less detail) by *in situ* analysis using spacecraft. Other forms of *in situ* analysis include mass spectroscopy and electrical and thermal conductivity measurements.

The compositions of the planets, asteroids and satellites show a dependence on heliocentric distance, with the objects closest to the Sun having the largest concentrations of dense materials (which tend to be **refractory**, i.e., have high melting and boiling temperatures) and the smallest concentration of ices (which are much more **volatile**, i.e., have much lower melting and boiling temperatures).

1.4.9 Surface Structure

The surface structure varies greatly from one planet or moon to another. There are various ways to determine the structure of a planet's surface:

- Structure on large scales (e.g., mountains) can be detected by imaging, either passively in the visible/infrared/radio or actively using radar imaging techniques. It is best to have imaging available at more than one illumination angle in order to separate tilt-angle (slope) effects from albedo differences.
- Structure on small scales (e.g., grain size) can be deduced from the radar echo brightness and the variation of reflectivity with **phase angle**, the angle between the illuminating Sun and the observer as seen from the body. The brightness of a body with a size much larger than the wavelength of light at which it is observed generally increases slowly with decreasing phase angle. For very small phase angles, this increase can be much more rapid, a phenomenon referred to as the **opposition effect**.

1.4.10 Atmosphere

Most of the planets and some satellites are surrounded by significant atmospheres. The giant planets Jupiter, Saturn, Uranus and Neptune are basically huge fluid balls, and their atmospheres are dominated by H_2 and He. Venus has a very dense CO_2 atmosphere, with clouds so thick that one cannot see its surface at visible wavelengths; Earth has an atmosphere consisting primarily of N_2 (78%) and O_2 (21%), and Mars has a more tenuous CO_2 atmosphere. Saturn's satellite Titan has a dense nitrogen-rich atmosphere, which is intriguing because it contains many kinds of organic molecules. Pluto and Neptune's moon Triton each have a tenuous atmosphere dominated by N_2, and the atmosphere of Jupiter's volcanically active moon Io consists primarily of SO_2. Mercury and the Moon each have an extremely tenuous atmosphere ($\lesssim 10^{-12}$ bar); Mercury's atmosphere is dominated by atomic O, Na and He, and the main constituents in the Moon's atmosphere are He and Ar. The gaseous components of cometary comae are essentially temporary atmospheres in the process of escaping.

The composition and structure (temperature–pressure profile) of an atmosphere can be determined from spectral reflectance data at visible wavelengths, thermal spectra and photometry at infrared and radio wavelengths, stellar occultation profiles, *in situ* mass spectrometers and attenuation of radio signals sent back to Earth by atmospheric/surface probes.

1.4.11 Interior

The interior of a planet is not directly accessible to observations. However, with help of the observable parameters discussed earlier, one can derive information on a planet's bulk composition and its interior structure.

The **bulk composition** is not an observable attribute, except for extremely small bodies, such as meteorites, that we can actually take apart and analyze (see Chapter 11). Thus, we must deduce bulk composition from a variety of direct and indirect clues and constraints. The most fundamental

constraints are based on the mass and the size of the planet. Using only these constraints together with material properties derived from laboratory data and quantum mechanical calculations, it can be shown that Jupiter and Saturn are composed mostly of hydrogen, simply because all other elements are too dense to fit the constraints (unless the internal temperature is much higher than is consistent with the observed effective temperature in a quasi-steady state). However, this method only gives definitive results for planets composed primarily of the lightest element. For all other bodies, bulk composition is best estimated from models that include mass and radius as well as the composition of the surface and atmosphere, the body's heliocentric distance (location is useful because it gives us an idea of the temperature of the region during the planet-formation epoch and thus which elements were likely to condense), together with reasonable assumptions of cosmogonic abundances (§1.5, Table 3.1 and Chapters 11 and 15).

The **internal structure** of a planet can be derived to some extent from its gravitational field and rotation rate. From these parameters, one can estimate the degree of concentration of the mass at the planet's center. The gravitational field can be determined from spacecraft tracking and the orbits of satellites or rings. Detailed information on the internal structure of a planet with a solid surface may be obtained if seismometers can be placed on its surface, as was done for the Moon by *Apollo* astronauts. The velocities and attenuations of seismic waves propagating through the planet's interior depend on density, rigidity and other physical properties, which in turn depend on composition, as well as on pressure, temperature and time. Reflection and refraction off internal boundaries provide information on layering. The free oscillation periods of gaseous planets can, in theory, also provide clues to internal properties, just as **helioseismology**, the study of solar oscillations, now provides important information about the Sun's interior. Evidence of volcanism and plate tectonics constrain the thermal environment below the surface. Energy output provides information on the thermal structure of a planet's interior.

The response of moons that are subject to significant time-variable tidal deformations depends on their internal structure. Repeated observations of such moons can reveal internal properties, including in some cases the presence of a subterranean fluid layer. Combining altitude and gravity field measurements could give indications about lateral inhomogeneities under the surface of icy moons and thus, for instance, indicate volcanic sources and tectonic structures.

Magnetic fields are produced by moving charges. Although a small magnetic field such as the Moon's may be the result of remanent ferromagnetism, substantial planetary magnetic fields are thought to require a conducting fluid region within the planet's interior. Whereas centered dipole fields are probably produced in or near the core of the planet, highly irregular offset fields are likely to be produced closer to the planet's surface.

1.5 Formation of the Solar System

The nearly planar and almost circular orbits of the planets in our Solar System argue strongly for planetary formation within a flattened circumsolar disk. Astrophysical models suggest that such disks are a natural byproduct of star formation from the collapse of rotating cores of molecular clouds. Observational evidence for the presence of disks of Solar System dimensions around young stars has increased substantially in recent years, and infrared excesses in the spectra of young stars suggest that the lifetimes of protoplanetary disks range from 10^6–10^7 years.

Our galaxy contains many molecular clouds, most of which are several orders of magnitude larger than our Solar System. **Molecular clouds** are the coldest and densest parts of the interstellar medium. They are inhomogeneous, and the densest parts of molecular clouds are referred to as **cores**. These are the sites in which star formation occurs at the current epoch. Even a very slowly rotating molecular cloud core has far too much spin angular momentum to collapse down to an object of stellar dimensions, so a significant fraction of the material in a collapsing core falls onto a rotationally supported disk orbiting the pressure-supported (proto)star. Such a disk has the same initial elemental composition as the growing star. At sufficient distances from the central star, it is cool enough for ~1%–2% of this material to be in solid form, either remnant interstellar grains or condensates formed within the disk. This dust is primarily composed of rock-forming compounds within a few AU of a $1\,M_\odot$ star, but in the cooler, more distant regions, the amount of ices (e.g., H_2O, CH_4, CO) present in solid form is comparable to that of rocky solids.

During the infall stage, the disk is very active and probably highly turbulent as a result of the mismatch of the specific angular momentum of the gas hitting the disk with that required to maintain Keplerian rotation. Gravitational instabilities and viscous and magnetic forces may add to this activity. When the infall slows substantially or stops, the disk becomes more quiescent. Interactions with the gaseous component of the disk affect the dynamics of small solid bodies, and the growth from micrometer-sized dust to kilometer-sized planetesimals remains poorly understood. Meteorites (see Chapter 11), minor planets and comets (see Chapter 12), most of which were never incorporated into bodies of planetary dimensions, best preserve a record of this important period in Solar System development.

The dynamics of larger solid bodies within protoplanetary disks are better characterized. The primary perturbations on the Keplerian orbits of kilometer-sized and larger planetesimals in protoplanetary disks are mutual gravitational interactions and physical collisions. These interactions lead to accretion (and in some cases erosion and fragmentation) of planetesimals. Eventually, solid bodies agglomerated into the terrestrial planets in the inner Solar System and into planetary cores several times the mass of the Earth in the outer Solar System. These massive cores were able to gravitationally attract and retain substantial amounts of gaseous material from the solar nebula. In contrast, terrestrial planets were not massive enough to attract and retain such gases, and the gases in their current thin atmospheres are derived from material that was incorporated in solid planetesimals.

The planets in our Solar System orbit close enough to one another that the final phases of planetary growth could have involved the merger or ejection of planets or planetary embryos on unstable orbits. However, the low eccentricities of the orbits of the outer planets imply that some damping process, such as accretion/ejection of numerous small planetesimals or interactions with residual gas within the protoplanetary disk, must also have been involved.

As researchers learn more about the individual bodies and classes of objects in our Solar System, and as simulations of planetary growth become more sophisticated, theories about the formation of our Solar System are being revised and (we hope) improved. The detection of planets around other stars has presented us with new challenges to develop a unified theory of planet formation that is more generally applicable. We discuss these theories in more detail in Chapter 15.

Key Concepts

- Planets are the wanderers of the night sky, changing in position relative to the 'fixed' stars.
- The study of the motions of the planets dates back thousands of years, but most of our knowledge about planets and smaller bodies within our Solar System has been obtained during the space age.
- The Sun dominates our Solar System in most respects, followed by Jupiter, then Saturn and after that the pair Uranus and Neptune.
- A wide variety of techniques are used to observe the properties of planetary bodies. However, some planetary characteristics, such as interior composition, cannot at present be directly observed and can only be deduced from theoretical modeling.
- When a molecular cloud core collapses, the inner portion becomes a star. Molecular cloud material with high angular momentum falls into a disk around that star and is available for planet formation.
- Whereas planets grow by accretion of small bodies into larger ones, stars form via the collapse of large clouds into smaller objects.

Further Reading

More extensive and technical accounts of most of the topics presented in this book (other than those connected to life) can be found in our graduate-level textbook:

de Pater, I., and J.J. Lissauer, 2010. *Planetary Sciences*, 2nd Edition. Cambridge University Press, Cambridge. 647pp.

A good nontechnical overview of our planetary system, complete with many beautiful color pictures, is given by:

Beatty, J.K., C.C. Peterson, and A. Chaikin, Eds., 1999. *The New Solar System*, 4th Edition. Sky Publishing Co., Cambridge, MA and Cambridge University Press, Cambridge. 421pp.

A terse but detailed overview, including reproductions of paintings of various Solar System objects by the authors, is provided by:

Miller, R., and W.K. Hartmann, 2005. *The Grand Tour: A Traveler's Guide to the Solar System*, 3rd Edition. Workman Publishing, New York. 208pp.

An overview of the Solar System emphasizing atmospheric and space physics is given by:

Encrenaz, T., J.-P. Bibring, M. Blanc, M.-A. Barucci, F. Roques, and Ph. Zarka, 2004. *The Solar System*, 3rd Edition. Springer-Verlag, Berlin. 512pp.

Two good overview texts aimed at college students not majoring in science are:

Morrison, D., and T. Owen, 2003. *The Planetary System*, 3rd Edition. Addison-Wesley Publishing Company, New York. 531pp.

Hartmann, W.K., 2005. *Moons and Planets*, 5th Edition. Brooks/Cole, Thomson Learning, Belmont, CA. 428pp.

Short summaries of a multitude of topics, ranging from mineralogy to black holes, at a level of sophistication a bit higher than that of this book, are provided by:

Cole, G.H.A., and M.M. Woolfson, 2002. *Planetary Science: The Science of Planets Around Stars*, Institute of Physics Publishing, Bristol and Philadelphia. 508pp.

Chemical processes on planets and during planetary formation are covered in some detail by:

Lewis, J.S., 2004. *Physics and Chemistry of the Solar System*, 2nd Edition. Elsevier, Academic Press, San Diego. 684pp.

The following encyclopedia forms a nice complement to this book:

McFadden, L., P. R. Weissman, and T.V. Johnson, Eds., 2007. *Encyclopedia of the Solar System*, 2nd Edition. Academic Press, San Diego. 982pp.

Extensive planetary data tables can be found in:

Yoder, C.F., 1995. Astrometric and geodetic properties of Earth and the Solar System. In *Global Earth Physics: A Handbook of Physical Constants*. AGU Reference Shelf 1, American Geophysical Union, 1–31.

For updated information, see http://ssd.jpl.nasa.gov.

A collection of beautiful images of planets and astrobiology can be found at https://fettss.arc.nasa.gov/collection/.

Problems

1-1. Because the distances between the planets are much larger than planetary sizes, very few diagrams or models of the Solar System are completely to scale. However, imagine that you are asked to give an astronomy lecture and demonstration to your niece's second-grade class, and you decide to illustrate the vastness and near emptiness of space by constructing a scale model of the Solar System using ordinary objects. You begin by selecting a (1-cm-diameter) marble to represent the Earth.
(a) What other objects can you use, and how far apart must you space them?
(b) Proxima Centauri, the nearest star to the Solar System, is 4.2 light years distant; where, in your model, would you place it?

1-2. The satellite systems of the giant planets are often referred to as 'miniature solar systems'. In this problem, you will make some calculations comparing the satellite systems of Jupiter, Saturn and Uranus with the Solar System.

(a) Calculate the ratio of the sum of the masses of the planets with that of the Sun and similar ratios for the jovian, saturnian and uranian systems using the respective planet as the primary mass.

(b) Calculate the ratio of the sum of the orbital angular momenta of the planets to the rotational angular momentum of the Sun. You can assume circular orbits at zero inclination for all planets and ignore the effects of planetary rotation and the presence of satellites. The Sun rotates differentially, with a mean rotation period of 25.4 days.

(c) Repeat the calculation in (b) for the jovian, saturnian and uranian systems using the respective planet as the primary mass.

(d) Calculate the orbital semimajor axes of the planets in terms of solar radii and the orbital semimajor axes of Jupiter's moons in jovian radii. How would a scale model of the jovian system compare with the model of the Solar System in Problem 1-1?

1-3. (a) Standing on the surface or floating in the atmosphere of which Solar System body would you see the brightest object in the nighttime sky? Justify your answer.

(b) Same question but assume that you are standing on a body with a solid surface and a significant atmosphere.

Hint: Calculate the angular area of the body being observed, multiply by the body's albedo and divide by the square of the distance to the Sun to account for flux of light reflected off of the observed body. The sizes, distances and albedos of Solar System objects are provided in Appendix E.

1-4. A planet that keeps the same hemisphere pointed towards the Sun must rotate once per orbit in the prograde direction.

(a) Draw a diagram to demonstrate this fact. Whereas the rotation period (in an inertial frame) or **sidereal day** for such a planet is equal to its orbital period, the length of a **solar day** on such a planet is infinite.

(b) Earth rotates in the prograde direction. How many times must Earth rotate per orbit for there to be 365.24 solar days per year? Verify your result by comparing the length of Earth's sidereal rotation period (Table E.2) with the length of a mean solar day.

(c) If a planet rotated once per orbit in the retrograde direction, how many solar days would it have per orbit?

(d)* Determine a general formula relating the lengths of solar and sidereal days on a planet. Use your formula to compute the lengths of solar days on Mercury, Venus, Mars and Jupiter.

(e)* For a planet on an eccentric orbit, the length of either the solar day or the sidereal day varies on an annual cycle. Which one varies, and why? Calculate the length of the longest such day on Earth. This longest day is how much longer than the mean day of its type?

(Note: The Earth's obliquity causes variations in the rate of apparent motion of the Sun along the equator, which also produce variations in the length of the day. The **equation of time** accounts for both types of variations and enables accurate calculation of the time using a sundial.)

1-5. A **total solar eclipse** occurs when the Moon blocks the entire disk of the Sun, allowing the observer to view only the Sun's extended atmosphere, the **corona**. An **annular eclipse** occurs when the Moon obscures the central portion of the Sun but a narrow annulus of the Sun's photosphere can be seen surrounding the Moon.

(a) Using the data in Tables E.4, E.5 and C.5 and equation (2.1), calculate the minimum and maximum separations between the Earth and the Moon and between the Earth and the Sun. Use this information to show that the eccentricities of the orbits of Earth about the Sun and the Moon about the Earth make it possible for both types of eclipse to be viewed from the surface of Earth. Ignore the finite size of the Earth, i.e., assume that the observer is located at the center of the Earth.

(b) How does the non-zero size of Earth affect your answer to part (a)?

CHAPTER 2

Dynamics

The Planets move one and the same way in Orbs
concentric, some inconsiderable Irregularities excepted,
which may have arisen from the mutual Actions of Planets
upon one another, and which will be apt to increase, till
this System wants a Reformation.

Isaac Newton, *Opticks*

Dynamical studies of planetary bodies characterize their motions, including rotation and deformation of bodies resulting from tidal distortions. Dynamics is the oldest of the planetary sciences. Gravitational interactions determine how the distance of a planet from the Sun varies with time and thus how much solar radiation the planet intercepts. Rotation rates determine the length of the day; obliquity influences pole-equator temperature differences and seasonal variations. Tidal heating produces extensive volcanism on bodies such as Jupiter's moon Io (see Fig. 10.4).

The history of observational studies of, and kinematical models for, planetary motions dates back to antiquity. Modern planetary dynamics began in the seventeenth century. In the first decades of that century, Johannes Kepler conducted an extensive analysis of planetary observations that had been made in the previous decades by Tycho Brahe. Towards the end of the seventeenth century, Isaac Newton provided a firm basis for dynamical studies by discovering physical laws that govern the motions of objects on Earth as well as in the heavens. Albert Einstein's (twentieth-century) theory of relativity fundamentally modified the underlying theories of motion and gravity, but the magnitude of relativistic corrections to planetary motions is generally quite small (Problems 2-4 and 2-5).

In 1687, Newton showed that the relative motion of two spherically symmetric bodies resulting from their mutual gravitational attraction is described by simple conic sections: ellipses for bound orbits and parabolas and hyperbolas for unbound trajectories. However, the introduction of additional gravitating bodies produces a rich variety of dynamical phenomena even though the basic interactions between pairs of objects can be straightforwardly described.

In this chapter, we describe the basic orbital properties of Solar System objects (planets, moons, minor bodies and dust) and their mutual

interactions. We also provide several examples of important dynamical processes that occur in the Solar System and lay the groundwork for describing some of the phenomena that are considered in other chapters of this book.

We begin in §2.1 with an overview of the **two-body problem**, i.e., the relative motion of an isolated pair of spherically symmetric objects that are gravitationally attracted to one another. Our discussion introduces Kepler's laws, Newton's laws and the terminology used to describe planetary orbits. In the next three sections, we discuss the consequences of gravitational interactions among larger numbers of bodies. We consider the dynamics of spherically symmetric objects of finite size in §2.5. We relax the assumption of spherical symmetry in §2.6 to analyze the dynamics of rotating planets and of orbits about them, and we consider the effects of tidal forces on deformable bodies in §2.7. Although gravity is the dominant force on the motions of large bodies in the Solar System, electromagnetic forces such as radiation pressure substantially affect the motions of small objects, which have larger surface area to mass ratios than do large objects; we discuss such forces in §2.8. We conclude the chapter with a brief overview of orbits about a mass-losing star, which may be important for very young and very old planetary systems.

2.1 The Two-Body Problem

All bodies in the Universe are subject to the gravitational attraction of all other bodies. But for many planetary science applications, the trajectory of one body is well approximated by considering just the gravitational force exerted on it by a single other body. We describe the analysis of this elementary yet nontrivial problem and various applications in this section.

(a)

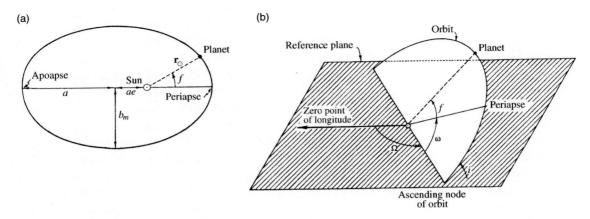

(b)

Figure 2.1 (a) Geometry of an elliptical orbit. The Sun is at one focus, and the vector $\mathbf{r}_\odot$ denotes the instantaneous heliocentric location of the planet (i.e., $r_\odot$ is the planet's distance from the Sun). The semimajor axis of the ellipse is a, e denotes its eccentricity and b_m is the ellipse's semiminor axis. The true anomaly, f, is the angle between the planet's perihelion and its instantaneous position. (b) Geometry of an orbit in three dimensions; i is the inclination of the orbit, Ω is the longitude of the ascending node and ω is the argument of periapse. (Adapted from Hamilton 1993)

2.1.1 Kepler's Laws of Planetary Motion

By careful analysis of the observed orbits of the planets, Kepler deduced his three 'laws' of planetary motion:

(1) All planets move along elliptical paths with the Sun at one focus. We can express the **heliocentric distance**, $r_\odot$ (i.e., the planet's distance from the Sun), as

$$r_\odot = \frac{a(1 - e^2)}{1 + e\cos f},$$
(2.1)

with a the **semimajor axis** (average of the minimum and maximum heliocentric distances). The **eccentricity** of the orbit, $e \equiv (1 - b_m^2/a^2)^{1/2}$, where $2b_m$ is the minor axis of the ellipse. The **true anomaly**, f, is the angle between the planet's **perihelion** (where it is closest to the Sun) and its instantaneous position. These quantities are displayed graphically in Figure 2.1a.

(2) A line connecting any given planet and the Sun sweeps out area, $\mathcal{A}$, at a constant rate:

$$\frac{d\mathcal{A}}{dt} = \text{constant}.$$
(2.2)

The value of this constant rate differs from one planet to the next. Kepler's second law is illustrated in Figure 2.2.

(3) The square of a planet's orbital period about the Sun (in years), P_{yr}, is equal to the cube of its semimajor axis (in AU), a_{AU}:

$$P_{yr}^2 = a_{AU}^3.$$
(2.3)

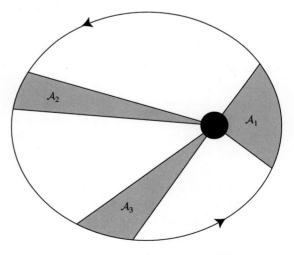

Figure 2.2 Schematic illustration of Kepler's second law. (Murray and Dermott 1999)

2.1.2 Newton's Laws of Motion and Gravity

Isaac Newton developed the first physical model that explained the motion of objects on Earth and in the heavens using a single, unified theory. Newton's theory includes four 'laws', three explaining motion and the fourth quantifying the gravitational force.

Newton's first law concerns inertia: A body remains at rest or in uniform motion unless a force is exerted upon it.

Consider a body of mass m_1 at instantaneous location $\mathbf{r}_1$ with instantaneous velocity $\mathbf{v}_1 \equiv d\mathbf{r}_1/dt$ and hence momentum $m_1\mathbf{v}_1$. The acceleration produced by a net force $\mathbf{F}_1$ is given by Newton's second law of motion:

$$\frac{d(m_1\mathbf{v}_1)}{dt} = \mathbf{F}_1. \tag{2.4}$$

Newton's third law states that for every action there is an equal and opposite reaction; thus, the force on each object of a pair due to the other object is equal in magnitude but opposite in direction:

$$\mathbf{F}_{12} = -\mathbf{F}_{21}, \tag{2.5}$$

where $\mathbf{F}_{ij}$ represents the force exerted by body j on body i.

Newton's universal law of gravity states that a second body of mass m_2 at position $\mathbf{r}_2$ exerts an attractive force on the first body given by

$$\mathbf{F}_{g12} = -\frac{Gm_1m_2}{r^2}\hat{\mathbf{r}}, \tag{2.6}$$

where $\mathbf{r} \equiv \mathbf{r}_1 - \mathbf{r}_2$ is the vector distance from particle 2 to particle 1, G is the gravitational constant and $\hat{\mathbf{r}} \equiv \mathbf{r}/r$.

Although Kepler's laws were originally deduced from careful observation of planetary motion, they were subsequently shown to be derivable from Newton's laws of motion together with his universal law of gravity. We present portions of this derivation (using modern mathematics and notation) below.

2.1.3 Reduction of the Two-Body Problem to the One-Body Problem

The equation for the relative motion of two mutually gravitating bodies can be derived from Newton's laws. Consider two mutually gravitating bodies of masses m_1 and m_2 and positions $\mathbf{r}_1$ and $\mathbf{r}_2$. Newton's second law of motion (eq. 2.4) can be combined with his law of gravitation (eq. 2.6) to yield the following two equations that govern the motion of these bodies:

$$m_1\frac{d^2\mathbf{r}_1}{dt^2} = -\frac{Gm_1m_2}{|\mathbf{r}_1 - \mathbf{r}_2|^3}(\mathbf{r}_1 - \mathbf{r}_2), \tag{2.7}$$

$$m_2\frac{d^2\mathbf{r}_2}{dt^2} = -\frac{Gm_1m_2}{|\mathbf{r}_2 - \mathbf{r}_1|^3}(\mathbf{r}_2 - \mathbf{r}_1). \tag{2.8}$$

To separate the motion of the center of mass from the relative motion of the two bodies, we apply the coordinate transformation $\mathbf{x} \equiv (m_1\mathbf{r}_1 + m_2\mathbf{r}_2)/(m_1 + m_2)$, $\mathbf{r} \equiv \mathbf{r}_1 - \mathbf{r}_2$. Substitution and simple algebraic manipulation yields:

$$(m_1 + m_2)\frac{d^2\mathbf{x}}{dt^2} = \mathbf{0}, \tag{2.9}$$

and

$$\frac{d^2\mathbf{r}}{dt^2} = -\frac{GM}{r^2}\hat{\mathbf{r}}, \tag{2.10}$$

where $M \equiv m_1 + m_2$. Equation (2.9) implies that the center of mass of the system does not accelerate and therefore moves at constant velocity. Equation (2.10) describes the acceleration of the relative position of the two bodies, $\mathbf{r}$.

Thus, the relative motion of the two bodies is completely equivalent to that of a particle orbiting a *fixed* central mass M. This reduces the two-body problem to an equivalent one-body problem.

2.1.4* Generalization of Kepler's Laws

Having reduced the two-body problem to an equivalent one-body problem, we proceed with the derivation of (Newton's generalization of) Kepler's laws. As $\mathbf{v} \equiv d\mathbf{r}/dt$, vector calculus manipulation implies that

$$\frac{d}{dt}(\mathbf{r} \times \mathbf{v}) = \mathbf{r} \times \frac{d\mathbf{v}}{dt} + \frac{d\mathbf{r}}{dt} \times \mathbf{v}$$

$$= \mathbf{r} \times \frac{d^2\mathbf{r}}{dt^2} + \mathbf{v} \times \mathbf{v} = 0, \qquad (2.11)$$

where the first two equalities in equation (2.11) are valid in general and the last equality uses the force law given by equation (2.6). Equation (2.11) implies that the angular momentum, $\mathbf{L}$, which is given by:

$$\mathbf{L} \equiv \mathbf{r} \times m\mathbf{v}, \qquad (2.12)$$

is conserved, i.e.,

$$\frac{d\mathbf{L}}{dt} = 0. \qquad (2.13)$$

In polar coordinates, the expression for the magnitude of the angular momentum is just $L = mrv_\theta$. The rate of sweeping is

$$\frac{d\mathcal{A}}{dt} = \frac{rv_\theta}{2} = \frac{L}{2m}. \qquad (2.14)$$

Conservation of angular momentum (eq. 2.12) thus yields the Newtonian generalization of Kepler's second law:

(2) A line connecting two bodies (as well as lines from each body to the center of mass) sweeps out area at a constant rate. The value of this constant is given by equation (2.14).

The derivation of the generalized versions of Kepler's first and third laws is mathematically straightforward but rather tedious. The details of these derivations are presented in many books and are available on the web. We therefore only sketch the procedure and quote the results below.

To derive Kepler's first law, take the dot product of $\mathbf{v}$ with equation (2.10) to derive the equation of conservation of energy per unit mass. Integrate your result to determine an expression for the specific energy of the system, E. Express your answer in polar coordinates and solve for dr/dt. Take the

reciprocal; multiply both sides by $d\theta/dt$; and then use the magnitude of the specific angular momentum, L, to eliminate the angular velocity from your expression, yielding the following purely spatial relationship for the orbit:

$$\frac{d\theta}{dr} = \frac{1}{r}\left(\frac{2Er^2}{L^2} + \frac{2GMr}{L^2} - 1\right)^{-1/2}. \qquad (2.15)$$

Integrate equation (2.15) and solve for r. Set the constant of integration equal to $-\pi/2$, define $r_0 \equiv L^2/(GM)$ and use the relationship $e = \left(1 + (2EL^2)/(G^2M^2)\right)^{1/2}$ to obtain:

$$r = \frac{r_0}{1 + e\cos\theta}. \qquad (2.16)$$

For $0 \le e < 1$, equation (2.16) represents an ellipse in polar coordinates. Thus, Kepler's first law is also precise in the two-body Newtonian approximation, although the Sun itself is not fixed in space. Note that if $E = 0$, then $e = 1$ and equation (2.16) describes a parabola, and if $E > 0$, then $e > 1$ and the orbit is hyperbolic. The generalized form of Kepler's first law reads:

(1) The two bodies move along elliptical paths, with one focus of each ellipse located at the center of mass (CM) of the system,

$$\mathbf{r}_{CM} = \frac{m_1\mathbf{r}_1 + m_2\mathbf{r}_2}{M}. \qquad (2.17)$$

To derive Kepler's third law, begin by showing that the semimajor and semiminor axes of the ellipse given by equation (2.16) are $a = r_0/(1 - e^2)$ and $b = r_0/(1 - e^2)^{1/2}$, respectively. Determine the orbital period, P, by setting the integral of $d\mathcal{A}/dt$ equal to the area of the ellipse, πab. The resulting generalized form of Kepler's third law is:

(3) The orbital period of a pair of bodies about their mutual center of mass is given by

$$P_{orb}^2 = \frac{4\pi^2 a^3}{GM}. \qquad (2.18)$$

Note that the result given in equation (2.18) differs from Kepler's third law by replacing the Sun's mass, m_1, by the sum of the masses of the Sun and the planet, M.

2.1.5 Orbital Elements

The Sun contains more than 99.8% of the mass of the known Solar System. The gravitational force exerted by a body is proportional to its mass (eq. 2.6), so to an excellent first approximation we can regard the motion of the planets and many other bodies as being solely influenced by a fixed central pointlike mass. For objects such as the planets, which are bound to the Sun and hence cannot go arbitrarily far from the central mass, the general solution for the orbit is the ellipse described by equation (2.1).

The orbital plane, although fixed in space, can be arbitrarily oriented with respect to whatever reference plane we have chosen. This reference plane is usually taken to be either the Earth's orbital plane about the Sun, which is called the **ecliptic**, or the equatorial plane of the largest body in the system, or the **invariable plane** (the plane perpendicular to the total angular momentum of the system). The Solar System's invariable plane is nearly coincident with the plane of Jupiter's orbit, which is inclined by 1.3° relative to the ecliptic. In this book, we follow standard conventions and measure inclinations of heliocentric orbits with respect to the ecliptic plane and inclinations of planetocentric orbits relative to the planet's equator.

The terminology and variables used to describe orbits are shown in Figure 2.1. The **inclination**, i, of the orbit is the angle between the reference plane and the orbital plane; i can range from 0° to 180°. Conventionally, secondaries orbiting in the same direction as the primary rotates are defined to have inclinations from 0° to 90° and are said to be on **prograde** (or **direct**) orbits. Secondaries orbiting in the opposite direction are defined to have 90° < i ≤ 180° and said to be on **retrograde** orbits. For heliocentric orbits, the Earth's orbital plane rather than the Sun's equator is usually taken

as the reference. The intersection of the orbital and reference planes is called the **line of nodes**, and the orbit pierces the reference plane at two locations – one as the body passes upward through the plane (the **ascending node**) and one as it descends (the **descending node**). A fixed direction in the reference plane is chosen, and the angle to the direction of the orbit's ascending node is called the **longitude of the ascending node**, Ω.

The angle between the line to the ascending node and the line to the direction of **periapse** (the point on the orbit when the two bodies are closest, which is referred to as **perihelion** for orbits about the Sun and **perigee** for orbits about the Earth) is called the **argument of periapse**, ω. For heliocentric orbits, Ω and ω are measured eastward from the vernal equinox. The **vernal equinox** is the great circle through the celestial poles that crosses the equator at the location of the Sun on the first day of spring. Finally, the true anomaly, f, specifies the angle between the planet's periapse and its instantaneous position. Thus, the six **orbital elements**, a, e, i, Ω, ω and f, uniquely specify the location of the object in space (Fig. 2.1). The first three quantities, a, e and i, are often referred to as the **principal orbital elements** because they describe the size, shape and tilt of the orbit.

For two bodies with known masses, specifying the elements of the relative orbit and the positions and velocities of the center of mass is equivalent to specifying the positions and velocities of both bodies. Alternative (sets of) orbital elements are often used for convenience. For example, the **longitude of periapse**,

$$\varpi \equiv \Omega + \omega, \qquad (2.19a)$$

can be used in place of ω. The time of perihelion passage, t_ϖ, is commonly used instead of f as an alternative way by which to specify the location of the particle along its orbital path. The **mean motion** (average angular speed),

$$n \equiv \frac{2\pi}{P_{\text{orb}}}, \qquad (2.19b)$$

and the **mean longitude**,

$$\lambda = n(t - t_\varpi) + \varpi, \qquad (2.19c)$$

are also used to specify orbital properties.

2.1.6 Bound and Unbound Orbits

For a pair of bodies to travel on a circular orbit about their mutual center of mass, they must be pulled towards one another enough to balance inertia. Quantitatively, gravity must balance the centrifugal pseudoforce that is present if the problem is viewed as a steady state in the frame rotating with the angular velocity of the two bodies, n. The **centripetal force** necessary to keep an object of mass m in a circular orbit of radius r with speed v_c is

$$\mathbf{F_c} = mn^2\mathbf{r} = \frac{mv_c^2}{r}\hat{\mathbf{r}}. \qquad (2.20)$$

Equating this to the gravitational force exerted by the central body of mass M, we find that the speed of a circular orbit is

$$v_c = \sqrt{\frac{GM}{r}}. \qquad (2.21)$$

The total energy of the system, E, is a conserved quantity:

$$E = \frac{1}{2}mv^2 - \frac{GMm}{r} = -\frac{GMm}{2a}, \qquad (2.22)$$

where the first term in the middle expression is the kinetic energy of the system and the second term is potential energy. For circular orbits, the second equality in equation (2.22) follows immediately from equation (2.21).

If $E < 0$, the absolute value of the potential energy of the system is larger than its kinetic energy, and the system is **bound**: The body orbits the central mass on an elliptical path. Simple manipulation of equation (2.22) yields an expression for the velocity along an elliptical orbit at each radius r:

$$v^2 = GM\left(\frac{2}{r} - \frac{1}{a}\right). \qquad (2.23)$$

Equation (2.23) is known as the **vis viva equation**. If $E > 0$, the kinetic energy is larger than the absolute value of the potential energy, and the system is **unbound**. The orbit is then described mathematically as a hyperbola. If $E = 0$, the kinetic and potential energies are equal in magnitude, and the orbit is a parabola. By setting the total energy (eq. 2.22) equal to zero, we can calculate the **escape velocity** (alternatively referred to as the **escape speed**) at any separation:

$$v_e = \sqrt{\frac{2GM}{r}} = \sqrt{2}\, v_c. \qquad (2.24)$$

As noted earlier, the orbit in the two-body problem is an ellipse, parabola or hyperbola corresponding to the energy being negative, zero or positive, respectively. These curves are known collectively as **conic sections** and are illustrated in Figure 2.3. The generalization of equation (2.1) to include unbound as well as bound orbits is

$$r = \frac{\zeta}{1 + e\cos f}, \qquad (2.25)$$

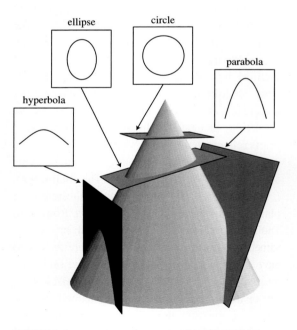

Figure 2.3 Conic sections. (Murray and Dermott 1999)

where r and f have the same meaning as in equation (2.1), e is the **generalized eccentricity** and ζ is a constant. Bound orbits have $e < 1$ and $\zeta = a(1 - e^2)$, but the generalized eccentricity can take any non-negative value. For elliptical orbits, the generalized eccentricity is no different from the eccentricity defined in §2.1.1. For a parabola, $e = 1$ and $\zeta = 2q$, where q is the **pericentric separation**, i.e., the distance of closest approach. For a hyperbola, $e > 1$ and $\zeta = q(1 + e)$; $e \gg 1$ signifies a hyperbola with only a slight bend, nearly a straight line. For all orbits, the three orientation angles i, Ω and ω are defined as in the elliptical case.

Whereas the energy of an orbit is uniquely specified by its semimajor axis (eq. 2.22), the angular momentum also depends on the orbit's eccentricity:

$$|\mathbf{L}| = m\sqrt{GMa(1 - e^2)}. \tag{2.26}$$

As with energy, the angular momentum of a circular orbit follows immediately from equation (2.21). For a given semimajor axis, a circular orbit contains the maximum possible amount of angular momentum (eq. 2.26). This occurs because when $r = a$ for an eccentric orbit, the magnitude of the velocity is the same as that for a circular orbit (by conservation of energy), but not all of this velocity is directed perpendicular to the line connecting the two bodies.

2.2 The Three-Body Problem

Gravity is not restricted to interactions between the Sun and the planets or individual planets and their satellites, but rather all bodies feel the gravitational force of one another. The motion of two mutually gravitating bodies is **completely integrable** (i.e., there exists one independent integral or constraint per degree of freedom), and the relative trajectories of the two bodies are given by simple conic sections, as discussed earlier. However, when more bodies are added to the system, additional constraints are needed to specify the motion; not enough integrals of motion are available, so the trajectories of even three gravitationally interacting bodies cannot be deduced analytically except in certain limiting cases. The general three-body problem is quite complex, and little progress can be made without resorting to numerical integrations. Fortunately, various approximations based on large differences between the masses of the bodies and nearly circular and coplanar orbits (which are quite accurate for most Solar System applications) simplify the problem sufficiently that some important analytic results may be obtained.

If one of the bodies is of negligible mass (e.g., a small asteroid, a ring particle or an artificial satellite), its effects on the other bodies may be ignored; the simpler system that results is called the **restricted three-body problem**, and the small body is referred to as a **test particle**. If the relative motion of the two massive particles is a circle, we refer to the situation as the **circular restricted three-body problem**. An alternative to the restricted three-body problem is **Hill's problem**, in which the mass of one of the bodies is much greater than the other two, but there is no restriction on the masses of the two small bodies relative to one another. An independent simplification is to assume that all three bodies travel within the same plane, the **planar three-body problem**. Various, but not all, combinations of these assumptions are possible.

Most of the results presented in this section are rigorously true only for the circular restricted three-body problem. However, they are valid to a good approximation for many configurations that exist in the Solar System.

2.2.1 Jacobi's Constant and Lagrangian Points

Our study of the three-body problem begins by considering an idealized system in which two massive bodies move on circular orbits about their common center of mass. A third body is introduced that is much less massive than the smaller of the first two, so that, to good approximation, it has no effect on the orbits of the other bodies. Our analysis is performed in a noninertial frame that rotates about the z-axis at a rate equal to the orbital frequency of the two massive bodies. We choose units such that the distance between the two bodies, the sum of the masses and the gravitational constant are all equal to one; this implies that the angular frequency of the rotating frame also equals unity (Problem 2-6). The origin is given by the center of mass of the pair, and the two bodies remain fixed at points on the x-axis, $\mathbf{r}_1 = (-m_2/(m_1 + m_2), 0)$ and $\mathbf{r}_2 = (m_1/(m_1 + m_2), 0)$. By convention, $m_1 \geq m_2$; in most Solar System applications, $m_1 \gg m_2$. The (massless) test particle is located at $\mathbf{r}$, so $|\mathbf{r} - \mathbf{r}_i|$ is the distance from mass m_i to the test particle. The velocity of the test particle in the rotating frame is denoted by v.

By analyzing a modified energy integral in the rotating frame, Carl Jacobi deduced the following constant of motion for the circular restricted three-body problem:

$$C_J = x^2 + y^2 + \frac{2m_1}{|\mathbf{r} - \mathbf{r}_1|} + \frac{2m_2}{|\mathbf{r} - \mathbf{r}_2|} - v^2. \quad (2.27)$$

The first two terms on the right-hand side of equation (2.27) represent twice the centrifugal potential energy, the next two twice the gravitational potential energy and the final one twice the kinetic energy; C_J is known as **Jacobi's constant**. Note that a body located far from the two masses and moving slowly in the inertial frame has small C_J because the gravitational potential energy terms are small and the centrifugal potential almost exactly cancels the kinetic energy of the test particle's motion viewed in the rotating frame.

For a given value of Jacobi's constant, equation (2.27) specifies the magnitude of the test particle's velocity (in the rotating frame) as a function of position. Because v^2 cannot be negative, surfaces at which $v = 0$ bound the trajectory of a particle

with fixed C_J (note that the allowed region need not be finite). Such **zero-velocity surfaces**, or in the case of the planar problem **zero-velocity curves**, are quite useful in discussing the topology of the circular restricted three-body problem.

Joseph Lagrange found that in the circular restricted three-body problem there are five points where test particles placed at rest would feel no net force in the rotating frame. The locations of three of these so-called **Lagrangian points** (L_1, L_2 and L_3) lie along a line joining the two masses m_1 and m_2. Zero-velocity curves intersect at each of the three collinear Lagrangian points, which are saddle points of the total (centrifugal + gravitational) potential in the rotating frame. The other two Lagrangian points (L_4 and L_5) form equilateral triangles with the two massive bodies. All five Lagrangian points are in the orbital plane of the two massive bodies. Figure 2.4 illustrates the positions of the Lagrangian points as well as trajectories and zero-velocity curves of various orbits that are close to these equilibrium positions.

Particles displaced slightly from the three collinear Lagrangian points will continue to move away; hence, these locations are unstable. The triangular Lagrangian points are potential energy maxima, but the Coriolis force stabilizes them for $m_1/m_2 \gtrsim 25$, which is the case for all known examples in the Solar System that are more massive than the Pluto–Charon system. If a particle at L_4 or L_5 is perturbed slightly, it will start to **librate** about these points (i.e., oscillate back and forth, without circulating past the secondary).

The L_4 and L_5 points are important in the Solar System. For example, the **Trojan asteroids** are located near Jupiter's triangular Lagrangian points, more than a dozen asteroids are known to librate about Neptune's L_4 and L_5 points and several small asteroids, including 5261 Eureka, are martian Trojans. There are also small moons in the saturnian system near the triangular Lagrangian points of Tethys and Dione (Table E.4). The L_4 or L_5 points

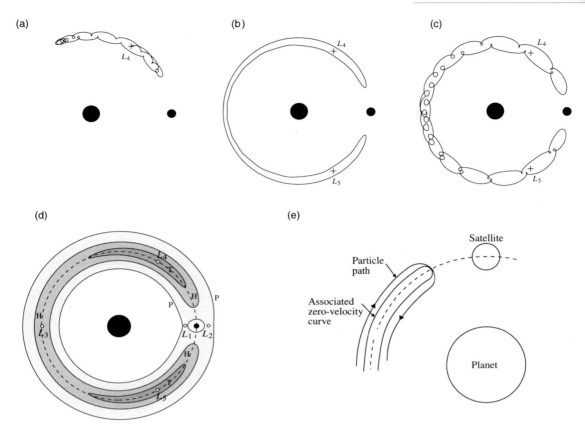

Figure 2.4 Schematic diagrams illustrating various properties of orbits in the circular restricted three-body problem. All cases are shown in the frame that is centered on the primary and rotating at the orbital frequency of the two massive bodies (corotating with the secondary). (a) Example of a tadpole orbit of a test particle viewed in the rotating frame. (b) Similar to (a) but for a horseshoe orbit with small eccentricity. (c) As in (b) but the particle has a larger eccentricity. (Panels a–c adapted from Murray and Dermott 1999) (d) The Lagrangian equilibrium points and various zero-velocity curves for three values of the Jacobi's constant, C_J. The mass ratio $m_1/m_2 = 100$. The locations of the Lagrangian equilibrium points L_1–L_5 are indicated by *small open circles*. The *white region* centered on the secondary is the secondary's Hill sphere. The *dashed line* denotes a circle of radius equal to the secondary's semimajor axis. The letters T (tadpole), H (horseshoe), and P (passing) denote the type of orbit associated with the curves. The regions enclosed by each curve (*shaded*) are excluded from the motion of a test particle that has the corresponding C_J. The largest horseshoe curve actually passes through L_2, and the largest tadpole curve passes through L_3. Horseshoe orbits can exist between these two extremes. (Courtesy Carl Murray) (e) Schematic diagram showing the relationship between a horseshoe orbit and its associated zero-velocity curve. The particle's velocity in the rotating frame drops as it approaches the zero-velocity curve, and it cannot cross the curve. (Adapted from Dermott and Murray 1981)

in the Earth–Moon system have been suggested as possible locations for a future space station.

2.2.2 Horseshoe and Tadpole Orbits

Consider a moon on a circular orbit about a planet. A particle just interior to the moon's orbit has a higher angular velocity and moves with respect to the moon in the direction of corotation. A particle just outside the moon's orbit has a smaller angular velocity and moves relative to the moon in the opposite direction. When the outer particle approaches the moon, the particle is pulled towards

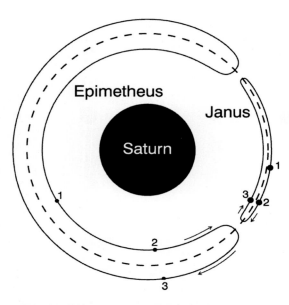

Figure 2.5 Diagram of the librational behavior of the Janus and Epimetheus coorbital system in a frame rotating with the average mean motion of both satellites. The system is shown to scale, apart from the radial extent of the librational arcs being exaggerated by a factor of 500 and the radii of the moons inflated by a factor of 50. The ratio of the radial widths (as well as the azimuthal extents) of the arcs is equal to the Janus/Epimetheus mass ratio (~0.25). The numbered points represent a temporal sequence of positions of the two moons over approximately one-quarter of a libration cycle. (Tiscareno et al. 2009)

the moon and consequently loses angular momentum. Provided the initial difference in semimajor axis is not too large, the particle drops to an orbit lower than that of the moon. The particle then recedes in the forward direction. Similarly, the particle on the lower orbit is accelerated as it catches up with the moon, resulting in an outward motion towards a higher, and therefore slower, orbit. Orbits like these encircle the L_3, L_4 and L_5 points and appear shaped like horseshoes in the rotating frame (Fig. 2.4b); thus they are called **horseshoe orbits**. Saturn's small moons Janus and Epimetheus execute just such a dance, changing orbits every 4 years, as illustrated schematically in Figure 2.5. Because Janus and Epimetheus are comparable in

mass, Hill's approximation is more accurate than is the restricted three-body formalism used earlier, but the dynamical interactions are essentially the same.

Because the Lagrangian points L_4 and L_5 are stable, material can librate about these points individually; such orbits are called **tadpole orbits** after their asymmetric elongated shape in the rotating frame (Fig. 2.4a). The Trojan asteroids librate about Jupiter's L_4 and L_5 points. The tadpole libration width at L_4 and L_5 is proportional to $(m_2/m_1)^{1/2}r$, and the horseshoe width varies as $(m_2/m_1)^{1/3}r$, where m_1 is the mass of the primary, m_2 the mass of the secondary and r the distance between the two objects. For a planet of Saturn's mass, $M_\hbar = 5.7 \times 10^{26}$ kg, and a typical moon of mass $m_2 = 10^{17}$ kg (a 30-km-radius object with density of ~1000 kg m^{-3}) at a distance of 2.5 R$_\hbar$, the tadpole libration half-width is ~3 km and the horseshoe half-width ~60 km.

2.2.3 Hill Sphere

The approximate limit to a secondary's (e.g., planet's or moon's) gravitational dominance is given by the extent of its **Hill sphere**,

$$R_H = \left(\frac{m_2}{3(m_1 + m_2)} \right)^{1/3} a, \qquad (2.28)$$

where m_2 is the mass of the secondary and m_1 the primary's (e.g., Sun's or planet's) mass. The Hill sphere stretches out to the L_1 point and essentially circumscribes the Roche lobe (§13.1) in the limit $m_2 \ll m_1$. Planetocentric orbits that are stable over long periods of time are those well within the boundary of a planet's Hill sphere; all known natural satellites lie in this region. As illustrated in Figure 2.6, stable heliocentric orbits are always well outside the Hill sphere of any planet. Comets and other bodies that enter the Hill sphere of a planet at very low velocity can remain gravitationally bound to the planet for some time as **temporary satellites**, an example of which is shown in Figure 2.7.

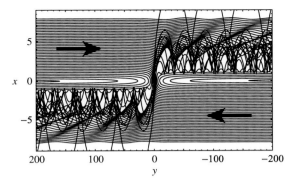

Figure 2.6 The trajectories of 80 test particles in the vicinity of a secondary of mass $m_2 \ll m_1$ are shown in the frame rotating with the secondary's (circular) orbit about the primary. The scale of the plot is expanded in the radial (x) direction relative to that in the azimuthal (y) direction, with numerical values in both directions given in units of the radius of the secondary's Hill sphere. The secondary mass is located at the origin and the L_1 and L_2 points are at $y = 0$, $x = \pm 1$. The particles were all started with $dx/dt = 0$ (i.e., circular orbits) at $y = \pm 200$. The *arrows* indicate their direction of motion before encountering the secondary. The primary is located at $y = 0, x = -\infty$. In an inertial frame, the secondary and the test particles all move from right to left. (Adapted from Murray and Dermott 1999)

The orbits of moons that lie in the inner part of a planet's Hill sphere are classified as prograde if the moons move in the sense that the planet rotates and retrograde if they travel in the opposite sense. However, for very distant satellites, the more important dynamical criterion is whether they travel in the same direction as the planet orbits the Sun (prograde) or in the opposite sense (retrograde). Retrograde orbits are stable to larger distances from a planet than are prograde ones, and moons on retrograde orbits are found at greater distances (Table E.4).

2.3 Perturbations and Resonances

Within the Solar System, one body typically produces the dominant gravitational force on any given object, and the resultant motion can be thought of as a Keplerian orbit about a primary,

(a)

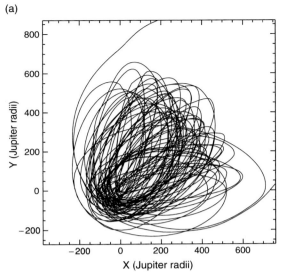

(b)

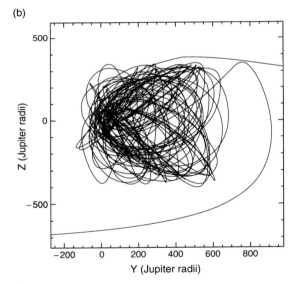

Figure 2.7 Trajectory relative to Jupiter of a test particle initially orbiting the Sun that was temporarily captured into an unusually long duration (140 years) unstable orbit about Jupiter. (a) Projected into the plane of Jupiter's orbit about the Sun. (b) Projected into a plane perpendicular to Jupiter's orbit. (Kary and Dones 1996)

subject to small perturbations by other bodies. Although perturbations on a body's orbit are often small, they cannot always be ignored. They must be included in short-term calculations if high accuracy is required, e.g., for predicting stellar occultations or targeting spacecraft. Most long-term perturbations are periodic in nature, their directions oscillating with the relative longitudes of the bodies or with some more complicated function of the bodies' orbital elements. Small perturbations can produce large effects if the forcing frequency is commensurate or nearly commensurate with the natural frequency of oscillation of the responding elements. Under such circumstances, perturbations add coherently, and the effects of many small tugs can build up over time to create a large-amplitude, long-period response. This is an example of **resonant forcing**, which occurs in a wide range of physical systems. In this section, we consider some important examples of the effects of these perturbations on the orbital motion.

2.3.1 Resonant Forcing

An elementary example of resonant forcing is given by the one-dimensional forced harmonic oscillator, for which the equation of motion is

$$m\frac{d^2x}{dt^2} + m\omega_o^2 x = F_f \cos \omega_f t, \qquad (2.29)$$

where x is the displacement, m is the mass of the oscillating particle, F_f is the amplitude of the driving force, ω_o is the natural frequency of the oscillator and ω_f is the forcing frequency. The solution to equation (2.29) is

$$x = \frac{F_f}{m(\omega_o^2 - \omega_f^2)} \cos \omega_f t + C_1 \cos \omega_o t + C_2 \sin \omega_o t, \qquad (2.30)$$

where C_1 and C_2 are constants determined by the initial conditions. Note that if $\omega_f \approx \omega_o$, a large-amplitude, long-period response can occur even if F_f is small. Moreover, if $\omega_o = \omega_f$, equation (2.30) is invalid. In this (resonant) case, the solution is given by

$$x = \frac{F_f}{2m\omega_o} t \sin \omega_o t + C_1 \cos \omega_o t + C_2 \sin \omega_o t. \qquad (2.31)$$

The t in the middle of the first term at the right-hand side of equation (2.31) leads to secular (i.e., steady rather than periodic) growth. Often this linear growth is moderated by the effects of nonlinear terms that are not included in the simple example provided above. However, some perturbations have a secular component.

2.3.2 Mean Motion Resonances

The simplest celestial resonances to visualize are so-called **mean motion resonances**, in which the orbital periods of two bodies are commensurate, i.e., can be written as the ratio of two integers. **First-order resonances**, which have the form $N/(N + 1)$, where N is an integer, are usually the strongest. Some examples of the consequences of mean motion resonance are given below. Almost exact orbital commensurabilities exist at many places in the Solar System. As illustrated in Figure 2.8, Io orbits Jupiter twice as frequently as Europa does, and Europa in turn orbits Jupiter in half of the time that Ganymede takes. **Conjunction** (the moons being at the same longitude in their orbits about the planet) between Io and Europa always occurs when Io is at its perijove (the point in its orbit that is closest to Jupiter). How can such commensurabilities exist? After all, the rational numbers form a set of measure zero on the real line, which means that the probability of randomly picking a rational from the real number line is nil! The answer lies in the fact that **orbital resonances** may be held in place by stable 'locks' that result from nonlinear effects not represented in the simple mathematical example of the harmonic oscillator. Differential tidal recession (§2.7.2) brings moons into resonance, and nonlinear interactions between the moons can keep them there.

Other examples of resonance locks include the Hilda and Trojan asteroids with Jupiter,

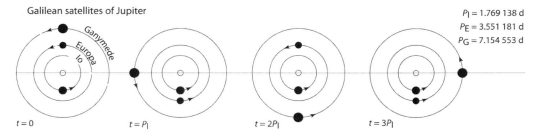

Galilean satellites of Jupiter

$P_I = 1.769\,138$ d
$P_E = 3.551\,181$ d
$P_G = 7.154\,553$ d

$t = 0$ $t = P_I$ $t = 2P_I$ $t = 3P_I$

Figure 2.8 Schematic illustration of the orbital resonances between the three inner Galilean satellites of Jupiter. Successive views represent the system at times separated by one orbital period of the moon Io, P_I. The configuration at $4\,P_I$ is identical to that at $t = 0$. (From Perryman 2011)

Neptune–Pluto and several pairs of moons orbiting Saturn, such as Janus–Epimetheus, Mimas–Tethys and Enceladus–Dione. Resonant perturbations can force bodies into eccentric and/or inclined orbits, which may lead to collisions with other bodies; this is thought to be the dominant mechanism for clearing the Kirkwood gaps in the asteroid belt (see below). Several moons of Jupiter and Saturn have significant resonantly produced **forced eccentricities**, which are denoted by the symbol f in Table E.4.

Spiral waves can be produced in a self-gravitating disk of particles by resonant perturbations of a satellite. Spiral density waves resonantly excited by moons are observed in Saturn's rings (§13.4.2). Analogous waves in protoplanetary disks can alter the orbits of young planets (§15.7.1).

2.3.3 Secular Resonances

Although many interactions among planets depend on their relative azimuthal positions, some important long-term effects are produced by the shapes and orientations of their orbits (eccentricities, apse locations, inclinations and nodes). **Secular perturbation theory** averages over orbital timescales and treats planets as elliptical wires of nonuniform density, with density corresponding to the time spent at a given phase of the orbit according to Kepler's second law (§2.1.1). Secular perturbations can alter

the eccentricity and inclination of an orbit but not its semimajor axis. A **secular resonance** occurs when the apses or nodes of two orbits precess at the same rate.

In the restricted circular three-body problem, secular perturbations can change the small body's eccentricity and its inclination relative to the orbit of the two massive bodies, but the quantity $\sqrt{1 - e^2}\cos i$ remains constant. Orbital inclination can thus be traded for eccentricity. For high values of inclination, $\cos^2 i < 3/5$, the **Kozai mechanism** forces the argument of periapse to remain fixed, and large periodic variations in eccentricity and inclination are produced. The Kozai mechanism causes some asteroids and comets to approach closely to and even collide with the Sun (§12.3.5) and highly inclined irregular satellites to collide with their planets. It is also likely to be one of the mechanisms responsible for the high observed eccentricities of some extrasolar planets (Fig. 14.25) and the high inclinations some of the hot jupiter planets' orbits have relative to the planes of their star's equator (§14.3.4).

2.3.4 Resonances in the Asteroid Belt

There are obvious patterns in the distribution of asteroidal semimajor axes that appear to be associated with mean motion resonances with Jupiter (Fig. 12.2a). At these resonances, a particle's period of revolution about the Sun is a small integer ratio multiplied by Jupiter's orbital period. The Trojan asteroids travel in a 1:1 mean motion resonance with Jupiter. Trojan asteroids execute small-amplitude (tadpole) librations about the L_4 and L_5 points 60° behind or ahead of Jupiter and therefore never have a close approach to Jupiter. Another example of a protection mechanism provided by a resonance is the Hilda group of asteroids at Jupiter's 3:2 mean motion resonance and the asteroid 279 Thule at the 4:3 resonance. The Hilda asteroids have a libration about 0° of their **critical argument** (the combination of orbital elements that signifies the resonant configuration), $3\lambda'-2\lambda-\varpi$, where λ' is Jupiter's longitude, λ is the asteroid's longitude and ϖ is the asteroid's longitude of perihelion. In this way, whenever the asteroid is in conjunction with Jupiter ($\lambda = \lambda'$), the asteroid is close to perihelion ($\lambda' \approx \varpi$) and well away from Jupiter.

Most orbits starting with small eccentricity in the general vicinity of the 3:1 mean motion resonance with Jupiter appear regular and show very little variation in eccentricity or semimajor axis over timescales of 5×10^4 yrs. However, orbits near the resonance can maintain a low eccentricity ($e < 0.1$) for nearly a million years and then have a 'sudden' increase in eccentricity to $e > 0.3$. Asteroids that begin on near-circular orbits in the gap acquire sufficient eccentricities to cross the orbits of Mars and the Earth and in some cases become so eccentric that they hit the Sun, so the perturbative effects of the terrestrial planets are probably capable of clearing out the 3:1 gap in a time equivalent to the age of the Solar System.

The ν_6 secular resonance occurs where the periapse angle of an asteroid precesses at the rate of the sixth secular frequency of our Solar System, which is essentially the same as the precession rate of Saturn's periapse. Perturbations resulting from the ν_6 resonance can excite asteroidal eccentricities to such high values that the ν_6 resonance is largely responsible for the inner edge of the asteroid belt near 2.1 AU.

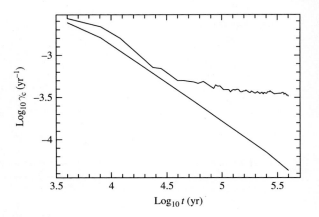

Figure 2.9 Distinction between regular (*lower curve, nearly straight*) and chaotic trajectories (*upper curve*) as characterized by the Lyapunov characteristic exponent, γ_c. Both trajectories are near the 3:1 resonance with Jupiter, and they have been integrated using the elliptic restricted three-body problem. For chaotic trajectories, a plot of $\log \gamma_c$ versus $\log t$ eventually levels off at a value of γ_c that is the inverse of the Lyapunov timescale for the divergence of initially adjacent trajectories. For regular trajectories, $\gamma_c \to 0$ as $t \to \infty$. (Adapted from Duncan and Quinn 1993)

2.3.5 Regular and Chaotic Motion

Direct integrations of multi-body systems on computers demonstrate that for some initial conditions, the trajectories are **regular** with variations in their orbital elements that seem to be well described by the perturbation series, but for other initial conditions, the trajectories are found to be **chaotic** and are not as confined in their motions. The evolution of a system that is chaotic depends so sensitively on the system's precise initial state that the behavior is in effect unpredictable even though it is strictly determinate in a mathematical sense.

Figure 2.9 shows a key feature of chaotic orbits that we use here as a definition of **chaos**: Two trajectories that begin arbitrarily close in phase space (which can be defined using coordinates such as positions and velocities, or a more complicated set of orbital elements) within a chaotic region typically diverge exponentially in time. Within a given chaotic region, the timescale for this divergence does not typically depend on the precise values of the initial conditions! The distance, $d(t)$, between two particles having an initially small separation, $d(0)$, increases slowly for regular orbits, with $d(t) - d(0)$ growing as a power of time t (typically linearly). In contrast, for chaotic orbits,

$$d(t) \sim d(0)e^{\gamma_c t}, \qquad (2.32)$$

where γ_c is the **Lyapunov characteristic exponent** and γ_c^{-1} is the **Lyapunov timescale**. From this definition of chaos, we see that chaotic orbits show such a sensitive dependence on initial conditions that the detailed long-term behavior of the orbits is lost within several Lyapunov timescales. Even a fractional perturbation as small as 10^{-8} in the initial conditions will result in a 100% discrepancy in about 20 Lyapunov times. However, one of the interesting features of much of the chaotic behavior seen in simulations of the orbital evolution of bodies in the Solar System is that the timescale for large changes in the principal orbital elements is often many orders of magnitude longer than the Lyapunov timescale.

In dynamical systems such as the Solar System, chaotic regions do not appear randomly; rather, many of them are associated with trajectories in which the ratios of characteristic frequencies of the original problem are sufficiently well approximated by rational numbers, i.e., near resonances. Figure 2.10 shows that the outer boundaries of the chaotic zone coincide well with the boundaries of the 3:1 Kirkwood gap.

The above discussion applies to orbits that do not closely approach any massive secondaries. Close approaches can lead to highly chaotic and unpredictable orbits, such as the possible future behaviors of the giant, distant cometary centaur Chiron shown in Figure 2.11 (see §12.2.2.). These planet-crossing trajectories do not require resonances to be unstable and generally are not well characterized by a constant Lyapunov exponent.

For nearly circular and coplanar orbits, the strongest mean motion resonances occur at

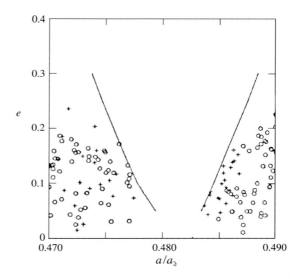

Figure 2.10 The outer boundaries of the chaotic zone surrounding Jupiter's 3:1 mean motion resonance in the a–e plane are shown as *lines*. Locations of numbered asteroids are shown as *circles* and Palomar–Leiden survey asteroids (whose orbits are less well determined) are represented as *plus signs*. Note the excellent correspondence of the observed 3:1 Kirkwood gap with theoretical predictions. (Adapted from Wisdom 1983)

locations where the ratio of test particle orbital periods to the massive body's period is of the form $N{:}(N{\pm}1)$, where N is an integer. At these locations, conjunctions (closest approaches) always occur at the same phase in the orbit, and tugs add coherently. (The locations of these strong resonances are shifted slightly when the primary is oblate; see §§2.6 and 13.4 for details.) The strength of these first-order resonances increases as N grows because the magnitudes of the perturbations are larger closer to the secondary. First-order resonances also become closer to one another near the orbit of the secondary (Problem 2-7). Sufficiently close to the secondary, the combined effects of greater strength and smaller spacing cause resonance regions to overlap; this overlapping can lead to the onset of chaos as particles shift between the nonlinear perturbations of various resonances. The region of overlapping resonances is approximately

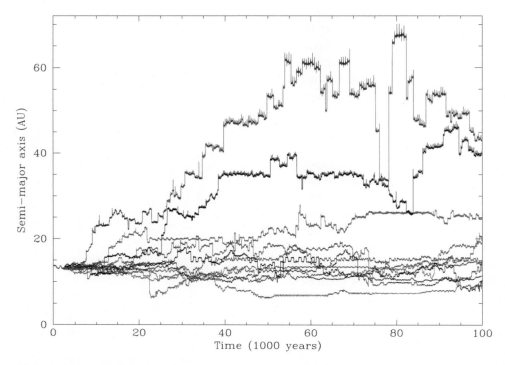

Figure 2.11 COLOR PLATE The future evolution of the semimajor axis of P/Chiron's orbit according to 11 numerical integrations. The initial orbital elements of the simulated bodies differed by about 1 part in 10^6. The orbit of Chiron currently crosses the orbits of both Saturn and Uranus, and is not protected from close approaches with either planet by any resonance. Chiron's orbit is highly chaotic, with gross divergence of trajectories in $<10^4$ years. (Courtesy L. Dones)

symmetric about the planet's orbit and has a half-width, $\Delta a_{\rm ro}$, given by

$$\Delta a_{\rm ro} \approx 1.5 \left(\frac{m_2}{m_1}\right)^{2/7} a, \qquad (2.33)$$

where a is the semimajor axis of the planet's orbit. Whereas the functional form of equation (2.33) has been derived analytically, the coefficient 1.5 is a numerical result.

2.4 Stability of the Solar System

We turn now to one of the oldest problems in dynamical astronomy: whether or not the planets will continue indefinitely in almost circular, almost

coplanar orbits. From an astronomical viewpoint, stability implies that the system will remain bound (no ejections), that no mergers of planets will occur for the possibly long but finite period of interest and that this result is robust against (most if not all) sufficiently small perturbations.

2.4.1 Orbits of the Eight Planets

Figure 2.12 shows the behavior of the eccentricities of all eight planets for 3 million years into the past as well as into the future. Mercury's eccentricity reaches higher values on 10^8-year timescales, as can be seen in Figure 2.13, but the eccentricities of the other planets do not extend much beyond their range shown in Figure 2.12 over this time interval. Variations in the semimajor axis of Earth's

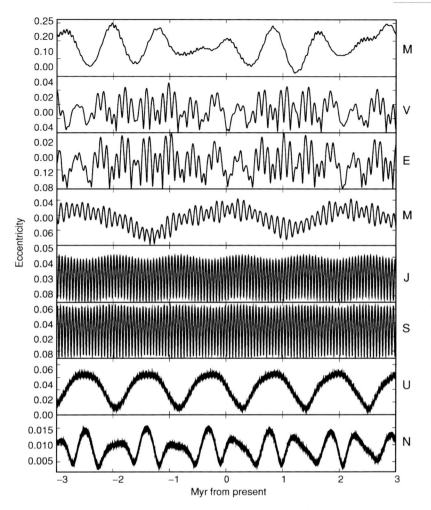

Figure 2.12 The eccentricities of the eight major planets are shown for 6×10^6 years centered on the present epoch. Mercury's eccentricity is displayed in the *top panel* followed by that of each of the other planets in order of their heliocentric distance. Note the relatively large amplitudes of the variations of the two smallest planets, Mercury and Mars, and the correlated oscillations of $e_\oplus$ with those of $e_\venus$ and of $e_\jupiter$ with those of $e_\saturn$. (Courtesy Tom Quinn; see Laskar et al. 1992 for an explanation of the integration used to compute these values)

orbit over ± 3 Myr are shown in Figure 2.14. The small fractional changes in semimajor axis relative to the variations in eccentricity evident for Earth are characteristic of all eight planets.

Long-duration numerical integrations show a surprisingly high Lyapunov exponent, $\sim (5\,\mathrm{Myr})^{-1}$. Such large Lyapunov exponents certainly suggest chaotic behavior. However, the apparent regularity of the motion of the Earth, and indeed the fact that the Solar System has survived for 4.5 billion years, implies that any pathways through phase space that might lead to (highly chaotic) close

approaches must be narrow. Nonetheless, the exponential divergence seen in all long-term integrations implies that the accuracy of the deterministic equations of celestial mechanics to predict the future positions of the planets will always be limited by the accuracy with which their orbits can be measured. For example, even if the position of Earth along its orbit were to be known to within 1 cm today and all other planetary masses, positions and velocities were known exactly, the exponential propagation of errors that is characteristic of chaotic motion implies that we would still have no

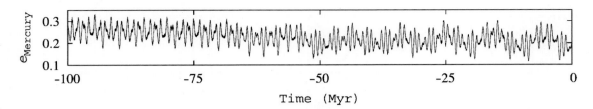

Figure 2.13 Variations in the eccentricity of Mercury's orbit over the past 100 million years. Integrations included the Sun, all eight planets and first-order post-Newtonian effects of general relativity; the eccentricities of all of the planets over the past 3 million years look the same in the integration used to produce this figure as those shown in Figure 2.12. (Courtesy Julie Gayon)

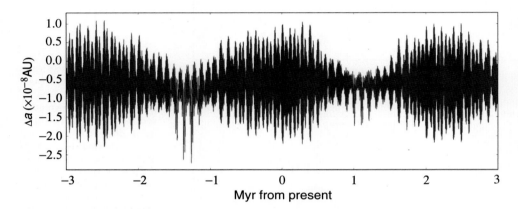

Figure 2.14 Variations in the semimajor axis of Earth's orbit (more precisely, the semimajor axis of the center of mass of the Earth–Moon system) over a time interval of 6 million years centered on the present epoch. Note the scale of the vertical axis, which indicates that the Earth's semimajor axis varies by only a few kilometers over timescales of millions of years. These data were taken from the integrations used to produce the plot of eccentricities of the planets shown in Figure 2.12. (Courtesy Tom Quinn)

knowledge of Earth's orbital longitude 200 million years in the future.

The situation is even less predictable when the gravitational influence of smaller bodies is accounted for. Asteroids exert small perturbations on the orbits of the major planets. These perturbations can be accounted for and do not adversely affect the precision to which planetary orbits can be simulated on timescales of tens of millions of years. However, unlike the major planets, asteroids suffer close approaches to one another and thus are subject to the types of chaos depicted in Figure 2.11. Close approaches between the two largest and most massive asteroids, 1 Ceres and 4 Vesta

(Tables E.6 and E.8), lead to exponential growth in uncertainty for backwards integrations of planetary orbits with doubling times of $<10^6$ years prior to 50–60 million years ago.

It is also worth bearing in mind the lessons learned from integration of test particle trajectories, namely that the timescale for macroscopic changes in the system can be many orders of magnitude longer than the Lyapunov timescales. Thus, the apparent stability of the current planetary system on billion-year timescales may simply be a manifestation of the fact that the Solar System is in the chaotic sense a dynamically young system. Indeed, there is a small but nontrivial chance that

Mercury's orbit will become so eccentric that it will cross the orbit of Venus before the Sun evolves off the main sequence 6 billion years from now.

Because planetary perturbations appear to be capable of bringing the Solar System to the verge of instability on geological timescales, the planets within our Solar System may be about as closely spaced as can be expected for a mature planetary system containing planets as massive as those orbiting the Sun. Although somewhat more crowded configurations can be long lived, it may well be that the planet formation process (see Chapter 15) is unlikely to produce more densely packed systems of similar planets that survive on gigayear timescales.

2.4.2 Survival Lifetimes of Small Bodies

Interplanetary space is vast, yet few bodies orbit within this great expanse. And those few bodies are far from randomly distributed. Rather, minor planets are concentrated within a few regions (§12.2): the Kuiper belt beyond Neptune's orbit, the main asteroid belt between the orbits of Mars and Jupiter, the regions surrounding the triangular Lagrangian points of the Sun–Jupiter system (§2.2.1) and probably around the regions surrounding the triangular Lagrangian points of the Sun–Neptune system. Dynamical analyses show that orbits within these regions remain stable for far longer than trajectories passing through most other locations in the Solar System. What causes the removal of bodies from other regions of the Solar System? How rapidly are they removed?

Trajectories crossing the paths of one or more of the major planets are rapidly destabilized by scatterings resulting from close planetary approaches unless they are protected by some type of resonance (as is Pluto). Small bodies can remain on orbits between a pair of terrestrial planets or a pair of giant planets for much longer, but most are perturbed into planet-crossing paths in less than the age of the Solar System by the same resonance

overlap-induced chaos that makes planetary orbits unpredictable on long timescales. Lifetimes of orbits vary greatly, and collections of test particles spread randomly over even fairly small regions of phase space last for quite diverse amounts of time. Figure 2.15 illustrates stability times for test particles located exterior to 5 AU; note the logarithmic scale for the time axis. Loss rates are rapid early on, but as particles near the stronger resonances are removed, it takes longer and longer for a given fraction of the remaining bodies to be destabilized. This decay rate is more gradual than that of other natural processes, such as radioactivity (§3.4), in which the population drops exponentially with time (§11.6.1).

2.5* Dynamics of Spherical Bodies

Thus far we have approximated Solar System bodies as point masses for the purpose of calculating their mutual gravitational interactions. Self-gravity causes most large celestial bodies to be approximately spherically symmetric.

2.5.1 Moment of Inertia

The moment of inertia, I, of a body about a particular axis is defined as

$$I \equiv \int\int\int \rho(\mathbf{r}) r_c^2 d\mathbf{r}, \tag{2.34}$$

where r_c is the distance from the axis and the integral is taken over the entire body. The rotational angular momentum of a simply rotating rigid body is given by

$$\mathbf{L} = \int\int\int \rho(\mathbf{r})(\mathbf{r} \times \mathbf{v})\, d\mathbf{r}$$
$$= \int\int\int \rho(\mathbf{r}) r_c^2 \omega_{rot} d\mathbf{r} = I\omega_{rot}, \tag{2.35}$$

where ω_{rot} is its spin angular velocity. Orbital angular momentum is given by equations (2.12)

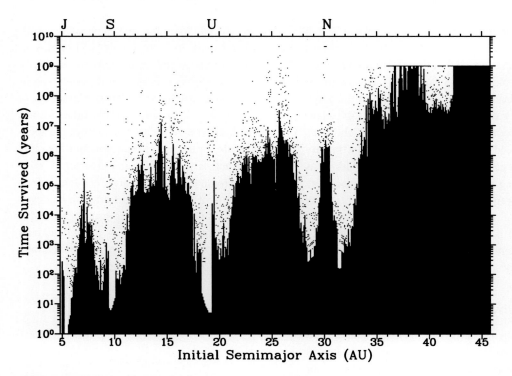

Figure 2.15 Stability map for test particles in the outer Solar System based on numerical integrations that include the Sun and the four giant planets. The time that each particle survived is plotted as a function of particle initial semimajor axis. For each semimajor axis bin, six particles were started at differing longitudes. The *vertical bars* mark the minimum of the six termination times. The *points* mark the termination times of the other five particles. The *scatter of points* gives an idea of the spread in particle lifetimes at each semimajor axis. The locations of the planets are denoted on the top of the figure; the spikes in particle lifetimes near these semimajor axes represent particles initially in tadpole or horseshoe orbits. The integrations extend to 4.5×10^9 years for particles initially interior to Neptune and to 10^9 years for those farther out. Only a few particles initially interior to Neptune survived the entire integrations, but many particles exterior to 33 AU and all particles beyond about 43 AU remained on non–planet-crossing orbits for the entire time interval simulated. (Courtesy Matt Holman; see Holman 1997 for details on the calculations)

and (2.26). Analogously, the kinetic energy of rotation is given by

$$E_{\text{rot}} = \frac{1}{2}I\omega_{\text{rot}}^2 = \frac{1}{2}\omega_{\text{rot}}L. \tag{2.36}$$

The moment of inertia of a uniform density sphere of radius R and mass m about its center of mass can be computed directly by performing the integration specified in equation (2.34) over the sphere. However, a more elegant and less tedious method exploits the symmetry of the sphere as follows. Note that to compute I about the z-axis, we

have $r_c^2 = x^2 + y^2$; about the x-axis, $r_c^2 = y^2 + z^2$; and about the y-axis, $r_c^2 = x^2 + z^2$. By symmetry, these three integrals are equal, so adding them gives:

$$3I = \iiint \rho(2x^2 + 2y^2 + 2z^2)d\mathbf{r}. \tag{2.37}$$

The quantity in parentheses in equation (2.37) is equal to $2r^2$, so the integral can most easily be performed using spherical coordinates. Dividing both sides by 3 and noting that the integrations in the angular coordinates give the surface area of a spherical shell, $4\pi r^2$, yields:

$$I = \frac{8\pi\rho}{3} \int_0^R r^4 dr = \frac{2}{5} mR^2. \tag{2.38}$$

Centrally condensed bodies have moment of inertia ratios $I/(mR^2) < 2/5$ (Problem 2-10). The moment of inertia ratios for the planets are listed in Table E.15.

2.5.2 Gravitational Interactions

Newton showed that the gravitational force exerted by a spherically symmetric body exterior to its surface is identical to the gravitational force of a pointlike particle of the same mass located at the body's center. We derive this result below using multiple integration of the gravitational potential, which is defined in the next paragraph.

For many applications, it is convenient to express the gravitational field in terms of a potential, $\Phi_g(\mathbf{r})$, defined as:

$$\Phi_g(\mathbf{r}) \equiv - \int_\infty^{\mathbf{r}} \frac{\mathbf{F}_g(\mathbf{r}')}{m} \cdot d\mathbf{r}'. \tag{2.39}$$

By inverting equation (2.39), one can see that the gravitational force is the gradient of the potential and

$$\frac{d^2\mathbf{r}}{dt^2} = -\nabla\Phi_g. \tag{2.40}$$

In general, $\Phi_g(\mathbf{r})$ satisfies Poisson's equation:

$$\nabla^2\Phi_g = 4\pi\rho G. \tag{2.41a}$$

In empty space, $\rho = 0$, so $\Phi_g(\mathbf{r})$ satisfies Laplace's equation:

$$\nabla^2\Phi_g = 0. \tag{2.41b}$$

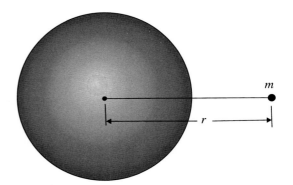

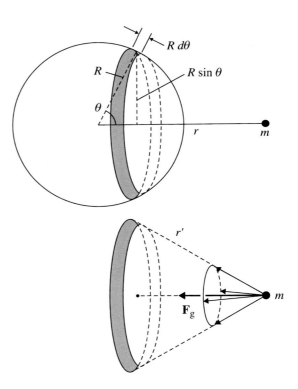

Figure 2.16 Diagram showing the notation used in the calculation of the gravitational potential exterior to a uniform sphere.

A spherically symmetric body can be viewed as the sum of thin concentric shells, each of uniform surface density. Without loss of generality, consider a shell of radius R and surface density unity centered at the origin and evaluate the potential at a location $\mathbf{r}$, where $r > R$. Subdivide the

shell into rings that are oriented perpendicular to the direction from the center of the sphere to the point at which the potential is being evaluated. As illustrated in Figure 2.16, let θ denote the angle between lines from the origin to a point on the ring and to **r**. The mass of the ring is given by $2\pi R \sin\theta$. The potential of the shell at the point under consideration is given by:

$$\Phi_g = \int_0^\pi 2\pi R \sin\theta \frac{1}{r'} R d\theta, \qquad (2.42)$$

where r' denotes the distance from the point at which the potential is being evaluated. The law of cosines gives the square of r' as:

$$r'^2 = R^2 + r^2 - 2rR\cos\theta. \qquad (2.43)$$

Because r and R are constant, we may differentiate equation (2.43) and rearrange the terms to obtain:

$$\frac{\sin\theta}{r'}d\theta = \frac{dr'}{Rr}. \qquad (2.44)$$

Substituting equation (2.44) into equation (2.42) yields:

$$\Phi_g = 2\pi R^2 \int_{r-R}^{r+R} \frac{dr'}{Rr} = \frac{4\pi R^2}{r}. \qquad (2.45)$$

The value of Φ_g given in equation (2.45) is equal to the area of the sphere (and thus to its mass if surface density of unity is assumed) divided by the distance from its center to the point at which the potential is being evaluated. This value is identical to the potential of a point particle of the same mass located at the center of the sphere, completing the proof.

2.6 Orbits about an Oblate Planet

Several forces act to produce distributions of mass that deviate from spherical symmetry. In the Solar System, rotation, physical strength and tides produce important departures from spherical symmetry in some bodies. The gravitational field near an aspherical body differs from that near a point-mass. The gravity field can be determined to quite high accuracy by tracking the orbits of spacecraft close to the body or from the rate of precession of the periapses of moons and rings orbiting the planet. The gravity field of a planet or moon contains information on the body's internal density structure.

Most planets are very nearly axisymmetric, with the major departure from sphericity being due to a rotationally induced equatorial bulge. Thus, in this section, we analyze the effects of an axisymmetric body's deviation from spherical symmetry on the gravitational force that it exerts and on its response to external torques.

2.6.1* Gravity Field

The analysis of the gravitational field of an axisymmetric planet is most conveniently done by using the Newtonian gravitational potential, $\Phi_g(\mathbf{r})$, which is defined in equation (2.39). Because $\Phi_g(\mathbf{r})$ in free space satisfies Laplace's equation (2.41b), the gravitational potential exterior to a planet can be expanded in terms of **Legendre polynomials**:

$$\Phi_g(r,\phi,\theta) = -\frac{Gm}{r}\left[1 - \sum_{n=2}^{\infty} J_n P_n(\cos\theta)\left(\frac{R}{r}\right)^n\right].$$
$$(2.46)$$

Equation (2.46) is written in standard spherical coordinates, with ϕ the longitude and θ representing the angle between the planet's symmetry axis and the vector to the particle (i.e., the colatitude). The terms $P_n(\cos\theta)$ are the Legendre polynomials, given by the formula:

$$P_n(x) = \frac{1}{2^n n!}\frac{d^n}{dx^n}\left(x^2 - 1\right)^n. \qquad (2.47)$$

The **gravitational moments**, J_n, are determined by the planet's mass distribution (§6.2.2). The origin is chosen to be the center of mass, thus the gravitational moment $J_1 = 0$. For a nonrotating fluid body in hydrostatic equilibrium, the potential is spherically symmetric and the moments $J_n = 0$. If the planet's mass is distributed symmetrically about the planet's equator, then the J_n are zero for all odd n.

Let us consider a small body, e.g., a moon or ring particle, that travels around a planet on a circular orbit in the equatorial plane ($\theta = 90°$) at a distance r from the center of the planet. The centripetal force must be provided by the radial component of the planet's gravitational force (eq. 2.20), so the particle's angular velocity n satisfies:

$$rn^2(r) = \left.\frac{\partial \Phi_g}{\partial r}\right|_{\theta=90°}. \tag{2.48}$$

If the particle suffers an infinitesimal displacement from its circular equatorial orbit, it will oscillate freely in the horizontal and vertical directions about the reference circular orbit with radial (epicyclic) frequency $\kappa(r)$ and vertical frequency $\mu(r)$, respectively, given by

$$\kappa^2(r) = r^{-3}\frac{\partial}{\partial r}[(r^2 n)^2], \tag{2.49}$$

$$\mu^2(r) = \left.\frac{\partial^2 \Phi_g}{\partial z^2}\right|_{z=0}. \tag{2.50}$$

2.6.2 Precession of Particle Orbits

Using the equations (2.46–2.50), one can show that the orbital, epicyclic and vertical frequencies can be written as

$$n^2 = \frac{Gm}{r^3}\left[1 + \frac{3}{2}J_2\left(\frac{R}{r}\right)^2 - \frac{15}{8}J_4\left(\frac{R}{r}\right)^4 + \cdots\right], \tag{2.51}$$

$$\kappa^2 = \frac{Gm}{r^3}\left[1 - \frac{3}{2}J_2\left(\frac{R}{r}\right)^2 + \frac{45}{8}J_4\left(\frac{R}{r}\right)^4 + \cdots\right], \tag{2.52}$$

$$\mu^2 = 2n^2 - \kappa^2. \tag{2.53}$$

For a perfectly spherically symmetric planet, $\mu = \kappa = n$. Because planets are oblate, μ is slightly larger than the orbital frequency, n, and κ is slightly smaller. The oblateness of a planet therefore causes periapse longitudes of particle orbits in and near the equatorial plane to precess in the direction of the orbit and lines of nodes of nearly equatorial orbits to regress. Orbits about

oblate planets are thus not Keplerian ellipses. However, because the trajectories are nearly elliptical, they are often specified by instantaneous Keplerian orbital elements. Note that

$$\frac{d\varpi}{dt} = n - \kappa, \tag{2.54}$$

$$\frac{d\Omega}{dt} = n - \mu. \tag{2.55}$$

2.6.3 Torques on an Oblate Planet

The nonspherical distribution of mass within an oblate planet allows the planet to exert torques upon its satellites, changing their orbital angular momenta and thereby precessing the planes of their orbits, as discussed earlier. Other bodies can exert torques on an oblate planet via a corresponding back force, thereby changing the direction of its rotational pole in inertial space; this spinning of the rotation axis is referred to as **precession**.

The strongest torques on an oblate planet are exerted by the Sun. In some cases, such as Earth, large satellites act as intermediaries, and their presence can affect the magnitude of the torque. Solar torques cause a planet's rotation axis to precess, resulting in a difference between the length of a tropical year (periodicity of the seasons) and the planet's orbital period, as well as changing the position of the pole on the celestial sphere. For example, the Sun, Moon and other planets produce a net torque on the Earth's equator, which leads to a precession of the Earth's rotation axis with a period of $\sim 26\,000$ years. Thus, we use a different northern pole star than did the ancient Greeks and Romans.

The torques on a planet's equatorial bulge resulting from gravitational forces of other planets are, of course, much smaller than are the corresponding solar torques. Nonetheless, interplanetary torques can be more fundamentally important than are solar torques because they can cause planetary obliquities (axial tilts) to vary. Figure 2.17 shows past values of the obliquity of Mars. Mars's obliquity varies chaotically as a result of planetary perturbations, reaching values as high

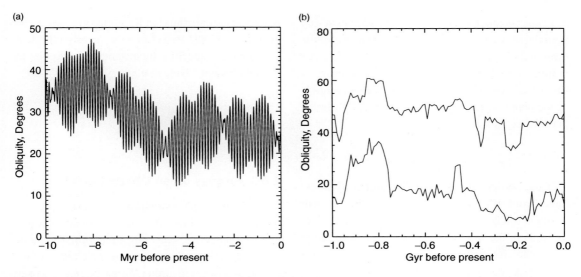

Figure 2.17 Torques by other planets on the martian equatorial bulge produce variations in the obliquity of Mars on a variety of timescales. (a) The martian obliquity over the past 10 Myr. (b) Maxima and minima of the martian obliquity during 10-Myr time intervals are shown for the past 1 Gyr. Because the Lyapunov time for this system is far less than 1 Gyr, the obliquity of Mars cannot be calculated accurately for this timescale, but this integration yields one possible, randomly selected realization. (Courtesy John Armstrong; see Armstrong et al. 2004 for details on the calculation)

as 60°. For obliquity in the range 54°−126°, the seasonally averaged flux of solar energy reaching a planet's pole is greater than that reaching its equator.

Earth's obliquity is stabilized by the Moon; without the Moon, Earth's obliquity would also vary considerably, resulting in substantial variations in climate. As it is, the small variations in Earth's obliquity that have occurred over the past few million years, shown in Figure 2.18, are correlated with ice ages. The quasi-periodic climate variations associated with changes in the Earth's obliquity and with our planet's eccentricity are known as **Milankovich cycles**.

2.7 Tides

The gravitational force arising from the pull of external objects varies from one part of a body to another. These differential tugs produce what is known as the **tidal force**. The net force on a body determines the acceleration of its center of mass. Figure 2.19 illustrates how tidal forces can alter the shape of a body. Tidal deformation can lead to torques that alter the body's rotation state. Time-variable tidal forces such as those experienced by moons on eccentric orbits can result in flexing, which leads to internal heating.

Tidal forces are important to many aspects of the structure and evolution of planetary bodies. For example, on short timescales, temporal variations in tides (as seen in the frame rotating with the body under consideration) cause stresses that can move fluids with respect to more rigid parts of the planet, such as the ocean tides with which we are familiar. These stresses can even cause seismic disturbances. (Although the evidence that the Moon causes some earthquakes is weak and disputable, it is clear that the tides raised by the Earth are a major cause of moonquakes.) On long timescales, tides change the orbital and spin properties of planets

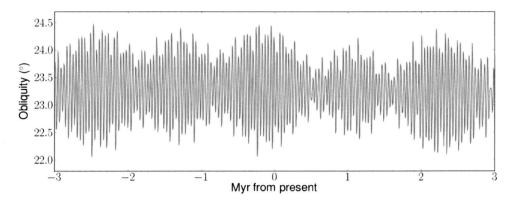

Figure 2.18 Variations in the obliquity of Earth over a time interval of 6×10^6 years centered on the present epoch. These data were taken from the same integrations used to produce Figures 2.12 and 2.14. (Courtesy Tom Quinn)

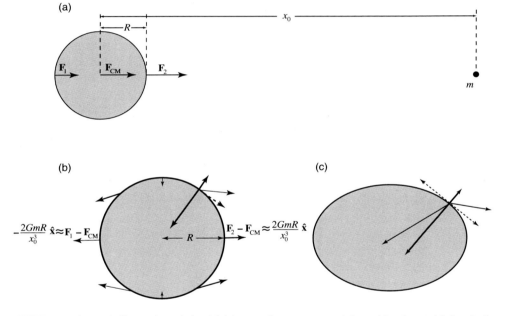

Figure 2.19 Schematic illustration of the tidal forces of a moon on a deformable planet. (a) Gravitational force of the moon on different parts of the planet. (b) Plain solid arrows indicate the differential force of the moon's gravity relative to the force on the planet's center of mass. (c) Response of the planet's figure to the moon's tidal pull. (de Pater and Lissauer 2010)

and moons. Tides, along with rotation, determine the equilibrium shape of a body located near any massive object; note that many materials that behave as solids on short timescales are effectively fluids on very long geological timescales, e.g., the Earth's mantle. In some cases, tidal forces are so strong that they exceed a body's cohesive force, and the body fragments.

2.7.1 The Tidal Force and Tidal Bulges

Consider a nearly spherical body of radius R, centered at the origin, which is subject to the gravitational force of a point mass, m, at $\mathbf{r_o}$, where $r_o \gg R$ (Fig. 2.19 with $r_o = x_o$). At a point $\mathbf{r} = (x, y, z)$, the specific (per unit mass) tidal force is the difference between the pull of m at $\mathbf{r}$ and the pull of m at the origin:

$$\mathbf{F_T}(\mathbf{r}) = \frac{Gm}{|\mathbf{r_o} - \mathbf{r}|^3}(\mathbf{r_o} - \mathbf{r}) - \frac{Gm}{r_o^3}\mathbf{r_o}. \qquad (2.56)$$

For points along the line joining the center of the body to the point mass (which we take to be the x-axis), equation (2.56) reduces to

$$F_T(x) = \frac{Gm}{(x_o - x)^2} - \frac{Gm}{x_o^2} \approx \frac{2xGm}{x_o^3}. \qquad (2.57)$$

The tidal approximation used for the last part of equation (2.57) can be derived by Taylor expansion of the first term in the middle expression and retaining only the first two terms. Equation (2.57) states that, to lowest order, the tidal force varies proportionally to the distance from the center of the stressed body and inversely to the cube of the distance from the perturber. The portion of the body with positive x coordinate feels a force in the positive x-direction, and the portion at negative x is tidally pulled in the opposite direction (Fig. 2.19).

Note from Figure 2.19 and equation (2.56) that material off the x-axis is tidally drawn towards the x-axis. If the body is deformable, it responds by becoming elongated in the x-direction. For a perfectly fluid body, the degree of elongation is that necessary for the body's surface to be an equipotential, when self-gravity, centrifugal force due to rotation and tidal forces are all included in the calculation (§6.2.2).

The gravitational attraction of, for example, the Moon and Earth on one another thus causes tidal bulges that rise along the line joining the centers of the two bodies. The near-side bulge is a direct consequence of the greater gravitational attraction closer to the other body; the bulge on the opposite side results from the weaker attraction at the far side of the object than at its center. The differential centrifugal acceleration across the body also contributes to the size of the tidal bulges.

The Moon spins once per orbit, so the same face of the Moon always points towards the Earth and the Moon is always elongated in that direction. The Earth, however, rotates much faster than the Earth–Moon orbital period. Thus, different parts of the Earth point towards the Moon and are tidally stretched. Water responds much more readily to these varying forces than does the 'solid Earth', resulting in the tidal variations in the water level seen at ocean shorelines. Because the combined effects of terrestrial rotation and the Moon's orbital motion imply that the Moon passes above a given place on Earth approximately once every 25 hours (Problem 2-14), there are almost two tidal cycles per day, and the principal tide that we see is known as the **semidiurnal tide**. The Sun also raises semidiurnal tides on Earth, with a period of 12 hours and an amplitude just under half those of lunar tides (Problem 2-15). Tidal amplitudes reach a maximum twice each (astronomical) month, when the Moon, Earth and Sun are approximately aligned, i.e., when the Moon is 'new' or 'full'. Tides are also larger when the Moon is near perigee and when the Earth is near perihelion (the latter occurs in early January).

Strong tides can significantly affect the physical structure of bodies. Generally, the strongest tidal forces felt by Solar System bodies (other than Sun-grazing or planet-grazing asteroids and comets) are those caused by planets on their closest satellites. Near a planet, tides are so strong that they can rip apart a fluid (or weakly aggregated solid) body. In such a region, large moons are unstable, and even small moons, which could be held together by material strength and friction, are unable to accrete because of tides. The boundary of this region is known as **Roche's limit**. Interior

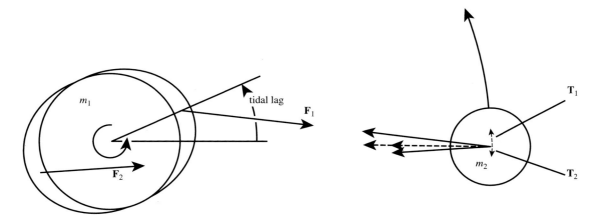

Figure 2.20 Schematic illustration of the tidal torque that a planet exerts on a moon orbiting in the prograde direction with a period longer than the planet's rotation period. Dissipation within the planet causes the tidal bulges that the moon raises on the planet to be located at places on the planet that were nearest to and farthest from the moon at a slightly earlier time. Because there is a temporal lag in the tidal bulges, for a moon on a slow prograde orbit, the bulges lead the position of the moon. The asymmetries in the planet's figure imply that its gravity is not a central force, and thus it can exert a torque on the moon. The far side bulge exerts a retarding torque, T_2, on the moon, but the near-side bulge exerts a larger positive torque, T_1, on the moon, so the moon receives a net positive torque, and its orbit evolves outward. (de Pater and Lissauer 2010)

to Roche's limit, solid material remains in the form of small bodies, and we see rings instead of large moons. The derivation of Roche's limit is outlined in §13.1.

2.7.2 Tidal Torque

Tidal dissipation causes secular variations in the rotation rates and orbits of moons and planets. Although the total angular momentum of an orbiting pair of bodies is conserved in the absence of an external torque, angular momentum can be transferred between rotation and orbital motions via tidal torques.

If planets were perfectly fluid, they would respond immediately to varying forces, and tidal bulges raised by a satellite would point directly towards the moon responsible. However, the finite response time of a planet's figure causes the tidal bulges to lag 'behind' at locations on the planet that pointed towards the moon at a slightly earlier time. This process is diagrammed in Figure 2.20.

Provided the planet's rotation period is shorter than the moon's orbital period and the moon's orbit is prograde, this tidal lag causes the nearer bulge to lie in front of the moon, and the moon's greater gravitational force on the near-side bulge than on the far side bulge acts to slow the rotation of the planet. The reaction force on the moon causes its orbit to expand. Satellites in retrograde orbits (e.g., Triton) and satellites whose orbital periods are less than the planet's rotation period (e.g., Phobos) spiral inwards towards the planet as a result of tidal forces.

The tidal torque depends on the size of the tidal bulge and the lag angle. The size of the tidal bulge varies as $m_2 r^{-3}$. The torque on a bulge of given size and lag angle is also a tidal effect and is also proportional to $m_2 r^{-3}$. Thus, the torque on a satellite from the tidal bulge that it raises on its primary varies as:

$$\dot{L} \propto \frac{m_2^2}{r^6}. \tag{2.58}$$

The above arguments remain valid if 'moon' is replaced by 'Sun', or if 'moon' and 'planet' are interchanged. Indeed, the stronger gravity of planets means that they have a much greater effect on the rotation of moons than vice versa. Most, if not all, major moons in our Solar System have been slowed to a synchronous rotation state in which the same hemisphere of the moon always faces the planet; thus, no tidal lag occurs for these moons.

Evidence exists for the tidal slowing of Earth's rotation on a variety of timescales. Growth bands observed in fossil bivalve shells and corals imply that there were 400 days per year approximately 350 million years ago. Eclipse timing records imply that the day has lengthened slightly over the past two millennia. Precise measurements using atomic clocks show variations in the Earth's rotation rate during the past few decades; however, care must be taken to separate secular tidal effects from the short-term periodic influences. Most of the secular decrease in Earth's rotation rate is caused by tides raised by the Moon, but at the present epoch, ~20% is due to solar tides.

The Pluto–Charon system has evolved even further. Charon is between one-ninth and one-eighth as massive as Pluto, a much larger secondary to primary mass ratio than observed for any of the more massive bodies in the Solar System. The semimajor axis of their mutual orbit is just 19 636 km, which is only 5% of the Earth–Moon distance and smaller than any planet–moon separation apart from Mars–Phobos (Table E.4). The Pluto–Charon system has reached a stable equilibrium configuration, in which each of the bodies spins on its axis in the same length of time that they orbit about their mutual center of mass. Thus, the same hemisphere of Pluto always faces Charon, and the same hemisphere of Charon always faces Pluto.

Solar tides have resulted in a stable spin-orbit lock for nearby Mercury, but one that is more complicated than the synchronous state that exists for most planetary satellites. Mercury makes three rotations around its axis every two orbits about the

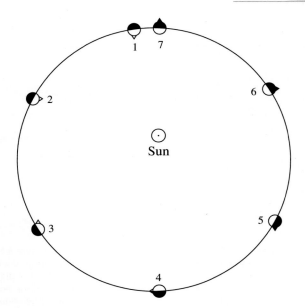

Figure 2.21 Mercury's rotation and solar day. Mercury's 3:2 spin-orbit resonance implies that the same axis is always aligned with the direction to the Sun at perihelion and that the mercurian (solar) day lasts twice as long as the mercurian year. (de Pater and Lissauer 2010)

Sun. The reason that equilibrium exists at exactly one and one-half rotations per orbit is that Mercury has a small permanent (nontidal) deformation and a highly eccentric orbit. It is energetically most favorable for Mercury's long axis (the axis with smallest moment of inertia) to point towards the Sun every time the planet passes perihelion, a configuration consistent with the observed 3:2 spin-orbit resonance. Figure 2.21 shows how Mercury's slow prograde rotation leads to a mercurian day lasting twice as long as a mercurian year.

Solar gravitational tides are probably the principal reason that Venus rotates very slowly, but they do not explain why our sister planet spins in the retrograde direction. Solar heating produces asymmetries in Venus's massive atmosphere that are known as **atmospheric tides**, and the Sun's gravitational pull on these atmospheric tides probably prevents Venus's solar day from becoming longer than it is at present. Tidal forces slow the rotation rates of the other planets but at rates too small to be significant, even over geologic time.

2.7.3 Tidal Heating

Temporal variations in tidal forces can lead to internal heating of planetary bodies. The locations of the tidal bulges of a moon having nonsynchronous rotation vary as the planet moves in the sky as seen from the moon. A synchronously rotating moon on an eccentric orbit is subjected to two types of variations in tidal forces. The amplitude of the tidal bulge varies with the moon's distance from the planet, and the direction of the bulge varies because the moon spins at a constant rate (equal to its mean orbital angular velocity), but the instantaneous orbital angular velocity varies according to Kepler's second law. Because planetary bodies are not perfectly rigid, these variations in tidal forces change the shape of the moon. Because bodies are not perfectly fluid either, moons dissipate energy as heat while they change in shape. Internal stresses caused by variations in tides on a body that is on an eccentric orbit or that is not rotating synchronously with its orbital period can therefore result in significant tidal heating of some bodies, most notably in Jupiter's moon Io.

If no other forces were present, the dissipation that results from variations in tides raised on Io by Jupiter would lead to a decay of Io's orbital eccentricity. Io's orbit would approach circularity, the lowest energy state (smallest semimajor axis) for a given angular momentum (eq. 2.26), with the dissipated orbital energy being converted into thermal energy. As Io orbits exterior to the semimajor axis at which its orbital period would be synchronous with Jupiter's spin, the tides raised on Jupiter by Io transfer some of Jupiter's rotational energy to Io's orbit, causing Io to spiral outward. These torques do not directly affect the eccentricity of Io's orbit. However, there exists a 2:1 mean motion resonance lock between Io and Europa (Table E.4 and §2.3.2). Io passes on some of the orbital energy and angular momentum it receives from Jupiter to

Europa, and because $n_{Io} > n_{Europa}$, Io's eccentricity is increased as a consequence of this transfer. This forced eccentricity maintains a high tidal dissipation rate, and consequently there is a large internal heating of Io, which displays itself in the form of active volcanism (§10.2.1).

The rate at which tides convert orbital energy to heat depends on a complicated combination of orbital and physical properties (e.g., rigidity) of the moon. For the Galilean moons, the key factors determining average energy dissipation rates are those that control the amount of energy that goes into the moon's orbital eccentricity. However, for non-equilibrium situations such as those that occurred in the distant past for Neptune's moon Triton, which was subjected to a major infusion of tidal heat after being captured into a highly eccentric orbit about Neptune, the strong dependences of tidal forces on eccentricity and periapse location dominated, and most of the heating occurred quite rapidly.

2.8 Dissipative Forces and the Orbits of Small Bodies

The gravitational interactions between the Sun, planets and moons were described in §§2.1–2.7. In this section, we consider the effects of solar radiation, the solar wind and gas drag. Whereas gravity exerts itself on the entire volume of a body, these other forces act only at the surface. Therefore, these nongravitational forces most significantly affect the orbits of small bodies, which have the largest surface-to-volume ratios.

Solar radiation affects the motions of small bodies in three ways:

(1) **Radiation pressure**, which pushes particles (primarily micrometer-sized dust) outward from the Sun
(2) **Poynting–Robertson drag**, which causes centimeter-sized particles to spiral inward towards the Sun

(3) The **Yarkovsky effect**, which changes the orbits of meter-to ten-kilometer-sized objects because of uneven temperature distributions across their surfaces

The solar wind produces a **corpuscular drag** similar in form to the Poynting–Robertson drag; corpuscular drag is most important for submicrometer particles. We discuss each of these processes in the next four subsections and then examine the effect of gas drag on orbital motion in §2.8.5. Nongravitational forces resulting from asymmetric mass loss by comets are considered in §12.2.4.

2.8.1 Radiation Pressure (Micrometer Grains)

The Sun's radiation exerts a repulsive force, $\mathbf{F}_{rad}$, on all bodies in our Solar System. This force is given by

$$\mathbf{F}_{rad} \approx \frac{\mathcal{L}_\odot A}{4\pi c r_\odot^2} Q_{pr}\hat{\mathbf{r}}, \qquad (2.59)$$

where A is the particle's geometric cross-section, $\mathcal{L}_\odot$ is the solar luminosity, $r_\odot$ is the heliocentric distance, c is the speed of light and Q_{pr} is the dimensionless **radiation pressure coefficient**. The radiation pressure coefficient accounts for both absorption and scattering and is equal to unity for a perfectly absorbing particle. For large particles, Q_{pr} is typically of order unity, but $Q_{pr} \ll 1$ for grains much smaller than the wavelength of the impinging radiation.

The radiation pressure exerted on a body that is large compared with the wavelength of the light impinging on it varies in proportion to its projected area, πR^2, but the gravitational force is proportional to the body's mass, $4\pi\rho R^3/3$, so the ratio of the two goes as $(\rho R)^{-1}$. This ratio is expressed numerically for grains in heliocentric orbit using the dimensionless parameter β, which is defined as the ratio between the forces caused by the radiation pressure and the Sun's gravity:

$$\beta \equiv \left|\frac{F_{rad}}{F_g}\right| = 5.7 \times 10^{-4}\frac{Q_{pr}}{\rho R}, \qquad (2.60)$$

with the particle's radius, R, in meters and its density, ρ, in kg m^{-3}. Because both radiation pressure and gravity fall off as $r_\odot^{-2}$, β is independent of heliocentric distance. The Sun's effective gravitational attraction is given by

$$F_{g,eff} = \frac{-(1-\beta)GmM_\odot}{r_\odot^2}, \qquad (2.61)$$

that is, particles 'see' a Sun of mass $(1-\beta)M_\odot$. It is clear that small particles with $\beta > 1$ are repelled more strongly by the Sun's radiation than they are attracted by solar gravity and thus quickly escape the Solar System unless they are gravitationally bound to one of the planets. Some 'large' bodies that orbit at the Keplerian velocity shed dust (Fig. 12.1); dust released from large bodies that are on circular orbits about the Sun is ejected from the Solar System if $\beta > 0.5$.

Figure 2.22 shows the relative strengths of radiation pressure and gravity for grains of different size and compositions. Solar radiation pressure is only important for micrometer- and submicrometer-sized particles. Extremely small particles are not strongly affected by radiation pressure because Q_{pr} decreases as the particle radius drops below the (visible wavelength) peak in the solar spectrum.

The importance of solar radiation pressure can, for example, be seen in comets (§12.7.4): Cometary tails always point away from the Sun. The ion tails point near the antisolar direction because the ions are dragged along with the solar wind (§12.7.5), which moves rapidly compared to orbital velocities. The dust tails also lie farther from the Sun than the nucleus but are curved. The dust grains initially have the same Keplerian orbital velocities as the comet nucleus. But as a consequence of radiation pressure, they feel a smaller net attraction to the Sun (eq. 2.61), so they drift slowly outward relative to the nucleus.

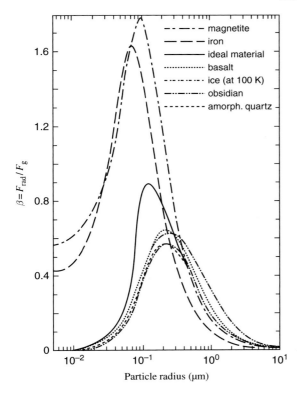

Figure 2.22 The relative radiation pressure force, $\beta = |F_{rad}/F_g|$, as a function of particle size for six cosmically significant substances and a hypothetical ideal material that absorbs all radiation of wavelength $\lambda < 2\pi R$ but is completely transparent to longer wavelengths and has a density $\rho = 3000$ kg m^{-3}. Most solar energy is radiated in the form of photons of wavelength 0.2–4 μm (see Fig. 4.2). For grains much larger than the wavelengths of these photons, the curves are inversely proportional to particle radius, with the constant of proportionality depending on particle reflectivity. Grains interact weakly with photons whose wavelength is much larger than the grain size, so β values decline precipitously for grains smaller than $\sim$0.1 μm. Note that these values are for particles in orbit about the Sun. Grains orbiting stars of different mass, luminosity and/or spectral type would have different values of β. (Adapted from Burns et al. 1979)

2.8.2 Poynting–Robertson Drag (Small Macroscopic Particles)

A particle in orbit around the Sun absorbs solar radiation and reradiates the energy isotropically in its own frame. The particle thereby preferentially

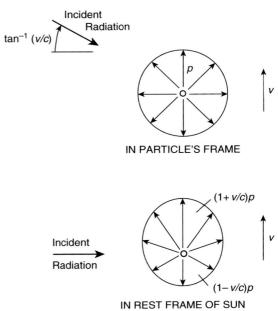

Figure 2.23 A particle in heliocentric orbit that reradiates the solar energy flux isotropically in its own frame of reference, preferentially emits more momentum, p, in the forward direction as seen in the solar frame because the frequencies and momenta of the photons emitted in the forward direction are increased by the particle's motion. (Adapted from Burns et al. 1979)

radiates (and loses momentum) in the forward direction when viewed from the inertial frame of the Sun, as illustrated in Figure 2.23. This leads to a decrease in the particle's energy and angular momentum and causes dust in bound orbits to spiral sunward. This effect is called **Poynting–Robertson drag**.

Radiation pressure removes small dust grains from the interplanetary medium, with (sub)micrometer-sized grains being rapidly blown out of the Solar System (see eq. 2.61 and surrounding discussion). In contrast, Poynting–Robertson drag causes millimeter- and centimeter-sized particles to slowly spiral inward towards the Sun. Poynting–Robertson drag also damps grain eccentricities. Typical decay times (in years) for particles on circular orbits are given by

$$t_{\mathrm{pr}} \approx 400 \frac{r_{AU}^2}{\beta}. \qquad (2.62)$$

The **zodiacal light** is a band centered near the ecliptic plane that appears almost as bright as the Milky Way on a dark night. It is visible in the direction of the Sun just after sunset or before sunrise. Particles that produce the bulk of the zodiacal light (at infrared and visible wavelengths) are between 20 and 200 μm, so their lifetimes at Earth's orbit are of the order of 10^5 years, which is much less than the age of the Solar System. Dust grains responsible for the zodiacal light are resupplied primarily from the asteroid belt, where numerous collisions occur between countless small asteroids, and from comets. Some dust grains released by Kuiper belt objects spiral inwards as a result of Poynting–Robertson drag and provide a small additional contribution to the zodiacal light.

2.8.3 Yarkovsky Effect (1–10⁴-Meter Objects)

Consider a rotating body heated by the Sun. The afternoon/evening hemisphere is typically warmer than the morning hemisphere by an amount $\Delta T \ll T$. Let us assume that the temperature of the morning hemisphere is $T - \frac{\Delta T}{2}$ and that of the afternoon/evening hemisphere is $T + \frac{\Delta T}{2}$. The excess emission on the evening side alters the body's orbit. This process is referred to as the diurnal **Yarkovsky effect**. The diurnal Yarkovsky force is positive (expands orbits) for an object that rotates in the prograde direction, $0° \le \psi < 90°$, with ψ the particle's obliquity, i.e., the angle between its rotation axis and orbit pole. It is negative (causes orbital decay, as does Poynting–Robertson drag) for an object with retrograde rotation, $90° \le \psi \le 180°$. There is also an analogous seasonal Yarkovsky effect that is produced by temperature differences between the spring/summer and the autumn/winter hemispheres.

The Yarkovsky effect significantly alters the orbits of bodies in the meter to 10-km size range. The first direct observational evidence of Yarkovsky forcing was obtained using measurements of the deviations in the orbit of the ∼300 m radius near-Earth asteroid 6489 Golevka from purely gravitational models. Another observed consequence of the Yarkovsky effect is the size-dependent distribution of orbital elements of objects within the Karin cluster of main belt asteroids (MBAs), which was produced by a disruptive collision 6 million years ago. The Yarkovsky effect plays an important role in transporting most meteorite parent bodies from the asteroid belt to Earth by helping to sweep them into resonances.

Asymmetric outgassing of comets produces a nongravitational force similar to the Yarkovsky force. Nongravitational forces on comets are discussed in §12.2.4.

2.8.4 Corpuscular Drag (Submicrometer Dust)

Particles with sizes much smaller than one micrometer are also subjected to a significant corpuscular 'drag' by solar wind particles. This effect can be calculated in a manner similar to the Poynting–Robertson drag. **Corpuscular drag** is more important than radiation drag for particles $\lesssim$ 0.1 μm in size, which couple poorly to solar radiation, and it is the primary force behind these very small dust grains' inward spiral towards the Sun. Figure 2.24 shows a graph of the relative magnitudes of corpuscular drag and Poynting–Robertson drag, β_{cp}, as a function of particle radius.

2.8.5 Gas Drag

Although for most purposes interplanetary space can be considered to be a vacuum, there are certain situations in which interactions with gas can significantly alter the motion of solid particles. Two prominent examples of this process are planetesimal interactions with the gaseous component of the protoplanetary disk during the formation of the

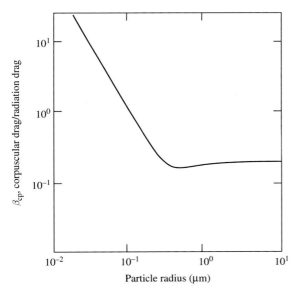

Figure 2.24 Ratio of corpuscular drag (caused by the solar wind) to that resulting from solar radiation, β_{cp}, is plotted against radius for grains composed of obsidian. (Adapted from Burns et al. 1979)

Solar System and orbital decay of ring particles as a result of drag caused by extended planetary atmospheres.

In the laboratory, gas drag slows solid objects down until their positions remain fixed relative to the gas. In the planetary dynamics case, the situation is more complicated. For example, a body on a circular orbit about a planet loses mechanical energy as a result of drag with a static atmosphere, but this energy loss leads to a decrease in semimajor axis of the orbit, which implies the body actually speeds up! Other, more intuitive, effects of gas drag are the damping of eccentricities and, when there is a preferred plane in which the gas density is the greatest, the damping of inclinations relative to this plane.

Objects whose dimensions are larger than the mean free path of the gas molecules experience **aerodynamic drag**:

$$F_D = -\frac{C_D A \rho_g v^2}{2}, \tag{2.63a}$$

where v is the velocity of the body with respect to the gas; ρ_g is the gas density; A is the projected surface area of the body; and C_D is a dimensionless drag coefficient, which is of order unity unless the **Reynolds number**, which is a measure of the ratio of inertial forces to viscous forces, is very small. Smaller bodies are subject to **Epstein drag**:

$$F_D = -A\rho_g v v_o, \tag{2.63b}$$

where v_o is the mean thermal velocity of the gas. Note that as the drag force is proportional to surface area and the gravitational force is proportional to volume (for constant particle density), gas drag is usually most important for the dynamics of small bodies.

The gaseous component of the protoplanetary disk was partially supported against the gravity of the Sun by a negative pressure gradient in the radial direction. Thus, less centrifugal force was required to maintain equilibrium, and consequently the gas orbited less rapidly than the Keplerian velocity. The 'effective gravity' felt by the gas was

$$g_{eff} = -\frac{GM_\odot}{r_\odot^2} - \frac{1}{\rho_g}\frac{dP}{dr_\odot}. \tag{2.64}$$

For circular orbits, the effective gravity must be balanced by centrifugal acceleration, $r_\odot n^2$. For estimated protoplanetary disk parameters, the gas rotated $\sim 0.5\%$ slower than the Keplerian speed. The implications of gas drag for the accretion of planetesimals are discussed in §15.4.1.

Drag induced by a planetary atmosphere is substantially more effective for a given gas density than is drag within a disk that is primarily centrifugally supported. Because atmospheres are almost entirely pressure supported, the relative velocity between the gas and orbiting particles is large. As atmospheric densities drop rapidly with height, particle orbits decay slowly at first, but as they reach lower altitudes, their decay can become very rapid. Gas drag is the principal cause of orbital decay of artificial satellites in low Earth orbit.

2.9 Orbits about a Mass-Losing Star

At the present epoch, the Sun's mass is decreasing as it expels a bit more than 10^{-14} $M_\odot$ per year in matter through the solar wind, emits about five times as much via photon luminosity (Problem 2-20a) and loses a relatively small amount through emission of neutrinos. Post-main-sequence stars can shed substantially more mass through enormous stellar winds during their distended red giant phases as well as via supernova explosions. Very young stars both accrete and eject substantial quantities of matter. The dynamical consequences of interactions of orbiting bodies with the photons and massive particles escaping from the Sun are discussed in §2.8. In this section, we consider the direct effects of stellar mass loss on planetary orbits.

The response of planetary orbits to stellar mass loss depends qualitatively on the timescale over which the mass loss occurs. If the mass is lost (passes beyond the orbit of the planet) in a time short compared with the planet's orbital period, then the planet's instantaneous position and velocity are unchanged, but the diminished mass of the star affects the subsequent motion of the planet. Provided this loss occurs symmetrically, the velocity of the star itself does not change. Then the situation is dynamically analogous to the release of a small dust grain from a larger body on a Keplerian orbit, with the fractional decrease in the star's mass being the parameter analogous to β in equation (2.61). For example, an initially circular orbit becomes unbound if the star suddenly loses more than half of its mass. A smaller amount of 'instantaneous' stellar mass loss results in eccentric bound orbits.

If the star loses mass on a timescale long compared with planetary orbital periods, as does the Sun, then planetary orbits expand gradually. No torque is exerted on the planet, so its orbital angular momentum (eq. 2.26) is conserved. The shape (eccentricity) of the planet's orbit also remains unchanged, so its semimajor axis increases according to the formula:

$$\frac{\dot{a}}{a} = -\frac{\dot{M}_\star}{M_\star}. \tag{2.65}$$

Key Concepts

- Kepler's three 'laws' of planetary motion state:
  (i)   Each planet travels along an elliptical path with the Sun at one focus.
  (ii) An imaginary line from the Sun to a given planet sweeps out area at a constant rate.
  (iii) The square of the orbital period of a planet is proportional to the cube of the long axis of its elliptical trajectory.

- Generalized versions of Kepler's laws can be derived from Newton's laws of motion and of gravity. Newton's laws are applicable on Earth as well as in the heavens.

- A Keplerian orbit of two objects with known masses is uniquely specified by six quantities. These quantities can be the relative positions and velocities of the bodies in space at a given time, but are more conveniently expressed in terms of orbital elements that do not change as rapidly with time.

- Gravitational potential energy is a negative quantity; kinetic energy is positive. Orbits with positive total (kinetic + potential) energy remain bound. At a given separation, the velocity required for escape to infinity equals $\sqrt{2}$ times that for a circular orbit.

- The three-body problem can only be solved analytically for certain special cases. When one of the three bodies is massless and the others travel on a circular orbit, there are five locations at which the massless body can remain stationary relative to the two massive bodies. All five of these Lagrangian points lie in the orbital plane of the massive bodies. Whereas the three Lagrange points that lie along the line connecting the two bodies are unstable, the two that make equilateral triangles with the massive bodies are stable.

- A planet dominates the motion of objects within its Hill sphere, whose size is proportional to one-third the power of the planet's mass.
- Planets perturb the orbits of one another. Strong perturbations resulting from resonances and from close approaches can cause trajectories to be chaotic.
- Resonant perturbations from Jupiter have cleared the Kirkwood gaps within the asteroid belt.
- The orbits of the planets in our Solar System are chaotic, but major orbital changes are unlikely to occur on a 10^{10}-year timescale.

- Orbits about nonspherical bodies are not Keplerian ellipses.
- The differential gravitational pull of the Moon across the Earth stretches out our planet along the line from the Earth to the Moon, producing tidal bulges near both the sub-Moon and anti-Moon points. Analogous elongation occurs between any two deformable bodies.
- Tidal torque slows the rotation of the Earth and increases the distance between the Earth and Moon, so both the day and the month are getting longer.
- Orbits of small bodies can be significantly altered by electromagnetic forces such as radiation pressure, Poynting–Robertson drag and the Yarkovsky effect.
- When a star loses mass, the orbits of its planets expand.

Further Reading

A good introductory text:

Danby, J.M.A., 1988. *Fundamentals of Celestial Mechanics*, 2nd Edition. Willmann-Bell, Richmond, VA. 467pp.

An excellent overview of many important aspects of planetary dynamics is presented by:

Murray, C., and S. Dermott, 1999. *Solar System Dynamics*. Cambridge University Press, Cambridge. 592pp.

Legendre expansions and spherical harmonics are covered in:

Jackson, J.D., 1999. *Classical Electrodynamics*, 3rd Edition. John Wiley and Sons, New York. 641pp.

A detailed discussion of the effects of solar radiation and the solar wind on the motion of small particles is given by:

Burns, J.A., P.L. Lamy, and S. Soter, 1979. Radiation forces on small particles in the Solar System. *Icarus*, **40**, 1–48.

Other useful individual articles include:

Duncan, M.J., and T. Quinn, 1993. The long-term dynamical evolution of the Solar System. *Annu. Rev. Astron. Astrophys.*, **31**, 265–295.

Peale, S.J., 1976. Orbital resonances in the Solar System. *Annu. Rev. Astron. Astrophys.*, **14**, 215–246.

Problems

2-1. A baseball pitcher can throw a fastball at a speed of ∼150 km/hr. What is the largest size spherical asteroid of density $\rho = 3000$ kg m^{-3} from which he can throw the ball fast enough that it:

(a) escapes from the asteroid into heliocentric orbit?

(b) rises to a height of 50 km?

(c) goes into a stable orbit about the asteroid?

2-2.* **(a)** Calculate the minimum velocity at which a rock must leave Mars's atmosphere for it to be on a trajectory that intersects Earth's orbit. You should take into account the velocity needed to escape from the martian gravity but not drag within the planet's atmosphere; you may also neglect planetary rotation and assume that the planets' orbits are circular and coplanar.

(b) At what speed does this rock impact the (upper atmosphere of) Earth?

(c) Under the same assumptions, calculate the minimum velocity for transfer from the top of Earth's atmosphere to Mars.

(d) At what speed does this rock impact the (upper atmosphere of) Mars?

(e) Discuss how planetary rotation and orbital eccentricities and inclinations affect your results.

2-3. Five of the spacecraft that flew past Jupiter and up to three other giant planets are escaping from the Solar System on unbound trajectories with velocities of several AU/year. Design the trajectory of a spacecraft to reach great distances from the Sun as quickly as possible given the limits of a specified chemical rocket.

(a) Suppose that your rocket could bring you into a circular low Earth orbit (200-km elevation) and subsequently provide an additional velocity change $\Delta v = X$ km/s in one quick burst. Calculate the maximum escape velocity from the Solar System on a direct trajectory (in AU/year). You may neglect the nonsphericity of Earth and the eccentricity of its orbit from the Sun.

(b) Repeat your calculation but allow the spacecraft to pass by Jupiter for a gravity assist similar to those given the spacecraft that have encountered the planet in the past. Assume that you have tiny rockets to correct trajectory errors, but your main rocket can only be used once, to escape Earth orbit.

(c) Repeat your calculation, this time with a rocket that can burn two times, once to escape Earth orbit and the other time near Jupiter, with a total $\Delta v = X$ km/s.

(d) How could you increase the escape speed by using a rocket that can burn multiple times with the same total $\Delta v = X$ km/s? Think of creative trajectories!

2-4. Einstein's **general theory of relativity** is conceptually quite different from Newton's theory of gravity, but the predictions of general relativity reduce to those of Newton's model in the low-velocity (relative to the speed of light, c), weak gravitational field (relative to that required for a body to collapse into a black hole) limit.

(a) Calculate v^2/c^2 for the following bodies:

(i) Mercury using its circular orbit velocity

(ii) Mercury using its velocity at perihelion

(iii) Earth using its circular orbit velocity

(iv) Neptune

(v) Io using its velocity relative to Jupiter

(vi) Metis using its velocity relative to Jupiter

(vii) A Sun-grazing comet (§12.3.5) on a parabolic orbit just above the Sun's photosphere

(b) The **Schwarzchild radius** of a body,

$$R_{\text{Sch}} = \frac{2Gm}{c^2}, \qquad (2.66)$$

is the radius inside of which light cannot escape from the body. Calculate the Schwarzchild radii of the following Solar System bodies and the ratios of these radii to the sizes of the bodies in question and to the semimajor axes of their nearest (natural) satellites:

(i) Sun

(ii) Earth

(iii) Jupiter

2-5. As demonstrated in the previous problem, relativistic corrections to Newton's theory of gravity are quite small for most Solar System situations. The most easily observable effect of general relativity is the precession of orbits because it is nil in the Newtonian two-body approximation. The first-order (weak field) general relativistic corrections to Newtonian gravity imply a precession of the periapse of orbit of a small body ($m_2 \ll m_1$) about a primary of mass m_1 at the rate

$$\dot{\varpi} = \frac{3(Gm_1)^{3/2}}{a^{5/2}(1 - e^2)c^2}. \qquad (2.67)$$

Calculate the general relativistic precession of the periapse of the following objects:

(a) Mercury

(b) Earth

(c) Io (in its orbit about Jupiter)

Quote your answers in arcseconds per year. Note: The average observed precession of Mercury's periapse is $56.00'' \text{ yr}^{-1}$, all but $5.74'' \text{ yr}^{-1}$ of which is caused by the observations not being done in an inertial frame far from the Sun. Newtonian gravity of the other planets accounts for $5.315'' \text{ yr}^{-1}$. Your answer should be close to the difference between these two values, $0.425'' \text{ yr}^{-1}$. The small difference between this rate and the value calculated by the procedure outlined above is within the uncertainties of the observations and calculations and may also be affected by a very small contribution resulting from solar oblateness (§2.6.2).

2-6. Show that if the two-body problem is analyzed using units in which the gravitational constant, the separation between the bodies and the sum of their masses are all set equal to one, then the orbital period of the two bodies about their mutual center of mass equals 2π.

2-7. A small planet travels about a star with an orbital semimajor axis equal to 1 in the units chosen. Calculate the locations of the 2:1, 3:2, 99:98 and 100:99 resonances of test particles with the planet.

2-8.* One of the greatest triumphs of dynamical astronomy was the (nineteenth-century) prediction of the existence and location of the planet Neptune on the basis of irregularities observed in the orbit of Uranus. The motion of Uranus could not be accurately accounted for using only (the Newtonian modification of) Kepler's laws and perturbations of the then-known planets. Estimate the maximum displacement in the position of Uranus caused by the gravitational effects of Neptune as Uranus catches up and passes this slowly moving planet. Quote your results both in kilometers along Uranus's orbital path and in seconds of arc against the sky as observed from Earth. For your calculations, you may neglect the effects of the other planets (which can be and were accurately estimated and factored into the solution) and assume that the unperturbed orbits of Uranus and Neptune are circular and coplanar. Note that although the displacements of Uranus in radius and longitude are comparable, the longitudinal displacement produces a much larger observable signature because of geometric factors.

(a) Obtain a very crude result by assuming that the potential energy released as Uranus gets closer to Neptune increases Uranus's semimajor axis and thus slows down Uranus. You may assume that Uranus's average semimajor axis during the interval under consideration is halfway between its semimajor axis at the beginning and the end of the interval and neglect the influence of Uranus on Neptune. Remember to use the synodic (relative) period of the pair of planets rather than just Uranus's orbital period.

(b) Obtain an accurate result using numerical integration of Newton's equations on a computer.

2-9. **(a)** Explain how the Lyapunov characteristic exponent, γ_c, is used to distinguish between regular and chaotic trajectories.

(b) What is the value of γ_c for regular trajectories?

2-10. **(a)** Compute the moment of inertia of a planet of mass M_p and radius R_p consisting of a core of radius $R_p/2$ that is twice as dense as the surrounding mantle and crust.

(b) Compare your result with the moment of inertia of a uniform density sphere of the same mass and radius and comment.

2-11. Saturn is the most oblate major planet in the Solar System, with gravitational moments $J_2 \approx 1.63 \times 10^{-2}$ and $J_4 \approx -9 \times 10^{-4}$ (Table E.15). Calculate the orbital periods, apse precession rates and node regression rates for particles on nearly circular and equatorial orbits at 1.5 $R_\hbar$ and 3 $R_\hbar$:

(a) Neglecting planetary oblateness entirely

(b) Including J_2, but neglecting higher order moments

(c) Including J_2 and J_4

2-12. If the Earth–Moon distance was reduced to half its current value then:
(a) Neglecting solar tides, how many times as large as at present would the maximum tide heights on Earth be?
(b) Including solar tides, how many times as large as at present would the maximum tide heights on Earth be?

2-13. **(a)** If a moon were on a synchronous orbit around Mars, what would its semimajor axis be? Express your answer in martian radii (R_σ) from the center of the planet.
(b) Compare this distance with the orbits of Phobos (at 2.76 R_σ) and Deimos (at 6.9 R_σ).
(c) Describe the motion of each of these three moons (two real, one hypothetical) on the sky as seen from Mars.

2-14. Calculate the mean **synodic period** (time it takes for a relative configuration to repeat) between Earth's rotation and the Moon's orbit. Note that this is equal to the average interval between successive moonrises.

2-15. Compute the ratio of the height of tides raised on Earth by the Moon to those raised by the Sun.

2-16. If the Moon's mass was reduced to half its current value, then:
(a) Neglecting solar tides, maximum tide heights on Earth would be how many times as large as at present?
(b) Including solar tides, maximum tide heights on Earth would be how many times as large as at present?

2-17. Calculate the orbital period around the Sun for a dust grain with $\beta = 0.3$ and semimajor axis $a = 1$ AU.

2-18. A grain with $\beta = \beta_0$ travels on a circular orbit about the Sun. The grain splits apart into smaller grains with $\beta = \beta_n$. What are the eccentricities and semimajor axes of these new grains?

2-19. A giant parasol at the L_1 inner Lagrangian point of the Sun–Earth system has been proposed to block some of the Sun's photons heading for Earth and thus counter anthropogenic greenhouse warming.
(a) Calculate the distance between Earth and L_1. You may ignore the eccentricity of Earth's orbit and neglect the perturbations of the Moon and the other planets.
(b) To minimize the weight of material launched from Earth or transported from elsewhere to L_1, this parasol would have a large surface area to mass ratio. Thus, radiation pressure could affect its orbit significantly. Would the L_1 equilibrium point of such a significantly non-zero β parasol be closer to or farther from Earth's surface than is the purely gravitational L_1 point? Explain your reasoning.

n.b. Such a parasol could only be a partial solution to greenhouse warming. By reducing energy coming in, it could counteract the first-order planet-averaged warming, but the lower inward and outward flows would have different latitudinal dependences. Thus, the poles would warm up, causing melting of the ice caps, and the atmospheric circulation might change greatly. Additionally, the amount of useful (e.g., for photosynthesis) solar energy arriving at Earth would be reduced.

2-20. **(a)** Calculate the rate at which the Sun is losing mass as a result of radiating photons by using Einstein's famous formula,

$$E = mc^2. \tag{2.68}$$

(b) Estimate the Sun's mass loss integrated over the past 4 billion years. You may assume that the solar luminosity and solar wind have remained constant and that the Sun's neutrino luminosity is negligible.

2-21. Derive equation (2.65) using conservation of angular momentum.

CHAPTER 3

Physics and Astrophysics

Thermodynamics is a funny subject. The first time you go through it, you don't understand it at all. The second time you go through it, you think you understand it, except for one or two small points. The third time you go through it, you know you don't understand it, but by that time you are so used to it, it doesn't bother you any more.

Arnold Sommerfeld, Physicist, 1868–1951

The focus of this textbook is on planets and smaller objects, but some background in physics and astrophysics beyond what is common to most undergraduate science majors is required to understand the structure of planets and planetary systems, how they formed and planetary habitability. We discuss planetary dynamics in Chapter 2 and address energy transport in Chapter 4. In this chapter, we provide other important background and concepts that are often not covered in introductory physics courses.

A planet is composed of $\sim 10^{50}$ molecules. This number is so large that it is impossible to model planets on a molecule by molecule basis. Rather, properties of matter are analyzed in a statistical sense to reveal the bulk behavior of matter in planetary bodies. Many of these properties are described by thermodynamics, so we begin with an introduction to those aspects of thermodynamics relevant to material presented in this book in §3.1.

The radial structure of planets and stars is determined primarily by a balance between the downwards force of gravity and the resistance to collapse provided by the pressure gradient. This balance plays a major role in determining the size and density of a body, whether it will remain bound as a single object, and internal energy transport. We describe this analysis in §3.2.

The Sun is the dominant body in our Solar System. It provides us with warmth and holds the system together with its gravity. Without it, there would be no life on Earth's surface. The properties of other stars are key to habitability of planets orbiting about them. Section 3.3 presents a brief introduction to stellar physics to inform students who have not taken a college-level astronomy course of those stellar properties that have implications for planets and habitability.

The properties of a planetary system depend on the mix of matter of which it is composed. Most elements were formed in the early universe just minutes after the big bang or in the interiors of stars; radioactive decay and various processes in interstellar space also affect the balance of elements present in our Solar System. We summarize key aspects of the formation of elements, as well as the relative abundances of the elements, in §3.4.

3.1 Thermodynamics

Thermodynamics is the study of heat and of transfers of energy between systems and between different states of matter. Systems are viewed in a macroscopic rather than microscopic sense, i.e., mean properties of ensembles are analyzed rather than motions of constituent molecules. For example, discontinuous phase transitions in a material brought about by melting, boiling, sublimation or condensation are familiar examples wherein **latent heat of transformation** is absorbed or liberated during the change of a material's phase. Thermodynamics is important in almost every branch of science, including atmospheric physics, rock formation and biology.

3.1.1 Laws of Thermodynamics

In **classical thermodynamics**, one deals with **closed systems**, i.e., systems without mass transfer to and from the external environment. Energy transfer, however, can take place across the boundary of a closed system. In contrast, both mass and energy are freely exchanged between an **open system** and its environment. In classical thermodynamics, systems (which are closed) are in equilibrium; this is known as the **zeroth law of thermodynamics**. If two systems with different temperatures are combined into one closed system, over time they will reach the same temperature and be in thermodynamic equilibrium. A system does not necessarily have to be homogeneous to be in equilibrium. A mix of ice and water at a pressure of 1 bar and temperature of 0°C is in equilibrium because the relative amounts of ice and water do not change unless energy is added to or removed from the system.

The **thermodynamic state** of a system refers to the system's physical properties (e.g., density and dielectric constant) in equilibrium. This state can be fully characterized by a few variables, usually temperature and pressure, although for more complex systems additional variables may be needed. In the water/ice system referred to above, the state is completely defined by the temperature and pressure of the system; the proportion of the two phases does not affect the system's thermodynamic state.

The **first law of thermodynamics** is that energy is conserved:

$$dQ = dU + P\,dV. \tag{3.1}$$

In equation (3.1), dQ is the amount of heat absorbed by the system from its surroundings, dU the change in internal energy, P the pressure and dV the change in volume V. The last term in equation (3.1), $P\,dV$, is the work done by the system on its environment via an expansion of the system. Many types of internal energy exist, including gravitational (potential) energy, kinetic energy (motion of atoms, molecules), chemical energy (e.g., in atomic/molecular bonds), electrical energy (moving charges) and radiant energy (photons). The first law of thermodynamics in essence constrains trade-offs between various kinds of energy.

Isolated systems that are not in equilibrium redistribute energy while evolving towards equilibrium; this redistribution leads to an increased 'disorder'. **Entropy**, S, is a quantitative measurement of a system's disorder. Unlike energy, entropy is not conserved. Its behavior is governed by the **second law of thermodynamics**, which states that for any process, the overall change in entropy is always positive:

$$\Delta S > 0. \tag{3.2}$$

For example, two gases in a closed system will spontaneously mix; if we add some dye to water, the dye will spread. In both cases, the entropy (disorder, or degree of mixing) increases. If the system is open, i.e., in contact with its surroundings, then the entropy of the system plus its environment must increase over time. The second law of thermodynamics thus requires that the net entropy of the Universe increases.

The **third law of thermodynamics** states that all processes virtually cease at absolute zero temperature, i.e., at 0 K. The entropy of a system at absolute zero is a well-defined (and usually small) constant (§3.1.3). This third law, just like the first and second laws, remains a postulate, i.e., these laws are derived **empirically** from experiments. Their plausibilities can be demonstrated by using statistical and/or quantum mechanical arguments.

3.1.2 Enthalpy

The **enthalpy**, H, is defined as the sum of the internal energy, U (potential energy stored in the interatomic bonding plus kinetic energy of the atomic vibrations) and the work done on the system, PV, with P the pressure and V the volume:

$$H \equiv U + PV. \tag{3.3}$$

In physical terms, we can think of enthalpy being simply heat exchanged at constant pressure:

$$dH = (dQ)_P. \tag{3.4}$$

Although the enthalpy of pure elements is defined to be zero, the enthalpy of changing the phase of a material or forming a mineral from its individual elements is not zero: energy is either required or released. The reaction is referred to as **endothermic** if energy is used, in which case $dH > 0$. If the reaction frees up energy, the reaction is **exothermic** and $dH < 0$.

The **formation enthalpy**, dH, is related to the **thermal heat capacity** of the system, C_P, defined as the amount of heat needed to raise the temperature of 1 **mole** of material by 1 K while keeping the pressure constant:

$$C_P \equiv \left(\frac{dH}{dT}\right)_P = \left(\frac{dQ}{dT}\right)_P, \qquad (3.5)$$

where T is the temperature. One mole contains $N_A \equiv 6.022 \times 10^{23}$ molecules, with N_A being **Avogadro's number**. The enthalpy of the system at temperature T_1 is thus equal to

$$H = H_0 + \int_0^{T_1} C_P dT, \qquad (3.6)$$

with H_0 the enthalpy at $T = 0$ K. The change in enthalpy,

$$\Delta H = \int_0^{T_1} C_P dT, \qquad (3.7)$$

is the heat required to raise the temperature of 1 mole of material to T_1 without changing the pressure. While heating the material, it may undergo a phase transition. For example, when heating 1 mole of liquid water at a constant pressure of 1 bar, it starts to boil and vaporize at 100°C. The temperature of the water does not change, and the heat is used to vaporize the liquid; the (formation) enthalpy, ΔH, is referred to as the **latent heat of vaporization**. Similarly, when melting ice, the heat is used for melting, and the temperature of the ice does not change (i.e., at $P = 1$ bar, the temperature of the ice stays at $T = 0°$C). The reverse processes generate heat (exothermic reactions).

3.1.3 Entropy

From a macroscopic point of view, the change in entropy is equal to the ratio of the amount of heat absorbed by the system, ΔQ, and the temperature, T:

$$\Delta S = \frac{\Delta Q}{T}. \qquad (3.8)$$

The entropy of a system at temperature T_1 is then given by

$$S = S_0 + \int_0^{T_1} \frac{C_P}{T} dT, \qquad (3.9)$$

with S_0 the entropy at $T = 0$ K, which in most cases is small.

From a microscopic point of view, **Boltzmann's postulate** relates entropy to the number of **equivalent microstates**, Ξ, which can be derived with the help of statistical mechanics:

$$S = k \ln \Xi, \qquad (3.10)$$

where k is Boltzmann's constant ($k = R_{gas}/N_A$), with R_{gas} the universal gas constant ($R_{gas} = N_A k$, with N_A Avogadro's number). All microstates that give rise to the same macrostate are defined as being equivalent.

In mineral physics, entropy measures the change in a mineral's state of order when it changes from one phase or structure to another. For a perfect crystal with $S_0 = 0$, all atoms are in the ground state. Upon heating the sample, disorder (and hence entropy) increases. If the heating is thermodynamically reversible, then upon cooling the sample back down to its original temperature, one might expect the total change in entropy to be zero. However, this would violate the second law of thermodynamics. Hence, if a mineral becomes more ordered in a transformation process (i.e., $dS < 0$), then the heat liberated in the process must increase disorder in the environment, so that in total: $dS > 0$.

3.1.4 Gibbs Free Energy

Gibbs free energy, G, is a thermodynamic potential that depends on the temperature and pressure of the system. The Gibbs free energy of a system is defined as:

$$G \equiv H - TS \qquad (3.11a)$$

and the change in Gibbs free energy, ΔG, as:

$$\Delta G \equiv \Delta H - T\Delta S. \qquad (3.11b)$$

Gibbs free energy has a minimum value for a system in equilibrium. Minimization of G is therefore often used to study processes such as phase

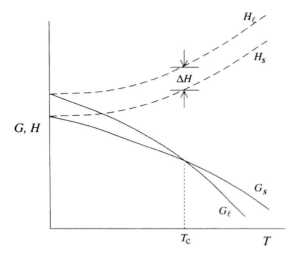

Figure 3.1 A graph of the variation in the Gibbs free energy, G, and the enthalpy, H, of the liquid (ℓ) and solid (s) phase of a solution as a function of temperature. The phase with the lowest free energy is the stable phase. Whereas at temperatures $T > T_c$, the mixture is in a liquid phase, at $T < T_c$ the solution is solid. (Adapted from Putnis 1992)

changes and chemical equilibrium. It is also *the* tool to assess the suitability or potential for life in an environment because life requires 'usable' energy to survive.

3.1.5 Material Properties: Phase Changes

Gibbs free energy, or more precisely the change therein, drives the evolution of a system, regardless of its constituency, living or not living. It is a particularly useful tool to examine near-isothermal systems at constant pressure because it predicts spontaneous changes, such as phase transitions in materials. Figure 3.1 shows a graph of the Gibbs free energy as a function of temperature for the liquid, ℓ, and solid, s, phases of a melt. The phase with the lowest free energy is the stable phase. At the critical temperature, T_c, the curves cross, and upon further heating or cooling of the melt, a phase transformation takes place. The phase transition brings about a change in the enthalpy, ΔH, which is the **latent heat of transformation**:

$$\Delta H = T\Delta S. \tag{3.12}$$

We are all familiar with this concept in daily life when thinking of water: melting, freezing, evaporization and sublimation. When you are hot and wet, you cool off because the water on your skin evaporates using your body heat. When clouds form, the heat generated in the process warms the environment (e.g., §5.3).

3.2 Barometric Law and Hydrostatic Equilibrium

The large-scale structure of a planetary body or star is governed by a balance between gravity and pressure; this balance is referred to as **hydrostatic equilibrium**. Figure 3.2 illustrates the pressure difference across a 'slab' of material of thickness Δz and density ρ. The z-coordinate is taken to be positive going outwards (decreasing pressure). This slab exerts a force because of its weight on the slabs below it. Per unit area, this force becomes a pressure. So the increase in pressure across the slab, ΔP, is simply how much a column of height Δz and density ρ weighs:

$$\Delta P = -g_p \rho \Delta z, \tag{3.13a}$$

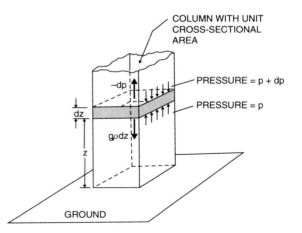

Figure 3.2 Sketch to help visualize the concept of hydrostatic equilibrium.

where g_p represents the gravitational acceleration. The equation of hydrostatic equilibrium in differential form is

$$\frac{dP}{dz} = -g_p(z)\rho(z). \qquad (3.13b)$$

An **equation of state** describes the relationship between two or more state variables such as pressure, density and temperature of a substance. At low to moderate pressures in planetary atmospheres and stars, the pressure is well approximated by the **ideal gas law** (alternatively referred to as the **perfect gas law**). The ideal gas law is written as:

$$P = NkT = \frac{\rho R_{gas} T}{\mu_a} = \frac{\rho kT}{\mu_a m_{amu}}, \qquad (3.14)$$

where N is the particle number density (m^{-3}); μ_a the mean molecular mass (in atomic mass units); and $m_{amu} \approx 1.67 \times 10^{-27}$ kg the mass of an atomic mass unit, which is slightly smaller than the mass of a hydrogen atom.

Using the ideal gas law (eq. 3.14) to eliminate ρ from equation (3.13b) results in:

$$\frac{dP}{dz} = -g_p(z)P\frac{\mu_a m_{amu}}{kT}. \qquad (3.15)$$

We define the **scale height**, $H(z)$, by

$$H(z) \equiv \frac{kT(z)}{g_p(z)\mu_a(z)m_{amu}}, \qquad (3.16)$$

where $g_p(z)$ is the acceleration due to gravity at altitude z and $\mu_a m_{amu}$ is the molecular mass.

In the following, we assume T, g_p and μ_a to be independent of z. We can now rearrange equation (3.15) to yield:

$$\frac{dP}{P} = -\frac{dz}{H}. \qquad (3.17)$$

Integrating equation (3.17) results in the **barometric law**:

$$P(z) = P(0)e^{-z/H}, \qquad (3.18)$$

with P_0 the pressure at the surface (or zero-reference level). Thus for a constant pressure scale

height, H is equal to the distance over which the pressure decreases by a factor e. Small values of H imply a rapid decrease of atmospheric pressure with altitude. Figure 3.3 illustrates the pressure and density profiles for constant H in an incompressible medium (panel a) and in an ideal gas (panel b). The scale height, however, usually varies with altitude. If g_p, T and H do vary with altitude, the barometric law can be written as:

$$P(z) = P(0)e^{-\int_0^z dr/H(r)}. \qquad (3.19)$$

Let us return to the general equation of hydrostatic equilibrium, as provided in equation (3.13b). The example given above is applicable only for atmospheres because we adopted the ideal gas law to relate temperature, pressure and density. In its more general form, the equation of hydrostatic equilibrium can also be applied to the internal structure of a planet because this structure is also determined by a balance between gravity and pressure:

$$P(r) = \int_r^R g_p(r')\rho(r')dr'. \qquad (3.20)$$

Equation (3.20) can be used to calculate the pressure throughout the body provided $\rho(r)$ is known. If the density is constant throughout a planet's interior, then the pressure at its center, P_c, is given by (Problem 3-3b):

$$P_c = \frac{3GM^2}{8\pi R^4}. \qquad (3.21)$$

Equation (3.21) provides a lower limit to the central pressure because the density usually decreases with distance r. This method yields good estimates for relatively small bodies with a nearly uniform density, such as the Moon, which has a central pressure of (only) 45 kbar.

An alternative quick estimate of P_c can be obtained by assuming the body consists of one slab of material, in which case the central pressure is a factor of two larger than the value obtained in the previous estimate (Problem 3-3a). Because

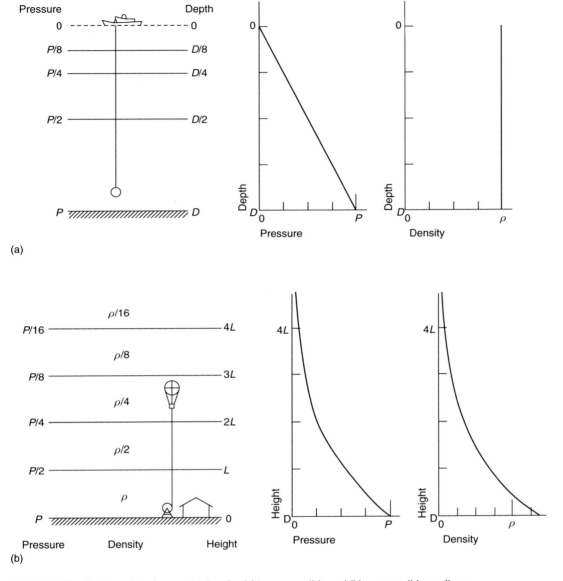

Figure 3.3 Visualization of the barometric law for (a) incompressible and (b) compressible medium.

the single-slab technique overestimates the gravity over most of the region of integration, the actual pressure at the center of a rocky planet usually lies between these two values. On the other hand, if the body is highly compressible, as are gaseous planets and stars, it may be extremely centrally condensed. For such bodies, the density increases sharply towards the center of the planet, and the pressure calculated using the single-slab model may still be too low compared with the actual value. We find that the central pressure of Earth calculated according to the single-slab model agrees quite well with the actual value of 3.6 Mbar (Problem 3-3). The Earth is differentiated (not homogeneous, see §6.1.4), and the increase in density towards the center just about compensates for our overestimate in gravity. Using the single-slab model, Jupiter's central pressure is still underestimated by a factor of ~4 because this planet is very dense near its center.

3.3 Stellar Properties and Lifetimes

Planets are intimately related to stars, their larger and much more luminous companions. Stellar gravity dominates planetary motions (see Chapter 2). The luminosity of our star, the Sun, is the primary source of energy for most planets (see Chapter 4), and solar energy inputs dominate planetary weather (see Chapter 5). The solar wind affects and controls planetary magnetospheres (see Chapter 7), and solar heating is responsible for cometary activity (see Chapter 12). Moreover, past generations of stars produced most of the elements out of which terrestrial planets are composed (§3.4.2), and stars and planets form together (§1.5 and Chapter 15). Thus, understanding a few basic properties of stars is quite useful for learning about planets.

3.3.1 Virial Theorem

The **virial theorem** states that the time-averaged potential energy of a bound self-gravitating system is twice as large as the negative of the time-average of the kinetic energy of the system,

$$\langle E_G \rangle = -2\langle E_K \rangle, \tag{3.22}$$

where the brackets denote the time-average of the variable that they surround.

The general proof of the virial theorem is quite involved, but it is straightforward to demonstrate that equation (3.22) holds for a particle on a circular orbit about a much more massive body. The proof goes as follows: A particle at rest an infinite distance from the massive body has neither kinetic nor potential energy, so a particle with just enough kinetic energy to escape to infinity has no total energy, i.e., $E_G + E_K = 0$. From equation (2.24), the kinetic energy of a circular orbit is equal to half of the kinetic energy of a particle at the same location that is moving at escape velocity, completing the proof.

The virial theorem provides a powerful tool for the analysis of various astrophysical problems. However, it is not applicable if significant amounts of energy in the system reside in forms apart from kinetic energy and gravitational potential energy. For example, it cannot be used for planetary interiors, where molecular potential energy is important.

3.3.2 Luminosity

Stars are huge balls of gas and **plasma** (ionized gas) that radiate energy from their surfaces and liberate energy via thermonuclear fusion reactions in their interiors. The internal structure of a star is determined primarily by a balance between gravity and pressure. Fusion reaction rates are extremely sensitive to temperature (§3.4.2), so a very small warming of the interior enables the star to release nuclear energy much more rapidly. A quasi-equilibrium state is maintained because if the interior gets too cool, the star's core contracts and heats up, but if it becomes too hot, pressure builds, and the core expands and cools.

During the star's long-lived **main sequence** phase, hydrogen in its core gradually 'burns up' (fuses into helium) to maintain the pressure balance. High-mass stars are much more luminous than low-mass stars because greater pressure and hence higher temperature are required to balance their larger gravity. Along the main sequence, stellar luminosity, $\mathcal{L}_\star$, is roughly proportional to the fourth power of a star's mass, $M_\star$:

$$\mathcal{L}_\star \propto M_\star^4. \tag{3.23}$$

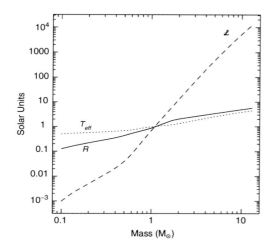

Figure 3.4 Logarithm of the radius (*solid line*), radiating temperature (*dotted line*) and luminosity (*dashed line*) of a solar composition zero-age main sequence star as functions of the star's mass. All quantities are ratioed to those of the Sun at the present epoch. (Courtesy Jason Rowe)

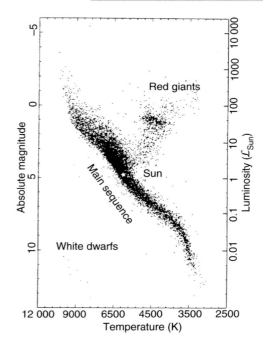

Figure 3.5 Hertzsprung–Russell (H–R) **diagram** for nearby, bright, single stars. (Adapted from Perryman et al. 1995)

Figure 3.4 shows the relationship between stellar mass and luminosity more accurately. As the amount of hydrogen fuel in the core increases approximately linearly with the star's mass, stellar lifetime varies inversely with the cube of the star's mass.

Stars range in mass from $0.08\,M_\odot$ (solar masses) to a little more than $100\,M_\odot$. Smaller objects cannot sustain sufficient fusion in their cores to balance gravitational contraction, but radiation pressure resulting from high luminosity would blow away the outer layers of more massive bodies. Low-mass stars are much more common than high-mass stars, but because high-mass stars are vastly more luminous, they can be seen from much farther away, and the majority of stars visible to the naked eye are more massive than the Sun.

The absolute magnitude of a star is equal to the apparent magnitude (§1.2.5) if it were observed from a distance of 10 parsecs ~ 32.6 light-years. Figure 3.5 shows that although stellar luminosities vary by many orders of magnitude, in other respects stars are a relatively homogeneous class of objects. Stars are supported against gravitational collapse by thermal pressure that is maintained by fusion reactions in their interiors; they range in mass by a little over 3 orders of magnitude. In contrast, even the most conservative definition of planets encompasses a more diverse family of objects within our Solar System alone. These include Mercury, a condensed body composed primarily of iron and other heavy elements, as well as Jupiter, a fluid object that is almost 4 orders of magnitude more massive than Mercury and is made mostly of hydrogen and helium. And some definitions of planets include Pluto and extrasolar objects more than ten times as massive as Jupiter (see Chapter 14), extending the range to over 6 orders of magnitude in mass.

Figure 3.6 shows the track that the Sun makes in the (photospheric) temperature-luminosity plane as it ages. In general, a star's luminosity grows slowly during the star's main sequence phase because fusion increases the mean particle mass in the core

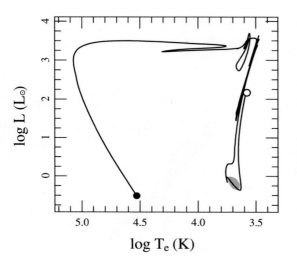

Figure 3.6 Evolutionary track of the Sun. The open circle denotes the beginning of the track and the filled circle the end. The Sun lies in the small highlighted region near the lower right of the figure for most of its life as a star. (Courtesy Marc Pinsonneault)

and greater temperature is required for pressure to balance gravity. When hydrogen in the core is completely used up, the core shrinks. Hydrogen burning occurs in a shell of material surrounding the helium-rich core. The star's luminosity increases significantly, and its outer layers swell and cool, turning the star into a **red giant**. If the star is at least ~60% as massive as our Sun, the core gets hot and dense enough for helium to fuse into carbon and oxygen. A helium-burning equilibrium analogous to the main sequence exists, but it lasts a much shorter time because the star is more luminous (more thermal pressure is needed to balance gravity in the star's denser core) and helium burning liberates far less energy than does hydrogen burning (see Fig. 3.9). Thus, the helium fuel is exhausted far more rapidly than is hydrogen.

In solar mass stars, electron degeneracy pressure prevents the star from attaining temperatures required for fusion to produce elements more massive than carbon and oxygen. **White dwarfs** are remnants of small- and medium-sized stars in which electron degeneracy pressure provides

the primary support against gravitational collapse. But in very massive stars, fusion continues until iron, the most stable nucleus, is produced in the core. No energy can be liberated by nuclear fusion beyond iron, so the core collapses, rapidly releasing an immense amount of gravitational energy. This energy can fuel a **supernova** explosion, freeing some of the heavy elements that the star has produced to be incorporated into subsequent generations of stars and planets and leaving a neutron star or black-hole remnant. Stellar mass objects such as white dwarfs and **neutron stars** that have exhausted their nuclear fuel are properly referred to as **stellar remnants** rather than stars.

Gravitational contraction is a major source of the energy radiated by giant planets and brown dwarfs. These objects shrink and (after some initial warming) cool as they age (Fig. 3.7), so there is not a unique relationship between luminosity and mass.

3.3.3 Size

Nuclear reactions maintain the temperature in the cores of low-mass stars close to $T_{nucl} \approx 3 \times 10^6$ K (§3.4.2) because the fusion rate is roughly proportional to T^{10} near T_{nucl}. The virial theorem (eq. 3.22) can be used to show that the radii of such stars must be roughly proportional to mass. In equilibrium, the thermal energy and the gravitational potential energy are in balance:

$$\frac{GM_\star^2}{R_\star} \sim \frac{M_\star k T_{nucl}}{m_{amu}}. \tag{3.24a}$$

Therefore

$$R_\star \propto M_\star, \tag{3.24b}$$

and the star's mean density

$$\rho_\star \propto M_\star^{-2}. \tag{3.24c}$$

More-massive stars are larger than low-mass stars (eq. 3.24b), but an equally important reason that they are able to radiate much more energy (eq. 3.23) is that they are hotter (bluer) than low-mass stars. The physics of electromagnetic radiation is discussed in §4.1.1.

At low densities, the hydrostatic structure of a star is determined primarily by a balance between gravity and thermal pressure. At sufficiently high densities, another source of pressure becomes significant. Electrons, because they have half-integer spins, must obey the **Pauli exclusion principle** and are accordingly forbidden from occupying identical quantum states. The electrons thus successively fill up the lowest available energy states. Electrons that are forced into higher energy levels contribute to **degeneracy pressure**. The degeneracy pressure scales as $\rho^{5/3}$ and is important when it is comparable in magnitude to or larger than the ideal gas pressure (which scales as ρT; see eq. 3.14). Near T_{nucl}, the degeneracy pressure dominates when densities exceed a few hundred grams per cubic centimeter.

Degeneracy pressure dominates over thermal pressure in cool brown dwarfs, neutron stars and white dwarfs; in these objects, it provides the primary upward force to balance the downward pull of gravity. (Nothing balances gravity in black holes, which are singularities in the space–time continuum.) Bodies supported primarily by degeneracy pressure are referred to as **compact objects**. In compact objects, the virial theorem implies that the energy of the degenerate particles (electrons in the case of brown dwarfs) is comparable to the gravitational potential energy:

$$\rho^{5/3}R^3 \sim \frac{GM^2}{R}. \tag{3.25a}$$

Therefore,

$$R \propto M^{-1/3}, \tag{3.25b}$$

(Problem 3-5), and compact objects shrink if more mass is added to them. The most massive cool brown dwarfs are indeed expected to have slightly smaller radii than their lower mass brethren. Young brown dwarfs can be hot and distended, depending on their age and formation circumstances; see Figure 3.7.

3.3.4 Sizes and Densities of Massive Planets

For bodies of planetary mass, **Coulomb pressure**, provided by the electromagnetic repulsion of electrons in one molecule from those in another, plays a larger role relative to degeneracy pressure. Coulomb pressure is characterized by constant density, which implies that the radius of such bodies scales as:

$$R \propto M^{1/3}. \tag{3.26}$$

The combination of Coulomb and degeneracy pressures results in radii similar to that of Jupiter for all cool brown dwarfs and giant planets of solar composition, as well as for the very lowest mass stars (Fig. 3.8). The largest size cool planets are expected to have $M_{\text{p}} \approx 4\,\text{M}_{\text{2+}}$.

A **polytrope** obeys an equation of state of the form

$$P = K_{\text{po}}\rho^{\frac{n_{\text{po}}+1}{n_{\text{po}}}}, \tag{3.27}$$

where K_{po} is the **polytropic constant**, and n_{po} is the **polytropic index**.

To derive this maximum radius for a sphere of cool pure hydrogen gas, consider the equation of state as given by equation (3.27) with a polytropic index $n_{\text{po}} = 1$, so that $P = K_{\text{po}}\rho^2$. In this particular case, the radius is independent of mass. Thus, when more mass is added to the body, the material gets compressed such that its radius does not change. With $n_{\text{po}} = 1$, the equations can be solved analytically, and the results are in good agreement with calculations based on more detailed pressure–density relations. Integration of the equation of hydrostatic equilibrium (eq. 3.20) leads to the following density profile:

$$\rho = \rho_{\text{c}}\left(\frac{\sin(C_K r)}{C_K r}\right), \tag{3.28}$$

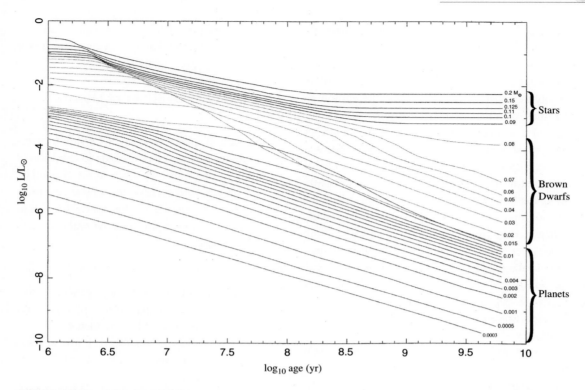

Figure 3.7 Evolution of the luminosity (in $\mathcal{L}_\odot$) of initially hot and distended very low mass stars and substellar objects plotted as functions of time (in years) after formation. All of these objects are isolated and have solar-metallicity. The stars, brown dwarfs and planets are shown as the upper, middle and lower sets of curves, respectively. The curves are labeled by the object's mass in units of $M_\odot$; the lowest curve corresponds to the mass of Saturn. All of the substellar objects become less luminous as they radiate away the energy released by their gravitational contraction from large objects to much more compact bodies with sizes of order $R_{2\!\!\!\downarrow}$. Stars ultimately level off in luminosity when they reach the hydrogen burning main sequence. In contrast, the luminosities of brown dwarfs and planets decline indefinitely. Objects with $M \gtrsim 0.012\ M_\odot$ exhibit plateaus between 10^6 and 10^8 years as a result of deuterium burning. The luminosities of the younger objects, especially giant planets $\lesssim 10^8$ years old, may be substantially smaller than the values shown here if they radiate away substantial portions of their accretion energy while they are growing. (Burrows et al. 1997)

with ρ_c the density at the center of the body and

$$C_K = \sqrt{\frac{2\pi G}{K_{\text{po}}}}. \qquad (3.29)$$

The radius of the body, R, is defined by $\rho = 0$, thus $\sin(C_K R) = 0$, and $R = \pi/C_K$. With a value for the polytropic constant $K_{\text{po}} = 2.7 \times 10^5\ \text{m}^5$ $\text{kg}^{-1}\ \text{s}^{-2}$, as obtained from a fit to a more precise equation of state, we find that the planet's radius $R = 7.97 \times 10^4$ km. This number is thus independent of the planet's mass. The radius of this

hydrogen sphere is slightly larger than Jupiter's mean radius, 6.99×10^4 km, and significantly larger than Saturn's mean radius, 5.82×10^4 km. This suggests that the Solar System's two largest planets are composed primarily, but not entirely, of hydrogen.

Using experimental data at low pressure and theoretical models at very high pressure, it can be shown that the maximum radius for cold self-gravitating spheres is approximately

$$R_{\max} \approx \frac{Z \times 10^5}{\mu_a m_{\text{amu}} \sqrt{Z^{2/3} + 0.51}}\ \text{km}, \qquad (3.30)$$

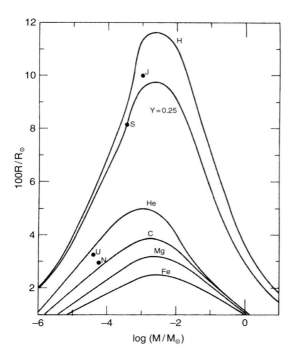

Figure 3.8 The mass–density relation for spheres of different materials at zero temperature, as calculated numerically using precise (empirical) equations of state. The *second curve* from the *top* is for a mixture of 75% H, 25% He by mass; all other curves are for planets composed entirely of one single element. The approximate locations of the giant planets are indicated on the graph. (Stevenson and Salpeter 1976)

where Z is the atomic number of the material composing the planet and $\mu_a m_{amu}$ the atomic mass. This equation results in $R_{max} \approx 82\,600$ km for a cool pure hydrogen planet and $R_{max} \approx 35\,000$ km for pure helium. For heavier material, R_{max} is smaller. Figure 3.8 graphs the mass–density relation for spheres of zero-temperature matter for different materials, as calculated numerically using precise (empirical) equations of state. The approximate locations of the giant planets are indicated on the graph, where the radius is taken as the distance from a planet's center out to the average 1 bar level along the planet's equator.

The observed atmospheric species form a boundary condition on the choice of elements to include in models of the interior structure of the giant planets. The atmospheric composition of Jupiter and Saturn is close to a solar composition, and the location of these planets on the radius–mass graph in Figure 3.8 shows that the composition of Jupiter and Saturn's interiors must indeed be close to solar. Calculations for more realistic (nonzero temperature) models that take contraction and cooling during a planet's formation into account are shown in Figure 3.7.

3.4 Nucleosynthesis

The nucleus of an ordinary hydrogen atom, ^{1}H, consists of a single proton. Nuclei of all other elements, as well as heavier isotopes of hydrogen, include both protons and neutrons (these two elementary particles are collectively referred to as **nucleons**). These nucleons are held together by the **strong nuclear force**. The combined mass of such a nucleus is less than that of the sum of the individual nucleons. This difference can be expressed as an energy using Einstein's famous relationship between mass and energy (eq. 2.68), and it is referred to as the **binding energy** of the nucleus. A higher value of the binding energy per nucleon means that the nucleus is more tightly bound. The binding energies per nucleon of many nuclides are shown in Figure 3.9. Although fusion of light elements typically releases energy, electromagnetic repulsion between nuclei dominates the strong nuclear force unless the nuclei are very close. This repulsion is referred to as the **Coulomb barrier** and is illustrated in Figure 3.10.

The formation of atomic nuclei is known as **nucleosynthesis**. The nuclei of the atoms that compose stars, planets, life, etc. formed in a variety of astrophysical environments. Models of nucleosynthesis, together with observational data from meteorites and other bodies, yield clues about the history of the material that was eventually incorporated into our Solar System. The two most

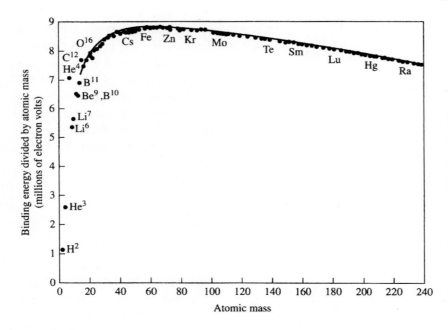

Figure 3.9 The nuclear binding energy per nucleon is shown as a function of atomic weight. For most elements, only the most stable isotope is plotted. Note that ^{4}He, ^{12}C and to a lesser extent somewhat heavier α-particle multiples lie above the general curve, indicating greater stability. The peak occurs at ^{56}Fe, indicating that iron is the most stable element. (Lunine 2005)

important environments for nucleosynthesis are the very early Universe and the interiors of stars. However, other environments are important for some isotopes; e.g., energetic cosmic rays can split nuclei with which they collide, and this **spallation** process is a major producer of the light elements lithium, beryllium and boron. Also, many isotopes form via radioactive decay (§ 11.6.1). The relative abundances of elements found in the Solar System are listed in Table 3.1.

3.4.1 Primordial Nucleosynthesis

The Universe began in an extremely energetic **hot big bang** roughly 13.7 billion years ago. The very young Universe was filled with rapidly moving particles. There were untold numbers of protons (ordinary hydrogen nuclei, ^{1}H, or more precisely ^{1}p$^+$) and neutrons (^{1}n). Free neutrons are unstable with a **half-life**, the time required for half of a sample to decay, of $t_{1/2} = 10.3$ minutes and decay via the reaction

$$^1\text{n} \xrightarrow[t_{1/2}=10.3\,\text{m}]{} {}^1\text{p}^+ + \text{e}^- + \bar{\nu}_e, \qquad (3.31)$$

where e$^-$ represents an electron, and $\bar{\nu}_e$ an (electron) antineutrino. Protons and neutrons collided and sometimes fused together to form deuterium (^{2}H) nuclei, but during the first few minutes, the cosmic background (blackbody) radiation field (§4.1.1) was so energetic that deuterium nuclei were photodissociated very soon after they formed. After about three minutes, the temperature cooled

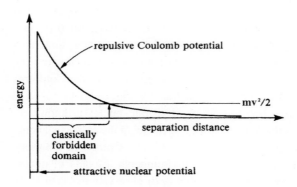

Figure 3.10 The electromagnetic repulsion between nuclei dominates the strong (but short-ranged) nuclear force unless the nuclei are very close. (Adapted from Shu 1982)

Table 3.1 **Elemental Abundances**

Element		Solar System[a] (atoms/10^6 Si)	CI Chondrites (mass fraction)	Element		Solar System[a] (atoms/10^6 Si)	CI Chondrites (mass fraction)
1	H	2.431×10^{10}	21.0 mg/g	44	Ru	1.90	692 ng/g
2	He	2.343×10^9	56 nL/g	45	Rh	0.37	141 ng/g
3	Li	55.5	1.46 μg/g	46	Pd	1.44	588 ng/g
4	Be	0.74	25.2 ng/g	47	Ag	0.49	201 ng/g
5	B	17.3	713 ng/g	48	Cd	1.58	675 ng/g
6	C	7.08×10^6	35.2 mg/g	49	In	0.18	78.8 ng/g
7	N	1.95×10^6	2.94 mg/g	50	Sn	3.73	1.68 μg/g
8	O	1.41×10^7	458.2 mg/g	51	Sb	0.33	152 ng/g
9	F	841	60.6 μg/g	52	Te	4.82	2.33 μg/g
10	Ne	2.15×10^6	218 pL/g	53	I	1.00	480 ng/g
11	Na	5.75×10^4	5.01 mg/g	54	Xe	5.39	31.3 pL/g
12	Mg	1.02×10^6	95.9 mg/g	55	Cs	0.37	185 ng/g
13	Al	8.41×10^4	8.50 mg/g	56	Ba	4.35	2.31 μg/g
14	Si	1.00×10^6	106.5 mg/g	57	La	0.44	232 ng/g
15	P	8370	920 μg/g	58	Ce	1.17	621 ng/g
16	S	4.45×10^5	54.1 mg/g	59	Pr	0.17	92.8 ng/g
17	Cl	5240	704 μg/g	60	Nd	0.84	457 ng/g
18	Ar	1.03×10^5	888 pL/g	62	Sm	0.25	145 ng/g
19	K	3690	530 μg/g	63	Eu	0.095	54.6 ng/g
20	Ca	6.29×10^4	9.07 mg/g	64	Gd	0.33	198 ng/g
21	Sc	34.2	5.83 μg/g	65	Tb	0.059	35.6 ng/g
22	Ti	2420	440 μg/g	66	Dy	0.39	238 ng/g
23	V	288	55.7 μg/g	67	Ho	0.090	56.2 ng/g
24	Cr	1.29×10^4	2.59 mg/g	68	Er	0.26	162 ng/g
25	Mn	9170	1.91 mg/g	69	Tm	0.036	23.7 ng/g
26	Fe	8.38×10^5	182.8 mg/g	70	Yb	0.25	163 ng/g
27	Co	2320	502 μg/g	71	Lu	0.037	23.7 ng/g
28	Ni	4.78×10^4	10.6 mg/g	72	Hf	0.17	115 ng/g
29	Cu	527	127 μg/g	73	Ta	0.021	14.4 ng/g
30	Zn	1230	310 μg/g	74	W	0.13	89 ng/g
31	Ga	36.0	9.51 μg/g	75	Re	0.053	37 ng/g
32	Ge	121	33.2 μg/g	76	Os	0.67	486 ng/g
33	As	6.09	1.73 μg/g	77	Ir	0.64	470 ng/g
34	Se	65.8	19.7 μg/g	78	Pt	1.36	1.00 μg/g
35	Br	11.3	3.43 μg/g	79	Au	0.20	146 ng/g
36	Kr	55.2	15.3 pL/g	80	Hg	0.41	314 ng/g
37	Rb	6.57	2.13 μg/g	81	Tl	0.18	143 ng/g
38	Sr	23.6	7.74 μg/g	82	Pb	3.26	2.56 μg/g
39	Y	4.61	1.53 μg/g	83	Bi	0.14	110 ng/g
40	Zr	11.3	3.96 μg/g	90	Th	0.044	30.9 ng/g
41	Nb	0.76	265 ng/g	92	U	0.0093	8.4 ng/g
42	Mo	2.60	1.02 μg/g				

[a] Protosolar material after radioactive decay to the present epoch.
Source: Data from Lodders (2003), who also gives uncertainties.

to the point that deuterium was stable for long enough to merge with protons, neutrons and other deuterium nuclei. Within the next few minutes, about one-quarter of the baryonic matter (nucleons) in the Universe agglomerated into alpha particles (^{4}He); most of the baryonic matter remained as protons, with small amounts forming deuterium, light helium (^{3}He) and tritium (^{3}H, which decays into ^{3}He with a half-life of 12 years), as well as very small but astrophysically significant amounts of the rare light elements lithium, beryllium, and boron, and minute amounts of heavier elements. Big bang nucleosynthesis did not proceed much beyond helium because by the time the blackbody radiation was cool enough for nuclei to be stable, the density of the Universe had dropped too low for fusion to continue to form heavier nuclei. After about 700 000 years, the blackbody radiation had cooled sufficiently for electrons to join the remaining protons and the larger nuclei that had formed in the early minutes of the Universe, producing atoms.

3.4.2 Stellar Nucleosynthesis

Most nuclei heavier than boron, as well as a small but still significant fraction of the helium nuclei, were produced in stellar interiors. Main sequence stars, such as the Sun, convert matter into energy via nuclear reactions that ultimately transform hydrogen nuclei into alpha particles. In normal stars, thermal pressure acts to counter gravitational compression. Protostars and young stars contract as they radiate away their thermal energy, and this contraction leads to an increase in pressure and density in the stellar core. Contraction continues until the core becomes hot enough to generate energy from **thermonuclear fusion**. The rates of fusion reactions increase steeply with temperature because only the tiny minority of nuclei in the high-velocity tail of the Maxwell–Boltzmann distribution (eq. 5.12) that describes the spread in velocities in equilibrium situations possess enough

kinetic energy to have a noninfinitesimal probability of quantum-mechanically tunneling through the barrier that is produced by Coulomb repulsion. If fusion proceeds too rapidly, the core expands and cools; if not enough energy is supplied by fusion, the core shrinks and heats up; in this fashion, equilibrium can be maintained. Deuterium fusion requires a lower temperature than fusion of ordinary hydrogen, so it occurs first, and stars rapidly deplete their supply of deuterium, although a significant amount of deuterium can remain in the outer (cooler) portion of a star if it is not convectively mixed with the lower hot regions. The cores of very low mass objects (brown dwarfs, §§1.3 and 3.3.3) get so dense that they are stopped from collapse by degenerate electron pressure before they reach a temperature high enough for fusion to occur at a significant rate.

In main sequence stars of solar mass and smaller, the primary reaction sequence is the **pp-chain**. The principal branch of the pp-chain occurs as follows:

$$2(^1\mathrm{H} + {}^1\mathrm{H} \rightarrow {}^2\mathrm{H} + \mathrm{e}^+ + \nu_e), \tag{3.32a}$$

$$2(^2\mathrm{H} + {}^1\mathrm{H} \rightarrow {}^3\mathrm{He} + \gamma), \tag{3.32b}$$

$$^3\mathrm{He} + {}^3\mathrm{He} \rightarrow {}^4\mathrm{He} + 2\,{}^1\mathrm{H} + 2\gamma, \tag{3.32c}$$

where e^+ represents a **positron** (the anti-particle of the electron), ν_e an (electron) neutrino and γ a photon. The reaction rate for the pp-chain becomes significant near $T_{\mathrm{nucl}} = 3 \times 10^6$ K. At temperatures close to that of the Sun's core, $T \approx 15 \times 10^6$ K, the fusion rate is roughly proportional to T^4. Although the rate of fusion in a 15-million-degree plasma is not as sensitive to temperature change as the T^{10} dependence near T_{nucl}, energy generation in the solar core still varies steeply with temperature. This steep temperature dependence implies that fusion acts as an effective thermostat: If the core gets too hot, it expands and cools, and energy production drops; if the core is too cold, it shrinks until adiabatic compression heats it enough for fusion rates to generate sufficient energy to balance the energy transported outwards. Moreover, the steep temperature dependence of fusion rates implies that more massive main sequence stars only require a slightly higher core temperature in order to generate a substantially higher luminosity than their smaller brethren.

In main sequence stars more massive than the Sun, the core temperature is somewhat higher, and the even more temperature-sensitive catalytic **CNO cycle** predominates. The principal branch of the CNO cycle is

$$^{12}C + {}^1H \rightarrow {}^{13}N + \gamma, \tag{3.33a}$$

$$^{13}N \underset{t_{1/2}=10\,m}{\longrightarrow} {}^{13}C + e^+ + \nu_e, \tag{3.33b}$$

$$^{13}C + {}^1H \rightarrow {}^{14}N + \gamma, \tag{3.33c}$$

$$^{14}N + {}^1H \rightarrow {}^{15}O + \gamma, \tag{3.33d}$$

$$^{15}O \underset{t_{1/2}=2\,m}{\longrightarrow} {}^{15}N + e^+ + \nu_e, \tag{3.33e}$$

$$^{15}N + {}^1H \rightarrow {}^{12}C + {}^4He. \tag{3.33f}$$

Note that although the half-lives are given for both of the **inverse β-decays** (emissions of positrons) in equation (3.33), the timescales for the four fusion reactions within the CNO cycle depend on the temperature and the abundances (densities) of the nuclei involved.

No nuclide with atomic mass 5 or 8 is stable, so to produce carbon from helium requires two fusions in immediate succession: first a pair of alpha particles combine to produce a (highly unstable, $t_{1/2} = 2 \times 10^{-16}$ s) beryllium 8 nucleus, and then another alpha particle is added before this nucleus decays:

$$^4He + {}^4He \leftrightarrow {}^8Be \tag{3.34a}$$

followed immediately by

$$^8Be + {}^4He \rightarrow {}^{12}C + \gamma. \tag{3.34b}$$

This **triple alpha process** requires much higher densities than do the pp-chain and CNO process described earlier. Helium fusion occurs when a sufficiently massive star ($\gtrsim 0.25\ M_\odot$) has exhausted the supply of hydrogen in its core, so the thermostat that maintained equilibrium during the star's main sequence phase is no longer active. Hydrogen fusion occurs in a shell surrounding the hydrogen-depleted core, and total stellar energy production greatly exceeds that during the star's main sequence phase, so its outer layers expand and cool, and the star becomes a **red giant**.

Nuclear growth beyond carbon does not require two reactions in immediate succession (as does the triple alpha process) and thus could occur in a lower density environment, but the increased Coulomb barrier implies that even higher temperatures and thus larger stellar masses are required. Growth can proceed by successive addition of alpha particles:

$$^{12}C + {}^4He \rightarrow {}^{16}O + \gamma, \tag{3.35a}$$

$$^{16}O + {}^4He \rightarrow {}^{20}Ne + \gamma, \tag{3.35b}$$

$$^{20}Ne + {}^4He \rightarrow {}^{24}Mg + \gamma, \tag{3.35c}$$

or at somewhat higher temperatures by reactions such as

$$^{12}C + {}^{12}C \rightarrow {}^{24}Mg + \gamma. \tag{3.35d}$$

The most stable nucleus is ^{56}Fe (Fig. 3.9), so fusion up to this mass can release energy. However, fusion of alpha particles (helium nuclei) into heavier nuclei (up to $Z = 28$) requires higher temperatures in order to overcome the Coulomb barrier. Nuclei composed of three to ten alpha particles are quite stable and easy to produce, so they are relatively abundant. Larger nuclei of this form are too proton-rich and rapidly inverse β-decay (emit positrons), thereby transforming themselves into more neutron-rich nuclei. Nonetheless, heavy elements with even atomic numbers tend to be more abundant than odd-numbered elements (Table 3.1). The proton and neutron numbers of all stable nuclides are shown in Figure 3.11.

Large quantities of elements up to the iron binding-energy peak can be produced by reactions

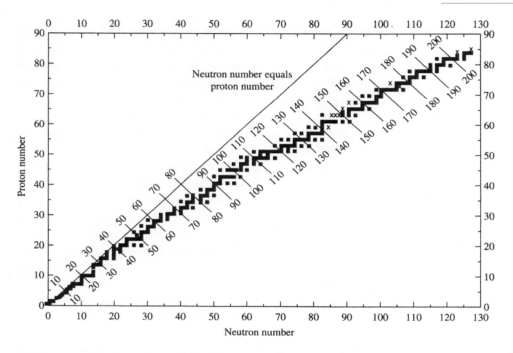

Figure 3.11 Distribution of stable nuclei, plotted as atomic number versus number of neutrons. The *diagonal line* representing equal numbers of protons and neutrons is plotted for reference. The *short lines* perpendicular to this diagonal represent nuclides with the same atomic weight. Long-lived but unstable isotopes are represented by *crosses*. (Adapted from Lunine 2005)

of the type discussed above, but the Coulomb barrier (Fig. 3.10) is too great for significant quantities of substantially more massive elements such as lead and uranium to be generated in this manner. Such massive nuclei are produced primarily by the addition of free neutrons, which are uncharged and thus do not need to overcome electrical repulsion. Free neutrons are released by reactions such as

$$^4He + {}^{13}C \rightarrow {}^{16}O + {}^1n, \qquad (3.36a)$$

and

$$^{16}O + {}^{16}O \rightarrow {}^{31}S + {}^1n. \qquad (3.36b)$$

Neutron addition does not produce a new element directly, but if enough neutrons are added, nuclei can become unstable and β-decay into elements of higher atomic number.

The mix of nuclides produced by neutron addition depends on the flux of neutrons. When the time between successive neutron absorptions is long enough for most unstable nuclei to decay, the mixture of nuclides produced lies deep within the **valley of nuclear stability**, where the mixture of neutrons and protons leads to the greatest binding energy for a nucleus with a given total number of nucleons; this 'slow' type of heavy element nucleosynthesis is referred to as the **s-process**. Nuclei with atomic masses as large as 209 may be formed via the s-process. The rapid **r-process** chain of nuclear reactions occurs during explosive nucleosynthesis, such as core-collapse supernovae and mergers of neutron stars with other neutron stars and with black holes. Such explosions provide a very high flux of neutrons, and explosive nucleosynthesis yields a more neutron-rich distribution of elements. Uranium and other very heavy naturally occurring elements are produced via r-process nucleosynthesis. Rare proton-rich heavy nuclei are formed by the poorly understood **p-process nucleosynthesis**.

Most of the elements produced via stellar nucleosynthesis are never released from their parent stars; only material ejected by stellar winds, nova outbursts and supernova explosions is available to enrich the interstellar medium and to form subsequent generations of stars and planets. The distributions of elements and isotopes found in individual interstellar grains and in the Solar System as a whole are indicative of the various environments in which stellar nucleosynthesis occurs and the conditions under which material is released from stars.

3.4.3 Radioactive Decay

Many naturally occurring nuclides are **radioactive**, that is, they spontaneously decay into nuclides of other elements that are usually of lesser mass. Radioactive decay rates can be accurately measured; thus, the abundances of decay products provide precise clocks that can be used to reconstruct the history of many rocks (§11.6). Every radioactive decay process releases energy, and the resultant heating can lead to differentiation of planetary bodies. The most common types of radioactivity are β-decay, whereby a nucleus emits an electron, and α-decay, in which a helium nucleus (composed of two protons and two neutrons) is emitted.

When an atomic nucleus undergoes β-decay, a neutron within the nucleus is transformed into a proton, so the atomic number increases by one; the total number of nucleons (protons plus neutrons) remains fixed, so the atomic mass number of the nucleus does not change, and the actual mass of the nucleus decreases very slightly. An example of β-decay is the transformation of an isotope of rubidium into one of strontium: $^{87}_{37}\text{Rb} \rightarrow {}^{87}_{38}\text{Sr}$; the number shown to the upper left of the element symbol is the atomic weight of the nuclide (number of nucleons), and the number to the lower left (which

is usually omitted because it is redundant with the name of the element) is the atomic number (number of protons). Proton-rich nuclei can undergo inverse β-decay, decreasing their atomic number by one; a closely related process is **electron capture**, whereby an atom's inner electron is captured by the nucleus; both of these processes convert a proton into a neutron. An example of a decay that can occur via positron emission or electron capture is $^{40}_{19}\text{K} \rightarrow {}^{40}_{18}\text{Ar}$; however, potassium 40 (β-) decays into calcium ($^{40}_{20}\text{Ca}$) eight times more frequently than it decays into argon.

When a nucleus undergoes α-decay, its atomic number decreases by two, and its atomic mass decreases by four; an example of α decay is the transformation of uranium to thorium: $^{238}_{92}\text{U} \rightarrow {}^{234}_{90}\text{Th}$. Some heavy nuclei decay via **spontaneous fission**, which produces at least two nuclei more massive than helium, as well as smaller debris. Spontaneous fission is an alternative decay mode for $^{238}_{92}\text{U}$ (and of the now almost extinct $^{244}_{94}\text{Pu}$), leading typically to xenon and lighter byproducts.

Key Concepts

- Laws of thermodynamics:
  - (0) 0th: Closed systems are in thermodynamic equilibrium.
  - (1) 1st: Energy is conserved.
  - (2) 2nd: The entropy of any closed system increases with time.
  - (3) 3rd: All processes virtually cease at 0 K.

- Gibbs free energy has a minimum value for a system in equilibrium.
- The relationship between temperature, pressure and density in planets and stars is governed by hydrostatic equilibrium.
- Mass is the principal differentiating factor between stars. More massive stars are much more luminous and shorter lived.

- Stars are objects that are supported against gravitational collapse by thermal pressure maintained by sustained thermonuclear fusion. Brown dwarfs are massive enough to fuse deuterium but not massive enough to become stars. Planets do not have enough mass to fuse deuterium.

- The Sun and other stars spend long periods of time on the main sequence, during which thermonuclear fusion of hydrogen into helium within the core provides the energy to replace that lost via electromagnetic radiation from the surface and allows pressure balance to be maintained. The luminosity of a star increases slowly while it remains on the main sequence. Changes are more rapid after the star exhausts the hydrogen in its core.

- A few minutes after the big bang, about three-quarters of the matter was in the form of hydrogen and one-quarter in helium; trace amounts of lithium, beryllium, boron and heavier elements were also present. Heavier elements have since been produced, mostly within stars and supernova explosions.

Further Reading

Thermodynamics is covered in great detail in the following books:

Anderson, G., 2005. *Thermodynamics of Natural Systems*. Cambridge Univ. Press, 2nd Edition. 648pp.

Douce, P., 2011. *Thermodynamics of the Earth and Planets*. Cambridge Univ. Press. 709pp.

A (mostly still current) thorough review of nucleosynthesis within stars is given in:

Clayton, D.D., 1983. *Principles of Stellar Evolution and Nucleosynthesis*. University of Chicago Press. 612pp.

An excellent popular account of big bang nucleosynthesis is provided by:

Weinberg, S., 1988. *The First Three Minutes*. Basic Books, New York. 198pp.

Problems

3-1. Show that enthalpy is simply heat exchanged at constant pressure, i.e., show that equation (3.4) is valid if under constant pressure. (Hint: Differentiate equation (3.3) and substitute for dU using equation (3.1)).

3-2. Show that isothermal expansion is accompanied by an increase in entropy.

3-3. Use the equation of hydrostatic equilibrium to estimate the pressure at the center of the Moon, Earth and Jupiter.

(a) Take the simplest approach, approximating the planet to consist of one slab of material with thickness R, the planetary radius. Assume the gravity $g_p(r) = g_p(R)$ and use the mean density $\rho(r) = \rho$.

(b) Assume the density of each planet to be constant throughout its interior and derive an expression for the pressure in a planet's interior as a function of distance r from the center. (Hint: You should get equation (3.21).)

(c) Although the pressure obtained in (a) and (b) is not quite right, it will give you a fair estimate of its magnitude. Compare your answer with the more sophisticated estimates given in Table E.14 and comment on your results.

3-4. Use the formulas specifying the kinetic and potential energy of a particle on a circular orbit about a much more massive body to verify the virial theorem (eq. 3.22) for the case of circular orbits.

3-5. For small bodies, the relationship between mass and size is given by equation (3.26). When more mass is added to a planet, the material gets compressed. When the internal pressure becomes very large, the matter becomes degenerate, as is the case for white dwarf stars. Consider the central pressure, P_c, of a white dwarf, which can be calculated from equation (3.20), as you did in Problem 3-3. The polytropic constant n in the equation of state (eq. 3.27) is 2/3 in the limit of high pressure. Show that for a white dwarf, $M \propto R^{-3}$.

3-6. **(a)** Calculate the number of hydrogen nuclei originally in the Sun using the data in Tables 3.1, C.2 and C.5.

(b) What is the total energy supply available from hydrogen fusion if this process liberates 0.75% of the mass of hydrogen and all of the hydrogen is available for fusion?

(c) Using the Sun's present luminosity, how long would it take to convert all of the hydrogen in the Sun's core (10% of the Sun's mass) into helium? Quote your answer in years.

3-7. Table 3.1 gives the relative cosmic abundances of the elements. Use the plot of binding energy per nucleon (Fig. 3.9) plus the astrophysics of nucleosynthesis to briefly explain the reasons that:

(a) H is the most abundant element, and He is the second most abundant element.

(b) C and O are far more abundant than are Li, Be and B.

(c) Fe is more abundant (by mass) than any other element heavier than O.

(d) Elemental abundances drop precipitously with increasing atomic number above 26 (the atomic number of iron).

(e) Apart from H, elements with odd atomic number are generally less abundant than neighboring elements of even atomic numbers.

CHAPTER 4

Solar Heating and Energy Transport

Like all other arts, the Science of Deduction and Analysis is one which can only be acquired by long and patient study, nor is life long enough to allow any mortal to attain the highest perfection in it. Before turning to those moral and mental aspects of the matter which present the greatest difficulties, let the inquirer begin by mastering more elementary problems.

Sherlock Holmes, *A Study in Scarlet*, Sir Arthur Conan Doyle

Temperature is one of the most fundamental properties of planetary matter, as is evident from everyday experience such as the weather and cooking a meal, as well as from the most basic concepts of chemistry and thermodynamics. For example, H_2O is a liquid between 273 K and 373 K (at standard pressure), a gas at higher temperatures and a solid (ice) when it is colder. We are familiar with all three phases here on Earth, but the liquid phase of water is not seen on the surface of any other planet in our Solar System. Rocks undergo similar transitions at substantially higher temperatures. For example, liquid rock is present in the Earth's mantle (magma) and can be seen on the surface of Earth during a volcanic eruption (e.g., lava flows in Hawaii). Some other substances are only in vapor form on Earth's surface. For example, carbon dioxide (CO_2) consists in gas form, and when cooled to 194.7 K, it freezes into a substance known as 'dry ice'. No liquid phase of CO_2 exists on Earth's surface because liquid CO_2 is only stable at pressures exceeding 5.1 bar. All three phases of methane exist at standard pressure, but methane condenses and freezes at lower temperatures.

Most substances expand when heated, with gases increasing in volume the most; the thermal expansion of liquid mercury allowed it to be the 'active ingredient' in most thermometers from the seventeenth century through the twentieth century. The equilibrium molecular composition of a given mixture of atoms often depends on temperature (as well as on pressure), and the time required for a mixture to reach chemical equilibrium generally increases rapidly as temperature drops. Gradients in temperature and pressure are responsible for atmospheric winds (and, on Earth, ocean currents) as well as convective motions that can mix fluid material within planetary atmospheres and interiors. Earth's solid crust is dragged along by convective currents in the mantle, leading to continental drift (§6.3.1).

Temperature, T, is a measure of the random kinetic energy of molecules, atoms, ions, etc. The energy, E, of a perfect gas is given by

$$E = \frac{3}{2}NkT, \tag{4.1}$$

where N is the number of particles and k is Boltzmann's constant. The temperature of a given region of a body is determined by a combination of processes. Solar radiation is the primary energy source for most planetary bodies, and radiation to space is the primary loss mechanism. In this chapter, we summarize the mechanisms for solar heating and energy transport. We then use this background in our discussions of planetary atmospheres, surfaces and interiors in subsequent chapters.

We begin in §4.1 with a discussion of temperature and the effects of absorbing and reflecting (sun)light. Conduction, convection and radiation are the three primary ways by which energy is transported within, to and from planetary bodies. We introduce these mechanisms in §4.2, and elaborate on each in the subsequent three sections. In §4.5.3, the basic equation of radiative transfer is developed and described; this knowledge is used in §4.6 to explain the greenhouse effect in an atmosphere.

4.1 Energy Balance and Temperature

Planetary bodies are heated primarily by absorbing radiation from the Sun, and they lose energy via radiation to space. While a point on the surface of a body is illuminated by the Sun only during the day, it radiates both day and night. The amount of energy incident per unit area depends both on the distance from the Sun and the local elevation angle of the Sun. As a consequence, most locales are coldest around sunrise and hottest a little after local noon, and the polar regions are (on average) colder than the equator for bodies with obliquity $\psi < 54°$ (or $\psi > 126°$).

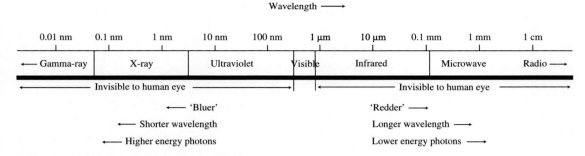

Figure 4.1 The electromagnetic spectrum. (Adapted from Hartmann 1989)

Over the long term, most planetary bodies radiate almost the same amount of energy to space as they absorb from sunlight; were this not the case, planets would heat up or cool off. (The giant planets Jupiter, Saturn and Neptune are exceptions to this rule. These bodies radiate significantly more energy than they absorb because they are still losing heat produced during the time of their formation.) Although long-term global equilibrium is the norm, spatial and temporal fluctuations can be large. Energy is stored from day to night, perihelion to aphelion and summer to winter, and can be transported from one location on a planet to another. We begin our discussion with the fundamental laws of radiation in §4.1.1 and factors affecting global energy balance in §4.1.2.

4.1.1 Thermal (Blackbody) Radiation

Electromagnetic radiation consists of photons at many wavelengths; a spectrum of the electromagnetic emission is shown in Figure 4.1. The frequency, ν, of an electromagnetic wave propagating in a vacuum is related to its wavelength, λ, by

$$\lambda \nu = c, \tag{4.2}$$

where c is the **speed of light in a vacuum**, 2.998×10^8 m s^{-1}.

Most objects emit a continuous spectrum of electromagnetic radiation. This **thermal emission** is well approximated by the theory of 'blackbody' radiation. A **blackbody** is defined as an object that absorbs all radiation that falls on it at all frequencies and all angles of incidence; i.e., no radiation is reflected or scattered. A body's capacity to emit radiation is the same as its capability of absorbing radiation at the same frequency. The radiation emitted by a blackbody is described by **Planck's radiation law**:

$$B_\nu(T) = \frac{2h\nu^3}{c^2} \frac{1}{e^{h\nu/(kT)} - 1}, \tag{4.3}$$

where $B_\nu(T)$ is the **brightness** (W m^{-2} Hz^{-1} sr^{-1}) and h is Planck's constant. Figure 4.2a shows a graph of brightness as a function of frequency for various blackbodies with temperatures ranging from 40 to 30 000 K. A spectrum of our Sun is shown in Figure 4.2b, with superposed a blackbody curve at a temperature of 5777 K. These two figures show that the Sun's brightness peaks at optical wavelengths. In contrast, those of the planets ($\sim$40–700 K) peak at infrared wavelengths. The brightness of most Solar System objects near their spectral peaks can be approximated quite well by blackbody curves.

Two limits of Planck's radiation law can be derived:

(1) **Rayleigh–Jeans law**: When $h\nu \ll kT$ (i.e., at radio wavelengths for temperatures typical of planetary bodies), the term $(e^{h\nu/(kT)} - 1) \approx h\nu/(kT)$ and equation (4.3) can be approximated by

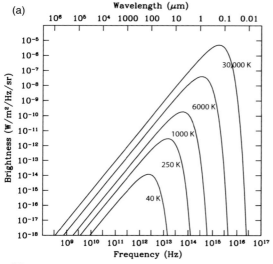

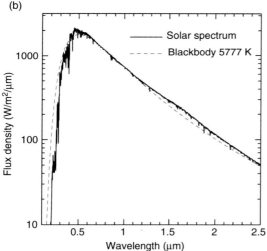

Figure 4.2 (a) Blackbody radiation curves shown as a function of frequency, $B_\nu(T)$, at various temperatures ranging from 40 K up to 30 000 K. The 6000 K curve is representative of the solar spectrum (eq. 4.3). (b) Solar spectrum as a function of wavelength between 0.1 and 2.5 μm. A blackbody spectrum at 5777 K is superposed. (Adapted from de Pater and Lissauer 2010)

$$B_\nu(T) \approx \frac{2\nu^2}{c^2}kT. \tag{4.4}$$

(2) **Wien's law**: When $h\nu \gg kT$:

$$B_\nu(T) \approx \frac{2h\nu^3}{c^2}e^{-h\nu/(kT)}. \tag{4.5}$$

Equations (4.4) and (4.5) are simpler than equation (4.3), and thus they can be quite useful in the regimes in which they are applicable.

The frequency, ν_{max}, at which the peak in the brightness, $B_\nu(T)$, occurs can be determined by setting the derivative of equation (4.3) equal to zero, $\partial B_\nu/\partial \nu = 0$. The result is known as **Wien's displacement law**:

$$\nu_{max} = 5.88 \times 10^{10}T, \tag{4.6}$$

with ν_{max} in Hz and T in degrees Kelvin. With

$$B_\lambda = B_\nu \left| \frac{d\nu}{d\lambda} \right|, \tag{4.7}$$

the blackbody spectral peak in wavelength can be found by setting $\partial B_\lambda/\partial \lambda = 0$:

$$\lambda_{max} = \frac{2.9 \times 10^{-3}}{T}, \tag{4.8}$$

with λ_{max} in fractions of a meter (Problem 4-2) and T in Kelvin. Note that $\lambda_{max} = 0.57\, c/\nu_{max}$, i.e., the brightness peak measured in terms of wavelength is blueward of the brightness peak measured in terms of frequency.

The **flux density**, $\mathcal{F}_\nu$ (W m^{-2} Hz^{-1}), of radiation from an object is given by:

$$\mathcal{F}_\nu = \Omega_s B_\nu(T), \tag{4.9}$$

where Ω_s is the solid angle subtended by the object. Just above the 'surface' of a planet of brightness B_ν, the flux density is equal to

$$\mathcal{F}_\nu = \pi B_\nu(T). \tag{4.10}$$

The **flux**, $\mathcal{F}$ (J s^{-1} m^{-2}), is defined as the flux density integrated over all frequencies:

$$\mathcal{F} \equiv \int_0^\infty \mathcal{F}_\nu d\nu = \pi \int_0^\infty B_\nu(T)d\nu = \sigma T^4, \tag{4.11}$$

where σ is the **Stefan–Boltzmann constant**. Equation (4.11) is known as the **Stefan–Boltzmann law**.

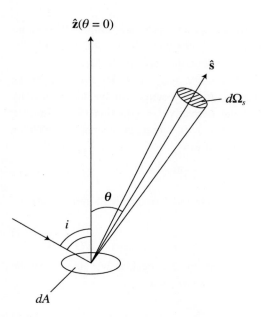

Figure 4.3 Sketch of the geometry of a surface element dA: $\hat{\mathbf{z}}$ is the normal to the surface, $\hat{\mathbf{s}}$ is a ray along the line of sight and θ is the angle the ray makes with the normal to the surface (de Pater and Lissauer 2010)

4.1.2 Albedo

When an object is illuminated by the Sun, it reflects part of the energy back into space (which makes the object visible to us), while the remaining energy is absorbed. In principle, one can determine how much of the incident radiation is reflected into space at each frequency; the ratio between incident and reflected + scattered energy is called the **monochromatic albedo**, A_ν. Integrated over frequency, the ratio of the total radiation reflected or scattered by the object to the total incident light from the Sun is called the **Bond albedo**, A_b. The energy or flux absorbed by the object determines its temperature, as discussed further in §4.1.3.

It is also important to consider how a surface scatters light. The Sun's light is scattered off a planet and received by a telescope. The three angles of relevance are indicated in Figures 4.3 and 4.4: The angle that incident light makes with the normal to the planet's surface, i, and the angle that

the reflected ray received at the telescope (i.e., the ray along the line of sight) makes with the normal to the surface, θ, are shown in Figure 4.3; the **phase angle** or angle of reflectance, ϕ, as seen from the object is shown in Figure 4.4. For purely backscattered radiation, the phase angle $\phi = 0$; in the case of forward scattered light, $\phi = 180°$. Light is often scattered in a preferred direction, which can be expressed by the **scattering phase function**. In particular, for particles similar in size to, or slightly larger than, the wavelength of light, the preferred angle of scattering is in the forward direction. For particles much smaller than the wavelength of light, scattering is more isotropic.

From Earth, all phase angles between 0° and 180° can only be measured for planets with heliocentric distances less than 1 AU (Mercury, Venus) as well as for the Moon. The outer planets are observed from Earth at phase angles close to 0°. At phase angle $\phi = 0$, the albedo or head-on

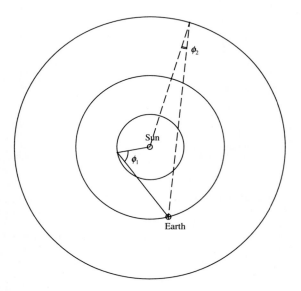

Figure 4.4 Scattering of light by a body that is illuminated by the Sun, with radiation received on Earth. For purely backscattered radiation, the phase angle $\phi = 0$; in the case of forward scattered light, $\phi = 180°$. Two planets are indicated: one inside Earth's orbit, with phase angle ϕ_1, and one outside Earth's orbit, with phase angle ϕ_2. (de Pater and Lissauer 2010)

reflectance, is often referred to as the **geometric albedo**, A_0:

$$A_0 = \frac{r_{\odot AU}^2 \mathcal{F}(\phi = 0)}{\mathcal{F}_\odot},$$

(4.12)

where $\mathcal{F}(\phi = 0)$ is the flux reflected from the body at phase angle $\phi = 0$. The heliocentric distance, $r_{\odot AU}$, is expressed in AU, and $\mathcal{F}_\odot$, the solar constant, is defined as the solar flux at $r_{\odot AU} = 1$:

$$\mathcal{F}_\odot \equiv \frac{\mathcal{L}_\odot}{4\pi r_\odot^2} = 1366 \text{ W m}^{-2},$$

(4.13)

with $\mathcal{L}_\odot$ the solar luminosity. The combination $(\mathcal{F}_\odot/r_{\odot AU}^2)$ is equal to the incident solar flux at heliocentric distance $r_{\odot AU}$ AU.

4.1.3 Temperature

Several types of 'temperature' are relevant to planetary sciences. In addition to the physical temperature of a body or a gas, we use the terms *brightness temperature*, *effective temperature* and *equilibrium temperature*. These temperatures are defined and described below.

Brightness and Effective Temperatures

One can determine the temperature of a blackbody using Planck's radiation law by measuring a small part of the object's radiation (Planck) curve. This is usually not practical because most bodies are not perfect blackbodies, but rather exhibit spectral features that complicate temperature measurements. It is common to relate the observed flux density, $\mathcal{F}_\nu$, to the **brightness temperature**, T_b, which is the temperature of a blackbody that has the same brightness at this particular frequency (i.e., replace T in eq. 4.3 by T_b). Conversely, if the total flux integrated over all frequencies of a body can be determined, the temperature that corresponds to a blackbody emitting the same amount of energy or flux $\mathcal{F}$ is referred to as the **effective temperature**, T_e:

$$T_e \equiv \left(\frac{\mathcal{F}}{\sigma}\right)^{1/4}.$$

(4.14)

The wavelength range at which the object emits most of its radiation can be estimated via Wien's displacement law (eq. 4.8). This is typically at mid-infrared wavelengths (10–20 μm) for objects with temperatures of 150–300 K (inner Solar System) and far-infrared wavelengths ($\sim$60–70 μm) for $\sim$40–50 K bodies in the outer Solar System.

Equilibrium Temperature

Provided the incoming solar radiation (insolation), $\mathcal{F}_{in}$, is balanced, on average, by reradiation outwards, $\mathcal{F}_{out}$, one can calculate the temperature of the object. This temperature is referred to as the **equilibrium temperature**. If indeed the temperature of the body is completely determined by the incident solar flux, the equilibrium temperature equals the effective temperature. Any discrepancies between the two numbers contain valuable information on the object. For example, the effective temperatures of Jupiter, Saturn and Neptune exceed the equilibrium temperature, which implies that these bodies possess internal heat sources (§§5.1 and 6.2.3). Venus's surface temperature is far hotter than the equilibrium temperature of the planet, a consequence of a strong greenhouse effect in this planet's atmosphere (§5.1). The effective temperature of Venus, which is dominated by radiation emitted from that planet's cool upper atmosphere, is equal to the equilibrium temperature, implying that Venus has a negligible internal heat source. We next discuss the average effect of insolation and reradiation for a rapidly rotating spherical object of radius R using approximate equations.

The sunlit hemisphere of a (spherical) body receives solar radiation in the amount of:

$$\mathcal{P}_{in} = (1 - A_b)\frac{\mathcal{L}_\odot}{4\pi r_\odot^2}\pi R^2,$$

(4.15)

with πR^2 the projected surface area for intercepting solar photons. A rapidly rotating planet reradiates energy from its entire surface (i.e., an area of $4\pi R^2$):

$$\mathcal{P}_{\text{out}} = 4\pi R^2 \epsilon \sigma T^4. \qquad (4.16)$$

Note that whereas the incoming solar radiation is primarily at optical wavelengths (Fig. 4.2), thermal emission from planets is radiated primarily at infrared wavelengths. The emissivity, ϵ_ν, depends on wavelength and size of the object. For objects that are large compared with the wavelength, ϵ_ν is usually close to 0.9 at infrared wavelengths, but it can differ substantially from unity at radio wavelengths. Objects much smaller than the wavelength ($R \lesssim 0.1\lambda$) do not radiate efficiently, i.e., $\epsilon_\nu < 1$.

From a balance between insolation and reradiation, $\mathcal{P}_{\text{in}} = \mathcal{P}_{\text{out}}$, one can calculate the equilibrium temperature, T_{eq}:

$$T_{\text{eq}} = \left(\frac{\mathcal{F}_\odot}{r^2_{\odot\text{AU}}} \frac{(1 - A_{\text{b}})}{4\epsilon\sigma} \right)^{1/4}. \qquad (4.17)$$

Even though this simple derivation has many shortcomings, the disk-averaged equilibrium temperature from equation (4.17) gives useful information on the temperature $\sim$1 m below a planetary surface. If ϵ is close to unity, it corresponds well with the actual (physical) temperature of subsurface layers that are below the depth where diurnal (day/night) and seasonal temperature variations are important, typically a meter or more below the surface. These layers can be probed at radio wavelengths, and the brightness temperature observed at these long wavelengths can be compared directly with the equilibrium temperature. In the derivation of equation (4.17), we omitted latitudinal and longitudinal effects of the insolation pattern. The magnitude of these effects depends on the planet's rotation rate, obliquity and orbit. Latitudinal and longitudinal effects are large, for example, on airless planets that rotate slowly, have small axial obliquities and/or travel on very eccentric orbits about the Sun.

In another limit for equilibrium temperatures, one can consider the subsolar point of a slowly rotating body. In this case, the surface areas πR^2 in equation (4.15) and $4\pi R^2$ in equation (4.16) should both be replaced by a unit area dA. It follows that the equilibrium temperature at the subsolar point of a slowly rotating body is $\sqrt{2}$ times the disk-average equilibrium temperature for a rapidly rotating body. The subsolar temperature calculated in this way corresponds well with the measured subsolar surface temperature of airless bodies.

4.2 Energy Transport

The temperature structure in a body is governed by the efficiency of energy transport. There are three principal mechanisms to transport energy: **conduction**, **radiation** and **convection**. Usually, one of these three mechanisms dominates and determines the thermal profile in any given region. Whereas energy transport in a solid is usually dominated by conduction, radiation typically dominates in gases that are neither too dense nor too tenuous. Convection is important in fluids and dense gases.

All three energy transport mechanisms are experienced in everyday life, e.g., when boiling water on a stove: The entire pan, including the handle (in particular if metal), is heated by conduction, as is the water very close to the surface of the pan. In contrast, most of the water in the pan is primarily heated through 'convection', up and down motions in the water. These motions are visible in the form of bubbles of vaporized water, which rise upward because they are lighter than the surrounding water. Heat is transported from the Sun to planets, moons, etc. via radiation. Although in these examples it is obvious which transport mechanism dominates, this is not always easy to determine; in some parts of a planet's interior, energy transport is dominated by convective motions, but in other parts, conduction is by far the most efficient. In a planet's atmosphere, we typically encounter all three mechanisms, although a particular mechanism usually dominates in a certain altitude range. Almost all of the energy transport to and from planetary bodies occurs via radiation. (Jupiter's moon Io is an exception to this rule; Io receives a substantial amount of energy from its orbit via tidal dissipation; see §2.7.3 and §10.2.1.)

In the next sections, we discuss all three principal mechanisms for energy transport and give equations for the thermal profile in an atmosphere under the assumption that a particular heat transport mechanism is dominant. Throughout this chapter, we assume atmospheres to be perfect (ideal) gases (eq. 3.14) in hydrostatic equilibrium (§3.2).

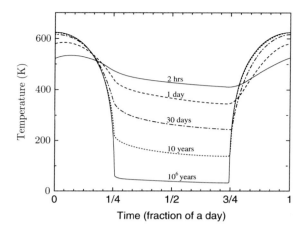

Figure 4.5 The surface temperature of a solid coherent rocky body that orbits the Sun in a circular orbit at a heliocentric distance of 0.4 AU and has zero obliquity. Curves are shown for bodies with rotation periods, P_{rot} (defined with respect to the Sun), of 2 hours, 1 day, 30 days, 10 years and 10^6 years. (Courtesy David L. Mitchell)

4.3 Conduction

Conduction, i.e., the transfer of energy via collisions between molecules, is important in a solid body, as well as in the tenuous upper part of an atmosphere (the upper thermosphere). In the latter situation, the mean free path is so long that atoms exchange locations very rapidly and the conductivity is therefore large. The high conductivity tends to equalize temperatures in this part of an atmosphere.

Sunlight heats a planet's surface during the day, and the heat is transported downwards from the surface mainly by conduction. The rate of flow of heat, the **heat flux**, Q (W m^{-2}), is determined by the **temperature gradient**, ∇T, and the **thermal conductivity**, K_T:

$$Q = -K_T \nabla T. \tag{4.18}$$

The thermal conductivity is a measure of the material's physical ability to conduct heat. The amplitude of diurnal temperature variations is largest at the surface and decreases exponentially into the subsurface. Moreover, because it takes time for the heat to be carried downwards, there is a

phase lag in the diurnal heating pattern of the subsurface layers. The peak temperature at the surface is reached at noon or soon thereafter, and the subsurface layers reach their peak temperature later in the afternoon. At night, the surface cools off, becoming cooler than the subsurface layers. Heat is then transported upward from below. Because the conductivity depends on the temperature, the surface may act as an insulator at night, preventing the subsurface from cooling off very rapidly.

Figure 4.5 shows the surface temperature as a function of local time (LT) for a hypothetical large rocky body in a circular orbit about the Sun at a heliocentric distance of 0.4 AU. Curves for various rotation periods from 2 hours up to 10^6 years are shown. This figure demonstrates that whereas the peak temperature is primarily determined by the heliocentric distance, the night-side temperature also depends on the planet's rotation rate. Note the time delay in peak temperature from local noon when $P_{rot} = 2$ hours. This delay is caused by the rapid rotation rate. Another factor that determines the time delay is the **thermal inertia** of the surface, which measures the ability of the surface to store energy.

4.4 Convection

In dense atmospheres, molten interiors of planets and protoplanetary disks, heat is often transported primarily by large-scale fluid motions, particularly by convection. Convection is the motion in a fluid caused by density gradients that result from temperature differences. Consider a parcel of air in a planet's atmosphere that is slightly warmer than its surroundings. To reestablish pressure equilibrium, the parcel expands, and thus its density decreases below that of its surroundings. This causes the parcel to rise. Because the surrounding pressure decreases with height, the rising parcel expands and cools. If the temperature of the environment drops sufficiently rapidly with height, the parcel remains warmer than its surroundings and thus continues to rise, transporting heat upward. This process is an example of convection. For convection to occur, the temperature must decrease with decreasing pressure (thus outward in a planetary environment) at a sufficiently rapid rate that the parcel remains buoyant.

The temperature structure of an atmosphere in which energy transport is dominated by convection follows an **adiabatic lapse rate**. Equations for both the dry and wet adiabatic lapse rate are derived in the next subsection.

4.4.1 Adiabatic Gradient

We now derive the thermal structure of an atmosphere in which energy transport is dominated by convection. The atmosphere is composed of an ideal gas in hydrostatic equilibrium, so equations (3.13b) and (3.14) are valid.

We assume the parcel of air moves **adiabatically**, i.e., no heat is exchanged between the parcel of air and its surroundings: $dQ = 0$. The first law of thermodynamics, i.e., conservation of energy (eq. 3.1), then reads:

$$dQ = dU + P\,dV = 0. \tag{4.19}$$

We define the thermal (or molar) heat capacity, C_P, in §3.1.2 (eq. 3.5). Similarly, we define the molar heat capacity, C_V, as the amount of heat, Q, necessary to raise the temperature of one mole of matter by one degree Kelvin without changing the volume. The **specific heat**, c_P (or c_V), is the amount of energy necessary to raise the temperature of 1 g of material by 1 K without changing the pressure (or volume):

$$m_{gm}c_P \equiv C_P \equiv \left(\frac{dQ}{dT}\right)_P, \tag{4.20a}$$

$$m_{gm}c_V \equiv C_V \equiv \left(\frac{dQ}{dT}\right)_V, \tag{4.20b}$$

where m_{gm} is a **gram-mole**, defined as the mass of one mole of molecules in units of grams. The mass of a gram-mole is numerically equal to its weight in atomic mass units (amu). Thus, a mole of the lightest and most common isotope of carbon atoms has a mass of 12 g.

Substituting equation (4.19) for Q in equations (4.20a) and (4.20b), we find:

$$C_V = \left(\frac{dU}{dT}\right)_V, \tag{4.21a}$$

$$C_P = \left(\frac{dU}{dT}\right)_P + P\left(\frac{dV}{dT}\right)_P. \tag{4.21b}$$

Combining equations (4.19) and (4.21a) shows that the following relationship holds for a parcel of air that moves adiabatically:

$$C_V dT = -P dV. \tag{4.22a}$$

The **specific volume** is the amount of space occupied by 1 g of molecules. We rewrite equation (4.22a) in terms of the specific heat, c_V:

$$c_V dT = -P dV. \tag{4.22b}$$

The ideal gas law then gives:

$$V = \frac{N_0 kT}{P},$$ (4.23)

with $N_0 = 1/(\mu_a m_{amu})$, where $\mu_a m_{amu}$ is the molecular mass.

Differentiating the ideal gas law gives

$$dV = \frac{N_0 k}{P} dT - \frac{N_0 kT}{P^2} dP.$$ (4.24)

In an ideal gas, the difference between the two thermal heat capacities (J mole^{-1} K^{-1}) and the specific heats (J kg^{-1} K^{-1}) is given by

$$C_P - C_V = R_{gas},$$ (4.25a)

$$c_P - c_V = \frac{R_{gas}}{m_{gm}} = N_0 k.$$ (4.25b)

Using equations (4.22b), (4.24) and (4.25b), we can write:

$$c_P \, dT = -P dV + N_0 k dT$$

$$= -P\left(\frac{N_0 k}{P} dT - \frac{N_0 kT}{P^2} dP\right) + N_0 k dT$$

$$= \frac{N_0 kT}{P} dP = V dP = \frac{dP}{\rho}.$$ (4.26)

Dividing equation (4.26) by dz, and using hydrostatic equilibrium (eq. 3.13b), we obtain the **dry adiabatic lapse rate**:

$$\frac{dT}{dz} = \frac{1}{c_P}\frac{1}{\rho}\frac{dP}{dz} = -\frac{g_P}{c_P}$$

$$= -\frac{\gamma - 1}{\gamma}\frac{g_P \mu_a m_{amu}}{k},$$ (4.27)

where γ is defined as the **ratio of the specific heats**:

$$\gamma \equiv c_P/c_V = C_P/C_V.$$ (4.28)

Typical values for γ are 5/3, 7/5, 4/3, for monatomic, diatomic and polyatomic gases, respectively. The dry adiabatic lapse rate in Earth's lower atmosphere is roughly 10 K km^{-1}.

In a moist atmosphere, the first law of thermodynamics is altered slightly by the inclusion of the release of latent heat, L_s,

$$c_V dT = -P dV - L_s dw_s,$$ (4.29a)

$$c_P dT = \frac{1}{\rho} dP - L_s dw_s,$$ (4.29b)

with w_s the mass of water vapor that condenses out per gram of air. The temperature gradient in saturated air becomes

$$\frac{dT}{dz} = -\frac{g_P}{c_P + L_s dw_s/dT}.$$ (4.30)

The latent heat is effectively added to the specific heat c_P, resulting in a decrease from the dry adiabatic lapse rate. The lapse rate in the presence of clouds is commonly referred to as the **wet adiabatic lapse rate**. On Earth (in the tropics), the wet lapse rate is 5–6 K km^{-1}, slightly more than half the dry rate. Note that the wet adiabatic gradient can never exceed the dry lapse rate.

Convection is extremely efficient at transporting energy whenever the temperature gradient or lapse rate is **superadiabatic** (larger than the adiabatic lapse rate). Energy transport via convection thus effectively places an upper bound on the rate at which temperature can increase with depth in a planetary atmosphere or fluid interior. Substantial superadiabatic gradients are only possible when convection is suppressed by gradients in mean molecular mass or by the presence of a flow-inhibiting boundary, such as a solid surface.

4.5 Radiation

The transport of heat in a planetary atmosphere is typically dominated by radiation in regions where the optical depth of the gas (eq. 4.38) is neither too large nor too small (of order $0.01 < \tau < 1$), so that absorption and re-emission of photons are efficient. This is usually the case in a planet's upper troposphere and stratosphere (§5.1). The radiation efficiency depends critically on the (photon) emission and absorption properties of the material involved. To develop equations of radiative transfer, one needs to be familiar with atomic structure and energy transitions in atoms and molecules and with the radiation 'vocabulary' such as specific intensity, flux density (§4.1.1) and mean intensity. We summarize the basic physics that one needs to know to understand the equations of radiative transfer in this section.

4.5.1 Photons and Energy Levels in Atoms

The energy and momentum of a photon are given by

$$E = h\nu, \tag{4.31a}$$

$$\mathbf{p} = \frac{E}{c}\hat{\mathbf{s}}, \tag{4.31b}$$

where h is Planck's constant, c the speed of light, ν the frequency and $\hat{\mathbf{s}}$ a unit vector pointing in the direction of propagation, as depicted in Figure 4.3.

Emission and absorption of photons by atoms or molecules involve a change in energy state. Each atom consists of a nucleus (protons plus neutrons) surrounded by a 'cloud' of electrons. A detailed description requires a full quantum mechanical analysis, but Bohr's semiclassical theory is sufficient for our purposes. In Bohr's model, the electrons orbit the nucleus such that the centrifugal force is balanced by the Coulomb force:

$$\frac{m_e v^2}{r} = \frac{Zq^2}{r^2}, \tag{4.32}$$

where m_e and v are the mass and velocity of the electron, respectively; r the radius of the electron's orbit (assumed circular); Z the atomic number; and q the electric charge. Electrons are in orbits such that the angular momentum,

$$m_e v r = n_{qm}\hbar, \tag{4.33}$$

and the radius,

$$r = \frac{n_{qm}^2 \hbar^2}{m_e Z q^2}, \tag{4.34}$$

where n_{qm}, an integer, is the **principal quantum number**, and $\hbar \equiv h/2\pi$. The radius of the lowest energy state ($n_{qm} = 1$) for the hydrogen atom ($Z = 1$) is called the **Bohr radius**:

$$r_{Bohr} = \hbar^2/(m_e q^2). \tag{4.35}$$

The energy of orbits n_{qm} can be derived from the kinetic and potential (Coulomb) energy:

$$\begin{aligned} E_e &= \frac{1}{2}m_e v^2 - \frac{Zq^2}{r} = \frac{Zq^2}{2r} - \frac{Zq^2}{r} \\ &= -\frac{Zq^2}{2r} \approx -\frac{\mathcal{R}Z^2}{n_{qm}^2}, \end{aligned} \tag{4.36}$$

where $\mathcal{R}$ is the **Rydberg constant** in the case of the hydrogen atom.

The frequency of various transitions can be calculated using equations (4.31a), (4.31b) and (4.36). An example of energy levels in the hydrogen atom is given in Figure 4.6. The transitions between the ground state and higher levels in a hydrogen atom are called the **Lyman series**, where Ly α is the transition between levels 1 and 2, Ly β between levels 1 and 3, etc. The **Balmer, Paschen** and **Brackett series** indicate transitions between levels 2, 3 and 4 with higher levels, respectively. If the electron is unbound, the atom is **ionized**. For hydrogen in the ground state, photons with energies ≥ 13.6 eV (1 eV $= 1.6 \times 10^{-19}$ J) or wavelengths shorter than 91.2 nm (the **Lyman limit**) may **photoionize** the atom from its ground state (Problem 4-16).

The energy levels of molecules are more numerous than those of isolated atoms because rotation and vibration of the nuclei with respect to one another require energy. This multiplicity leads to numerous molecular lines.

Transitions between energy levels may result in the absorption or emission of a photon with

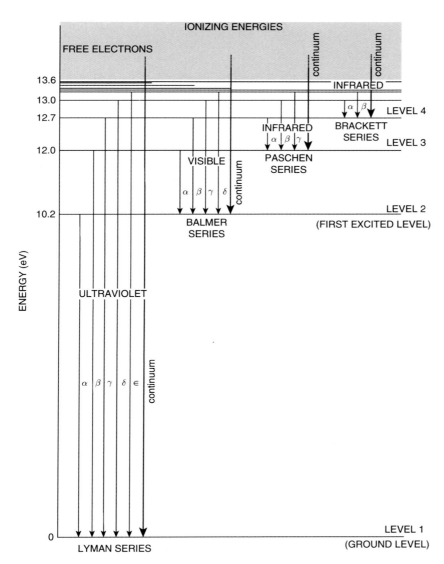

Figure 4.6 The energy levels of hydrogen and the series of transitions among the lowest of these energy levels. (Adapted from Pasachoff and Kutner 1978)

an energy ΔE_{ul} equal to the difference in energy between the two levels u and l. However, transitions are only possible between certain levels: They follow specific selection rules.

The energy difference between electron orbits and therefore the frequency of the photon associated with the transition decrease with increasing n_{qm} (Fig. 4.6). Whereas electronic transitions involving the ground state ($n_{qm} = 1$) may be observed at ultraviolet or optical wavelengths, transitions at high n_{qm}, (hyper)fine structure in atomic spectra and molecular rotation and rotation–vibration transitions all occur at infrared or radio wavelengths because the spacing between energy levels is much smaller. Because each atom/molecule has its own unique set of energy transitions, one can use measurements of absorption/emission spectra to identify particular species in an atmosphere or surface.

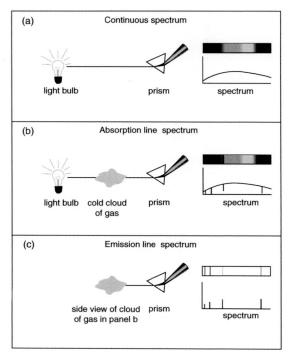

Figure 4.7 COLOR PLATE Illustration of Kirchhoff's laws. (a) The light from the light bulb is refracted through a prism. Because blue light is refracted much more than red light and the bulb produces light with a continuous spectrum, a smooth continuous rainbow of light is observed. (b) When a cloud of gas lies between the light bulb and the prism, we see dark absorption lines at the wavelengths where the gas has absorbed photons; such an absorption line spectrum reveals the composition of the gas. (c) When the cloud of gas from panel b is viewed from a different direction, i.e., without the light bulb behind the cloud (but still illuminating the cloud), we see an emission line spectrum. The emission lines are at the same wavelengths as the absorption lines in panel b.

4.5.2 Spectroscopy

Spectroscopy pertains to the dispersion of light as a function of wavelength. We distinguish among three types of spectra, as illustrated in Figure 4.7. These spectra represent **Kirchhoff's laws** of radiation:

1. A solid, liquid or high-density gas produces (glows with) a continuous thermal spectrum (panel a).

2. A low-density cloud of gas between the observer and a hotter continuous spectrum source absorbs light at specific wavelengths (panel b).

3. A low-density cloud of gas produces emissions at specific wavelengths. Such an emission spectrum can be used to deduce the cloud's composition and temperature (panel c). If the composition of this gas is the same as the gas in panel b, the wavelengths of the absorption and emission lines are equal.

For material that is in **thermodynamic equilibrium** with the radiation field, the amount of energy emitted via thermal excitation, j_ν, must be equal to the amount of energy absorbed:

$$j_\nu = \kappa_\nu B_\nu(T), \qquad (4.37)$$

where κ_ν is the mass absorption coefficient. Planck's function, $B_\nu(T)$, describes the radiation field in thermodynamic equilibrium.

Spectra

In astrophysics, one generally sees absorption lines when atoms or molecules absorb photons at a particular frequency from a beam of broadband radiation and emission lines when they emit photons; this is illustrated in Figure 4.7. The intensity, $\mathcal{F}_{\nu_0}$, at the center of an absorption line is less than the intensity from the background continuum level, $\mathcal{F}_c$: $\mathcal{F}_{\nu_0} < \mathcal{F}_c$, as shown graphically in Figure 4.8. For emission lines: $\mathcal{F}_{\nu_0} > \mathcal{F}_c$. For planets, we see the effects of atomic and molecular line absorption both in spectra of reflected sunlight (at ultraviolet, visible and near-infrared wavelengths) and in thermal emission (at infrared and radio wavelengths) spectra, as exemplified in Figure 4.9.

Planets, moons, asteroids and comets are visible because sunlight is reflected off their surfaces, cloud layers or atmospheric gases, as illustrated in Figure 4.10. Sunlight itself displays a large number of absorption lines, the **Fraunhofer absorption spectrum**, because atoms in the outer layers of

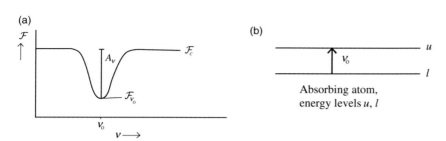

Figure 4.8 (a) Sketch of an absorption line profile. The flux density at the continuum level is $\mathcal{F}_c$; at the center of the absorption line at frequency ν_0, the flux density is $\mathcal{F}_{\nu_0}$. The absorption depth is A_ν. (b) Sketch of the upper, u, and lower, ℓ, energy levels in an atom giving rise to the absorption line in (a).

the Sun's atmosphere (photosphere) absorb part of the sunlight coming from the deeper, hotter layers. If all of the sunlight hitting a planetary surface is reflected back into space, the planet's spectrum is shaped like the solar spectrum (aside from an overall Doppler shift induced by the planet's motion, eq. 4.39); the spectrum thus exhibits the solar Fraunhofer line spectrum. Atoms and molecules in a planet's atmosphere or surface may absorb some of the Sun's light at specific frequencies, producing additional absorption lines in the planet's spectrum. For example, Uranus and Neptune are greenish-blue because methane gas, abundant in these planets' atmospheres, absorbs in the red part of the visible spectrum, so the sunlight reflected back into space is primarily bluish.

As in the case of the Sun, most of the thermal emission from a planetary atmosphere comes from deeper warmer layers, and some of these photons may be absorbed by gases in the outer layers. The temperature decreases with altitude in the Sun's photosphere, and the Fraunhofer absorption lines are visible as a decrease in the line intensity. Similarly, spectral lines formed in a planet's troposphere (lower atmosphere) are also visible as absorption profiles. Lines formed up in a planet's

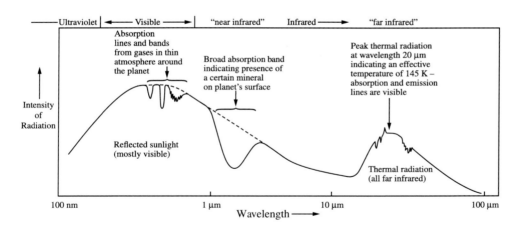

Figure 4.9 Example of a spectrum from a hypothetical planet with an effective temperature of 145 K. The spectrum is shown from ultraviolet through far-infrared wavelengths. At the shorter wavelengths, the Sun's reflected spectrum is shown. The *dashed line* shows the spectrum if there were no absorption lines and bands. The spectrum is already corrected for the Sun's Fraunhofer line spectrum. At infrared wavelengths, the planet's thermal emission is detected, where both absorption and emission lines might be present. Note the hyperfine structure of the molecular bands. (Adapted from Hartmann 1989)

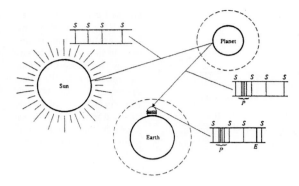

Figure 4.10 A sketch to help visualize the various contributions to an observed planetary spectrum. Sunlight, with its absorption spectrum (indicated by lines S) is reflected off a planet, where the planet's atmosphere may produce additional absorption/emission lines, P. Finally, additional absorption may occur in the Earth's atmosphere before the spectrum is recorded at a telescope, indicated by lines E. (Adapted from Morrison and Owen 2003)

stratosphere, however, are visible as emission profiles. To understand the cause of this difference, we first introduce the concept of **optical depth**, τ_ν, which is defined as the integral of the extinction coefficient along the line of sight, s:

$$\tau_\nu \equiv \int_{s_1}^{s_2} \alpha_\nu(s)\rho(s)\,ds, \qquad (4.38)$$

where α_ν is the mass extinction coefficient. In a spectrum of a planet's atmosphere, the optical depth at the center of the line is always largest, (much) larger than in the far wings or with respect to the continuum background. The line profile of a planet's thermal emission spectrum reveals the temperature and pressure at the altitudes probed. In the troposphere, the temperature decreases with altitude, so that lines forming in the troposphere are seen in absorption against the warm continuum background. In contrast, if a line is formed above the tropopause, where the temperature is increasing with altitude, the line is seen in emission against the cooler background. Thus, rather than speaking about emission or absorption lines as in (astro)physics, in atmospheric science, spectral lines are seen **in emission** ($\mathcal{F}_{\nu_0} > \mathcal{F}_c$) or **in**

absorption ($\mathcal{F}_{\nu_0} < \mathcal{F}_c$) depending on whether the temperature is increasing or decreasing with altitude in the region of line formation.

The Sun, planets and the largest satellites in our Solar System are usually well resolved on modern photographs. In such images, one can discern an effect referred to as **limb darkening** or **limb brightening**, where the former refers to a gradual (or sometimes more abrupt) decrease in the intensity of the light with distance from the center of the object towards the limb (edge) and the latter refers to an increase in intensity. If we observe the thermal emission from a giant planet, this effect can easily be explained using the concept of optical depth, discussed earlier, combined with the temperature profile in an atmosphere. If the temperature in the atmosphere is decreasing with altitude, we see limb darkening because of the greater path length through the cooler outer portion of the atmosphere near a planet's limb than at its center. If instead the temperature were increasing, one would expect to see limb brightening. Using the same concept of optical depth, one can also explain these effects on images of bodies with atmospheres seen in reflected sunlight. For example, near-infrared (1–2 µm) images of Titan show a limb-brightened disk. At these wavelengths, the satellite is seen in reflected sunlight. Haze layers in the atmosphere, which reflect light, cause a limb brightening because the haze optical depth increases towards the limb.

Doppler Shift and Line Broadening

In §4.5.1, we show that photons emitted or absorbed when an atom gets excited or de-excited have a specific energy (frequency), which corresponds to the difference in the energy levels between the upper, u, and lower, l, states, ΔE_{ul}. However, according to the **Heisenberg uncertainty principle**, a photon can be absorbed/emitted with an energy slightly different from ΔE_{ul}, which results in a finite width for the observed line profile,

Φ_ν. Usually a line profile is broader than this **natural broadened** profile, as discussed below.

The frequency of the object's emission and absorption lines is Doppler shifted by the amount

$$\Delta\nu = \frac{\nu v_r}{c}, \tag{4.39}$$

with v_r the velocity of the object along the line of sight. The **Doppler shift** is positive (**blue shifted**) if the atom moves towards the observer and negative (**red shifted**) if the atom moves in the opposite direction. This phenomenon has been used, for example, to determine the rotation rate of the planet Mercury using radar techniques. Upon transmitting a radar pulse at a specific frequency, the signal is blue shifted by the approaching and red shifted by the receding limb. These Doppler shifts broaden the radar signal upon reflection by an amount proportional to the planet's rotation rate.

In an atmosphere, atoms and molecules move in all directions, and hence the observed line profile is **Doppler broadened**, i.e., much broader than the species' natural line profile.

In a dense gas, collisions between particles perturb the energy levels of the electrons such that photons with a slightly lower or higher frequency can cause excitation/deexcitation. This also leads to a broadening of the line profile. The higher the pressure, the broader the line profile; this is referred to as **pressure broadening**, and pressure broadened profiles tend to be broader than Doppler broadened profiles. The detailed shape of a line profile is thus determined by the abundance of the element or compound producing the line, as well as the pressure and temperature of the environment.

4.5.3 Radiative Energy Transport

When the primary mechanism of energy transport in an atmosphere is the absorption and re-emission of photons, the temperature–pressure profile is governed by the equations of radiative energy transport. The change in intensity, dI_ν, caused by absorption and emission within a cloud of gas is equal to the difference in intensity between emitted and absorbed radiation:

$$dI_\nu = j_\nu \rho \, ds - I_\nu \alpha_\nu \rho \, ds, \tag{4.40}$$

where I_ν is the **specific intensity**, which has the dimensions of W m^{-2} sr^{-1} Hz^{-1}.

In equation (4.40), j_ν is the emission coefficient due to scattering and/or thermal excitation: $j_\nu = j_\nu(\text{scattering}) + j_\nu(\text{thermal excitation})$. The quantity α_ν is the mass extinction coefficient. Absorption and scattering both contribute to the extinction: $\alpha_\nu = \kappa_\nu + \sigma_\nu$, where κ_ν and σ_ν are the mass absorption and mass scattering coefficients, respectively.

The specific intensity of radiation at frequency ν emitted by a blackbody is

$$I_\nu = B_\nu(T). \tag{4.41}$$

The intensity is emitted into a solid angle $d\Omega_s$, as depicted in Figure 4.3. The solid angle, expressed in steradians (sr, i.e., radians^2), is defined such that, integrated over a sphere, it is

$$\oint d\Omega_s = \int_0^{2\pi} \int_0^\pi \sin\theta \, d\theta \, d\phi = 4\pi \text{ sr}. \tag{4.42}$$

The **mean intensity**, J_ν, of the radiation field is equal to

$$J_\nu \equiv \frac{\oint I_\nu \, d\Omega_s}{\oint d\Omega_s} = \frac{1}{2} \int_{-1}^1 I_\nu \, d\cos\theta. \tag{4.43}$$

Denoting the coordinate in the direction of the normal to the planet's surface by **z** and the angle between the line of sight **s** and **z** by θ, as in Figure 4.3, we get $ds = dz/\cos\theta$. Equation (4.40) can then be rewritten:

$$\frac{\cos\theta}{\rho} \frac{dI_\nu}{dz} = j_\nu - \alpha_\nu I_\nu. \tag{4.44}$$

Using $\mu_\theta \equiv \cos\theta$, and the definition of optical depth (eq. 4.38), equation (4.44) becomes

$$\mu_\theta \alpha_\nu \frac{dI_\nu}{d\tau_\nu} = j_\nu - \alpha_\nu I_\nu. \qquad (4.45)$$

The **source function**, S_ν, is defined as

$$S_\nu \equiv \frac{j_\nu}{\alpha_\nu}. \qquad (4.46)$$

Dividing equation (4.45) by α_ν and using equation (4.46), we obtain the **general equation of radiative transport**:

$$\mu_\theta \frac{dI_\nu}{d\tau_\nu} = S_\nu - I_\nu. \qquad (4.47)$$

If S_ν does not vary with optical depth, the solution to equation (4.47), assuming $\mu_\theta = 1$, is:

$$I_\nu(\tau_\nu) = S_\nu + e^{-\tau_\nu}(I_\nu(0) - S_\nu), \qquad (4.48)$$

where $I_\nu(0)$ is the 'background' radiation that gets attenuated when propagating through the absorbing medium. In radiative transfer calculations of planetary atmospheres, where one probes down into the atmosphere, one usually defines optical depth as increasing into the atmosphere, i.e., $\tau_\nu = 0$ at the top of the atmosphere.

In the remainder of this subsection, we examine the equation of radiative transfer, or, more explicitly, the source function S_ν for four 'classic' cases. These examples help to elucidate the theory of radiative transfer.

(1) Consider a cloud of gas along the line of sight, as sketched in Figure 4.11. Assume that the cloud itself does not emit radiation, thus $j_\nu = 0$. Suppose that there is a source of radiation behind the cloud. The incident light $I_\nu(0)$ is reduced in intensity according to equation (4.48), and the resulting observed intensity becomes

$$I_\nu(\tau_\nu) = I_\nu(0)e^{-\tau_\nu}. \qquad (4.49)$$

This relationship is called **Lambert's exponential absorption law**. If the gas cloud is optically thin ($\tau_\nu \ll 1$), equation (4.49) can be

$I_\nu(0)$ $I_\nu(\tau_\nu)$

Figure 4.11 Illustration of a cloud of gas along the line of sight.

approximated by: $I_\nu(\tau_\nu) = I_\nu(0)(1 - \tau_\nu)$. For very small τ_ν, $I_\nu \rightarrow I_\nu(0)$; i.e., the intensity of the radiation is defined by the incident radiation. If the cloud is optically thick ($\tau_\nu \gg 1$), the radiation is reduced to near zero.

(2) If the material in the cloud is in **local thermodynamic equilibrium** (LTE), the radiation field obeys:

$$I_\nu = J_\nu = B_\nu(T). \qquad (4.50)$$

If there is no scattering, i.e., $\sigma_\nu = 0$, then $\kappa_\nu = \alpha_\nu$. Because the material is in equilibrium with the radiation field, the amount of energy emitted must be equal to the amount of energy absorbed, as described by Kirchhoff's law, equation (4.37). The source function is therefore given by

$$S_\nu = B_\nu(T). \qquad (4.51)$$

(3) Let us assume that j_ν is due to scattering only: $j_\nu = \sigma_\nu I_\nu$. We receive sunlight that is reflected towards us (see Fig. 4.4). In general, scattering removes radiation from a particular direction and redirects or introduces it into another direction. If photons undergo only one encounter with a particle, the process is referred to as **single scattering**; multiple scattering refers to multiple encounters. The **single scattering albedo**, ϖ_ν, represents the fraction of radiation lost due to scattering:

$$\varpi_\nu \equiv \frac{\sigma_\nu}{\alpha_\nu}. \qquad (4.52)$$

The single scattering albedo is equal to unity if the mass absorption coefficient, κ_ν, is equal to zero. In this case

$$S_\nu = \varpi I_\nu = I_\nu. \qquad (4.53)$$

(4) In the situation of LTE ($I_\nu = J_\nu$) and isotropic scattering, the fraction of radiation that is absorbed by the medium, $\kappa_\nu/\alpha_\nu = 1 - \varpi_\nu$. Using Kirchoff's law (eq. 4.37), we can write the source function

$$S_\nu = \varpi_\nu J_\nu + (1 - \varpi_\nu)B_\nu(T). \qquad (4.54)$$

Problems (4-11) and (4-12) contain exercises related to radiative transfer in an atmosphere. In Problem (4-11), the student is asked to calculate the hypothetical brightness temperature of Mars at different frequencies. The temperature depends on the optical depth in the planet's atmosphere; if the optical depth is near zero or infinity, the problem essentially simplifies to case (1) discussed above. If the optical depth is closer to unity, the radiation from the planetary disk as well as the atmosphere contributes to the observed intensity.

4.5.4 Radiative Equilibrium

Energy transport in a planet's stratosphere, the region above the tropopause (§5.1), is usually dominated by radiation. If the total radiative flux is independent of height, the atmosphere is in **radiative equilibrium**:

$$\frac{d\mathcal{F}}{dz} = 0. \qquad (4.55)$$

If an atmosphere is in both hydrostatic and radiative equilibrium and its equation of state is given by the perfect gas law, then its temperature–pressure relation can be derived (§4.6.2):

$$\frac{dT}{dP} \approx -\frac{3}{16}\frac{T}{g_p}\left(\frac{T_e}{T}\right)^4 \alpha_R. \qquad (4.56)$$

with α_R a mean absorption coefficient. Both T and $B_\nu(T)$, as well as the abundances of many of the absorbing gases, vary with depth in a planetary atmosphere. The best approach to solving for the temperature structure is to solve the transport equation (4.47) at all frequencies, together with the requirement that the flux, $\mathcal{F}$, is constant with depth (eq. 4.55).

4.6 Greenhouse Effect

The surface temperature of a planet can be raised substantially above its equilibrium temperature if the planet is overlain by an atmosphere that is optically thick at infrared wavelengths, a situation sketched in Figure 4.12, and referred to as the **greenhouse effect**. Sunlight, which has its peak intensity at optical wavelengths (Fig. 4.2 and eq. 4.3 for a blackbody of temperature $\sim$5700 K), enters an atmosphere that is relatively transparent at visible wavelengths and heats the surface. The warm surface (perhaps augmented by geothermal heat) radiates its heat at infrared wavelengths. This radiation does not immediately escape into interplanetary space but is absorbed by air molecules, especially CO_2, H_2O and CH_4. When these molecules deexcite, photons at infrared wavelengths are emitted in a random direction. The net effect of this process is that the atmospheric (and surface) temperature is increased until equilibrium is reached between solar energy input and the emergent planetary flux (Fig. 4.12). The greenhouse effect raises the (near) surface temperature of Earth, Venus and Titan considerably, and that of Mars to a lesser extent.

To compute the temperature profile of an actual planetary atmosphere, one needs to account for the energy absorbed from incident light as well as the emitted thermal flux. Globally averaged, almost one-quarter of the solar radiation that reaches the top of Earth's atmosphere is absorbed by the atmosphere, almost half is absorbed at the surface and almost one-third is reflected back to space without being absorbed. Nonetheless, the optical depth of Earth's atmosphere (and those of other planets) is far larger to thermal infrared radiation than it is to (visible wavelength) sunlight. Strong positive correlations between temperature and CO_2 abundance in Earth's atmosphere during the past 400 000 years are illustrated in Figure 5.13.

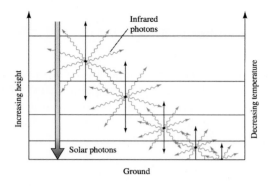

Figure 4.12 Schematic view of the transmission and scattering of radiation by an atmosphere that contains greenhouse gases. The short-wavelength sunlight passes through the atmosphere, delivering energy to the surface. In contrast, energy is reradiated by the planet at longer wavelengths, and much of this radiation is scattered back downwards by the greenhouse gases in the atmosphere. (Lunine 2005)

4.6.1 Quantitative Results

In §4.6.2, we show that the temperature in an atmosphere at an optical depth to outgoing radiation τ, i.e., $T(\tau)$, can be calculated from its temperature at the top of the atmosphere, T_0, and τ:

$$T^4(\tau) = T_0^4 \left(1 + \frac{3}{2}\tau \right). \qquad (4.57)$$

We further show that the effective temperature, T_e, is related to T_0:

$$T_e^4 = 2T_0^4. \qquad (4.58)$$

The temperature profile in an idealized gray atmosphere, assuming $T_0 = 200$ K and the temperature $T(\tau)$ to follow equation (4.57), is shown in Figure 4.13. We indicate the values of T_0 and T_e on the graph. If the temperature of a body is determined exclusively by the incident solar flux, the effective temperature is equal to the equilibrium temperature, i.e., $T_e = T_{eq}$. At the top of the atmosphere, where the optical depth is zero, the temperature $T_0 \approx 0.84\, T_e$. By combining equations (4.57) and (4.58), we see that the temperature, $T(\tau)$, is equal to the effective temperature, T_e, at an optical depth $\tau(z) = 2/3$. Thus, an external observer detects continuum radiation that emerged from an effective depth in the atmosphere where $\tau = 2/3$ (Problem 4-17).

We denote the air temperature just above the ground by $T(\tau_g)$, where τ_g is the optical depth (again of outgoing, typically IR, radiation) to the ground. The value of $T(\tau_g)$ can be computed from equations (4.57) and (4.58). The actual ground or surface temperature, T_g, is larger:

$$T_g^4 = T^4(\tau_g) + \frac{1}{2}T_e^4 = T_e^4 \left(1 + \frac{3}{4}\tau_g \right). \qquad (4.59)$$

The surface temperature T_g is indicated on Figure 4.13 for $\tau_g = 100$. Equation (4.59) and Figure 4.13 show that the surface temperature in a radiative atmosphere can be very high if the infrared opacity of the atmosphere is high.

The greenhouse effect is particularly strong on Venus, where the surface temperature reaches a

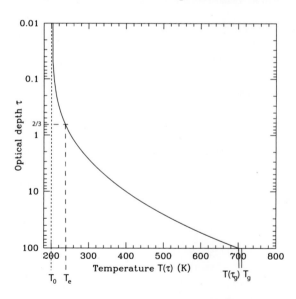

Figure 4.13 Sketch of a temperature profile as a function of optical depth in an idealized gray atmosphere. We adopted a temperature at the top of the atmosphere, $T_0 = 200$ K, and calculated the temperature profile with equation (4.57). T_0, the effective temperature T_e, the temperature just above the surface $T(\tau_g)$ and the surface temperature T_g, assuming an optical depth at the ground $\tau_g = 100$, are indicated along the horizontal axis.

value of 733 K, almost 500 K above the equilibrium temperature of ∼240 K. The greenhouse effect is also noticeable on Titan and Earth, where the temperature is raised by 21 K and 33 K, respectively, compared with their equilibrium temperatures. On Mars, the temperature is raised by ∼6 K.

The greenhouse warming on Titan, however, is partially compensated by cooling produced by small haze particles in the stratosphere that block short-wavelength sunlight but are transparent to long-wavelength thermal radiation from Titan; this process is known as the **anti-greenhouse** effect. Similar effects are observed on Earth after giant volcanic eruptions, such as the 1991 explosion of Mount Pinatubo in the Philippines, which injected huge amounts of ash into the stratosphere.

Icy material allows sunlight to penetrate several centimeters or more below the surface but is mostly opaque to reradiated thermal infrared emission. Thus, the subsurface region can become significantly warmer than the equilibrium temperature would indicate. In analogy with atmospheric trapping of thermal infrared emission, this process is known as the **solid-state greenhouse effect**. This process may be important on icy bodies, such as the Galilean satellites and comets. The subsurface Lake Untersee in Antarctica is maintained by a special case of the solid-state greenhouse effect known as the **ice-covered greenhouse effect** that is sketched in Figure 4.14. The ice-covered greenhouse effect may induce habitable liquid water environments just below the surface of a body that is too cold (and/or does not have sufficient atmospheric pressure) to have liquid water on its surface.

4.6.2* Derivations

Thermal Profile

The temperature structure in an atmosphere that is in radiative equilibrium can be obtained from the **diffusion equation**, an expression for the radiative

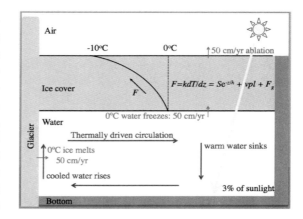

Figure 4.14 Schematic illustration of the ice-covered greenhouse effect. Ice is more transparent to visible radiation than to IR photons, so energy from sunlight penetrates below the surface and little is able to escape, enabling the subsurface water to warm and melt. Earth's atmosphere ablates ice from the surface, and the top of the lake freezes to maintain sufficient ice above the liquid for energy balance to be maintained. (Courtesy Chris McKay).

flux at altitude z. In the following, we derive the diffusion equation in an optically thick atmosphere that is approximately in LTE: $I_\nu \approx S_\nu \approx B_\nu(T)$. We assume the atmosphere to be in monochromatic radiative equilibrium: $d\mathcal{F}_\nu/dz = 0$.

Integration of equation (4.47) over a sphere yields:

$$\frac{d\mathcal{F}_\nu}{d\tau_\nu} = 4\pi(B_\nu - J_\nu), \qquad (4.60)$$

where $\mathcal{F}_\nu$ is the flux density across a layer in a stratified atmosphere. Multiplying equation (4.47) by μ_θ and integrating over a sphere yields the following relationship between J_ν and $\mathcal{F}_\nu$:

$$\frac{4\pi}{3}\frac{dJ_\nu}{d\tau_\nu} = -\mathcal{F}_\nu. \qquad (4.61)$$

Setting $d\mathcal{F}_\nu/d\tau_\nu = 0$ in equation (4.60), and using equation (4.61), we find:

$$\frac{dB_\nu}{d\tau_\nu} = -\frac{3}{4\pi}\mathcal{F}_\nu. \qquad (4.62)$$

Integrating over frequency yields the total radiative flux, or the **radiative diffusion equation**:

$$\mathcal{F}(z) = -\frac{4\pi}{3\rho}\frac{\partial T}{\partial z}\int_0^\infty \frac{1}{\alpha_v}\frac{\partial B_v(T)}{\partial T}dv. \qquad (4.63)$$

Equation (4.63) can be simplified by the use of a wavelength-averaged absorption coefficient, such as the **Rosseland mean absorption coefficient**, α_R:

$$\frac{1}{\alpha_R} \equiv \frac{\int_0^\infty \frac{1}{\alpha_v}\frac{\partial B_v}{\partial T}dv}{\int_0^\infty \frac{\partial B_v}{\partial T}dv}. \qquad (4.64)$$

With this simplification, we write the radiative diffusion equation as

$$\mathcal{F}(z) = -\frac{16\,\sigma T^3}{3\,\alpha_R\rho}\frac{\partial T}{\partial z}. \qquad (4.65)$$

Note that flux travels upward in an atmosphere if the temperature gradient dT/dz is negative (i.e., where the temperature decreases with altitude).

Applying equations (4.14) and (4.65), the atmospheric temperature profile is

$$\frac{dT}{dz} = -\frac{3}{16}\frac{\alpha_R\rho}{T^3}T_e^4. \qquad (4.66)$$

If an atmosphere is in both hydrostatic and radiative equilibrium and its equation of state is given by the perfect gas law, then its temperature–pressure relation is given by equation (4.56).

Greenhouse Effect

For this analysis, it is convenient to use the **two-stream approximation**:

$$I_v = (I_v^+ + I_v^-), \qquad (4.67)$$

where I_v^+ is the upward and I_v^- the downward radiation at frequency v. The net flux density across a layer becomes

$$\mathcal{F}_v = \pi(I_v^+ - I_v^-). \qquad (4.68)$$

We consider an atmosphere in monochromatic radiative equilibrium ($d\mathcal{F}_v/dz = 0$) and LTE, which is heated from below (i.e., $I_v^- \equiv 0$ at the top of the atmosphere). We are thus treating the incoming solar radiation as a source of energy at the ground, and modelling the outgoing radiation in the thermal IR, which is a reasonable approximation if $\tau_{IR} \ll \tau_{vis}$. The upward intensity at the ground, I_{vg}^+, can be expressed with help of equations (4.60), (4.67) and (4.68):

$$I_{vg}^+ \equiv B_v(T_g) = B_v(T(\tau_g)) + \frac{1}{2\pi}\mathcal{F}_v. \qquad (4.69)$$

The downward and upward intensities at the top of the atmosphere are

$$I_{v0}^- \equiv 0 = B_v(T_0) - \frac{1}{2\pi}\mathcal{F}_v, \qquad (4.70a)$$

$$I_{v0}^+ = B_v(T_0) + \frac{1}{2\pi}\mathcal{F}_v = 2B_v(T_0), \qquad (4.70b)$$

where T_0 is the temperature of the upper boundary, usually referred to as the **skin temperature**. Thus, the upward intensity at the top of the atmosphere is twice as large as that emitted by an opaque blackbody at temperature T_0. We can derive the brightness, and therefore temperature, at τ by solving equation (4.62) and using equation (4.70a):

$$B_v(\tau) = B_v(T_0)\left(1 + \frac{3}{2}\tau_v\right). \qquad (4.71)$$

Integration over frequency and conversion to temperature via the Stefan–Boltzmann law (eq. 4.11) yields equation (4.57) given in the previous section.

The total radiant flux from a body can be obtained by integrating equation (4.70b) over frequency. This flux translates into the effective temperature, T_e, given in equation (4.58).

The ground or surface temperature, T_g, can be obtained from equation (4.69) as the form given by the first equality within equation (4.59). Note that there is a discontinuity: The surface temperature, T_g, is higher than the air temperature, $T(\tau_g)$, just above it. In a real planetary atmosphere, conduction reduces this difference. Application of equations (4.71) and (4.58) results in the second equality within equation (4.59).

Key Concepts

- Any object with a temperature above absolute zero emits a continuous spectrum of electromagnetic radiation at all frequencies, which is its thermal or blackbody radiation, given by Planck's radiation law.
- The total emitted flux from a blackbody is proportional to the fourth power of its temperature, and is given by the Stefan–Boltzmann law.
- The (monochromatic) albedo is the ratio of reflected + scattered to the total incident radiation at frequency ν. When integrated over frequency, this is referred to as the Bond albedo.
- The equilibrium temperature of a body is that temperature the body would have if incoming sunlight is balanced by reradiation outwards.
- The effective temperature is the temperature that corresponds to a blackbody emitting the same amount of flux, integrated over all frequencies.
- Energy can be transported via conduction, radiation and convection.
- A parcel of air moves adiabatically when no heat is exchanged between the parcel of air and its environment.
- The temperature profile in an atmosphere that is marginally unstable to convection is given by the adiabatic lapse rate. The temperature in such a dry atmosphere follows a dry adiabat. If air is saturated, the temperature follows a wet adiabat.
- Spectral lines are seen in emission or in absorption depending on whether the temperature is increasing or decreasing with altitude in the region of line formation.
- Spectral lines are Doppler shifted towards longer (red) wavelengths when the object is moving away from the observer, and towards shorter (blue) wavelengths when the object is moving towards the observer.
- When energy transport is dominated by absorption and re-emission of photons, the temperature structure is governed by the equations of radiative transport.
- If a planet has an atmosphere that is optically thick at infrared wavelengths, its surface temperature can be raised considerably as a result of the greenhouse effect.

Further Reading

A classical book on spectroscopy:

Herzberg, G., 1944. *Atomic Spectra and Atomic Structure*. Dover Publications, New York. 257pp.

Books that discuss radiative transfer in detail:

Chandrasekhar, S., 1960. *Radiative Transfer*. Dover, New York. 392pp.

Thomas, G.E., and K. Stamnes, 1999. *Atmospheric and Space Science Series: Radiative Transfer in the Atmosphere and Ocean*. Cambridge University Press, Cambridge. 517pp.

Problems

4-1. Write Planck's radiation law (eq. 4.3) in terms of wavelength λ rather than frequency v.

4-2. Use your expression from Problem 4.1 to derive equation (4.8).

4-3. A dust grain is heated by the interstellar radiation field, which has a temperature T_{ISM}. If the grain were a perfect blackbody, its temperature would also be T_{ISM}. Do you expect the grain to be warmer or colder than this value, and why?

4-4. Sedna is a minor planet that orbits far beyond the orbit of Pluto. It is difficult to measure the radius of a small, distant Solar System object because optical imaging only constrains AR^2, the product of the square of the radius and the albedo (so for a given brightness, you do not know if the object is big but dark or small and bright). However, Sedna was also observed with the *Spitzer Space Telescope*, which measured the total flux emitted by Sedna in the infrared. Explain, using the appropriate equation, why this measurement allows the radius to be measured.

4-5. Calculate the equilibrium temperature for the Moon:
(a) Averaged over the lunar surface. (Hint: Assume the Moon to be a rapid rotator.)
(b) As a function of solar elevation, assuming the Moon to be a slow rotator.

4-6. Is the spectrum of emitted thermal radiation from a planet broader or narrower than the blackbody spectrum? Why?

4-7. A body emits and absorbs radiation of any given frequency with the same efficiency. Given this fact, why does the temperature of a planet depend on its albedo?

4-8. Thermal emissions by solid bodies can be analyzed to provide information on the temperature at a few wavelengths below the surface. Observations of the night hemisphere of Mercury longward of ~ 0.1 m yield temperatures close to the diurnal equilibrium temperature. Because radiation is obviously able to escape directly from the regions being observed, why does it not get much colder than this during the long mercurian night?

4-9. **(a)** Neglecting internal heat sources and assuming rapid rotation, calculate the average equilibrium temperature for all eight planets using the data provided in Tables E.1, E.9 and E.10.
(b) At which wavelength would you expect the blackbody spectral peak of each planet?

4-10. Jupiter's effective temperature is observed to be 125 K. Compare the observed temperature with the equilibrium temperature calculated in the previous problems. What could be responsible for the difference? (Hint: Consider the assumptions that are involved in the derivation of the equilibrium temperature.)

4-11. The solid surface of Mars is opaque at all wavelengths and is surrounded by an optically thin atmosphere. The atmosphere absorbs in a narrow spectral region: Its absorption coefficient is large at a wavelength $\lambda_0 = 2.6$ mm and negligibly small at other radio wavelengths; thus, for most radio wavelengths, λ_1: $\alpha_{\lambda_0} \gg \alpha_{\lambda_1}$. Assume that the temperature of Mars's surface $T_s = 230$ K and of its atmosphere $T_a = 140$ K. (Note: For both parts of this problem, you may use approximations appropriate for radio wavelengths, even though they are not fully correct at mm wavelengths.)

(a) What are the observed brightness temperatures at λ_1 and at λ_0?

(b) What is the observed brightness temperature at λ if the optical depth of the atmosphere at this wavelength is equal to 0.5?

4-12. Consider a hypothetical planet of temperature T_p, surrounded by an extensive atmosphere of temperature T_a, where $T_a < T_p$.

The atmosphere absorbs in a narrow spectral line; its absorption coefficient is large at frequency ν_0 and is negligibly small at other frequencies, such as ν_1: $\alpha_{\nu_0} \gg \alpha_{\nu_1}$. The planet is observed at frequencies ν_0 and ν_1. Assume that the Planck function does not change much between ν_0 and ν_1.

(a) When observing the center of the planet's disk, at which frequency, ν_0 or ν_1, is the brightness temperature higher? Is the same true when observing near the limb, where the limb is defined as the atmosphere beyond the outer edge of the solid planet? Make a sketch of the 'observed' spectra.

(b) Repeat for $T_a > T_p$.

4-13. (a) Using your answers to Problems 4-9 and 4-10 and assuming that all energy transport inside Jupiter is by conduction, estimate the temperature at the interior of Jupiter. The thermal conductivity of dense hydrogen (the main constituent of Jupiter) is about 1000 J m^{-1} s^{-1} K^{-1}.

(b) We know from the size of Jupiter that the typical interior temperatures must be below 10^5 K. Compare your answer to this number and discuss whether or not your result is plausible. If not, which assumption that led to this result is likely in error?

4-14. During the ice ages, much more of the Earth was covered by ice.

(a) Would that increase or decrease Earth's albedo?

(b) Estimate an 'ice age' albedo for Earth and compute the equilibrium temperature for the planet.

(c) What if the entire Earth were covered by ice? Again estimate a planetary albedo and equilibrium temperature.

(d) Thinking just about the solar energy balance, would temperature changes tend to be stable or unstable?

4-15. Calculate the expected increase in the global average temperature of the Earth at a full Moon compared with a new Moon (neglecting eclipses). Note: Two qualitatively different factors must be taken into account.

4-16. (a) Calculate the wavelength and energy of photons corresponding to the Lyman α, Balmer β and Brackett α emission from a hydrogen atom.

(b) Calculate the wavelength and energy of photons necessary to ionize a hydrogen atom from the electronic ground state.

4-17. Consider a rapidly rotating planet with an atmosphere in radiative equilibrium. The planet is located at a heliocentric distance $r_\odot = 2$ AU; its Bond albedo $A_b = 0.4$ and emissivity $\epsilon = 1$ at all wavelengths. Assume that the planet is heated exclusively by solar radiation.

(a) Calculate the effective and equilibrium temperatures.

(b) Calculate the temperature at the upper boundary of the atmosphere, where $\tau = 0$.

(c) Show that continuum radiation from the planet's atmosphere is received from a depth where $\tau = 2/3$.

(d) If the optical depth of the atmosphere $\tau_g = 10$, determine the surface temperature of this planet.

4-18. Other things equal, the temperature near Earth's surface cools off more rapidly on a humid night than on a dry night.

(a) Assuming it is not so humid that condensation occurs, what is responsible for this difference?

(b) What two additional factors are important if the humidity is high enough that water droplets form?

CHAPTER 5

Planetary Atmospheres

Blow, winds, and crack your cheeks! rage! blow!
You cataracts and hurricanoes, spout
Till you have drench'd our steeples,
drown'd the cocks!

William Shakespeare, *King Lear*, Act III, Scene II

The **atmosphere** is the gaseous outer portion of a planet. Atmospheres have been detected around all eight planets in our Solar System, several planetary satellites, the dwarf planet Pluto and several extrasolar planets. Atmospheric properties exhibit tremendous diversity. Some atmospheres are very dense, and those of the giant planets gradually blend into fluid envelopes that contain most of the planet's mass. Other atmospheres are extremely tenuous, so tenuous that even the best laboratory vacuum seems dense in comparison. The composition of planetary atmospheres varies from the solar-like hydrogen/helium envelopes of the giant planets to atmospheres dominated by nitrogen and/or carbon dioxide for the three largest terrestrial planets, and a diverse ensemble, including esoteric gases such as sulfur dioxide or sodium, for smaller bodies.

The more substantial atmospheres were formed while the planets accreted. In contrast, the most tenuous atmospheres are continuously produced by, for example, volcanic activity (Io, Enceladus), sublimation of ices (Mars, Io, Pluto, Triton) and **sputtering** (Mercury and the Moon, many of the icy satellites and Saturn's rings). Sputtering is a process wherein atoms and molecules from a planet's surface are 'kicked' up by energetic particles (solar wind, magnetospheric plasma) and micrometeorites.

Despite their diverse masses and compositions, all atmospheres are governed by the same physical and chemical processes. The thermal structure of a planet's atmosphere results primarily from the efficiency of energy transport, and the primary source of energy is the Sun (star). The various mechanisms by which energy is transported are discussed in §4.2. For most bodies with substantial atmospheres, temperature decreases with altitude from the surface (or the deep interior for giant planets) upwards; this region is referred to as the **troposphere**. Near the 0.1 bar level there is a temperature minimum; this is the **tropopause**. Temperature

increases with height above the tropopause. Clouds form in the troposphere of many atmospheres, but their compositions differ vastly from one body to another because the gases available to condense differ. The upper layers of an atmosphere are modified by photochemistry, with the particulars depending on atmospheric composition. Variations in temperature and pressure within an atmosphere lead to winds, which can be steady or turbulent, strong or weak.

In this chapter, we discuss the various processes operating in planetary atmospheres and relate these to sample observations in our Solar System. We start in §5.1 with discussions of an atmosphere's thermal structure, the various sources of heat and how energy is transported. This section concludes with a short summary of observed temperature–pressure profiles. The composition of atmospheres is discussed in §5.2. Clouds and basic concepts of meteorology are summarized in §5.3 and §5.4, respectively. Section 5.5 explains photochemistry and photoionization using ozone chemistry and Earth's ionosphere as examples. Diffusion processes are discussed in §5.6, and atmospheric escape is summarized in §5.7. The formation and subsequent evolution of the atmospheres and climates of terrestrial planets are described in §5.8. The atmospheres of individual planets and moons within the Solar System are discussed in more detail in Chapters 8–10. Exoplanet atmospheres are considered in §14.3.3.

5.1 Thermal Structure

The relationships between temperature, pressure and density in a planetary atmosphere are governed by a balance between gravity and pressure. To first approximation, atmospheres are in hydrostatic equilibrium, and the temperature, pressure and density of the gases are related to one another via the ideal gas law. In §3.2, we derive the barometric law, equation (3.19), with the scale height, $H(z)$, given by equation (3.16). Approximate pressure scale heights for the planets, Titan, the Moon, Triton and Pluto are shown in Tables E.9–E.11. It is interesting to note that H is of the order of 10–25 km for most planets because the ratio $T/(g_p\mu_a)$ for the giant and terrestrial planets is similar. Only in the tenuous atmospheres of Mercury, Pluto and various moons is the scale height larger.

The **thermal structure** of a planet's atmosphere, dT/dz, is primarily governed by the efficiency of energy transport. The various mechanisms by which energy is transported depend largely on the optical depth in an atmosphere and are discussed in §4.2–4.5. To determine an atmosphere's thermal structure, one has to consider the energy source(s) and all possible processes that may, directly or indirectly, affect the temperature:

(1) The top of the atmosphere is irradiated by the Sun. Some of this radiation is absorbed and scattered in the atmosphere. This process, together with radiative losses and conduction, are the primary factors defining the temperature profile in the upper part of the atmosphere.

(2) Energy from internal heat sources and reradiation of absorbed sunlight by a planet's surface or dust in its atmosphere modify (in some cases dominate) the temperature profile.

(3) Chemical reactions in an atmosphere change its composition, which leads to changes in opacity and hence thermal structure.

(4) Clouds and/or photochemically produced haze layers scatter incident light, affecting the energy balance. Clouds and hazes also increase the atmospheric opacity and change the temperature locally through release (cloud formation) or absorption (evaporation) of latent heat.

(5) Volcanoes and geyser activity on some planets and satellites may modify their atmospheres substantially.

(6) Atmospheres of terrestrial planets and satellites are affected by chemical interactions between the atmosphere and the crust or ocean.

(7) The Earth's atmospheric composition, opacity and thermal structure are influenced by biochemical and anthropogenic processes.

Even though the atmospheric composition varies drastically from one planet/satellite to another, the temperature structure of all but the most tenuous atmospheres is qualitatively similar, as shown in Figure 5.1. Moving upwards from the surface or, for the giant planets, from the deep atmosphere, the temperature decreases with altitude; this part of the atmosphere is called the **troposphere**. It is in the troposphere that condensable gases, usually trace elements, form clouds. The atmospheric temperature typically reaches a minimum at the **tropopause**, near a pressure level of ~0.1 bar. Above the tropopause, in the **stratosphere**, the temperature increases going upwards. At higher altitudes, the **mesosphere** is characterized by temperature decreasing with altitude. The **stratopause** forms the boundary between the stratosphere and mesosphere. On Earth, Titan and perhaps Saturn, the **mesopause** forms a second temperature minimum. Above the mesopause, in the **thermosphere**, the temperature increases with altitude, up to the **exosphere**, which is the outermost part of an atmosphere. Collisions between gas molecules in the exosphere are rare, and the rapidly moving molecules have a relatively large chance to escape into interplanetary space. The **exobase**, at the bottom of the exosphere (~500 km on Earth), is the altitude above which the mean free path length exceeds the atmospheric scale height H (eq. 5.11).

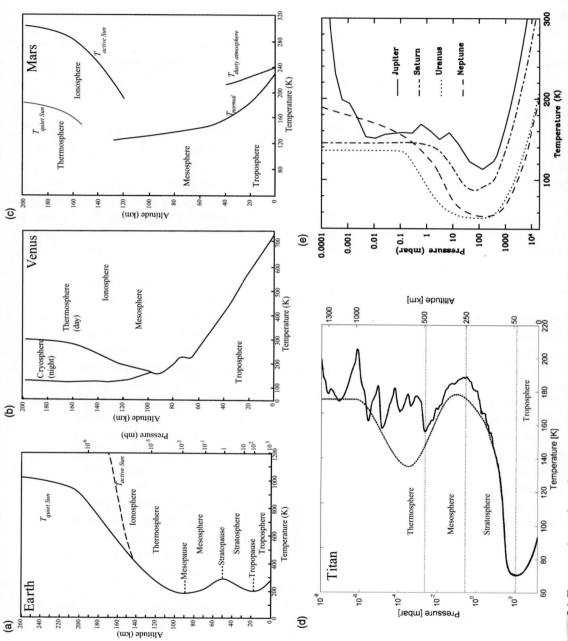

Figure 5.1 The approximate thermal structure of the atmospheres of (a) Earth; (b) Venus; (c) Mars; (d) Titan; and (e) Jupiter, Saturn, Uranus and Neptune. The temperature–pressure profile in each planet's atmosphere is shown as a function of altitude (Venus, Mars), pressure (gaseous planets), or both (Earth, Titan). (de Pater and Lissauer 2010)

5.1.1 Sources and Transport of Energy

Heat Sources

All planetary atmospheres within our Solar System are subject to the Sun's radiation, which heats an atmosphere through absorption of solar photons. Because the solar 5700 K blackbody curve peaks near 500 nm, most of the Sun's energy output is in the visible wavelength range. These photons heat a planet's surface (terrestrial planets) or layers in the atmosphere where the optical depth is moderately large (typically near the cloud layers). Re-radiation of sunlight by a planet's surface or atmospheric molecules, dust particles or cloud droplets occurs primarily at infrared wavelengths and forms a source of heat within or below the atmosphere. Internal heat sources may also heat the atmosphere from below; this is important for the giant planets.

Solar heating of the upper atmosphere is very efficient at extreme ultraviolet (EUV) wavelengths even though the number of photons in this wavelength range (100–10 nm) is very low. Typical EUV photons have energies between 10 and 100 eV, which is enough to ionize several of the atmospheric constituents (§5.5.2). The excess energy from ionization is carried off by electrons freed in the process.

In addition to heating processes triggered by sunlight, an upper atmosphere can be heated substantially by **charged particle precipitation**: charged particles that enter the atmosphere from above (i.e., from the solar wind or a planet's magnetosphere). On planets with intrinsic magnetic fields, charged particle precipitation is confined to high magnetic latitudes, the **auroral zones**; aurora are discussed in more detail in §7.3.2.

Energy Transport

The temperature profile in an atmosphere is governed by energy transport. There are three distinct mechanisms to transport energy: Conduction, mass motion (e.g., convection) and radiation. Each of these mechanisms is discussed in detail in §4.3–4.5.

Conduction is important in the very upper part of the thermosphere and in the exosphere, as well as very near the surface if one exists. Collisions tend to equalize the temperature distribution, resulting in a nearly isothermal profile in the exosphere; the temperature of the atmosphere just above a surface tends towards that of the surface.

In the troposphere, energy transport is usually driven by convection, and the temperature profile is therefore close to an adiabat. The dry adiabatic lapse rate is given in §4.4 (eq. 4.27); the formation of clouds reduces the temperature gradient through the latent heat of condensation, as discussed in the same section and given by equation (4.30). Convection thus effectively places an upper bound to the rate at which the temperature can decrease with height.

When the most efficient way for energy transport is via absorption and re-emission of photons (i.e., radiation) the thermal profile is governed by the equations of radiative energy transport. The temperature structure for an atmosphere in radiative equilibrium is given in §4.6 (eq. 4.56).

The temperature profile in any part of an atmosphere is governed by the most efficient mechanism to transport energy. In Chapter 4, the thermal structure is given for an atmosphere in which energy transport is by either convection (eq. 4.27, 4.30) or radiation (eq. 4.56). Which process is most efficient depends on the temperature gradient, dT/dz. In the tenuous upper parts of the thermosphere, energy transport is dominated by conduction, and the atmosphere is isothermal. At deeper layers, down to a pressure of ~ 0.5 bar, an atmosphere is usually in radiative equilibrium, and below that, convection dominates. On Mars, the combination of conduction (collisions) near/with the surface and radiation from the surface leads to a superadiabatic layer just above ($\lesssim 100$ m) the surface during the day and an inversion layer at night (Problem 5-20).

5.1.2 Observed Thermal Profiles

A body's temperature can be determined from observations of its thermal energy flux (eq. 4.11). It is interesting to compare these effective (observed) temperatures (eq. 4.14) with their equilibrium values (eq. 4.17), discussed in §4.1.3. Tables E.9–E.11 show a comparison of these numbers for many bodies in our Solar System. The observed effective temperatures of Jupiter, Saturn and Neptune are substantially larger than the equilibrium values, which implies the presence of internal heat sources. The observed surface temperatures of Venus, Earth and Mars exceed the equilibrium value because of a greenhouse effect (§4.6).

The thermal structure in an atmosphere can be measured remotely via observations at different wavelengths. Different wavelengths probe different depths in a planet's atmosphere because opacity is a strong function of wavelength. Although the altitudes probed differ from planet to planet, we can make a few generic statements here. At optical and infrared wavelengths, the radiative part of an optically thick atmosphere is probed. Convective regions at $P \gtrsim 0.5$–1 bar can be investigated at infrared and radio wavelengths. The tenuous upper levels, at $P \lesssim 10$ µbar, are typically probed at ultraviolet (UV) wavelengths, or via stellar occultations at UV, visible and infrared (IR) wavelengths. Atmospheric profiles for several bodies are shown in Figure 5.1. The profiles of the terrestrial planets and Titan have been derived from *in situ* measurements by probes and/or landers, as well as inversion of IR and microwave spectra. For the giant planets, the temperature–pressure profiles have been derived via inversion of IR spectra combined with UV and radio occultation profiles from *Voyager* and other spacecraft. At deeper levels in the atmosphere, typically at pressures $P > 1$–5 bar, where no direct information on the temperature structure can be obtained via remote observations, the temperature is usually assumed to follow an adiabatic lapse rate. *In situ* observations by the Galileo probe showed the temperature lapse rate in Jupiter's atmosphere to be close to that of a dry adiabat (§8.1.1).

Earth

The average temperature just above Earth's surface is 288 K, and the average pressure at sea level is 1.013 bar. This temperature is 33 K above the equilibrium value, a difference that can be attributed to the greenhouse effect (§4.6), most notably caused by the presence of water vapor, carbon dioxide (CO_2) and a variety of trace gases, including ozone (O_3), methane (CH_4) and nitrous oxide (N_2O).

Earth's troposphere extends up to an altitude of ∼20 km at the equator, decreasing to ∼10 km above the poles. Above the tropopause, the temperature in the stratosphere increases with altitude as a result of the formation and presence of ozone, which absorbs at both UV and IR wavelengths. Decreased O_3 production and increased CO_2 cooling to space lead to a decrease in temperature with increasing altitude above the stratopause at ∼50 km. This region is referred to as the mesosphere. A second temperature minimum is found at the mesopause at altitudes of ∼80–90 km. The temperature structure in Earth's stratosphere – mesosphere is unusual; massive atmospheres other than Earth's and Titan's (and perhaps Saturn's) show a single temperature minimum. Above the mesosphere lies the thermosphere. Thermospheres are usually hot because they are dominated by atoms or molecules that do not radiate efficiently. In Earth's thermosphere, the temperature increases with altitude, partly caused by absorption of UV sunlight (O_2 photolysis and ionization), but primarily because there are too few atoms/molecules to cool the atmosphere efficiently through emission of IR radiation. Most of the IR emission originates from O and NO, molecules that radiate less efficiently than CO_2. At the base of the thermosphere, there is enough CO_2 gas to cool the atmosphere.

The upper thermosphere heats up to 1200 K or more during the day and cools to ~800 K at night.

Venus

Venus's lower atmosphere, or troposphere, extends from the ground up to the level of the visible cloud layers, at ~65 km, which is also the altitude of the tropopause. The surface temperature and pressure are 737 K and 92 bar, respectively. Venus has a very strong greenhouse effect primarily because of the planet's massive CO_2 atmosphere. Venus's middle atmosphere, the mesosphere, extends from the top of the cloud layers up to ~90 km.

At higher altitudes, in the thermosphere, there is a distinct difference in temperature between the day and night sides. The temperature rises above about 100 km on the day side, reaching about 300 K at 170 km. The night side is much colder, 100–130 K, and is often referred to as the **cryosphere**.

Mars

The average surface pressure on Mars is 6 mbar, and the mean temperature is ~215 K. At mid-latitudes, Mars's surface temperature varies from ~200 K at night to ~300 K during the day. At the winter pole, the temperature is ~130 K, and the temperature at the summer pole may reach ~190 K. Similar to Venus, Mars lacks a stratosphere. Mars's thermosphere has a temperature of ~200 K at altitudes above ~200 km. As on Venus, the low temperature can be explained by the efficiency of CO_2 as a cooling agent.

Titan

The thermal structure in Titan's atmosphere has been measured by the *Huygens* probe. At the surface, the temperature and pressure were measured at 93.65 K and 1.467 bar, respectively. This temperature results from the competing greenhouse and anti-greenhouse effects (§4.6).

Giant Planets

The various parameters that characterize the thermal structure of the giant planets' atmospheres are summarized in Table E.9. The observed effective temperatures of Jupiter, Saturn and Neptune are substantially higher than expected from solar insolation alone. This excess emission implies that Jupiter, Saturn and Neptune each emit roughly twice as much energy as they receive from the Sun. The excess heat escaping from these planets is attributed to a slow cooling of the planets since their formation combined with energy release by He differentiation (§6.2.3). For Uranus, the upper limit to excess heat is 14% of the solar energy absorbed by the planet. It is not known why Uranus's internal heat source is so different from those of the other three giant planets.

5.2 Atmospheric Composition

The composition of a planetary atmosphere can be measured either via remote sensing techniques or *in situ* using mass spectrometers on a probe or lander. In a mass spectrometer, the atomic weight and number density of the gas molecules are accurately measured. However, most molecules are not uniquely specified by their mass (unless it is measured far more accurately than is currently possible using spacecraft), and isotopic variations further complicate the situation. Hence, atmospheric composition is deduced from a combination of *in situ* measurements, observations via remote sensing techniques and/or theories regarding the most probable atoms or molecules to fit the mass spectrometer data.

In situ measurements have been made in the atmospheres of Venus, Mars, Jupiter, the Moon, Titan, Mercury and Enceladus (and, of course, Earth). These data contain a wealth of information on atmospheric composition because trace elements as well as atoms and molecules that do not exhibit observable spectral features, such as nitrogen and the noble gases, can be measured with great accuracy. A drawback of such measurements, besides the cost, is that they are performed only along the path of the probe at one specific moment in time. Landers can measure the composition for a longer time, although again at only one location, and rovers can measure the composition over longer timescales and over a range of positions, although such a range is still limited because rovers move very slowly. So *in situ* data, although extremely valuable, may not be representative of the atmosphere as a whole at all times.

Spectral line measurements are performed either in reflected sunlight or from a body's intrinsic thermal emission. Whereas the central frequency of a spectral line is indicative of the composition of the gas (atomic and/or molecular) producing the line, the shape of the line contains information on the abundance of the gas, as well as the temperature and pressure of the environment. Background on spectra, spectral line profiles and the basic principles of atomic and molecular line transitions is presented in §4.5. The strongest spectral lines can be used to detect small amounts of trace gases (volume mixing ratios of $\lesssim 10^{-9}$ in the giant planet atmospheres) and the composition of extremely tenuous atmospheres (e.g., Mercury, whose surface pressure $P \lesssim 10^{-12}$ bar). The line profile may contain information on the altitude distribution of the gas (through its shape) and the wind velocity field (through Doppler shifts).

The most common species in the atmospheres of the planets and Titan are listed in Tables E.12 and E.13. We specify **abundances** as **volume mixing ratios**, that is, the fractional number density of particles (or mole fraction) of a given species per unit volume.

Earth's atmosphere consists primarily of N_2 (78%) and O_2 (21%). The most abundant trace gases are H_2O, Ar and CO_2, but many more have been identified. The atmospheres of Mars and Venus are dominated by CO_2, roughly 95%–97% on each planet; nitrogen (N_2) contributes approximately 3% by volume; the most abundant trace gases are Ar, CO, H_2O and O_2. On Venus, we also find a small amount of SO_2. Ozone, abundant in Earth's stratosphere, has also been identified on Mars. The differences in atmospheric composition among these three planets must result from differences in their formation and evolutionary processes, such as differences in temperature, volcanic and tectonic activity (§6.3) and biogenic evolution (§§16.6.1 and 16.8).

Titan's atmosphere, similar to Earth's, is dominated by N_2 gas. The second major species is methane gas. The *Huygens* probe measured composition while descending through Titan's atmosphere; these results are discussed in detail in §10.3.1.

The compositions of the giant planets' atmospheres, together with the elemental protosolar mixing ratios, are shown in Table E.13. If the giant planets had, like the Sun, formed via a gravitational collapse in the primitive solar nebula, one would expect these planets to have a composition similar to the protosolar values. They are indeed composed primarily of molecular hydrogen (~80%–90% by volume) and helium (~15%–10% by volume). However, helium appears to be depleted on Jupiter and Saturn, and carbon, in the form of methane gas, is enhanced compared with the protosolar value by a factor increasing with heliocentric distance. Accurate measurements of the abundances of these species would help refine models of the formation of the giant planets (§15.6).

Sample Spectra

To help the reader develop a better understanding of planetary spectra and how to interpret and analyze them, we show a few sample spectra in Figures 5.2 and 5.3. Figure 5.2 shows coarse (low-resolution) thermal infrared spectra of Earth, Venus and Mars between 5 and 100 μm. Each spectrum

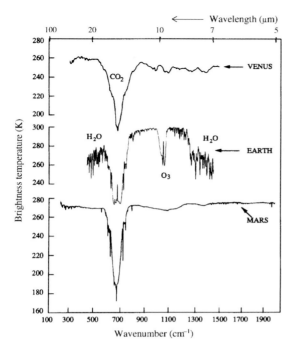

Figure 5.2 Thermal infrared emission spectra of Venus, Earth and Mars. The Venus spectrum was recorded by *Venera 15*, the spectrum of the Earth by *Nimbus 4* and that of Mars by *Mariner 9*. (Adapted from Hanel et al. 1992)

displays a broad CO_2 absorption band at $\sim$15 μm. The width of the absorption profile is similar for the three planets despite vast differences in pressure, since a molecular absorption band consists of numerous transitions (Fig. 4.9). These bands can be used to investigate different depths in the atmospheres. Under clear conditions, when no other absorbers are present, the surfaces of Earth and Mars are probed in the far wings (continuum) of the band. For Venus, the cloud deck rather than surface is probed. Higher altitudes are probed closer to the center of the band. Because the profile is seen in absorption, the temperature must decrease with altitude on all three planets. Therefore, CO_2 must be present in their troposphere. Earth's spectrum has a small emission spike at the center of the CO_2 absorption profile, indicative of some CO_2 in Earth's stratosphere, where, in contrast to the troposphere, the temperature increases with altitude. Other prominent features in the terrestrial spectrum are ozone at 9.6 μm and methane at 7.66 μm. Note the emission spike at the center of the ozone profile, as in the CO_2 absorption band. Numerous water lines are visible in the spectrum. These

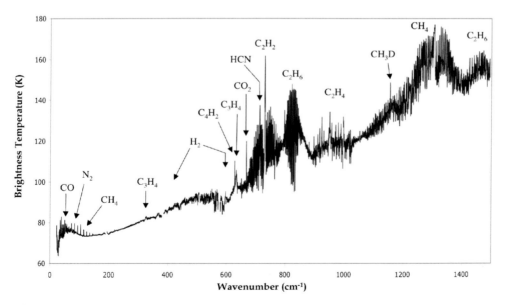

Figure 5.3 Thermal infrared spectrum of Titan, obtained with *Cassini/CIRS*. Note the numerous emission (i.e., stratospheric) lines from hydrocarbons and nitriles, superposed on a smooth continuum. (Coustenis et al. 2007)

lines make the Earth's atmosphere almost opaque in some spectral regions. Note that the CO_2 band prevents transmission near 15 µm. Water lines, although of lesser strength, are also visible in the spectra of Mars and Venus.

Because the emission and absorption lines in planetary atmospheres depend on the local temperature profile, spectra taken at different locations on a planet may appear very different even if the concentrations of the absorbing gases are similar. For example, the CO_2 absorption band on Mars is seen in emission above the martian poles and thus indicates that the atmospheric temperature must be higher than that of Mars's surface at the poles. As the poles are covered by CO_2 ice, such observations can be readily understood.

Figure 5.3 shows a thermal spectrum of Titan, revealing numerous hydrocarbons and nitriles in emission. Because these lines show up in emission, they must form in Titan's stratosphere, where the temperature rises with altitude.

Thermal and reflection spectra for the giant planets are shown in Chapter 8.

5.3 Clouds

Earth's atmosphere contains a small (~1%) and highly variable amount of water vapor (Table E.12). The air is said to be **saturated** if the abundance of water vapor (or, in general, of any condensable species under consideration) is at its maximum vapor partial pressure. Under equilibrium conditions, air cannot contain more water vapor than indicated by its **saturation vapor pressure curve**. The saturation vapor pressure of H_2O is sketched in Figure 5.4a, and given by the **Clausius–Clapeyron equation of state**:

$$P = C_L e^{-L_s/(R_{gas}T)}, \tag{5.1}$$

where L_s is the latent heat, R_{gas} the gas constant and C_L a constant.

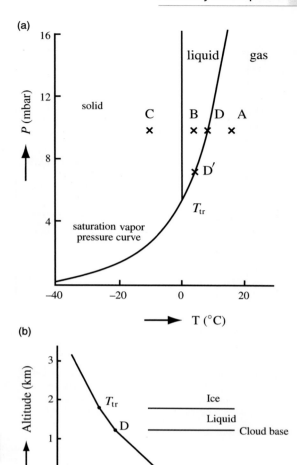

Figure 5.4 (a) Saturation vapor pressure curve for water. The *vertical axis* indicates the partial pressure for H_2O vapor at the temperature (in °C) indicated along the *horizontal axis* (see text for a detailed discussion). (b) Idealized sketch of the temperature structure in the Earth's atmosphere. In the lower troposphere, the air follows a dry adiabat. A wet air parcel rising up through the atmosphere starts to condense when the water vapor inside the air parcel exceeds the saturated vapor curve (at point D in panel a). The temperature profile in the atmosphere follows the wet adiabat between D and T_{tr} and changes again at T_{tr} when the ice line is crossed.

Water at a partial pressure of $\sim$10 mbar in a parcel of air to the right of the solid curves (e.g., point A in Figure 5.4a) is all in the form of vapor, liquid water is present in parcels between the two solid curves (e.g., at point B) and we find water-ice on the left side of the solid curves (e.g., at point C). The solid lines indicate the saturated vapor curves for liquid (on the right) and ice (on the left). Along these lines, evaporation (called **sublimation** if ice transforms directly into gas) is balanced by condensation. The symbol T_{tr} indicates the **triple point** of water, where ice, liquid and vapor coexist.

Consider a parcel of air at point A, with a vapor pressure of 10 mbar and a temperature of 15 °C. If the parcel is cooled, condensation starts when the solid line is first reached, at point D. Upon further cooling, the partial vapor pressure decreases along the curve $D - D' - T_{tr}$. At 3 °C (point D′), the water vapor pressure is 7.6 mbar. Further chilling to below 0 °C results in the formation of ice, where ice first forms upon crossing the second solid line. As an example, the vapor pressure at −10 °C (point C) is 2.6 mbar.

Figure 5.4b shows an idealized sketch of the temperature structure in the Earth's troposphere. The temperature gradient in the lower troposphere is $\sim$10 K km^{-1}; the pressure drops according to equation (3.18). The symbols A, D and T_{tr} correspond roughly to the same points as in Figure 5.4a. A moist parcel of air rising upward in the Earth's troposphere cools adiabatically as it rises from A to D. At point D, the air parcel is saturated, and liquid water droplets condense out. The condensation process releases heat: the **latent heat of condensation**. This decreases the atmospheric lapse rate, as shown by the change in slope from D to T_{tr}. At T_{tr} the atmospheric temperature is 0 °C (273.16 K), and water-ice forms, reducing the lapse rate even more because the latent heat of fusion is added to that of condensation. We have seen in §4.4 that this latent heat of condensation changes the adiabatic lapse rate from a dry to a wet profile (assuming 100% humidity; humidity is explained later).

The numerous water droplets and ice crystals that form this way make up **clouds**. Clouds on other planets are composed of various condensable gases, e.g., in addition to H_2O, we find NH_3, H_2S and CH_4 clouds on the giant planets and CO_2 clouds on Mars. Clouds on Venus are made up of H_2SO_4 droplets. Clouds may modify the surface temperature and atmospheric structure considerably by changing the radiative energy balance. Clouds are highly reflective; thus, they reduce the amount of incoming sunlight, cooling off the surface. Clouds absorb incoming sunlight, heating the immediate environment. Clouds can also block the outgoing infrared radiation, increasing the greenhouse effect. The thermal structure of an atmosphere is influenced by cloud formation through these types of radiative effects and through the release of latent heat of condensation. Clouds further play a major role in the meteorology of a planet, particularly in the formation of storm systems (§5.4).

To account for the effect of clouds on a planet's climate, and hence habitability, one needs to know the cloud coverage, particle sizes, cloud altitudes, etc. This is very difficult, and evolutionary models of a planet's climate therefore have substantial uncertainties.

Relative humidity is the ratio of the partial pressure of the vapor to that in saturated air. The relative humidity in terrestrial clouds is usually 100% ± 2%, although considerable departures from this value have been observed. The humidity can be as low as 70% at the edge of a cloud caused by turbulent mixing or entrainment of drier air. In the interior layers, the humidity can be as high as 107%.

5.4 Meteorology

Everyone is familiar with weather. On Earth, we have different seasons, and each season is associated with particular weather patterns, which vary with geographic location. One sometimes experiences long periods of dry sunny weather, but at other times, we are subjected to long cold spells, periods of heavy rain, huge thunderstorms, blizzards, hurricanes or tornadoes. What causes this weather, and what can we infer about weather on other planets? In this section, we summarize the basic motions of air as caused by pressure gradients (induced by, e.g., solar heating) and the rotation of the planetary body. We further discuss the vertical motion of air that leads its temperature to change via adiabatic expansion or contraction.

5.4.1 Coriolis Effect

Winds are induced by gradients in atmospheric pressure. Because planets rotate, winds cannot blow straight from a high-pressure region to an area of low pressure but rather follow a curved path. This phenomenon can be visualized with help of a turntable, as sketched in Figure 5.5. Draw a line along a ruler held fixed in inertial space while the platform rotates: the line comes out curved in the direction opposite to the platform's rotation. According to the same principle, winds on

Earth (or any other prograde rotating planet) are deflected to the right on the northern hemisphere and to the left on the southern hemisphere. (The opposite pairing occurs on planets rotating in the retrograde direction.) This is called the **Coriolis effect**, and the 'fictitious force' causing the wind to curve is referred to as the **Coriolis force**.

The Coriolis effect follows from the conservation of angular momentum about the rotation axis. A parcel of air at latitude θ has angular momentum per unit mass:

$$L = (\omega_{\text{rot}} R \cos\theta + u)R\cos\theta, \qquad (5.2)$$

where R is the planet's radius and u is the wind velocity along the x coordinate. If an air parcel initially at rest relative to the planet moves poleward while conserving angular momentum, then u must grow in the direction of the planet's rotation to compensate for the decrease in $\cos\theta$. Hence, a planet's rotation deflects the wind perpendicular to its original direction of the motion. The direction of the wind is changed, but because the acceleration is always perpendicular to the wind direction, no work is done, and the speed of the wind is not altered.

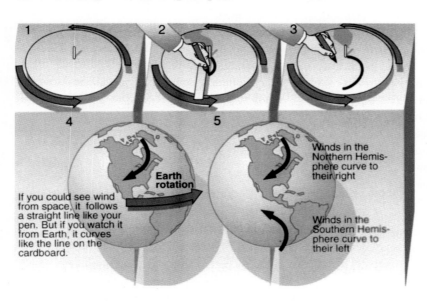

Figure 5.5 A schematic explanation of the Coriolis force: (1) A turntable that rotates counterclockwise. (2) Hold a ruler at a fixed position in inertial space and draw a 'straight' line on the turntable. (3) Even though you drew a straight line, the line on the turntable is curved. This is caused by the 'Coriolis' force. (4, 5) The Coriolis force on the rotating Earth. The rotation of the Earth is indicated by the *thick arrow*. (Williams 1992)

5.4.2 Winds Forced by Solar Heating

Differential solar heating induces pressure gradients in an atmosphere, which trigger winds. Some examples of wind flows triggered directly by solar heating are the Hadley circulation, eddies and vortices, thermal tidal winds and condensation flows. Each of these topics is discussed below. The effects of planetary rotation on the winds are discussed in §5.4.1.

Hadley Circulation

If a planet's rotation axis is approximately perpendicular to its orbital plane, the planet's equator receives more solar energy than do other latitudes. Hot air near the equator rises and flows towards regions with a lower pressure, thus toward the poles. The air then cools, subsides and returns back to the equator at low altitudes. This pattern of atmospheric motion is called the **Hadley cell circulation**.

For a slowly rotating or nonrotating planet, such as Venus, there is one Hadley cell per hemisphere. If the planet rotates rapidly, the meridional winds are deflected (Fig. 5.5; §5.4.1), and the circulation pattern breaks up. As diagrammed in Figure 5.6,

Earth has three mean-meridional overturning cells per hemisphere, and the cell closest to the equator is called the Hadley cell. The middle, or **Ferrel, cell** in each hemisphere circulates in a thermodynamically indirect sense: the air rises at the cold end and sinks at the warm end of the pattern. The third cell, closest to the poles, is referred to as the **polar cell**. The giant planets rotate very rapidly, and latitudinal temperature gradients lead to a large number of zonal winds. If the planet's rotation axis is not normal to its orbital plane, the Hadley cell circulation is displaced from the equator, and weather patterns can vary with season. Moreover, a planet with a large obliquity on an eccentric orbit (e.g., Mars) may have large orbit-averaged differences between the two polar regions. Topography and other surface properties also affect the atmospheric circulation.

The Hadley cells on Earth cause the well-known **easterly** (from the east) trade winds in the tropics, as the return Hadley cell flow near the surface is deflected to the west by the Coriolis force. Similarly, one might expect **westerlies** at mid-latitudes on the low-altitude return flow in the Ferrell cell, as indicated in Figure 5.6; however, in reality, the situation is more complex. On the giant planets, the large gradient in the Coriolis force with latitude leads to a large number of **zonal winds**, where the winds flow along lines of constant pressure, (i.e., **isobars**), as depicted in Figures 5.7 and 8.3.

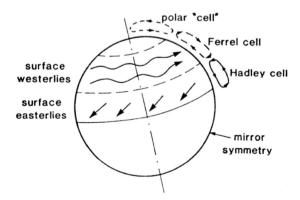

Figure 5.6 Sketch of the Hadley cell circulation on Earth. Three cells are indicated, with the surface winds caused by the Earth's rotation. The winds are indicated as easterly and westerly winds, that is, winds blowing from the east and west, respectively. (Ghil and Childress 1987)

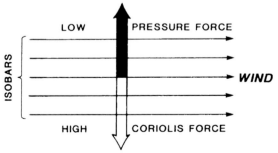

Figure 5.7 Geostrophic balance: the pressure and Coriolis forces balance each other and the wind flows along isobars. (Kivelson and Schubert 1986)

(a)   (b)

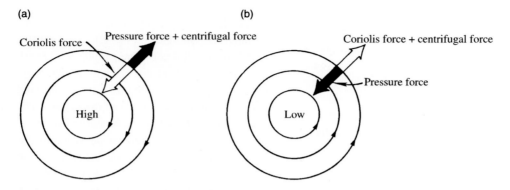

Figure 5.8 Isobars and wind flows around (a) a high-pressure region (anticyclone) in the Earth's northern hemisphere and (b) a low-pressure region (cyclone) in the Earth's northern hemisphere. (Kivelson and Schubert 1986)

The winds continue to blow if the pressure gradient and the Coriolis force just balance each other, a situation referred to as **geostrophic balance**.

It is possible that the centrifugal force of zonal winds, which is directed outward from the planet's 'surface', may balance the force induced by a meridional pressure gradient. Such a situation may be encountered if the zonal winds are very fast; this balance, seen e.g., on Venus (§9.3.2), is known as **cyclostrophic balance**.

Eddies and Vortices

Local topography on a terrestrial planet may induce **stationary eddies**, which are storms that do not propagate in an atmosphere. Stationary eddies are seen over mountains on Earth and Mars and on Earth at the interface between oceans and continents where large differences in temperature exist.

Eddies often form in transition layers between two flows. The most common of these are the so-called **baroclinic eddies**, which form in an atmosphere with geostrophic flows and where the density varies along isobaric surfaces. Only **prograde** (rotating in the direction of flow motion) baroclinic eddies survive. The winds in such eddies flow along isobars, where the pressure force is balanced by the combined Coriolis and centrifugal forces, as shown graphically in Figure 5.8. Whereas in

cyclones, the wind blows around a region of low pressure, in an **anticyclone**, the wind blows around a high-pressure region. Cyclones and anticyclones are observed in the atmospheres of Earth, Mars and the giant planets. The Great Red Spot on Jupiter is an anticyclone.

Thermal Tides

If there is a large difference in temperature between the day and night hemispheres of a planet, air flows from the hot day side to the cool night side. Such winds are called **thermal tidal winds**. A return flow occurs at lower altitudes. The presence of such winds thus depends on the fractional change in temperature over the course of a day, $\Delta T/T$. To estimate this number, we compare the solar heat input, $\mathcal{P}_{in}$ (eq. 4.15), with the heat capacity of the atmosphere.

The fractional change, $\Delta T/T$, is typically less than 1% for planets with substantial atmospheres, such as Venus and the giant planets. But it can be large for planets that have tenuous atmospheres. For example, it is ~20% for Mars. We therefore expect the strong thermal winds near the surface only on Mars and on those planets and satellites with even more tenuous atmospheres. On Venus and Earth, thermal tides are strong in the thermosphere, well above the visible cloud layers, where the density is low and the day–night temperature difference is large.

Condensation Flows

On several bodies, such as Mars, Triton and Pluto, gas condenses out at the winter pole and sublimes in the summer. Such a process drives **condensation flows**. At the martian summer pole, CO_2 sublimes from the surface, thus enhancing the mass of the atmosphere. At the winter pole, it condenses, either directly onto the surface or onto dust grains that then fall down because of their increased weight. Mars's atmospheric pressure varies by $\sim20\%$ from one season to the next (Mars's eccentric orbit contributes to this large annual variation). The condensable gases nitrogen and methane may induce condensation flows on Triton and Pluto. Such flows may explain why a fresh layer of ice overlaid most of Triton's (cold) equatorial regions during the *Voyager* flyby (1989; §10.5), but no ice cover was seen in the warmer areas. The observed decrease in Pluto's albedo as it approached perihelion in 1989 may also be evidence for evaporation of a substantial amount of ground frost (e.g., §12.5). Sulfur dioxide on Io sublimates on the day side and condenses at night, which may drive fast (supersonic) day-to-night winds.

5.5 Photochemistry

All planetary atmospheres are subjected to solar irradiation, which both heats the atmosphere and changes its composition. Typically, absorption of photons at far-IR and radio wavelengths ($\lambda \gtrsim 100$ μm) induces excitation of a molecule's lowest quantum states (i.e., of the rotational levels), whereas photons at IR wavelengths ($\lambda \sim 2$–20 μm) can excite vibrational levels, and photons at visible and UV wavelengths can excite electrons to higher quantum states within atoms and molecules. Photons with $\lambda \lesssim 1$ μm can break up molecules, a process referred to as **photodissociation**. Photons

at higher energies, $\lambda \lesssim 100$ nm, can **photoionize** atoms and molecules. The solar blackbody curve (Fig. 4.2b) peaks at visible wavelengths, and the number of UV and higher energy photons drops significantly with decreasing wavelength. The penetration depth of any solar photon into an atmosphere depends on the optical depth at the particular wavelength of radiation, which is affected by, for example, clouds, hazes, Rayleigh scattering (scattering by atmospheric molecules) and absorption by molecules and atoms. Because the optical depth is particularly large for high-energy photons, most of the photochemical reactions occur at high altitudes. If the production rate of a particular species created via photochemistry is balanced by its loss rate, there is **photochemical equilibrium**. In the subsections below, we assume the species to be in photochemical equilibrium, and under these conditions, we derive information on the altitude distribution of such species.

5.5.1 Photolysis and Recombination

Photodissociation typically takes place at high altitudes; the reverse reaction, **recombination**, proceeds faster at lower altitudes. Therefore, the balance between dissociation and recombination may well be affected by vertical transport. In this section, we discuss the formation of ozone on Earth in detail because the detection of ozone on other planets may indicate life.

Oxygen Chemistry on Earth

The reactions for photodissociation (reaction 1) and recombination (reactions 2–4) for oxygen in the Earth's atmosphere can be written as:

Photodissociation:

(1) $O_2 + h\nu \rightarrow O + O$, $\lambda < 175$ nm.

The production rate at altitude z:

$$\frac{d[O]}{dt} = 2[O_2]J_1(z), \tag{5.3}$$

where an atom or a compound in square brackets, e.g., $[O_2]$, refers to the number of O_2 molecules per unit volume and the subscript 1 in $J_1(z)$ refers to reaction 1. The photodissociation rate, $J(z)$, is given by

$$J(z) = \int \sigma_{x_v} \mathcal{F}_v e^{-\tau_v(z)/\mu_\theta} dv, \qquad (5.4)$$

where σ_{x_v} is the photon absorption cross-section at frequency v, μ_θ is the cosine of the angle between the solar direction and the local vertical and $\mathcal{F}_v$ is the solar flux density impinging on the atmosphere (expressed in photons m^{-2} s^{-1} Hz^{-1}). The mean photodissociation rate for oxygen at an altitude of 20 km is $J_1(20 \text{ km}) = 4.7 \times 10^{-14}$ s^{-1}; at an altitude of 60 km: $J_1(60 \text{ km}) = 5.7 \times 10^{-10}$ s^{-1}. Because the number density of solar photons decreases exponentially with optical depth (thus with decreasing altitude) and the number of oxygen molecules decreases with increasing altitude according to the barometric law (eq. 3.19), the concentration of oxygen atoms increases with altitude in the Earth's lower atmosphere.

Recombination: the direct, two-body reaction:

(2) $\quad O + O \rightarrow O_2 + hv$,

is very slow, with a reaction rate k_r: $k_{r2} < 10^{-26}$ m^3 s^{-1} (the subscript 2 stands for reaction 2). Oxygen recombination is therefore dominated by the three-body processes:

(3) $\quad O + O + M \rightarrow O_2 + M$
(4) $\quad O + O_2 + M \rightarrow O_3 + M$,

where M is an atmospheric molecule that takes up the excess energy liberated in the reaction. Because the abundance of M and O_2 follows the barometric altitude distribution, reactions (3) and (4) are most effective at low altitudes, provided atomic oxygen is present. In reaction 4, **ozone**, O_3, is produced, with a reaction rate $k_{r4} = 6 \times 10^{-46} \left(\frac{T}{300}\right)^{-2.3}$ m^6 s^{-1}.

Ozone is very important for life on Earth because it effectively blocks penetration of UV

sunlight to the ground. To deduce the vertical distribution of ozone in our atmosphere, we must consider the processes that lead both to its formation and to its destruction. In a pure oxygen atmosphere, the relevant processes are the **Chapman reactions** (reactions 1, 4, 5 and 6), where ozone is formed in reaction (4) and destroyed by:
Photodissociation:

(5) $\quad O_3 + hv \rightarrow O_2 + O$,
$\quad\quad \lambda \lesssim 310$ nm, $J_5(60 \text{ km}) = 4.0 \times 10^{-3}$ s^{-1},
$\quad\quad J_5(20 \text{ km}) = 3.2 \times 10^{-5}$ s^{-1},

or the reaction:

(6) $\quad O + O_3 \rightarrow O_2 + O_2$,
$\quad\quad k_{r6} = 8.0 \times 10^{-18} e^{-2060/T}$ m^3 s^{-1}.

The net changes in the atomic oxygen and ozone number densities are:

$$\frac{d[O]}{dt} = 2J_1(z)[O_2] + J_5(z)[O_3] - k_{r6}[O][O_3]$$
$$- k_{r4}[O][O_2][M], \qquad (5.5)$$

$$\frac{d[O_3]}{dt} = k_{r4}[O][O_2][M] - k_{r6}[O][O_3] - J_5(z)[O_3],$$
$$(5.6)$$

where the subscripts to J and k_r refer to reactions (1)–(6). In chemical equilibrium, the net changes of [O] and $[O_3]$ are equal to zero. This leads to the altitude profiles of atomic oxygen and ozone (Problem 5-11):

$$[O] = \frac{J_1(z)[O_2]}{k_{r6}[O_3]}, \qquad (5.7)$$

$$[O_3] = \frac{k_{r4}[O][O_2][M]}{k_{r6}[O] + J_5(z)}. \qquad (5.8)$$

Figure 5.9 shows the altitude distributions of atomic oxygen, molecular oxygen and ozone in our atmosphere. As expected from photodissociation arguments, the number of oxygen atoms increases with increasing altitude up to a certain altitude, and the number of oxygen molecules decreases

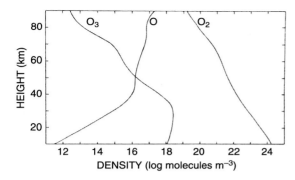

Figure 5.9 Graph of calculated densities of O, O_2 and O_3 in the Earth's atmosphere. (Chamberlain and Hunten 1987)

with altitude. Ozone peaks in number density at altitudes near 30 km.

Our atmosphere is not composed of pure oxygen, so **catalytic** destruction of ozone takes place in addition to reactions (5) and (6) given above. Given that ozone protects life on Earth from harmful solar UV photons, much research is currently devoted to the chemistry of production and destruction of ozone. Free hydrogen atoms, nitric oxides (NO_x), chlorine and halomethanes are important **catalysts** for the destruction of ozone. These molecules react with O_3 and then are almost immediately regenerated through a reaction with atmospheric oxygen; hence, the abundance of these molecules is constant. Nitric oxides are generated by biological activities, lightning storms, engine exhausts and industrial fertilizers close to the ground. Before these nitric oxides can destroy ozone, they must be brought up into the stratosphere. It is therefore not clear how harmful the production of NO_x in the troposphere is; it actually leads to a local production of ozone in the troposphere through the formation of atomic oxygen. However, O_3 is not desirable in the lower troposphere because it is highly reactive and corrosive.

5.5.2 Photoionization: Ionospheres

UV photons at $\lambda \lesssim 100$ nm can ionize atoms and molecules. Radiative recombination of atomic

ions (e.g., $O^+ + e^- \rightarrow O + h\nu$) is very slow compared with molecular recombination. Atomic ions are usually converted to molecular ions by ion–neutral reactions, and the molecular ions recombine. Photolysis in a tenuous atmosphere therefore leads to the formation of the **ionosphere**, a region characterized by the presence of free electrons. The electron density in the ionosphere is determined by both the ionization rate and how rapidly the ions recombine, whether directly or indirectly via charge exchange. Each planet with a substantial atmosphere is expected to have an ionosphere.

On Earth, there are four distinct ionospheric layers: the D, E, F_1 and F_2 layers, as shown in Figure 5.10. The ionosphere starts at the D layer, which covers the range from $\sim$50 km up to 90 km; at 90 km, the electron density is $\sim 10^9$ m^{-3}. The E layer lies between 90 and 130 km, with electron densities 1–2 orders of magnitude larger than that in the D layer. Above the E layer are the F_1 and F_2

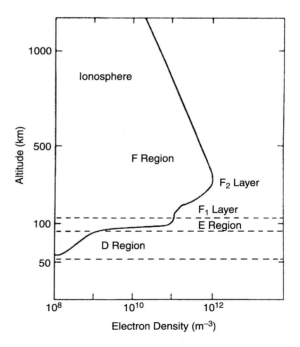

Figure 5.10 Sketch of the electron density in Earth's dayside atmosphere, with the approximate locations of the ionospheric layers. (Russell 1995)

layers. The electron density in the F_2 layer peaks at $\sim$300 km at $\sim$$10^{12}$ m^{-3}. The heights and electron densities of the various layers are highly variable in time because they depend sensitively on the solar UV flux, which varies strongly during the day. The D and F_1 layers are usually absent at night, with the electron densities in the E and F_2 layers being smaller at night than during the day. Typical neutral densities at these altitudes can be calculated from the barometric law (4.19) with allowance for diffusive separation (§5.6) and are many orders of magnitude larger than the electron densities.

The ionospheric layers are distinct from each other because the ionization and recombination processes are different for each layer. This is caused by variations in the composition and absorption characteristics of the atmosphere with altitude.

All planets with a substantial atmosphere have an ionosphere. The electron density in Mars's ionosphere reaches a maximum of $\sim$$10^{11}$ m^{-3} at an altitude of $\sim$140 km, and it is a factor of 3–5 higher in Venus's ionosphere. Interestingly, the dominant ambient ion is not CO_2^+ but O_2^+ (and O^+ at higher altitudes), which forms quickly through various reactions.

Direct photoionization of molecular hydrogen, the main constituent of a giant planet atmosphere, produces H_2^+. This ion undergoes rapid charge transfer interactions with molecular hydrogen, which ultimately results in an ionosphere dominated by H^+ ions.

5.6 Molecular and Eddy Diffusion

In the previous section, we discussed how the atmospheric composition is changed by photochemical reactions. We showed how to calculate the altitude distribution of photochemically derived constituents if the reaction and photolysis rates of the relevant reactions are known. Theoretical altitude profiles for these constituents, however, seldom agree exactly with observations. This is, at least in part, caused by vertical movements of air parcels, known as **eddy diffusion**, and of individual molecules within the air, referred to as **molecular diffusion**.

5.6.1 Eddy Diffusion

Eddy diffusion is a macroscopic process and may occur if an atmosphere is unstable against turbulence. Eddy diffusion dominates the atmosphere below the **homopause**, which is the boundary between the lower atmosphere, wherein constituents are well mixed, and the upper atmosphere, in which different constituents have different scale heights. On Earth, the homopause is located at an altitude of $\sim$100 km.

In (super)adiabatic atmospheres, convection is usually the primary mechanism for vertical mixing. In subadiabatic regions, eddy diffusion may be driven by internal gravity waves or tides. The eddy diffusion coefficient, $\mathcal{K}$, is usually estimated from observed altitude distributions of trace gases. Typical values for $\mathcal{K}$ are of the order of 10^{10}–10^{14} m^2 s^{-1}.

5.6.2 Molecular Diffusion

Density gradients in individual molecular species drive **molecular diffusion**, resulting in a mixing ratio that is constant with altitude. For example, the O/O_2 mixing ratio in the Earth's atmosphere is calculated in §5.5 and shown in Figure 5.9. Molecular diffusion is most effective above the homopause, where O atoms can be carried downwards to equalize the O/O_2 ratio with height. At lower altitudes, the oxygen atoms combine into molecules. This process causes the O/O_2 ratio at $z > 100$ km to be less than expected from the chemistry described in §5.5.

In contrast, molecular diffusion caused by buoyancy drives an atmosphere towards a barometric height distribution for individual constituents. Because the scale height varies with $1/\mu_a$, heavy molecules tend to concentrate at lower altitudes.

Collisions between particles slow the diffusion process, so diffusion is only effective at high altitudes. Hence, in contrast to molecular diffusion, **molecular buoyancy diffusion** enhances the O/O_2 ratio at high altitudes, above that expected from local photochemical equilibrium considerations alone. The net effect of the two processes is that the O/O_2 mixing ratio is less than expected from photochemical considerations alone at altitudes around 100 km but larger at altitudes above this level.

The net result from molecular diffusion is a vertical flux, Φ_i, for each species, which depends on a net molecular diffusion coefficient, D_i, which is inversely proportional to the atmospheric number density N. The maximum rate of diffusion occurs for complete mixing, $\partial(N_i/N)/\partial z = 0$, which leads to the concept of **limiting flux**, Φ_ℓ. For an atmosphere with small or no temperature gradients, where a light gas flows through the background atmosphere, the limiting flux can be written:

$$\Phi_\ell \approx \frac{N_i D_i}{H}, \tag{5.9}$$

with H the atmospheric scale height. Hence, the net outward flux from a planet's atmosphere is limited by the diffusion rate and cannot exceed the limiting flux. The limiting flux depends only on the mixing ratio of constituent i and the pressure scale height. To calculate the limiting flux for hydrogen atoms in Earth's atmosphere (Problem 5-14), one needs to consider the upward flux of all hydrogen-bearing molecules (H_2O, CH_4, H_2) just below the homopause (i.e., $\sim$100 km). The mixing ratio of all hydrogen-bearing molecules at the homopause is $\sim 10^{-5}$, which results in a limiting flux of $\sim 2 \times 10^{12}$ m^{-2} s^{-1}.

5.7 Atmospheric Escape

A particle may escape a body's atmosphere if its kinetic energy exceeds the gravitational binding energy *and* it moves along an upward trajectory without intersecting the path of another atom or molecule. The region from which escape can occur is referred to as the **exosphere**, and its lower boundary is called the **exobase**.

The mean free path for collisions between molecules, ℓ_{fp}, is given by:

$$\ell_{fp} = 1/(\sigma_x N), \tag{5.10}$$

where σ_x is the molecular cross-section and N is the local particle (number) density. The exobase is located at the altitude z_{ex} at which

$$\int_{z_{ex}}^{\infty} \sigma_x N(z)dz \approx \sigma_x N(z_{ex})H = 1. \tag{5.11}$$

If the scale height, H, is constant in the exosphere, then the exobase is located at the altitude z_{ex} where $\ell_{fp}(z_{ex}) = H$. Within the exosphere the mean free path for a molecule or atom is thus comparable to, or larger than, the atmospheric scale height, so an atom with sufficient upward velocity has a reasonable chance of escaping. In addition to the thermal or Jeans escape, various nonthermal processes can also lead to atmospheric escape.

5.7.1 Thermal (Jeans) Escape

For a gas in thermal equilibrium, the velocities follow a **Maxwellian distribution function**:

$$f(v)dv = N\left(\frac{2}{\pi}\right)^{1/2}\left(\frac{m}{kT}\right)^{3/2}v^2 e^{-mv^2/(2kT)}dv. \tag{5.12}$$

At and below the exobase, collisions between particles drive the velocity distribution into a Maxwellian distribution. Above the exobase, collisions are essentially absent, and particles in the tail of the Maxwellian velocity distribution that have a velocity $v > v_e$ may escape into space. A Maxwellian distribution formally extends up to infinite velocities, but because of the steep dropoff, there are practically no particles with velocities larger than about four times the thermal or most probable velocity $v_o = \sqrt{2kT/m}$.

The ratio of the potential to kinetic energy is referred to as the **escape parameter**, λ_{esc}:

$$\lambda_{esc} = \frac{GMm}{kT(R+z)} = \frac{(R+z)}{H(z)} = \left(\frac{v_e}{v_o}\right)^2. \quad (5.13)$$

Integrating the upward flux in a Maxwellian velocity distribution above the exobase results in the **Jeans formula** for the rate of escape (atoms m^{-2} s^{-1}) by thermal evaporation:

$$\Phi_J = \frac{N_{ex}v_o}{2\sqrt{\pi}}(1 + \lambda_{esc})e^{-\lambda_{esc}}, \quad (5.14)$$

where the subscript 'ex' refers to the exobase and λ_{esc} is the escape parameter at the exobase. Typical parameters for Earth are $N_{ex} = 10^{11}$ m^{-3} and $T_{ex} = 900$ K. For atomic hydrogen $\lambda_{esc} \approx 8$, and $\Phi_J \approx 6 \times 10^{11}$ m^{-2} s^{-1}, which is a factor of 3–4 smaller than the limiting flux (eq. 5.9) for hydrogen atoms on Earth (Problem 5-14). Note that lighter elements and isotopes are lost at a much faster rate than heavier ones. **Jeans escape** can thus produce a substantial **isotopic fractionation**. To first approximation, calculations of Jeans escape can be used to predict whether or not an object has an atmosphere (Problem 5-15).

5.7.2 Nonthermal Escape

Jeans escape gives a lower limit to the escape flux; **nonthermal processes** often dominate the escape rate. Examples of such nonthermal processes include dissociation of molecules and **charge exchange** reactions. In the latter process, a fast ion may exchange a charge (usually an electron) with a neutral, where the ion loses its charge but retains its kinetic energy. The new neutral (former ion) may have sufficient energy to escape the body's gravitational attraction.

Another important process is **sputtering**, where an atmospheric atom may gain sufficient energy to escape the body's gravitational attraction when it is hit by a fast atom or ion. Because it is much easier to accelerate an ion than an atom, sputtering is usually caused by fast ions.

5.7.3 Hydrodynamic Escape and Impact Erosion

In the early Solar System, atmospheric losses on many bodies were dominated by impact erosion and hydrodynamic escape. In the present era, hydrodynamic escape can be important for some bodies, such as Pluto.

Hydrodynamic escape occurs when a planetary wind composed of a light gas (e.g., H) entrains heavier gases, which by themselves would not escape according to the Jeans equation (Problem 5.14). Hydrodynamic escape requires a large input of energy to the upper atmosphere. Solar energy is usually not large enough to produce hydrodynamic escape in any present-day terrestrial-type atmosphere. The early atmospheres of Venus, Earth and Mars may have experienced periods of hydrodynamic escape, triggered by intense solar UV radiation and a strong solar wind.

Impact erosion of the atmosphere of a terrestrial planet can occur during or immediately after a large impact. The underlying concept is that atmospheric gases can be swept into space by the momentum of a hydrodynamically expanding vapor plume produced upon impact with the planet (see §6.4.3). Atmospheric escape occurs if this vapor plume attains escape velocity and if it has enough extra momentum to carry the intervening atmosphere with it.

For an impactor that is smaller than the atmospheric scale height, shock-heated air flows around the impactor, and the energy is dispersed over a relatively large volume of the atmosphere. If the impactor is larger than an atmospheric scale height, however, a large fraction of the shock-heated gas can be blown off because for such impactors, the impact velocity exceeds the escape velocity from the planet. The mass of the atmosphere blown into space, M_e, is given by

$$M_e = \frac{\pi R^2 P_0 \mathcal{E}_e}{g_p}, \quad (5.15)$$

where R is the radius of the impactor and P_0/g_p is the mass of the atmosphere per unit area. The atmospheric mass that can escape is thus the mass intercepted by the impactor multiplied by an enhancement factor $\mathcal{E}_e$:

$$\mathcal{E}_e = \frac{v_i^2}{v_e^2(1 + \mathcal{E}_v)}, \tag{5.16}$$

where v_i and v_e are the impact and escape velocities, respectively, and $\mathcal{E}_v$ is the **evaporative loading parameter**. This parameter accounts for the burden imposed on escape because in addition to atmospheric gas, the plume also carries the vapor of impactor and target material. $\mathcal{E}_v$ is proportional to the energy imparted on the gas ($\sim 0.5v_i^2$ per unit mass) and the latent heat of evaporation, L_s:

$$\mathcal{E}_v = C_H \frac{v_i^2}{2L_s}, \tag{5.17}$$

where C_H is the heat transfer coefficient, which is of order 0.1–0.2 near the surface on Earth. A typical value of L_s for meteorites is $\sim 8 \times 10^6$ J kg^{-1}.

Significant escape occurs when $\mathcal{E}_e > 1$. If $\mathcal{E}_e < 1$, evaporative loading is much larger than the energy gained by impact heating, and the gas does not have enough energy left to escape into space. In the case of a colossal cratering event, the ejecta from the crater may also be large enough and contain sufficient energy to accelerate atmospheric gas to escape velocities. Such large impactors can remove all of the atmosphere above the horizon at the location of the impact.

5.8 History of Secondary Atmospheres

5.8.1 Formation

The initial stages of planetary growth involve the accumulation of solid materials, but gas can be trapped within some solids; chemical alteration can produce volatiles, and radioactive nuclides can decay into volatiles. Moreover, if a planet becomes massive enough, it may gravitationally trap gases.

The formation of planets and their atmospheres is discussed in detail in Chapter 15.

The atmospheres of the giant planets are composed primarily of hydrogen and helium, with traces of C, O, N, S and P in the form of CH_4, H_2O, NH_3, H_2S and PH_3, respectively. In contrast, the atmospheres of the terrestrial planets and satellites are dominated by CO_2, N_2, O_2, H_2O and SO_2. The main difference between the giant and terrestrial planets is gravity, which allowed the giant planets to accrete large quantities of common species (e.g., H_2, He; see Table 3.1) that remain gaseous at Solar System temperatures. The light elements H and He (if present originally) would have escaped the shallow gravitational potential wells of the terrestrial planets.

Below we argue that the atmospheres of the terrestrial planets cannot be remnants of gravitationally trapped **primordial atmospheres** but must have formed from outgassing of bodies accreted as solids. In §5.8.2, we discuss the subsequent evolution of the 'climate' on Earth, Mars and Venus.

The following chemical reactions can occur in an atmosphere between H_2 and other volatiles:

$$CH_4 + H_2O \longleftrightarrow CO + 3H_2$$
$$2NH_3 \longleftrightarrow N_2 + 3H_2$$
$$H_2S + 2H_2O \longleftrightarrow SO_2 + 3H_2$$
$$8H_2S \longleftrightarrow S_8 + 8H_2$$
$$CO + H_2O \longleftrightarrow CO_2 + H_2$$
$$CH_4 \longleftrightarrow C + 2H_2$$
$$4PH_3 + 6H_2O \longleftrightarrow P_4O_6 + 12H_2.$$

A loss of hydrogen shifts the equilibrium towards the right, hence oxidizing material. We refer to an atmosphere as **reducing** if a substantial amount of hydrogen is present, as on the giant planets, and as **oxidizing** if little hydrogen is present, as on the terrestrial planets.

If the atmospheres of the terrestrial planets were primordial in origin (accreted from a solar composition gas, similar to the formation of the giant planets; §15.6) and all of the hydrogen and helium subsequently escaped, then the most abundant gases in these atmospheres would be CO_2 ($\sim 63\%$),

Ne ($\sim$22%) and N_2 ($\sim$10%), with a small fraction of carbonyl sulfide (OCS, $\sim$4%; this molecule does not form via one simple reaction as those listed above). In addition, one would expect solar concentrations for Ar, Kr and Xe. These abundances are quite different from what is observed. In particular, neon on Earth is present in minuscule amounts, about ten orders of magnitude less than predicted from this model. Similarly, nonradiogenic Ar, Kr and Xe are present but at abundances over six orders of magnitude less than expected for a solar composition atmosphere. These latter three noble gases are too heavy to escape via thermal processes if initially present and could not be chemically confined to the condensed portion of the planet, as CO_2 is on Earth (see below). The observed small abundances of the noble gases provide strong evidence that the atmospheres of the terrestrial planets are secondary in origin. A **secondary atmosphere** could have been produced (*i*) during the accretion phase of the planet when impacts caused intense heating and/or by (late) accreting volatile-rich asteroids and comets, (*ii*) at the time of core formation when the entire planet was molten and/or (*iii*) through 'steady' outgassing via volcanic activity.

The measured ratio Argon isotopes, $^{40}Ar/^{36}Ar$, in Earth's atmosphere and volcanic glasses can be used to deduce when gases were released into the atmosphere. ^{36}Ar is a primordial isotope, incorporated into planetesimals only at extremely low temperatures ($\lesssim$ 30 K). ^{40}Ar, in contrast, originates from radioactive decay of potassium, ^{40}K, which has a half-life of 1.25 Gyr (Table 11.1). Both the stable and radioactive isotopes of potassium are incorporated in rock-forming minerals. Upon decay of ^{40}K, the resulting argon gets released only when the mineral melts. The $^{40}Ar/^{36}Ar$ ratio seen in bubbles within volcanic glasses is about one hundred times as large as the atmospheric value of 300. This implies that the vast majority of the ^{36}Ar now in the atmosphere either never resided in the mantle or was outgassed from the mantle within the first few tens of million years of Earth's accretion.

5.8.2 Climate Evolution

As discussed in Chapter 4, a planet's surface temperature is determined mainly by solar insolation, its Bond albedo and atmospheric opacity. The Sun's luminosity has slowly increased during the past $\sim$4.5 Gyr. Early on the solar luminosity was $\sim$25% smaller than it is nowadays, implying lower surface temperatures for the terrestrial planets. In contrast, although the Sun's total energy output was less than it is at the present time, its X-ray and UV emission were much larger, and the solar wind was stronger. In addition to changes in the solar flux, the planetary albedos may have fluctuated as a result of variations in a planet's cloud deck, ground-ice coverage and volcanic activity. Changes in atmospheric composition, particularly with regard to greenhouse gases, also have a profound effect on climate.

Variations in albedo that result in changes in ice cover and in the abundance of greenhouse gases provide **positive feedback** mechanisms, i.e., the effect of a change in the primary input is amplified, as sketched in Figures 5.11 and 5.12, respectively. Figure 5.11 shows that a decrease in a planet's

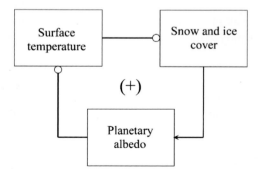

Figure 5.11 A change in a planet's albedo provides a positive feedback loop on climate, as sketched here for a snow and ice cover. An increase in ice cover increases the planet's albedo, which leads to a decrease in temperature and further freezing of water. This could lead to a phenomenon known as snowball Earth. (Catling and Kasting 2007)

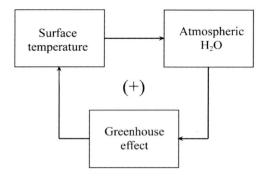

Figure 5.12 A change in the complement of greenhouse gases provides a positive feedback loop on climate, as sketched here for the amount of water vapor in a planet's atmosphere. A larger concentration of water vapor leads to an increase in temperature, which further increases atmospheric H_2O. (Catling and Kasting 2007)

albedo will lower its temperature as more sunlight is reflected; this in turn will increase the snow and ice cover, as more H_2O freezes out, etc. This can eventually lead to a phenomenon known as **snowball Earth**. Alternatively, a decrease in albedo leads to a warming of the planet and hence evaporation/sublimation of snow and ice, thereby decreasing the albedo even more. Because water vapor is a greenhouse gas, this scenario is closely coupled to the feedback loop sketched in Figure 5.12, where an increase in the atmospheric water vapor content leads to a warming of the atmosphere and surface and hence an increase in evaporation/sublimation and so on. This could ultimately lead to a **runaway greenhouse effect**. Small changes in input parameters in which positive feedback occurs can have large effects on a planet's climate.

In addition to these feedback mechanisms, periodic variations (e.g., the Milankovitch cycles discussed in §2.6.3) or sudden modifications in a planet's orbital eccentricity and the obliquity of its rotation axis (e.g., caused by a large impact) also play an important role in climate evolution. In particular, the ~40 000- and 100 000-year cycles of Earth's ice ages (Fig. 5.13a) have been attributed to the Milankovitch cycles.

Earth

Even though the luminosity of the young Sun was smaller, the presence of sedimentary rocks and the possible absence of glacial deposits on Earth about 4 Gyr ago suggest that the Earth may have been even warmer than today. If true, this could be attributed to an increased greenhouse effect because the atmospheric H_2O, CO_2, CH_4 and possibly NH_3 content was likely larger during the early stages of outgassing.

One might expect that an increase in CO_2 gas, a greenhouse gas, would lead to a continuous increase in a planet's temperature. However, the long-term cycling of CO_2 may be regulated in such a way that the Earth's surface temperature does not change too much over *long* (many thousands of years) time periods. Carbon dioxide is removed from the atmosphere–ocean system via the **Urey weathering reaction**, a chemical reaction between CO_2, dissolved in water (e.g., rain), with silicate minerals in the soil. This cycling of CO_2 between the atmosphere and other reservoirs is illustrated in Figures 5.14 and 5.15. The reaction releases Ca and Mg ions and converts CO_2 into bicarbonate (HCO_3^-). The bicarbonate reacts with the ions to form other carbonate minerals. An example of such a chemical reaction for calcium is given by

$$CaSiO_3 + 2CO_2 + H_2O \rightarrow Ca^{++} + SiO_2 + 2HCO_3^-$$
(5.18a)

$$Ca^{++} + 2HCO_3^- \rightarrow CaCO_3 + CO_2 + H_2O. \quad (5.18b)$$

We note that most of the calcium carbonates on Earth are produced by organisms in the oceans. Carbonate sediments on the ocean floor are carried downwards by plate tectonics (§6.3.1) and are transformed back into CO_2 in the high-temperature and high-pressure environment of the Earth's mantle:

$$CaCO_3 + SiO_2 \rightarrow CaSiO_3 + CO_2. \quad (5.18c)$$

Volcanic outgassing returns the CO_2 to the atmosphere. Whereas the weathering rate increases

(a)

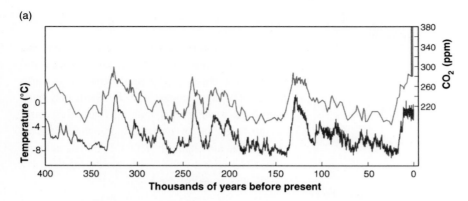

(b)

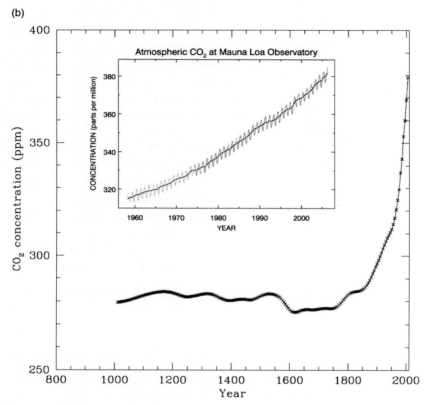

Figure 5.13 (a) Long-term variations in temperature (*bottom graph*) and in the atmospheric concentration of carbon dioxide (*top graph*) over the past 400 000 years as inferred from Antarctic ice-core records. (Fedorov et al. 2006) (b) CO_2 concentration in the Earth's atmosphere over the past 1000 years. (Adapted from Etheridge et al. 1996). The *insert* is the CO_2 concentration from data obtained at a single location in Hawaii. This graph reveals a steady increase of the CO_2 concentration, in addition to the seasonal oscillations. (NOAA Earth System Research Laboratory)

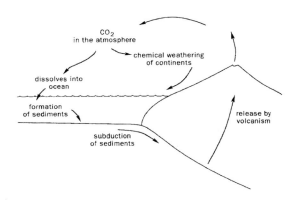

Figure 5.14 Schematic of the CO_2 cycle on Earth. Carbon dioxide is removed from the atmosphere by the Urey weathering reaction, transported down into the mantle via plate tectonics and recycled back into the atmosphere by volcanic activity. (Jakosky 1998)

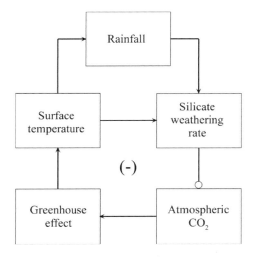

Figure 5.15 The negative feedback loop involving the silicate weathering rate, the CO_2 abundance in the atmosphere and climate. (Catling and Kasting 2007)

when the surface temperature is higher, the surface temperature is related to the CO_2 content of the atmosphere through the greenhouse effect. This effectively causes a self-regulation in the atmospheric CO_2 abundance on Earth. The role of this carbonate–silicate weather cycle during the recovery from glaciations is well established.

The abundance of O_2 in the Earth's atmosphere is primarily attributable to photosynthesis of green plants together perhaps with a small contribution

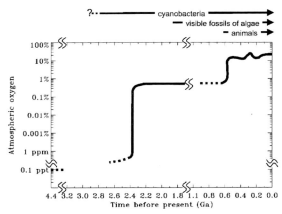

Figure 5.16 The abundance of oxygen in the Earth's atmosphere is shown as a function of time. Oxygenation of our planet's atmosphere appears to have occurred in a highly nonuniform manner. Although the general trend of increasing oxygen is well established, quantitative estimates of the abundance are quite uncertain, especially in the more distant past. The best constrained epochs are represented by the *solid portions* of the curve, *gaps in the curve* denote epochs with no or very weak constraints and *dashes* represent values that are moderately uncertain. (Courtesy David Catling)

from past photodissociation of H_2O and subsequent escape of the hydrogen atoms. As illustrated in Figure 5.16, the O_2 in our atmosphere rose to significant levels about 2.2 Gyr ago. The presence of iron carbonate ($FeCO_3$) and uranium dioxide (UO_2) in sediments that date back to more than 2.2 Gyr ago and its absence in younger sediments provide evidence for the presence of free oxygen in Earth's atmosphere over the past 2.2 Gyr because oxygen destroys these compounds today. The most striking evidence of a low oxygen abundance on Earth is provided by **banded iron formations** (BIFs) on the ocean floor, a sediment that consists of alternating (few centimeters thick) layers of iron oxides (e.g., hematite and magnetite) and sediments (e.g., shale and chert) (§6.1.2). An example is shown in Figure 5.17. Banded iron formations, which were formed by precipitation, imply that iron deposits built up in seawater, something that cannot occur in today's oxygen-rich oceans. Because BIFs are common in sediments laid down prior to 1.85 Gyr ago but

Figure 5.17 COLOR PLATE Photograph of a 3.8-Gyr-old rock from the Isua Formation in Greenland showing a banded iron formation. The minerals comprising this rock cannot form in equilibrium with Earth's current oxygen-rich atmosphere, but banded iron formations are abundant in the geologic record of rocks more than 2.4 Gyr old. (Courtesy Minik Rosing)

very rare in more recently formed rocks, free oxygen must have been a rare commodity more than 2 Gyr ago. The rise of oxygen coincided with a large ice age. The increase in O_2 may have eliminated much of the methane gas, hypothesized to have been a major greenhouse gas on early Earth, by reducing its photochemical lifetime and constraining the environments in which methanogens (methane-producing archaebacteria) could survive. The influence of the biosphere on Earth's atmosphere is discussed more thoroughly in §§16.6.1 and 16.8.

The influence of humans on the evolution of the Earth's atmosphere at the present time should not be underestimated. The CO_2 levels in our atmosphere are rising at an alarming rate (Fig. 5.12), which is leading to a global warming through enhancement of the greenhouse effect. Although the increased temperature will also increase the weathering rate, these geophysical processes occur on much longer timescales than the present (largely human-induced) rapid rate of CO_2 accumulation in the atmosphere. Increased absorption (dissolution) of CO_2 in seawater, however, has been measured already and causes the ocean to become less basic, i.e., this decreases the ocean's pH. This process is referred to as **ocean acidification** and may have serious consequences for marine ecosystems. In addition, chemical reactions between atmospheric gases and pollutants may also influence the atmospheric composition, the consequences of which are difficult to predict with any certainty. We have embarked on a giant inadvertent experiment with our home planet's atmosphere, which may have dire consequences for the future of life on Earth.

Mars

Mars's small size may be as important a cause of the difference in climate between Mars and Earth as is the difference in heliocentric distance of these two planets. Although liquid water cannot exist in significant quantities on the surface of Mars at the present time, the numerous channels on the planet, layers of sandstone and minerals that can only form in the presence of water are suggestive of running water in the past (§9.4). This implies that Mars's atmosphere must have been denser and warmer. Because the runoff channels are confined to the ancient, heavily cratered terrain, the warm martian climate did not extend beyond the end of the heavy bombardment era, about 3.8 Gyr ago. Models of Mars's early atmosphere suggest a mean surface pressure of the order of 1 bar and temperature close to 300 K. Widespread volcanism, impacts by planetesimals and tectonic activity must have provided large sources of CO_2 and H_2O, whereas impacts by very large planetesimals may also have led to (repeated) losses of atmospheric gases through impact erosion. In addition to atmospheric escape into space, Mars has probably lost most of its CO_2 via carbonaceous (weathering) processes, adsorption onto the regolith and/or condensation onto the surface. Because Mars does not show current tectonic activity, the CO_2 cannot be recycled back into the atmosphere. Without liquid water on the surface, weathering has ceased, and Mars has retained a small fraction of its CO_2 atmosphere. The present abundance of H_2O on Mars is not accurately known. Most of the H_2O might have escaped and/or there may be large amounts of subsurface water-ice on the planet (§9.4). A potential problem with the weathering theory is the apparent lack of carbonates on the martian surface, although this can be reconciled if the water was very acidic.

Climatic changes may also be caused by changes in Mars's orbital eccentricity and obliquity. On Earth, these parameters vary periodically on timescales of $\sim 10^4$–10^5 years (see Figs. 2.12 and 2.18) and may be responsible for the succession of ice ages and ice-free epochs during the past million years. For Mars, these parameters have periodicities about ten times larger than for Earth, and departures from the mean values are also much larger (see Figs. 2.12 and 2.17). The polar regions receive more sunlight when the obliquity is large, and large eccentricities increase the relative amount of sunlight falling on the summer hemisphere at perihelion. The layered deposits in Mars's polar region and glaciations in the tropics and at mid-latitudes (§9.4) suggest that such periodic changes have taken place on Mars.

The Tharsis region of Mars contains many volcanoes that appear to be roughly the same age. The eruptions of these volcanoes must have enhanced the atmospheric pressure and, via the greenhouse effect, the surface temperature. However, the sparsity of impact craters implies that the volcanic eruptions occurred well after the formation of the runoff channels on Mars's highlands.

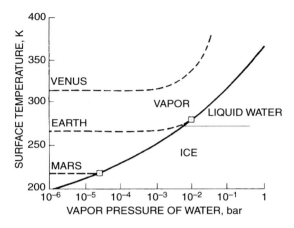

Figure 5.18 Evolution of the surface temperatures of Venus, Earth and Mars for a pure water-vapor atmosphere. (Adopted from Goody and Walker 1972)

Venus

Venus is very dry, with an atmospheric H_2O abundance of only 100 parts per million. This totals about 10^{-5} as much H_2O as in the Earth's oceans. Various theories have been offered to explain the lack of water on Venus. It could have simply formed with very little water because the minerals that condensed in this relatively warm region of the solar nebula lacked water. However, mixing of planetesimals between the accretion zones and asteroid/cometary impacts may have provided similar amounts of volatiles to Venus and Earth, in which case it seems probable that there was an appreciable fraction of a terrestrial ocean on early Venus. The D/H ratio on Venus is ~ 100 times larger than on Earth, which is persuasive evidence that Venus was once much wetter than it is now. But where did the water go? Water can be dissociated into hydrogen and oxygen, by either photodissociation or chemical reactions, and the hydrogen then escapes into space.

The classical explanation for Venus's loss of water is via the runaway greenhouse effect (Fig. 5.12). Figure 5.18 shows the evolution of the surface temperatures for Earth, Mars and Venus, where each of these planets was assumed to have outgassed a pure water-vapor atmosphere, starting from an initially airless planet. Although the vapor pressure of water in the atmosphere is increasing, the surface temperature increases because of the enhanced greenhouse effect. There is thus a positive feedback between the increasing temperature and increasing opacity. On Venus, the temperature stayed well above the saturation pressure curve, which led to a runaway greenhouse effect, and all of Venus's water accumulated in the atmosphere as steam. If one postulates an effective mixing process, the water is distributed throughout the atmosphere and is photodissociated at high altitudes, with subsequent escape of the hydrogen atoms from the top of the atmosphere.

An alternative model to explain Venus's loss of water is the **moist greenhouse effect**. This model, in contrast to the runaway greenhouse model, relies on moist convection. Venus may have had a surface temperature near $100\,°C$. Convection transported the saturated air upward. Because the temperature near the surface decreases with altitude, water condensed out, and the latent heat of condensation led to a decrease in the atmospheric lapse rate and increased the altitude of the tropopause. In this scenario, water vapor naturally reached high altitudes, where it was dissociated and escaped into space. With the accumulation of water vapor in the atmosphere, the atmospheric pressure increased, preventing the oceans from boiling. Because of the rapid loss rate of water from the top of the atmosphere, the oceans continued to evaporate. As long as there was liquid water on the surface, CO_2 and O_2 were removed from the atmosphere by weathering processes. When the liquid water was gone, CO_2 could not form carbonate minerals and accumulated in the atmosphere.

5.8.3 Summary of Secondary Atmospheres

Although the atmospheres of the terrestrial planets and satellites are all different in detail, common processes probably led to their formation. These atmospheres were formed predominantly via outgassing, but the mass (gravity) of the individual bodies, together with their atmospheric temperature and composition, is key to retaining the atmosphere. Subsequent evolution led to large variations in composition and atmospheric pressure. The carbon–silicate weathering process probably played a crucial role on Venus, Earth and Mars, and surface temperature together with tectonic activity ultimately led to the observed differences. Late-accreting planetesimals may have both eroded atmospheres and supplied the planets with additional volatiles.

Satellites around the giant planets formed within the planets' local subnebulas, where temperatures were well above those where planetesimals could have trapped N_2 and noble gases. The nitrogen in these planetesimals was therefore incorporated in the form of nitrogen compounds, particularly ammonia-ice. Outgassing of this constituent long ago probably led to Titan's dense nitrogen atmosphere, whereas methane gas must be supplied continuously or episodically via cryovolcanism or perhaps via a meteorological cycle akin to the water cycle on Earth. Because of the large kinetic energies involved, any late impacting planetesimals would have eroded any pre-existing atmosphere around the large jovian satellites Ganymede and Callisto. Analogous impacts on Titan, being less energetic because typical collision speeds were smaller, might have led to an increase, rather than a decrease, in the satellite's volatile budget.

Key Concepts

- The thermal structure of an atmosphere is primarily governed by the efficiency of energy transport.
- Spectroscopy can be used to study a planet's atmospheric composition and thermal structure.
- The formation of clouds reduces the atmospheric lapse rate through release of latent heat of condensation.

- Winds are triggered by pressure gradients in an atmosphere. A planet's rotation causes the winds to curve (the Coriolis effect).
- The composition of the upper layers of an atmosphere can be significantly altered through photochemistry (photodissociation and ionization).
- Particles can escape a planet's atmosphere through Jeans escape or via nonthermal processes.
- The giant planet atmospheres formed via accretion from the solar nebula. In contrast, the atmospheres of the terrestrial planets and Titan formed later via outgassing.
- Positive and negative feedback mechanisms may have played important roles in climate evolution.
- Life was responsible for the oxygenation of Earth's atmosphere.
- Solar heating led to the runaway greenhouse on Venus.
- On Earth, the Urey weathering reaction acts as a thermostat over long time periods.

Further Reading

A good book on weather for nonscience majors is:

Williams, J., 1992. *The Weather Book.* Vintage Books, New York. 212pp.

A few books on atmospheric sciences that we recommend are:

Salby, M.L., 1996. *Fundamentals of Atmospheric Physics.* Academic Press, New York. 624pp.

Seinfeld, J.H., and S.N. Pandis, 2006. *Atmospheric Chemistry and Physics: From Air Pollution to Climate Change*, 2nd Edition. John Wiley and Sons, New York. 1203pp.

Chamberlain, J.W., and D.M. Hunten, 1987. *Theory of Planetary Atmospheres.* Academic Press, New York. 481pp.

Irwin, P., 2009. *Giant Planets of our Solar System*, 2nd Edition. Springer-Praxis, 2009. 436pp.

Papers reviewing the origin and evolution of planetary atmospheres can be found in:

Atreya, S.K., J.B. Pollack, and M.S. Matthews, Eds., 1989. *Origin and Evolution of Planetary and Satellite Atmospheres.* University of Arizona Press, Tucson, Arizona. 881pp.

Descriptions of the atmospheres of individual planets are in:

McFadden, L., P. R. Weissman, and T.V. Johnson, Eds., 2007. *Encyclopedia of the Solar System*, 2nd Edition. Academic Press, San Diego. 982pp.

Problems

5-1. Estimate the pressure scale height near the surfaces of Earth, Venus, Mars, Pluto and Titan and at the 1-bar levels of Jupiter and Neptune. Assume each of the atmospheres to be composed of the two most abundant gases in the measured ratios. Comment on similarities and differences.

5-2. Estimate the pressure scale height of the atmosphere of the planet Kepler-11d. Use the data presented in Table 14.1 and assume that the atmosphere is composed primarily of H and He in the cosmic abundance ratio given in Table 3.1.

5-3. If you were to observe Jupiter's thermal emission at radio wavelengths, where one probes down to well below the planet's tropopause, would you expect limb brightening, darkening or no change in intensity when you scan the planet from the center to the limb? Explain your reasoning.

5-4. Assume that you observe Saturn at infrared wavelengths in the line transitions of C_2H_2 and PH_3. The C_2H_2 line is seen in emission, and the PH_3 line in absorption.
(a) Where in the atmosphere are these gases located?
(b) Do you expect the limb of the planet to be brighter or darker than the center of the planet in each of these lines?

5-5. Consider a parcel of dry air in the Earth's atmosphere. Show that if you replace some portion of the air molecules (79% N_2, 21% O_2) by an equivalent number of water molecules, the parcel of air becomes lighter and rises.

5-6. Calculate the dry adiabatic lapse rate (in $K\ km^{-1}$) in the atmospheres of the Earth, Jupiter, Venus and Mars. Approximate the atmospheres of Venus and Mars to consist entirely of CO_2 gas; that of Earth by 20% O_2 and 80% N_2; that of Jupiter as 90% H_2 and 10% He. Make a reasonable guess for the value of γ in each atmosphere. (Hint: See §4.4.1.)

5-7. Estimate (crudely) the wet adiabatic lapse rate (in $K\ km^{-1}$) in the Earth's lower troposphere following steps (a)–(d) below.
(a) Determine c_P from the dry adiabatic lapse rate (see previous problem).
(b) Set $T = 280$ K and $P = 1$ bar. The saturation vapor pressure of water near 280 K is roughly approximated by the Clausius–Clapeyron relation (eq. 5.1),

with $C_L = 3 \times 10^7$ bar, and $L_s = 5.1 \times 10^4$ J mole^{-1}. Calculate the partial pressure of H_2O in a saturated atmosphere at 280 K.
(c) As the concentration of water in a saturated atmosphere decreases with height much more rapidly than the total pressure, you may estimate the value of w_s (fraction of a kilogram of water per kilogram of air) by multiplying the value of the partial pressure of water in bars by the ratio of the molecular mass of water to the mean molecular mass of air. Determine the value of w_s.
(d) Estimate the wet adiabatic lapse rate. Note: Watch your units. Whereas the latent heat in equation (5.1) is given in J mole^{-1}, that in equation (4.30) is in J kg^{-1}.

5-8. The saturated vapor pressure curve for NH_3 gas is given by equation (5.1) with $C_L = 1.34 \times 10^7$ bar and $L_s = 3.12 \times 10^4$ J mole^{-1}.
(a) Calculate the temperature at which ammonia gas condenses out if the NH_3 volume mixing ratio is 2.0×10^{-4}. Assume the atmosphere to consist of 90% H_2 and 10% He and that the pressure is 1 bar. (Hint: Convert the volume mixing ratio to partial pressure.)
(b) Calculate the temperature at which ammonia gas condenses out if the NH_3 volume mixing ratio is 1.0×10^{-3}.

5-9. Explain why the temperature within Earth's troposphere falls with increasing altitude to a minimum value at the tropopause and then increases with height in the stratosphere. Discuss how ultraviolet, visible and infrared light is transmitted and/or absorbed in each layer. In the case of absorption, explain what the important absorbing gasses are.

5-10. Explain briefly, qualitatively, why the density of ozone in Earth's atmosphere peaks near an altitude of 30 km. List the relevant reactions.

5-11. **(a)** Using the Chapman reactions (reactions 1, 4, 5 and 6 in §5.5.1), derive equations (5.5) and (5.6).
(b) Assuming chemical equilibrium, derive the altitude profiles given by equations (5.7) and (5.8).
(c) Calculate the number density of O_3 molecules in Earth's atmosphere at altitudes of 20 and 60 km using the number densities of [O] and [O_2] from Figure 5.9. The number density for [M] at $z = 0$ km on Earth is referred to as **Loschmidt's number** (Table C.3).
(d) Compare the value of number density of ozone that you've calculated in part (c) to that plotted at $z = 20$ km in Figure 5.9. Explain which simplifying assumptions made for your calculations may have led to this difference.

5-12. Explain why the NH_3 mixing ratio in the stratospheres of Saturn and Jupiter is well below the mixing ratio based on the saturated vapor curve.

5-13. Cloud or haze layers of hydrocarbons are observed in the stratospheres of Uranus and Neptune. Explain why these hazes are present above rather than below the tropopause.

5-14. The limiting flux is the maximum diffusion rate through a planetary atmosphere. The limiting flux can be calculated by assuming the same value for N_i/N for the part of the atmosphere that is well mixed (i.e., below the homopause, which for Earth is at $z < 100$ km). Consider hydrogen atoms in the Earth's atmosphere in all forms (H_2O, CH_4, H_2) at a fractional abundance $N_i/N \approx 10^{-5}$.
(a) Calculate the limiting flux of hydrogen-bearing molecules (and thus hydrogen) from the Earth's atmosphere. You can approximate the diffusion parameter, $D_i = 1.41 \times 10^{30}/N$ m^{-4} s^{-1}. (Hint: Calculate the limiting flux for an altitude $z = 100$ km; why?)
(b) Calculate the Jeans rate of escape for hydrogen atoms from Earth.
(c) Compare your answers from (a) and (b) and comment on the results.

5-15. The presence or absence of an atmosphere is to first order determined by Jeans escape. To verify this statement, plot the escape parameter for atomic hydrogen, λ_{esc}, as a function of heliocentric distance for all eight planets; Io; Ganymede; Titan; Enceladus; the asteroid Ceres; and the Kuiper belt objects Pluto, Eris and Varuna. To calculate relative temperatures, simply use the equilibrium temperature (see Chapter 4) at perihelion, with an albedo of zero, and emissivity of unity. Assume the exobase altitude $z = 0$. Discuss your findings. (Hint: You may want to use Tables E.1–E.8.)

5-16. Suppose a body with a radius of 15 km hits a planet at a velocity of 30 km s^{-1}. Assume the evaporative loading parameter (§5.7.3) is ~20.
(a) Calculate the mass of the atmosphere blown into space if the impactor hits the Earth.
(b) Calculate the mass of the atmosphere blown into space if the impactor hits Venus.

(c) Calculate the mass of the atmosphere blown into space if the impactor hits Mars.

(d) Express the masses of escaping gas calculated in (a)–(c) as a fraction of each planetary atmosphere's mass. Comment on your results.

5-17. Why is the $^{15}N/^{14}N$ ratio larger in Mars's atmosphere than it is in Earth's atmosphere?

5-18. (a) Explain how feedback effects in the carbon cycle can act on climatic variations on Earth.

(b) What geologic processes might perturb the carbon cycle on geologic timescales?

5-19. Firestorms sparked by a major nuclear war would release large amounts of soot into the atmosphere.

(a) Explain qualitatively why surface temperatures would be expected to drop well below freezing. This scenario is widely known as **nuclear winter**.

(b) Consider the case of an extreme nuclear winter, where smoke prevents any sunlight from penetrating deeply into Earth's atmosphere. In such a situation, the temperature of the top of the soot layer is equal to the equilibrium temperature, T_{eq}, because not enough atmosphere lies above the soot to provide significant greenhouse warming. The atmosphere below the soot layer is essentially isothermal, because no sunlight penetrates to this region. If the albedo of Earth remains unchanged at $A_b = 0.3$, what is the surface temperature?

(c) Repeat your calculation in part (b) for the case in which Earth's albedo is zero.

5-20. We find a superadiabatic layer in the martian atmosphere just above the surface during the day, which changes into an inversion layer at night. Explain this effect qualitatively. Do you expect much turbulence in this layer during the day and/or night?

5-21. The primary constituent of Pluto's atmosphere is N_2. Assume that this molecular nitrogen atmosphere is in equilibrium with the surface. What is the scale height of such an atmosphere? Compare this to the radius of Pluto. Does Pluto's nitrogen atmosphere run the risk of escaping? Discuss.

CHAPTER 6

Surfaces and Interiors

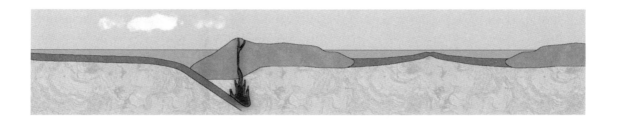

One of the most aggravating, restricting facets of lunar surface exploration is the dust and its adherence to everything no matter what kind of material, whether it be skin, suit material, metal . . . and its restrictive, friction-like action to everything it gets on.

Eugene A. Cernan, *Apollo 17* Commander, at the crew technical debriefing

The four largest planets in our Solar System are gas-rich giants with very deep atmospheres and no detectable solid 'surface'. All of the smaller bodies, the terrestrial planets, asteroids, moons and comets, have solid surfaces. Bodies with a solid surface display geological features that yield clues about past and current geological activity.

Each Solar System body that has been studied in detail has its own peculiarities. Every solid surface has its own geology, which often looks quite different from that of any other known body. The surface reflectivity varies dramatically from one body to another. Some surfaces have very low albedos; examples include the maria on the Moon, carbonaceous asteroids and comet nuclei. Others, such as the surfaces of Europa and Enceladus, are highly reflective. Our Moon and Mercury are almost completely covered by impact craters, whereas Io, Europa and Earth display little or no sign of impacts.

The terrestrial planets and many of the larger moons show clear evidence of past volcanic activity. Earth, Io, Enceladus and Triton are volcanically active at the present epoch. Past volcanic activity may manifest itself in the form of volcanoes of different shapes and size, such as those seen on Earth, Mars and Venus, or in the form of large solidified lava lakes, as on e.g., our Moon. Most bodies, even small asteroids, display linear features such as faults, ridges and scarps that are suggestive of past tectonic activity. However, only Earth displays the motion of tectonic plates. Why are planetary surfaces so different superficially, and what similarities do they share?

Surfaces and atmospheres are readily accessible for study, but we cannot observe the inside of a planet directly. We have seismic data for the Earth and the Moon. Such data reveal the propagation of waves deep below the surface, thereby providing information on the interior structure. The interior structure of every other body is deduced through less direct means. Observations of the body's mass and size give us the density, providing a first clue on its composition. The body's rotational period and geometric oblateness, gravity field, characteristics of its magnetic field (or absence thereof), the total energy output and the composition of its atmosphere and/or surface provide additional information that can be used to constrain the interior structure of bodies as diverse as large giant planets to small rocky objects.

In this chapter, we review planetary geological and geophysical processes. We start in §6.1 with discussions of rocks and minerals and of the behavior of different constituents under high temperature and pressure. This section concludes with a summary of the crystallization of **magma** (molten rock at depth). We cover planetary interiors in §6.2. A brief overview of the interior structure of Earth is followed by a more general discussion of gravity fields and equipotential surfaces, including the effects of mass anomalies and isostatic equilibrium. Processes that 'shape' the surfaces, such as gravity, volcanism and tectonics, are discussed in §6.3. Impacts are covered in §6.4. Details on the surface morphology and interior structure of individual planets and satellites are presented in Chapters 8–10, and surfaces of individual minor planets are discussed in §12.5.

6.1 Mineralogy and Petrology

Rocks are made up of different minerals, and the study of the composition, structure and origin of rocks is known as **petrology**. We begin this section with a short summary of the basics of mineralogy before reviewing the various types of rocks, how rocks are formed and where they are found.

6.1.1 Minerals

Minerals are solid chemical compounds that occur naturally and that can be separated mechanically

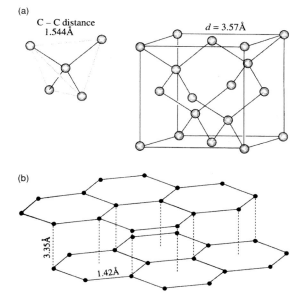

(a)

C – C distance
1.544Å

$d = 3.57$Å

(b)

3.35Å

1.42Å

Figure 6.1 Two very different solids consisting exclusively of carbon atoms are illustrated. (a) Mineral structure of diamond. Each carbon atom is surrounded by four others in a regular tetrahedron (shown by the *dashed line, left figure*). The diamond (*right figure*) is built up from these tetrahedra. The distance between atoms, d, is indicated in units of Angstrom (1 Å $= 0.1$ nm $= 10^{-10}$ m). (Putnis 1992) (b) The structure of graphite is made up of layers in which the carbon atoms lie at the corners of a hexagonal mesh. Within the layers, the atoms are strongly bonded, but the layers are weakly bonded to one another. (Putnis 1992)

from other components of a rock. Each mineral is characterized by a specific chemical composition and a specific regular architecture of the atoms. Several thousand minerals are known, each with its own unique set of properties.

A different spatial arrangement of the atoms that make up one material can lead to a very different mineral even if the chemical composition is the same. A classic example is **graphite** versus **diamond**, each of which consists exclusively of carbon (C) atoms. As shown in Figure 6.1a, in diamond, each C atom is surrounded by four others in a regular tetrahedron, and these atoms are held together by a **covalent bond**, i.e., the atoms share electrons in their outer shells. As a result,

diamond is an extremely hard mineral. In contrast, as shown in Figure 6.1b, graphite is made up of layers within which the C atoms form a hexagonal mesh. Bonding here is caused by **Van der Waals forces**, a weak electrical attractive force that exists between all ions and electrons in a solid. Graphite, therefore, is very soft.

6.1.2 Rocks

Planetary surfaces are composed of solid material, which is generally referred to as 'rock', the assemblages of different minerals. Rocks are classified on the basis of their formation history. We distinguish four major groups, each of which is discussed in more detail below; these are primitive, igneous, metamorphic and sedimentary rocks. Within these groups, the rocks can be further subdivided on the basis of the minerals of which they are composed or on the basis of their texture, such as the size of the grains that make up the rock. Some rocks, such as breccias, can include material from various groups. Figure 6.2 shows images of many different types of rocks.

Primitive Rocks

Primitive rocks are formed directly from material that condensed in the protoplanetary disk. These rocks have not undergone transformations in interiors of objects like the planets and larger moons and asteroids, where materials are altered significantly because of the high temperatures and pressures prevailing there. These primitive rocks have never been heated much, although some of their constituents (e.g., chondrules) may have been quite hot early in the history of our Solar System. Primitive rocks are common on the surfaces of many asteroids, and the majority of meteorites are primitive rocks (§11.1).

Igneous Rocks

Igneous rocks are the most common rocks on Earth and other bodies that have undergone melting.

Figure 6.2 COLOR PLATE Examples of different types of rocks and minerals. This collection is by no means complete but gives an idea of the differences between important rock types. An approximate scale (1 cm bar) is indicated on each frame. The rocks are grouped by type (i, igneous; s, sedimentary; m, metamorphic) and grain size. In some cases, we show two examples, e.g., where colors may differ substantially (e.g., a white and pink granite, a dark-gray and a reddish hematite). Individual crystals (e.g., quartz, biotite, muscovite, plagioclase) can be identified in the granite samples, but in rhyolite, only small specks of grains are visible. Pumice is a very light and froth-like rock. In gabbro, individual crystals are large, as in granite, but the crystals are smaller in basalt and andesite. No crystal structure is present in obsidian, a glassy rock. Sandstones, sedimentary rocks, can be clearly recognized from their granular appearance (a gray sandstone and an Arizona pinkish one are shown); conglomerates are much coarser grained. Chert is a relatively hard sediment. Halite and gypsum are representative samples of evaporites. Magnetite is an iron oxide. Limonite and hematite are sediments formed from clay minerals. Minerals are sometimes present in the form of large crystals, such as the quartz and garnet crystals shown. Several metamorphic rocks are displayed (*bottom row*): marble, gneiss, schist and slate. (Photographs taken by Floris van Breugel using rock samples provided by K. Ross, Museum of Geology, UC Berkeley. Figure adapted from de Pater and Lissauer 2010)

Igneous rocks are formed when a **magma**, i.e., a large amount of hot molten rock, cools. The physics and chemistry of a cooling magma and the crystallization of minerals therein are discussed in §6.1.4. Here we describe the end products of the melt, i.e., the various types of rocks that result from the cooling process. The magma can cool either underground, producing **intrusive** rocks, or above ground, forming **extrusive** or **volcanic** rocks. Magma deep underground cools slowly, and crystals have plenty of time to grow. The resulting intrusive rocks are therefore coarse grained, and the minerals can easily be distinguished with the naked eye (e.g., common granite, Fig. 6.2). When magma erupts through the planetary crust, it cools rapidly via radiation into space. Volcanic rocks thus show a fine-grained structure in which individual minerals can only be seen through a magnifying glass. In cases of extremely rapid cooling, the rock may 'freeze' into a glassy material. Obsidian is a volcanic rock that cooled so fast that it shows no crystalline structure (Fig. 6.2). The minerals in these rocks show an amorphous, glassy structure. The texture of the rock thus depends on how rapidly the magma cools, and the composition of the rock depends on the minerals that crystallize from the melt.

Although the classification of the major rock groups is based on their chemical and mineralogical composition, for practical purposes, one can simply use the silica content of the rock. The two basic rock types are basalts and granites; basalts contain 40%–50% silica (by weight) and granites much more ($\sim$70% by weight). In addition to silica, basalts consist largely of heavy minerals, such as pyroxenes and olivines. Whereas most granites are light colored, basalts are dark. Basaltic rocks are probably the most common rocks on planetary surfaces because they make up the lava (solidified magma) flows on bodies such as the Earth and Moon. A lunar basalt is shown in Figure 6.3a. Although abundant on Earth, granite is less common on other planetary bodies.

Extrusive rocks are directly correlated with volcanic activity. The type of eruption is determined by the **viscosity** (a measure of resistance to flow in a fluid) of the magma and depends on the temperature, composition (in particular silica content) and gas content of the melt. Typically, magma and lava with a high silica content have a high viscosity, which is increased even more if the gas content is high. Such eruptions are explosive, with typical temperatures of $\sim$1050–1250 K, and they form thick local deposits. In contrast, basaltic melts are very fluid, erupt with temperatures of 1250–1500 K and flow fast and far. Basaltic melts can form large lava beds and fill in lowlands, such as the maria on the Moon and the lava beds on Hawaii. These lava flows are usually dark in color, in contrast to the lighter felsic deposits.

Metamorphic Rocks

Because there has been much reworking of the Earth's surface by great forces from within the interior, many rocks have been altered when subjected to high temperature and pressure or when introduced to other chemically active ingredients. Rocks that have been so altered are called **metamorphic** rocks. Particular types of rocks are often named for a mineral constituent that is predominant in the rock, such as marble (from limestone or other carbonate rock) and quartzite (from quartz; see Fig. 6.2 for some examples). Metamorphism can act on either a **regional** or a **local** scale. On the regional scale, rocks (igneous, sedimentary, metamorphic) are transformed many kilometers below the surface by extremely high temperatures and pressures. Large areas or regions of rock can be metamorphosed this way. On a local scale, rocks are transformed near an igneous intrusion, largely by heat. Magma forces its way into layered rocks or penetrates cracks or cavities. If the stress on the rocks and the temperature are high enough, the rock is altered or metamorphosed. Although not as common on Earth as regional or contact

(a)

(b)

(c)

Figure 6.3 COLOR PLATE (a) Basaltic rock from the lunar mare brought to Earth by *Apollo 15*. This rock, sample 15016, crystallized 3.3 billion years ago. The numerous vesicles (bubbles) were formed by gas that had been dissolved in the basaltic magma before it erupted. (NASA/Johnson) (b) A breccia from the lunar highlands, sample 67015, collected by *Apollo 16* astronauts. This rock is termed **polymict** because it contains numerous fragments of preexisting rocks, some of which were themselves breccias. It was compressed into a coherent rock about 4.0 billion years ago. (NASA/Johnson) (c) Petrographic thin section (2–3 mm across) of the sample in panel (b) in polarized transmitted light. (Courtesy Paul Spudis)

metamorphic rocks, rocks may also get altered as a result of impact-induced shocks. Shocked quartz has been found at several impact sites.

Sedimentary Rocks

On planets that possess an atmosphere, material may be transported by winds, rain and liquid (e.g., water) flows. Sedimentation is the final stage of this process. These sediments may form new **sedimentary** rocks (Fig. 6.2). While the rock fragments are transported, the minerals are sorted by size and weight with a variable efficiency. Through the sorting process, rocks form with different textures, varying from coarse to very fine grained. Coarse-grained fragments, such as gravel, form **conglomerates** when cemented together. Medium-grained sands form **sandstone**, and fine-grained clays and silt may be cemented together into **mudstone** or **shale**. The individual grains in shale and sandstones are fairly round as a result of the erosion. Shale and sandstone are the two most abundant sedimentary rocks on Earth. Shale makes up ~70% of the sediments on Earth, and sandstone comprises ~20%.

The composition of rocks may be altered through the interaction with other chemical constituents, such as those present in an atmosphere. Such sediments are referred to as **chemical sediments**. A common example is calcium carbonate or calcite, $CaCO_3$. Although most carbonates on Earth are from biological deposits – fossils of animal shells produced by organisms in the oceans – carbonates are also formed when atmospheric CO_2, dissolved in water, reacts with silicate minerals in the soil (§5.8.2). **Evaporite** is a rocky material from which liquid evaporated, leaving behind sediments such as halite or common salt and sulfate minerals, such as gypsum. Evaporites may bond other rocks together into a loose, crumbly rock. Another type of sediment is formed by clay minerals, hydrous aluminum silicates such as hematite and limonite. They are abundant in erosion products on both Earth and Mars, as well as in carbonaceous material on asteroids and bodies in the outer Solar System.

Breccias

Breccias are 'broken rocks' that consist of sharp angular fragments that are cemented together. A lunar breccia is shown in Figure 6.3. Brecciated rocks may originate from meteoroid impacts, where the pieces are 'glued together' under the high temperature and pressure during and immediately after the impact. Therefore, breccias cover the bottom of many impact craters.

6.1.3 Material under High Temperature and Pressure

Whether a material is in a solid, liquid or vapor phase depends on both the temperature and the pressure of the environment. Thus, the state of a material in a planet's interior may be quite different from that at the standard temperature of 273 K and pressure of 1 bar. The relationship of a material's state to the temperature and pressure is given by the **constituent relations**. For modeling purposes, one also needs to be able to relate the pressure, density, temperature and composition of a material in the form of an equation, $P = P(\rho, T, f_i)$. Such relations are called the **equation of state**. A well-known example is the perfect (ideal) gas law, given by equation (3.14), which is valid in planetary atmospheres at pressures below ~50 bar. At higher temperatures and pressures, this simple equation is no longer adequate.

Both the constituent relations and the equation of state are largely determined empirically. Although such experiments may be relatively easy to conduct at low pressure, it is quite difficult to determine such relations at the high temperatures and pressures encountered in a planet's interior, so relationships in these regimes are not as well known.

We next discuss the various phases of materials that make up planets and relate our findings to what is known about planetary interiors (§6.2). We describe individual planetary and satellite interiors in Chapters 8–10.

Hydrogen and Helium

Typical temperatures and pressures in the giant planets range from about 50–150 K at a planet's tropopause up to 7000–20 000 K at their centers, and the pressure varies from near zero in the outer atmosphere up to 20–80 Mbar at the planet's center. At Mbar pressures, the fluid is so densely packed that the separation between the molecules becomes comparable to the size of the molecules, so that their electron clouds start to overlap and an electron can hop or percolate from one molecule to the next. Shock wave experiments show that at $T \gtrsim 2000$ K and $P \gtrsim 1.4$ Mbar, where hydrogen is a fluid, this fluid behaves like a metal, a

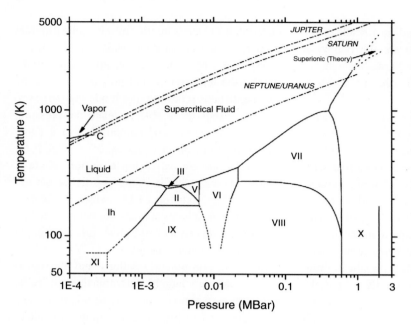

Figure 6.4 Phase diagram for water at temperatures and pressures relevant for the icy satellites as well as the giant planets (adiabats for the giant planets are superposed). The various crystal forms of ice are indicated by roman numerals I–XII, where 'h' in Ih refers to the hexagonal crystal form of ordinary ice (all natural snow and ice on Earth are in the Ih form). The metastable ices IV and XII are not shown (they fall in the top corner of II and middle of V, respectively). The critical point of water is indicated by a C. (Diagram courtesy of Stephen A. Gramsch, Carnegie Institution of Washington; adiabats for the planets were provided by William B. Hubbard)

phase referred to as **fluid metallic hydrogen**. In Jupiter, this transition occurs at a radius of ~0.90 $R_{2\!\!\!+}$. Convection in this fluid is thought to create the magnetic fields observed to exist around Jupiter and Saturn.

Helium transforms into a liquid metallic state only at pressures much higher than encountered in the giant planets. The temperatures and pressures in the giant planets are also not high enough for hydrogen and helium to fully mix. When they separate, helium, being heavier, sinks. The observed depletions of helium (compared with the He/H ratio on the Sun) in the atmospheres of Jupiter and Saturn have been attributed to He separation within the metallic hydrogen region.

Ices

Water-ice is a major constituent of bodies in the outer parts of our Solar System. Depending on the temperature and pressure of the environment, water-ice can take on at least 15 different crystalline forms, many of which are shown in the water phase diagram in Figure 6.4. Although the

water molecule does not change its identity, the molecules are more densely packed at higher pressures, so that the density of the various crystalline forms varies from 920 kg m^{-3} for common ice (form I) up to 1660 kg m^{-3} for ice VII and near the triple point with ices VI and VIII. Temperatures and pressures expected in the interiors of the icy satellites range from ~50–100 K at the surface up to several hundred K at pressures of up to a few tens kbar in their deep interiors. So one might expect a wide range of ice forms in these satellites.

At the higher temperatures, above 273 K at moderate pressures, water is a liquid. The **critical point** of water ($T = 647$ K, $P = 221$ bars) is indicated by a C; above this temperature, there is no first-order phase transition between gaseous and liquid H$_2$O. Water becomes a supercritical fluid, characterized as a substance that is neither gas nor liquid, with properties that are very different from ambient water. The adiabats of the giant planets are superposed on Figure 6.4.

In addition to water-ice, one would expect the outer planets and moons to contain substantial amounts of other 'ices', such as ammonia, methane and hydrogen sulfide. Phase diagrams of these ices might be as complex as those for water. Although experiments up to 0.5–0.9 Mbar have been carried out for some of these constituents, they are not as well studied under high pressures as is water-ice, and mixtures of these various ices are even less understood.

Rocks and Metals

The phase diagrams of magmas, i.e., molten rocks, are very complex. The various elements and compounds interact in different ways, depending on the temperature, pressure and composition of the magma. Experiments have shown that at higher pressure, the molecules get more closely packed, which leads to a reorganization of the atoms in the lattice structure. This rearrangement of atoms and molecules into a more compact crystalline structure involves an increase in density, and characterizes a phase change. A common example is carbon, which at low pressure is present in the form of graphite, and at high pressure as diamond (Fig. 6.1).

6.1.4 Cooling of a Magma

The composition, pressure and temperature of a magma determine which minerals ultimately form when the magma cools. The states that the magma goes through as it cools and crystallizes can be shown on a **phase diagram**, such as discussed in §3.1 (Fig. 3.1). Phase diagrams for a single constituent are usually relatively straightforward. When multiple mineral components are involved, phase diagrams can become very complicated. For example, the solid and/or melt may become **immiscible** (insoluble in each other) at certain temperatures, solid phases may exist in different forms at different temperatures, the melting point of a substance may be lowered in the presence of a particular melt (**eutectic** behavior), intermediate products may be formed, etc.

Magmas or molten 'rocks' crystallize over a large range of pressures and temperatures. While the melt is cooling, the composition of the crystallizing material as well as that of the magma changes continuously. The minerals that crystallize from the magma and the sequence in which they crystallize depend on pressure, temperature and composition and on how these vary as the system changes from a liquid to a solid phase. While the magma is cooling, existing crystals or nucleation seeds grow. The rate of growth is regulated because the latent heat of crystallization (§3.1) warms the local environment. This heat has to be removed from the crystal to allow it to grow. On the other hand, if the temperature is reduced too rapidly, the magma becomes very viscous and does not provide the crystal with enough material to continue to grow.

To complicate matters further, a magma usually does not cool under equilibrium conditions. The crystallized matter may not equilibrate with the magma, or heavy crystals may sink down in the magma, a process referred to as **differentiation**. In the latter case, the crystals are essentially removed from the melt. In the former case, if the crystals do not equilibrate, zoned crystals may form. For example, in a two-component melt, A + B, zoned crystals consist of a core rich in composition A, which we assume to have the higher melting temperature, surrounded by a mantle that gradually grades to a composition richer and richer in B, which has the lower melting temperature.

Starting with a high-temperature magma, olivine crystals are the first that condense out. Because these crystals are heavy, they sink to the bottom of the magma chamber and are thus removed from the melt through magmatic differentiation. The remaining melt is of basaltic composition. Upon further cooling, pyroxenes condense and differentiate out, leaving a melt of a more silica-rich composition. Then amphiboles and biotite micas appear, and the magma left behind is more and more silicic in composition. If the crystals had remained in the cooling melt, they may have reacted with the magma and consequently been changed. For example, if the olivine crystals had not settled out, they would have been converted to pyroxenes through interaction with the cooling melt, in which case there would not be much olivine on Earth.

Upon cooling a more silica-rich melt, crystallization starts with calcium-rich plagioclase (anorthite), and at lower temperatures, the crystals gradually become more and more sodium rich (albite). Because feldspars are relatively light, they tend to float on top of the magma. The cooling magma becomes gradually more and more silicic, ending with granitic material (potassium feldspar, muscovite mica and quartz). At any time during the cooling process, if the melt reaches the surface, the resulting rocks have the composition of the melt 'frozen'. So, if the melt surfaces early on in the cooling process, basaltic rocks are formed. If the magma surfaces later during the cooling process, it contains relatively more silica.

The reaction series of fractional crystallization and magmatic differentiation is based on laboratory experiments of melting and subsequent cooling of igneous rocks. The sequence of reactions taking place in a cooling magma in the Earth's mantle, however, is often far more complicated, and research is still going on. Usually there is partial rather than complete melting of rocks, and there is a wide range of temperatures even within one magma chamber. The differences in temperature may create chemical separation. Although convective motions may mix magmas of different composition, some melts are immiscible, so that there may be melts with different compositions within one magma chamber. These melts each give rise to their own crystallization products.

6.2 Planetary Interiors

As discussed in §1.4.11, the interior structure of bodies other than the Earth and the Moon can only be deduced from remote observations.

Measurements of a body's mass and size together yield an estimate for the average density, which provides some first-order estimates of a body's composition. For example, a density $\rho \lesssim 1000$ kg m^{-3} for a small body implies an icy and/or porous object, but giant planets of this density consist primarily of hydrogen and helium. For objects comparable to Earth in size or smaller, a density $\rho \approx 3000$–3500 kg m^{-3} suggests a rocky composition, and higher densities indicate the presence of heavier elements, in particular iron, one of the most abundant heavy elements in the cosmos (Table 3.1). Note that we refer here to the **uncompressed density**, i.e., the density that a solid or liquid planet would have if material was not compressed by the weight of overlying layers.

In addition to a body's mass and size, its shape, rotational period and gravity field are key to constraining its interior structure. Knowledge regarding internal energy sources provides information on the thermal structure of its interior, and the presence or absence of a magnetic field adds information regarding the conductivity of its interior.

Because it is handy for comparative planetology to be familiar with Earth, we start this section with a brief overview of the Earth's interior. This is followed by a longer discussion of a body's shape and gravity field. We then consider sources, transport and losses of a body's internal heat.

6.2.1 Interior Structure of the Earth

The interior structure of the Earth has been deduced from seismological data. Figure 6.5 shows a sketch of the Earth's structure. The Earth consists of a solid iron–nickel **inner core** surrounded by a fluid metallic **outer core**. The core extends over roughly half the Earth's diameter. Earth's outer ∼3000 km is the primarily silicate **mantle**, which itself is divided into a lower and upper mantle, separated by a transition zone. A cool elastic **lithosphere** sits on a hot, highly viscous 'fluid', the **asthenosphere**. Although the asthenosphere is

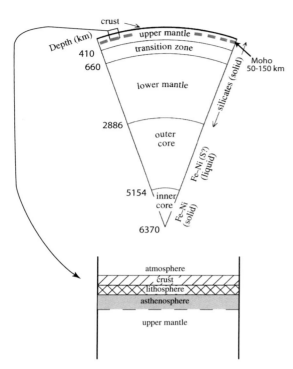

Figure 6.5 Sketch of the interior structure of the Earth. The lower lithosphere and asthenosphere are parts of the upper mantle. (Adapted from Putnis 1992)

highly viscous, the mantle below it is even more so, and the asthenosphere, therefore, is like a 'lubricating' layer between the rigid lithosphere and highly viscous mantle below it.

Earth's outer 'skin' is the **crust**, a rather brittle layer. Earth's crust has a mean thickness of ~6 km under the oceans and a mean thickness of ~35 km under the continents. Oceanic crust, primarily of basaltic composition, is denser than the more silicic (granitic) continental crust.

6.2.2 Shape and Gravity Field

Gravity is ubiquitous. It wants to pull everything 'down', shaping a nonrotating 'fluidlike' planet into a perfect sphere, i.e., the shape that corresponds to its lowest energy state. The term 'fluidlike' in this context means deformable over

geologic time (i.e., $\gtrsim$ millions of years). Thus, an object is approximately spherical if the weight of the mantle and crust exerted on its inner parts is large enough to deform the body over time into a **spheroid**. Later we discuss how a body's shape depends on its size, composition, material strength, rotation rate, temperature and history (including tidal interactions for moons).

On the largest scales, the force that best competes with gravity is rotation, through the centrifugal force. Rotation flattens a deformable object somewhat, changing its figure to an **oblate spheroid**, the equilibrium shape under the combined influence of gravity and centrifugal forces. Polar flattening caused by rotation is the largest deviation from sphericity for planet-sized (radii exceeding a few hundred km) bodies.

In the following, we discuss the equipotential surface of a rotating body. We then proceed with a short discussion of how the internal density distribution for rotating bodies in hydrostatic equilibrium can be deduced from the body's moment of inertia. Mass anomalies may have profound effects on the orientation and spin of a body. Precession and nutation are discussed in §2.6.3; below we address a phenomenon known as polar wander. We conclude this subsection with a brief summary of isostatic equilibrium.

Equipotential Surface

A body's gravity field can be measured to quite high accuracy by tracking the orbits of spacecraft close to the body or from the rate of precession of the periapses of moons and rings orbiting the planet. Mathematical expressions for the gravity field in terms of its gravitational moments, J_n, are provided in §2.6.1. For a nonrotating fluid body in hydrostatic equilibrium, the moments $J_n = 0$, and the gravitational potential reduces to $\Phi_g(r, \phi, \theta) = -GM/r$. Rotating fluid bodies in hydrostatic equilibrium have $J_n = 0$ for all odd n. This is a very good approximation for the giant planets.

For rotating bodies, the effective surface gravity is less than the gravitational attraction calculated for a nonrotating planet because the centrifugal force induced by rotation is directed outward from the planet (§2.1.6). The **equipotential surface** on a planet must thus be derived from the sum of the gravitational potential, Φ_g, and the rotational, or centrifugal, potential Φ_c:

$$\Phi_g(r,\phi,\theta) + \Phi_c(r,\phi,\theta) = \text{constant}, \qquad (6.1)$$

where $\Phi_g = -GM/r$ and Φ_c is defined as

$$\Phi_c = -\frac{1}{2}r^2\omega_{rot}^2 \sin^2\theta. \qquad (6.2)$$

On Earth, the equipotential surface is referred to as the **geoid**, which is measured at the mean sea level.

The precise shape or **figure** of a planet depends on the **rheology** of the material in addition to the rotation rate. Rheology is an empirically derived quantity that determines the stress/strain relation of a material. **Strain** refers to deformation. **Stress**, like pressure, is defined as force per unit area. **Elastic** material responds to stress, and if the stress is removed, an elastic material will regain its original properties. If a stress is applied to a viscous material, the material deforms or flows in a slow, smooth way as long as the stress is exerted. When the stress is removed, the flow stops because of the material's intrinsic resistance against deformation. Only small bodies, where the gravity field is weak, can have very irregular shapes (e.g., Figs. 1.4, 1.7 and 10.21). Minimum radii for a body to be spherical can be estimated from its density and material strength. A rocky body with a typical density $\rho = 3500 \text{ kg m}^{-3}$ and material strength $S_m = 2 \times 10 \text{ Pa}$ is approximately round if $R \gtrsim 350$ km; the maximum radius is ~220 km for iron bodies to be oddly shaped (Problem 6-1).

Moment of Inertia Ratio, $I/(MR^2)$

For axisymmetric bodies in hydrostatic equilibrium, the body's moment of inertia, I, its rotation rate and the gravitational moment J_2 are related.

The moment of inertia ratio, $I/(MR^2)$, contains valuable information on a body's internal density structure. If the density ρ is uniform throughout the planet, $I = 0.4 MR^2$ (§2.5.1). The moment of inertia ratio is less than 0.4 if ρ increases with depth in the planet, which is usually the case. Table E.15 shows the J_n and $I/(MR^2)$ values for the planets and several large moons. The relatively small $I/(MR^2)$ values for the giant planets, bodies that are rapid rotators in hydrostatic equilibrium, suggest a pronounced increase in density towards their centers. The situation is different for Mercury and Venus. These planets rotate very slowly, so non-hydrostatic effects (e.g., mantle convection) have a much larger contribution to the J_2 than does the rotational effect. This complicates interpretation of the $I/(MR^2)$ values.

Our Moon and the Galilean satellites are in synchronous rotation with their orbital period, so that in addition to rotation, tidal forces affect their shape and hence gravity fields (§2.7.1). A synchronously rotating fluidlike satellite takes on a (nearly) triaxial shape with axes $A > B > C$, where the long axis is directed along the planet–satellite line and the short axis is parallel to the rotation axis.

Polar Wander

If a body has a large positive mass anomaly (excess of mass) that is not at the equator (or a deficiency in mass that is not at the pole), the planet will reorient itself in space while the rotational axis stays fixed with respect to distant stars. The rotation axis appears to 'move' across the globe, a phenomenon known as **polar wander**. Large mass anomalies can lead to a complete reorientation of the polar axis with respect to the globe.

Isostatic Equilibrium

Deviations in the measured gravity with respect to the geoid provide information on the structure of

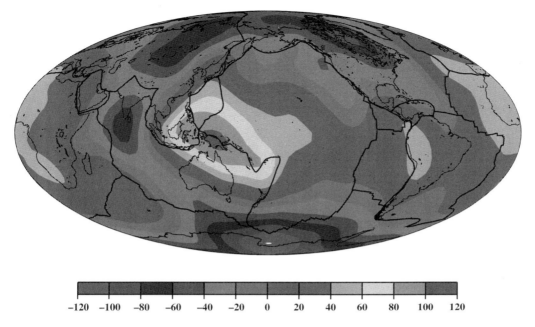

Figure 6.6 COLOR PLATE The observed geoid (degrees 2–15) superposed on a map of the planet Earth showing the boundaries of land masses and crustal plates. Contour levels are from −120 m (near the South Pole) to +120 m (just north of Australia). (Lithgow-Bertelloni and Richards 1998)

the crust and mantle. The measured surface gravity field of Earth does not deviate substantially from an oblate spheroid even in the proximity of high mountains, despite the large land masses that make up the mountains. This observation led to the concept of **isostatic equilibrium**, which is based on the **Archimedes principle** and the theory of hydrostatic equilibrium (§3.2).

The Archimedes principle states that a floating object displaces its own weight of the substance on which it floats. Similar to an iceberg floating in water, a mountain in isostatic equilibrium is compensated by a deficiency of mass underneath because the part of the mountain that is submerged in the upper mantle is lighter than the mantle material displaced. Similarly, ocean and impact basins in isostatic equilibrium have extra mass deeper down.

Figure 6.6 shows the geoid shape for Earth superposed on a map of the Earth. This is a relatively low order (J_2–J_{15}) gravity map,

comparable to those obtained for other planets. Although there clearly is structure in the geoid, there is no correlation with the topography. However, the structure does seem to correlate well with tectonic features, such as mid-ocean ridges and subduction zones, and must be caused by mantle convection and deep subduction (dynamic isostacy). This poor correlation between topography and low-order harmonic gravity maps is unique to Earth. Most other bodies reveal a correlation between gravity field and topography, as discussed in §9.1.3, §9.3.3 and §9.4.2

6.2.3 Internal Heat: Sources, Losses and Transport

The temperature profile in a planet's interior depends on its internal energy sources, transport of energy and loss of heat. In §5.1.2, we compare the equilibrium temperatures of the giant planets, i.e., the temperature the planet would have if heated by solar radiation only, with their observed effective temperatures (Table E.9). This comparison showed that Jupiter, Saturn and Neptune are warmer than can be explained from solar heating alone, which led to the suggestion that these planets possess internal heat sources. The Earth must also have an internal source of heat, as deduced from its measured heat flux of 0.075 J m^{-2} s^{-1}. Measured planetary heat-flow parameters are summarized in Table E.16.

Heat Sources

Accretion of material during formation is likely one of the largest sources of heat for large bodies (§15.5.2; Problem 15-16). Bodies hit the forming planet with roughly the escape velocity, yielding an energy for heating of GM/R per unit mass. The internal heat sources of the giant planets Jupiter, Saturn and Neptune are attributed to this **gravitational energy**, either from gradual escape of primordial heat generated during the planet's formation or from previous or ongoing differentiation. In particular, detailed models of Saturn's interior structure show that, in addition to primordial heat, differentiation of helium from hydrogen adds to its internal heat source. Helium differentiation is important to a lesser extent for Jupiter. This process also explains the observed depletion in the helium abundance in both of these giants' atmospheres compared with solar values.

The intrinsic luminosity, L_i, of the three most massive giant planets in our Solar System is comparable to the amount of incident sunlight absorbed by the planet and re-emitted at infrared wavelengths, L_{ir}. For the terrestrial planets, L_i is very small. The effective temperature T_e is obtained by integrating the emitted energy over all infrared wavelengths, and it thus consists of both L_{ir} and L_i. The equilibrium temperature T_{eq} is the temperature the planet would have in the absence of internal heat sources (§4.1.3). It is interesting to note that Uranus does not show any excess heat; the reasons for the difference between Uranus and Neptune are not understood.

Gravitational heating alone cannot account for the heat budget of the terrestrial planets and smaller bodies. An important source of heat in the interiors of the terrestrial planets, satellites, asteroids and the icy bodies in the outer Solar System is **radioactive decay** (§11.6.1) of elements with long half-lives, roughly of the order of a billion years. Important sources are ^{235}U, ^{238}U, ^{232}Th and ^{40}K. Temporal variations in tidal forces can lead to internal heating of planetary bodies, as discussed in detail in §2.7.3. For most objects, tidal dissipation is a much smaller source of energy than gravitational contraction plus radioactive heating. For the moons Io, Europa and likely Enceladus, however, tidal heating is a major source of energy (see §§10.2.1, 10.2.2 and 10.3.3).

Energy Transport and Loss of Heat

Energy transport determines the temperature gradient in a planet's interior, just like it does in a planetary atmosphere. The temperature gradient is determined by the process that is most effective in transporting heat. The three mechanisms by which heat is transported are conduction, radiation and mass motion, primarily convection (see §4.2–4.5). Conduction and convection are important in planetary interiors, whereas radiation is important in transporting energy from a planet's surface into space and in a planet's atmosphere where the opacity is small.

Solid bodies generally lose heat by conduction upwards through the crust and radiation from the surface into space. At deeper levels, for planets that are large enough that solids 'flow' over geological timescales, the energy may be transported by convection, a process that, when it occurs, typically transports more heat than does conduction. In the upper layers, heat transport is again by conduction upwards through the rigid lithosphere and crust. However, conduction alone may not be sufficient to 'drain' the incoming energy from below. On Earth, additional heat is lost through tectonic activity along plate boundaries and via hydrothermal circulation along the mid-ocean ridge. Heat may also be lost during episodes of high volcanic activity, either in volcanic eruptions or through vents or hot spots. The latter source of heat loss is dominant on Io (and Enceladus). The total heat outflow through Io's hot spots may, at the present time, even exceed the generation of heat due to tidal dissipation.

In the giant planets, heat is transported by convection throughout the mantle and most of the troposphere, while radiation to space plays an important role at higher altitudes in the stratosphere and lower thermosphere.

6.3 Surface Morphology

The surfaces of planets, asteroids, moons and comets show distinct morphological features, such as mountain chains, volcanoes, craters, basins, (lava) lakes, canyons, faults, scarps, etc. Such features can result from **endogenic** (within the body itself) or **exogenic** (from outside) processes. In this section, we summarize endogenic processes that are common on planetary bodies. In §6.4, we discuss exogenic processes, in particular, impact cratering.

The surface **topography** of a planet is measured with respect to the planet's geoid. The ability of local structures on planetary surfaces to survive the gravitational pull towards the planet's center depends on the density and strength of the material. Although downhill movements of material are induced by gravity, whether or not such movements occur is determined by the steepness of the slope compared with the **angle of repose**, which is the greatest slope that a particular material can support. The angle of repose depends mainly on friction. If one piles up sand in a sand box, the slope of the resulting hill is the same for small and large hills, but the slope is different if the mound is built out

of fine sand, gravel or pebbles. The angle of repose is determined by the type of material, the size and shapes of the 'granules', water and air content and temperature. If the slope of a hill is steeper than the angle of repose, **mass wasting** such as landslides, mudflows or rockslides will occur. But even on slopes less steep than the angle of repose, material can migrate downhill in **slumping** motions (e.g., landslides, avalanches) or as a slow, continuous **creeping** motion (e.g., glaciers, lava flows). Such downhill migrations can be triggered by seismic activity (earthquakes on Earth, moonquakes on the Moon, etc.) caused by either internal or external processes. Precipitation and the presence of a liquid, such as water on Earth and methane on Titan, also play major roles in downhill motions. Spacecraft observations have revealed that mass wasting is a widespread surface process, and the characteristics of landslides (e.g., length/height ratios) seem to be remarkably insensitive to gravity and the material properties.

Stratigraphy is the study of the temporal sequence of geological events and of dating the events with respect to one another. The word *stratigraphy* originates from the sequence and correlation of stratified rock layers. On some bodies, uplifts of strata (e.g., the walls of an impact crater) yield information not only on the dating of the event but also on the crustal layers and the forces involved. As shown in Figure 6.7, features produced by an endogenic process sometimes cross or dissect a preexisting crater, but others appear partially obliterated by an impact crater. The stratigraphy of these features yields information on the chronology of the various events.

6.3.1 Tectonics

Any crustal deformation caused by motions of the surface, including those induced by extension or compression of the crust, is referred to as **tectonic** activity. Many planetary bodies (terrestrial planets and most major satellites) show evidence of

(a)

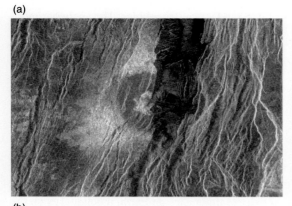

(b)

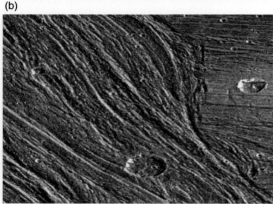

Figure 6.7 Examples of stratigraphy. (a) A *Magellan* radar image of a 'half crater' on Venus, located in the rift between Rhea and Theia Montes in Beta Regio. The crater is 37 km in diameter and has been cut by many fractures or faults. (NASA/*Magellan*, PIA00100) (b) The Nippur Sulcus region, an example of bright terrain on Ganymede, shows a complex pattern of multiple sets of ridges and grooves. The intersections of these sets reveal complex age relationships. The Sun illuminates the surface from the southeast (*lower right*). In this image, a younger sinuous northwest–southeast trending groove set cuts through and has apparently destroyed the older east–west trending features on the right of the image. The area contains many impact craters; the large crater at the bottom of the image is about 12 km in diameter. (NASA/*Galileo Orbiter*, PIA01086)

crustal motions from shrinking and/or expanding of the surface layers, commonly caused by heating or cooling of the crust.

Tectonic activity is controlled by the rheology of a planet's materials. Consider a forming planet as a hot ball of fluid magma. The outer layers are in direct contact with cold outer space and thus cool off first by radiating the heat away, so that a thin crust forms over the hot magma. While the crust cools, it shrinks. Convection in the mantle may move 'hot plumes' around and heat the crust locally, leading to a local expansion of the crust. The interior cools off through convection and conduction, with volcanic eruptions at places where the crust is thin enough that the hot magma can burst through. The added weight of the magma on the crust may lead to local depressions, such as the 'coronae' on Venus shown in Figure 9.16.

Extensional and compressional forces on the crust result in folding and faulting. Common tectonic deformations are illustrated in Figure 6.8. **Folding** refers to an originally planar structure that has been bent. **Faulting** involves fractures, such as the **faults** seen when two tectonic plates slide alongside each other (e.g., as across the San Andreas fault in California). Fault displacements can be of order ∼30 m after a very large

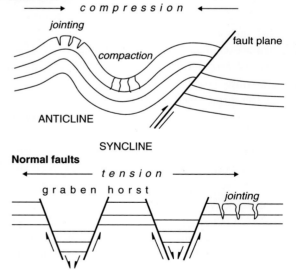

Figure 6.8 Schematic showing common tectonic deformations such as faults, and the formation of grabens and horsts. (Greeley 1994)

Figure 6.9 Vertical basalt columns in Devils Postpile National Monument (California). (Courtesy Cooper, Wikimedia Commons)

earthquake. If there is no crustal motion involved, cracks are called **joints**. Along such joints, rocks are relatively weak and therefore particularly susceptible to erosion and weathering. A spectacular example is the columnar basalts that are shown in Figure 6.9.

Folding and faulting help shape a planet's surface, and the effects are visible as distinct geological features. Typical examples are **grabens** and **horsts**. A graben is an elongated fault block that has been lowered in elevation relative to surrounding blocks (Fig. 6.8), and a horst is a fault block that has been uplifted. **Scarps** are steep cliffs that can be produced by faulting or by erosion processes. Many bodies display **rilles**, which are elongated trenches, either sinuous in shape or relatively linear. Rilles can be tectonic (from, e.g., faulting) or volcanic (e.g., collapsed lava tubes) in origin. Figure 6.10 shows the volcanically produced Hadley Rille on the Moon. Folding and faulting processes can also lead to the formation of mountain ridges, such as those found on many continents on Earth.

Plate Tectonics

A study of the shapes and motions of the continents on Earth has led to the concept of **plate tectonics**. Figure 6.11 shows how the various continents seem to fit together as a jigsaw puzzle, and current theories imply that roughly 200 Myr ago there was only one large landmass, **Pangaea**. Since that time, the continents have moved away from one another, a process known as **continental drift**. This motion is induced by plate tectonics. The lithosphere consists of 15 large plates, which move with respect to each other by a few, in some cases up to nearly 20, centimeters per year. Before the existence of the supercontinent Pangaea, there were other continental assemblages and breakups. This cycle of opening and closing of ocean basins (complementary to the formation and breakup of supercontinents) is known as the **Wilson cycle**. The Wilson cycle has the longest period of the geologic variations on Earth – at present roughly 300–500 million years. Continents built up gradually over time; there were no large land masses on Earth 4 billion years ago.

Plate tectonics is caused by convection in the mantle, which induces a large-scale circulation pattern, where the plates 'ride' on top. This circulation is sometimes compared to a conveyor belt. The

Figure 6.10 Hadley Rille, a typical sinuous rille on the Moon, close to the base of the Apennine mountains. The rille starts at a small volcanic crater and 'flows' downhill. The picture is approximately 130 × 150 km. (NASA/*Lunar Orbiter IV-102H3*)

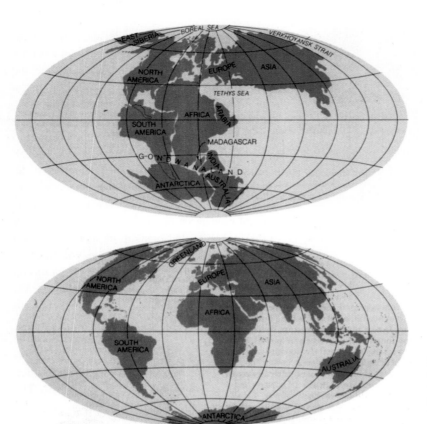

Figure 6.11 Continental drift on Earth: 200 million years ago the continents fit together as a jigsaw puzzle, a supercontinent named Pangaea (*top panel*). (Press and Siever 1986)

plates recede from one another at the mid-oceanic rift, where hot magma rises and fills the void, as illustrated in Figure 6.12. When this magma reaches the surface, it solidifies and becomes part of the oceanic plates; i.e., new ocean floor is created at mid-oceanic ridges. The recession of plates is referred to as **sea floor spreading**. Where plates meet, they bump into each other (similar to a river full of logs) or slip past one another at **transform** faults. These 'collisions' result in **earthquakes**. The energy, E, released by earthquakes can be quantified using the **Richter magnitude** scale, $\mathcal{M}_R$:

$$\log_{10} E = 6.24 + 1.44 \mathcal{M}_R. \qquad (6.3)$$

A (large) earthquake of magnitude 7.7 on the Richter scale releases 10^{17} J.

Oceanic plates (lithosphere + crust) are thinner (0–100 km) than continental plates ($\sim$200 km) and also denser because they are composed of basalt; the continental crust is made of lighter (more granitic) material. When an oceanic and continental plate are pushed against each other, the oceanic plate will therefore **subduct**, or dive under, the continental plate, at least at active margins. At passive margins, oceanic and continental plates are juxtaposed without subduction. Mountain ranges and volcanoes form on the continental side of a subduction zone and ocean trenches on the oceanic side (Fig. 6.12b). While an oceanic plate descends, sediments on its 'surface' (e.g., bones and shales from marine animals, sands from rivers) are squeezed and heated, so that new metamorphic rocks may form, and at greater depths this

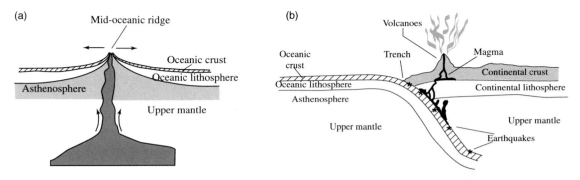

Figure 6.12 Schematic presentations of plate tectonics. Note that the layers labeled 'oceanic lithosphere', 'continental lithosphere' and 'asthenosphere' are parts of the upper mantle, and the crust is part of the lithosphere. (a) Sea floor spreading: Plates recede from each other at the mid-oceanic rift, where magma rises and fills the void. (b) Subduction zones: At convergent boundaries between two lithospheric plates, at least one of which is oceanic, the heavier oceanic plate is subducted. Volcanoes form near such subduction zones. (de Pater and Lissauer 2010)

material, which has a low melting temperature, melts. This forms a continuous source of magma for the volcanoes (which it formed) along the fault line. Because water lowers the melting temperature of rocks, solidification of the rising magma (Fig. 6.12b) results in more granitic-type rocks. Recycling of oceanic lithosphere typically occurs on timescales of $\sim 10^8$ years (Problem 6-8).

Plate tectonics is unique to Earth and is not seen on any other body in our Solar System. The smaller planets and large satellites, such as Mercury, Mars and the Moon, cooled off rapidly and developed one thick lithospheric plate. Tectonic features on these planets involve primarily vertical movements with features such as grabens and horsts. Venus shows evidence of local lateral tectonic movements, but not of plates. The absence of plate tectonics has been attributed to a lack of water on Venus, a consequence of its high surface temperature, making its outer shell too stiff to break into plates. Water in Earth's interior is thought to play a major role in plate tectonics by reducing the strength of rocks, leading to a breakup of the lithosphere. Based on the geology, petrology and magnetic field measurements, it has been suggested that Mars may have had plate tectonics early in its history, although these theories are highly controversial. Perhaps the jovian satellite

Europa shows features that most closely resemble the mid-oceanic ridges seen on Earth (Fig. 10.8). The bands on Europa, however, formed in a different way and by other forces (tidal), when ice or water rose to fill fractures created by regional stresses in the surface.

6.3.2 Volcanism

Many planets and several moons show signs of past volcanism, and a few bodies, especially Earth, Io and Enceladus, are still active today. Volcanic eruptions can change a planet's surface drastically, both by covering up old features and by creating new ones. Volcanism can also affect, and even produce, atmospheres (§10.2.1). In this section, we discuss what volcanic activity is, where it is found and how it changes the surface. We focus our discussion on Earth, a volcanically active planet that has been studied in detail. Figures 6.13, 6.16 and 6.17 show examples of volcanism.

A prerequisite for volcanic activity is the presence of buoyant material, like magma, below the crust. There are several sources of heat to accomplish this: (1) Heat can be generated from accretion during the era of the planet's formation, as discussed in §15.5.2. This is referred to as **primordial heat**. Processes of ongoing differentiation of

(a)

(b)

(c)

Figure 6.13 COLOR PLATE Examples of volcanic activity on Earth. (a, b) Photographs are shown of Mount St. Helens before (a) and after (b) the big explosion in 1980. This volcano is located in Washington state, along the boundary of the Pacific and North American plates. (Courtesy USGS/Cascades Volcano Observatory) (c) Eruption of the Kilauea volcano on Hawaii on 6 September 1983. (Courtesy J.D. Griggs and USGS/Hawaii Volcano Observatory)

heavy and light material, such as the phase separation between hydrogen and helium discussed in §§6.1.3 and 6.2.3, also generate heat. (2) Tidal interactions between bodies can lead to substantial heating, such as is the case for Jupiter's moon Io (§§2.7.3, 10.2.1). (3) Radioactive nucleides form an important source of heat for all of the terrestrial planets.

Earth's upper mantle consists of hot, primarily unmolten, rock under pressure, and it behaves as a highly viscous fluid. That is, when considering timescales of at least a few hundred years, (upper) mantle material flows, but on shorter timescales, it does not move much (Problem 6-10). The solid lithosphere and crust overlying the mantle can be compared to a lid on a pressure cooker or espresso machine. Magma formed within the hot rock, being less dense than the solid rocks surrounding it, is buoyant and hence rises. It is pushed out through any cracks or weakened structures in the surface.

On Earth, we distinguish three types of volcanism. Two of these three types are associated with plate tectonics: eruptions along the mid-oceanic ridges and in subduction zones (§6.3.1; Fig. 6.12). A third type of volcanism is found above hot thermal mantle 'plumes' at places where the crust is weak and the magma can break through. Such magma is very hot – the plumes are thought to originate close to the core–mantle boundary and do not appear to be connected to plate tectonics. As tectonic plates move over the hot magma plume, the magma forms islands in the ocean, such as the chain of Hawaiian islands.

The style of volcanic eruptions is governed by the (chemical) composition and the physical properties of the magma. Volatile compounds such as water, carbon dioxide and sulfur dioxide dissolve in magma at high temperatures and pressures. These volatiles come out of solution when the temperature and pressure drop, i.e., when the magma approaches the surface. The type of volcanism, **explosive** versus **effusive** (the nonexplosive extrusion of magma at the surface), depends primarily

Figure 6.14 COLOR PLATE Lava pours down a well-developed lava channel near the erupting vents (background) in Hawaii (on 28 March 1984). (Courtesy R.W. Decker and USGS)

on the viscosity of the magma, which is determined by its composition, especially its silica content. The higher the silica content, the lower the melting temperature and the higher the viscosity. In low-viscosity basaltic melts, volatiles can rise and escape freely into space as gas bubbles, but in high-viscosity silicic magma bubbles cannot rise, so volatiles are carried upwards with the magma, resulting in explosive eruptions at the surface.

Basaltic magma erupts at temperatures of 1250–1500 K. Because of its low viscosity, basaltic magma is very fluid and can cover vast (many square kilometers) areas within hours. Photographs of low-viscosity lava flows on Earth are shown in Figures 6.14 and 6.15. In the first image, the low-viscosity fast-flowing hot lava comes down like a river. The second image shows such lava after it has cooled down; the glistening smooth surface is referred to as **pahoehoe** (Hawaiian word for ropy).

This pahoehoe is in part covered by **'a'a**, a very rough, jagged and broken lava flow, painful on feet. Extensive basaltic lava flows are also seen on the Moon and the planets Venus and Mars.

Figure 6.15 COLOR PLATE Lava flows on Hawaii: A glowing 'a'a flow front advancing over pahoehoe on the coastal plain of Kilauea Volcano, Hawaii. (USGS Volcano Hazards program)

of $\sim$100 km. Relative sizes of volcanoes on Earth, Mars and Venus are depicted on Figure 6.18.

In contrast, highly viscous silicic magma erupts at temperatures of 1050–1250 K. This lava flows very slowly, oozing out of a vent like toothpaste from a tube. Such volcanic events create **domes**, which may largely consist of obsidian. Such 'glass mountains' can, for example, be seen near Mono Lake and Mount Shasta in California.

Geysers are another form of volcanic activity. Groundwater that is heated by magma can produce hot springs and geysers, such as shown in Figure 6.19. Such springs and geysers are found in volcanic areas on Earth, such as Yellowstone National Park. They have been observed on Saturn's moon Enceladus (Fig. 10.19), and may occur on Europa, one of Jupiter's Galilean moons. The *Voyager* spacecraft imaged geysers of liquid nitrogen on Triton, Neptune's largest moon (Fig. 10.25).

One can find **craters** on the summits of most volcanoes. A volcanic crater is centered over a vent and is produced when the central area collapses as the pressure that caused the eruption dissipates.

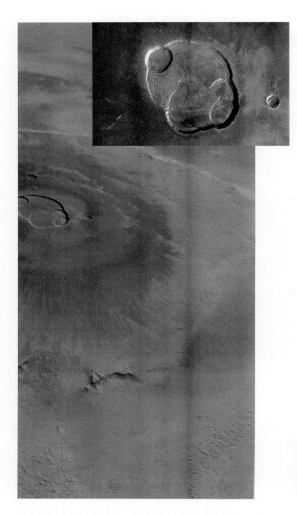

Figure 6.16 The largest shield volcano in our Solar System: Olympus Mons on the planet Mars. (NASA/*Mars Global Surveyor*) The *inset* shows a *Mars Express* image of the caldera at the summit. (ESA/DLR/FU Berlin; G. Neukum)

Basaltic magma eruptions can build up a **shield volcano**, a gently sloping volcanic mountain, such as those shown in Figures 6.16 and 6.17. These volcanoes may be very large; the largest known shield volcano is Olympus Mons on Mars, with a height of $\sim$25 km and base diameter $\sim$600 km. The largest shield volcano on Earth, Mauna Loa on Hawaii, measures about 9 km from its peak to the bottom of the ocean floor and has a base diameter

Figure 6.17 Three-dimensional, computer-generated view of the surface of Venus showing Maat Mons, an 8-km-high volcano. This view is based upon radar data obtained with the *Magellan* spacecraft. Lava flows extend for hundreds of kilometers across the fractured plains in the foreground. The vertical scale has been exaggerated 22.5 times. (NASA/JPL, PIA00254)

Heights of mountains on Mars, Venus, Earth

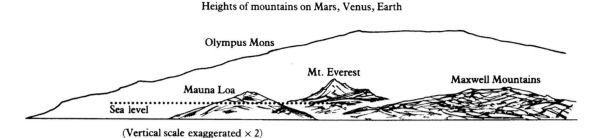

(Vertical scale exaggerated × 2)

Figure 6.18 A comparison of volcanoes and mountains on Mars (Olympus Mons), Earth (Mauna Loa and Mount Everest) and Venus (Maxwell Mountains). (Morrison and Owen 1996)

Because the original crater walls are steep, they usually cave in after an eruption, enlarging the crater to several times the vent diameter. Craters can be hundreds of meters deep. A volcano in Italy, Mount Etna, has a central vent that is 300 m in diameter and more than 850 m deep. **Calderas** are large basin-shaped volcanic depressions, varying in size from a few kilometers up to 50 km in diameter. Such volcanic depressions are caused by collapse of the underlying magma chamber. After a volcanic eruption, when the magma chamber is empty, its roof, which is the crater floor, may collapse. Over time, the crater walls erode, and lakes may form within the depression. Many years later (this may take 10^5–10^6 years), new magma may enter the chamber and push up the crater floor, and the entire process may start again. Several calderas show evidence of multiple eruptions. Examples of such **resurgent calderas** are the Yellowstone caldera in Wyoming, the Valles caldera in Mexico, Crater Lake in Oregon and the Aniakchak Caldera in Alaska that is shown in Figure 6.20.

After an explosion, when the lava flow cools and contracts, shrinkage cracks may form. Sometimes when the source of a lava flow is cut off, and the outer layers have solidified, the lava drains out, and **lava tubes** or **caves** result. Such caves are found, for example, in Hawaii and northern California. The sinuous Hadley Rille in Mare Imbrium on the Moon, shown in Figure 6.10, might be a lava tube where the roof has collapsed. Such volcanic structures, characterized as steep-sided troughs, are also known as **lava channels**.

6.3.3 Atmospheric Effects on Landscape

An atmosphere can profoundly alter the landscape of a planetary body. If the atmospheric pressure and temperature at the surface are high enough

Figure 6.19 One of the numerous geysers in Yellowstone National Park, USA. (Courtesy Wil van Breugel)

Figure 6.20 The Aniakchak Caldera in Alaska is an example of a resurgent volcanic caldera. The caldera formed about 3500 years ago; it is 10 km in diameter and 500–1000 m deep. Subsequent eruptions formed domes, cinder cones and explosion pits on the caldera floor. (Courtesy M. Williams, National Park Service 1977)

In this section, we summarize the most common morphological features caused by water and wind on Earth. These features serve as a basis for comparative studies with other terrestrial planets (see Ch. 9) and Titan (see §10.3.1).

Water and Other Liquids

The atmospheric temperature and pressure on Earth are close to the triple point of water, so H_2O exists as a vapor, a liquid and ice. Although at present Earth is unique in this respect, water may have flowed freely over Mars's surface during the first 0.5–2 Gyr after its formation, and Titan's surface shows clear evidence of lakes and fluvial features, although produced by liquid hydrocarbons rather than water. Europa most likely has a large water ocean under its ice crust. Liquids on planetary surfaces, whether water, hydrocarbons or lava, tend to flow downhill at a velocity determined by the flow's viscosity, the terrain and the planet's gravity. In addition to the liquid itself, the flow transports solid materials, such as sediments eroded from rocks. The faster a river flows, the larger the particles or rocks it can transport. The largest particles usually stay close to the bottom and may roll and slide over the surface; the finest particles (clay in the case of water flows) may be suspended throughout the flow. The finest particles are carried the farthest, a sifting process that produces the various sedimentary rocks described in §6.1.2.

for liquids, such as water, to exist, there may be oceans, rivers and precipitation, which modify the landscape through both mechanical and chemical interactions. In 'dry' areas, e.g., currently on Mars and in deserts on Earth, winds displace dust grains and erode rocks. Over time, these processes 'level' a planet's topography: high areas are gradually worn down, and low areas filled in. This process is called **gradation**, and material is displaced by **mass wasting**. The main driving force for gradation is gravity. Although gradation and mass wasting occur on all solid bodies, the presence of an atmosphere and liquids on the surface enhance these processes and give rise to particular surface features, as discussed in more detail below. In addition to mass wasting, there are numerous chemical interactions between the crust and the atmosphere. These vary from planet to planet, because they depend on atmospheric and crustal composition, temperature and pressure, as well as the presence of life. In addition to the changes a massive atmosphere can make to surface morphology, it also protects a planet's surface from impacting debris, especially small and/or fragile projectiles, as well as from cosmic rays and ionizing photons.

In addition to running water on the surface, groundwater just below the surface may leave profound marks as well. Some rocks are dissolved in water (e.g., limestone, gypsum, salt), leading to **karst topography**, which may show its appearance in the form of, e.g., sinkholes. Groundwater seeping up from below also leads to particular drainage patterns. Finally, dry lake beds or **playas** are associated with former lakes, swamps or oceans, and sea cliffs and beaches mark the shorelines of oceans.

The temperature on most bodies in our Solar System is well below freezing, which makes ice an important constituent of planetary surfaces. The thawing and freezing of permafrost leave a characteristic pattern of polygons, such as the patterns on Earth and Mars shown in Figure 6.21. This pattern is caused by a contraction of the ground when it is freezing cold (in the winter), which creates spaces that fill with melt water in the summer. During the winter, this water freezes, thereby widening the cracks.

On Earth we find **valley glaciers** and **ice sheets**, as well as the morphological features of past glaciation, such as U-shaped valleys, grooves and striations parallel to the flow and amphitheater-shaped **cirques** at the head of the flow. Usually the ice contains dust and rocks, which are left behind when the ice melts or sublimes. The dust and rocks may be deposited, carried away by melt water or blown away by winds. The morphology of the deposits contains information on the glaciers and hence the surface topography and past climate.

In the outer reaches of our Solar System, ices of methane, ammonia or carbon dioxide may have similar roles as water-ice on Earth. For example, whereas carbon dioxide on Mars freezes out above the winter pole, it sublimes during the summer. It is cold enough above Mars's poles for permanent water-ice caps; water vapor freezes near this planet's equator at night, subliming again during the day. The temperature on Titan's surface is near the triple point of methane, so methane vapor, liquid and ice coexist. On Pluto and Triton, nitrogen-ice forms during the winter and sublimes in the summer. Ice may be considered a 'pseudo-plastic fluid' that moves downhill.

Winds

Planets that have significant atmospheres in contact with solid surfaces show the effects of **aeolian** or wind processes. Examples from Earth, Mars and Venus are shown in Figure 6.22. On Earth, aeolian processes are most pronounced in desert and

(a)

(b)

Figure 6.21 (a) *Phoenix Mars Lander* image of the arctic region of Mars at a latitude of 68° N. The flat landscape is strewn with tiny pebbles and shows a pattern of polygons, similar to the patterns seen in permafrost terrains on Earth. (NASA/JPL-Caltech/Univ. of Arizona) (b) Polygon permafrost pattern on northeastern Spitsbergen (Norway) shows a striking similarity to that seen on Mars. Although this photo shows a large amount of surface water, the process could presumably occur beneath the surface with far less water. (NASA/Visible Earth, O. Ingolfsson)

coastal areas. The winds transport material. The smallest particles, such as clay and silt ($\lesssim$60 μm), are **suspended** in the atmosphere. Larger dust and sand grains (~60–2000 μm) are transported via

(a)   (b)

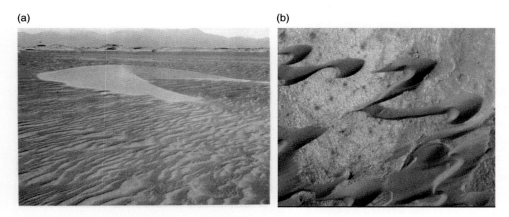

Figure 6.22 (a) Small sand dunes on Earth in Peru. Prevailing wind is from the left to the right. (USGS Interagency Report, 1974) (b) The dark sand dunes of Nili Patera, Syrtis Major, on Mars. The shape of these dunes indicates that the wind has been steadily transporting dark sand from the right or upper right toward the lower left. The width of the picture is 2.1 km. (NASA/*Mars Global Surveyor*, MOC2-88)

saltation, an intermittent 'jumping' and 'bounding' motion of the dust grains. Still larger grains are transported via **surface creep**, where particles are rolled or pushed over the ground. The amount of dust that the winds can displace depends on the atmospheric density, viscosity, temperature and surface composition and roughness. When the wind blows over a large sandy area, it first ripples the surface and then builds dunes. Because the turbulence created by the wind increases with increasing surface roughness, the winds get stronger during this process (positive feedback).

The amount of dust that winds can transport depends on the atmospheric density and wind strength. On Earth, half a ton of sand can be moved per day over a meter-wide strip of sand if the wind blows at ∼50 km per hour; the amount of sand transported by stronger winds increases more rapidly than the increase in wind speed. In large terrestrial dust storms, one cubic kilometer of air may carry up to 1000 tons of dust, and thus many millions of tons of dust can be suspended in the air if the dust storm covers thousands of square kilometers. On planets with low-density atmospheres, the winds need to be much stronger to transport

material. The winds on Mars need to be about an order of magnitude stronger than on Earth to transport the same amount of material.

Winds both erode and shape the land. They erode the land through removal of loose particles, thereby lowering or **deflating** the surface. When winds are loaded with sand, they can wear away and shape rocks through **sandblasting**. This causes erosion and rounding of rocks. The best known example of wind-blown landforms are **dunes**, such as those shown in Figure 6.22. Any obstacle to the wind, such as a large rock, can start the formation of a dune, where sand grains are deposited at the lee-side of the obstacle. The observed shape of dunes can be used to deduce the local wind patterns. Dunes are found in desert and coastal areas, where winds are strong and there is an abundance of particulate material. Dunes have also been found on Mars, Venus and Titan and **wind streaks** on Mars and Venus.

Chemical Reactions

The interaction between a planet's atmosphere and its surface can lead to **weathering**, a process that depends on the composition of both the atmosphere and surface rocks. Weathering on Earth is usually a two-part process, consisting of **mechanical weathering** or fragmentation of rocks together with **chemical weathering** or decay of the rock fragments. Many iron silicates, such as pyroxene, weather or oxidize slowly through the interaction with oxygen and water and get a rusty-iron color. **Hydration** is a more general process on planetary surfaces because it only requires the presence of water. It is therefore not too surprising that hydrated minerals have been detected on several asteroids. Some minerals, such as feldspar, partially dissolve when in contact with water and leave behind a layer of clay. Calcite and some other minerals may completely dissolve away.

The presence of life may have profound effects on the surface morphology of a planet, as we well know from our own planet, Earth. Plants cover large fractions of continental crust, changing the albedo (even varying with season), atmospheric and soil composition as well as the local climate. Plants add matter to the soil and influence erosion. Humankind has large effects on the surface morphology, e.g., through building and mining projects and by altering the composition of the atmosphere. Even micro-organisms change the atmospheric and soil composition (locally), e.g., through metabolism.

6.4 Impact Cratering

The distances between bodies within our Solar System are quite large compared with their sizes. Nonetheless, when considering geological timescales, collisions are frequent events. Collision speeds are typically high enough that impacts are very violent, with **ejecta** thrown outwards from the location of the impact, often leaving a long-lasting depression in the surface. **Impact craters** are produced on every body in the Solar System that has a solid surface and have been observed on the four terrestrial planets, many moons, asteroids and comets. Impact craters, their ejecta blankets, and on small asteroids **seismic shaking** caused by impacts, produce the dominant landform on geologically inactive bodies that lack substantial atmospheres. The lunar surface is covered by craters, which were formed by meteoroids that hit the Moon over the past 4.4 Gyr. Earth has been subjected to a somewhat higher flux of impacts (due to gravitational focusing, see §15.4.2), but most craters have disappeared from our planet as a result of plate tectonics and erosion.

Impact cratering involves the nearly instantaneous transfer of energy from the impactor to the target. If the target has a substantial atmosphere, as does Earth, the impactor is seen as a fireball, or **bolide**, before impact. High-velocity collisions are common, with impact energy being provided from the kinetic energy of the relative orbital motion of the two colliding bodies augmented by the gravitational potential energy released as they approach one another. Typical impact velocities of large meteoroids (which are not significantly slowed by the atmosphere) on Earth are 11–40 km s^{-1}, although long-period comets may impact at speeds up to 73 km s^{-1} (Problem 11-1). These speeds imply characteristic impact energies of the order $\sim 10^8$ J kg^{-1}. This is much larger than the specific energy of chemical explosives (TNT releases $\sim 4 \times 10^6$ J kg^{-1}) and that of the foods that we consume, but about six orders of magnitude smaller than the specific energy of nuclear explosives. For example, a nickel–iron meteoroid, 15 m in radius moving at 15–20 km/s, would impart an energy a few times 10^{16} J, or the equivalent of several million tons (megatons) of TNT. This energy is comparable to that of a (large) earthquake with Richter magnitude 7 (eq. 6.3); but note that the impact energy is not fully transferred into kinetic energy of material motion within the target body.

(a)

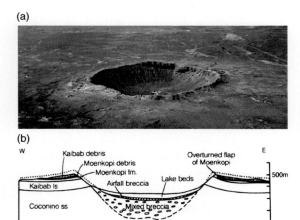

(b)

Figure 6.23 (a) Meteor Crater in Arizona. The crater has a diameter of 1 km and is 200 m deep. (Courtesy D. Roddy, USGS/NASA) (b) A cross-section through Meteor Crater. (Melosh 1989)

Nonetheless, collisions can be very energetic, and the impacting body typically creates a hole (crater) that is much larger than its own size.

Figure 6.23 shows an aerial photograph and a geological cross-section of Meteor Crater in Arizona. This crater has a diameter of about 1 km and a depth of 200 m. It was formed (in less than one minute!) by the impact of an $R \approx 15$ m nickel–iron meteoroid.

Impact cratering has been studied using astronomical and geological data pertaining to the crater morphology seen on our planet, on the Moon and on more remote Solar System objects. **Hypersonic** (much faster than the speed of sound) impact experiments using small (millimeter to centimeter) projectiles have been conducted in the laboratory. Studies of craters produced by conventional and nuclear explosions sample a broad range of energies. Numerical computer simulations are of great value in studies concerning impacts and impact craters. Impact theories were tested and refined after astronomers witnessed a series of large impacts of Comet D/Shoemaker–Levy 9 on Jupiter in 1994 (§8.1.2).

6.4.1 Crater Morphology

Craters can be 'grouped' according to their morphology into four classes:

(1) **Microcraters**, often referred to as 'pits', such as the one shown in Figure 6.24, are tiny (sub-centimeter) craters caused by impacts of micrometeoroids or high-velocity cosmic dust grains on rocky surfaces. Microcraters are only found on airless bodies. The central hole is often covered with glass.

(2) Small or **simple craters**, typically up to several kilometers across, are bowl shaped. Figure 6.25 is an image of a simple crater on the Moon. The depth (bottom to rim) of a simple crater is approximately one fifth of its diameter, although variations do occur, depending on the strength of the surface material and the surface gravity.

(3) Large craters are more complex. They usually have a flat floor and a central peak, and the inside of the rim is characterized by terraces. These features are readily apparent in Figure 6.26. **Complex craters** have diameters of a few tens up to a few hundred kilometers. The transition size between simple and complex craters is ∼12 km on the Moon and scales inversely with the gravitational acceleration, g_p, because gravity

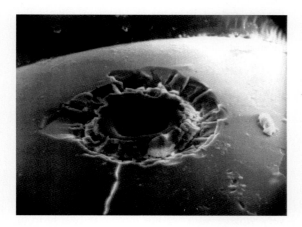

Figure 6.24 Microcrater with a diameter of 30 μm. This is a scanning electron microprobe photograph of a glass sphere from the Moon, brought to Earth by *Apollo 11*. (Courtesy D. McKay, NASA S70-18264)

Figure 6.25 This photograph shows the 2.5-km-diameter simple crater Linne in western Mare Serenitatis on the Moon. (NASA/*Apollo* panoramic photo AS15-9353)

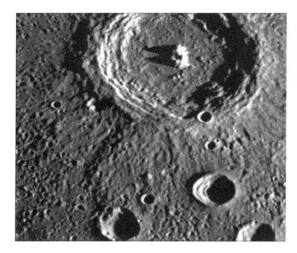

Figure 6.26 A close-up of a 98-km-diameter complex mercurian crater, characterized by a relatively flat crater floor, a central peak and terraced walls. Note that the smaller craters in the foreground (25 km diameter) also are terraced. This image (FDS 80) was taken during *Mariner 10's* first encounter with Mercury. (NASA/JPL/Northwestern University)

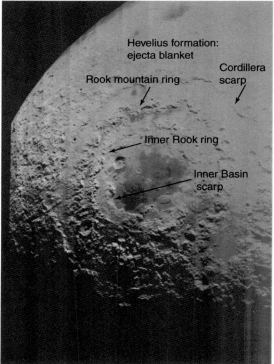

Figure 6.27 A photograph of the lunar multiring basin Mare Orientale. (NASA/*Lunar Orbiter IV 194 M*)

is responsible for the modifications that convert transient simple craters into complex craters. The minimum size for a crater to acquire a complex morphology also depends on the strength of the target's surface material. Craters with dimensions between 100 and 300 km on the Moon, Mars and Mercury show a concentric ring of peaks rather than a single central peak. The inner ring diameter is typically half the rim-to-rim diameter. The crater size at which the central peak is replaced by a peak ring scales in the same way as the transition diameter between small and complex craters. More details on the attributes of complex craters, together with a discussion of their formation, are provided in §6.4.2.

(4) **Multiring basins** are systems of concentric rings, which cover a much larger area than the complex craters mentioned above. Figure 6.27 shows an annotated image of the Mare Orientale basin on the Moon. The inner rings often consist of hills in a rough circle, and the crater floor may be partly flooded by lava. In some cases, it is not clear which of the outer rings is the true crater rim.

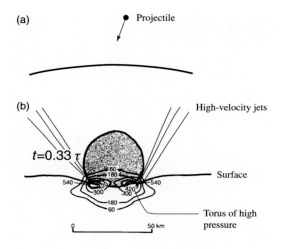

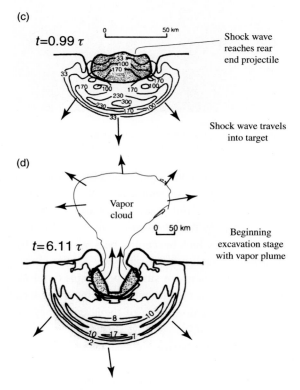

Craters on icy satellites show systematic differences from craters on rocky bodies, such as the minimum size of a crater when central peaks form. Also, no peak-ring craters have been seen on icy satellites. These differences are presumably caused by the difference in material properties.

6.4.2 Crater Formation

The formation of a crater consists of a rapid sequence of phenomena, which starts when the impactor first hits the target and ends when the last debris around the crater has fallen down. It helps to understand the process by identifying three stages: An impact event begins with the **contact and compression stage**, is followed by an **excavation stage** and ends with a **collapse and modification stage**. These three stages are sketched in Figure 6.28 and are discussed below for an airless planet. The influence of an atmosphere is summarized in §6.4.3.

Contact and Compression Stage

Upon collision of a meteoroid with a planet, the relative kinetic energy is transferred to the bodies in the form of shock waves, one of which propagates into the planet and the other into the projectile. The impact velocity of a typical meteoroid with the Moon and terrestrial planets is of the order of 10 km s^{-1}. Because the velocity of seismic waves in rocks is only a few km s^{-1}, the impact velocity is hypersonic.

Rocks are typically compressed to pressures well over a few Mbar. The shock waves originate at the point of first contact and compress the target and projectile material to extremely high pressures.

Figure 6.28 Schematic presentation of a hypervelocity (15 km/s) impact of a 23.2 km radius iron projectile onto a rocky surface. Times subsequent to impact are given in units of τ, the ratio of the diameter of the projectile to its initial velocity, which is ~3 s in this case, and pressure contours are labeled in GPa (10^9 Pa). (a) Projectile approaching the target. (b) A torus of extra-high pressure is centered on the circle of contact between the projectile and the target (perpendicular impact). Heavily shocked material squirts or jets outwards at velocities of many km s^{-1}. (c) Shock waves propagating into the target and projectile. The latter wave has reached the rear end of the projectile. The projectile melts or vaporizes (depending on the initial pressure) when decompressed by rarefaction waves. (d) Beginning of the excavation stage, preceded by the vapor plume leaving the impact site. (Adapted from Melosh 1989)

A region of extra-high pressure (Fig. 6.28b) is centered on the point of contact between the projectile and target.

The geometry of the shock wave system is modified by the presence of free surfaces on the target and the meteoroid, i.e., the outer surface of the material that is in contact with air or the interplanetary medium. Free surfaces cannot sustain a state of stress, and therefore **rarefaction** (release) **waves** develop behind the shock wave. When the shock wave in the projectile reaches its rear surface, a rarefaction wave is reflected from the surface. This rarefaction wave travels at the speed of sound through the shocked projectile, thereby decompressing the material to near-zero pressure. The contact and compression phase lasts as long as it takes the shock wave and subsequent rarefaction wave to traverse the projectile, which is typically 1–100 ms for meteoroids with sizes between 10 m and 1 km (Problem 6-13).

As soon as the heavily shocked mixture of target and projectile material is decompressed by rarefaction waves, it squirts or **jets** outward at velocities of many km s^{-1}. This jetting occurs nearly instantaneously with the projectile hitting the target and is often finished by the time the projectile is fully compressed. The shock, propagating hemispherically into the target, can be detected as a seismic wave.

Most rocks vaporize when suddenly decompressed from pressures exceeding ~600 kbar. Because the initial pressure of the shocked material can be very high, the projectile may nearly completely melt or vaporize upon decompression, and shortly after the passage of the rarefaction wave, remnants of the projectile leave the crater as a **vapor plume**.

Excavation Stage

Upon decompression, the projectile and target region vaporize, provided the initial pressure was high enough. The vapor plume or fireball expands adiabatically upward and outward. Meanwhile, the shock wave expands and weakens as it propagates into the target. The shock wave gradually degrades to a stress wave, propagating at the local speed of sound. The rarefaction waves behind the shock decompress the material and initiate a subsonic **excavation flow**, which opens up the crater. The excavation of material may last for several minutes, depending on the size of the crater and the surface gravity. Target material below the excavation depth is pushed downwards, and the strata above this depth are flipped over, creating an **inverted stratigraphy**, including the crater's **rim**. The rim height on relatively small craters (diameter ≲15 km on the Moon) is typically ~4% of the crater diameter:

$$h_{\rm rim} \approx 0.04D \quad \text{(small craters)}, \qquad (6.4a)$$

and the depth (bottom to rim):

$$d_{\rm br} \approx 0.2D \quad \text{(small craters)}. \qquad (6.4b)$$

For larger craters, these relationships break down (depths and rim heights are smaller than given by eqs. 6.4a and 6.4b) because gravity causes various morphological changes, including rim and crater collapse.

Rocks and debris excavated from the crater are ejected at velocities that are much lower than those of the initial 'jets' of fluidlike material during the compression stage. These ejecta are thrown up and out along ballistic, nearly parabolic, trajectories. The excavation flow forms an outwardly expanding **ejecta curtain**, which has the shape of an inverted cone. A laboratory example is shown in Figure 6.29. The ejection velocities are highest (up to several km s^{-1}) early in the excavation process, thus near the impact site. The sides of the crater continue to expand until all of the impact energy is dissipated by viscosity and/or carried off by the ejecta. The resulting crater is many times as large as the projectile that produced it. The crater is nearly hemispherical until the maximum depth is reached, after which it only grows horizontally. The ejecta form an **ejecta blanket** around

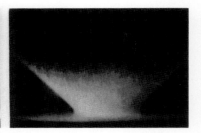

Figure 6.29 The excavation flow forms an outward expanding conical ejecta curtain. The excavation flow in a large planetary impact is thought to be very similar to that from this small impact in a laboratory. (Courtesy P.H. Schultz)

the crater (Fig. 6.32), up to one or two crater radii from the rim, covering up the old surface.

Some ejecta blankets on Mars look morphologically like mudflows, suggestive of a fluid substance. The most likely fluid is subsurface ice that liquefies when heated by an impact. Alternative possibilities include adsorbed CO_2 released by the impact and trapped air.

Some of the excavated rocks, when they hit the surface, create **secondary craters**. Because the ejecta follow ballistic trajectories, the secondaries are closer to the primary crater on more massive planets. The sizes of the secondary craters depend on the mass and impact velocity of the ejecta, as well as the properties of the material on or near the target's surface. Because the impact velocities of the ejecta are lower than those of the original impactors, the morphology of secondary craters is somewhat different from those of **primary craters**, but the differences are often subtle. The secondary craters are usually seen outside the ejecta blanket from the primary crater and may be found many crater radii away from the primary crater. Relatively young craters on the Moon ($\lesssim 10^9$ yr) display bright **rays** emanating outwards from the primary. As shown in Figure 6.30, these rays may be visible over a large surface area, extending to ten or more crater radii from the primary. Many, but not all, secondary craters are associated with bright rays. The rays are strings of secondary craters and their ejecta. Over time the rays disappear, probably because of radiation damage.

At the end of the excavation stage, the crater is referred to as the **transient crater**. The shape of this crater depends on the meteoroid's size, speed and composition; the angle at which it struck; the planet's gravity; and the material and structure of the surface in which the crater formed. The energy required for excavation varies approximately as D^4 because the mass of excavated material is proportional to the third power of the crater diameter, and the distance that it must be moved to be clear of the crater adds the fourth power. The crater dimension

Figure 6.30 Rays from the young lunar crater Tycho. The rays are visible over nearly an entire hemisphere. (Courtesy UCO/Lick Observatory)

thus scales approximately with the meteoroid's kinetic energy as:

$$D \propto (E/g_p)^{1/4}, \qquad (6.5)$$

where the subscript p signifies the planetary target. Equation (6.5) describes **energy scaling**, meaning that the size depends only on the energy of the event. Energy is, indeed, the most important factor in determining crater size, but other factors also matter.

A more general scaling law, derived empirically, is given by (in SI units)

$$D \approx 2\rho_i^{0.11}\,\rho_p^{-0.33}\,g_p^{-0.22}R^{0.12}E^{0.22}(\sin\theta)^{1/3}, \quad (6.6)$$

where ρ_p and ρ_i are the densities of the target and the impactor, respectively; R the physical radius of the projectile; E the impact (kinetic) energy; and θ the angle of impact from the local horizontal. Some key terms in equation (6.6) have simple physical interpretations beyond the simple energy scaling given in equation (6.5). For fixed projectile energy, crater size is larger for more massive (higher momentum) projectiles, leading to the ρ_i and R terms in equation (6.6), as well as the slightly smaller explicit dependence on E. The energy is required to work against the target's gravity, leading to the g_p term. The mass of material excavated varies as $D^3\rho_p$. The shallower the impact angle, the less effectively the kinetic energy of the impact is coupled to the surface; although only very oblique impacts (within $\sim 10°$ of horizontal) produce asymmetric craters, smaller departures from verticality can lead to significant reduction in crater size, as well as asymmetric ejecta blankets. It follows from equation (6.6) that a typical crater on Earth is roughly ten times as large as the size of the impacting meteoroid.

Collapse and Modification

After all of the material has been excavated, the crater is modified by geological processes induced by the planet's gravity, which tends to pull excess mass down, and by the relaxation of compressed material in the crater floor, pushing it upwards. The shape of the final crater depends on the original morphology of the crater, its size, the planet's gravity and the material involved. The four basic morphological classes of craters are summarized in §6.4.1. The transition size between simple and complex craters is inversely proportional to the planet's gravity, and depends on, e.g., the material strength, melting point and viscosity.

Complex craters are characterized by central peaks and terraced rims. Shortly after the excavation process ends, the debris remaining in the crater moves downwards and back towards the center, and the crater floor undergoes a rebound of the compressed rocks, which probably leads to the formation of the central peak or mountain rings. The formation of the central peak might be analogous to the impact of a droplet into a fluid, shown in Figure 6.31. After the initial crater has formed (panel d), the central peak comes up (panel e) and grows (panels e–g). Collapse of the peak (panel h) triggers the formation of a second concentric ring (the first being the crater rim), which propagates outwards. The possibly analogous process on a solid surface is sketched in Figure 6.32. The rebound starts before the crater has been completely excavated, and the central uplift 'freezes' to form the central peak. For larger craters, the peak becomes too high and collapses, triggering an outwardly propagating ring, which 'freezes' into a ring of mountains. Although the details of this process are not completely understood, the peak ring must form before the crater material comes to rest.

After material has been excavated, the crater rim collapses or **slumps**, moving outwards and thereby increasing the crater diameter, filling in the floor of the crater and shaping the wall in the form of terraces. Observations indicate that these terraces form before the rock, which acts like a liquid that because of the 'shaking' from the impact, has 'solidified'. The entire collapse process typically takes several minutes for a large complex crater.

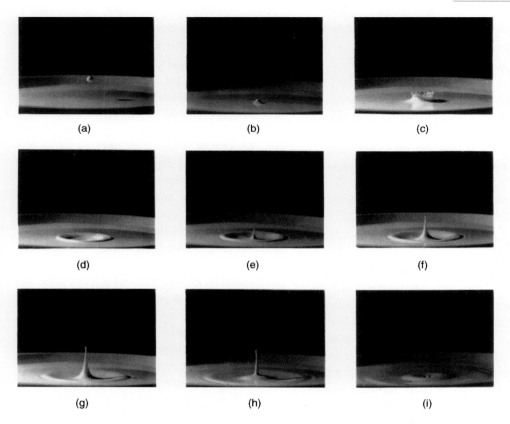

Figure 6.31 A series of photographs from the impact of a milk drop into a 50/50 mixture of milk and cream. An ejecta curtain forms immediately after impact (c), which creates the 'crater' wall. After the initial 'crater' forms (d), the central peak appears (e) and grows (f and g). Collapse of the peak (h) triggers the formation of a second, outwardly expanding ring (i). (Courtesy R.B. Baldwin; photographs by Gene Wentworth of Honeywell Photograph Products)

Characteristics of Craters

Some crater properties vary with crater size in a well-defined manner. For craters with sizes between about 15 and 80 km on the Moon, the height of the central peak, h_{cp} (in km), typically increases with crater diameter, D (in km), as

$$h_{cp} \approx 0.0006D^2. \tag{6.7}$$

For larger craters, the central peak on the Moon tops out at about 3 km, so the central peak is usually lower than the rim (eq. 6.4a). The central peak width is approximately 20% of the crater diameter. In lunar craters with $D > 140$ km, a ring of mountains develops in the crater, replacing the central peak. This ring develops about halfway between the center and crater rim. In some craters, both a central peak and peak ring are seen. Similar to the transition between simple and complex craters, the transition from craters with a central peak to peak rings is gravitationally induced, and the size at which it takes place differs from planet to planet in proportion to $1/g_p$. Thus, this transition also occurs at substantially smaller crater diameters on Venus and Earth than it does on the Moon.

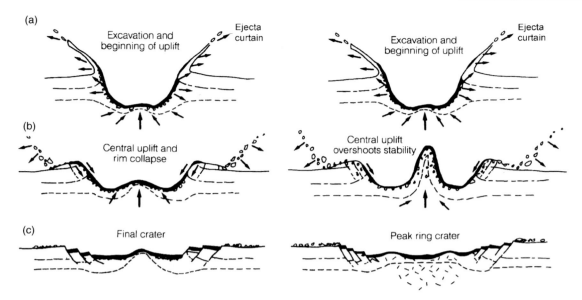

Figure 6.32 (a)–(c) Schematic illustration of the formation of central peaks (*left panels*) and peak rings (*right panels*). (Adapted from Melosh 1989)

Some complex craters have a **central pit** rather than a peak. These are seen on the ice-dominated surfaces of Ganymede and Callisto for craters larger than 16 km in diameter. The formation of central pits might be caused by the properties of icy surfaces and subsurface regions.

The Orientale basin on the Moon (Fig. 6.27) is the youngest and best preserved multiring basin, with four or five rings surrounding the original crater rim. Note that peak rings in complex craters always form inside the crater rim. Multiring basins on other planets and satellites are similar but differ in details. The Valhalla structure on Callisto has more than a dozen rings, and the steep scarps face outwards rather than inwards.

Further modification of craters happens on long timescales, i.e., months, years and eons. Erosion and micrometeoroid impacts slowly erode away the rim and smooth out or flatten the crater. The lifetime of a terrestrial 1-km impact crater against erosion is $\lesssim 10^6$ years. Isostatic adjustments

(§6.2.2) may be important in large craters, e.g., the crater floor may be uplifted to account for the deficiency in mass caused by the excavation of the crater. On icy satellites, craters are slowly flattened by plasticlike ice flows. Many large craters on Ganymede and Callisto may have disappeared and left vague discolored circular patches on the surface called **palimpsests**. Volcanism and tectonic forces within the general area of the crater can modify a crater at a (much) later stage. Many impact basins on the Moon have been flooded with basaltic lava.

Analysis of craters provides information regarding material on and below the surface. The most direct examples are the central peak and ejecta blankets around the craters, which were formed from material originally below the surface. The number, shape and size of craters also yield information regarding the surface composition, materials and the impacting bodies, as discussed later in this chapter and in Chapters 9 and 10.

Regolith

Whereas large impacts may fracture the crust down to many kilometers below the surface, small impacts affect only the upper few millimeters to centimeters of the crust. The cumulative effect of meteoritic bombardment over millions or billions of years pulverizes the bedrock so that a thick layer of rubble and dust is created on airless bodies. This layer is called **regolith** (the Greek word for rocky layer), although it is sometimes called 'soil' in a terrestrial analog. However, the properties of regolith are very different from those of terrestrial soil.

Most planetary bodies are covered by a thick layer of regolith. Because the population of meteoroids displays a steep size distribution, the rate of 'gardening' or turnover decreases rapidly with depth. With present-day meteoritic bombardment rates, half of the regolith on the Moon has been overturned to a depth of 1 cm within the past 10^6 years; in most locations, the uppermost millimeter has been turned over several tens of times. The thickness of the regolith depends on the age of the underlying bedrock. In the ~3.5-Gyr-old lunar maria, the regolith is typically several meters thick, but it is well over ten meters deep in the 4.4-Gyr-old lunar highlands. The ejecta from larger impacts have formed a 2- to 3-km-thick **mega-regolith**, which consists of (many) meter-sized boulders. The regolith may be bonded at depth as a consequence of the higher (past or present) pressure and temperature.

Focusing Effects

Large impacts can produce surface and body waves of sufficient amplitude that they may propagate through the entire planet. In planets with a seismic low-velocity core, the waves are 'focused' at the **antipode** (the point directly opposite the impact site). If the impact is very large, the waves still have enough energy to substantially modify terrain. Preexisting landforms in regions antipodal to

very large impact features on these rocky bodies are overprinted and modified with rugged equidimensional hills and narrow linear troughs. This effect is, for example, clearly seen on Mercury and the Moon. No unequivocal evidence of antipodal effects from impacts has yet been identified on any icy body.

Erosion and Disruption

Sufficiently energetic impacts can substantially erode or even catastrophically disrupt the target. The impact energy required to disrupt and disperse a planet-sized target body is substantially larger than the energy required to produce a crater whose diameter is about equal to the diameter of the target body.

Summary

The following features are recognized in connection with cratering events:

- Primary crater: the crater formed upon impact of a projectile from a distant source.
- Ejecta blanket: debris ejected from the crater up to roughly one crater diameter beyond the rim. The appearance of the ejecta blanket depends on the subsurface properties of the target. For example, ejecta blankets on the Moon consist largely of boulders; in contrast, some ejecta blankets on Mars display evidence of fluid motions.
- Secondary impact craters: these are caused by impacts of high-velocity chunks of target material ejected from the primary crater.
- Rays: bright linear features 'radiating' outward from the impact crater. They extend about ten crater diameters out. Rays are composed of secondary craters and their ejecta.
- Crater chains: linear arrays of secondary craters, often similar in size and overlapping, sometimes form along rays emanating out from a primary crater. Chains of primary craters form when a meteoroid is tidally disrupted by a planet

and crashes into a moon before the fragments disperse.

- Breccia and melt glasses: high temperature and pressure minerals, melt glasses and breccias (rocks of broken fragments cemented together, §6.1.2) form and line the inside of a crater. A lunar breccia is shown in Figure 6.3b and c.
- Regolith: the rocks that were broken or ground down by (micro-)meteoroids and secondary impacts.

6.4.3 Impact Modification by Atmospheres

Hitherto we discussed impacts on airless bodies. If the target is enveloped by a dense atmosphere, as are Earth and Venus, impacts may be modified extensively. Projectiles can be completely vaporized while plunging through the atmosphere and never hit the ground. Or a projectile might break up into many pieces. Such fragmentation is thought to have led to the formation of **crater clusters** on Venus. No analogous close groupings are observed on the lunar surface. Atmospheric drag can slow down a small meteoroid so that it hits the surface at merely the terminal velocity (eq. 11.3). Larger bodies can explode in the air, never creating an impact crater. These processes are described quantitatively in §11.3. In this section, we consider the effects of atmospheric transit on the cratering process.

Not only does the atmosphere affect the projectile but the projectile can also noticeably perturb the atmosphere during and after the explosion. Exploding meteors are a source of dust and vapor that are not in chemical equilibrium with surrounding atmospheric gases, and they can also release considerable amounts of energy, most of which is immediately converted into heat. Both dynamical and chemical interactions can be substantial.

Passage through the Atmosphere

A meteoroid is slowed down by atmospheric drag, with the aerodynamic force being proportional to the atmospheric ram pressure (ρv^2). Atmospheric drag may be substantial compared with a meteor's internal strength. Meteors, therefore, often break up and fall down in clusters. The effect of such a breakup is usually reflected in the crater morphology. A body that is broken apart by ram pressure is slowed very rapidly in the atmosphere; nonetheless, the cluster of remnant bodies can hit the ground and produce either a crater or a strewn field (Figure 11.15).

The mass displaced during a meteor's fall through the atmosphere is $\sigma_\rho A/\sin\theta$, where σ_ρ is the mass of the atmosphere per unit area, A is the body's cross-sectional area and θ is the angle that the trajectory makes with the surface. The meteor is slowed substantially when it encounters an atmospheric mass about equal to its own mass. Significantly smaller bodies are slowed close to the terminal velocity before hitting the surface. It follows that the radius of a typical iron meteor must be larger than 1 m for it to hit Earth in a hypersonic impact (Problem 6-15). Note that when a meteor breaks up, A is increased, typically several fold.

However, actual meteors do not remain constant in size as they pass through the atmosphere. As their exteriors are heated to melting and even vaporization temperatures, they shed material. This process is known as **ablation** and is discussed in greater detail in §11.3. Mechanical forces can also ablate a meteor by breaking off material.

Meteors that hit the surface at speeds $\lesssim 100$ m s^{-1} produce a small hole or pit equal in size to the diameter of the meteor itself. At somewhat higher speeds, the hole produced is larger than that of the impacting meteor; impacts at hypersonic speeds produce craters as discussed in §6.4.2.

When a meteor plunges through the atmosphere at supersonic speeds, a bow shock forms in front of it, and gases are considerably compressed. The shock waves of such meteors, even for projectiles that do not hit the surface, can be devastating and leave obvious marks. On 30 June 1908, an ~40-m radius meteor that disintegrated in an airburst produced a shock wave that flattened about 2000 km^2 of forest near the Tunguska river in Siberia (§16.7). On Venus, the *Magellan* spacecraft detected hundreds of radar-dark features, some of which have been connected to impact events. Such features probably were produced by the blast waves from meteors that never hit the ground.

Fireball

Immediately after a high-velocity impact, a hot plume of gases, known as the **fireball**[1], is formed at the impact site. The fireball expands almost adiabatically, and its radius can be calculated from its initial pressure, P_i, and volume, V_i, assuming PV^γ = constant, where γ, the ratio of specific heats, usually equals 1.5. The fireball is preceded by a shock wave in the lower atmosphere. Because the vapor is hotter and therefore less dense than its surroundings, the fireball rises, driven by buoyancy forces. Fine dust brought up in the plume or by ejecta may stay suspended in the Earth's atmosphere for many months to a year, which may have profound climatic consequences through blocking of sunlight and trapping of outgoing infrared radiation (§16.7.1). If the size of the impactor exceeds the atmospheric scale height (i.e., ~10 km on Earth; §5.7.3), a large fraction of the shock-heated air may be blown off into space.

6.4.4 Spatial Density of Craters

Although most airless bodies show evidence of impact cratering, the **crater density** (number of craters per unit area) varies substantially from object to object. The **size–frequency distribution** of craters quantifies the number of craters per unit area as a function of crater size. Examples of size–frequency distributions for various bodies are shown in Figure 6.33. Some surfaces,

[1] This term, unfortunately, is also used to describe a large meteor blazing its way through a planetary atmosphere on its way toward the surface.

including regions of the Moon, Mercury and Rhea, appear **saturated** with craters, i.e., the craters are so closely packed that, on average, each additional impact obliterates an existing crater. The surface has reached a 'steady state'. Other bodies, such as Io and Europa, have very few, if any, impact craters. Why is there such a large range in crater densities? The variations must be caused by the combined effect of impact frequency and crater removal, two topics that are discussed below.

Cratering Rate

The surface of the Moon shows the cumulative effect of impact cratering since the time that the Moon's crust solidified. The size–frequency distribution of lunar craters is shown in Figure 6.33a. For the Moon, we can determine the ages of some regions of the surface accurately using radioisotope dating (§11.6) on rock samples returned to Earth by the various Apollo and Luna landers. Nine different missions returned a total of 382 kg of rocks and soil to Earth (Table F.1). A comparison of these absolute ages with graphs such as shown in Figure 6.33a has been used to determine the historic cratering rate. The very oldest regions, the lunar highlands (~4.45 Gyr), are saturated with craters, but younger regions (lunar maria, ~3–3.5 Gyr) show a much lower crater density. This difference can be understood when considering planetary accumulation processes (see Chapter 15). Because there were many more stray bodies around during the early history of the Solar System, the impact frequency was highest during the planet-forming era. This period is referred to as the **early bombardment era**. The impact frequency dropped off rapidly during the first billion and a half years to a roughly constant cratering flux during the past 3 Gyr.

The present cratering rate on the Moon for craters with sizes $D > 4$ km is ~2.7×10^{-14} craters km^{-2} yr^{-1}, a rate that can be accounted for within a factor of ~3 by the observed and

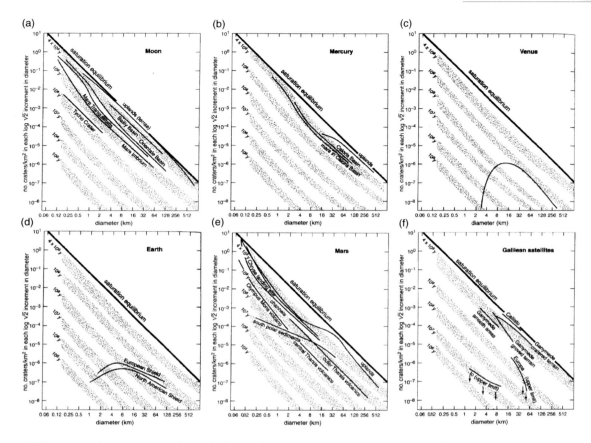

Figure 6.33 Graphs of the crater density (differential crater size–frequency) as a function of crater diameter for (a) the Moon, (b) Mercury, (c) Venus, (d) Earth, (e) Mars, (f) Galilean satellites. (Hartmann 2005)

extrapolated distributions of Earth-crossing asteroids and comets. This rate, scaled to values expected at Mars, agrees with the ~20 craters between 2 and 150 m in diameter that were observed (by *Mars Global Surveyor*) to have formed between 1999 and 2006 in a region of area ~20 million km².

The cratering rate was clearly larger prior to 3.2 Gyr ago than it is at present and much larger prior to 3.8 Gyr ago. But the data are not adequate to uniquely constrain the cratering rate in the first 700 Myr of the Solar System. One possibility is a monotonically decreasing rate of impacts, as represented by curve 'a' of Figure 6.34. However, most of the impact melts in rocks returned from

the Moon by the *Apollo* and *Luna* missions, as well as lunar meteorites, date from 3.8–3.9 Gyr before the present. These dates suggest that the Moon may have been subjected to a terminal cataclysm during this epoch, known also as the **late heavy bombardment**, an era when the cratering rate substantially exceeded that of the previous few hundred million years. The cratering rate as a function of time in this scenario is sketched in curve 'b' of Figure 6.34. Assuming that it occurred, the late heavy bombardment produced a spike in the cratering rate. The size distribution of craters (Fig. 6.33) also provides a historical record of the size–frequency distribution of impactors.

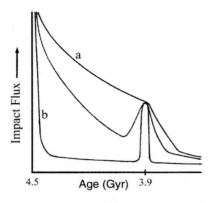

Figure 6.34 Schematic diagram showing possible temporal variations of the early impact flux at the Moon and Earth. Curve 'a' is for a relatively long period of heavy bombardment after planetary accretion. Curve 'b' shows a sharp decline in the impact cratering rate after planetary accretion followed by an intense cataclysmic period of bombardment 3.9 Gyr ago (known as the late heavy bombardment). Note that no scale is given for the vertical axis, and the variations shown may be orders of magnitude. Although it is generally agreed that there was a decrease in the impact rate ∼3.85 Gyr ago, the precise 'shape' of the variation in impact flux over time before ∼3.85 Gyr ago is not well known, as indicated by the variety of curves on this graph. (Adapted from Kring 2003)

Crater Removal

The gravity and material strength of the target body influence the size and shape of craters produced by impacts. **Viscous relaxation** sets an upper limit on how long a surface feature can be recognized as an impact crater, although this time can exceed the age of the Solar System. No impact craters have been seen on Jupiter's satellite Io; their absence has been attributed to Io's extremely active volcanism. A major removal process on Earth is plate tectonics, which 'recycles' the Earth's oceanic crust on timescales of $\sim 10^8$ years (§6.3.1). Non-plate tectonic processes have affected craters on various planets and moons. Atmospheric (and oceanic) weathering slowly erodes craters, both mechanically (e.g., water and wind flows) and through chemical interactions.

In addition to endogenic removal processes, craters can be destroyed and eroded by exogenic processes. Impacts may hit a preexisting crater and destroy it or can cover up old craters with ejecta blankets. Seismic shaking resulting from an impact can obliterate craters either locally or globally. Seismic shaking has the greatest influence on the smallest bodies.

Small craters are easier to obliterate than larger ones. The more densely cratered a surface is, the more craters are likely to be destroyed by a given impact. Eventually, the surface reaches a statistical equilibrium, wherein one new crater destroys, on average, one old crater. Further bombardment by the same population of projectiles does not cause any further secular changes in the size–frequency distribution of craters. Such a surface is saturated with craters. Because small craters are easier to destroy than large ones, saturation is usually first reached for the smallest craters. However, on some bodies, such as Jupiter's moon Callisto (§10.2.3), viscous relaxation is much more efficient at destroying small craters than large ones, leaving the surface saturated with moderate to large craters but with far less area covered by small craters.

Micrometeoritic impacts have an eroding effect on airless bodies, which is referred to as **sandblasting**. A similar effect, **sputtering**, is caused by low-energy ions (keV range) (solar wind, magnetosphere) (§5.7.2). Impacts by charged particles and UV photons also contribute to this process, although these are more important in changing the chemical composition of the surface locally through radiolysis and photolysis, respectively. The lightest atoms, such as H, may gain sufficient energy in this process to escape the gravitational attraction from the body, leaving the darker material behind. Icy surfaces, especially those containing carbon-bearing molecules, tend to darken over time, an effect clearly seen on the outer planets' satellites, most small to midsized Kuiper belt objects and comets.

Dating Cratered Surfaces

The simple counting of craters on a given surface yields a crude estimate of its age. Surfaces of bodies that are $\sim$4.5 Gyr old are saturated with craters. Because all large bodies in our Solar System formed $\sim$4.5 Gyr ago, we know that something happened to a body if its surface is less than 4.5 Gyr old. For example, when the plains on the Moon (maria) were flooded by lava, any preexisting craters were covered and disappeared. When the lava solidified, new craters started to accumulate, and the number of craters increased with time.

The relative ages of some rocks and planetary surfaces can be estimated by stratigraphic techniques, but the most common method to determine relative ages of planetary and satellite surfaces uses the spatial density of craters. Such cratering ages can be deduced from remote imaging observations.

The longer a surface has been solid, the greater the integrated flux of impactors that it has been exposed to and thus the larger the expected number of craters on its surface. However, many complications are inherent in estimating the age of a surface from the size–frequency distribution of craters observed thereupon. As discussed earlier, cratering has occurred at a very nonuniform rate over the history of the Solar System. Cratering rates can differ substantially from body to body and can even vary markedly across the surface of an individual object.

Assuming a model for the density and size distribution of impactors over time and throughout the Solar System, one can estimate a body's age from graphs such as those shown in Figure 6.33 and using the Moon as the 'ground-truth'. This **crater dating** technique is widely used because it typically provides the only means to assign an age to a body's surface.

6.4.5 Impacts on Earth

The well-preserved history of impacts on the Moon, together with the widely observed impact of Comet Shoemaker–Levy 9 with Jupiter, makes us uncomfortably aware of the dangers of being hit by meteoroids. Because the Earth's crust is continuously renewed, impact craters are relatively rare and difficult to find and to recognize. Yet studies of asteroids, comets and interplanetary dust suggest that Earth 'sweeps up' $\sim$10 000 tons of micrometeoritic material each year. Meteors, primarily centimeter-sized and smaller material falling through Earth's atmosphere, are a familiar sight, especially around 10 August (Perseids), 11 December (Geminids) and 1 January (Quadrantids). One calculation suggests that $\sim$7000 meteorites greater than 100 g in mass ($\gtrsim$2 cm in radius) make it to the ground each year, which translates into one fall per square kilometer every 100 000 years.

The smallest meteoroids to hit the Earth's atmosphere are slowed to benign speeds by gas drag or vaporized before they hit the ground. Extremely large impactors can melt a planet's crust and eliminate life entirely.

An airburst creates a shock wave whose destructive reach varies roughly as the one-third power of the energy released by the explosion. However, the size of the area devastated by an airburst depends on the burst's height as well as its energy because shock waves created by high airbursts can dissipate before reaching the ground and shocks from low airbursts approach the ground very obliquely aside from immediately below the explosion.

The Tunguska event involved an airblast of $\sim$ 4 MT and flattened about 2000 km^2 of forests (Fig. 16.6). This area is somewhat larger than the typical value for an object with this amount of kinetic energy because the explosion appears to have occurred near the optimal height for producing destructive effects.

A smaller meteor, about 10 meters in radius, exploded at an altitude of $\sim$20 km near the Russian city of Chelyabinsk on 15 February 2013. This explosion was less than one-tenth as energetic as that over Tunguska, but because it occurred in a more densely populated region, more than 1000 people were injured seriously enough to seek medical attention.

In addition to the direct collisional destruction, an impact can have dramatic effects on climate worldwide. The impact of a 10-km-sized asteroid 66 Myr ago altered the environment so strongly that a majority of the animal species on our planet, including all of the dinosaurs, went extinct (§16.7.1).

The Moon likely formed about 4.5 Gyr ago in a disk that was produced by an impact of a Mars-sized or larger body with the proto-Earth (§15.10.2). At that time, just after the formation of the planets, there still were many planetesimals around that battered the surfaces of planets, and moderately large impacts were probably not uncommon. Impact probabilities at the current epoch are far lower (Fig. 6.34) but nonetheless represent a significant hazard (§16.7.2).

Table 16.1 summarizes when an impactor of a given size last hit the Earth. The impact rate at the current epoch is shown in Figure 16.9. Based on the observed size distribution of asteroids and comets, one expects the present-day Earth to be hit by a body 5 km in radius roughly once every 10^8 years. Smaller bodies hit more frequently: an object with a 2.5 km radius impacts Earth on average once every $\sim 10^7$ years and a body with a radius of 200 m once every $\sim 10^5$ years.

Note that most impact frequencies shown in Figure 16.9 are less than the inverse time since the last impact of a body of a given size given in Table 16.1. A factor of two difference is to be expected, because typically we would expect to be halfway between two impacts, and some uncertainties are inherent in the estimation techniques used for the two compilations. However, examining the individual differences provides interesting insights. The K–T impactor that wiped out the dinosaurs 66 Myr ago (§16.7.1) seems to have been one of the most energetic impact events on Earth within the past couple of billion years, but this is difficult to confirm because the majority of Earth's crust is oceanic and therefore gets subducted into the mantle in ~ 200 Myr (§6.3.1). It is also possible that long-period comets, not included in the near-Earth object (NEO) data, make a significant contribution to impacts of this and higher energy. Finally, the impact rate during the first billion years of Solar System history must have been much larger than it is at the current epoch to account for the very large impacts that occurred $\gtrsim 3.8$ Gyr ago.

Key Concepts

- A mineral is characterized by a specific chemical composition and architecture of the atoms from which it is made.
- Rocks are made up of different minerals and are classified into four groups based on their formation history: primitive, igneous, sedimentary and metamorphic.
- The phase of a material depends upon the temperature and pressure of the environment. Materials behave very differently under high temperature and pressures.
- The shape of a planet-sized body is determined by gravity and rotation. A nonrotating fluid body will take on the shape of a sphere; spinning planets are flattened by rotation. Surfaces of bodies are further 'shaped' by endogenic (e.g., tectonics, volcanism, erosion) and exogenic (e.g., impacts) processes.
- The gravity field of a planet or moon contains information on the internal density structure.
- The degree to which surface topography and gravity are correlated can be interpreted in terms of how much or little isostatic compensation is present.
- The principal sources of internal heat are gravitational, radioactive and tidal heating.

- The interior structure of the Earth has been determined via seismology. For other bodies, models have been constrained based on observable parameters.
- Plate tectonics is unique to Earth. It is caused by convection in the mantle, which induces a large-scale circulation pattern. Tectonic plates recede from each other at the mid-oceanic rift, where hot magma rises and creates new ocean floor.
- An atmosphere can profoundly affect a body's surface through, e.g., water (or other liquids) and wind.
- Impact craters are the most ubiquitous features on planetary surfaces in our Solar System.
- The response of a surface to a hypervelocity impact is more similar to that of an explosion than to a low-velocity collision. The shock waves from the impact's 'explosive' release of energy propagate into the target, causing compression; a rarefaction wave follows, releasing pressure and ejecting material. The primary factors affecting crater size are the energy of the impact and the gravity (g_p) and density of surface material on the target body.
- Small craters tend to have the shape of a simple bowl with diameter about five times the depth. Larger craters have smaller depth/diameter ratios and increasingly more complex morphology at greater sizes, including central peaks or pits, rings, etc.
- Ejecta from craters cover nearby parts of the surface. In some cases, they produce rays and crater chains.
- Atmospheres can slow or disrupt small impactors, preventing small craters from forming. Atmospheres can also modify trajectories of crater ejecta.
- The cratering rate was extremely high soon after the Solar System formed. It dropped by orders of magnitude during the subsequent billion years.
- Graphs of crater density as a function of crater size provide information on the relative age of a surface; absolute dating, however, also depends on knowledge of the impact frequency.

Further Reading

Examples of books on general geology and geophysics of the Earth and other planetary bodies are:

Fowler, C.M.R., 2005. *The Solid Earth: An Introduction to Global Geophysics*, 2nd Edition. Cambridge University Press, New York. 685pp.

Greeley, R., 1994. *Planetary Landscapes*, 2nd Edition. Chapman and Hall, New York, London. 286pp.

Greeley, R., 2012. *Introduction to Planetary Morphology*. Cambridge University Press, New York. 240pp.

Grotzinger, J., T. Jordan, F. Press, and R. Siever, 2006. *Understanding Earth*, 5th Edition. W.H. Freeman and Company, New York. 579pp.

Melosh, H.J. 2011. *Planetary Surface Processes*. Cambridge University Press, Cambridge. 500pp.

A short nontechnical review of the Earth as a planet is given by:

Pieri, D.C., and A.M. Dziewonski, 2007. Earth as a planet: surface and interior. In *Encyclopedia of the Solar System*, 2nd Edition. Eds. L. McFadden, P.R. Weissman, and T.V. Johnson. Academic Press, San Diego. pp. 189–212.

A good book (although it is outdated at places) on the interior structure of all planets is:

Hubbard, W.B., 1984. *Planetary Interiors*. Van Nostrand Reinhold Company Inc., New York. 334pp.

An excellent monograph on impact cratering is:

Melosh, H.J., 1989. *Impact Cratering: A Geologic Process*. Oxford Monographs on Geology and Geophysics, No. 11. Oxford University Press, New York. 245pp.

A pleasant 'novel' about the development of the impact theory on the extinction of the dinosaurs is written by:

Alvarez, W., 1997. *T. Rex and the Crater of Doom*. Princeton University Press, Princeton, NJ. 185pp.

Problems

6-1. Assume that material can be compressed significantly if the pressure exceeds the material strength. If material can be compressed considerably over a large fraction of a body's radius, the body will take on a spherical shape (the lowest energy state of a nonrotating fluid body).

(a) Calculate the minimum radius of a rocky body to be significantly compressed at its center. Assume an (uncompressed) density $\rho = 3500$ kg m^{-3} and material strength $S_m = 10^8$ Pa.

(b) Calculate the minimum radius of the rocky body in (a) to be significantly compressed over a region containing about half the body's mass.

(c) Repeat for an iron body, which has an (uncompressed) density of $\rho = 8000$ kg m^{-3} and material strength $S_m = 2 \times 10^8$ Pa.

(d) Repeat for an icy body of density $\rho = 1000$ kg m^{-3} and material strength $S_m = 4 \times 10^6$ Pa.

6-2. **(a)** Show that the net gravitational plus centrifugal acceleration, $g_{\text{eff}}(\theta)$, on a rotating sphere is:

$$g_{\text{eff}}(\theta) = g_p(\theta) - \omega_{\text{rot}}^2 r \sin^2 \theta, \quad (6.8)$$

where ω_{rot} represents the spin angular velocity, g_p is the gravitational acceleration and θ is the planetocentric colatitude.

(b) Calculate the ratio of the centrifugal to the gravitational acceleration for Earth, the Moon and Jupiter.

6-3. **(a)** Calculate the total amount of internal energy lost per year from Earth, assuming a heat flux of 0.075 J m^{-2} s^{-1}.

(b) Calculate the temperature of Earth if the 'heat flux' had been constant over the age of the Solar System but none of the heat had escaped the planet. A typical value for the specific heat of rock is $c_P = 1.2 \times 10^3$ J kg^{-1} K^{-1}.

6-4. Compare the Earth's intrinsic luminosity (L_i) or heat flux (0.075 J m^{-2} s^{-1}) with the luminosity from reflection of sunlight (L_v) and from emitted infrared radiation (L_{ir}). Assume the Earth's albedo is 0.36. Note that you can ignore the greenhouse effect and that the value of the infrared emissivity is irrelevant. (Hint: Use §4.1.2.) Comment on your results and the likelihood of detecting Earth's intrinsic luminosity via remote sensing techniques from space.

6-5.* Calculate the ratio of internal radioactive heat to absorbed solar radiation for Pluto. You may assume that Pluto consists of a 60:40 mixture of chondritic rock to ice and that its albedo is 0.4.

6-6. The continents on Earth drift relative to each other at rates of up to a few centimeters per year; this drift is caused by convective motions in the mantle.
(a) Assuming typical bulk motions in the mantle of Earth are 0.01 m/yr, calculate the total energy associated with mantle convection.
(b) Calculate the kinetic energy associated with the Earth's rotation.
(c) Calculate the kinetic energy associated with the Earth's orbital motion about the Sun.

6-7. Suppose that the shield volcanoes on Mars are produced from a partial melting of martian rocks at a temperature of $\sim$1100 K. The density is then $\sim$10% less dense than that of the surrounding rocks, and the magma will rise up through the surface and build up the 20-km high Tharsis region. Assuming the pressure on the magma chamber below the magma column is equal to the ambient pressure at that depth, calculate the depth of the magma chamber below the martian surface.

6-8. If the lithospheric plates move, on average, at a speed of 0.06 m yr^{-1}, what would be a typical recycling time of terrestrial crust? (Hint: Calculate the recycling time based on the motion of one plate over the Earth's surface. How would your answer change if you have several plates moving over the surface?)

6-9. Suppose a child leaves a toy truck on a deserted sandy beach facing the ocean and returns many years later to retrieve it. The beach is characterized by strong winds, which usually blow inland. Sketch the dune formed around the toy truck.

6-10. Calculate a typical timescale over which material in the Earth's upper mantle flows. The dynamic viscosity is $\sim$10^{21} Pa s, and the shear modulus is $\sim$250 GPa.

6-11. Calculate the gas-to-dust ratio in a large terrestrial dust storm using the parameters given in §6.3.3. Comment on your result.

6-12. Geysers have been observed on several bodies within the Solar System. Water shot up by Old Faithful in Yellowstone Park reaches a height of up to 50 m. The Prometheus plume on Io is 60 km high, much of the water from Enceladus's south polar plume escapes the moon entirely and plumes on Triton reach a height of 8 km.
(a) Assuming that the heights of each of the above geysers are controlled only by ballistics, compute the exit velocity for each geyser as the erupted material leaves the vent. Express your answer both as a velocity and as a ratio to each world's escape velocity.
(b)* Discuss what your answers may be telling you about the fundamental mechanism behind each type of plume and consider where this simple ballistic assumption may be wrong.

6-13. **(a)** The compression stage typically lasts a few times longer than the time required for the impacting body to fall down a distance equal to its own diameter. Calculate the duration of the compression stage for $R = 10$ m and $R = 1$ km meteoroids that impact Earth at $v = 15$ km s^{-1}.

(b) Estimate the pressure involved in these collisions, assuming the meteoroids are stony bodies with density $\rho = 3000$ kg m^{-3}.

6-14. **(a)** Determine the diameter of the crater produced when a stony ($\rho = 3000$ kg m^{-3}) meteoroid, 1 km across, hits the Earth at a velocity of 15 km s^{-1}, at an angle of 45° (ignore the Earth's atmosphere). Use a density of 3500 kg m^{-3} for the Earth's surface layer.
(b) Determine the crater diameter from (a) if the meteoroid is 10 km across.
(c) Determine the diameter of craters produced by both meteoroids if they had hit the Moon ($v = 15$ km s^{-1}) instead of Earth.

6-15. **(a)** Determine the minimum radius of an iron meteoroid ($\rho = 8000$ kg m^{-3}) to impact the Earth at hypersonic speed. (Hint: See §11.3.) Note that as the value of the drag coefficient depends on the shape of the body, and the velocity at which the meteoroid hits the atmosphere isn't specified, you really can't calculate to an accuracy of better than a factor of ~ 2. You can get this accuracy by simply setting the mass of a column of air through which the meteoroid passes equal to the mass of the meteoroid.

(b) Determine the minimum radius for a similar meteoroid of similar composition to make it through Venus's atmosphere.
(c) Calculate the approximate crater size the meteoroids produce on both planets if the impacting velocity is equal to the escape velocity from the planet. Have craters smaller than this size been observed on Venus and Earth?

6-16. **(a)** Draw a 5 × 5 square grid on which 25 craters are uniformly distributed at the centers of each grid square. You may represent the craters as either circles (which are substantially smaller than the grid squares) or points.
(b) Draw the same grid but place the center of one crater at random within each grid square. Use a random number generator to compute the coordinates of each grid point.
(c) Place the 25 craters at random within the 5 × 5 square grid, to simulate the production of craters under most circumstances.
(d) Place the 25 craters in a clustered distribution as would be expected if one part of the surface is much older than another part.
(e) Comment on using the spatial distribution of craters to determine the relative ages of planetary surfaces.

Sun, Solar Wind and Magnetic Fields

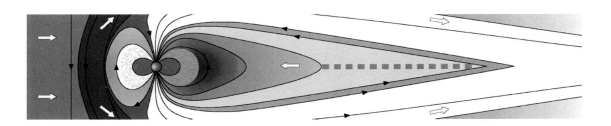

De Magnete, Magneticisque Corporibus, et de Magno Magnate Tellure (On the Magnet and Magnetic Bodies, and on That Great Magnet the Earth)

Title of a major scientific work on magnetism published in 1600 by the English physician and scientist, William Gilbert

Most planets are surrounded by huge magnetic structures known as **magnetospheres**. The magnetic cavity formed by the Sun, known as the **heliosphere**, resembles a bubble 'blown' into the interstellar medium by the solar wind. Within this bubble, the solar wind flows around and interacts with the magnetic fields surrounding the Earth, the giant planets and Mercury. The fields are generated in these planets' interiors via a dynamo process. Venus, Mars and comets are surrounded by magnetic structures that are induced by the interaction of the solar wind with their ionospheres. Mars, the Moon and some asteroids show evidence of large-scale remanent magnetism.

The shape and size of a planet's magnetosphere are determined by the strength and orientation of its magnetic field, the solar wind flow past the field and the motion of charged particles within the magnetosphere. Many planets have a magnetosphere that is more than 10–100 times as large as the planet itself. Thus, planetary magnetospheres form the largest structures in our Solar System other than the heliosphere.

Charged particles are present in all magnetospheres, although the density and composition vary from planet to planet. The particles may originate in the solar wind, in the planet's ionosphere or on satellites or ring particles whose orbits are partly or entirely within the planet's magnetic field. The motion of these charged particles gives rise to currents and large-scale electric fields, which in turn influence the magnetic field and the particles' motion through the field.

Although most of our information is derived from *in situ* spacecraft measurements, atoms and ions in some magnetospheres have been observed from Earth through the emission of photons at ultraviolet and visible wavelengths. Accelerated electrons emit photons at radio wavelengths observable at frequencies ranging from a few kilohertz to several gigahertz. Radio emissions at $\sim$10 MHz were detected from Jupiter in the early 1950s

and formed the first evidence that planets other than Earth might have strong magnetic fields.

In the first section of this chapter, we discuss our Sun because it is the center of our Solar System and the main source of energy for most life forms on Earth. The Sun is also the source of the solar wind, which separates us from the interstellar medium by forming the heliosphere. In §7.2, we discuss the solar wind and its interaction with bodies in our Solar System in more detail; we also touch on the phenomenon of 'space weather' in that section. In §7.3, we describe the planetary magnetospheres, including sources and losses of plasma located therein. The generation of and variability in magnetic fields are discussed in §7.4.

7.1 The Sun

Figure 7.1 shows an image of the Sun taken at visible wavelengths. This image shows the solar surface, referred to as the **photosphere**. It reveals several dark 'blemishes' on the surface, known as **sunspots**. Sunspots appear dark because their temperature is lower ($\approx$4500 K) than the average photospheric temperature ($\approx$5800 K). In contrast, the magnetic field strength in sunspots is higher (up to several thousand Gauss) than in the surrounding regions (a few Gauss), such that there is approximate pressure equilibrium between the spots and the surrounding regions. Each sunspot shows a dark center, the **umbra**, surrounded by a lighter, radially striated **penumbra**. Sunspots range in diameter from $\sim$3500 km up to 50 000 km, and their lifetimes are from about 1 week to several months.

The basic internal structure of the Sun is sketched in Figure 7.2. Similar to other stars, the Sun is composed primarily of hydrogen (volume mixing ratio of 92.1%) and helium (7.8%) outside of its core. The core of the Sun, where hydrogen

(a)   (b)

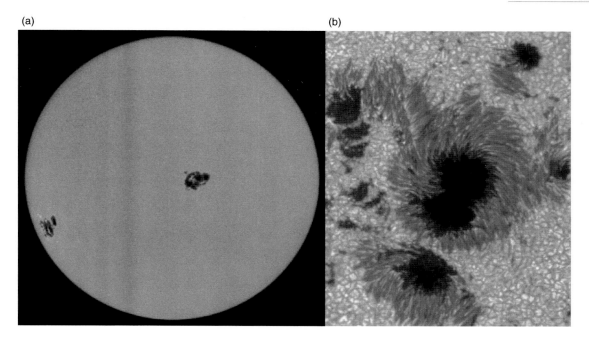

Figure 7.1 COLOR PLATE (a) Visible (red) light image of the entire Sun, with some large (planet-sized) sunspot groups on it. The image was taken in the continuum near the Ni I 676.8 nm line, and the contrast has been enhanced. (b) A close-up of some sunspots (different from those seen in panel a). The dark center of the spots is referred to as the umbra, with the surrounding lighter part the penumbra. The background shows the granulation on the Sun. (SOHO [NASA/ESA] and the Royal Swedish Academy of Sciences)

burns into helium through thermonuclear reactions (see §3.3), is enriched in helium and has a temperature of $\sim$15 million K. Energy is transported outwards via radiation in the **radiative zone** and via convection in the outer third of the Sun, the **convection zone**. It takes photons about 0.1–1 million years to reach the solar surface.

The convective motions just below the photosphere are visible in the form of a **granulation pattern**, as shown in Figure 7.1b. Typical granules have sizes of order 1000 km and lifetimes of several minutes. The subphotospheric gas rises at the bright centers of the granulation cells, cools via radiation and flows down the intergranular lanes. Typical velocities are of order 1 km s^{-1}. The convection pattern is also organized in larger structures (e.g., **supergranulation**) for granules at $\sim$20 000 km.

Sunspots usually come in pairs, and some sunspots occur in complex groups, called **active regions**. The two spots in each pair of sunspots show opposite magnetic polarity, with the field lines pointing either in or out of the spots, as sketched in Figure 7.3. Figure 7.4 shows an image of the looplike magnetic field structures as 'traced out' by the emission from Fe ions moving along the field lines.

When the number of sunspots is plotted as a function of time, as in Figure 7.5, one discerns an 11-year cycle. Throughout the cycle, the leading sunspot of a pair tends to show the same polarity. This polarity as well as the overall magnetic field of the Sun reverses every 11 years. The spatial pattern traced out by the sunspots over time is shown in Figure 7.6. At the beginning of the 11-year cycle, sunspots appear at high solar latitudes. New

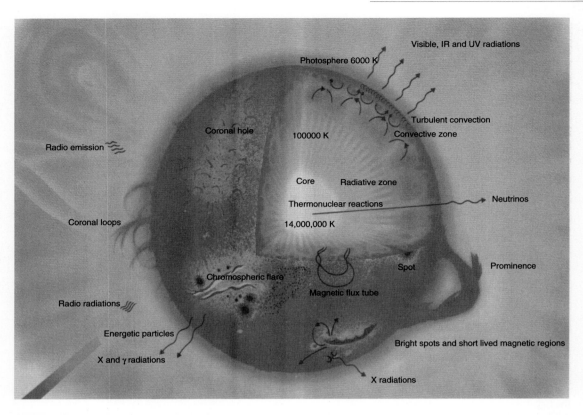

Figure 7.2 COLOR PLATE Cutaway view of the Sun. Energy is released via thermonuclear reactions in the core. The energy is transported via photons in the radiative zone and via convection in the convective zone. The photosphere represents the solar surface. Above the photosphere we find the chromosphere (lower solar atmosphere) and the corona (upper atmosphere). Many phenomena associated with the solar activity cycle are shown, such as filaments, prominences and flares. (NASA/ESA)

sunspots appear at lower and lower latitudes as the cycle progresses, so that over time a **butterfly pattern** is formed on the plot.

It is apparent from Figure 7.5 that the number of sunspots during **solar maximum** (i.e., maximum number of sunspots during a solar cycle) and **solar minimum** also varies over time. Because sunspots lower the total luminosity of the Sun by only ∼0.15% at solar maximum, this has a negligable effect on the climate on Earth. The extreme ultraviolet (EUV) radiation from the Sun, however, which also displays an 11-year solar cycle, may influence Earth's climate because this radiation affects the ionization in a planet's ionosphere.

Very few sunspots were seen between about 1645 and 1715, implying that the Sun was relatively inactive. This period, referred to as the **Maunder minimum**, coincides with a climatic period known as the **Little Ice Age** (§5.8).

Figure 7.7 shows the Sun during a solar eclipse. This image reveals the solar **corona** through sunlight scattered off electrons. The corona consists of highly variable magnetically controlled **loops** and **streamers**, as shown. Closer to the Sun's photosphere, in the low corona and **chromosphere**, are bright **prominences**, elongated clouds of material that are held aloft by magnetic fields.

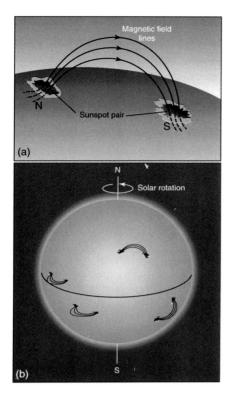

Figure 7.3 An artist-rendition of sunspots on the Sun and the magnetic field lines connecting them.

Figure 7.4 This image of prominences on the Sun was taken by the Transition Region and Coronal Explorer (TRACE) in emission lines of FeIX. The image clearly shows the magnetic loop structures such as visualized in Figure 7.3; the details in this picture are startling. (NASA)

7.2 The Interplanetary Medium

7.2.1 Solar Wind

The loops and streamers in the solar corona contain hot plasma that is visible at EUV and X-ray wavelengths. Figure 7.8 shows X-ray images of the Sun, which reveal a tremendous amount of structure. The luminosity from the X-ray bright areas is produced by a dense hot thermal plasma. The dark areas indicate regions largely devoid of hot X-ray-emitting plasma and are referred to as **coronal holes**. Magnetic field lines in these regions have opened up into interplanetary space, so that particles can freely escape into space, forming the **solar wind**.

The solar wind consists of a roughly equal mixture of protons and electrons, with a minor fraction of heavier ions. The density decreases roughly as

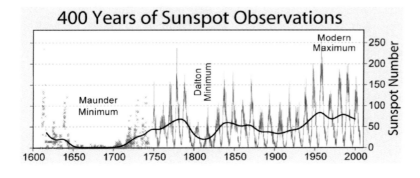

Figure 7.5 The number of sunspots as a function of time, from the early seventeenth century onwards. Note the 11-year periodicity and the variations in the minimum and maximum number of sunspots from period to period. In particular, the Maunder minimum is intriguing. (Courtesy Robert A. Rohde)

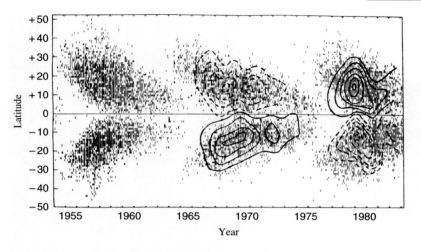

Figure 7.6 The spatial distribution of sunspots as a function of time, which shows the familiar butterfly diagram. The 11-year solar sunspot cycle is clearly visible. The contours outline the radial mean magnetic field at multiples of 0.27 μT. *Solid lines* indicate positive magnetic field values; *dashed lines* indicate negative values. (Stix 1987)

the inverse square of the heliocentric distance. A typical ion density at Earth's orbit is 6 to 7 million protons m^{-3}. The particles flow radially outward from the Sun, carrying the solar magnetic field as if it were 'frozen in'. The magnetic field strength at Earth's orbit is typically 5–7 nT, and the solar wind speed at Earth's orbit and beyond is on average ~400 km s^{-1}. During solar minima, there are large coronal holes near the Sun's poles from which the wind emanates; at these times, the speed is typically 750–800 km s^{-1} at higher solar latitudes.

A large filament or prominence sometimes erupts in connection with a **coronal mass ejection** (CME), such as shown in Figure 7.9. CMEs often occur in association with **solar flares**, events during which the coronal X-ray and ultraviolet (UV) emissions suddenly intensify by several orders of magnitude in a relatively localized region in the low corona and chromosphere. Major solar flares typically occur approximately once per week during years of maximum sunspot activity, with weak flares and CMEs happening a few times daily. During years of minimum solar activity, such weaker events may happen once a week.

7.2.2 The Parker Model

In 1958, Eugene Parker predicted the existence of the continuous solar wind flow, assuming that

particles flowed radially outward from the Sun, carrying the solar magnetic field as if it were 'frozen in'. The outward acceleration of the solar wind is primarily caused by the pressure difference between the corona and the interplanetary medium.

Whereas the expansion of the solar wind is radially outwards, the Sun rotates 'underneath'. Each fluidlike element of the wind effectively carries a specific magnetic field line, which is rooted at the Sun. Consequently, the solar wind magnetic field takes on the approximate form of an Archimedean spiral, as displayed in Figure 7.10. The radial and azimuthal components of the field are roughly equal at Earth's orbit, with a strength of a few nT each. Because the total magnetic flux through any closed surface around the Sun must be zero, inward and outward magnetic fluxes must balance each other. Spacecraft measurements have shown that the inward and outward fluxes are distributed in a systematic way such that there are interplanetary sectors with predominantly outward fluxes and others with predominantly inward fluxes. The different sectors are magnetically connected to different regions on the solar surface – generally different coronal holes. However, because of the predominantly dipolar nature of the Sun's largest scale magnetic field, the field in the solar

Figure 7.7 COLOR PLATE The total solar eclipse of 29 March 2006 viewed at visible wavelengths. The dark center is the Moon as it passes between Earth and the Sun. The bright white streamers extending outwards from the Sun are visible because sunlight is scattered to us by electrons in the streamers. (Courtesy Hana Druckmüllerová and Miloaslav Druckmüller)

wind or heliosphere is generally in the form of a positive (outwards) hemisphere and a negative (inwards) hemisphere. The **heliospheric current sheet** separates the hemispheres of opposite magnetic polarity. This current sheet is effectively the extension of the solar magnetic equator into the heliosphere.

The flows from different coronal holes often have different speeds and hence 'collide' and produce spiral-shaped compressions in the solar wind. The entire magnetic field and stream structure rotates with the Sun. The magnetic sector structure seen in 1963/1964 by the spacecraft *IMP-1* is illustrated in Figure 7.11. This sector structure changes over time as conditions on the Sun change. While the Sun rotates, the different magnetic sectors sweep by the various bodies in our Solar System. The sudden reversals of magnetic field direction and stream structure are possibly responsible for the 'disconnection' events seen in cometary ion tails (§12.7.5), as well as for certain magnetospheric disturbances.

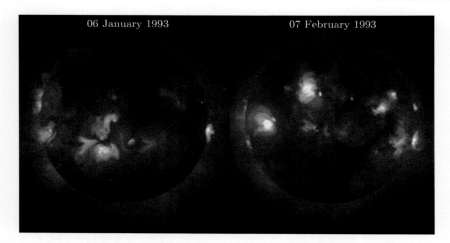

06 January 1993   07 February 1993

Figure 7.8 The Sun as observed by the *Yohkoh* spacecraft in soft X-rays. X-ray bright regions indicate heating to temperatures in excess of 2×10^6 K. These regions usually overlie sunspots or active regions. The very dark regions are the coronal holes, which are usually located above the solar poles. These coronal holes sometimes extend down to lower latitudes, as illustrated by these two images. (Yohkoh Science Team)

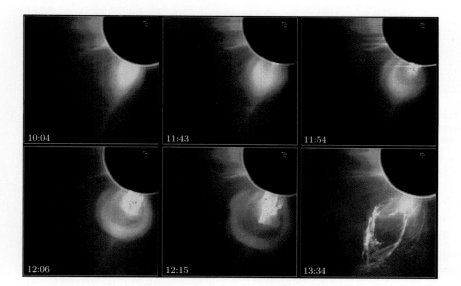

10:04   11:43   11:54

12:06   12:15   13:34

Figure 7.9 Sequence of images from 18 August 1980, showing the formation of a coronal mass ejection through the distortion of a coronal streamer, triggered by the disruption of a solar prominence beneath it. The prominence material is blown outward along with the original streamer material; the bright, filamentary structures on the 13:34 frame are the remnants of the prominence. The dark circle in the upper right corner of each frame is the occulting disk of the Solar Maximum Mission coronagraph, which is 60% larger than the solar disk. (Courtesy J. Burkepile, High Altitude Observatory, NCAR)

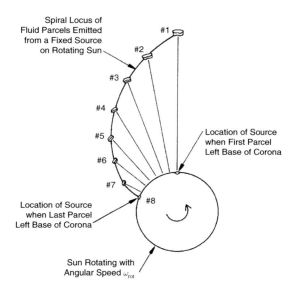

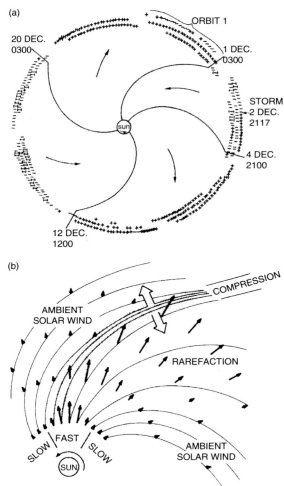

Figure 7.10 Archimedean (or Parker) spiral of solar wind particles streaming away from the Sun. (Hundhausen 1995)

As discussed in §1.2.6 and shown in Figure 1.5, the heliosphere forms a 'bubble' in the interstellar medium. Its boundary, the **heliopause**, is at the location where the solar wind pressure is balanced by that in the interstellar medium (analogous to the magnetopause, discussed in §7.2.4).

7.2.3 Space Weather

Interplanetary space is full of complicated and sometimes violent processes. Interplanetary shocks, preceding high-speed winds and coronal mass ejections accelerate particles locally to very high energies. Although it takes a typical solar wind particle a few days to reach the Earth, the most energetic cosmic ray-like particles injected through a solar flare may travel at a considerable fraction of the speed of light and thereby reach Earth less than 1 hour after the event. The response of Earth's environment to solar activity and the continuously varying interplanetary medium is known as **space weather**, and the effects of CMEs and solar flares on the interplanetary medium

Figure 7.11 (a) Sector structure in the interplanetary field as observed by *IMP-1* in 1963 and 1964. *Plus signs* indicate magnetic field directed outward from the Sun. (Wilcox and Ness 1965), (b) The interaction between a region of high-speed solar wind with the ambient solar wind. The high-speed solar wind is less tightly bound. (Hundhausen 1995)

and Earth's environment are referred to as **space weather storms**. Space weather, a plasma effect, should not be confused with space weathering, a material processing effect, which is discussed in §12.4.1.

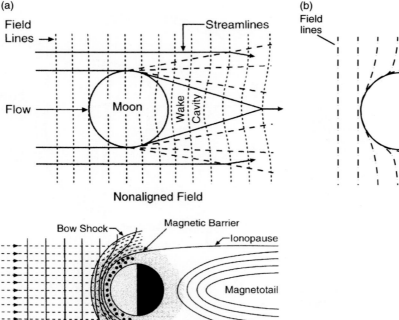

(a)

Field Lines →

Streamlines

Flow → Moon Wake Cavity

Nonaligned Field

(b)

Field lines

Bow Shock Magnetic Barrier

Ionopause

Magnetotail

Interplanetary Magnetic Field Draped Field Magnetosheath

Figure 7.12 Interaction of the solar wind with various types of planetary bodies that do not possess internal magnetic fields. (a) A nonconducting body. (b) A conducting body. (c) A body with an ionosphere. (Adapted from Luhmann 1995)

Space weather is of particular importance because of its effects on spacecraft operations and communication systems and on some ground-based electronics. Particles accelerated as a result of fast CMEs may damage exposed electronic equipment directly and may disrupt communications indirectly. The Earth's upper atmosphere is heated by enhanced auroral precipitation and expands, which increases the drag on near-Earth satellites and 'space junk', leading to changes in their orbits. Such changes may result in a (temporary) 'loss' of spacecraft. Radio communication systems on the ground rely on the reflection of radio waves by the Earth's ionosphere. Such communications are temporarily disrupted when the ionosphere is altered by a solar flare or CME. Transformers in power grids and long conductivity cables can be damaged by currents induced by the related changes in the magnetic field on the ground.

The most powerful space weather storm ever recorded was the **Carrington Event** in 1859, when a major CME traveled directly towards Earth. Aurorae were seen as far south as the Caribbean, and telegraph systems in Europe and North America failed. As described above, a similar event today would have far-reaching consequences on our entire society.

7.2.4 Solar Wind–Planet Interactions

All planetary bodies interact to some extent with the solar wind, as schematically indicated in Figure 7.12. For bodies without intrinsic magnetic fields, the interaction depends on the conductivity of the body, and/or its atmosphere. Rocky objects, such as the Moon and most asteroids, are poor conductors. In such circumstances, as shown in Figure 7.12a, the solar wind particles hit the body directly and are absorbed. The interplanetary magnetic field (IMF) lines simply diffuse through the body. The wake immediately behind the object is practically devoid of particles.

The interaction of the solar wind with a conducting body is more complicated. If a planetary body is highly conductive, the interplanetary magnetic field lines drape around the body because the plasma flows around the conductor, as shown in Figure 7.12b.

If a poorly conducting body has an atmosphere but no internal magnetic field, the solar wind interacts with the atmosphere. The interaction is mainly between charged particles, i.e., with a planet's or satellite's ionosphere, or in the case of comets, with the outflowing gases after they are ionized. Because interplanetary magnetic field is moving with the solar wind plasma, atmospheric and cometary ions exposed to it are accelerated and picked up by the solar wind. These charged particles 'drape' the field lines around the body, as depicted for a comet in Figure 7.13.

If a planet has an extensive ionosphere, currents are set up, which inhibit the magnetic field from diffusing through the body, as depicted in Figure 7.12c. This situation gives rise to a magnetic configuration very similar to the magnetic 'cavity' created by the interaction of the solar wind with a magnetized planet, discussed in §7.3. Venus, Mars and comets have such **induced magnetospheres**. However, all of these bodies show solar wind field penetration into the ionosphere when the solar wind pressure is larger than the ionospheric thermal plasma pressure, so the shielding is only partial and varies with the interplanetary conditions. Titan, in Saturn's magnetosphere, exhibits similar behavior.

As discussed in §12.7.5, the precise structure of a comet's ion tail depends on the interplanetary medium and its magnetic field. For example, Figure 7.13 shows the change in a comet's magnetic field when it encounters a sudden reversal in the interplanetary magnetic field direction. This may cause the comet to appear to 'lose' its tail, although it immediately starts forming a new one. Such phenomena are referred to as **disconnection events**.

7.3 Planetary Magnetospheres

If a body has an internal magnetic field, the solar wind interacts with the field around the object. It confines the magnetic field to a 'cavity' in the solar wind, referred to as a **magnetosphere**. The size of a planet's magnetosphere depends on the planet's **magnetic dipole moment**, $\mathcal{M}_B$, expressed in T m^3, and the local strength of the solar wind. Typical parameters for the Earth and other magnetospheres are summarized in Table E.18. Note that the 'surface' magnetic field strengths (the 1-bar level for the giant planets) for Earth, Saturn, Uranus and Neptune are all $\sim 3 \times 10^4$ nT; however, because the radii of the giant planets are much larger than that of Earth, their magnetic dipole moments are 25–500 times larger than the terrestrial moment. Jupiter has by far the strongest magnetic dipole moment, nearly 20 000 times stronger than that of the Earth. The standoff distance of the magnetopause for Earth and the giant planets is typically larger than 6–10 planetary radii. For Mercury, the standoff distance is 1.4–1.5 R$_{\mathrsfso{\male}}$. At times Mercury's magnetopause might be pushed down to the surface, in which case the solar wind interacts directly with the planet's surface. The north magnetic poles on Earth and Mercury are near the planets' geographic south poles; thus, field lines exit the planets in the south and enter in the northern hemisphere. The north magnetic poles of Jupiter and Saturn are in the northern hemispheres.

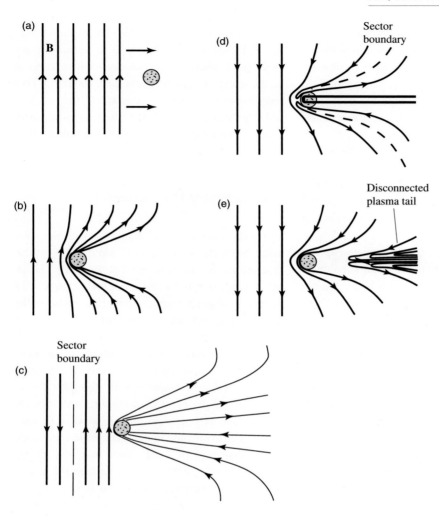

Figure 7.13 The field-draping model of Alfvén, in which interplanetary magnetic field lines are deformed by a comet's ionosphere. The sequence from (a) to (c) shows the gradual draping of interplanetary magnetic field lines around the comet into a magnetic tail. When the comet encounters a sector boundary in the interplanetary magnetic field (where the magnetic field reverses direction; dashed line), the tail becomes disconnected, as depicted in the sequence from (c) to (e). (de Pater and Lissauer 2010)

In §7.3.1, we summarize the topology of the Earth's magnetosphere, which provides an example of such fields. Many general characteristics are shared by all planetary magnetospheres, but there is also considerable diversity that results from different field strengths, orientations and shapes, as well as the diversity of interactions with the atmospheres or surfaces of the planet and its satellites. Aurora are discussed in §7.3.2 and magnetospheric plasmas in §7.3.3. This section concludes with a short summary of radio emissions in §7.3.4.

7.3.1 Earth's Magnetosphere

The Earth's magnetosphere is sketched in Figure 7.14. The Earth's magnetic field is approximately dipolar, similar to that generated by a bar magnet. At the Sun-facing side, the interaction of the supersonic solar wind with the Earth's magnetic field induces a **bow shock**. This shock is analogous to a bow wave in front of a speedboat on a lake. In both cases, the quiescent flow (solar wind or water) does not 'know' about the obstacle because the relative velocity between the obstacle

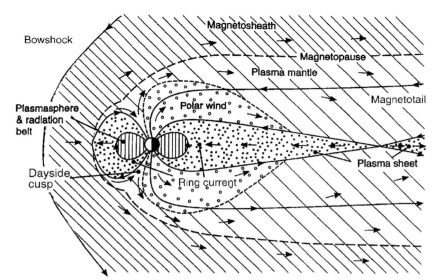

Figure 7.14 Sketch of the Earth's magnetic field. The *solid arrowed lines* indicate magnetic field lines, the *heavy long-dashed line* represents the magnetopause and the *arrows* represent the direction of the plasma flow. *Diagonal hatching* indicates plasma in, or directly derived from, the solar wind or magnetosheath. Outflowing ionospheric plasma is indicated by *open circles*; the *solid dots* indicate hot plasma accelerated in the tail, and the *vertical hatching* shows the corotating plasmasphere. (Adapted from Cowley 1995)

and medium is larger than the speed of the relevant waves (surface waves in the case of water and magnetosonic waves in the case of the solar wind). The bow shock is located at ~15 $R_\oplus$ on the side facing the Sun and is only ~20 km thick. The solar wind plasma is decelerated at the bow shock to subsonic speeds. The turbulent subsonic region behind the bow shock is the **magnetosheath**. The magnetosheath is shielded from the Earth's magnetic field by the **magnetopause**, which separates the solar wind plasma from the terrestrial magnetic field. At the subsolar point, the magnetopause is located where the ram pressure of the solar wind and the pressure inside the magnetosphere are balanced, which is at ~10 $R_\oplus$. The solar wind flows smoothly around the magnetospheric obstacle. The solar wind pressure shapes the 'nose' of the Earth's field; the flow around the magnetosphere drags the Earth's field lines back and stretches them out into the **magnetotail**. The magnetotail consists of two lobes of opposite polarity separated by a neutral sheet. Because countless shocks and discontinuities propagate through the interplanetary medium, the interaction between the solar wind and a magnetosphere is very dynamic.

Close to Earth's surface, where the magnetic field is hardly deformed by the solar wind, we find the **plasmasphere** and the **radiation belts** (the Earth's radiation belts are also known as the **Van Allen belts**). The plasmasphere is characterized by cold dense plasma from Earth's ionosphere, while the Van Allen belts are regions in the magnetosphere where energetic charged particles are trapped. These particles execute three types of motion, as shown in Figure 7.15: They rapidly gyrate around magnetic field lines; on somewhat longer timescales, they 'bounce' up and down magnetic field lines; and they slowly drift around the Earth. These particles cannot escape from the magnetosphere under normal equilibrium circumstances.

7.3.2 Aurora

The Earth's magnetic field largely protects life on Earth from high-energy cosmic rays, as charged particles are deflected at the magnetopause. Magnetic field lines in the **polar cusp** are connected to the interplanetary magnetic field, with the amount of interconnection depending on the orientation of the interplanetary magnetic field with respect

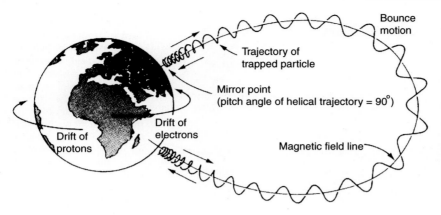

Figure 7.15 The motion of charged particles in the Earth's magnetic field is characterized by a gyro motion around the field lines and bounce motion along the field line. These motions together result in the corkscrew motion indicated in the figure. In addition, the particles also drift around the Earth, as indicated. (Lyons and Williams 1984)

to the Earth's field. These charged particles enter the atmosphere along magnetic field lines. Atmospheric atoms, molecules and ions are excited through interactions with the precipitating particles or by photoelectrons produced in the initial 'collision' with these particles. Upon deexcitation, the atmospheric species emit photons that can be observed at optical, infrared, UV and, on Jupiter and Earth, at X-ray wavelengths. Examples of these emissions, known as **aurora** and commonly referred to as **northern** or **southern lights**, are shown in Figure 7.16. Because the footpoints of field lines that thread through the 'storage place' of charged particles form an oval about the magnetic poles, the emissions occur in an oval-shaped region, referred to as the **auroral zone**.

As shown in Figure 7.16, aurora can be truly spectacular. On Earth, they are regularly seen in the night sky at high latitudes. Rapidly varying, colorful displays fill large fractions of the sky. The lights may appear diffuse, in arcs with or without ray patterns, or may resemble draperies. On Earth, auroral emissions are studied both from the ground and from space at wavelengths varying from X-rays up to radio wavelengths. Aurora are clearly related to disturbances of the geomagnetic field, usually induced by fluctuations in the solar wind, such as those triggered by solar flares or coronal mass ejections. The emissions generally vary rapidly in color, form and intensity.

The most prominent lines and bands in the terrestrial aurora are caused by molecular and atomic nitrogen and oxygen. The visible display is dominated by the green and red atomic oxygen lines at 557.7 nm and 630.0/636.4 nm. Blue emissions are produced by nitrogen (427.8 nm and 391.4 nm). Strong lines at UV wavelengths are also caused primarily by nitrogen and oxygen.

Aurora have also been detected on all four giant planets. Images of the aurora on Jupiter and Saturn are quite spectacular; examples are shown in Figures 7.17 and 7.18. As on Earth, the emissions are predominantly oval shaped. In addition, *the Hubble Space Telescope* (*HST*) image in Figure 7.17 shows faint UV emissions extending roughly 60° in the wake or plasma flow direction beyond Io's magnetic footprint, providing direct evidence of charged particles entering Jupiter's ionosphere along magnetic field lines connecting Io with Jupiter's ionosphere. Such currents at Io's orbit have been detected directly by the *Galileo* spacecraft.

7.3.3 Magnetospheric Plasmas

Magnetospheres are populated with charged particles: protons, electrons and ions. We find primarily oxygen and hydrogen ions in the Earth's magnetosphere; in Jupiter's magnetosphere, these are augmented with sulfur ions; in Saturn's

Figure 7.16 (a) COLOR PLATE Photographs of the northern (auroral) lights from Fairbanks, Alaska. One can see a double arc with a developing ray below the Big Dipper. A few minutes later (*right image*) the arcs have become unstable and develop curtains/draperies. (Courtesy J. Curtis, Geophysical Institute, UAF) 🖟 (b) Aurora Australis (Southern Lights) four days after a strong solar flare went off. (NASA) 🖟 (c) Aurora Australis as photographed from the space shuttle. (NASA)

magnetosphere, with ions derived from water; and in Neptune's magnetosphere, both H^+ and N^+ have been detected. There are large differences in ion densities: Whereas the maximum ion densities in the magnetospheres of Uranus and Neptune are only 2–3 million protons m^{-3}, the densities exceed a few $\times 10^9$ m^{-3} in some regions of the magnetospheres of Earth and Jupiter. All magnetospheres

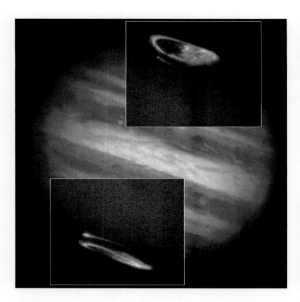

Figure 7.17 Composite *HST* image of Jupiter at visible wavelengths, with superposed northern and southern aurorae at ultraviolet wavelengths. Note the 'trail of light' produced by Io just outside the aurora. See also Figure 8.17. (Courtesy John Clarke, NASA/*HST*)

Sources of Plasma

There are several sources for magnetospheric plasma, and the relative contributions of each source vary from planet to planet. Charged particles can originate in cosmic rays, the solar wind or the planet's ionosphere or on satellites/rings that are partially or entirely embedded in the magnetosphere. Although ionospheric particles are usually gravitationally bound to the planet, some charged particles escape along magnetic field lines into the magnetosphere (§5.7.2). Sputtering by micrometeorites, charged particles and high-energy solar photons may cause ejection of atoms and molecules from moons/rings (§5.7.2); if such particles become ionized, they enrich the magnetospheric plasma.

A planetary magnetosphere is embedded in the solar wind; simple entry of solar wind particles into the magnetosphere would populate a planet's magnetic field with solar wind plasma. The detection of protons, electrons and helium nuclei in Earth's magnetosphere suggests that the solar wind is a rich source of plasma. Both solar and galactic cosmic rays can also enter the magnetosphere. Their access is mainly at high latitudes and is energy dependent.

Particle Losses

Moons, rings and atmospheres are both sources and sinks of magnetospheric plasma. Particles that hit the surface of a solid body are generally absorbed and lost from the magnetosphere. Similarly, if a particle enters the collisionally thick part of an atmosphere, it gets 'captured' and will not return to the magnetosphere. Charged particles carry out a helical bounce motion along magnetic field lines, as shown in Figure 7.15. The particle is reflected at the **mirror point**, and if this point lies in or below the ionosphere/atmosphere, the numerous collisions with atmospheric particles 'trap' the magnetospheric particle in the atmosphere. These particles are thus lost from the magnetosphere.

also contain large numbers of electrons; on average, magnetospheric plasma is approximately neutral in charge.

The spatial distribution of plasma is determined by the sources and losses of the plasma, as well as the motions of the particles in the planet's magnetic field. In equilibrium situations, a charged particle's motion is completely defined by the magnetic field configuration, the planet's gravity field, centrifugal forces, large-scale electric fields and the particle's charge-to-mass ratio, q/m. We note that diffusion of particles in and through a magnetosphere is an important process: Without particle diffusion, the radiation belts would be empty, unless there is an *in situ* source. **Radial diffusion** displaces particles across field lines and hence transports particles from their place of origin to other regions in a magnetosphere. Radial diffusion is driven by large-scale electric fields and is the primary mechanism by which particles are distributed throughout a planet's magnetosphere.

Figure 7.18 *HST* image of Saturn's ultraviolet (UV) aurora. The planet and rings were imaged at visible wavelengths, and the bright circle of light at the south pole is an image of the aurora at UV wavelengths. A sequence of three images taken over several days shows the day-to-day changes in brightness in response to large changes in the solar wind dynamic pressure. (Courtesy John Clarke and Z. Levay, NASA/ESA)

Another loss process is induced by charge exchange of magnetospheric ions (§5.7.2). The ions roughly corotate with the planet. Their velocity at large planetocentric distances is therefore much higher than the Keplerian velocity of a neutral particle. If such an ion undergoes a charge exchange with a neutral, the newly formed ion picks up the rotation speed of the ambient plasma and stays trapped in the magnetosphere. The former ion, however, becomes a fast neutral and escapes the system if the corotation speed exceeds the escape velocity.

7.3.4 Radio Emissions

All four giant planets and Earth are strong radio sources at low frequencies (kilometric wavelengths). The strongest planetary radio emissions usually originate near the auroral regions and are intimately related to auroral processes. Jupiter is the strongest low-frequency radio source followed by Saturn, Earth, Uranus and Neptune.

Periodic modulations in these radio emissions are thought to be the best indicator of the rotation of the deep interior of a planet because the electrons, producing the emissions, are tied to the magnetic field, which is rooted in a planet's interior. The rotation period of the interior is important because this provides a rotating coordinate system against which the atmospheric winds can be measured. Based on early radio observations,

in 1965, the International Astronomical Union defined Jupiter's internal rotation period to be 9 h 55 m 29.71 s. This was essentially confirmed later, adding measurements made by the *Ulysses* and the *Galileo* spacecraft.

For Saturn, even though the radio emission is highly variable over time, a clear periodicity at 10 h 39 m 24 s ± 7 s was derived from the *Voyager* data, which was adopted as the planet's rotation period. Because the emission is tied to Saturn's magnetic field, which is axisymmetric, the cause of the modulation remains a mystery, although it may be indirect evidence of higher order moments in Saturn's magnetic field. Even more mysterious, however, is that the modulation periods measured later by the *Ulysses* and *Cassini* spacecraft vary by 1% or more (several minutes) on timescales of a few years or less. This implies that Saturn's radio rotation rate does not accurately reflect the rotation of its interior, in contrast to the radio periods of the other giant planets. No consensus has yet been reached over the cause of these variations in the radio rotation rate.

7.4 Generation of Magnetic Fields

7.4.1 Variability of Earth's Magnetic Field

Earth's magnetic field is not a static phenomenon; it is continuously changing, although the timescales involved are long compared with human lifetimes. The magnetic polarity in rock samples of the oceanic ridge and volcanic lava flows implies that the Earth's magnetic field reverses its polarity at irregular intervals of between 10^5 and a few million years. The reversals take place so quickly, on a geological timescale, that it is difficult to find rocks preserving a record of such changes. From the (sparse) records, it seems that the intensity of the field decreases by a factor of three or four during the first few thousand years of a field reversal, while the field maintains its direction. The field then swings around a few times by $\sim30°$ and finally moves to the opposite polarity. Afterwards, the intensity increases again to its normal value. The details of the changes in magnetic field configuration during these episodes, however, are not clear. Whether or not similar changes occur in magnetic fields around other planets is not known. Despite short-term variability, paleomagnetic evidence from the magnetization of old rocks implies that the Earth's magnetic field has existed for at least $\sim3.5 \times 10^9$ years and that its strength has usually been within a factor of two from its present value (except during field reversals).

7.4.2 Magnetic Dynamo Theory

Magnetic fields around planets cannot be caused by permanent magnetism in a planet's interior. The **Curie point** for iron is near 800 K; at higher temperatures, iron loses its magnetism. Because planetary interiors are much warmer, all ferromagnetic materials deep inside a planet have lost their permanent magnetism. In addition, a permanent magnet with a characteristic length scale ℓ and conductivity σ_o would gradually decay away over the characteristic **Ohmic dissipation time**:

$$t_d = \sigma_o \mu_o \ell^2, \tag{7.1}$$

where μ_o is the permeability of free space.

According to equation (7.1), the Earth's magnetic field would dissipate away in 10^4–10^5 years

(Problem 7-5). The pattern of remanent magnetism observed near mid-ocean rifts implies that Earth's magnetic field has reversed its direction frequently during geologically recent times; a permanent magnet would not behave in this manner. Thus, the variability of Earth's magnetic field discussed earlier also implies that there must be a mechanism that continuously produces magnetic field.

Electric currents in the interior of a planet form the only plausible source of planetary magnetism. The outer core of the Earth is liquid nickel–iron and thus is highly conductive. Similarly, the interiors of Jupiter and Saturn are fluid and conductive (metallic hydrogen), and Uranus and Neptune have large ionic mantles. Electric currents are likely to exist in all of these planets, and it is generally assumed that planetary magnetic fields are generated by a **magnetohydrodynamic dynamo**, a process that reinforces an already present magnetic field. Because the field is generated by electric currents in the Earth's interior, variability in the strength and pattern can be understood, in principle.

Key Concepts

- The Sun's energy is created via thermonuclear reactions in its core and transported outward via radiation and convection.
- The Sun is a highly dynamic environment, as revealed, e.g., via sunspots, prominences and CMEs.
- The solar wind consists of charged particles emanating from coronal holes on the Sun.
- A planet's magnetic field can be generated in the interior of the planet or induced by the interaction of the solar wind with the body's ionosphere.
- If a body generates its own magnetic field, it must contain a fluid that is both conductive and convective.

- The shape of a planet's magnetosphere is determined by the strength of its magnetic field, the solar wind flow past the field and the motion of charged particles within the magnetosphere.

- Precipitation of charged particles from 'outer space' may excite atmospheric particles; upon deexcitation, these particles produce auroral emissions.

Further Reading

Good reviews of the solar wind, magnetospheres and space weather can be found in:

McFadden, L., P.R. Weissman, and T.V. Johnson, Eds., 2007. *Encyclopedia of the Solar System*, 2nd Edition. Academic Press, San Diego. 982pp.

We recommend the chapters by:

Aschwanden, M.J., The Sun.

Gosling, J.T., The solar wind.

Luhmann, J.G. and S.C., Solomon, The Sun–Earth connection.

Kivelson, M.G. and F. Bagenal, Planetary magnetospheres.

Problems

7-1. Compute the ratio of energy per unit area radiated from a sunspot at temperature $T = 4500$ K to that of the typical solar surface with temperature $T = 5800$ K.

7-2. Calculate the mean free path between collisions in the solar wind, assuming quiescent solar wind properties (Table E.17). Compare this number with the size of the Earth's magnetosphere and the thickness of the bow shock.

7-3. What two pressures are equal at the outer boundary of a magnetized planet's magnetosphere? Does the size of a planet's magnetosphere depend on its distance from the Sun?

7-4. Consider a proton in the Earth's magnetic field about 1 $R_\oplus$ above Earth's surface. The particle is confined to the magnetic equator, and has a kinetic energy of 1 keV.

(a) Calculate the proton's velocity and compare this with the Keplerian orbital velocity of a neutral particle at the same geocentric distance (2 $R_\oplus$; assume a circular orbit).

(b) Describe what would happen if the ion undergoes charge exchange and becomes neutral.

7-5. If Earth's magnetic field would not be regenerated, calculate how long it would take for the field to dissipate.

7-6. State two ways in which life in each of the following categories would be affected were Earth's magnetic field to disappear:

(a) Technological civilization

(b) Nontechnological life, short-term (years or less)

(c) Nontechnological life, long-term (millennia or longer)

CHAPTER 8

Giant Planets

There is not perhaps another object in the heavens that presents us with such a variety of extraordinary phenomena as the planet Saturn: a magnificent globe, encompassed by a stupendous double ring: attended by seven satellites: ornamented with equatorial belts: compressed at the poles: turning upon its axis: mutually eclipsing its ring and satellites, and eclipsed by them . . .

Sir William Herschel, *Philosophical Transactions of the Royal Society of London*, 1805

The four giant planets in our Solar System, Jupiter, Saturn, Uranus and Neptune, dominate the mass of the planetary system and the angular momentum of the known Solar System. All four have been observed extensively from the ground and with spacecraft. Their masses and bulk densities, given in Table E.3, imply that they consist primarily of light elements. Note that Saturn is less dense than water. The two largest planets in our Solar System are generally referred to as **gas giants** even though the elements that make up these giants are no longer gases at the high pressures prevailing in the interiors of Jupiter and Saturn. Analogously, Uranus and Neptune are frequently referred to as **ice giants** even though the astrophysical ices, such as H_2O, CH_4, H_2S and NH_3, that are thought to make up the majority of these planets' mass are in fluid rather than solid form.

All four giant planets have deep atmospheres composed primarily of molecular hydrogen ($\sim$80%–90% by volume) and helium ($\sim$15%–10% by volume). Solid surfaces, if they exist at all, are deep within these planets at crushingly high pressures. A summary and short discussion of the composition of the giant planets' atmospheres are provided in §5.2, and the compositions of these atmospheres are given in Table E.13. The interior structure of each of the giant planets is deduced from measurements of its mass, radius, rotation period, oblateness, internal heat source and gravitational moments J_2, J_4 and J_6. Some background was provided in §§6.1.3 and 6.2, and numerical values are listed in Table E.15.

In this chapter, we give brief descriptions of the atmospheres, interior structure and magnetic field configurations for each of the giant planets. The moons of these planets are discussed separately in Chapter 10. Rings have been observed around each of the giant planets and are covered in Chapter 13.

We begin in §8.1 with a discussion of Jupiter, the largest, closest and most colorful giant planet in our Solar System. We consider Jupiter's atmosphere, impacts observed thereon, the planet's interior and

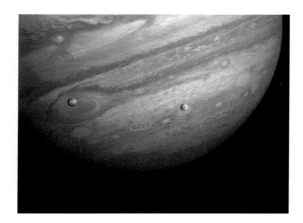

Figure 8.1 COLOR PLATE Image of Jupiter at visible wavelengths taken by the *Voyager* spacecraft in 1979. The white zones, brown belts and Great Red Spot (GRS) are prominent features, as well as the satellites Io (near the GRS) and Europa. Two large white ovals are visible at latitudes immediately south of the GRS. (*Voyager 1*/NASA, PIA00144)

its magnetic field and charged particle belts. Section 8.2 covers Saturn, which has been studied intensively by the *Cassini* orbiter since 2004. Far less is known about the smaller and more distant worlds Uranus and Neptune, which closely resemble one another in terms of bulk properties; these two planets are described and compared in §8.3.

8.1 Jupiter

8.1.1 Atmosphere

Zones, Belts and Ovals

Figure 8.1 shows a close-up image of Jupiter's southern hemisphere at visible wavelengths. The white **zones**, brown **belts** and the **Great Red Spot** (GRS) all feature prominently in this view. G.D. Cassini was the first to point out a large red feature on Jupiter, in 1665. There is some debate, however, whether or not the spot that Cassini saw persisted to become the present-day GRS because no red spots were seen during an extended period in the

(a) (b) (c)

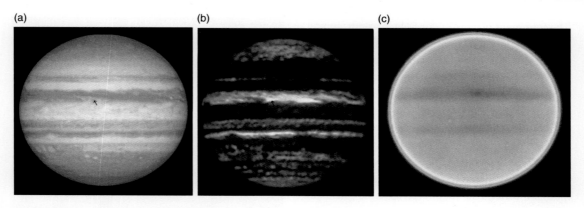

Figure 8.2 COLOR PLATE Images of Jupiter at (a) visible (HST), (b) thermal infrared (IRTF) and (c) radio wavelengths (Very Large Array, VLA). At visible wavelengths, the planet's zones and belts show up as white and brown regions, respectively. At infrared and radio wavelengths, we receive thermal radiation from the planet. The belts, as well as other brown regions, appear to be warmer than the zones. This suggests that the opacity in the belts is lower than in the zones, so deeper, warmer layers in the planet are probed. All images were taken within a few months of the time that the Galileo probe entered Jupiter's atmosphere (December 1995). The position of the probe entry is indicated by an *arrow* on the *HST* and IRTF images. The radio image is integrated over 6–7 hours, so longitudinal structure is smeared out. In panel (c), the 'colors' are inverted: the dark bands across the disk represent a high temperature, and the light 'ring' around the planet represents a low temperature (limb darkening). (*HST:* Courtesy R. Beebe and *HST*/NASA; IRTF: Courtesy G. Orton; VLA: de Pater et al. 2001)

nineteenth century. However, we know from observations throughout the 1980s and 1990s that the appearance of the GRS can change drastically in prominence, size and shape, so the system might be very long lived.

At visible wavelengths, such as the images displayed in Figures 8.1 and 8.2a, planets are seen in reflected sunlight. At longer wavelengths, a planet's thermal emission can be detected. Images of Jupiter's thermal emission at wavelengths of 5 μm and 2 cm are shown in Figure 8.2, panels b and c, respectively. A comparison of these images with the visible light image in panel a shows a strong correlation between optical color and temperature at 5 μm and radio wavelengths. The white zones are generally slightly colder than the brownish belts, suggesting more opaque clouds (and therefore emission being received from higher, colder levels of the atmosphere) in the zones than the belts. The opacity at radio wavelengths is mainly provided by ammonia gas, and the clouds

are (almost) transparent. Overall, the belts are radio-bright, implying a relatively low ammonia gas abundance that allows deep warm layers to be probed. All observations together indicate gas rising in the zones, with subsidence in the belts. As a parcel rises, condensable gases form clouds at the altitudes where their partial pressure exceeds their saturated vapor curve. Dry air subsides in the belts, and this forces a simple convection pattern. Because the air above the belts is relatively dry, the ammonia-ice cloud layer is either absent or rather thin and hence not seen at infrared wavelengths. The rising/subsiding air motions produce a latitudinal temperature gradient between belts and zones, which drives the zonal winds on the planet through geostrophic balance with the Coriolis force (§5.4.1). A schematic of the motions of the gas is shown in Figure 8.3, and Figure 8.4 shows the zonal winds on Jupiter.

Immediately south of the GRS are the three large white ovals, and in the South Equatorial Belt

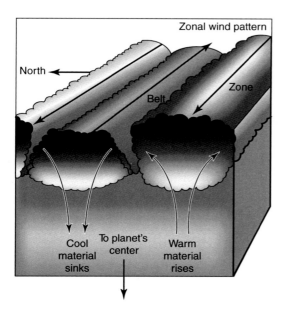

Figure 8.3 Sketch of the rising and sinking motions of the gas in Jupiter's zones and belts. The resulting zonal flow is also indicated.

one can discern a row of smaller white ovals. The three large white ovals trace back to the 1930s and merged during 1998–2000. Initially, the newly formed oval was no different from the white ovals in Figure 8.1, but in late 2005, this oval turned red, as red as the GRS. The two storm systems are shown side by side in Figure 8.5. The small black-and-white images on the side are taken at three different wavelengths (ultraviolet (UV), visible and near infrared). The contrast of the ovals varies drastically from image to image. Radiative transfer calculations show that this variation in contrast comes about because the storm systems rise ~8 km above the surrounding cloud deck.

The GRS and the ovals in Jupiter's atmosphere are giant storm systems. The circulation in almost all of these eddies is clockwise in the northern hemisphere and counterclockwise in the south, indicative of high-pressure systems. Although the white color of the white ovals can probably be attributed to ammonia ice, no one knows why the

GRS is red or why the belts are brownish. These colors are usually attributed to **chromophores**, coloring agents that contaminate the cloud particles. These chromophores may contain red phosphorus or sulfur compounds.

Composition

Spectroscopic observations of Jupiter have revealed the presence of numerous gaseous species, as shown in Figures 8.6 and 8.7; the main atmospheric species are summarized in Table E.13. The *Galileo* probe entered Jupiter's atmosphere in December 1995 and made *in situ* measurements down to pressures of 15–20 bar. At these deep levels, CH_4, NH_3, H_2S, Ar, Kr and Xe were detected at abundances several (3–6) times the protosolar value, and He and Ne were at below-solar values.

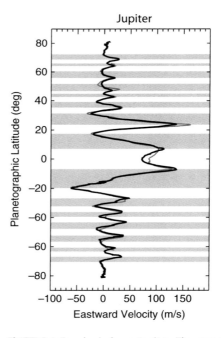

Figure 8.4 Zonal winds on Jupiter. The *narrow gray line* is the average zonal wind profile derived from *Voyager 2* images (1979), and the *heavy black line* is the profile derived from *Cassini* data (2001). The *gray regions* represent the belts; *white* regions the zones. (Adapted from Vasavada and Showman 2005)

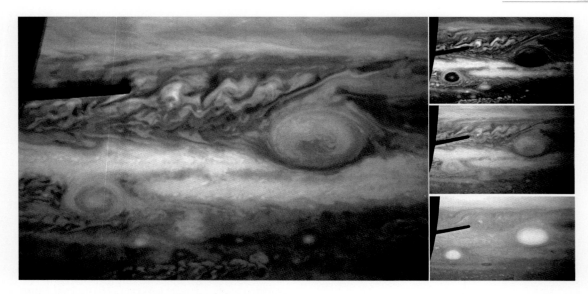

Figure 8.5 COLOR PLATE Deprojected image of the Great Red Spot and Red Oval on Jupiter obtained with the *Hubble Space Telescope* (*HST*) on 25 April 2006. This visual image is constructed from images in a red (F658N), green (F502N) and blue (F435W) filter. The small *black-and-white images* on the side are deprojected *HST* images at different wavelengths (330, 550 and 892 nm from top to bottom). These images can be used to derive the altitudes of the clouds. The black protrusion on the images is the occulting finger in the camera. (Adapted from de Pater et al. 2010)

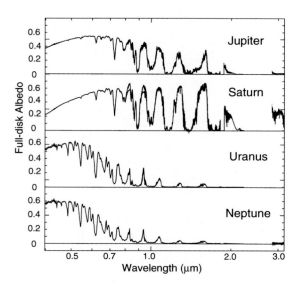

Figure 8.6 Full-disk albedo spectra of Jupiter, Saturn, Uranus and Neptune. All spectra show strong CH_4 absorption bands. (de Pater and Lissauer 2010)

Although images of Jupiter, such as those shown in Figures 8.1 and 8.5, clearly show dense clouds in the planet's atmosphere, the composition of the clouds is not easily determined by remote sensing techniques. Nor can we 'see' clouds directly below the upper cloud deck unless there is a clearing in the upper cloud deck or significant convective penetration of the clouds. Our knowledge of the cloud layers on Jupiter, as well as on the other giant planets, is therefore largely theoretical, based on the (usually) known composition of the atmospheres, together with laboratory measurements of the saturated vapor pressure curves of the condensable gases. Spectroscopic observations show Jupiter's upper cloud deck to be composed of ammonia ice, as predicted by theoretical calculations. Below the ammonia ice cloud, we expect clouds of ammonium hydrosulfide, water-ice and liquid water in a solution with minor species such as NH_3 and

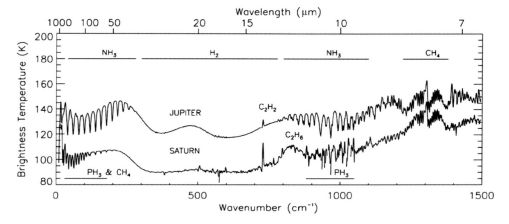

Figure 8.7 Thermal infrared spectra of Jupiter and Saturn taken with CIRS on *Cassini*. (Courtesy Conor Nixon and NASA/GSFC/UMCP)

H_2S. Sketches of the cloud layers on all four giant planets are shown in Figure 8.8.

Photochemistry

At altitudes where the pressure $P \lesssim 0.3$ bar, CH_4, NH_3, H_2S and PH_3 photodissociate and form complex compounds, some of which may be the red chromophores mentioned earlier. Indeed, the observed ammonia abundance in Jupiter and Saturn's atmospheres near and above the tropopause is well below saturation levels as a result of photochemistry.

Photolysis of methane gas, followed by a complex series of chemical reactions, results in the hydrocarbons acetylene (C_2H_2), ethylene (C_2H_4) and ethane (C_2H_6). Ethane is 1–2 orders of magnitude more abundant than the other hydrocarbons. These species have been detected on all four giant planets, as shown in Figure 8.7.

8.1.2 Impacts on Jupiter

In 1993, Carolyn and Gene Shoemaker, in collaboration with David Levy, discovered their ninth comet, Comet Shoemaker–Levy 9 (or SL9). Comet SL9 was very unusual: It consisted of more than 20 individual cometary fragments that orbited Jupiter rather than the Sun. An *HST* image is shown in Figure 8.9. Orbital calculations showed that the comet was captured into an unstable jovicentric orbit around 1930. In July 1992, the comet's perijove passage was ~ 1.3 $R_{2\!\!\!+}$ from Jupiter's center, well within Roche's limit of tidal stability (§13.1). As a consequence, SL9 was torn apart by Jupiter's strong tidal forces. Each fragment, less than 1 km across, developed into a small comet and continued its journey around the planet.

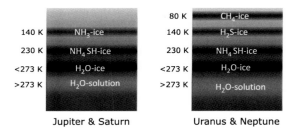

Figure 8.8 Illustration of the theoretically derived cloud layers on the giant planets.

Figure 8.9 *HST* image of Comet D/Shoemaker–Levy 9 (SL9) about 2 months before the comet crashed into Jupiter. Note that each fragment, A–W, is a small comet, with its own tail. (Courtesy Hal Weaver and T. Ed Smith, *HST*/NASA)

Two years later, approaching the subsequent perijove, each fragment crashed into the planet. The impacts occurred over a 6-day interval on the hemisphere facing away from Earth, a few degrees behind the morning limb. The *Galileo* spacecraft, at 1.6 AU from Jupiter on its way to the planet, had a direct view of the impact sites. These are the only large impacts that humankind has witnessed directly. Their interpretation, mostly through numerical models, taught us much about impacts in general.

By using both ground-based and *Galileo* data at different wavelengths, together with elaborate numerical simulations of the impacts, the sequence of events sketched as a cartoon in Figure 8.10 was identified. The first observed event was a short flash, lasting a few tens of seconds, interpreted as a meteor shower when the fragment's coma first impacted Jupiter's atmosphere, as illustrated in Figures 8.10a and 8.11b-2. The fragment disrupted and vaporized in the atmosphere. Most of its kinetic energy was deposited near its terminal atmospheric depth, probably at a few bars. *Galileo* measured a temperature of ~8000 K in the **fireball** immediately after the explosion, consistent with an explosion temperature of more than 10 000–20 000 K. As soon as the **plume** of hot material became visible from Earth, a second flash was observed (Fig. 8.10b, 8.11b-4). The plumes reached a height of ~3000 km above Jupiter's

cloud tops (Fig. 8.10c). When the plume collapsed, the splash-back of material onto the atmosphere was characterized by a dramatic brightening at infrared wavelengths, six minutes after the first flash (Figs. 8.10d and 8.11b-8), which lasted for about ten minutes.

Figure 8.12 shows that *HST* images revealed the post-impact sites to have a specific morphology. Each site showed a brown dot at the point of entry, surrounded by one or two dark rings propagating outwards at a velocity of 400–500 m s^{-1}. A large crescent of impact material is visible to the southwest. Although the impact phenomena appeared dark against Jupiter's clouds at visible wavelengths, they were bright at near-infrared wavelengths, indicative of material at high altitudes, well above the visible clouds. Much of this material has been attributed to submicron-sized dust, likely carbonaceous material or 'soot' generated by shock-heating atmospheric methane. The impact morphology can be accounted for by

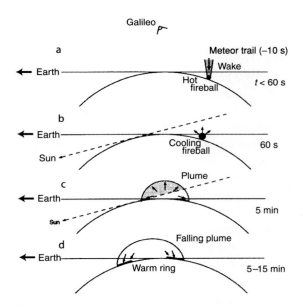

Figure 8.10 A cartoon illustrating the geometry during the SL9 impacts. For each panel, the time, t, after impact is indicated. (Adapted from Zahnle 1996)

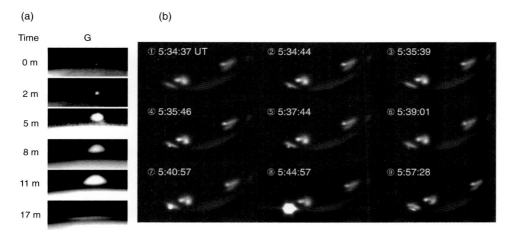

Figure 8.11 (a) A sequence of *HST* images showing the meteor shower (at time $t = 0$ min), thermal emission from the plume ($t = 2$ min), plume (in sunlight) rising ($t = 5$ min), spreading ($t = 8–11$ min) and collapsing ($t = 17$ min) after the impact of SL9 fragment G. The plume continued to slide after it had fully collapsed. (Adapted from Hammel et al. 1995; *HST*/NASA) (b) Impact of SL9 fragment R with Jupiter as observed at 2.3 μm by the Keck telescope. Each panel, labeled with UT time, is a frame from the movie. Only Jupiter's southern hemisphere is shown, with the former impact sites G (coming into view on the dawn limb), L and K indicated. The planet is dark at 2.3 μm because methane and hydrogen gases in the atmosphere absorb (incoming and reflected) sunlight. The impact sites are bright because some of the impact material is located at high altitudes above most of the absorbing gases. (Graham et al. 1995)

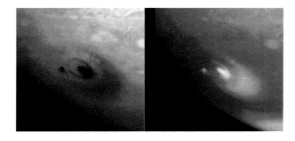

Figure 8.12 Two *HST* images of Jupiter on 18 July 1994, approximately 1.75 hours after the impact of SL9 fragment G, one of the largest fragments. The left image was taken through a green filter (555 nm) and the right image through a near-infrared methane filter (890 nm). Note that the impact sites appear bright at near-infrared wavelengths in methane absorption bands and dark at visible wavelengths. The G impact site has concentric rings around it, with a central spot 2500 km in diameter. The thick outermost ring's inner edge has a diameter of 12 000 km. The small spot to the left of the G impact site was created by the impact of the smaller sized fragment D, about 20 hours earlier. (Courtesy Heidi B. Hammel, *HST*/ NASA)

a combination of the ballistic trajectories of the plume material, the Coriolis force and a horizontal 'sliding' of material for about 20–30 minutes after impact.

Forensic evidence implies Jupiter was hit by an object of radius $\sim$100–200 meters in July 2009. The scar on Jupiter produced by this event had a morphology very similar to those left by medium-sized SL9 fragments, and it was spotted by amateur astronomers within a day after the impact (neither the impactor nor the impact event were detected). Analysis of spectroscopic observations revealed that the impacting body was most likely a small asteroid.

Occasionally, amateur astronomers spot flashes of small ($R \sim 10$ m) bolides entering Jupiter; an example is shown in Figure 8.13. Because subkilometer objects near and beyond Jupiter's orbit cannot be detected directly, analysis of such bolides

Figure 8.13 Color composite of images taken by amateur astronomer Anthony Wesley, revealing the flash (at 4 o'clock) from a bolide entering Jupiter's atmosphere in June 2010. (Adapted from Hueso et al. 2010)

may lead to more reliable estimates of the population of small bodies in the outer regions of our Solar System.

8.1.3 Interior Structure

Most models for Jupiter's interior structure predict a relatively small heavy element core of mass 5–10 $M_{\oplus}$. However, there are also models that predict no core and others that predict a much more massive core. Overall, Jupiter has a total of ~15–30 $M_{\oplus}$ high-Z ($Z > 2$) material that is distributed throughout its core and surrounding envelope, so that the planet, overall, is enriched in high-Z material by a factor of 3–5 compared with a solar composition planet. This value is consistent with the abundances observed in Jupiter's atmosphere (Table E.13). The core is probably composed of relatively large quantities of iron and rock, some of which was incorporated initially when the planet formed from solid body accretion, but more may have been added later via gravitational settling, surrounded by 'ices' of H_2O, NH_3, CH_4 and S-bearing

materials. A schematic of the interior structure of Jupiter is shown in Figure 8.14.

As discussed in §6.1.3, theories and laboratory experiments suggest that hydrogen in Jupiter's deep interior must be present in the form of liquid metallic hydrogen. Helium and neon are immiscible in metallic hydrogen. Both elements were measured at below-solar values by the *Galileo* probe, suggesting that they rained out of the metallic hydrogen mantle and are concentrated above Jupiter's heavy element–rich core.

Jupiter emits almost twice as much energy as it receives from the Sun. The excess energy has been attributed to an internal source of heat, which most likely originates from accretion and subsequent gravitational contraction. At present, the planet is still contracting by ~3 cm/year, and its interior cools by 1 K/Myr.

8.1.4 Magnetic Field

Jupiter possesses a strong magnetic field, the presence of which was first postulated in the late 1950s after the detection of nonthermal radio signals from the planet. As of 2018, seven spacecraft have flown past Jupiter (*Pioneer 10* and *11*, *Voyager 1* and *2*, *Ulysses*, *Cassini*, *New Horizons*), and two, *Galileo*,

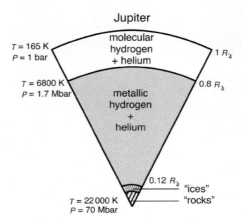

Figure 8.14 Model of the interior structure of Jupiter assuming a fully convective hydrogen–helium envelope (adiabatic models). (Adapted from Guillot et al. 1995)

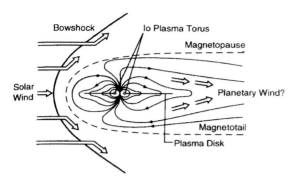

Figure 8.15 A sketch of Jupiter's magnetosphere. (Kivelson and Bagenal 1999)

which orbited the planet for almost 8 years, and *Juno* (Appendix G), is currently in orbit about Jupiter. These nine spacecraft studied the planet's magnetic field and the plasma environments of the Galilean satellites in detail through remote sensing techniques and *in situ*. Intensive monitoring programs from near-Earth and ground-based telescopes enrich the available database. According to magnetohydrodynamic dynamo theory (§7.4.2), one expects the generation of electromagnetic currents, and hence a magnetic field, in Jupiter's metallic hydrogen region.

The general form of Jupiter's magnetosphere resembles that of Earth, with a primarily dipole-like field tilted by $\sim$10° with respect to the rotation axis. But the dimensions of Jupiter's magnetosphere are over three orders of magnitude larger than that of Earth. If visible to the eye, Jupiter's magnetosphere would appear several times as large in the sky as the Moon. Jupiter's magnetotail extends to beyond Saturn's orbit; at times, Saturn is engulfed in Jupiter's magnetosphere. The magnetosphere is diagrammed in Figure 8.15, and quantities characterizing the field are listed in Table E.18.

Io's Neutral Clouds and Plasma Torus

Sublimation of SO_2 frost from Io's surface, volcanism and sputtering create a tenuous, yet collisionally thick, atmosphere around the satellite (§10.2.1). Ions corotating with Jupiter's

magnetosphere have typical velocities of 75 km s^{-1}, and hence readily overtake Io, which moves on a Keplerian orbit about Jupiter at a speed of 17 km s^{-1}. The ions interact with Io's atmosphere through collisions, 'throwing' some of the atmospheric molecules into Io's **neutral cloud**. The main constituents of the neutral cloud are oxygen and sulfur atoms. The cloud was first discovered, however, via emissions by the trace constituents sodium and potassium because Na and K are easily excited by resonant solar scattering.

When the atoms and molecules in Io's neutral cloud get ionized (e.g., via electron impact or charge exchange), they are accelerated to corotational velocities. Together these ions form Io's **plasma torus**. The torus surrounds Jupiter and is located near Io's orbit at 5.9 $R_{2\!+}$. Images of the neutral cloud and plasma torus are shown in Figure 8.16.

On a much larger scale, Jupiter is enveloped by a giant disk-shaped sodium cloud, extending out to a few hundred $R_{2\!+}$. This cloud is likely formed by fast neutrals flung out from the magnetosphere via charge exchange or elastic collisions between particles in the Io plasma torus and neutral cloud.

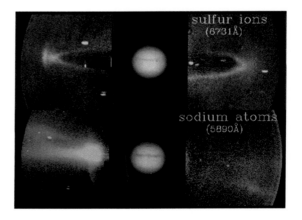

Figure 8.16 An image of Io's plasma torus imaged in S^{+} (*top*) and neutral sodium cloud (*bottom*). Jupiter is shown in the center of each image; the planet was imaged through a neutral density filter. (Courtesy Nick M. Schneider and John T. Trauger)

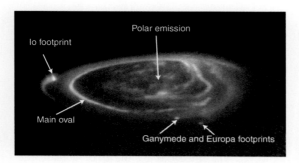

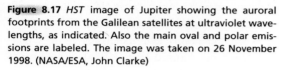

Figure 8.17 *HST* image of Jupiter showing the auroral footprints from the Galilean satellites at ultraviolet wavelengths, as indicated. Also the main oval and polar emissions are labeled. The image was taken on 26 November 1998. (NASA/ESA, John Clarke)

Radio Emissions

Jupiter's **decametric radio emission** is confined to frequencies below 40 MHz and has routinely been observed from the ground since its discovery in the early 1950s. The low-frequency radio spectrum has been observed by spacecraft from several kHz up to 40 MHz. The dynamic spectra in the frequency–time domain are extremely complex. The satellite Io appears to modulate some of these emissions, as charged particles travel along magnetic field lines connecting Io to Jupiter's ionosphere. Electrons that move down along the field lines emit radio emissions. Some of the particles on these magnetic flux tubes enter Jupiter's atmosphere, where locally they excite atmospheric atoms and molecules through collisions. De-excitation of the atmospheric molecules results in auroral emissions. These particular emissions trace the footpoints of Io's flux tube, as shown in Figures 8.17 and 7.17.

Because the dipole moment of Jupiter is tilted by ~10° from the rotational axis, most jovian radio emissions exhibit a strong rotational modulation. Given that Jupiter is a gas giant, this modulation is thought to be the best indicator of the rotation of the deep interior of the planet. The rotation period of the interior is important, for example, because this provides a rotating coordinate system against which the atmospheric winds can be measured.

At higher frequencies, we receive Jupiter's **synchrotron radiation**, emitted by relativistic electrons trapped in Jupiter's magnetic field. An image of Jupiter's emission at a wavelength of 20 cm (frequency of 1.4 GHz) is shown in Figure 8.18. The thermal emission from the planet itself is visible at the center of the image. This emission arises from its deep atmosphere (as in Figure 8.2c). The synchrotron emitting radiation belts on either side of the planet are indicated by the letters L (left) and R (right).

8.2 Saturn

8.2.1 Atmosphere

Saturn, similar to Jupiter, displays a large variety of atmospheric phenomena, although such features are usually not as prominent as those on Jupiter. Some spectacular photographs taken by the Cassini spacecraft are shown in Figures 8.19 and 8.20. Clear zonal bands can be discerned in Figure 8.19, together with a highly unusual storm in the planet's northern hemisphere. Although large storms on Saturn typically appear once every few decades,

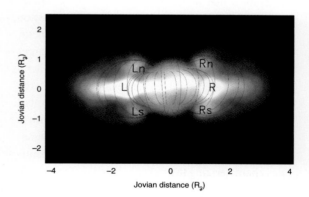

Figure 8.18 Image of Jupiter's radiation at a wavelength of 0.2 m, taken with the VLA. The spatial resolution is $0.3\ R_{\mathrm{2\!+}}$. Several magnetic field lines are superposed. The planet's main radiation belts are indicated by the letters L (left) and R (right); the subscripts n and s refer to the high-latitude radiation regions to the north and south of the main regions. (Adapted from de Pater et al. 1997)

Figure 8.19 COLOR PLATE Image of Saturn taken by the *Cassini* spacecraft on 25 February 2011 about 12 weeks after a powerful storm was first detected in Saturn's northern hemisphere. This storm is seen overtaking itself as it encircles the entire planet. This storm is the largest, most intense storm ever observed on Saturn. In addition, the banded structure on the planet is clearly visible. (NASA/JPL/Space Science Institute, PIA12826)

(a)

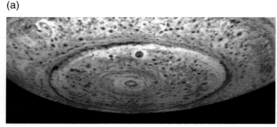

(b)

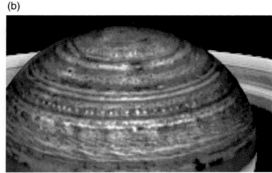

(c)

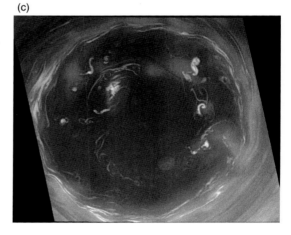

Figure 8.20 (a) Thermal emission from Saturn's south polar region, imaged by *Cassini*'s Visual and Infrared Mapping Spectrometer (VIMS) at 5 μm. (NASA/JPL/University of Arizona, PIA11214) (b) Thermal 5 μm emission from Saturn's northern hemisphere, imaged by *Cassini* VIMS, revealing a hexagon-shaped vortex near a latitude of 78°. Near 40°N, note the well-defined bright bands and the 'string of pearls' spaced every 3.5° in longitude over a 60 000-km stretch. (NASA/JPL, PIA01941) (c) High-resolution view of Saturn's south polar vortex constructed from *Cassini* images at 617 and 750 nm taken on 14 July 2008 at a resolution of 2 km/pixel. Vigorous convective storms are visible inside the brighter ring, or 'eyewall' of the hurricane, which spans ∼2500 km. (NASA/JPL/Space Science Institute, PIA11104)

a storm as dramatic as this one has never before been seen. The shadow cast by Saturn's rings has a strong seasonal effect, and the powerful storm in the northern hemisphere in 2011 may have been related to the change of seasons after the planet's August 2009 equinox.

Spectroscopic observations of Saturn have revealed the presence of numerous gaseous species, as shown in Figures 8.6 and 8.7; the main species are summarized in Table E.13. Similar to Jupiter, Saturn's composition is dominated by hydrogen and helium. Methane gas has been measured at an abundance 2–2.5 times that on Jupiter. The composition of the clouds on Saturn is expected to be quite similar to those on Jupiter, as depicted in Figure 8.8.

Thermal emission from Saturn is shown in Figure 8.20a and b. The patterns observed by *Cassini*'s Visual and Infrared Mapping Spectrometer in both hemispheres are striking. Each pole is

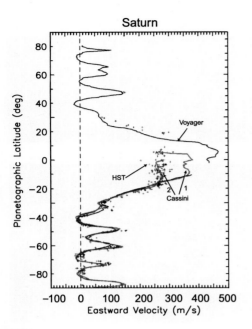

Saturn

Figure 8.21 Zonal winds on Saturn. One of the solid lines is the averaged profile as derived from *Voyager 1* and *2* images (1980–1981; wavelength: broadband green). The *crosses* mark the wind velocities derived from *HST* data between 1995 and 2002 (wavelength: broadband red). The two other *solid lines* are data as measured by *Cassini* (wavelengths: 1: methane band and 2: broadband red). The wind velocities are based on the *Voyager* period (*vertical dashed line*). (Adapted from Del Genio et al. 2009)

The velocity of the equatorial jet on Saturn is much stronger than that on Jupiter: $\sim$100 m s^{-1} for Jupiter versus several hundred meters per second for Saturn.

8.2.2 Interior Structure

Saturn is enriched, overall, in high-Z material by a factor of $\sim$10 compared with a solar composition mixture. Saturn's core is probably somewhat larger than Jupiter's.

Saturn emits almost twice as much energy as it receives from the Sun; in fact, relative to the solar input, the planet's excess energy exceeds that of Jupiter by about 10%. As for the larger planet, Saturn's excess energy has been attributed to an internal, most likely gravitational, source of heat. For Saturn, in addition to the overall contraction, a substantial part (roughly half) of the excess heat is attributed to the release of gravitational energy by helium sedimentation. Helium, being immiscible in and of higher density than metallic hydrogen, rains down to the bottom of the envelope. This slow 'drainage' of helium also explains the depletion of this element in Saturn's atmosphere compared with the solar value (Table E.13).

8.2.3 Magnetic Field

Saturn's magnetic field is weaker than that of Jupiter, probably because the liquid metallic hydrogen region on Saturn is not as extended as that on Jupiter. The planet's magnetosphere is intermediate in extent between those of Earth and Jupiter. The strength of the magnetic field at the equator is a little less than that found on the Earth's surface. However, Saturn is roughly 10 times larger than Earth, and it is roughly 10 times farther away from the Sun; both of these factors lead to a substantially larger magnetosphere around Saturn than around Earth. Most remarkable is the nearly perfect alignment between Saturn's magnetic and rotational axes. The center of its dipole field is slightly shifted towards the north, by 0.04 R_h.

characterized by a giant vortex. Around the north pole, the vortex is shaped like a hexagon. The south polar vortex resembles a hurricane (cyclone), with a clearly defined 'eyewall', which in analogy to Earth may consist of a ring of towering thunderstorms. The high-resolution *Cassini* image shown in Figure 8.20c reveals numerous small storm systems inside the eye, although the clear region within the vortex eye indicates that most of the air is descending.

Saturn's zonal wind flows, shown in Figure 8.21, are similar to those observed on Jupiter but with 3–4 jets in each hemisphere rather than 5–6. The equatorial jets on both of these planets move in the eastward direction (faster than the planet rotates).

(a)   (b)

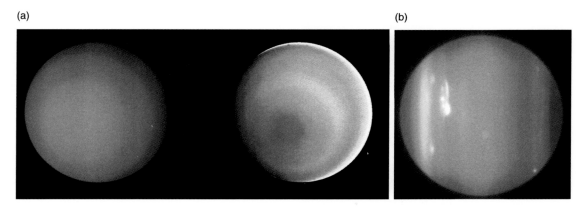

Figure 8.22 (a) COLOR PLATE Visible-light *Voyager 2* image of Uranus taken in 1986, with (on the right side) enhanced variations in albedo. The bull's-eye pattern is centered at Uranus's south pole. (NASA/JPL). (b) Image of Uranus at a wavelength of 1.6 μm taken in August 2007 (a few months before equinox) with the Keck telescope. Note the bright storm in the southern (*left side* of image) hemisphere. (Sromovsky et al. 2009)

Radio Emissions

Similar to Jupiter, Saturn emits a broad spectrum of nonthermal radio emissions. Saturn's kilometric radiation (SKR) resembles Jupiter's decametric radiation emissions. The intensity is strongly correlated with the solar wind ram pressure, suggesting a continuous transfer of the solar wind into Saturn's low-altitude polar cusps. A detailed comparison between high-resolution *HST* images of Saturn's aurora (Fig. 7.18) with SKR shows a strong correlation between the intensity of UV auroral spots and SKR.

Even though the SKR emission is highly variable over time, a clear periodicity at 10 h 39 m 24 s ± 7 s was derived from the *Voyager* data, which was adopted as the planet's rotation period. Because the emission is tied to Saturn's magnetic field, which is axisymmetric, the cause of the modulation remains a mystery but may be indirect evidence of higher order moments in Saturn's magnetic field. Even more mysterious, however, is that the SKR modulation period measured by *Ulysses* and *Cassini* varies by 1% or more (several minutes) on timescales of a few years or less. This implies that Saturn's radio rotation rate does not accurately

reflect the rotation of its interior, in contrast to the radio period of Jupiter.

Saturn electrostatic discharges (SEDs) are strong, unpolarized, impulsive events and are a counterpart of lightning flashes in Saturn's atmosphere. Some SED episodes have been linked directly to cloud systems observed in Saturn's atmosphere by the *Cassini* spacecraft, particularly events related to the storm system shown in Figure 8.19.

8.3 Uranus and Neptune

8.3.1 Atmospheres

Voyager images taken in 1986, such as the one in Figure 8.22a, showed Uranus as a rather bland planet without prominent cloud features or convective storms. Starting in the 1990s, however, prominent cloud features became visible in the northern hemisphere as it rotated into sunlight after spending decades in darkness. During the last few years before the 2007–2008 equinox, eye-catching clouds developed in the southern hemisphere. The

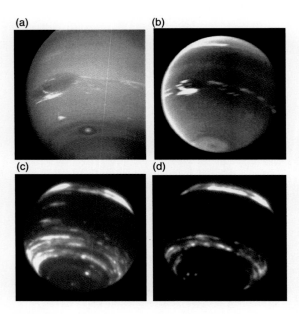

Figure 8.23 (a) *Voyager 2* image of Neptune at visible wavelengths, taken in 1989, revealing the Great Dark Spot (GDS), the Little Dark Spot and Scooter (the bright white cloud feature in between the dark spots). The GDS is located relatively deep in the atmosphere in contrast to the white hazes, which are at higher altitudes. (NASA/JPL) (b) A *Voyager 2* image in the methane band (890 nm) shows that the white clouds are at high altitudes. (NASA/JPL) (c) Keck adaptive optics image at a wavelength of 1.6 μm, in a methane absorption band, taken on 5 October 2003. (d) Keck adaptive optics image at a wavelength of 2.2 μm, where methane gas is strongly absorbing, taken on 3 October 2003. (Panels c, d from de Pater et al. 2005)

cloud complex featured in Figure 8.22b may have been around since 1994, but in 2005, it changed drastically in morphology. Around the same time, it started moving towards the equator, where it dissipated in 2009.

As shown in Figure 8.23, Neptune's appearance is strikingly different from that of Uranus. During the *Voyager* era (1989), Neptune was characterized by the Great Dark Spot (GDS), a smaller dark spot (DS2) in the south and a small bright cloud feature, referred to as Scooter, moving faster than either dark spot. When Neptune was imaged

several years later by *HST*, all three features had vanished. Since the late 1990s, the planet has been imaged regularly by *HST* and the 10-m Keck telescope equipped with adaptive optics. As shown in panels c and d of Figure 8.23, Neptune has changed dramatically since 1989.

On both Uranus and Neptune, the wind near the equator lags behind the planet's rotation. Profiles are shown in Figure 8.24. On Uranus, the wind speed is $\lesssim 100$ m s^{-1}, but on Neptune, speeds of ~ 350 m s^{-1} are reached. At higher latitudes ($\gtrsim 20°$ on Uranus, $\gtrsim 50°$ on Neptune), the winds blow in the direction of planetary rotation and reach velocities of ~ 200 m s^{-1}.

The hydrocarbons detected in the stratospheres of Uranus and Neptune have been attributed to photolysis of methane gas, just as for Jupiter and Saturn (§8.1). Prominent emission lines of CO and HCN have been discovered at (sub)millimeter wavelengths on Neptune; these data imply abundances ~ 1000 times higher than predicted from thermochemical equilibrium models. Disequilibrium species in a planet's stratosphere could have been brought up from below, via rapid vertical transport, or have fallen in from outside, just like the H_2O molecules that have been detected in the stratospheres of all four giant planets. Sources for such molecules are the rings and moons, as well as interplanetary dust and meteoritic material. It has been suggested that the extremely high abundances of CO and HCN in Neptune's stratosphere might result from a large cometary impact sometime in the recent (few hundred years ago) past.

8.3.2 Interiors

The positions of Uranus and Neptune on a mass-radius diagram (see Fig. 3.8) imply that these planets are quite different from Jupiter and Saturn. Although the total mass of $Z > 2$ elements is very similar for all four planets, the abundance

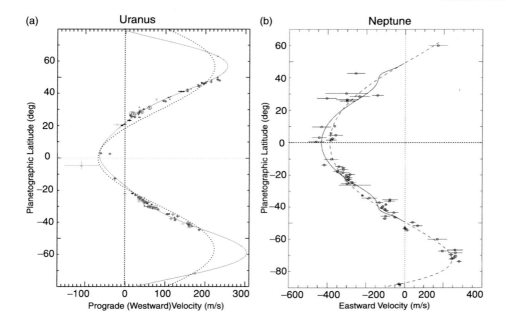

Figure 8.24 (a) Zonal winds on Uranus. A compilation of data derived from the *HST*, Keck and *Voyager* data from 1986 to 2008. The *dotted curve* is a symmetric fit to the *Voyager* zonal wind profile. The *solid line* is an average high-order polynomial fit to all of the data. (Adapted from Sromovsky et al. 2009) (b) Zonal winds on Neptune. The data are bin averaged wind velocities based on cloud features in *Voyager* images and the *Voyager* radio occultation results. The *dashed curve* is an empirical fit to the *Voyager* observations, and the *solid line* is an empirical fit to *HST* data from 1995–1998. (Adapted from Sromovsky et al. 2001)

of heavy elements relative to hydrogen is enhanced by factors of over 30 compared with solar elemental abundances for Uranus and Neptune. Uranus and Neptune each contain only a few $M_{\oplus}$ of hydrogen and helium, but these light gases occupy most of the volume in both planets. However, large uncertainties exist, with respect to both bulk composition and the extent of segregation and layering. Jupiter and Saturn must be primarily composed of H and He because no other elements can form cool planets with such low densities and large radii. In contrast, intermediate-density planets such as Uranus and Neptune can be composed primarily of astrophysical ices with a small amount of light gases or of a mixture of ices, rock and gas or even rock and gas with minimal amounts of ices.

A model for the interior structure of the ice giants is shown in Figure 8.25. Some models have a core subtending over 20% of the planet's radius, but others have no core at all – the mantle consists of 'ices' and comprises up to ~80% of the mass in these models. The outer 5%–15% of the planets' masses are likely composed primarily of a hydrogen- and helium-rich atmosphere.

Because Uranus and Neptune each have an internal magnetic field, their interiors must be both convective and electrically conductive. Although the pressure and temperature are high in the planets' interiors, the pressure is too low throughout most of the planet interiors for metallic hydrogen to form. The conductivity in Uranus's and Neptune's interiors must thus be attributed to other materials.

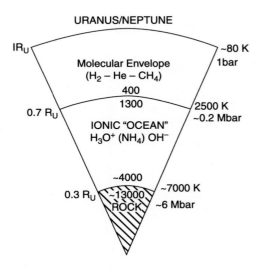

Figure 8.25 Schematic representation of the interiors of Uranus and Neptune. The approximate densities (in kg m⁻³) are indicated in each of the three layers. (After Stevenson 1982)

Because of the high temperature and pressure, the icy mantles are probably hot, dense, liquid, ionic 'oceans' of water or a brine, which are electrically conductive.

A comparison of the equilibrium and effective temperatures shows that whereas Neptune must have a large internal heat source, Uranus has only a very small one, if any. The observed upper limit for Uranus is consistent with the heat flow expected from radioactive decay alone. This difference in heat sources may explain the observed difference in atmospheric dynamics and hydrocarbon species in the planets' atmospheres.

8.3.3 Magnetic Fields

Our knowledge of the magnetospheres of Uranus and Neptune is limited to that obtained during brief encounters of these planets by the *Voyager 2* spacecraft in 1986 (Uranus) and 1989 (Neptune). The axis of Uranus's magnetic field makes an angle of ~60° with the planet's rotation axis. This is much larger than those of Mercury, Earth, Jupiter

and Saturn, each of which have axes aligned to within ~10°. The magnetic center is displaced by ~0.3 $R_{\circleddash}$ from the planet's center, resulting in the configuration sketched in Figure 8.26.

Neptune's magnetic axis makes an angle of 47° with its rotation axis, which has an obliquity of ~30°. The center of the magnetic dipole is displaced by 0.55 R_{Ψ} with respect to the planet's center, even more than in Uranus's case. The misalignment between Neptune's rotational and magnetic axes brings about a unique magnetic field configuration. While the field is rotating with the planet, two extreme situations are encountered, as sketched in Figure 8.27. At times the field is similar to that of the Earth, Jupiter and Saturn, where the magnetic field in the tail has two lobes of opposite polarity, separated by the plasma sheet. Half a rotation later the field topology is that of a 'pole-on' configuration, with the magnetic pole directed towards the Sun. The magnetic field topology in this case is very different, with a cylindrical plasma sheet, separating planetward field lines on the outside and field lines pointing away from the planet on the inside. The magnetic pole faces the Sun, and the solar wind flows directly into the planet's polar cusp.

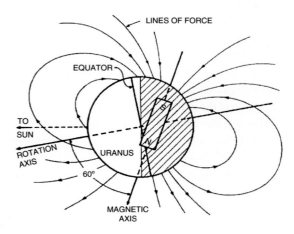

Figure 8.26 A sketch of Uranus's offset dipole magnetic field. (Ness et al. 1991)

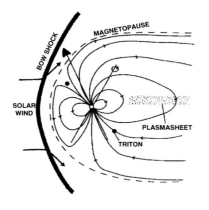 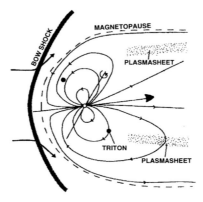

Figure 8.27 A sketch of two extreme situations of Neptune's magnetic field configuration at the epoch of *Voyager 2's* encounter with the planet in 1989. The *left* and *right* panels are separated by half a planetary rotation. (Bagenal 1992)

Radio emissions have been detected from both magnetospheres, and they are rather similar to the emissions detected from Saturn. The periodicity of radio emissions led to the estimated rotation periods of the planets of 17.24 ± 0.01 hr for Uranus and 16.11 ± 0.02 hr for Neptune, although as with Saturn the uncertainties in rotation rates are far larger than suggested by the formal error bars.

Key Concepts

- All four giant planets have deep atmospheres that are composed primarily of hydrogen and helium.
- Jupiter and Saturn have such low densities for their sizes that they must be composed primarily of the lightest gases, and they are referred to as gas giants.
- Uranus and Neptune are too dense to be primarily H_2/He and not dense enough to be primarily rocky compounds. They are probably made largely of astrophysical ices such as CH_4, H_2O and NH_3. Even though these substances would be in fluid form, the planets are called ice giants.

- The atmospheres of all four giant planets are characterized by banded structures with zonal winds.
- Zonal wind velocities are measured with respect to a planet's interior rotation rate. The rotation rates of the giant planets are taken to be equal to the rotation period derived from radio measurements, which signify the rotation period of the planet's magnetic field.
- The magnetic fields around the gas giants are produced in the metallic hydrogen region in their interiors. Magnetic fields of the ice giants are produced in their ionic ocean mantles.
- The Great Red Spot and white ovals on Jupiter are high-pressure storm systems.
- The collision of Comet Shoemaker–Levy 9 with Jupiter was the first impact observed on a body other than Earth and provided a wealth of information on impacts in general, comets and Jupiter.
- Huge storm systems develop occasionally (every 20–30 years) on Saturn and may depend on the season.
- Storms on Uranus have been observed in the years before and after equinox.

Further Reading

We recommend a number of chapters in the *Encyclopedia of the Solar System,* 2nd Edition, 2007. Eds. L. McFadden, P. Weissman, and T.V. Johnson, Academic Press, Inc.:

West, R.A., Atmospheres of the giant planets, pp. 383–402.

Marley, M.S., and J.J. Fortney, Interiors of the giant planets, pp. 403–418.

Kivelson, M.G., and F. Bagenal, Planetary magnetospheres, pp. 519–540.

Bhardwaj, A., and C.M. Lisse, X-rays in the Solar System, pp. 637–658.

Hendrix, A.R., R.M. Nelsomn, and D.L. Domingue, The Solar System at ultraviolet wavelengths, pp. 659–680.

de Pater, I., and W.S. Kurth, The Solar System at radio wavelengths, pp. 695–718.

Problems

8-1. **(a)** Calculate the kinetic energy and pressure involved when the Earth gets hit by a stony meteoroid ($\rho = 3400$ kg m^{-3}) that has a diameter of 10 km, and $v_\infty = 0$.
(b) Calculate the kinetic energy were the same meteoroid to hit Jupiter instead of Earth, assuming the body has zero velocity at a large distance from Jupiter.
(c) Calculate the kinetic energy involved when a fragment of Comet D/Shoemaker–Levy 9 ($\rho = 500$ kg m^{-3}, $R = 0.5$ km) hits Jupiter at the planet's escape velocity.
(d) Express the energies from (a)–(c) in magnitudes on the Richter scale and compare these with common earthquakes.

8-2. Approximate Jupiter and Saturn by pure hydrogen spheres, with an equation of state $P = K\rho^2$ and $K = 2.7 \times 10^5$ m^5 kg^{-1} s^{-2}.
(a) Determine the moment of inertia for the planets.
(b) Assume the planets each have a core of density $10\,000$ kg m^{-3} and the moment of inertia ratio is $I/(MR^2) = 0.254$ for Jupiter

and 0.210 for Saturn. Determine the mass of these cores.

8-3. **(a)** Use equation (3.21) to compute a lower bound on the central pressure for Jupiter, Saturn and an extrasolar planet of mass 5 $M_{\text{⚄}}$ and radius 1.1 $R_{\text{⚄}}$.
(b) Calculate the density of hydrogen at the pressures you computed in part (a) assuming a polytropic equation of state $P = K\rho^2$ and $K = 2.7 \times 10^5$ m^5 kg^{-1} s^{-2}, as in Problem 8-2. Compare your results with the mean densities of these planets and comment on how good an approximation the lower bound that you computed in part (a) is likely to be.
(c) Assume that the pressure at which hydrogen changes from its molecular to metallic phase is 2 Mbar. Estimate the highest mass planet for which hydrogen is in the molecular and not metallic phase all the way to its center. How might the magnetic field of such a planet be different from that of Jupiter?

8-4. (a) Why do we think that we can calculate fairly precisely the temperature versus altitude profiles well below the observable clouds for Jupiter and Saturn?

(b) Sketch one of these profiles and describe how it is derived.

(c) Why would the assumptions made in deriving this profile be questionable if applied to Uranus?

8-5. The base of the methane cloud in Uranus's atmosphere is at a pressure level of 1.25 bar and temperature of 80 K. The saturation vapor pressure curve is given by equation (5.1), with $C_L = 4.658 \times 10^4$ bar and $L_s = 9.71 \times 10^3$ J mole^{-1}. Derive the CH_4 volume mixing ratio in Uranus's atmosphere, assuming the composition of the atmosphere is 83% H_2 and 15% He. Compare your answer with the solar volume mixing ratio for carbon.

8-6. Calculate the rotation period for Neptune from the observed oblateness 0.017, equatorial radius of 24 766 km and $J_2 = 3.4 \times 10^{-3}$. Compare your answer with a typical rotation period of 18 hours for atmospheric phenomena and of 16.11 hours for Neptune's magnetic field. Comment on your results.

CHAPTER 9

Terrestrial Planets and the Moon

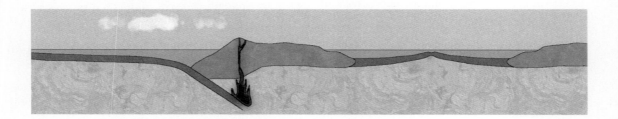

Those who are skeptical about carbon dioxide greenhouse warming might profitably note the massive greenhouse effect on Venus . . . The climatological history of our planetary neighbor, an otherwise Earthlike planet on which the surface became hot enough to melt tin or lead, is worth considering – especially by those who say that the increasing greenhouse effect on Earth will be self-correcting, that we don't really have to worry about it, or . . . that the greenhouse effect is a 'hoax'.

Astronomer Carl Sagan, 1934–1996, in *Pale Blue Dot*, 1994

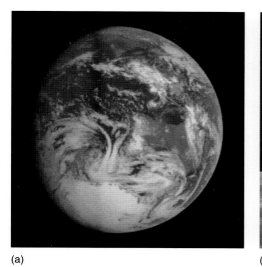

(a)   (b)

Figure 9.1 COLOR PLATE (a) Image of the Earth taken by the *Galileo* spacecraft on 11 December 1990, from a distance of ~2.5 × 10⁶ km. India is near the top of the picture, and Australia is to the right of center. The white, sunlit continent of Antarctica is below. (NASA/*Galileo*, PIA00122) (b) This photo of 'Earthrise' over the lunar horizon was taken by the Apollo 8 crew as their spacecraft orbited the Moon in December 1968. It showed humanity Earth as it appears from deep space for the first time. Note the contrast between the vibrant colors of Earth and the stark grayness of the Moon. (NASA)

Figure 9.1 shows Earth as viewed from space. From this perspective, Earth bears resemblance to other planetary bodies. Our planet's most distinctive features are the blue color of the oceans and the highly variable (but far from random) pattern of white clouds. The continents are visible as brown land masses, and white ice sheets dominate the polar regions.

In this chapter, we summarize key aspects of the three other terrestrial planets and our Moon. We often compare these bodies with our Earth, the planet we are most intimately familiar with, and the only planet where we know life thrives. We will not discuss the Earth here, because it has been described in previous chapters: The Earth's interior structure and surface morphology has been addressed in Chapter 6, its atmosphere in Chapter 5 and its magnetic field and magnetosphere in Chapter 7.

In terms of bulk properties, Venus is the planet most similar to Earth. It is slightly smaller and

less dense, so it is not an identical twin, but these two planets are far less different from one another than the chasm between the properties that they share and those of their brethren within our Solar System. However, Venus's massive CO_2 atmosphere produces a huge greenhouse heating of that planet's surface, making conditions completely uninhabitable for life as we know it.

Mars is far smaller than either Earth or Venus, and its density, even when discounting the effects of compression (Table E.14), differs by an amount that implies a more significant difference in bulk composition. Nonetheless, the conditions on the martian surface and in general habitability considerations imply that from a biological perspective, Mars is by far the most Earth-like of our home world's neighbors, and geological evidence reveals that it was even more earthlike in the distant past.

From a dynamical viewpoint, Earth's Moon is a planetary satellite and would best be discussed in Chapter 10. However, from a geological perspective, the Moon and the planet Mercury are far more analogous to one another than to any other bodies in our Solar System. The Moon's radius is 71% that of Mercury. Both are heavily cratered, nearly spherical worlds with rocky surfaces that undergo substantial temperature variations, and both have very tenuous atmospheres and small but nontrivial amounts of water and other volatiles concentrated in permanently shadowed regions of craters located near the poles. The uncompressed densities of the Moon and Mercury place them at opposite extremities of the five bodies that are being discussed in or compared with in this chapter, but interior models suggest that these differences are primarily accounted for by differences in the extents of the iron-rich cores, so similarities may extend well below the crusts of these two bodies. Furthermore, formation models suggest that the extreme densities of both the Moon and Mercury were produced by highly energetic collisions during the planet formation epoch (§§15.5.1 and 15.10.2).

In this chapter, we start our discussion with the Moon and follow with summaries of the Sun's three other rocky planets in order of heliocentric distance.

9.1 The Moon

9.1.1 Surface

One can discern two major types of geological units on the Moon with the naked eye: The bright **highlands** or **terrae**, which account for more than 80% of the Moon's surface area and have an albedo of 11%–18%, and the darker plains or **maria**, with an albedo of 7%–10%, that cover 16% of the lunar surface. These two types of terrain can easily be distinguished in Figure 9.2. The maria are concentrated on the hemisphere facing Earth.

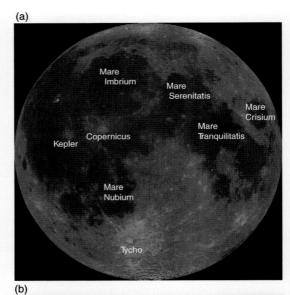

(a)

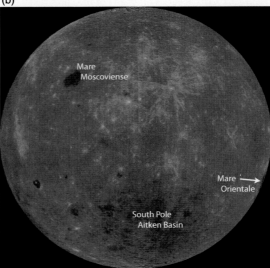

(b)

Figure 9.2 Images of the near (a) and far (b) sides of the Moon taken by the *Clementine* spacecraft. Some of the most prominent craters, maria and one basin are labeled. (Courtesy USGS)

The dominant landforms on the Moon are impact craters. The highlands appear saturated (§6.4.4) with craters, whose diameters range from micrometers (see Fig. 6.24) up to hundreds of kilometers in size (e.g., Orientale basin, see Fig. 6.27). Some of the large younger craters show the bright rays and patterns of secondary craters. The highlands clearly date back to the early bombardment era ~4.4 Gyr ago (Fig. 6.32a). In contrast, the maria are much less heavily cratered, and must therefore be younger.

Some maria cover parts of seemingly older impact basins, e.g., Mare Imbrium within the Imbrium basin. Radioisotope dating (§11.6) of rocks brought back by the various *Apollo* and *Luna* missions indicates that the maria are typically between 3.1 and 3.9 Gyr old. The lunar samples further indicate that the maria consist of fine-grained, sometimes glass-like basalt, rich in iron, magnesium and titanium. All of this together suggests that the basaltic lavas that formed the maria originated hundreds of kilometers below the surface and must have been brought up by volcanic activity, 3.1–3.9 Gyr ago. The lava lakes cooled and solidified rapidly, as evidenced by the glassiness and small grain size of the minerals in the rocks. A few small volcanic domes and cones can be discerned on the surface.

The most pronounced topographic structure on the Moon is the South Pole Aitken Basin, the oldest discernible impact feature. It is the largest (2500 km in diameter) and deepest (13 km from the rim crest to the crater floor) impact basin known in the entire Solar System. The highest point on the Moon, 8 km above the reference ellipsoid, is in the highlands on the far side of the Moon adjacent to the South Pole Aitken Basin.

Some locations within craters near the lunar poles are in permanent shadow, as shown in Figure 9.3. These regions may remain as cold as 40 K. Based on data by several spacecraft (especially the impact by the *Lunar Crater Observation and Sensing Satellite, LCROSS*), some of the shadowed craters contain tiny ice crystals mixed in with the soil at a concentration of a few percent by weight.

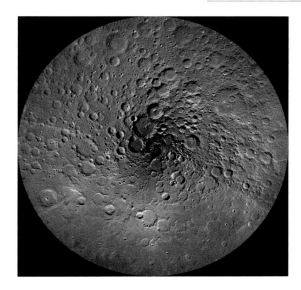

Figure 9.3 Mosaic of the Moon's north pole, constructed from images taken with the wide-angle camera on the *Lunar Crater Observation and Sensing Satellite, LCROSS*. A polar stereographic projection between latitudes of 60°N and 90°N is shown, including permanently shadowed craters. (PIA14024; NASA/GSFC/Arizona State University)

This ice must have been brought in by comets and asteroids well after the Moon had formed because the Moon, in general, is very dry.

The *Apollo* missions showed that the lunar crust in the highlands has been pulverized by numerous impacts. Continued micrometeoroid impacts on the broken rocks created a fine-grained layer of regolith, more than 15 m deep. The maria are also covered by regolith, but because the maria are younger, this layer is only 2–8 m deep. The lunar highlands generally lack rocks rich in siderophile elements, i.e., heavy minerals such as iron and titanium. The *Apollo* missions revealed an unusual chemical component in the lunar rocks, referred to as **KREEP**, named after its mineral components: potassium (K), rare-earth elements (REEs; elements with atomic numbers 57–70, see Appendix D) and phosphorus (P).

Although impact cratering is by far the most important geological process on the Moon, there is also clear evidence of volcanism and tectonics. Volcanism is most evident in the maria, but there are a few other features that are probably produced by volcanic flows. The sinuous (Hadley) rille shown in Figure 6.10 has been interpreted as a collapsed lava channel. The Japanese *Kaguya* spacecraft and NASA's *Lunar Reconnaissance Orbiter* (*LRO*) have imaged caverns on the Moon, which appear to be skylights to lava tubes. Tectonic features such as linear **rilles**, similar to graben faults, likely formed via crustal expansion or contraction.

9.1.2 Atmosphere

Mass and UV spectrometers on the *Apollo* spacecraft detected He and Ar on our Moon, with a surface density of a few billion atoms m^{-3} on the day side and an order of magnitude larger on the night side. Ground-based spectroscopy revealed Na and K at levels of a few tens of millions of atoms m^{-3}. The Moon's atmosphere is in part formed from sputtering by micrometeorites and energetic particles and by capturing particles from the solar wind.

9.1.3 Interior

Measurements of the Moon's moment of inertia show a value $I/(MR^2) = 0.3932 \pm 0.0002$, only slightly less than the value of 0.4 for a homogeneous sphere. A model that fits the measured $I/(MR^2)$ together with the lunar seismic data suggests the average density of the Moon to be 3344 ± 3 kg m^{-3}, with a lower density crust ($\rho = 2850$ kg m^{-3}) that is on average between 54 and 62 km thick and an iron core ($\rho = 8000$ kg m^{-3}) with a radius $R \lesssim 300\text{–}400$ km.

The interior structure of the Moon has largely been determined from seismic measurements of **moonquakes** at several *Apollo* landing sites. In contrast to the Earth, where quakes originate close to the surface, moonquakes originate both deep within the Moon (down to $\sim$1000 km) and close to its surface. Deep (700–1000 km) moonquakes are usually caused by tides raised by Earth, although such quakes can also occur closer to the surface. Meteoritic impacts are another common source of moonquakes. Attenuation of S waves below 1000 km implies the Moon has a liquid core with a radius of $\sim$350 km; there is some evidence for a solid inner core of the Moon. Hence, the Moon's core may resemble that of Earth in form, although it is much smaller in size. Seismic waves propagating within the Moon require a long time to damp, implying that the Moon has little water and lacks other volatiles.

Maps of the lunar topography, gravity and crustal thickness are shown in Figure 9.4. These maps show that the lunar highlands are gravitationally smooth, indicative of isostatic compensation such as seen over most regions on Earth (§6.2.2). The lunar basins, however, show a broad range of isostatic compensations. Some basins on the near side show clear gravity highs (e.g., the Imbrium basin), circular features called **mascons** (from mass concentrations). The gravity highs in some of the lunar basins suggest that the Moon's lithosphere here was very strong at and since the time the area was flooded by lava. The crust is generally thinner under the maria, with a minimum near 0 km at Mare Crisium; the thickest crust, 107 km, is on the far side beneath a topographic high.

Figure 9.5 shows a sketch of the interior structure of the Moon. The difference in crustal thickness produces an offset of the Moon's center of mass from its geometric center by 1.68 ± 0.05 km in the Earth–Moon direction, with the center of mass closer to Earth. This offset is probably caused by an asymmetry that developed during crystallization of the magma.

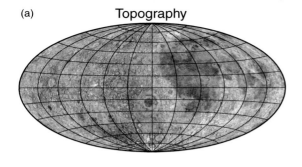

(a) Topography

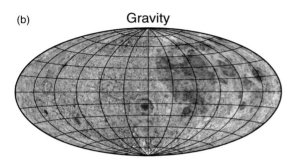

(b) Gravity

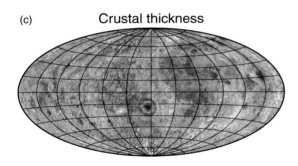

(c) Crustal thickness

Figure 9.4 COLOR PLATE A map of the lunar topography, free-air gravity anomaly and the crustal thickness. (a) Topography model based on *Clementine* data (Smith et al. 1997). (b) Gravity model that also includes data from *Lunar Prospector*. (Konopliv et al. 1998) (c) Crustal thickness model that takes the presence of mare basalt in the major impact basins into account. (Hood and Zuber 1999)

9.1.4 Magnetic Field

The Moon shows strong localized patches of surface magnetic fields with a few up to 250 nT. These patches appear to be correlated with the antipodal regions of large young impact basins, such as the

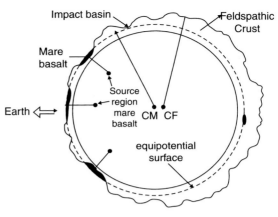

Figure 9.5 Sketch of the interior structure of the Moon in the equatorial plane showing the displacement toward the Earth of the center of mass (CM) (greatly exaggerated) relative to the center of the Moon's figure (CF). (Taylor 2007)

Crisium, Serenitatis and Imbrium basins. These data suggest the Moon to have had a global magnetic field, and hence a magnetic dynamo, 3.9–3.6 Gyr ago. Paleomagnetic data obtained from returned *Apollo* samples are also indicative of an early magnetic field with a surface field strength $\sim 10^4$–10^5 nT. Given recent estimates on the presence and size of the Moon's iron core, it appears as if a lunar magnetic dynamo almost 4 Gyr ago is quite plausible.

9.2 Mercury

9.2.1 Surface

Images of Mercury's surface, such as shown in Figure 9.6, resemble those of the Moon because craters are the dominant landform on both bodies. However, the two bodies differ significantly in detail. Craters on Mercury are shallower than like-sized craters on the Moon, and secondary craters and ejecta blankets are closer to the primary craters

Figure 9.6 Image of Mercury taken with the *MESSENGER* spacecraft. Most striking on this image are the large rays that appear to emanate from a relatively young crater in the far north. (NASA/JHU/CIW, PIA11245)

of a given size. Both of these differences are caused by the greater surface gravity on Mercury. Mercury's heavily cratered terrain is interspersed with smooth **intercrater plains**, resembling in some ways the maria on the Moon; the least cratered areas on Mercury, however, are bright, in contrast to the Moon's maria.

By far the largest feature observed on Mercury is the 1550-km-diameter Caloris basin. This huge ring basin, analogous to the large impact basins on the Moon, is shown in Figure 9.7. The basin is directly facing the Sun at perihelion of every other orbit, a consequence of Mercury's 2:3 spin-orbit resonance (§2.7.2).

Mercury shows unique scarps, or **rupes**, on its surface. These are linear features hundreds of kilometers long, which range in height from a few hundred meters up to several kilometers. An example is shown in Figure 9.8a. These lobate scarps are the most prominent tectonic features on the planet; they cut across all terrain at seemingly

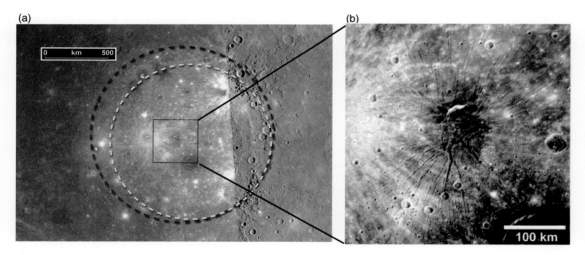

(a)

(b)

Figure 9.7 (a) A composite image of Caloris basin, the largest impact basin on Mercury. The eastern half of the basin was photographed in 1974 by *Mariner 10*; it was the only part in sunlight at that time. During its first flyby of Mercury, on 14 January 2008, the *MESSENGER* spacecraft imaged the western half of the basin. This composite image shows that Caloris basin is larger (outer *dashed circle*, diameter of 1550 km) than originally derived from the *Mariner 10* data (*inner dashed circle*, diameter of 1300 km). The *black box* at the center is enlarged in panel (b), which shows a detailed image of the center of Caloris basin. The radial troughs probably result from an extension of the floor materials that filled the Caloris basin after its formation. (NASA/*MESSENGER*, PIA10383; Murchie et al. 2008)

(a)

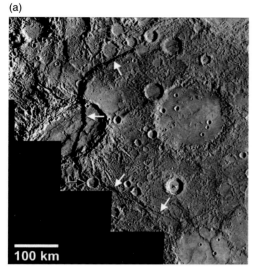

100 km

(b)

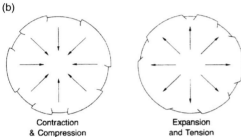

Contraction
& Compression

Expansion
and Tension

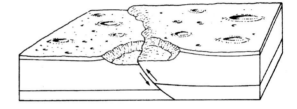

Figure 9.8 (a) Mercury shows prominent lobate scarps, such as Beagle Rupes, an ∼600-km-long scarp (*white arrows*) that offsets the floor and walls of an ∼220-km-diameter impact crater. Lava appears to have flooded this crater, which subsequently was deformed by wrinkle ridges before the scarp developed. In contrast, the *black arrow* points at an ∼30-km-diameter crater that must have formed afterwards. (NASA/*MESSENGER*, Solomon et al. 2008) (b) Schematic of the formation of scarps. (Hamblin and Christiansen 1990)

random orientations. These scarps may have been produced by planet-wide contraction caused by cooling, including partial core solidification, similar to the wrinkles on the skin of a dried-out apple. This model is sketched in Figure 9.8b. The sizes of the scarps suggest a global decrease of Mercury's radius by up to ∼4 km.

As on the Moon, the smooth plains formed after the heavy bombardment epoch, suggesting that they are not more than 3.8 Gyr old, but the cratered highlands stem from the early bombardment era.

Volcanism appears to be widespread on Mercury. Smooth plains seem to fill craters and embay crater rims. At places, volcanic plains are more than 1 km thick and appear to have occurred in multiple phases of emplacement in a flood-basalt style consistent with the measured surface compositions being more refractory than those of basalts. Features likely of volcanic origin, or at least indicative of subsurface magmatic events, include domelike features; pyroclastic vents; and pit craters, fractures and graben, sometimes radial, and at other

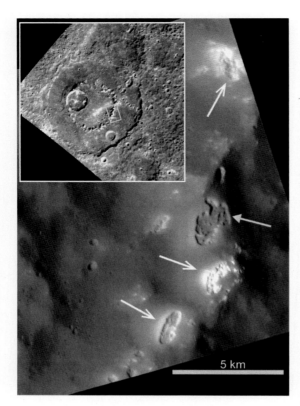

Figure 9.9 The *MESSENGER* spacecraft imaged an unexpected class of shallow, irregular depressions (*arrows*), referred to as hollows. Some hollows have bright interiors and halos (*white arrows*). This image shows hollows on the peak-ring mountains of an unnamed 170-km-diameter impact basin (*inset*). (Blewett et al. 2011)

places in concentric patterns; and irregular shallow depressions without a rim, referred to as **hollows**. Examples of these features are shown in Figure 9.9. Many hollows are associated with impact craters. These hollows likely involve a recent loss of volatiles through (some combination of) sublimation, outgassing, volcanic venting or space weathering.

Because Mercury's obliquity is ~0°, the poles receive very little sunlight. Crater floors in some regions are in permanent shadow, and the temperature stays well below 100 K. Radar echoes indicate an unusual large reflection from the poles, which

has been attributed to the presence of water-ice in permanently shaded craters near the poles. Confirmation of this hypothesis is shown in Figure 9.10, which displays a composite image with radar reflections obtained with the Arecibo Observatory superposed on a *MESSENGER* image of Mercury's south pole. It seems counterintuitive to have water-ice on a planet so close to the Sun. However, as on the Moon, comets and volatile-rich asteroids have continued to impact Mercury. Although the impactors' volatile material rapidly evaporates, some of it does not immediately escape Mercury's gravitational attraction. Water molecules might 'hop' over the surface until they hit the polar regions, where they freeze and remain stable for long periods of time. At temperatures <112 K, water-ice in a vacuum is stable to evaporation over billions of years.

In addition to the latitudinal variation in surface temperature, the average diurnal insolation varies significantly with longitude as a result of the combined effects of Mercury's large orbital eccentricity and the 3:2 spin-orbit resonance between Mercury's rotation and orbital periods with the Sun (§2.7.2). The longitudes that see the Sun at high noon when Mercury is at perihelion receive on average approximately 2.5 times as much sunlight as longitudes 90° away from these, which causes the peak (noon) surface temperature near Mercury's equator to vary between 700 and 570 K. (The nighttime surface temperature is approximately 100 K, independent of longitude.) This nonuniform heating pattern also produces longitudinal variations in the subsurface temperature, an effect that has been observed at radio wavelengths.

Mercury's surface composition appears to be very different from that of the other terrestrial planets and the Moon. Both ground-based microwave data and X-ray spectroscopy from the *MESSENGER* spacecraft reveal a surface that is depleted in Fe and Ti by roughly an order of magnitude. In contrast, the Mg/Si ratio is ~2–3 times as high,

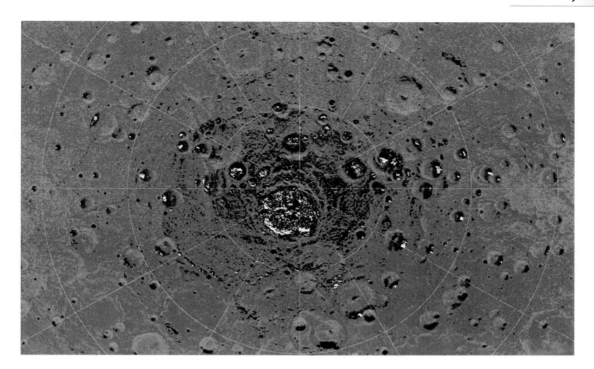

Figure 9.10 COLOR PLATE The highest-resolution radar image of Mercury's south polar region made from the Arecibo Observatory (Harmon et al. 2011) is shown in white on an image obtained with the *MESSENGER* spacecraft. The *MESSENGER* image is colorized by the fraction of time the surface is illuminated. Areas in permanent shadow are black. Radar-bright features in the Arecibo image all co-locate with craters that are in permanent shadow. This image is shown in a polar stereographic projection with every 5° of latitude and 30° of longitude indicated and with 0° longitude at the top. The large crater near Mercury's south pole, Chao Meng-Fu, has a diameter of 180 km. (PIA15533, NASA/Johns Hopkins University Applied Physics Laboratory/Carnegie Institution of Washington)

and the Al/Si and Ca/Si ratios are half as large as the ratios seen in terrestrial ocean and lunar mare basalts; the difference in these three ratios with lunar highland rocks is even larger (by another factor of 2) and hence rules out a lunarlike feldspar-rich crust. Perhaps even more intriguing is the $\gtrsim$ order of magnitude enhancement in the abundance of sulfur (S/Si), a rather volatile element, compared with that of the Earth, Moon and Mars.

Abundances of the radioactive elements potassium (K), thorium (Th) and uranium (U) were measured with the gamma-ray spectrometer on board the *MESSENGER* spacecraft. The K/Th ratio is similar to that measured on the other terrestrial planets (it is an order of magnitude lower on the Moon, indicative of the depletion of lunar volatiles compared to Earth). However, the absolute abundances of K, Th and U are more similar to that of martian meteorites, which are a factor of 3–4 lower than the average of the martian crust. The low values on Mercury's surface suggest differences in the magmatic and crustal evolution of Mars and Mercury.

The measured values of the radiogenic elements, which are the primary long-lived source of internal heat generation, indicate that heat production was about 4 times higher 4.5 Gyr ago. Calculations indicate that the internal heat production declined substantially since Mercury's formation, which is consistent with widespread volcanism shortly after the end of the late heavy bombardment 3.8 Gyr ago.

9.2.2 Atmosphere

Mercury has an extremely tenuous atmosphere, with a surface pressure $\lesssim 10^{-12}$ bar. This atmosphere is composed primarily of oxygen, sodium and helium, with number densities near Mercury's surface of a few $\times 10^9$ atoms m^{-3} for He and up to a few $\times 10^{10}$ atoms m^{-3} for O and Na. In addition, magnesium, potassium and hydrogen atoms have been detected. Many of the atoms heavier than He probably come from the planet's surface after having been kicked up into the atmosphere through sputtering (§5.7.2). H and He, major constituents of the solar wind, are probably captured therefrom (§5.7.2). Neutral species that are ejected from the surface with sufficient energy are accelerated by solar radiation pressure and form an extended tail in the antisolar direction, as observed.

9.2.3 Interior

Mercury has a high bulk density, $\rho = 5430$ kg m^{-3}, but most curious is its extremely high uncompressed density of 5300 kg m^{-3}, which implies that ~60% of the planet's mass consists of iron (Problem 9-4). Such a high iron abundance is twice the chondritic percentage. Models suggest the iron core extends out to 75% of the planet's radius. The outer 600 km is the mantle, composed primarily of rocky material.

The absence of a large rocky mantle has led to the theory that the planet was hit by one or more large objects towards the end of its formation. The impact(s) ejected and possibly vaporized much of Mercury's mantle, leaving an iron-rich planet. When the large iron core started to cool, it contracted, and the rigid outer crust collapsed and formed the unique scarps seen all over Mercury.

Variations in the planet's spin rate during Mercury's 88-day orbital period suggest that the planet's mantle is decoupled from the core, so Mercury must have a liquid outer core. The outer core likely consists of a mixture of Fe and FeS. A sulfur content of a few percent would lower the melting temperature of the outer core sufficiently to maintain it as a liquid yet permit the solidification of an inner core. In analogy with Earth, the solidification releases sufficient energy to keep the outer core convective, a necessary condition to sustain a magnetohydrodynamic dynamo system that could produce the magnetic field detected around Mercury. The presence of a liquid outer core has been confirmed by *MESSENGER* measurements of Mercury's gravity field. These measurements yield a moment of inertia ratio, $I/(MR^2) = 0.353$, which is in between that of Earth (0.331) and Mars (0.366). The gravity measurements together with the high density of the planet have led to an interior structure of Mercury as sketched in Figure 9.11.

9.2.4 Magnetic Field

Two close encounters by the *Mariner 10* spacecraft revealed Mercury to possess a small Earth-like magnetosphere. These observations, combined with data taken by the *MESSENGER* spacecraft, show that the planet's global magnetic field can be represented by a southward directed, spin-aligned (within 3° from the rotation axis), offset dipole that is centered on the spin axis, displaced almost 500 km towards the north. The field is very weak compared with that of Earth, with a surface field strength of 195 nT, a bit less than 1% of that on Earth. A sketch of Mercury's magnetosphere and plasma population based on the *MESSENGER* findings is shown in Figure 9.12.

Under quiescent solar wind conditions, Mercury's intrinsic field is strong enough to stand off the solar wind above its surface (1.2–1.8 R$_{\mercury}$). However, at times of increased solar wind pressure,

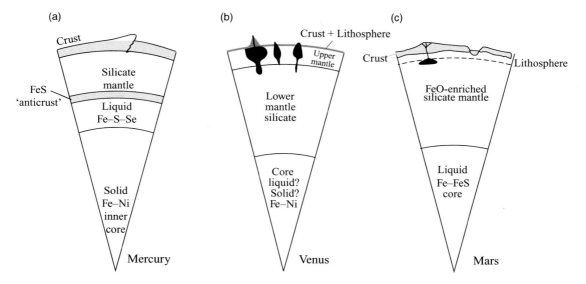

Figure 9.11 Sketch of the interior structure of (a) Mercury, (b) Venus and (c) Mars. These sketches are based on our 'best guess' models for the interior structure of the three planets, as discussed in the text.

the interplanetary particles may impinge directly onto Mercury's surface.

In contrast to the similarity in magnetosphere morphology between Earth and Mercury, both being shaped by the interaction with the solar wind, the differences in size make Mercury's field unique. Tiny Mercury occupies a much larger fractional volume of its magnetosphere than do Earth and the giant planets. This implies that the stable

trapping regions we see in other planetary magnetospheres, the radiation belts, cannot form, and indeed none have been detected.

The *MESSENGER* spacecraft detected sodium, oxygen and water ions throughout Mercury's magnetosphere. These ions are produced via surface sputtering effects. Enhancements in ion densities were seen in the nightside magnetosphere and cusp regions, as indicated on Figure 9.12.

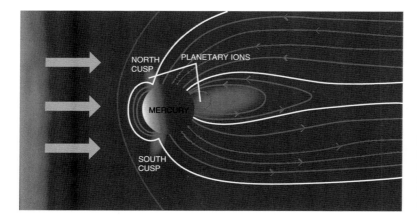

Figure 9.12 Sketch of Mercury's magnetic field and plasma population, as derived from *MESSENGER* data. Maxima in heavy ion fluxes are indicated. (Adapted from Zurbuchen et al. 2011)

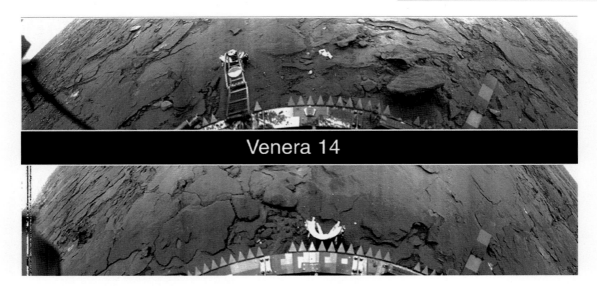

Figure 9.13 *Venera 14 Lander* images of the surface of Venus. The lander touched down at 13° S, 310° E on 5 March 1982. It transmitted from the surface for 60 minutes before succumbing to the planet's heat. Parts of the lander can be seen at the bottom of each picture (a mechanical arm in the upper picture, a lens cover on the lower one). The landscape appears distorted because *Venera 14*'s wide-angle camera scanned in a tilted sweeping arc. The horizon is seen in the upper left and right corners of both images. (Courtesy Carle Pieters and the Russian Academy of Sciences)

9.3 Venus

9.3.1 Surface

Although Venus is covered by a thick cloud deck, the surface can be probed at a few specific infrared wavelengths and at radio wavelengths longwards of a few centimeters. Global maps of Venus's surface have been constructed using radar. Several *Venera* spacecraft have landed on the surface and sent back photographs thereof, such as the ones displayed in Figure 9.13. The landscape in these images is orange colored because the dense, cloudy atmosphere scatters and absorbs the blue component of sunlight. The photographs reveal a dark surface and slightly eroded rocks; the rocks are not as smooth, however, as typical terrestrial rocks. The compositions of rocks at the landing sites are similar to various types of terrestrial basalts.

A global map of Venus is shown in Figure 9.14. A small fraction ($\sim$8%) of the surface is covered by four highlands, large continent-sized areas of volcanic origin that are well (3–5 km) above the average surface level. Roughly 20% of the surface consists of lowland plains and $\sim$70% of rolling uplands. Overall, most of the surface lies within a kilometer of the mean planetary radius. The difference in elevation between the highest and lowest features is $\sim$13 km, which is similar to the elevation contrast on Earth. However, histograms of surface area as a function of elevation, such as the ones in Figure 9.15, show very different distributions for the two planets: Earth has a bimodal distribution, reflecting the division between oceans and continents, and Venus has one peak centered near zero km altitude. Such a unimodal height distribution argues strongly against the presence of plate tectonics.

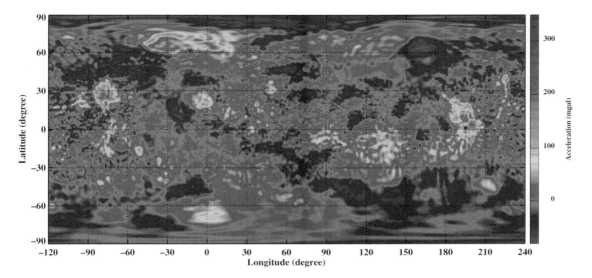

Figure 9.14 COLOR PLATE Mercator-projected view of Venus's surface as derived from *Magellan*'s radar altimeter data. Maxwell Montes, the planet's highest mountain region, rises 12 km above the mean elevation. (Courtesy Peter Ford, NASA/*Magellan*)

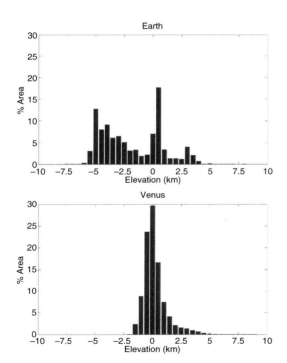

Figure 9.15 Histograms of the elevation (in 0.5-km bins) for Earth and Venus, normalized by area. Note the multiple peaks for Earth and the single peak for Venus. (Smrekar and Stofan 2007)

Volcanism

The *Magellan* spacecraft used radar to identify more than a thousand volcanic constructs on Venus. These include numerous small dome-like hills, which are probably shield volcanoes, many circular flattened domes with a small pit at their summit, as well as peculiar structures, including, e.g., **pancake-like domes** and **coronae**, examples of which are shown in Figure 9.16. Pancake-like domes typically have diameters of ~20–50 km, and their heights range from ~100 to ~1000 m. They must have formed by highly viscous lava flows, which makes them different from the basalt flows seen elsewhere on Venus. The complex fractures on top of these domes suggest that the outer layer cooled before magma activity below had completely stopped, resulting in a stretching (and hence fracturing) of the surface. The coronae, shown in panel b, are large circular or oval structures with concentric multiple ridges and diameters ranging from ~100 km to more than 1000 km. They are located primarily within the volcanic plains and are thought to form

(a)   (b)

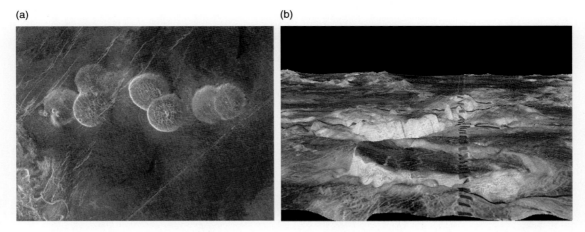

Figure 9.16 A variety of volcanic features on Venus. (a) Radar image showing seven pancake-shaped domes on Venus's surface, each averaging ~25 km in diameter with maximum heights of 750 m. (NASA/*Magellan*, PIA00215) (b) A perspective view of Venus with Atete Corona in the foreground, an approximately 600 × 450 km oval volcano-tectonic feature. (NASA/*Magellan*/JPL/USGS, PIA00096)

over hot upwellings of magma within the venusian mantle.

The low-lying plains on Venus are covered by volcanic deposits, likely caused by massive floods of basalt-like outpourings, volumetrically comparable to, e.g., the Deccan Traps in India. (The Deccan Traps are composed of a $\gtrsim$2-km-thick plateau that covers an area of about half a million km^2.)

Tectonics

Magellan radar images show that Venus's surface has undergone numerous episodes of volcanism and tectonics. The distribution of impact craters suggests that volcanic resurfacing is locally very efficient, but also quite episodic. Prominent tectonic features include long linear mountain ridges and strain patterns, which can be parallel to one another, or they may crosscut each other. Some of these tectonic deformations extend over hundreds of kilometers. They reflect the crustal response to dynamical processes in the mantle.

Erosion

The formation of sediments and erosion is less important on Venus than on Earth and Mars because of Venus's extremely dense and hot atmosphere. The atmosphere prevents small meteoroids from hitting the ground, a process that is the main source of erosion and regolith formation on airless bodies. The lack of water and thermal cycling on Venus limits weathering processes, and because there is little wind near the surface (wind velocities are $\lesssim 1$ m s^{-1}), there is little erosion by wind. However, even though the wind velocities near the surface of Venus are extremely low, because Venus's atmosphere is 90 times as dense as the Earth's atmosphere, even slow winds may be able to transport a considerable amount of sand.

Impact Craters

All impact craters seen on Venus's surface have diameters exceeding 3 km. The projectiles that would have created smaller craters must have been broken up or substantially slowed down in the atmosphere. The ejecta blankets around impact

Figure 9.17 Three impact craters, with diameters that range from 37 to 50 km, are visible in this radar image of Venus. Numerous domes, 1–12 km in extent and probably caused by volcanic activity, are seen in the lower right corner of the mosaic, which shows a region of Venus's fractured plains. (NASA/*Magellan*, PIA00214)

Venus, whose atmosphere is so thick that ejecta cannot travel very far. The ejecta patterns are often asymmetric as a result of oblique impacts, where the missing sector is in the uprange direction.

Craters seem to be randomly distributed over the planet's surface, and their number and size distribution suggest that Venus's surface is younger than that of Mars but older than most of Earth's (i.e., the rocks were emplaced later). Typical age estimates range from a few hundred million up to one billion years.

9.3.2 Atmosphere

At visible wavelengths, Venus appears as a bright yellow featureless disk. Distinct markings with an overall V-shaped morphology are discerned at UV wavelengths, as shown in Figure 9.18.

The composition of Venus's atmosphere is dominated by CO_2 gas, 96%–97%; nitrogen gas contributes approximately 3% by volume. We further find traces of, e.g., Ar, CO, H_2O and SO_2. At 0-km altitude (i.e., Venus's geoid), Venus's surface pressure is 92 bar, and its surface temperature is 737 K. These high numbers result from the strong greenhouse effect in Venus's atmosphere (§4.6). Venus's main cloud layers span the altitude range between 45 and 70 km, with additional hazes up to 90 km and down to 30 km. The cloud particles consist of

craters typically extend out to $\sim$2.5 crater radii. In radar images, such as the one shown in Figure 9.17, ejecta blankets look like bright patterns of flower petals. Such patterns are seen only on

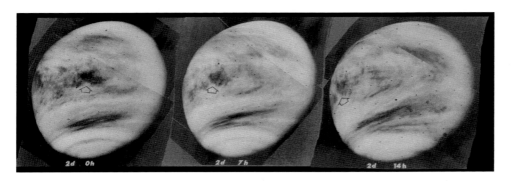

Figure 9.18 Three images of Venus taken in ultraviolet light 7 hours apart. A right-to-left motion of the cloud features can be seen (indicated by the *tiny arrow*). (*Mariner 10*/NASA: P14422)

sulfuric acid, H_2SO_4, with some contaminants. The droplets are formed at high altitudes (80–90 km), where solar UV light photodissociates SO_2, and chemical reactions (with, e.g., H_2O) lead to production of H_2SO_4. When the droplets fall, they grow. However, because the temperature increases at lower altitudes, the droplets tend to evaporate below 45 km. Below 30-km altitude, the temperature is too high for the droplets to exist, and the atmosphere is clear.

Wind patterns on Venus are characterized by a 'classical' Hadley cell circulation, i.e., one cell per hemisphere. Air rises at the equator and subsides at latitudes near $\sim 60°$. Strong westward (in the same direction as the planet's rotation) zonal winds are observed in Venus's cloud deck at altitudes of ~ 60 km. The winds circle the planet in 3–5 days (~ 100 m s^{-1}) and hence are **superrotating**, i.e., the upper atmosphere moves around the polar axis far faster than the bulk of the planet does. These winds decrease linearly in strength with decreasing altitude, and are only ~ 1 m s^{-1} at the surface. These superrotating zonal winds are in cyclostrophic balance (§5.4.2).

At higher altitudes, in the thermosphere, strong day-to-night winds prevail as a consequence of the large temperature gradient between Venus's thermosphere and cryosphere (Fig. 5.1b).

9.3.3 Interior

Venus is very similar to Earth in size and mean density, suggesting similar interior structures for the two planets. Venus's gravity field is highly correlated with its topography, suggestive of a lithosphere strong enough to support the topography. Although some highlands may, in part, be isostatically compensated, most are not and appear to be compensated by large mantle plumes.

In contrast to the Earth, Venus does not possess an intrinsic magnetic field, which implies the absence of a convective metallic region in its mantle and/or core. Another large difference between

the two planets is the absence of planet-wide tectonic plate activity on Venus (Fig. 9.15). Water on Earth plays a key role in driving plate tectonics. Water weakens a rock's rigidity, or strength, and lowers its melting temperature. The water in Earth's lithosphere is therefore thought to be essential to enable plate tectonics. Venus's atmosphere and rocks are extremely dry, which most likely is a consequence of the planet's very high surface temperature having driven off all the water. The dry rocks retain a high rigidity even at relatively high temperatures so that Venus's lithosphere does not break up.

On Earth, plate tectonics is a major avenue of heat loss. In the absence of plate motions, hot spot and volcanic activity may be more important on Venus, as indeed suggested by the numerous volcanic features on this planet. In addition to the volcanoes, some models suggest that the lithosphere may completely subduct, producing a **catastrophic resurfacing** event every few hundred million years.

9.4 Mars

The photograph of Mars in Figure 9.19a shows polar caps and the planet's typical red color. This unusual color is attributed to relatively large quantities of rust, Fe_2O_3, something not seen on any other planet. The detailed drawing of the surface of Mars from around 1900, shown in Figure 9.19b, reveals long straight linear features that were referred to as 'channels' or 'canals'. The martian polar caps were also discovered around 1900 and were observed to vary seasonally, reminiscent of the polar ice caps on our own planet. The ice caps, 'canals', and other seasonal changes convinced some scientists that life existed on Mars. Although we now know that the surface of the red planet is not currently inhabited, the questions of whether there is life underground or there has been life in the past are central to Mars exploration programs today.

(a)   (b)

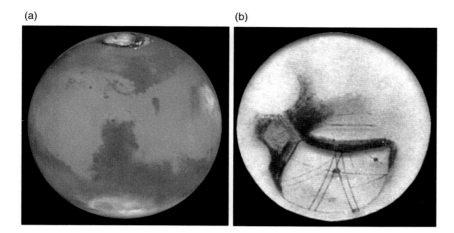

Figure 9.19 (a) *Hubble Space Telescope* (*HST*) image of Mars. The *dark feature* at the center is Syrtis Major. To the south of Syrtis, near the limb of the planet, a large circular feature, Hellas basin, is visible. On this picture, it is partly filled with surface frost and water-ice clouds. Towards the planet's right limb, late afternoon clouds have formed around the volcano Elysium. Note also the ring of dunes around Mars's north pole. (Steve Lee, Jim Bell, Mike Wolff and *HST*/NASA) (b) One of Percival Lowell's sketches of Mars, showing details of his 'canals'. Lowell thought that most canals were in pairs of two, as shown here.

9.4.1 Global Appearance

Mars has been mapped in detail by numerous orbiting spacecraft, and these maps, as shown in Figure 9.20, reveal a striking asymmetry between the northern and southern hemispheres. One half of the planet, mostly in the southern hemisphere, is heavily cratered and elevated 1–4 km above the 'nominal' surface level, the martian geoid, at a mean equatorial radius of 3396.0 ± 0.3 km. The other hemisphere is relatively smooth and lies at or below this level. The geologic division between these two hemispheres is referred to as the **crustal dichotomy**, characterized by complex geology and prominent scarps.

In addition to the global asymmetry, Mars's appearance is characterized by four massive shield volcanoes in the Tharsis region, including Olympus Mons and a giant canyon system, Valles Marineris. The Tharsis region is about 4000 km wide and rises 10 km above Mars's mean surface level. Three of the shield volcanoes rise another 15 km higher, and Olympus Mons, the largest volcano in our Solar System with a base of $\sim$600 km, rises 18 km above the surrounding high plains to a total height of 27 km above the martian geoid. Valles Marineris is a tectonically formed canyon system extending eastwards of Tharsis for 4000 km. The canyons of the Valles Marineris system are 2–7 km deep and more than 600 km wide at its broadest section.

Volcanism and tectonics have clearly been important in the planet's history. Although Mars is small, the scale of these martian features dwarfs similar structures on Earth. Figure 6.18 compares Olympus Mons with Mauna Loa, the largest volcano on Earth. Mars's relatively low surface gravity and cold, thick lithosphere enable the existence of such high mountains, which on Earth and Venus would have collapsed because of the larger surface gravity (and tectonic plate movements on Earth).

9.4.2 Interior

Figure 9.21 shows global maps of the topography, gravity and crustal thickness of Mars. The martian

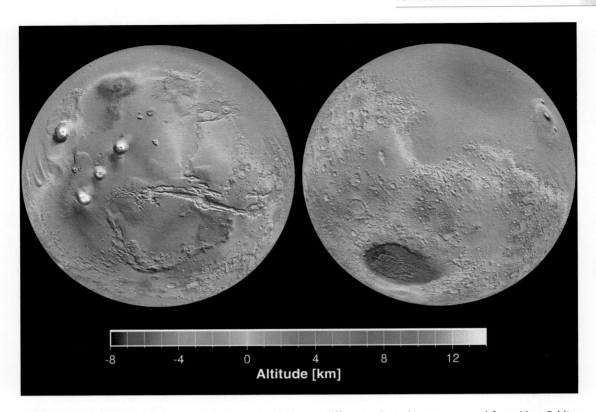

Figure 9.20 COLOR PLATE Global topographic views of Mars at different orientations, constructed from *Mars Orbiter Laser Altimeter* (*MOLA*) data. The *image on the right* features most strikingly the crustal dichotomy (division between the northern plains and heavily cratered southern hemisphere) and the Hellas impact basin (*dark blue*). The *left-hand image* shows the Tharsis topographic rise and Valles Marineris. (D. Smith, NASA/MGS-MOLA, PIA02820)

geoid appears to be highly correlated with topography, indicating that topographic features are not isostatically compensated. The Tharsis region, for example, shows a clear gravity high of $\gtrsim 1000$ mGal[1]. Localized mass concentrations appear centered at impact basins, and Valles Marineris shows a pronounced mass deficit. Such features are suggestive of a thick rigid lithosphere. Assuming that the gravity anomalies are directly correlated with variations in crustal thickness, Figure 9.21c shows that to first order such variations correlate well with Mars's topography. For example, the crust is thinnest in the northern hemisphere and at Hellas and thickest under the Tharsis region.

[1] A Gal is a unit of acceleration used in gravimetry. 1000 mGal $\equiv$ 1 Gal $\equiv$ 1 cm s^{-2}.

A model taking into account Mars's gravity field, moment of inertia and tidal deformation by the Sun suggests that Mars most likely has a (at least partially) liquid core with a radius between 1520 and 1840 km. Because there is no internal magnetic field, the core is probably completely fluid, as expected from laboratory measurements if it is composed of an Fe–FeS mixture.

9.4.3 Atmosphere

As on Venus, the primary atmospheric constituent on Mars is CO_2 ($\sim 95\%$) with $\sim 3\%$ N_2 gas. We further find traces of Ar, CO and H_2O. The average surface pressure on Mars is 6 mbar, and the mean temperature is ~ 215 K, which is raised by just a few degrees above the equilibrium value (§4.6).

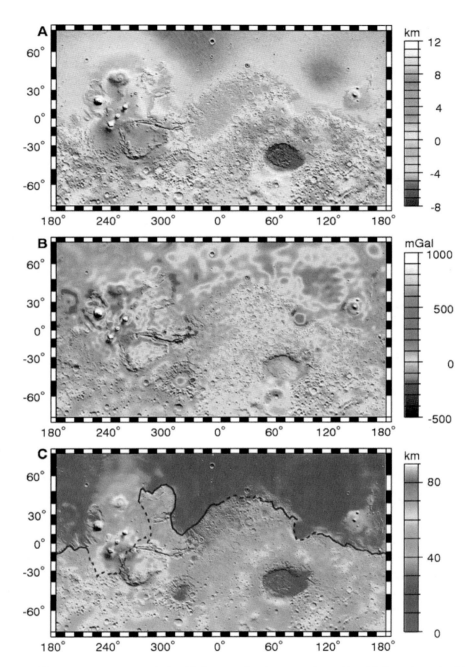

Figure 9.21 COLOR PLATE Relationship between local topography on Mars (a), free-air gravity (b) and crustal thickness (c). Note the good correlation between topography and gravity: topographic highs, such as the Tharsis region, show large positive gravity anomalies and a thick crust. Also the Hellas basin, a topographic low, shows a large negative gravity anomaly, and the crust is thin. (Zuber et al. 2000)

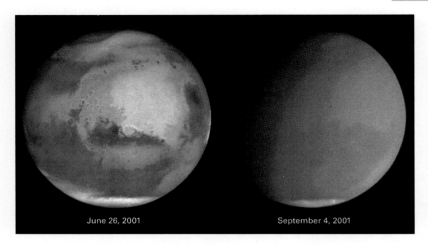

June 26, 2001   September 4, 2001

Figure 9.22 *HST* images of Mars at the onset of spring in the Southern Hemisphere. The *left image* shows the seeds of a storm brewing in the giant Hellas Basin and another storm near the north pole. Over the months following, surface features got obscured, and 75 days later (*right image*), most of the surface could no longer be distinguished. (*HST*/NASA, J. Bell, M. Wolff, STScI/AURA)

The low surface pressure on Mars, $\sim$6 mbar, is below the saturated vapor pressure curve for liquid water, so water is present either as a vapor or as ice. The small amount of water vapor in the martian air forms water-ice clouds at altitudes of $\sim$10 km above the equatorial regions. Such clouds are often seen near the martian volcanoes, as shown in Figure 9.19a. At higher altitudes, typically near $\sim$50 km, the temperature is low enough ($\sim$150 K) for CO_2–ice clouds to form.

Air rises over Mars's summer hemisphere and subsides above the winter hemisphere. Because the warmest latitude does not usually coincide with the equator, the Hadley cells are not confined to the northern and southern hemispheres but are displaced. Local topography with extreme altitude variations, from the deep Hellas basin up to the top of the Olympus and Tharsis ridge, leads to the formation of stationary eddies. Baroclinic eddies form over the winter hemisphere. Mars has substantial condensation flows, where CO_2 freezes out over the winter pole and sublimates above the summer pole.

Because Mars's atmosphere is tenuous, it responds rapidly to the solar heating, leading to large latitudinal, diurnal and seasonal variations in surface temperature. The temperature is as low as $\sim$130 K at the winter pole and peaks at $\sim$300 K at the subsolar point during the day. These large day-to-night temperature variations lead to strong winds across the terminator, the day-night line. These are the **thermal tide winds**, similar to the winds in Venus's thermosphere. On arid planets, when such winds exceed $\sim$50–100 m s^{-1}, they may start local dust storms, either initiated by **saltation**, where grains start hopping over the surface, or when the dust is raised up in **dust devils**, caused by convection in an atmosphere with a superadiabatic lapse rate. Once in the air, dust fuels the tidal winds, because the grains absorb sunlight and heat the atmosphere locally. Within just a few weeks, dust storms may grow so large that they envelop the entire planet, as shown in Figure 9.22. Such global storms may last for several months and have pronounced effects on Mars's climate.

Dust devils and their tracks have been photographed regularly both by Mars orbiters and landers and rovers; an example is shown in Figure 9.23. Some dust devils are hundreds of meters in diameter and several kilometers high. Dust devils occur most frequently during spring and summer. They usually leave a dark streak behind where dust has been removed (light-colored streaks have been seen, too). These streaks can be kinked and curved because of the swirly motion of the dust devils. Dust devils are also frequently seen in desert areas on Earth.

(a)

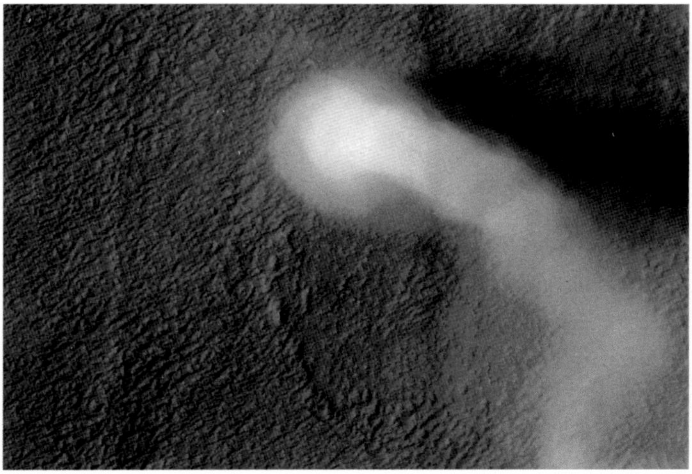

(b)

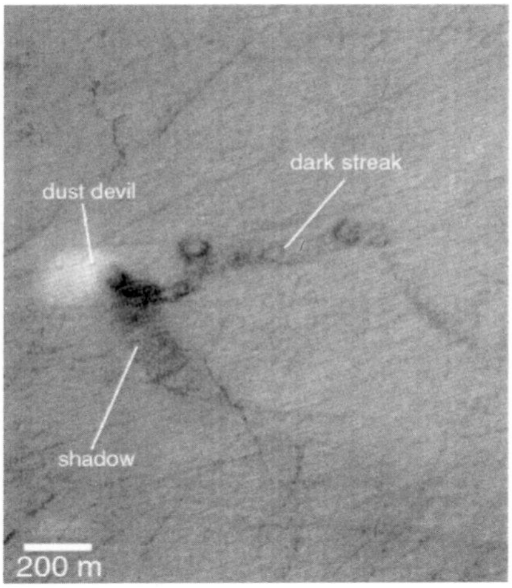

Figure 9.23 (a) COLOR PLATE This MRO image shows a martian dust devil roughly 20 kilometers high winding its way along the Amazonis Planitia region of northern Mars. Despite its height, the plume is only 70 meters wide. (NASA/JPL-Caltech/Univ. of Arizona) (b) Dust devil (~100 m diameter) and track observed with *Mars Global Surveyor*. Note the curlicue shape of the track, indicative of the path and spin of the dust devil. (MOC image M1001267, NASA/JPL/Malin Space Science Systems)

9.4.4 Frost, Ice and Glaciers

The temperature at the martian winter pole is only ~130 K; in the summer, it may rise to ~190 K, which is still well below the freezing temperature of water. Hence, water is permanently frozen at the martian poles. In the winter, the temperature above the poles drops below the freezing point of carbon dioxide, so CO_2 condenses out to form a (seasonal) polar cap of dry ice.

Figure 9.24 shows spacecraft images of the layered structure of dust and ice near the martian poles. These layers were produced by the sublimation and condensation of CO_2 over time. The dry ice in the northern ice cap sublimes completely away during the summer, leaving behind a permanent cap of water-ice ~1000 km in diameter. In the south, a permanent CO_2 ice cap ~350 km in diameter survives the summer. Mixed in with the permanent CO_2 ice is ~15% water-ice. This residual south polar cap displays a history indicative of depositional and ablational events unique to Mars's south pole. The difference between the ice caps on the two poles has been attributed to periodic variations in the orbital eccentricity, obliquity and season of perihelion of Mars. At the current epoch, the northern summer is hotter but shorter than the southern summer.

Although the presence of ice deposits in Mars's polar regions was never surprising, claims of glacial deposits at mid-latitudes were initially met with skepticism. However, evidence for past glaciers at mid-latitudes is accumulating. Figure 9.25 shows a 3.5- to 4-km-high massif with a viscous glacier-like flow of material from one crater down to the next one through a narrow notch. Periods of glaciation at such low latitudes may be caused by changes in the planet's obliquity (see Fig. 2.17, §5.8.2), where excursions from the nominal value can be much larger than for Earth (see Fig. 2.18). Additional ice reservoirs have been found at mid-latitudes, as exemplified in Figure 9.26. This figure shows that hydrogen, presumably in the form of water-ice, is common not only in the polar regions but also at a few locations at the equator and mid-latitudes.

(a)   (b)   (c)

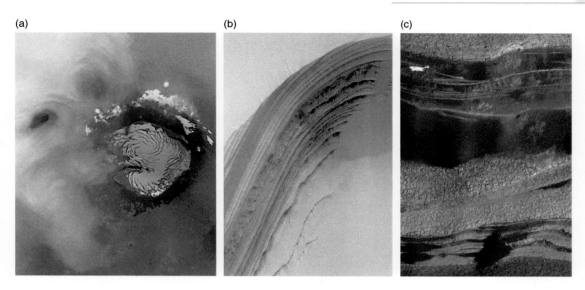

Figure 9.24 The martian north and south polar regions are covered by large areas of layered deposits that consist of a mixture of ice and dust. (a) This picture shows Mars's north polar cap in its entirety, surrounded by dunes. The image has a resolution of about 7.5 km/pixel. Annular clouds, visible in the upper left corner, are common here in mid-northern summer. They typically dissipate later in the day. The summer caps exhibit many exposed layers and steep cliffs, some of which are shown at higher resolution in adjacent panels. (b) A springtime view of frost-covered layers on an eroded scarp in the martian north polar cap. Some layers are known to be a source for the dark sand seen in nearby dunes. The picture covers an area about 3 km wide and is illuminated by the Sun from the lower left. (MGS-MOC, NASA/JPL/Malin Space Science Systems) (c) This image reveals the basal layers of Mars's north polar layered deposits at Chasma Boreale. The image is taken by the High-Resolution Imaging Science Experiment (HiRISE) on NASA's *Mars Reconnaissance Orbiter* (MRO). The resolution is 64 cm/pixel, and the imaged region is 568 m wide. (NASA/JPL/University of Arizona, PIA01925)

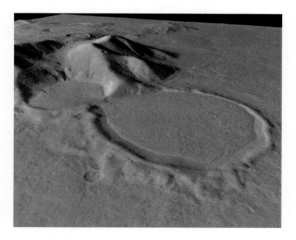

Figure 9.25 This *Mars Express* HRSC image shows a simulated perspective view of a 3.5- to 4-km-high massif in the Hellas region, with a viscous flow of material from one crater down to the next one, through a narrow notch. The image was taken from an altitude of 590 km, at a resolution of 29 m/pixel. (ESA/DLR/FU Berlin, G. Neukum)

9.4.5 Water on Mars

Much current research is focused on the search for water, liquid and frozen, on the surface or immediate subsurface of Mars. As mentioned earlier, under current climate conditions, liquid water cannot exist on Mars. Nonetheless, morphological features imply that water once flowed on this cold planet, and small-scale water flows may happen at the present epoch. We summarize below the morphological evidence that Mars was once much wetter than today.

Impact Craters

As evident in Figure 9.27, ejecta blankets of many martian craters appear to have 'flowed' to their current positions, rather than traveled through space

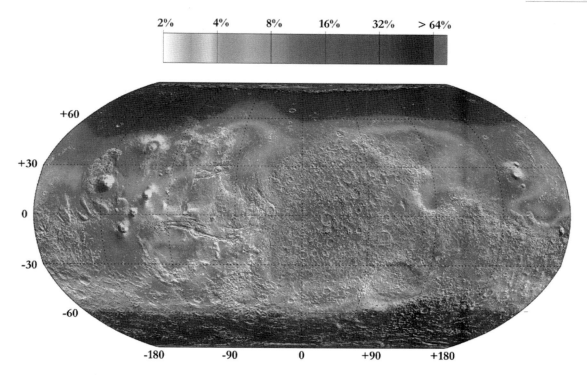

Figure 9.26 COLOR PLATE This map shows the estimated lower limit of the water content in Mars's surface. The estimates are obtained from the epithermal neutron flux as measured with the neutron spectrometer component of the gamma-ray spectrometer on *Mars Odyssey*. The epithermal neutron flux is sensitive to the amount of hydrogen in the upper meter of the soil. (NASA/JPL/Los Alamos National Laboratory)

along ballistic trajectories. This suggests that the surface was fluidized by the impacts. Craters with fluidized ejecta blankets are referred to as **rampart craters**. Hence, in contrast to the Moon and Mercury, Mars must have a significant fraction of water-ice in its crust or at least have had subsurface ice during the early bombardment era. In addition, most martian craters are shallower than those seen on the Moon and Mercury, and the craters (rocks and rim) show signs of atmospheric erosion, although not as much as on Earth. *Mars Global Surveyor* (MGS) images revealed evidence of seepage at the edge of some crater walls (see gullies below) and of (past) 'ponding', the accumulation of water in ponds on some crater floors. The observed polygon structures on Mars, shown in Figure 6.21a, are typical of ice-wedge

polygons on Earth that form via seasonal (or episodic) melting and freezing of water in and on the surface. A photograph of the latter is shown in Figure 6.21b.

Valleys and Outflow Channels

The oldest martian terrain contains numerous fluvial features, similar in appearance to **dendritic river systems** on Earth; an example is shown in Figure 9.28. In addition to these dendritic systems, there are also immense channel systems, or **outflow channels**, starting in the highlands and draining into the low northern plains. Some of these channels, as shown in Figure 9.29, are many tens of kilometers wide, several kilometers deep and hundreds to thousands of kilometers long. The

Figure 9.27 The ejecta deposits around this martian impact crater Yuty (18 km in diameter) consist of many overlapping lobes. This type of ejecta morphology is characteristic of many craters at equatorial and mid-latitudes on Mars but is unlike that seen around small craters on the Moon. (NASA/*Viking Orbiter* image 3A07)

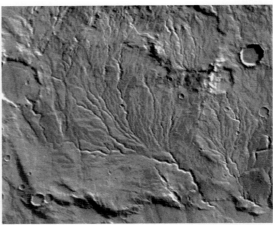

Figure 9.28 These valley systems on Mars resemble dendritic drainage patterns on Earth, where water acts at slow rates over long periods of time. The channels merge together to form larger channels. Because the valley networks are confined to relatively old regions on Mars, their presence may indicate that Mars possessed a warmer and wetter climate in its early history. The area shown is about 200 km across. (Courtesy Brian Fessler, image from the Mars Digital Image Map, NASA/*Viking Orbiter*)

presence of teardrop-shaped 'islands' in the outflow channels suggests that vast flows of water have flooded the plains.

Gullies

Figure 9.30 shows perhaps the most tantalizing images of groundwater flow, morphological features that likely result from fluid seepage and surface runoff. The 'head alcove', located just below the brink of a slope (e.g., on the wall of a crater, valley or hill), seems to be the 'source' of a depositional apron just below it. Most of these aprons show a main and some secondary channels emanating from the downslope apex of the alcove. The formation of these gullies likely involves a low-viscosity fluid, such as water, moving downhill. Most intriguing, though, is the observation that all gullies must be very young: There are no impact craters superposed on these features, and some features partly obscure aeolian landforms – indicative of the (geologically) very recent past.

9.4.6 Geology at Rover Sites

The Mars Exploration Rovers (MER) *Spirit* and *Opportunity* arrived on Mars in January 2004. Although the nominal mission was designed for 90 **sols** (martian days), *Spirit*'s mission continued until March 2010, and *Opportunity* is still roaming around in 2013. Both rovers were equipped with a full suite of instruments, among them a rock abrasion tool to grind into rocks and various spectrometers to analyze the rocks.

Spirit landed in Gusev crater, a flat-floored, 160-km-diameter crater that was most likely a lake ~4 Gyr ago, connected to the northern lowlands via a channel. Surprisingly, no sedimentary rocks were found. Instead, the rocks in Gusev crater are mostly basaltic in composition, with a texture that also points at a volcanic origin. A view from the rover is shown in Figure 9.31. The rocks are weathered primarily by impacts and wind. *Spirit* set course, via some small impact craters, to the 'Columbia Hills'. On approach, the rocks and soil

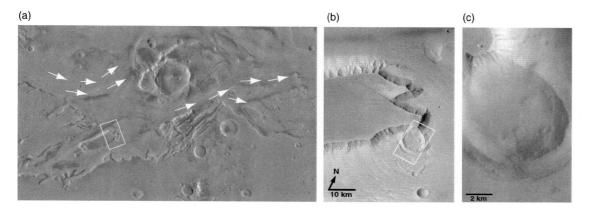

Figure 9.29 Image of a portion of the Kasei Vallis outflow channel system. (a) Large-scale view from *Viking* showing flow patterns (*arrows*) in a portion of the Kasei Vallis outflow channel that created 'islands'. The *large white box* shows the outline of the *Viking 1* image shown in panel (b), and the *small white box* outlines the area imaged by Mars Global Surveyor (MGS), shown in panel (c). The large crater in the *upper center* of this overview scene is 95 km in diameter. Panel (c) shows a 6-km-diameter crater that was once buried by about 3 km of martian 'bedrock'. This crater was partly excavated by the Kasei Vallis floods more than a billion years ago. The crater is poking out from beneath an 'island' in the Kasei Vallis. The mesa was created by a combination of the flood and subsequent retreat via small landslides of the scarp that encircles it. (USGS *Viking 1* mosaic; *Viking* 226a08; MOC34504)

changed. The rocks became largely granular in appearance, and both the rocks and soil in the hills are relatively rich in salts, suggestive of significant aqueous alteration compared with the rocks near the landing site.

Opportunity landed in Eagle crater on Meridiani Planum, a landing site that was selected because spectroscopic data from orbiting spacecraft revealed areas rich in the mineral hematite. Hematite can form in various ways, some involving

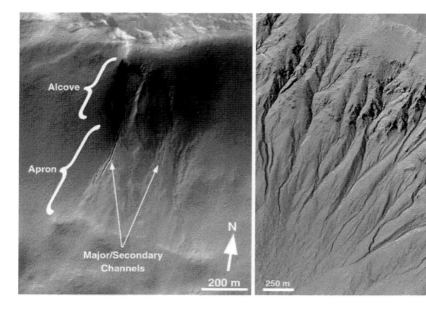

Figure 9.30 Examples of landforms that contain martian gullies. These features are characterized by a half-circle-shaped 'alcove' that tapers downslope, below which is an apron. The apron appears to be made of material that has been transported downslope through the channels or gullies on the apron. On the *right* is a larger scale view of some such channels. (M03_00537, M07_01873; Malin and Edgett 2000)

Figure 9.31 COLOR PLATE A view from Mars Exploration Rover *Spirit*, taken during its winter campaign in 2006. In the distance (850 m away) is 'Husband Hill' behind a dark-toned dune field and the lighter-toned 'home-plate'. In the foreground are wind-blown ripples along with a vesicular basalt rock. (NASA/JPL-Caltech/Cornell)

Figure 9.32 A panoramic view from Mars Exploration Rover *Opportunity* of the 'Payson' outcrop on the western edge of Erebus crater. One can see layered rocks in the ~1 m thick crater wall. To the left of the outcrop, a flat, thin layer of spherule-rich soil lies on top the bedrock. (NASA/JPL-Caltech/USGS/Cornell, PIA02696)

the action of liquid water. With the rovers' prime goal of searching for evidence of liquid water, in the past or present, this appeared to be an opportune area for closer investigation. *Opportunity* landed near a 30- to 50-cm high bedrock outcrop, shown in Figure 9.32. The bedrock is mostly sandstone composed of materials derived from weathering of basaltic rocks, with several tens of percent (by weight) sulfate minerals, as magnesium and calcium sulfates and the iron sulfate jarosite, as well as hematite. Scattered throughout the outcroppings and partly embedded within, *Opportunity* discovered small (4–6 mm across) gray/blue-colored spherules, 'blueberries', sometimes multiply fused, composed of >50% hematite by mass. An image of the blueberries is shown in

Figure 9.34 A false-color view of a mineral vein imaged with the panoramic camera (Pancam) on NASA's Mars Exploration Rover *Opportunity*. The vein is about 2 cm wide and 45 cm long. *Opportunity* found it to be rich in calcium and sulfur, possibly the calcium–sulfate mineral gypsum. (NASA/JPL/Cornell, PIA15034)

Figure 9.33 Small (millimeter-sized) spherules, dubbed 'blueberries', are scattered throughout the rock outcrop near rover *Opportunity*'s landing site. The rocks show finely layered sediments, which have been accentuated by erosion. The blueberries are lining up with individual layers, showing that the spherules are concretions, which formed in formerly wet sediments. (NASA/JPL/Cornell, PIA05584)

Figure 9.33. Blueberries are likely concretions that formed when minerals precipitated out of water-saturated rocks. In the same outcrops, small voids or **vugs** in the rocks also hint at the past presence of water; soluble materials, such as sulfates, dissolved within the rocks, leaving vugs behind. Although rocks partially dissolved or weathered away, the hematite concretions fell out of the bedrock, covering the plains. The sulfate-rich sedimentary rocks at Meridiani Planum, underneath a meter-thick layer of sand, preserve a historic record of a climate that was very different from the martian conditions we know today. Liquid water most likely covered Mars's surface, at least intermittently, with wet episodes being followed by evaporation and desiccation.

While traversing Meridiani Planum, *Opportunity* investigated several craters. It reached Victoria crater in September 2006 and ventured inside the crater a year later. In Summer 2008, after climbing out of Victoria crater, *Opportunity* set course to the 22-km-diameter Endeavour crater, where it arrived

in the summer of 2011. On its way, it investigated Santa Maria crater. Layers of bedrock exposed at Victoria and other locations revealed a sulfate-rich composition indicative of an ancient era when acidic water was present. After arriving at the rim of the 22-km-diameter Endeavour crater, the rover stumbled upon a vein, shown in Figure 9.34, rich in calcium and sulfur, possibly made of the calcium–sulfate mineral gypsum. This vein shows that water must have flowed through underground fractures in the rock, forming the chemical deposit gypsum.

On August 6, 2012, the rover *Curiosity* landed on Mars at Gale Crater. The HiRISE camera on MRO captured the image of *Curiosity* and its parachute shown in Figure 9.35. The overarching science goal of this mission is to assess whether the landing area has ever had or still has environmental conditions favorable to microbial life, both its habitability and its preservation.

9.4.7 Magnetic Field

Mars Global Surveyor detected surprisingly intense localized magnetic fields, shown in Figure 9.36. The strongest field measured ~0.16 nT at an altitude of 100 km, which, in combination with the ambient ionospheric pressure, is strong enough to stand off and deflect the solar wind at Mars. As

Figure 9.35 NASA's *Curiosity* rover and its parachute were photographed by HiRISE on MRO as *Curiosity* descended to the surface on August 6, 2012. The parachute and rover are seen in the center of the white box; the inset image is a cutout of the rover stretched to avoid saturation. (NASA/JPL-Caltech/Univ. of Arizona, PIA15978).

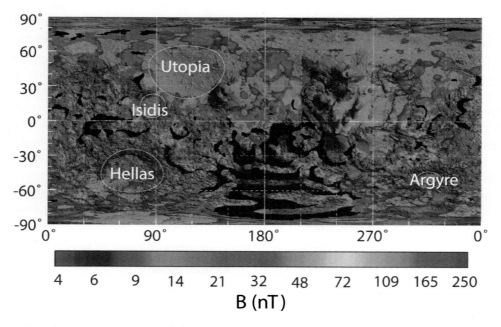

Figure 9.36 COLOR PLATE Smoothed magnetic map of Mars constructed from electron reflectometer data from *Mars Global Surveyor* (*MGS*). The logarithmic color scale represents the crustal magnetic field magnitude at an altitude of 185 km overlaid on a topography map as derived from laser altimeter data on MGS. The lower limit of the color scale is the threshold for unambiguously identified crustal features, and the scale saturates at its upper end. *Black* represents sectors with fewer than 10 measurements within a 100-km radius. These regions are areas where there is a closed crustal magnetic field and so the solar wind electrons cannot penetrate to the altitude of the spacecraft where they can be detected. The four largest visible impact basins are indicated (*dotted circles*). (Adapted from Lillis et al. 2008)

at Venus, solar wind magnetic field lines are compressed and drape around the planetary obstacle below the bow shock.

The localized magnetic fields on Mars are caused by remanent crustal magnetism. Most of the strong sources are located in the heavily cratered highlands south of the crustal dichotomy boundary. There is no evidence for crustal magnetization inside some of the younger giant ($\gtrsim$1000 km) impact basins (e.g., Hellas, Utopia and Argyre). These data suggest that early in the planet's history, Mars may have had a geodynamo with a magnetic moment comparable to, or larger than, Earth's dynamo at present.

Key Concepts

- The lunar surface is divided into two major types of geological units. The highlands are old, heavily cratered and relatively bright. The maria are younger, dark basaltic units with few large craters.
- Earth's Moon is substantially depleted in iron relative to all of the terrestrial planets and primitive meteorites. Nonetheless, it has a small Fe-dominated core.
- The Moon is also depleted in H_2O, but small reservoirs of H_2O-ice exist in permanently shadowed regions near the lunar poles. The polar regions of Mercury also host H_2O–ice.
- Mercury is substantially enriched in iron relative to all of the other terrestrial planets and primitive

meteorites. Mercury's excess iron appears to be concentrated in an Fe-dominated core. The outer core is fluid, and a dipolar magnetic field is generated in this region. Mercury's surface is depleted in Fe and Ti and enriched in the volatile element sulfur.
- Both the Moon and Mercury have very tenuous atmospheres. The constituents of these atmospheres escape rapidly and must be continually replenished from the solar wind or internal sources.
- Venus has a thick CO_2-dominated atmosphere that induces several hundred degrees of greenhouse warming at the surface.
- Venus is enshrouded by SO_2-rich clouds that give the planet a high albedo and obscure the view of the surface.
- Venus lacks plate tectonics and therefore has a single-peaked altitude distribution in contrast to the ocean–continent dichotomy seen on Earth.
- Mars's radius is half that of Earth, and its mountains and valleys are substantially higher because of the lower surface gravity.
- Mars has a thin CO_2-dominated atmosphere with a surface pressure less than 1% that of Earth.
- At present, Mars is cold and dry. But dry river beds imply that significant quantities of water flowed on the martian surface billions of years ago.

Further Reading

Excellent reviews of each of the planets, including Earth as a planet, are provided in:

Encyclopedia of the Solar System, 2nd Edition. Eds. L. McFadden, P.R. Weissman, and T.V. Johnson. Academic Press, San Diego. 482pp.

Problems

9-1. **(a)** Use the present-day lunar cratering rate given in §6.4.4 to estimate the average crater density (km^{-2}) for craters more than 4 km in size for a region that is 3.3 Gyr old.

(b) Explain why the same procedures cannot be used to provide a good estimate of the lunar maria.

9-2. The secondary craters related to a primary crater of a given size on Mercury typically lie closer to the primary crater than do the secondary craters of a similarly sized primary on the Moon. Presumably, this is the result of Mercury's greater gravity reducing the distance that ejecta travel.

(a) Verify this difference quantitatively by calculating the 'throw distance' of ejecta launched at a 45° angle with a velocity of 1 km s^{-1} from the surfaces of Mercury and the Moon.

(b) Typical projectile impact velocities are greater on Mercury than they are on the Moon. Why doesn't this difference counteract the surface gravity effect discussed earlier?

9-3. By examining the morphology of craters of various sizes in Figure 9.8, deduce:

(a) the direction to the Sun

(b) the form of the depth/diameter ratio for craters as a function of diameter

9-4. Mercury's mean density $\rho = 5430$ kg m^{-3}. This value is very close to the planet's uncompressed density. If Mercury consists entirely of rock ($\rho = 3300$ kg m^{-3}) and iron ($\rho = 7950$ kg m^{-3}), calculate the planet's fractional abundance of iron by mass.

9-5. Does the shaking last longer for moonquakes or for earthquakes? Why?

9-6. How cold can the inside of a shadowed crater on the Moon be? Follow the derivation of equilibrium temperature for a rapidly rotating planet in §4.1.3 but make a series of assumptions to make the problem more realistic:

(a) Compute the usual equilibrium temperature for the Moon.

(b) Instead of assuming direct overhead sunlight, adjust the incident light intensity to be appropriate for a very high latitude on the Moon. You will need to work out the geometry to relate latitude to incident flux and derive an equation relating equilibrium temperature to latitude. What is the equilibrium temperature at 89°S?

(c) Make a plot showing the equilibrium temperature as a function of latitude.

(d) What would be the surface temperature at a location that sees only 1 hour of sunlight per lunar day?

9-7. Estimate the temperature at the surface of Mercury at the following places and times. State the assumptions that you make for your calculations.

(a) At the subsolar point when Mercury is at perihelion

(b) At the subsolar point when Mercury is at apohelion

(c) 45° from the subsolar point when Mercury is at apohelion

9-8. Although in some respects Earth and Venus are 'twin planets', they have very different atmospheres. For example, the surface pressure on Venus is almost 2 orders of magnitude larger than that on Earth.
(a) Calculate the mass of each atmosphere; state your answer in kilograms.
(b) Recalculate these values for Earth, including Earth's oceans as part of its 'atmosphere'. (If all of the water above Earth's crust were spread evenly over the planet, this global ocean would be ~3 km deep.)
(c) Compare the values for the two planets and comment.

9-9. State and explain two pieces of evidence, one physical and the other chemical, that Mars was warmer and wetter in the distant past than it is at the present epoch.

9-10. **(a)** Estimate the typical collision velocity of asteroids with Mars.
(b) Calculate the size of a crater produced by the impact of a 10-km-radius stony asteroid onto Mars at this speed.

9-11. Contrast the differences between the northern lowlands and southern highlands on Mars (other than elevation) and give one possible explanation for the difference.

9-12. Consider the impact between an iron meteoroid ($\rho = 7\,000$ kg m^{-3}) with a diameter of 300 m and the Moon.
(a) Calculate the kinetic energy involved if the meteoroid hits the Moon at $v = 12$ km s^{-1}.
(b) Estimate the size of the crater formed by a head-on collision and one in which the angle of impact with respect to the local horizontal is 30°.
(c) If rocks are excavated from the crater with typical ejection velocities of 500 m s^{-1}, calculate how far from the main crater one may find secondary craters.

9-13. Repeat the same questions as in Problem 9-12 for Mercury. Comment on the similarities and differences.

9-14. After the Moon has been hit by the meteoroid from Problem 9-12, many rocks are excavated from the crater during the excavation stage.
(a) If the ejection velocity is 500 m s^{-1}, calculate how long the rock remains in flight if its ejection angle with respect to the ground is 25°, 45° and 60°.
(b) Calculate the maximum height above the ground reached by the three rocks from (a).

CHAPTER 10

Planetary Satellites

I had now decided beyond all question that there existed in the
heavens three stars wandering about Jupiter as do Venus and
Mercury about the Sun, and this became plainer than daylight
from observations on similar occasions which followed. Nor were
there just three such stars; four wanderers complete their
revolution about Jupiter . . .

Galileo, *The Starry Messenger*, 1610

Six of the eight major planets in our Solar System, as well as many minor planets, are orbited by smaller companion satellites, often referred to as **moons**. The largest moons, Jupiter's Ganymede and Saturn's Titan, are more voluminous than is the planet Mercury, albeit not as massive. Jupiter's Callisto is almost as large as the aforementioned three bodies, and Io and Europa, the other two moons discovered by Galileo four centuries ago, straddle Earth's Moon in size. In contrast, most known moons are tiny bodies, from a few kilometers to tens of kilometers in size. Objects classified as moons span a range of several thousand in radius and one hundred billion (10^{11}) in mass, so it should come as no surprise that this is a very heterogeneous category of celestial bodies.

Large moons are nearly spherical, whereas small ones can be quite oddly shaped; the dividing line is about 200 km in radius. Dynamically, moons fall into two classes, regular satellites traveling on low-inclination, near-circular orbits within a few dozen planetary radii of the planet and irregular satellites, most of which orbit at much greater distances and have large eccentricities and inclinations.

Most moons are airless, but Titan has a N_2/CH_4-dominated atmosphere that has a higher surface pressure than that which we experience on Earth. Neptune's Triton, the largest moon in our Solar System not mentioned above, has a surface pressure only 10^{-5} times that of Titan yet still orders of magnitude larger than that of any other known moon.

The vast majority of moons are geologically dead, and impact craters are the dominant features on most moons that are large enough to be roundish. A few moons, however, form dramatic exceptions to this general trend. Io is the most volcanically active body in the Solar System, and Saturn's moon Enceladus spews out gigantic geysers from its south pole. Europa's icy crust, which has solidified in the geologically recent past, lies above a still-liquid H_2O ocean. This liquid water, warmed by tidal heating, makes Europa a prime target for speculations on the possible existence of a variety of life forms. Conditions may be analogous to those near hot vents in the deep ocean on early Earth. Liquid hydrocarbon lakes have been discovered near Titan's poles, and numerous channel-like features on Titan's surface are indicative of liquid flows. Triton and the much smaller moon Miranda (which orbits Uranus) have varied and intriguing surfaces. The *Voyager 2* spacecraft discovered liquid nitrogen geysers on Triton.

In this chapter, we discuss the moons of the five planets orbiting exterior to our Earth. The two inner planets lack moons, although they may once have had satellites that were long ago lost to tidal decay (§2.7.2). Earth's Moon, more analogous in many ways to terrestrial planets than to the bodies considered here, is included in Chapter 9, and satellites of minor planets are discussed with their larger companions in Chapter 12.

Our treatment is organized by heliocentric distance, beginning with the moons of Mars and ending with those of Neptune. We concentrate on moons that are the most interesting from a geological, and in some cases astrobiological, perspective.

10.1 Moons of Mars: Phobos and Deimos

Mars has two small moons, Phobos and Deimos, traveling on nearly circular orbits close to the planet's equatorial plane. Both their visual albedos, $A_v \sim 0.07$, and their spectral properties are similar to those of primitive, carbon-rich asteroids. Their densities, $\sim 2000 \ kg \ m^{-3}$, suggest their composition to be either a mixture of rock and ice or primarily rock with significant void space.

Phobos, the larger of the pair with mean radius $R \approx 11$ km, orbits Mars at a distance of 2.76 R_{σ}, which is well inside the synchronous orbit, and tiny Deimos ($R \approx 6$ km) orbits Mars outside synchronous orbit at 6.92 R_{σ}. Both satellites are in

Figure 10.1 Image of Phobos, the inner and larger of the two moons of Mars, taken by *Mars Express* in 2004. The spatial resolution is 7 m/pixel. (ESA/DLR/FU Berlin, G. Neukum) The associated movie clip shows Phobos passing in front of the smaller and more distant Martian moon, Deimos, on 1 August 2013, from the perspective of NASA's Mars rover *Curiosity*. (NASA/JPL/Malin Space Science/Texas A&M)

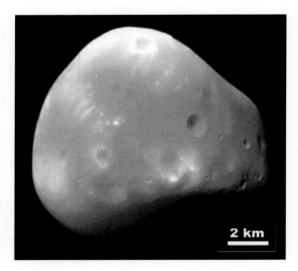

Figure 10.2 Image of Deimos taken 21 February 2009 at a spatial resolution of 20 m/pixel. (HiRISE/MRONASA/JPL/ University of Arizona, PIA11826)

synchronous rotation. Images of these two moons are shown in Figures 10.1 and 10.2. It is not surprising that both objects, being so small (Table E.5), are oddly shaped.

Phobos is heavily cratered, close to saturation. Its most unusual features are the linear depressions or grooves, typically 10–20 m deep, which are centered on the leading apex of Phobos in its orbit. These grooves may have formed as (secondary) crater chains from material ejected into space from impacts on the surface of Mars. Deimos's surface is rather smooth and shows prominent albedo markings, varying from 6%–8%. The images also show a concavity 11 km across, twice as large as the mean radius of the object.

10.2 Satellites of Jupiter

Jupiter's four largest moons, shown in Figure 10.3, range in size from Europa, which is slightly smaller than Earth's Moon, to Ganymede, the largest moon in our Solar System. They are collectively referred to as the Galilean satellites, named after Galileo Galilei, who discovered them in 1610.

10.2.1 Io

Io's mass and density are similar to those of Earth's Moon. However, in contrast to the Moon, no impact craters have been seen on Io, and hence its surface must be extremely young, less than a few Myr. Io's youthful surface and spectacular visual appearance result from the extreme volcanic activity on this moon. Examples of plumes and eruptions are shown in Figure 10.4.

Reflectance spectra, such as the one shown at the top of Figure 10.5, reveal a surface rich in SO_2 frost and other sulfur-bearing compounds. In addition, mafic minerals such as pyroxene and olivine have been identified in Io's dark (volcanic) calderas.

Figure 10.3 COLOR PLATE Galilean satellites: Io, Europa, Ganymede and Callisto, shown (*left* to *right*) in order of increasing distance from Jupiter. All satellites have been scaled to a resolution of 10 km/pixel. Images were acquired in 1996 and 1997. (NASA/*Galileo Orbiter* PIA01299)

Io's orbit is slightly eccentric and remains eccentric despite Jupiter's strong tidal forces because the satellite is locked in a 4:2:1 orbital resonance with the satellites Europa and Ganymede. Jupiter's strong tidal variations cause daily distortions in Io's shape that are many tens of meters in amplitude. Because Io is not perfectly elastic, this leads to dissipation of massive amounts of energy in its interior, so much that Io's global heat flux is ~20–40 times as large as the terrestrial value. This amount of heat is too large to be removed by conduction or solid-state convection. Melting therefore occurs, and lavas erupt through the surface via giant volcanoes. More than 400 volcanic calderas, up to $\gtrsim$200 km in size, are distributed over Io's surface. Some of the lava flows from the calderas are hundreds of kilometers long. Io further displays a variety of geological features, such as ridges, mountains and calderas, all of which are probably connected to the satellite's extreme volcanic activity.

Observations of Io at (near-) infrared wavelengths reveal a body covered by numerous **hot spots**, as exemplified in Figure 10.6. Hot spots are usually associated with low-albedo regions at visible and near-infrared (1–2.5 μm) wavelengths, as visualized in the top panel of Figure 10.6. When Io is in **eclipse** (i.e., in Jupiter's shadow) its near-IR luminosity is dominated by thermal emission from glowing hot spots, as shown in the lower panel of Figure 10.6. Hot spots appear at random times; because it takes time for the hot lava to cool off, a hot spot usually lasts for weeks to months and may stay 'active' for years. Blackbody fits to near-infrared (1–5 μm) spectra reveal temperatures in excess of 1000 K, indicative of silicate volcanism, as on Earth. Some observations suggest temperatures exceeding 1700 K, which would indicate volcanism driven by ultramafic magmas (e.g., komatiites), a style of volcanism that has not occurred for billions of years on Earth.

Some of the hot spots are associated with volcanic plumes, such as the Tvashtar eruption in 2007, shown in Figure 10.7a. Plumes are usually dominated by SO_2 gas and often also dust. These volcanic gases have probably led to large areas being covered by SO_2 ice. Sublimation of this ice, in addition to direct outgassing from the vents, produces Io's atmosphere, which is largely composed of sulfur dioxide. The spacecraft image of Io in eclipse shown in Figure 10.7b reveals hot spots, volcanic plumes and auroral glows.

The combination of Io's detailed shape and gravity field, both influenced by tidal and rotational

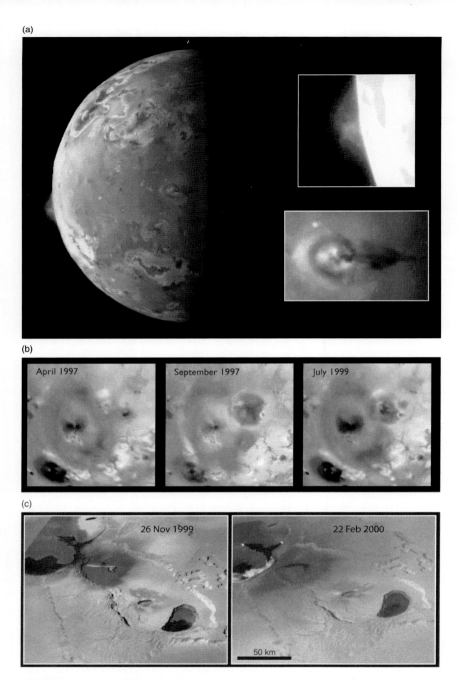

Figure 10.4 COLOR PLATE (a) A 140-km high plume rises above the bright limb of Io (see *inset* at *upper right*) on 28 June 1997. A second plume is located near the terminator (see *inset* at *lower right*). The shadow of the 75-km high plume extends to the right of the eruption vent near the center of the bright and dark rings. The blue color of the plumes is caused by light scattering off micron-sized dust grains, which makes the plume shadow reddish. (NASA/*Galileo Orbiter* PIA00703). (b) Images of Pele taken on 4 April 1997, 19 September 1997 and 2 July 1999 show dramatic changes on Io's surface. Between April and September 1997, a new dark spot, 400 km in diameter, developed surrounding Pillan Patera, just northeast of Pele. The plume deposits to the south of the two volcanic centers also changed, perhaps due to interaction between the two large plumes. The image from 1999 shows further changes, such as the partial covering of Pillan by new red material from Pele. A new eruption took place in Reiden Patera, northwest of Pillan, that deposited a yellow ring. (NASA/*Galileo*, PIA02501) (c) A pair of images taken of Tvashtar Patera. The eruption site has changed locations over a period of a few months in 1999 and early 2000. (NASA/*Galileo*, PIA02584)

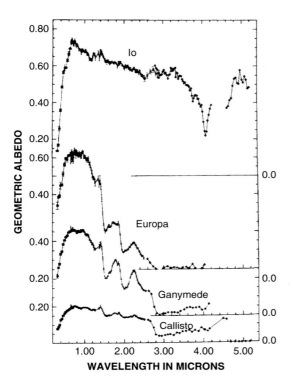

Figure 10.5 Spectra of the Galilean satellites. (Clark et al. 1986)

cloud of particles, including, for example, O, S, Na and K, around Io, referred to as the neutral cloud. Upon ionization, the newly formed ions move with Jupiter's magnetic field and form the Io plasma torus, a donut-shaped region of charged particles surrounding Jupiter located near Io's orbit. Io's neutral cloud and plasma torus are discussed in more detail in §8.1.4.

10.2.2 Europa

Europa is slightly smaller and less dense than the Moon. Its surface is very bright and has the spectral properties of nearly pure water-ice. Europa's moment of inertia ratio, $I/(MR^2) = 0.346$, implies a differentiated, centrally condensed body. Its mean density of 3010 kg m^{-3} is indicative of a rock/ice composition wherein the rocky mantle/core provides more than 90% of the mass. Europa is therefore best modeled as a mostly rocky body with perhaps a metal core overlain with an H$_2$O 'layer' $\sim$100–150 km thick. The top of the H$_2$O layer is a solid ice crust, whose thickness may be as small as a few kilometers or as large as a few tens of kilometers. Part of the lower portion of the H$_2$O layer must be liquid; an underground ocean is suggested by details in surface topology and, most convincingly, from measurements of Jupiter's magnetic field near Europa. This liquid ocean is maintained by tidal heating and decouples the ice shell from Europa's interior.

Europa's surface is relatively flat compared with the surfaces of, for example, Io and the Moon. Only a few tens of impact craters with radii over 2 km have been detected, implying a surface age of tens to at most a few hundred million years. Most of the geologic features that have been seen on Europa's surface were produced by diurnal tidal stresses. The oldest terrain is characterized by **ridged plains**, which are often criss-crossed by younger **bands**. In many places, two parallel ridges are separated by a V-shaped trough, as

forces, provides constraints on the satellite's internal properties. Io's core extends to $\sim$40%–50% of its radius. Overlying Io's core is a hot silicate mantle, topped off with a lower density crust and lithosphere that may be $\sim$30–40 km thick. The observed eruption temperatures, as well as measurements of variations in Jupiter's magnetic field near Io, are consistent with a partially molten global magma layer that is concentrated in a $\gtrsim$50-km-thick asthenosphere.

Sublimation of SO$_2$ frost from Io's surface, volcanism and sputtering create a tenuous yet collisionally thick atmosphere around the satellite. Ions corotating with Jupiter's magnetosphere have typical velocities of 75 km s^{-1} and hence readily overtake Io, which orbits Jupiter at the Keplerian speed of 17 km s^{-1}. The ions interact with Io's atmosphere, which leads to the formation of a

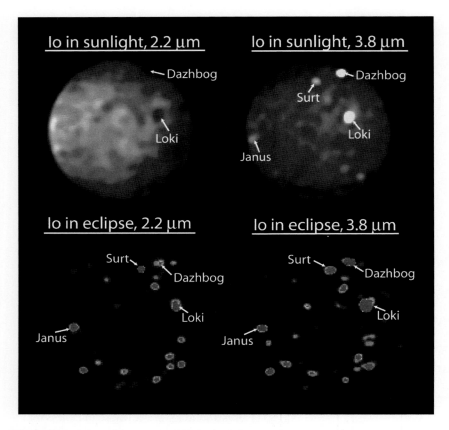

Figure 10.6 Images of Io at 2.2 μm (*left*) and 3.8 μm (*right*), taken with the Keck telescope, equipped with adaptive optics. (*top*): Io in sunlight. At both wavelengths, Io's emission is dominated by sunlight reflected off the satellite. Because the Sun's intensity is lower at 3.8 μm than at 2.2 μm and 3.8 μm is closer to the peak of a typical hot spot's blackbody curve, hot spots are easier to recognize at a wavelength of 3.8 μm than at 2.2 μm. Note that some volcanoes (Loki, Dazhbog) show up as hot spots at 3.8 μm but as low-albedo features at 2.2 μm. (*bottom*): Io in eclipse. Images of Io taken 2 hours later, after the satellite had entered Jupiter's shadow. Without sunlight reflecting off the satellite, even faint hot spots can be discerned by taking images with longer exposure times. The difference in brightness between the two wavelengths gives an indication of the temperature of the spot. Both Loki and Dazhbog, very bright at 3.8 μm, are low-temperature (~500 K) hot spots. Surt and Janus, on the other hand, are also very bright at 2.2 μm, indicative of higher temperatures (~800 K). (Adapted from de Pater et al. 2004a)

shown in Figures 10.8 and 10.9. These **wedge-shaped ridges** may have formed by an expansion of the crust or when two ice plates pulled slightly apart. Warmer, slushy or liquid material may have been pushed up through the crack, forming a ridge. The brownish color suggests that the slush consists in part of rocky material, hydrated minerals or clays or salts.

Although most ridges are linear in shape, some of them, the **cycloids** (Fig. 10.8a), are curved. The cycloidal shape results from the propagation of a crack in the surface caused by diurnal stresses, producing a curved rather than straight feature.

One of the youngest features on the satellite's surface is the **chaotic terrain**, displayed in Figure 10.9. The morphology of these features

(a)
(b)

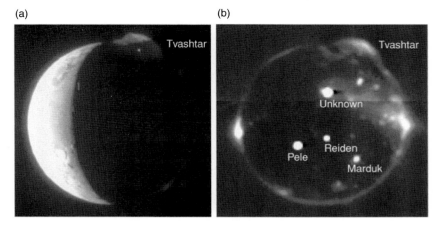

Figure 10.7 (a) An image of the 2006–2007 Tvashtar eruption, captured by the *New Horizons* spacecraft, 28 February 2007. Io's day side is overexposed to bring out faint details in the plumes and on the moon's night side. On the night side, at the 'center' of the eruption, the glow of the hot lava is visible as a bright point of light. Another plume is illuminated by Jupiter just above the lower right edge. (b) A LORRI *New Horizons* image of Io in eclipse, showing only glowing hot lava (the brightest points of light), as well as auroral displays in Io's tenuous atmosphere and the moon's volcanic plumes. The edge of Io's disk is outlined by the auroral glow produced as charged particles from Jupiter's magnetosphere bombard the (patchy) atmosphere. Both images are composites of images taken at wavelengths between 350 and 850 nm. (NASA/APL/SWRI, PIA09250, PIA09354)

(a)
(b)

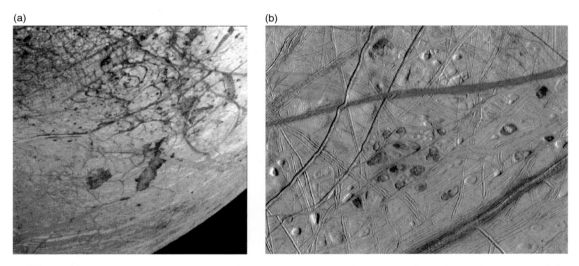

Figure 10.8 (a) Europa's southern hemisphere. The upper left portion of the image shows the southern extent of the 'wedges' region, an area that has undergone extensive disruption. The image covers an area approximately 675 by 675 km, and the finest details that can be discerned are about 3.3 km across. (NASA/*Galileo Orbiter*, PIA00875) (b) Reddish spots and shallow pits pepper the surface of Europa. The spots and pits on this image are about 10 km across. (PIA03878; NASA/JPL/University of Arizona/University of Colorado)

(a)

(b)

(c)

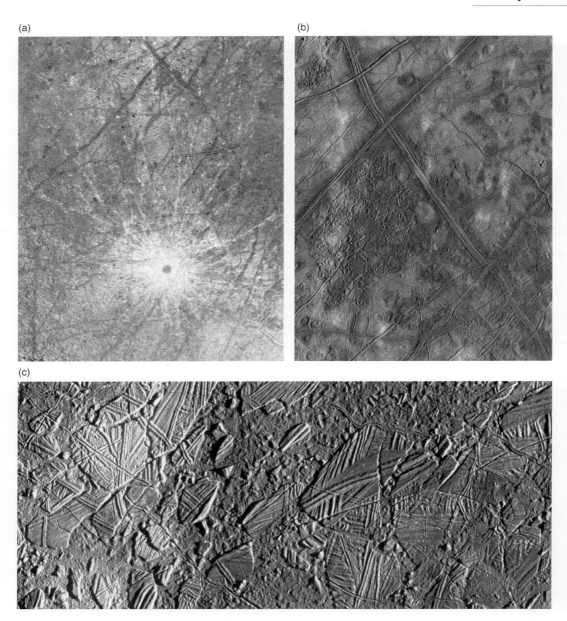

Figure 10.9 COLOR PLATE (a) The 26-km diameter impact crater Pwyll, just below the center of the image, is likely one of the youngest major features on the surface of Europa. The central dark spot is ~40 km in diameter, and bright white rays extend >1000 km in all directions from the impact site. One can also discern several dark lineaments, called 'triple bands' because they have a bright central stripe surrounded by darker material. The order in which these bands cross each other can be used to determine their relative ages. The image is 1240 km across. (NASA/*Galileo Orbiter*, PIA01211) (b) A close-up of the X-shaped ridges north of the Pwyll crater. The area covered in this panel is ~250 × 200 km. Surface features such as domes and ridges, as well as 'disrupted terrain', can be distinguished. (NASA/*Galileo Orbiter*, PIA01296) (c) Amplified view of a small region of the thin, disrupted ice crust in the Conamara region of Europa, the disrupted terrain displayed in panel b. The *white* and *blue* colors outline areas that have been blanketed by a fine dust of ice particles ejected at the time of formation of the large crater Pwyll. The image covers an area of 70 × 30 km; north is to the right. (NASA/*Galileo Orbiter*, PIA01127)

resembles kilometer-scale blocks or sheets of ice 'floating' on softer or slushy ice below. Some of these broken-up plates have been rotated, tilted or moved. They can be reassembled like a jigsaw puzzle. In these areas, ocean water may have reached the surface and produced new crust.

Other intriguing extremely young features are **lenticulae**, the Latin term for freckles (Fig. 10.8b). Their morphology suggests that they originate from convective upwelling of warm buoyant ice in **diapirs**, somewhat analogous to lava lamps. This could lead to **domes** when the diapirs reach the surface and depressions where the diapir does not break through to the surface but instead weakens or melts the ice above it, which subsequently sags down.

Because Europa's surface is covered by water-ice, it is not surprising that this satellite has a tenuous oxygen atmosphere. Sputtering processes knock water molecules off Europa's surface; upon dissociation, this H_2O breaks up into hydrogen and oxygen. Hydrogen escapes the low gravity field of Europa, leaving an oxygen-rich atmosphere behind. In addition to O, both Na and K have been detected in Europa's extremely tenuous atmosphere.

10.2.3 Ganymede and Callisto

The outer two Galilean satellites represent a fundamentally different type of body from the inner two. Their low densities, $\rho = 1940$ kg m^{-3} for Ganymede and $\rho = 1830$ kg m^{-3} for Callisto, are suggestive of a mixture of comparable amounts (by mass) of rock and ice. Global views of these two huge moons are shown in Figure 10.3, and Figures 10.10 and 10.11 show closeup images of specific regions on Callisto and Ganymede, respectively. Ice has been observed in the spectra of both bodies (Fig. 10.5), but it is far more contaminated than the ice on Europa's surface.

Impact craters are ubiquitous on both satellites. Callisto's surface is essentially saturated with medium-sized craters, although there is a dearth

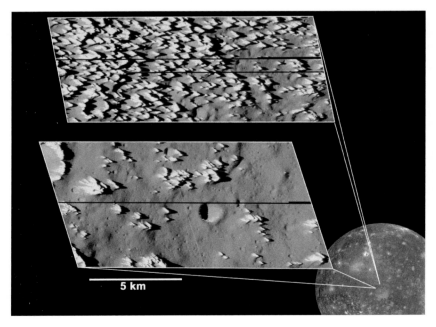

5 km

Figure 10.10 A region just south of the multiring impact crater Asgard on Callisto reveals numerous bright, sharp knobs, approximately 80–100 m high. They may consist of material thrown outward from a major impact billions of years ago. These knobs, or spires, are very icy but also contain some darker dust. As the ice erodes, the dark material appears to slide down and accumulate in low-lying areas. The lower close-up image shows somewhat older terrain, judging from the number of impact craters. Comparison of these images suggests that the spires erode away over time. (NASA/JPL/Arizona State University, PIA03455)

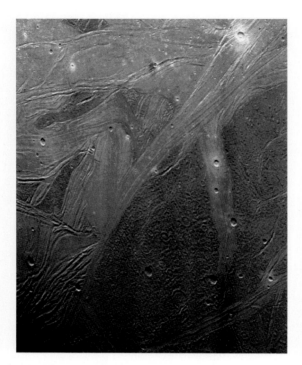

Figure 10.11 View of the Marius Regio and Nippur Sulcus area on Ganymede showing the dark and bright grooved terrain, which is typical on this satellite. The older, more heavily cratered dark terrain is rutted with furrows, shallow troughs perhaps formed as a result of ancient giant impacts. Bright grooved terrain is younger and was formed through tectonics, probably combined with icy volcanism. The image covers an area ~664 × 518 km at a resolution of 940 m/pixel. (NASA/*Galileo Orbiter*, PIA01618)

relief; such relief was probably present initially but was erased by relaxation caused by flowing subsurface ice.

Callisto's surface shows signs of weakness or crumbling at small scales, which can be seen in Figure 10.10. This crumbling or degradation may be produced by sublimation of a volatile component of the crust and may bury or destroy craters, leading perhaps to the above-mentioned lack of (sub-) kilometer-sized craters.

The geology of Ganymede is more diverse than that of Callisto. At low resolution Ganymede resembles the Moon, in that both dark and light areas are visible (Fig. 10.3). However, in contrast to the Moon, the dark areas on Ganymede's surface are the oldest regions, being heavily cratered, nearly to saturation. The lighter terrain is less cratered, albeit more than the lunar maria, so it must be younger than the dark terrain, although probably still quite old. The light-colored terrain shown in Figure 10.11 is characterized by a complex system of parallel ridges and grooves, up to tens of kilometers wide and maybe a few hundred meters high. These features are clearly of endogenic origin.

Ganymede's low moment of inertia ratio, $I/(MR^2) = 0.312$, implies that its mass is heavily concentrated towards the center. Most intriguing was the discovery of a magnetic field of intrinsic origin around this satellite, and hence Ganymede must have a liquid metallic core, most likely with a small solid inner core. Best fits to the gravity, magnetic field and density data are obtained with a three-layer internal model in which each layer is ~900 km thick. The innermost layer is a metallic core surrounded by a silicate mantle, which is topped off by a thick H_2O-ice shell. The water may be liquid at a depth of ~150 km (2 kbar), where the temperature (253 K) corresponds to the minimal melting point of water.

Callisto's moment of inertia, $I/(MR^2) = 0.355$, is slightly less than expected for a pressure compressed, yet compositionally homogeneous,

of (sub-) kilometer-sized craters. Most craters on both moons are flatter than those on our Moon, and some show unique features. These characteristics have been attributed to the relatively low (compared with rock) viscosity of the icy crust, even at low temperatures. Craters $\lesssim 2$–3 km in diameter reveal the classic bowl-shaped morphology (§6.4), and at larger sizes, central peaks appear. In some cases, all that can be distinguished from an impact crater is a large bright circular patch, some with and others without concentric rings around them. These features, called **palimpsests**, are similar to large impact basins but lack any topographic

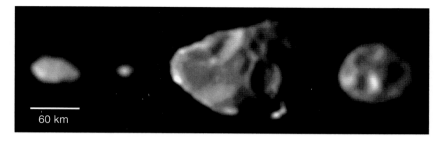

60 km

Figure 10.12 The four small, irregularly shaped 'ring' moons that have orbits within Jupiter's ring system. The moons are shown in their correct relative sizes. From left to right, arranged in order of increasing distance from Jupiter, are Metis, Adrastea, Amalthea and Thebe. (NASA/*Galileo Orbiter*, PIA01076)

mixture of ice and rock. The data are not conclusive, but it has been speculated that Callisto may be partially differentiated, with an icy crustal layer (a few hundred kilometers) and an ice/rock mantle that is slightly denser towards the center of the satellite. The magnetometer on board the *Galileo* spacecraft discovered magnetic field disturbances that suggest the presence of a salty ocean within Callisto. As on Ganymede, such an ocean may exist at a depth of $\sim$150 km.

10.2.4 Jupiter's Small Moons

Jupiter has dozens of known moons, but the others are all very small compared with the Galilean satellites. Their combined mass is about 1/1000 that of Europa, the smallest of the Galilean satellites.

Inner Satellites

Four moons have been detected inside the orbit of Io; images of all four are shown in Figure 10.12. The largest, **Amalthea**, with a mean radius of 83.5 km, is distinctly nonspherical in shape, dark and red, and heavily cratered. Its low density, 860 ± 100 kg m^{-3}, combined with a presumed rocky composition, implies substantial voids, suggestive of a 'rubble pile' composition. Shock waves from impacts on such a porous body are quickly damped, which might explain how Amalthea can have several craters almost half its

size. The low density itself hints at a violent collisional history. The other inner satellites of Jupiter, **Thebe**, **Metis** and **Adrastea**, are also dark and red. All four of these moons are obviously associated with, and likely provide most of the particles in, Jupiter's dusty rings (§13.3.1).

Irregular Satellites

Jupiter's outer moons are much farther away from Jupiter than are the Galilean satellites. Their orbits are highly eccentric and inclined, in many cases retrograde. Collectively they are referred to as the **irregular satellites**. As of late 2018, seven jovian irregular moons had been discovered on prograde orbits, and several dozen were known on retrograde orbits.

The orbital elements of Jupiter's irregular satellites are not randomly distributed but reveal the presence of dynamical groupings. Five such 'families' have been identified. Each of these families most likely resulted from the breakup of a body (likely an asteroid, judging from spectra) subsequent to capture by Jupiter. The giant planet is also known to have captured Jupiter-family comets in the past. Some such bodies orbited Jupiter for decades before being ejected from the system or colliding with the planet (e.g., Comet D/Shoemaker–Levy 9; §8.1.2) or with a satellite.

10.3 Satellites of Saturn

Saturn has more than 60 known satellites, many of which are discussed in the following subsections. Titan is by far the largest satellite, with a mass more than 20 times that of all other saturnian satellites combined (Table E.5). In addition to Titan, we also devote an entire subsection (§10.3.3) to Enceladus, which is a most enigmatic small moon.

10.3.1 Titan

Titan, Saturn's largest satellite, was discovered in 1655 by Christiaan Huygens. With a radius of 2575 km, Titan is similar in size to Ganymede, Callisto and Mercury. Its mean density of 1880 kg m^{-3} puts Titan in the 'icy' satellite class ($\sim$50% rock, 50% ice by mass).

Titan's most distinctive attribute is its dense atmosphere, with a surface pressure of 1.44 bar. This atmosphere is composed primarily of nitrogen and a small but significant amount of methane. Methane is easily dissociated by solar photons; photolysis would destroy all CH_4 in Titan's atmosphere within $\sim$20–30 million years. Hence, its (presumably) continued presence of methane in Titan's atmosphere is a puzzle. Methane must somehow be resupplied to Titan's atmosphere, perhaps via a methane cycle analogous to the hydrological cycle on Earth (i.e., via evaporation of oceans, formation of clouds and precipitation).

As on the giant planets, photolysis of methane gas should lead to the production of numerous hydrocarbons, including acetylene (C_2H_2), ethylene (C_2H_4) and ethane (C_2H_6). Because of the low stratospheric temperature, C_2H_6 and other such photochemically produced complex molecules condense to form a dense layer of smog in Titan's atmosphere. Laboratory measurements by Carl Sagan and coworkers in a simulated Titan atmosphere show the formation of such 'gunk', a reddish-brown powder referred to as **tholins**. The smog particles ultimately sediment out and fall to the ground, where they might have built up, over the eons, a few-hundred-meter thick layer of hydrocarbons.

Titan is covered globally by an optically thin methane-ice cloud at altitudes of 25–35 km. A persistent light methane drizzle below this cloud is present, at least near the Xanadu mountains. During the summer, distinct clouds were seen near Titan's south pole. In addition, cloud features have been seen regularly at southern mid-latitudes, near $-40°$. Although the former clouds may have been triggered by the high surface temperature during the summer, the latter ones may be confined latitudinally by Titan's geography and/or by a global atmospheric circulation pattern. Large, but short-lived, clouds were observed over tropical latitudes near the epoch of the saturnian equinox. The *Cassini* spacecraft identified a large cloud of ethane over Titan's north (winter) pole in the upper troposphere, and some smaller (presumably methane) clouds at lower altitudes, which have been hypothesized as lake effects. At much higher altitudes, in the stratosphere and mesosphere, distinct haze layers are present.

The dense (photochemical) smog layer in Titan's atmosphere makes it impossible to remotely probe the satellite's surface at visible wavelengths. The smog is transparent, however, at longer wavelengths, so the surface can be imaged at infrared wavelengths outside of the methane absorption bands. An example of such an infrared image is shown in Figure 10.13a, with a view of the *Huygens* probe landing site near the Xanadu mountains taken by the *Cassini Orbiter* presented in panel b. Both images show significant surface albedo variations.

The combined observations of the *Cassini Orbiter* and photographs such as those shown in Figure 10.14 that were taken below the haze by the *Huygens* probe reveal a surface that has been etched by fluids. *Cassini* radar images show numerous channels that cut across different types of terrain. Radar-bright rivers may be filled with

(a)

(b)

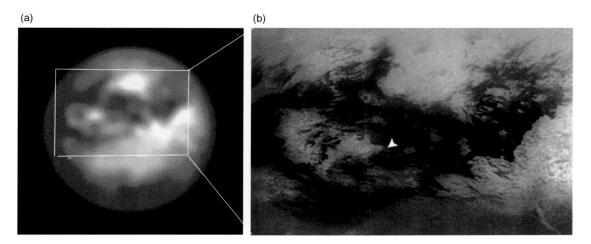

Figure 10.13 (a) An image of Titan's surface at a wavelength of 2.06 μm, obtained with the adaptive optics system on the W.M. Keck telescope one day after the descent of the *Huygens probe* on 14 January 2005. (Adapted from de Pater et al. 2006c) (b) Map of the region on Titan's surface as outlined (approximately) in panel (a), taken by the *Cassini Orbiter* at a wavelength of 938 nm. The *Huygens probe* landing site is indicated by an *arrow*. (NASA/JPL *Cassini Orbiter*, PIA08399)

boulders, and radar-dark channels suggest the presence of either liquids or smooth deposits. Lakes filled with hydrocarbon liquids have been discovered at high latitudes. One area with lakes is shown in Figure 10.15. The depth and extent of these lakes have been observed to vary over time.

However, even if all of the radar-dark features over both poles, which combined cover more than $600\,000$ km^2 (about 1% of Titan's total surface area), are filled with liquids, it may not be enough to explain Titan's methane cycle in a manner analogous to the hydrological cycle on Earth.

(a)

(b)

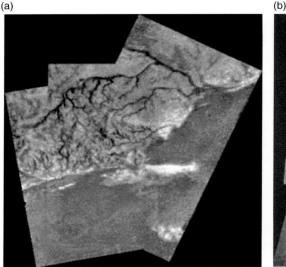

Figure 10.14 (a)   Mosaic of three frames from the *Huygens probe* shows a remarkable view of a 'shoreline' and channels, from an altitude of 6.5 km. The bright 'island' is about 2.5 km long. (NASA/ JPL/ ESA/ University of Arizona, PIA07236) (b) After landing, the *Huygens probe* obtained this view of Titan's surface, including 0.1–0.15 mm sized rocks, presumably made of ice. (ESA/NASA/JPL/University of Arizona, PIA06440)

Figure 10.15 *Cassini* radar images of lakes of liquid hydrocarbons near Titan's north pole. The lakes are darker than the surrounding terrain, indicative of regions of low backscatter. The strip of radar imagery is foreshortened to simulate an oblique view of the highest latitude region, seen from a point to its west. (NASA/JPL/USGS, PIA09102)

Only a handful of craters have been detected on Titan, indicative of a geologically young surface, perhaps a few hundred million years old. Many radar-bright 'flows' may be cryovolcanic lava flows, resurfacing Titan at a rapid rate. Such volcanic activity would also supply methane gas to the atmosphere. One ~180-km-wide feature may be a shield volcano, with an ~20-km diameter caldera at its center and sinuous channels and/or ridges radiating away from the caldera. Numerous longitudinal dunes, shown in Figure 10.16, dark both in radar echoes and at infrared wavelengths, are present in the equatorial region. These dunes are all oriented in the east–west direction and are up to thousands of kilometers long. The orientation

of the dunes has been used to derive the wind direction, which contrary to expectation is towards the east rather than the west.

10.3.2 Midsized Saturnian Moons

Saturn's six midsized moons range in radius from a little under 200 km (Mimas) up to 750 km (Rhea). Their densities range from just under 1000 kg m^{-3} (Tethys) up to 1600 kg m^{-3} (Enceladus), implying that they are ice-rich bodies with different amounts of rock. All of these moons are relatively spherical, suggestive of relatively low viscosities in their interiors at some point in their histories. Apart from one hemisphere of Iapetus, Saturn's midsized satellites are quite bright, with albedos, A_v, ranging from about 0.3 up to 1.0. All show water-ice in their surface spectra.

Detailed images taken by various spacecraft show that each satellite has its own unique characteristics. We defer discussion of Enceladus to the next subsection and show images of Saturn's four other midsized satellites that orbit interior to Titan in Figure 10.17. The surfaces of Mimas, Tethys and Rhea are heavily cratered. **Mimas** is characterized by one gigantic crater near the center of its leading hemisphere, about 140 km in diameter, one-third the moon's own size. The crater is about 10 km deep, and the central peak is ~6 km high. The impacting body must have been ~10 km across. **Tethys** displays an ~2000-km-long complex of valleys or troughs, Ithaca Chasma, which stretches three-quarters of the way around the satellite. **Dione** exhibits variations in surface albedo of almost a factor of two, which is much larger than those seen on **Rhea** but much less extreme than Iapetus's hemispheric asymmetry.

Iapetus, shown in Figure 10.18, is a bizarre body, with its trailing hemisphere ~10 times as bright as the leading hemisphere ($A_v \approx 0.5$ vs. $A_v \approx 0.05$). The black material on its

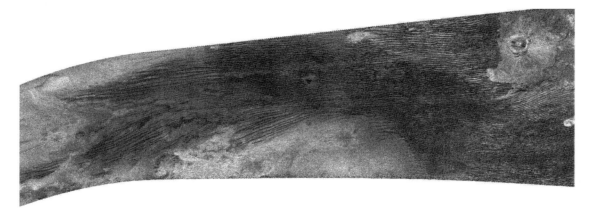

Figure 10.16 Part of Titan's surface as mapped by the *Cassini* radar instrument. Three of Titan's major surface features – dunes, craters and the Xanadu mountains – are shown. The hazy bright area at the *left* that extends to the *lower center* of the image marks the northwest edge of Xanadu. In the *upper right* is the crater Ksa, and the *dark lines* between these two features are linear dunes, similar to sand dunes on Earth in Egypt and Namibia. These longitudinal dunes make up most of Titan's equatorial dark regions. These ~100-km-long features run east–west on the satellite, are 1–2 km wide and spaced similarly, and are roughly 100 m high. They curve around the bright features in the image – which may be high-standing topographic obstacles – following the prevailing wind pattern. Unlike Earth's (silicate) sand dunes, these may be solid organic particles or ice coated with organic material. The image covers an area 350 km × 930 km, with a resolution of about 350 m/pixel. (NASA/JPL *Cassini Orbiter*, PIA14500)

leading hemisphere is reddish and might consist of organic, carbon-bearing compounds. Iapetus may just sweep up low albedo 'dirt' from Saturn's magnetosphere, such as dust from the dark satellite Phoebe. Iapetus's most remarkable topographic feature is a mysterious ~1300-km-long ridge, up to 20 km high at places, that coincides almost exactly with its geographic equator, as shown in Figure 10.18. Crater counts suggest the ridge to be ancient. Isolated peaks are observed at many of the places where segments of the ridge are absent.

10.3.3 Enceladus

Enceladus, with a radius of only 250 km, is a most remarkable satellite. Parts of this moon are heavily cratered, but large regions on the surface show virtually no impact craters at all. The youngest parts are probably no more than one million years old, and the oldest terrain solidified billions of years ago. Enceladus's surface reflectivity is very high, implying fresh, uncontaminated ice. With a bulk

density of 1600 kg m^{-3}, the satellite probably has a rocky core ($R \approx 170$ km, $\rho \approx 3000$ kg m^{-3}) and an ~80-km-thick icy crust. As on Ganymede and Europa, the crust displays regions of grooved terrain, indicative of tectonic processes, and smoother parts, possibly resurfaced by water flows.

Enceladus orbits Saturn between Mimas and Tethys. *Cassini*'s discovery of giant plumes of vapor, dust and ice emanating from Enceladus's south pole was unexpected. The first evidence for active geysers came from *Cassini*'s magnetometer data, which found a bending of field lines around the moon, indicative of mass-loading processes such as observed on Io (§§8.1.4 and 10.2.1). The plumes emanate from 'cracks' in the satellite's south polar region. Figure 10.19 shows these cracks, dubbed **tiger stripes**. *Cassini* detected temperatures of at least 180 K along some of the brightest tiger stripes, well above the 72-K background temperatures at other places in the south polar region. When flying through the plume, *Cassini* measured a gas composition similar to that

Mimas

Tethys

Dione

Rhea

Figure 10.17 Images (not to scale) of Saturn's inner midsized satellites other than Enceladus. The cratered surface of Mimas shows its 140-km-diameter crater Herschel (PIA06258). Tethys shows its anti–Saturn-facing hemisphere. The rim of the 450-km-diameter impact basin Odysseus lies on the eastern limb, making the limb appear flatter than elsewhere. Other large craters seen here are Penelope (*left of center*) and Melanthius (*below center*) (PIA08870). The trailing hemisphere of Dione shows many bright cliffs. At lower right is the feature called Cassandra, exhibiting linear rays extending in multiple directions (PIA08256). Rhea's crater-saturated surface shows a large bright blotch and radial streaks, which were likely created when a geologically recent impact sprayed out bright, fresh ice ejecta (PIA08189). (All images taken by the *Cassini* spacecraft; NASA/JPL/SSI)

Figure 10.18 The leading side of Iapetus, displayed in this image, is about ten times darker than its trailing side. An ancient, 400-km-wide impact basin shows just above the center of the disk. Along the equator is a conspicuous, 20-km-wide topographic ridge that extends from the western (*left*) side of Iapetus almost to the day/night boundary on the *right*. On the left horizon, the peak of the ridge rises at least 13 km above the surrounding terrain. (*Cassini*, NASA/JPL/SSI, PIA06166)

seen in comets. These plumes are likely the source of most of the material in Saturn's E ring. An image of Enceladus in Saturn's E ring is displayed in Figure 10.20.

The observed geyser activity on Enceladus requires a substantial heat source, the cause of which is still a puzzle. Primordial heat or radioactive decay is not sufficient, and tidal heating resulting from orbital eccentricities excited by its 2:1 orbital resonance with Dione may only be marginally adequate. Because the plume is composed primarily of water, it may erupt from chambers of liquid water just below the surface. The jets may also be 'driven' by **diapirs**, where warmer buoyant material (mushy ice or liquid) moves upwards through the ice shell to 'explode' into jets on the surface.

10.3.4 Small Regular Satellites of Saturn

All of Saturn's small inner moons are oddly shaped, heavily cratered and as reflective as Saturn's larger satellites. **Hyperion**, shown in Figure 10.21, is ~400 × 250 × 200 km and saturated with craters that appear to be deeply eroded. Its jagged and decidedly nonspherical shape implies that it is a collisional remnant of a larger body. Hyperion is the only satellite that displays a chaotic rotation.

The small inner moons of Saturn are shown to scale in Figure 10.22. Two of these moons, **Janus** and **Epimetheus**, share the same orbits and change places every four years (§2.2.2). **Calypso** and **Telesto** are located at the L_4 and L_5 Lagrangian points of Tethys's orbit, and **Helene** and **Polydeuces** reside in Dione's Lagrangian points. **Atlas** is a small moon orbiting just outside the A ring. **Prometheus** and **Pandora** are the inner and outer shepherds of Saturn's F ring and play a key role in shaping the kinky appearance of this ring (Fig. 13.24). The *Cassini* spacecraft discovered the satellites **Pallene** and **Methone** between the orbits of Mimas and Enceladus. Pallene is embedded within a faint ring of material. **Pan** and **Daphnis** orbit within the Encke and Keeler gaps within Saturn's A ring, respectively. The densities of these inner moons are very low, less than that of water (Table E.5). Such low densities imply that the moons are very porous.

10.3.5 Saturn's Irregular Moons

A large number of smaller irregular satellites orbit Saturn at relatively large distances ($\lesssim 20 \times 10^6$ km). These moons typically move in highly eccentric and inclined orbits, many of which are retrograde, suggestive of captured objects rather than formation within Saturn's subnebula.

Phoebe is by far the largest of Saturn's irregular moons and the only one for which we have resolved images. The moon is very dark ($A_v \approx 0.06$), similar to that of C-type asteroids and comets. Phoebe's density (1600 kg m^{-3}) suggests

(a)

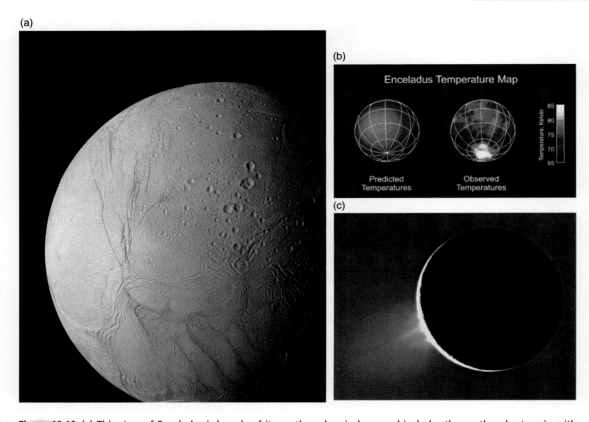

Figure 10.19 (a) This view of Enceladus is largely of its southern hemisphere and includes the south polar terrain with the blue 'tiger stripes' at the *bottom* of the image. The south polar region is encircled by a conspicuous and continuous chain of folds and ridges, a near absence of craters, and the presence of large blocks or boulders, presumably made of ice. Other parts of the disk are heavily cratered, where ancient craters appear somewhat pristine in some areas but have clearly relaxed in others. (NASA/JPL/SSI, PIA07800) (b) *Left:* Model of Enceladus's surface temperature as it would be if sunlight were the only source of energy. *Right:* A global temperature map made from measurements of Enceladus's radiation at wavelengths between 9 and 16.5 μm. The spatial resolution is 25 km. As expected, temperatures near the equator peak at 80 K, but the south pole reaches 85 K, ~15 K warmer than expected. The composite infrared spectrometer data suggest that small areas of the pole exceed 180 K. (NASA/JPL/GSFC/SWRI, PIA09037) (c) *Cassini* detected plumes of vapor, dust and ice emanating from the tiger stripes on Enceladus. The plumes are backlit by the Sun. The jets are geysers erupting from pressurized subsurface reservoirs of liquid water. (NASA/JPL/SSI, PIA07758)

a body composed of ice and rock. Water-ice was detected via ground-based spectroscopy, and *Cassini* identified CO_2-ice and organic materials on its surface. These findings, together with Phoebe's retrograde orbit, point at a capture origin. *Cassini* images of the moon, such as Figure 10.23, show an unusual variation in brightness, where some crater slopes and floors display bright

material – probably ice – on what is otherwise an extremely dark body.

10.4 Satellites of Uranus

Uranus's five largest moons are shown in Figure 10.24. They orbit in or near the plane of the planet's

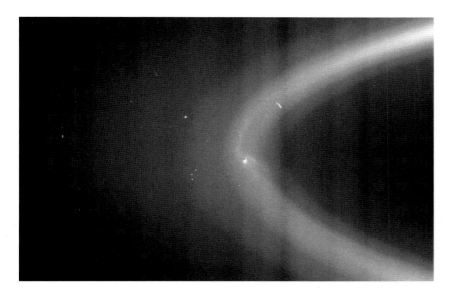

Figure 10.20 Wispy streaks of bright, icy material reach tens of thousands of km outward from Saturn's moon Enceladus into the E ring, and the moon's active south polar jets continue to fire away. The Sun is almost directly behind the Saturn system from *Cassini*'s vantage point, so small particles are 'lit up', but Enceladus itself is dark. Tethys is visible to the left of Enceladus. The image was taken in visible light when *Cassini* was at ~2.1 million km from Enceladus. (NASA/JPL/SSI, PIA08321)

Figure 10.21 Chaotically tumbling and seriously eroded by impacts, Hyperion is one of Saturn's more unusual satellites. The moon is quite porous; in this view it resembles a sponge rather than a solid rocky object. Its color is unusual as well, being rosy tan, perhaps from debris eroded from moons farther out. (*Cassini*, NASA/JPL/SSI, PIA07740)

equator, which is tilted by 98° with respect to Uranus's orbit around the Sun. Their radii range from 235 km for Miranda to almost 800 km for Titania.

The innermost of these five moons, **Miranda**, has a bizarre surface, which is readily apparent in Figures 10.24a and b. Some areas of Miranda are extremely heavily cratered, but other regions have only a few craters and a surprising endogenic terrain characterized by almost parallel sets of bright and dark bands, scarps and ridges. The boundaries between differing types of terrain are very sharp. There is no good explanation for this great diversity in terrain types.

Next in distance from Uranus is **Ariel**. Ariel shows clear signs of local resurfacing but nothing as striking as Miranda. The age of terrain varies significantly, but the entire surface appears to be younger than the oldest terrains on each of the three outer major moons. Although **Umbriel** is similar in size to Ariel, this moon is heavily cratered and appears to have the oldest surface in the uranian system. There is little or no evidence of tectonic activity. Most of **Titania** is heavily cratered; however, some patches of smoother material with fewer craters imply local resurfacing. An extensive network of faults cuts the surface of Titania. **Oberon**'s surface is dominated by craters, but

Figure 10.22 Saturn's smallest regular satellites, from the innermost one (Pan) outward, all to scale. All satellite images were taken by the *Cassini* spacecraft. (NASA/JPL/SSI; Courtesy Peter Thomas)

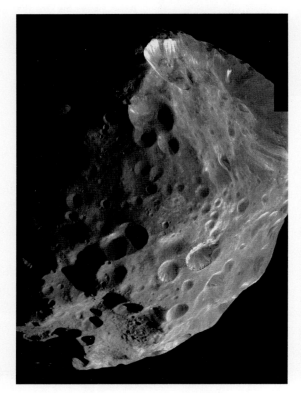

Figure 10.23 This mosaic of Saturn's moon Phoebe reveals the satellite's topography. Unusual variations in brightness are visible on some crater slopes and floors, showing evidence of layered deposits of alternating bright and dark material. (*Cassini*, NASA/JPL/SSI, PIA06073)

there are several high-contrast albedo features and signs of faulting.

Thirteen small moons are known to orbit Uranus interior to Miranda's orbit. The largest of the small satellites is **Puck**, with $R \approx 80$ km. Puck is darker than any of the five large satellites, slightly irregular in shape and heavily cratered. Nine moons, collectively referred to as the 'Portia group' after their largest member, have orbits between 59 200 km and 76 400 km from Uranus. Orbital calculations show that this family of satellites is chaotic and dynamically unstable; they may be remnants of a larger satellite. **Mab** is a particularly intriguing satellite because its orbit is centered in the outermost ring of Uranus, the μ ring, a ring which shows some similarities to Saturn's E ring (§13.3.3). **Cordelia** and **Ophelia** are two ring shepherds that control the inner and outer edges of the ϵ ring, the brightest and outermost of the planet's main rings (§13.3.3).

Irregular satellites have been discovered up to $\sim 20 \times 10^6$ km from Uranus. Similar to the irregular satellites of the other giant planets, many are on retrograde and/or highly eccentric orbits. These moons are most likely captured objects.

10.5 Satellites of Neptune

Before the 1989 *Voyager* flyby, only two moons of Neptune, Triton and Nereid, were known. Both of these bodies occupy 'unusual' orbits. **Triton** orbits Neptune at 14.0 R$_\Psi$, and has a very small eccentricity ($e < 0.0005$), but its orbit is inclined

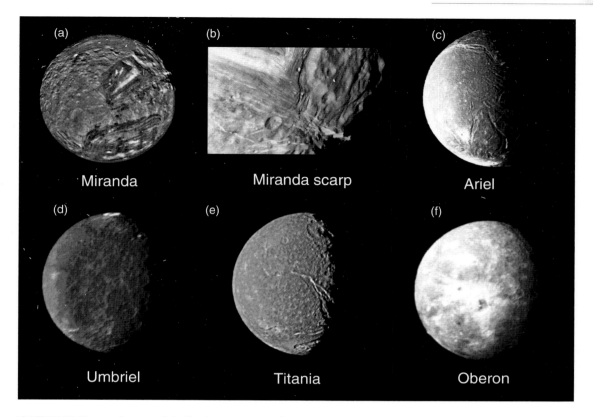

Figure 10.24 *Voyager* images of the five largest moons of Uranus, not shown to scale. (a) Miranda displays two strikingly different types of terrain. An old, heavily cratered rolling terrain with relatively uniform albedo, contrasted by a young, complex terrain that is characterized by sets of bright and dark bands, scarps, ridges and cliffs up to 20 km high, as seen most distinctly in the 'chevron' feature displayed in panel (b). (PIA01490) (c) Most of Ariel's visible surface consists of relatively heavily cratered terrain transected by fault scarps and fault-bounded valleys, as shown on this mosaic. (PIA01534) (d) Umbriel is the darkest of Uranus's larger moons and the one that appears to have experienced the least geological activity. Note the bright ring at top, ~140 km across, which lies near the satellite's equator. This may be a frost deposit, perhaps associated with an impact crater. Just below this feature, on the terminator, is an ~110-km diameter crater that has a bright central peak. (PIA00040) (e) Titania, Uranus's largest satellite, is heavily cratered and displays prominent fault valleys up to 1500 km long and 75 km wide. In valleys seen at right-center, the sunward-facing walls are very bright, suggestive of younger frost deposits. A prominent impact crater is visible at top. (PIA00039) (f) Oberon's icy surface is covered by impact craters, many of which are surrounded by bright rays. Near the center of the disk is a large crater with a bright central peak and a floor partially covered with very dark material. A large mountain, about 6 km high, is visible on the lower left limb. (PIA00034, NASA.JPL)

159° with respect to Neptune's equator, so that the satellite circles the planet in the retrograde direction. This odd orbit is the reason that it is generally accepted that Triton must have been captured from the Kuiper belt. With Neptune's rotation axis inclined by 28.8° relative to the planet's orbit about the Sun and Triton's inclined orbit precessing about Neptune's equatorial plane, Triton undergoes a complicated cycle of seasons within seasons, which lasts about 600 years.

Nereid is a small moon, with a radius of ~170 km, but nonetheless is reasonably round.

(a)

(b)

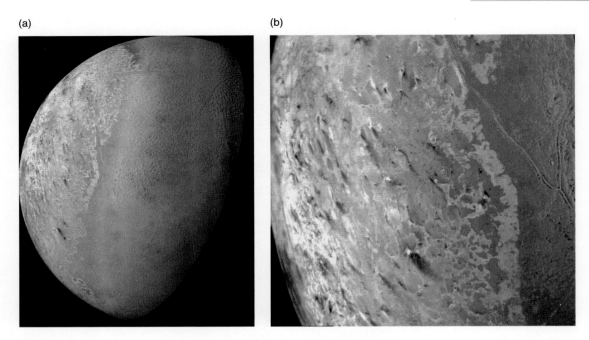

Figure 10.25 (a) COLOR PLATE A color mosaic of Neptune's largest moon, Triton. Triton's surface is covered by nitrogen-ice, while the pinkish deposits on the south polar cap (on the left) may contain methane-ice, which would have reacted under sunlight to form pink or red compounds. The dark streaks may be carbonaceous dust deposited from huge geyser-like plumes. The bluish-green band extends all the way around Triton near the equator; it may consist of relatively fresh nitrogen-frost deposits. The greenish areas include what is called the cantaloupe terrain and a set of 'cryovolcanic' landscapes. (NASA/*Voyager 2*, PIA00317) (b) This image of the south polar terrain of Triton reveals about 50 dark 'wind streaks'. A few plumes are observed to originate at dark spots; these are several km in diameter and some are more than 150 km long. The spots may be vents or geysers where gas has erupted from beneath the surface, carrying dark particles into Triton's atmosphere. Southwesterly winds then transported this dust, which formed gradually thinning deposits to the northeast of most vents. (NASA/*Voyager 2*, PIA00059)

Nereid's orbit about Neptune is prograde, and has a semimajor axis of 219 R_Ψ. The orbit has the largest eccentricity of any known moon, with $e = 0.76$. Nereid's orbit is inclined by $\sim27°$ to Neptune's equator and, more importantly, by 7.2° to its Laplace plane (essentially Neptune's orbital plane).

Triton, shown in Figure 10.25, is by far the largest moon in the neptunian satellite system, with a size somewhat smaller than Jupiter's Europa. It has a very tenuous atmosphere of nitrogen, with a trace of methane gas (mixing ratio $\sim10^{-4}$). Triton has the lowest observed surface temperature of any planetary satellite, 38 ± 4 K. The polar cap on the southern hemisphere is bright, with an albedo of

~0.9, but the equatorial region is somewhat darker and redder. Most of the surface is covered with a thin layer of nitrogen and methane ice that does not completely hide the underlying terrain. The western (trailing) hemisphere of Triton looks like a cantaloupe, a dense concentration of pits or dimples criss-crossed by ridges or fracture systems. The leading hemisphere consists of a smoother surface, with large calderas and/or lava lakes. The polar regions are covered with N_2-ice, which evaporates in the spring. The ice has a slightly reddish tint, indicative of organic compounds. A large number of relatively dark (10%–20% lower albedos than the surroundings) streaks are seen near the illuminated pole. At least two of these streaks appear to be active geyser plumes, likely driven by liquid nitrogen. The plumes rise ~8 km and then are swept westwards by the winds. Debris streaks produced by such plumes reach lengths exceeding 100 km. Although the heating mechanism for the geysers has not yet been understood, sunlight and the solid-state greenhouse effect (§4.6) might play roles (Problem 10-6) because all four geysers detected are in the south polar region, where the surface was continuously illuminated by the Sun.

Voyager discovered six satellites within and near Neptune's ring system. The largest of these, **Proteus**, with a radius of 200 km, is slightly bigger than Nereid. The radii of the other satellites are between 27 and 100 km. All of these satellites are dark, and those that were resolved in *Voyager* images are nonspherical.

Several irregular satellites have been found as far as 50×10^6 km from the planet. Little is known about these small moons, which, as with the irregular satellites of Uranus, were all discovered well after the *Voyager* encounter.

Key Concepts

- All four giant planets have both regular and irregular satellites. The regular satellites formed with the planet in the planetary subnebula. The irregular satellites were captured from heliocentric orbits.
- The surfaces of the large regular moons that are closest to the planet are typically younger (more endogenic activity) than the moons that are farther away.
- All giant planets have small regular inner satellites that 'interact' with the planets' ring systems.
- Io is the most volcanically active body in our Solar System. Tidal heating is the dominant source of energy for Io.
- Europa has a young ice surface and a massive subsurface liquid water ocean.
- Ganymede is the only moon with its own internally generated magnetic field.
- Titan has some similarities to Earth that other bodies lack, including atmospheres of similar density and large amounts of liquid falling as rain, flowing as rivers to shape the landscape and collecting into lakes.
- Titan's atmosphere may resemble the atmosphere of Earth before life developed.
- Enceladus has active H_2O geysers near its south pole; a reservoir of underground liquid water probably exists below. Tidal heating is likely the dominant, but probably not the only, source of energy.
- Triton appears to have geysers composed of liquid nitrogen.

Further Reading

Excellent reviews of many of the satellites are provided in *The Encyclopedia of the Solar System*, 2nd Edition. Eds. L. McFadden, P.R. Weissman, and T.V. Johnson. Academic Press, San Diego:

Planetary Satellites, by B.J. Buratti and P.C. Thomas, pp. 365–382.

Io: the Volcanic Moon, by R.M.C. Lopes, pp. 419–430.

Europa, by L.M. Prockter and R.T. Pappalardo, pp. 431–448.

Ganymede and Callisto, by G. Collins and T.V. Johnson, pp. 449–466.

Titan, by A. Coustenis, pp. 467–482.

Triton, by W.B. McKinnon and R.L. Kirk, pp. 483–502.

Problems

10-1. The *Voyager 1* spacecraft detected nine active volcanoes on Io. If we assume that, on average, there are nine volcanoes active on Io and that the average eruptive rate per volcano is 50 km^3 yr^{-1}, calculate:
(a) the average resurfacing rate on Io in m yr^{-1}
(b) the time it takes to completely renew the upper kilometer of Io's surface

10-2. **(a)** Calculate the energy involved when an ion in the Io plasma torus impacts Io. (Hint: Calculate the orbital velocity of Io and the ion, assuming both orbit Jupiter at a planetocentric distance of 6 R$_{2\!+}$.)
(b) Calculate the size of Io's neutral sodium cloud if the lifetime for sodium atoms is a few hours.
(c) Explain in words why the banana-shaped neutral sodium cloud is pointed in the forward direction.

10-3. **(a)** Calculate the ratio of the rate of collisions per unit area at the **apex** (the center of Callisto's leading hemisphere) to that at the **antapex** for a population of impactors that approaches Jupiter with $v_\infty = 5$ km s^{-1}. You may assume that Callisto is on a circular orbit and neglect the moon's gravity, but do not neglect Jupiter's gravitational pull on the impactors.
(b) Calculate the ratio of kinetic energies per unit mass at impact for the situation considered in (a).
(c) Repeat your calculations for Io in place of Callisto.
(d) Repeat your calculations in (a) and (b) for long-period comets that approach Jupiter with $v_\infty = 15$ km s^{-1}.

10-4. Methane (CH_4) is photolyzed by solar vacuum ultraviolet light (uv), chiefly the Lyman-α line at 121.6 nm. When a CH_4 molecule absorbs a Lyman-α photon, the molecule is broken up. Methane photolysis produces methyl (CH_3) radicals and hydrogen atoms:

$$CH_4 + uv \rightarrow CH_3 + H.$$

Assume that every CH_3 radical combines with another CH_3 radical to make a molecule of ethane (C_2H_6) and every H atom combines with another H atom to make a molecule of hydrogen (H_2):

$$CH_3 + CH_3 \rightarrow C_2H_6,$$

$$H + H \rightarrow H_2.$$

Hence it takes two Lyman-α photons to create a molecule of ethane from methane.
(a) If the Solar Lyman-α radiation is on average 1×10^{16} photons/m^2/s at Earth, what is the incident flux at Titan, which orbits Saturn at 9.5 AU? What is the flux of Lyman-α averaged over the entire globe?
(b) If every H_2 molecule escapes to space, what is the average H_2 escape flux? What is the C_2H_6 creation rate? What is the CH_4 destruction rate?
(c) At this rate, how long will the methane currently in Titan's atmosphere last? Titan's atmosphere is 1.5 bars, it has a mean molecular weight of 28, and it is 5% methane by number. The surface pressure is the weight of a column of atmosphere, $P = N\mu_a g_p$, where g_p is the surface gravity, N is the number of moles in

the column (think of a column as the atmosphere directly above a square centimeter of surface, extending from the surface out to space) and μ_a is the mean molecular weight of the atmosphere. Titan's surface gravity is 1.4 m/sec^2. Compute the number of moles of gas in a column of Titan's atmosphere. Use Avogadro's number to express the column density (the number of molecules in a column) of methane. Use the CH_4 loss rate calculated earlier to estimate the lifetime of the current reservoir of CH_4 in Titan's atmosphere.

(d) How many moles of ethane accumulate at the surface over this period of time?

(e) The density of liquid ethane is 550 kg m^{-3}. How much liquid ethane would accumulate over the age of the Solar System at the current predicted rate?

(f) What is the boiling point of CH_4 at 1 bar pressure? What is the boiling point of C_2H_6 at 1 bar pressure? What are the freezing points of CH_4 and C_2H_6?

(g) Comment on why the European Space Agency designed the *Huygens* *probe* to float and comment on what was actually found and why the predictions were incorrect.

10-5. **(a)** Under what conditions do moons tidally evolve inwards towards their planet? Explain.

(b) Name two moons that are tidally evolving inwards.

10-6. Geysers have been observed to erupt from the subsurface of Neptune's moon Triton. These eruptions may imply that a solid state greenhouse effect, in which solid molecular nitrogen plays the role that carbon dioxide does in the atmospheres of some terrestrial planets, is operating on Triton. Discuss the characteristics that a slab of frozen nitrogen would need to have in order for solar heating to produce a higher temperature underneath the slab than on top.

10-7. To calculate both the radius and the albedo of an unresolved moon, one needs observations at infrared as well as visible wavelengths. Explain in one sentence why this is the case.

CHAPTER 11

Meteorites

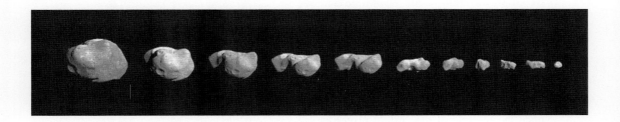

On the seventh of November, 1492, a singular miracle happened: for between eleven and twelve in the forenoon, with a loud crash of thunder and a prolonged noise heard afar off, there fell in the town of Ensisheim a stone weighing 260 pounds. It was seen by a child to strike the ground in a field of the canton of Gisguad, where it made a hole more than five feet deep.

Document in the town archives of Ensisheim, Germany

Von dem donnerstein gefallē jm xcij.iar:vor Ensißein.

Figure 11.1 Woodcut depicting the fall of a meteorite near the town of Ensisheim, Alsace, on 7 November 1492. A literal translation of the German caption reads 'of the thunder-stone (that) fell in 92 year outside of Ensisheim'. This meteorite is the oldest recorded fall from which material is still available.

A **meteorite** is a rock that has fallen from the sky. It was a **meteoroid** (or, if it was large enough, an asteroid) before it hit the atmosphere and a **meteor** while heated to incandescence by atmospheric friction. A meteor that explodes while passing through the atmosphere is termed a **bolide**. Meteorites that are associated with observations before or during the impact are called **falls**, and those simply recognized in the field are referred to as **finds**.

The study of meteorites has a long and colorful history. Meteorite falls have been observed and recorded for many centuries. The oldest recorded fall is the Nogata meteorite, which fell in Japan on 19 May 861. Figure 11.1 illustrates the fifteenth-century fall in Ensisheim, Germany. Iron meteorites were an important raw material for some primitive societies. However, even during the Enlightenment, it was difficult for many people (including scientists and other natural philosophers) to accept that stones could possibly fall from the sky, and reports of meteorite falls were sometimes treated with as much skepticism as UFO 'sightings' are given today. The extraterrestrial origin of meteorites became commonly acknowledged after the study of some well-observed and documented falls in Europe around the year 1800. The discovery of the first four asteroids, celestial bodies of subplanetary size, during the same period added to the conceptual framework that enabled

scientists to accept extraterrestrial origins for some rocks.

Meteorites provide us with samples of other worlds that can be analyzed in terrestrial laboratories. The overwhelming majority of meteorites are pieces of small asteroids that never grew to anywhere near planetary dimensions. Primitive meteorites, which contain moderate abundances of iron, come from planetesimals that never melted. Most of the iron-rich meteorites are thought to have once resided within the deep interiors of planetesimals that differentiated (§6.2.3) before being disrupted, whereas iron-poor meteorites are samples from the outer layers of differentiated bodies. Because small objects cool more rapidly than do large ones, the parent bodies of most meteorites either never got very hot or cooled and solidified early in the history of the Solar System. Many meteorites thus preserve a record of early Solar System history that has been wiped out on geologically active planets such as the Earth. The clues that meteorites provide to the formation of our planetary system are the focus of most meteorite studies and also of this chapter.

Meteorites are a diverse lot, and the properties and abundances of the major classes of meteorites are described in §11.1. Section 11.2 summarizes the relationships between meteorites and the bodies from which they come. We next discuss the passage of meteoroids through Earth's atmosphere

(§11.3). The remainder of this chapter concentrates on meteorite characteristics that provide clues to the formation of meteorites in particular and our Solar System in general. This discussion introduces the reader to models for various physical processes and, in §11.6, to the techniques that are used to determine the ages of meteorites and other rocks.

11.1 Classification

The traditional classification of meteorites is based on their gross appearance. Many people think of meteorites as chunks of metal because metallic meteorites appear quite different from ordinary terrestrial rocks. Museums also tend to specialize in metal meteorites because most people find these odd-shaped pieces of nickel–iron interesting to look at. Metallic meteorites are made primarily of iron, with a significant component of nickel and smaller amounts of several other **siderophile** elements (elements that readily combine with molten iron, such as gold, cobalt and platinum); thus, metal meteorites are referred to as **irons**. Meteorites that do not contain large concentrations of metal are known as **stones**. Many stony meteorites are difficult for the untrained eye to distinguish from terrestrial rocks. Meteorites that contain comparable amounts of macroscopic metallic and rocky components are called **stony-irons**.

A more fundamental classification scheme is based on the history of meteorite parent bodies. Most irons and stony-irons, as well as some stones known as **achondrites** (in contrast to chondrites, which are described later), come from **differentiated** parent bodies (i.e., bodies that have undergone density-dependent phase separation, §§6.1.4 and 6.2). Differentiated bodies experienced an epoch in which they were mostly molten, and much of their iron sank to the center, taking with it siderophile elements. (A small fraction of the analyzed irons was probably produced by impact-induced localized melting.) The bulk compositions of achondrites are enriched in **lithophile** and/or **chalcophile** elements. Lithophile elements tend to concentrate in the silicate phases of a melt, and chalcophile elements tend to concentrate in the sulfide phases of a melt. The abundances of both lithophiles and chalcophiles are also enhanced in the Earth's crust. Relative to a solar mixture of refractory elements, achondrites are significantly depleted in iron and siderophile elements. **Primitive meteorites** have not been differentiated; they are composed of material that formed directly from solar nebula condensates and surviving interstellar grains, modified in some cases by aqueous processing (implying that liquid water was once present in the parent body) and/or thermal processing (implying that the body was quite warm at some time).

Figure 11.2 shows photographs of three primitive meteorites that have been cut to produce a flat face, thereby revealing their internal structures. Greatly magnified views of cut primitive meteorites are displayed in Figure 11.3. Silicates, metals and other minerals are found in close proximity within primitive meteorites. Primitive meteorites are called **chondrites** because most of them contain small, nearly spherical, igneous inclusions known as **chondrules**, which solidified from melt droplets. Some chondrules are glassy, implying that they cooled extremely rapidly.

Figure 11.4 shows that apart from the most volatile elements, the composition of all chondrites is remarkably similar to that of the solar photosphere. Because meteorite compositions are easier to measure than are abundances in the Sun, analysis of chondrites provides the best estimates of the average Solar System composition of all but the most volatile elements (see Table 3.1). Densities of meteorites vary from 1700 kg m^{-3} for the primitive Tagish Lake carbonaceous chondrite meteorite (which fell in Canada on 18 January 2000) to between 7000 and 8000 kg m^{-3} for irons.

Most chondrites fit into one of three different **classes** that have been cataloged on the basis

(a)   (b)   (c)

Figure 11.2 Photographs of various chondritic meteorites. The scale bars are labeled in both centimeters and inches. (a) Brownfield H3.7 ordinary chondrite, which fell in Texas in 1937. Very small chondrules, plus highly reflective metal and sulfide grains, can be picked out. (b) Parnallee LL3 ordinary chondrite, which fell in India in 1857. The cut surface clearly shows well-delineated chondrules and slightly larger clasts. Parnallee is of very similar metamorphic grade to Brownfield, but it has a much coarser texture. (c) Vigarano CV3 carbonaceous chondrite, which fell in Italy in 1910. Vigarano has beautifully delineated chondrules. It also contains large calcium-aluminum inclusions (CAIs), which appear white in this photograph.

of composition and mineralogy; these classes are subdivided into various **groups**. Members of the most volatile-rich class of chondrites contain up to several percent carbon by mass and are known as **carbonaceous chondrites**. The carbonaceous chondrite class is divided into eight major groups that differ slightly in composition and are denoted CI, CM, CO, CV, CR, CH, CB and CK. The most common primitive meteorites are the **ordinary chondrites**, which are subclassified primarily on the basis of their Fe/Si ratio: H (high Fe), L (low Fe) and LL (low Fe, low metal; i.e., most of the iron that is present is oxidized). The third class of primitive meteorites, **enstatite chondrites**, is named after their dominant mineral ($MgSiO_3$). These highly reduced chondrites are also divided on the basis of their iron abundance, and groups are denoted EH and EL.

Although chondritic meteorites have never been melted, they have been processed to some extent in 'planetary' environments (i.e., in asteroid-like parent bodies) via thermal metamorphism, shock, brecciation (breaking up and reassembly) and chemical reactions often involving liquid water. Chondrites are assigned a **petrographic type** ranging from 1 in the most volatile-rich primitive meteorites to 6 in the most thermally equilibrated chondrites (Fig. 11.3). Type 3 chondrites (Figs. 11.2 and 11.3) appear to be the least altered in planetary environments and provide the best data on the conditions within the protoplanetary disk. Types 1 and 2 show progressively more aqueous alteration; all known aqueously altered chondrites are carbonaceous. Type 1 chondrites are devoid of chondrules, which either never were present or have been completely destroyed by aqueous processing. In contrast, the degree of metamorphic alteration increases in higher numbered types above type 3.

Iron meteorites are classified primarily on the basis of their abundance of nickel and of the moderately volatile trace elements germanium and

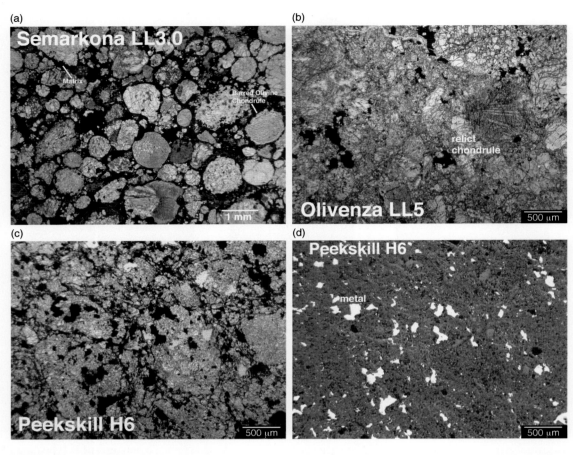

Figure 11.3 COLOR PLATE Thin-section photomicrographs of representative examples of chondrites showing petrographic variations among chondrite groups. (a) The Semarkona L3.0 ordinary chondrite in plane-polarized light, showing a high density of chondrules. Semarkona fell in India in 1940. (b) The Olivenza LL5 ordinary chondrite showing the texture of a recrystallized LL chondrite in which some relict chondrules are discernible. Olivenza fell in Spain in 1924. (c) The Peekskill H6 chondrite in plane-polarized light. Peekskill is a recrystallized ordinary chondrite. Peekskill fell in upstate New York on 9 October 1992. (d) Peekskill shown in reflected light, which emphasizes the 7% (by volume) metal content of this meteorite. (Weisberg et al. 2006)

gallium. Compositional differences are correlated to observed differences in structure. Crystallization textures of irons, such as the **Widmanstätten pattern** shown in Figure 11.5, also depend on the rate at which the meteorite cooled.

There are two classes of stony-irons: **pallasites**, which are closely related to irons, and **mesosiderites**, which are related to achondrites. Pallasites, an example of which is pictured in Figure 11.6, consist of networks of iron–nickel alloy

surrounding tiny nodules of olivine. Pallasites are of igneous origin, and they probably formed at a core–mantle interface. Mesosiderites, such as the meteorite pictured in Figure 11.7, contain a mixture of metal and magmatic rocks similar to **eucrite** achondrites.

Several different types of achondrites exist, presumably from different regions of differentiated (or at least locally melted) asteroids of a variety of compositions and sizes. A small percentage of

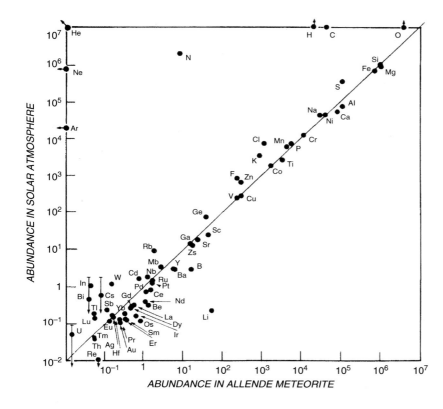

Figure 11.4 The abundance of elements in the Sun's photosphere plotted against their abundance in the Allende CV3 chondrite. Most elements lie very close to the curve of equal abundance (normalized to silicon). Several volatile elements lie above this curve, presumably because they are depleted in meteorites (rather than being enriched in the Sun). Only lithium lies substantially below the curve; lithium is depleted in the solar photosphere because it is destroyed by nuclear reactions near the base of the Sun's convective zone.

known achondrites are from two larger bodies: the Moon and Mars (§11.2).

The total mass of cosmic debris impacting the Earth's atmosphere in a typical year is 10^7–10^8 kg. Dust and micrometeorites $\sim$1–100 μm in radius account for most of the mass during most years, although the infrequent impacts of kilometer-sized and larger bodies dominate the flux averaged over very long timescales. The overwhelming majority of meteorites collected after being observed to fall are stones, most of which are chondrites.

Figure 11.5 A cut acid-etched surface of the Maltahhe iron meteorite. This surface shows a Widmanstätten pattern, an intergrowth of several alloys of iron and nickel that formed by diffusion of nickel atoms into solid iron during slow cooling within an asteroid's core. (Courtesy Jeff Smith)

11.2 Source Regions

Meteorites are identified by their extraterrestrial origin. The overwhelming majority (>90%) of

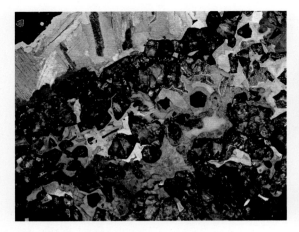

Figure 11.6 Photograph of the Seymchan pallasite, which was found in Siberia in 1967. A continuous network of iron-nickel metal acts as a frame holding grains of Mg-rich olivine. The imaged section is 83 mm across. (Courtesy Laurence Garvie)

Figure 11.7 Photograph of the Esterville mesoderite, which fell in Iowa in 1879. Note Esterville's brecciated structure, with large pods of iron-nickel metal mixed with a seemingly random jumble of stony clasts. The scale bar is labeled both in centimeters and inches, and the size of the meteorite is ~0.3 m.

meteorites are from bodies of subplanetary size in the asteroid belt. Comparison of the spectra of reflected light from several types of meteorites with asteroid spectra yields many close correspondences. Four spectral matches are shown in Figure 11.8. Firm identification of individual meteorites with specific asteroids is difficult, but strong cases can be made for some meteorite classes. The spectrum of 4 Vesta is unique among large asteroids and very similar to those of HED achondrite meteorites. Vesta's orbit in the inner asteroid belt (Table E.6) makes delivery to Earth of debris excavated by impacts easier than is the case for most asteroids. Thus, Vesta is the likely source of these interesting and well-studied differentiated meteorites (§11.7.1).

Pre-impact orbits have been determined for only a tiny fraction of meteorite falls, most of which are ordinary chondrites. The first four meteorite orbits to be determined are diagrammed in Figure 11.9. All of these orbits had perihelia near Earth's orbit; most penetrated the asteroid belt but remained well interior to Jupiter's orbit.

The typical time for a large object to reach Earth from the main asteroid belt is much longer than the age of the Solar System except near certain strong resonances. Poynting–Robertson drag (§2.8.2) can move very small meteoroids to the vicinity of resonances that can transport them into Earth-crossing orbits with characteristic lifetimes of 10^7 years; the Yarkovsky effect (§2.8.3) can transport ~1-m–10-km bodies into the same resonances. Earth-crossing asteroids provide another source of meteoroids with typical lifetimes as small space rocks of ~10^7 years. These short intervals are consistent with cosmic-ray exposure ages (§11.6.4), but note that a source is required for

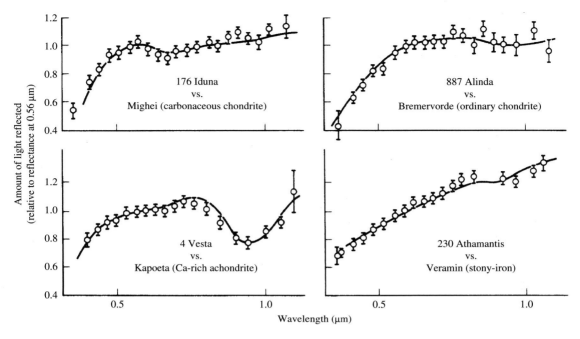

Figure 11.8 A comparison of the reflection spectra of four asteroids (*points with error bars*) with meteorite spectra as determined in the laboratory (*solid curves*). (Morrison and Owen 1996)

Earth-crossing asteroids, whose dynamical lifetime is far less than the age of the Solar System (§12.2).

Based on a comparison with *Apollo* samples, several dozen achondrite meteorites, many of which are **anorthositic breccias**, are clearly of lunar origin. A similar number of achondrite meteorites, including four falls (representing a total of ~0.4% of all known meteorite falls), are in the

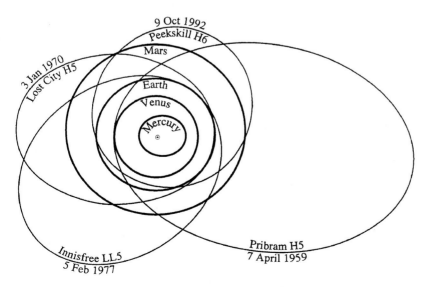

Figure 11.9 Pre-impact orbits of the first four recovered meteorites that were photographed well enough to allow for the calculation of accurate trajectories. All four are metamorphically processed chondrites, with petrographic types shown on the figure. The deduced orbit of Neuschwanstein, which fell on 6 April 2002, is almost identical to that of Pribram, so it was very surprising when laboratory analysis revealed that Neuschwanstein is an EL6 enstatite chondrite. (Adapted from Lipschutz and Schultz 2007)

Figure 11.10 A cut surface of the martian meteorite ALH 77005. This rock contains dark olivine crystals and light-colored pyroxene crystals plus some patches of impact melt, which were probably produced when it was ejected from the surface of Mars. The cube (W) is 1 cm on a side. (NASA/JSC)

SNC class (shergottites, nakhlites and chassignites, named after the places where the first examples were found), which are of martian origin. Figure 11.10 shows an SNC meteorite found in Antarctica. These rocks are young, with most having crystallization ages of $<1.3 \times 10^9$ years. (Meteorite ages are measured using radiometric dating techniques, which are described in detail in §11.6.) Convincing evidence for the martian origin of the SNC meteorites came from the similarity between their noble gas abundances and isotopic ratios of both noble gases and nitrogen to those measured in the martian atmosphere by the *Viking* landers. A 4.1×10^9-year-old non-SNC martian meteorite was recovered from Antarctica. This rock, ALH84001, has been extensively studied because it possesses several intriguing characteristics (including magnetite similar to that produced biologically on Earth) that were initially interpreted as evidence for ancient life on Mars (§16.13.1). Theoretical studies indicate that it is much easier for impacts to eject unvaporized rocks from Mars than from Earth because the velocity required to escape from Mars is less than half of Earth's escape velocity.

Meteorites are identified by their extraterrestrial origin. In theory, a rock could be knocked off the Earth, escape the Earth's gravity and orbit the Sun for a period of time before reimpacting Earth. Such a body would be a meteorite despite its terrestrial origin. However, it would be very difficult to identify such rocks as meteorites, and no such meteorite has ever been identified. Nonetheless, **tektites**, which are rocks formed from molten material knocked off the Earth's surface by impacts, have been found in many locations. The geographic distribution of tektite finds and the lack of cosmic ray tracks in tektites imply that these rocks did not escape but rather quickly fell back to Earth's surface.

11.3 Fall Phenomena

Meteoroids encounter the Earth's atmosphere at speeds ranging from 11 km s^{-1} to 73 km s^{-1}, with typical velocities of $\sim$15 km s^{-1} for bodies of asteroidal origin and $\sim$30 km s^{-1} for cometary objects (Problem 11-1). At such velocities, meteoroids have substantial kinetic energy per unit mass, enough energy to completely vaporize if it was converted to heat.

In the rarefied upper portion of a planet's atmosphere, gas molecules independently collide with the rapidly moving meteoroid. Interactions with this tenuous gas are dynamically insignificant for large meteoroids but are able to retard the motion of tiny bodies substantially (see eq. 11.2). **Micrometeoroids** smaller than $\sim$10–100 μm are able to radiate the heat they acquire from this drag rapidly enough that they can reach the ground. These very small particles are known as **cosmic spherules** if they are fully melted objects and **micrometeorites** if not.

Most of the highly porous, friable aggregates collected in the stratosphere, such as the one pictured in Figure 11.11, come from interplanetary dust particles (IDPs) that are likely to have been

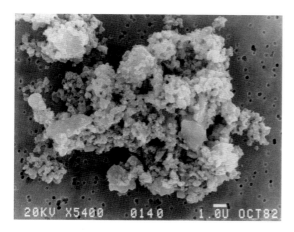

Figure 11.11 Scanning electron microscope image of a micrometeorite that was once an interplanetary dust particle. Note the fluffy fractal-like structure. (Courtesy Donald Brownlee)

more coherent than those collected in the stratosphere, and most come from IDPs of asteroidal origins. Micrometeorites of the types related to carbonaceous chondrites appear to be several times as common as those related to ordinary chondrites, but ordinary chondrites are much more abundant than carbonaceous chondrites in meteorite collections.

The surface of a meteor is heated by radiation from the atmospheric shock front that it produces. Meteors rarely get significantly hotter than 2000 K because this temperature is sufficient to cause iron and silicates to melt. The liquid evaporates or just falls off the meteor, so ablation provides an effective thermostat. Most visible meteors are from millimeter- and centimeter-sized bodies. The initial mass necessary for part of a meteoroid to make it to the ground depends on its initial speed, impact angle and composition. The surface of a meteor can become hot enough to melt rock to a depth of ~1 mm (~10 mm for irons, which have a higher thermal conductivity). However, interior to this hot outer skin the

incorporated in cold, volatile-rich bodies such as comets. Figure 11.12 shows eight micrometeorites (including cosmic spherules) that were collected on the Earth's surface. Micrometeorites found on the surface (typically the bottoms of lakes in Greenland or Antarctica) tend to be larger and

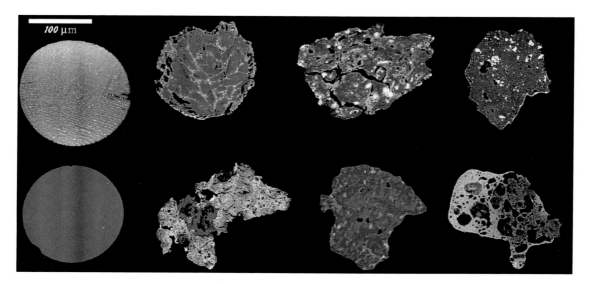

Figure 11.12 A selection of micrometeorites and cosmic spherules recovered from Antarctic aeolian sediments. Particles were mounted in epoxy and polished to expose a cross-section. All images are backscattered electron micrographs, where brighter gray means a higher atomic number. (Courtesy Ralph Harvey)

Figure 11.13 The 0.8-kg Lafayette (Indiana) nakhlite (martian meteorite) shows an exquisitely preserved fusion crust. During the stone's rapid transit through the atmosphere, air friction melted its exterior. The lines trace beads of melted rock streaming away from the apex of motion. (Courtesy Smithsonian Institution)

meteor is subject to an average pressure, P, given approximately by the formula:

$$P \approx \frac{C_D \rho_g v^2}{2}, \tag{11.1}$$

where v is the velocity of the meteor, C_D is the drag coefficient (which is approximately unity for a sphere) and ρ_g is the local density of the atmosphere. The meteor is slowed by this pressure, but it continues to be accelerated by the planet's gravity, so its velocity varies as

$$\frac{d\mathbf{v}}{dt} = -\frac{C_D \rho_g A v}{2m} \mathbf{v} - g_p \hat{\mathbf{z}}, \tag{11.2}$$

where m is the mass of the meteor, g_p is the gravitational acceleration and $\hat{\mathbf{z}}$ is the unit vector pointing in the upwards direction. Equation (11.2) accounts for aerodynamic drag (eq. 2.63a) in a uniform gravitational field. A meteor loses a substantial fraction of its initial kinetic energy if it passes through a column of atmosphere with a mass equal to its own.

The fate of centimeter- to meter-sized meteoroids depends primarily on their vertical velocity at atmospheric entry, v_{z_0}. Rapidly moving ($v_{z_0} >$ 15 km s^{-1}) meteoroids of this size tend to be ablated away, but similar objects that are slowly

Figure 11.14 Carbonaceous chondrite meteorite ALH 77307, which was found in Antarctica. The meteorite's rounded surface was sculpted during its passage through Earth's atmosphere. The cracked surface was produced by subsequent cooling. The cube is 1 cm on a side. (NASA/JSC)

temperature stays close to 260 K (Problem 11-2), so a meteorite is cold a few minutes after hitting the surface of Earth. The crust is melted and greatly altered, as can be seen in Figures 11.13 and 11.14, but the interior of the meteorite is basically undisturbed.

Larger cosmic intruders continue to travel at **hypersonic** (substantially supersonic) speeds as they penetrate into denser regions of the planet's atmosphere, and therefore they induce a shock in the gas in front of them. The leading face of a

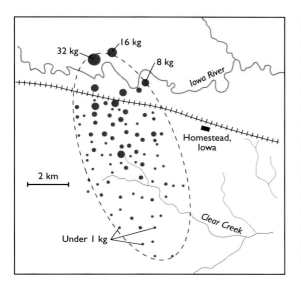

Figure 11.15 Illustration of the strewn field in the Homestead, Iowa, meteorite shower, which occurred on 12 February 1875. The meteor traveled in the upwards direction, slightly towards the left, in the geometry of the figure. Note that larger objects, which are less susceptible to atmospheric drag, landed at the far end of the ellipse, in the upper left of the figure. (Adapted from O. Farrington, 1915)

moving ($v_{z_0} < 10$ km s^{-1}) tend to be aerobraked to the **terminal velocity**, v_∞, at which gravitational acceleration balances atmospheric drag:

$$v_\infty = \sqrt{\frac{2g_p m}{C_D \rho_g A}}. \tag{11.3}$$

If the pressure given by equation (11.1) exceeds the compressive tensile strength of the meteor, the meteor is likely to fracture and disperse (Problem 11-4), with the resulting debris scattered over a **strewn field**. The strewn field of the Homestead meteorite is diagrammed in Figure 11.15.

Stony meteors in the $\sim$1–10-m size range tend to break up in the atmosphere; the fragments are aerobraked to terminal velocity (eq. 11.3) and are often recovered in many pieces spread out over strewn fields. Stony bodies between $\sim$10 and 100 m in size continue at high speed to deeper and denser levels in the atmosphere, where they can

be disrupted by ram pressure of the dense atmosphere. The giant ($\sim$5 $\times$ 10^{16} J $\approx$ 10 megatons of TNT) explosion that occurred over Tunguska, Siberia, in 1908 was probably produced by the disruption of a stony bolide roughly 100 m in diameter about 5–10 km above the Earth's surface (§6.4.5). Meteoroids larger than $\sim$100 m generally reach the surface with high velocity because even if they are flattened out by ram pressure from their interactions with the atmosphere, they still collide with an amount of gas totaling much less than their own mass. In contrast, iron meteorites are very cohesive, and iron meteors over a broad size range can reach Earth's surface moving with sufficient velocity to produce impact craters, such as the famous Meteor Crater in Arizona (Fig. 6.23). Cratering of planetary surfaces by hypervelocity impacts is described in detail in §6.4. Impact erosion of planetary atmospheres is discussed in §5.7.3.

11.4 Chemical and Isotopic Fractionation

Meteorites provide the oldest and most primitive rocks available for study in terrestrial laboratories. Analysis of meteorites yields important clues as to how, when and of what type of materials the planets themselves formed. The primary information on early Solar System conditions obtained from meteorites is given by their chemical and isotopic composition, and variations thereof between different parts of individual meteorites and among meteorites as a group. In this section, we provide the basic geochemistry background necessary to understand these results. In §11.6, we discuss how observed variations can be used to determine the ages of meteorites. The mineralogical structure of meteorites also yields information on conditions in the early Solar System, as does remanent ferromagnetism detected in some meteorites; these subjects are reviewed in §11.5 and §11.7.

11.4.1 Chemical Separation

In a sufficiently hot gas or plasma, atoms become well mixed. Provided diffusion has had sufficient time to erase initial gradients and turbulence is large enough to prevent gravitational settling of massive species, a gas is generally well mixed at the molecular level. Solid bodies, however, tend to mix very little, retaining the molecular composition that they acquired when they solidified. When solids form from a gas or from a melt, molecules tend to group with mineralogically compatible counterparts, and distinct minerals are formed. Such minerals have elemental compositions that can be considerably different from the bulk composition of the mixture (§6.1). Condensation from a gas can produce small grains, which mix heterogeneously with one another. Crystallization from a melt allows greater separation of materials, producing samples with large-scale heterogeneities. On an even larger scale, the combination of chemical separation in a melt and density-dependent settling results in planetary differentiation (§6.2). However, under most circumstances, the isotopic composition of each element usually remains uniform across mineral phases.

Analysis of the most primitive meteorites known implies that to a good approximation the material from which the planets formed was well mixed over large distance scales on both the isotopic and elemental level. Gross chemical differences result primarily from temperature variations within the protoplanetary disk. Exceptions to isotopic homogeneity have been used to determine the age of the Solar System (§11.6) and to show that at least a small amount of presolar grains survived intact and never melted or vaporized before being incorporated into planetesimals (§§11.5, 11.7 and 12.7.4).

11.4.2 Isotopic Fractionation

Although different isotopes of the same element are almost identical chemically, several physical and nuclear processes can produce isotopic inhomogeneities. Sorting out these different processes is essential to use the isotope data obtained

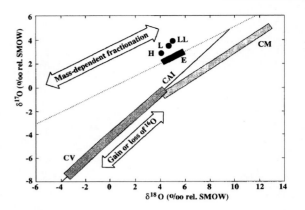

Figure 11.16 Plot showing the distribution of the three stable oxygen isotopes in various Solar System bodies. Isotope abundance ratios are shown relative to the standard (terrestrial) mean ocean water (SMOW), with units being parts per thousand variations. The *dotted line* represents the mass-dependent fractionation pattern observed in terrestrial samples. (Kerridge 1993)

from meteorites. One explanation for isotopic differences between grains is that they came from different reservoirs that were never mixed (e.g., interstellar grains that formed in distinct parts of the galaxy from material with different nucleosynthesis histories). This explanation can have profound consequences for our understanding of planetary formation, so other processes must also be considered.

Isotopes can be separated from one another by mass-dependent processes. These processes can rely on gravitational forces, such as the preferential escape of lighter isotopes from a planet's atmosphere (§5.7), or be the result of molecular forces, such as the preference for deuterium (as opposed to ordinary hydrogen) to bond with heavy elements, which is a consequence of a slightly lower energy resulting from deuterium's greater mass. **Mass-dependent fractionation** is easy to identify for elements such as oxygen, which have three or more stable isotopes, because the degree of fractionation is proportional to the difference in mass. Figure 11.16 shows a plot of the differences in the $^{17}O/^{16}O$ ratio against the corresponding difference in $^{18}O/^{16}O$ for various objects. Mass-dependent fractionation of oxygen leads to points along a line of slope 0.52; the fractionation slope is slightly larger than 1/2 because 17/16 > 18/17. The deviations of meteorite data from this line may result from **self-shielding** of the abundant (and thus optically thick) isotope $^{12}C^{16}O$ from photodissociation (§5.5). An alternative explanation involves **chemical mass-independent fractionation**, which can occur because the stability of some gaseous compounds containing multiple oxygen atoms differs between the symmetric case (all ^{16}O atoms) and the asymmetric one (i.e., containing a ^{17}O or ^{18}O atom).

Nuclear processes can also lead to isotopic variations. Paramount among these is radioactive decay, which transforms radioactive parents into stable daughter isotopes, thereby producing isotopic differences. Cosmic rays produce a variety of nuclear reactions. Energetic particles from local radioactivity may also induce nuclear transformations.

11.5 Main Components of Chondrites

Chondrites contain very nearly (within a factor of two) solar abundance ratios of refractory nuclides (Fig. 11.4). CI carbonaceous chondrites are the most similar to the Sun in elemental composition. Note, however, that even the most volatile-rich carbonaceous chondrites are depleted relative to the Sun in the highly volatile elements oxygen, carbon and nitrogen (as well as, of course, the extremely volatile noble gases and hydrogen). Isotopic ratios are even more strikingly regular; almost all differences can be accounted for by radioactive decay (excesses of daughter nuclides), cosmic ray-induced *in situ* nucleosynthesis or mass fractionation (§11.4.2). However, slight deviations from this rule show that the material within the solar nebula was not completely mixed at the atomic level.

Small condensates that clearly predate the protosolar nebula, some of which formed in outflows from stars and others that may have accreted within the interstellar medium, represent a volumetrically insignificant but scientifically critical component of chondrites. Many **presolar grains** are carbon-rich stardust, which occurs as **nanodiamonds**, graphite and silicon carbide (SiC). Other common types of presolar grains include silicates and the oxides corundum (Al_2O_3), hibonite ($CaAl_{12}O_{19}$) and spinel ($MgAl_2O_4$).

Chondrites have not been melted since their original accretion $\sim 4.56 \times 10^9$ years ago. Although these primitive meteorites represent well-mixed isotopic and elemental (except for volatiles) samples of the material in the protoplanetary disk, they are far from uniform on small scales. In addition to chondrules, many chondrites contain **CAIs**, which are refractory inclusions that are rich in calcium and aluminum. Chondrules and CAIs are embedded within a dark fine-grained **matrix** that is present in all chondrites. Chondrites formed with different percentages of inclusions (CAIs, chondrules) and differing amounts of moderately volatile elements. Up to $\sim 20\%$ of the mass of some chondrites is composed of Fe–Ni metal.

Chondrules, such as those shown in Figure 11.3, are small (typically ~ 0.1–2 mm), rounded igneous rocks (i.e., they solidified from a melt) composed primarily of refractory elements. They range from 0%–80% of the mass of a chondrite, with abundances depending on compositional class (CI chondrites contain neither chondrules nor CAIs) and petrographic type. Chondrules are totally absent in petrographic type 1 (they may have been destroyed by aqueous processes) and are substantially degraded by recrystallization resulting from thermal metamorphism in types 5 and 6; the most pristine chondrules are found in type 3 chondrites (Figs. 11.2 and 11.3). Mineralogical properties imply that chondrules cooled very quickly, dropping from a peak temperature of ~ 1900 K to ~ 1500 K over a period ranging from 10 minutes to a few hours. Chondrules are diverse, with a wide variety of compositions, mineralogies and sizes. However, strong correlations of chondrule properties (size and compositions) are observed within individual meteorites. These correlations, combined with the compositional complementarity of chondrules and matrix within individual primitive meteorites (together they are nearly solar in composition apart from volatiles, but separately they differ substantially), imply that chondrules were not well mixed within the protoplanetary disk before incorporation into larger bodies. Many chondrules have melted rims, providing evidence for multiple heating events.

CAIs are light-colored inclusions, typically 1–10 mm in size (Fig. 11.2c). They are composed of very refractory minerals, including substantial amounts of Ca and Al and abundances of high-Z elements that are greatly enhanced relative to bulk chondrites. They are among the oldest objects formed in the Solar System. Many CAIs have melted rims that formed up to 300 000 years after the core CAI formed. These rims solidified in a more oxygen-rich environment than their hosts. The interiors of the CAIs were not heated to anywhere near melting during this secondary processing, implying very short duration heating. Clearly, the protoplanetary disk was an active and sometimes violent place!

The characteristics of the fine-grained (10 nm–5 μm) matrix material that makes up the bulk of most chondrites vary with petrographic type. Chondrite matrices appear to contain material from a wide variety of sources, including presolar grains, direct condensates from the protoplanetary disk and dust from fragmented chondrules and CAIs. In most chondrites, many of these grains have been altered by post-accretional aqueous or thermal processing.

11.6 Radiometric Dating

Several different ages may be assigned to a given meteorite. All of these meteorite ages are determined by **radionuclide dating**, which relies on the existence of naturally occurring radioactive nuclides (§11.6.1).

The most fundamental age of a meteorite is its **formation age**, which is often referred to simply as the age. The formation age is the length of time since the meteorite (or its components) solidified from a molten or gaseous phase. Most meteorites have ages of 4.55–4.57×10^9 years, but a small fraction are much younger. Techniques for determining meteorite formation ages are based on radioactive decay of long-lived radioactive isotopes. These techniques are presented in §11.6.2. The relative ages of individual chondrites can be measured more precisely than their absolute ages using short-lived extinct radioactive isotopes (§11.6.3).

Some isotopes of noble gases such as helium, argon and xenon are produced by radioactive decay. These isotopes build up in rocks as time progresses but can be lost if the rock is fractured or heated. The abundance of such radiogenic gases can be used to determine the **gas retention age** of the rock. Usually this gas retention age is less than the formation age, but for some rocks, these two ages are equal. Lighter noble gases diffuse more readily than the heavy ones, implying a lower **closure temperature** to remain bound. So some events lead to the loss of most of a rock's helium but little of its argon, resulting in a meteorite with a helium retention age younger than its argon retention age.

Finally, nuclear reactions produced by cosmic rays can be used to determine how long the meteorite existed as a small body in space and the time at which it reached Earth (§11.6.4).

11.6.1 Decay Rates

The time required for an individual radioactive nucleus to decay is not fixed; however, there is a characteristic lifetime for each radioactive nuclide. The probability that a nucleus will decay in a specified interval of time does not depend on the age of the nucleus, so the number of atoms of a given radioactive nuclide remaining in a sample drops exponentially if no new atoms of this nuclide are

Table 11.1 Half-Lives of Selected Nuclides

Parent	Measurable Stable Daughter(s)	Half-Life $(t_{1/2})$
Long-lived radionuclides		
^{40}K	^{40}Ar, ^{40}Ca	1.25 Gyr
^{87}Rb	^{87}Sr	49 Gyr
^{147}Sm	^{143}Nd, ^4He	106 Gyr
^{176}Lu	^{176}Hf	36 Gyr
^{187}Re	^{187}Os	42 Gyr
^{190}Pt	^{186}Os, ^4He	500 Gyr
^{232}Th	^{208}Pb, ^4He	14 Gyr
^{235}U	^{207}Pb, ^4He	0.704 Gyr
^{238}U	^{206}Pb, ^4He	4.47 Gyr
Extinct radionuclides		
^{10}Be	^{10}B	1.4 Myr
^{22}Na	^{22}Ne	2.6 yr
^{26}Al	^{26}Mg	0.72 Myr
^{36}Cl	^{36}Ar, ^{36}S	0.30 Myr
^{41}Ca	^{41}K	0.10 Myr
^{44}Ti	^{44}Sc	52 yr
^{53}Mn	^{53}Cr	3.6 Myr
^{60}Fe	^{60}Ni	2.4 Myr
^{92}Nb	^{92}Zr	35 Myr
^{99}Tc	^{99}Ru	0.21 Myr
^{107}Pd	^{107}Ag	6.5 Myr
^{129}I	^{129}Xe	16 Myr
^{146}Sm	^{142}Nd	68 Myr
^{182}Hf	^{182}W	9 Myr
^{244}Pu	$^{131-136}Xe$	82 Myr

abundance of a 'parent' species at time t is related to its abundance at t_0 as

$$N_p(t) = N_p(t_0)e^{-(t-t_0)/t_m}. \qquad (11.4)$$

Alternatively, the **half-life** of the nuclide, $t_{1/2}$, which represents the time required for half of a given sample to decay, can be used to quantify the decay rate. The relationship between these quantities is

$$t_{1/2} = \ln 2\, t_m. \qquad (11.5)$$

Trees and other organisms that were alive during the past few dozen millennia are accurately dated by measuring abundances of the heavy isotope of carbon ^{14}C ($t_{1/2} = 5730$ yr), which is continually produced by cosmic ray interactions within Earth's atmosphere. The half-lives of nuclides commonly used to date events in the early Solar System are given in Table 11.1.

Many nuclides produced by radioactive decay are themselves unstable. In many cases, these 'daughter' nuclides have shorter half-lives than their 'parents'. A sequence of successive radioactive decays leading to a stable or nearly stable nuclide is referred to as a **decay chain**. Two decay chains that are important in meteorite evolution are shown in Figure 11.17. Note that the first decay in each of these chains takes of order 10^9 years, but subsequent decays are much more rapid.

produced. The timescale over which this process occurs can be characterized by the **mean lifetime** of a species, t_m, or the **decay constant**, t_m^{-1}. The

$$^{238}_{92}U \xrightarrow[t_{1/2}=4.47\times10^9\,y]{\alpha} {}^{234}_{90}Th \xrightarrow[21.4d]{\beta} {}^{234}_{91}Pa \xrightarrow[6.75h]{\beta} {}^{234}_{92}U \xrightarrow[2.47\times10^5\,y]{\alpha} {}^{230}_{90}Th \xrightarrow[8\times10^4y]{\alpha} {}^{226}_{88}Ra \xrightarrow[1600y]{\alpha} {}^{222}_{86}Rn \xrightarrow[3.8d]{\alpha} {}^{218}_{84}Po$$

$$\xrightarrow[3m]{\alpha} {}^{214}_{82}Pb \xrightarrow[27m]{\beta} {}^{214}_{83}Bi \xrightarrow[19.9m]{\beta} {}^{214}_{84}Po \xrightarrow[1.64\times10^{-4}s]{\alpha} {}^{210}_{82}Pb \xrightarrow[21y]{\beta} {}^{210}_{83}Bi \xrightarrow[5d]{\beta} {}^{210}_{84}Po \xrightarrow[183d]{\alpha} {}^{206}_{82}Pb \text{ (stable)},$$

$$^{235}_{92}U \xrightarrow[t_{1/2}=7.04\times10^8\,y]{\alpha} {}^{231}_{90}Th \xrightarrow[25.5h]{\beta} {}^{231}_{91}Pa \xrightarrow[3.25\times10^4y]{\alpha} {}^{227}_{89}Ac \xrightarrow[21.6y]{\beta} {}^{227}_{90}Th \xrightarrow[18.5d]{\alpha} {}^{223}_{88}Ra \xrightarrow[11.43d]{\alpha} {}^{219}_{86}Rn \xrightarrow[4s]{\alpha} {}^{215}_{84}Po$$

$$\xrightarrow[1.8\times10^{-3}s]{\alpha} {}^{211}_{82}Pb \xrightarrow[36m]{\beta} {}^{211}_{83}Bi \xrightarrow[2.15m]{\alpha} {}^{207}_{81}Tl \xrightarrow[4.8m]{\beta} {}^{207}_{82}Pb \text{ (stable)}.$$

Figure 11.17 The principal decay chains of the two long-lived isotopes of uranium. Most uranium atoms radioactively decay on the paths indicated, but some elements/isotopes have multiple possible decay paths, leading to small amounts of other isotopes produced.

11.6.2 Dating Rocks

With the passage of time, the abundance of radioactive 'parent' nuclides in a rock, $N_p(t)$, decreases, as these atoms decay into 'daughter' species (or 'granddaughter' etc. nuclides if the initial decay products are unstable with short half-lives, e.g., the decay chains shown in Fig. 11.17). The abundance of the daughter species can be expressed as

$$N_d(t) = N_d(t_0) + \xi(1 - e^{-(t-t_0)/t_m})N_p(t_0), \quad (11.6)$$

where the **branching ratio**, $0 < \xi \leq 1$, represents the fraction of the parent nuclide that decays into the daughter species under consideration. The branching ratio is a fundamental property of a given nuclide. In most cases, $\xi = 1$.

The current abundances, $N_d(t)$ and $N_p(t)$, are measurable quantities. The initial abundance of the parent, $N_p(t_0)$, can be expressed in terms of the measured abundance and the age of the rock, $t - t_0$, using equation (11.4). However, this combination of equation (11.4) and the decay chains in Figure 11.17 yields a single equation for two unknowns, $(t - t_0)$ and $N_d(t_0)$. If we could determine independently the 'initial' (nonradiogenic) abundance of the daughter nuclide (its abundance when the rock solidified), $N_d(t_0)$, then we could determine both the initial abundance of the parent and the age of the rock, $t - t_0$.

Chemical separation during a rock's solidification epoch can create an inhomogeneous sample that may be analyzed to determine both initial abundances and the age of the rock. Analysis of two samples within the rock containing different ratios of the parent element to the daughter element provides the two equations needed to solve for both age and initial abundance of the parent isotope. In practice, several samples are usually analyzed, and solutions are obtained graphically using an **isochron diagram**, such as the one shown in Figure 11.18.

Radiometric dates for chondritic meteorites cluster tightly around 4.56×10^9 years. The majority of differentiated meteorites are of similar

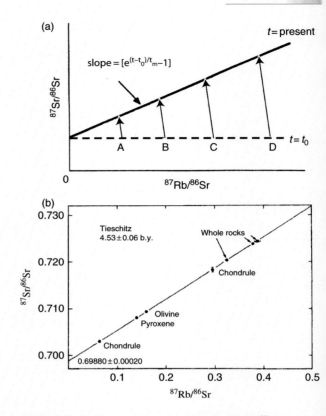

Figure 11.18 (a) Schematic isochron diagram of the ^{87}Rb–^{87}Sr system. Phases A, B, C and D have identical initial ^{87}Sr/^{86}Sr ratios at $t = 0$, but differing ^{87}Rb/^{86}Sr ratios. Assuming that the system remains closed, these ratios evolve as shown by the arrows to define an isochron for which t is the age of the rock. (de Pater and Lissauer 2010) (b) ^{87}Rb–^{87}Sr isochron for the Tieschitz unequilibrated H3 chondrite meteorite. (Taylor 1992)

age, but many are younger, in some cases much younger. These results and their implications are discussed in more detail in §11.7.

11.6.3 Extinct-Nuclide Dating

Absolute radiometric dating requires that a measurable fraction of the parent nuclide remains in the rock. For rocks that date back to the formation of the Solar System, this implies long-lived parents, which because of their slow decay rates cannot give highly accurate ages. The **relative ages** of rocks that formed from a single well-mixed reservoir of material can be determined more precisely using the daughter products of short-lived radionu-clides that are no longer present in the rocks. Additionally, extinct nuclei can provide estimates of the time between nucleosynthesis and rock formation.

Correlations between the Al/Mg ratio and $^{26}Mg/^{24}Mg$ excess have been detected within chondritic meteorites. This excess cannot be the result of mass-dependent fractionation because the relative abundance of the nonradiogenic isotopes of magnesium, $^{25}Mg/^{24}Mg$, is normal. The stable isotope of aluminum, ^{27}Al, is a common nuclide, but the lighter isotope, ^{26}Al, decays on a timescale much shorter than the age of the Solar System ($t_{1/2} = 720\,000$ years for inverse β decay into ^{26}Mg). If a meteorite or piece thereof contained ^{26}Al when it solidified eons ago, then decay of this isotope will have produced excess ^{26}Mg, and the amount of this excess is proportional to the local aluminum abundance. The constant of pro-portionality is the fractional abundance of ^{26}Al at the time of solidification. Provided the ^{26}Al was of presolar origin and the isotopes of aluminum were well mixed within the protoplanetary disk, the rel-ative ages of different samples can be determined with an uncertainty significantly smaller than the half-life of ^{26}Al.

11.6.4 Cosmic-Ray Exposure Ages

Galactic cosmic rays are extremely energetic par-ticles that can produce nuclear reactions in parti-cles with which they collide. Most galactic cos-mic rays are protons ($\sim$87%) or alpha particles ($\sim$12%), with $\sim$1% being heavier nuclei. Cosmic rays and the energetic secondary particles that they produce have a mean interaction depth of $\sim$1 m in rock; thus, they do not affect the bulk of material in any sizable asteroid. The amount of cosmic rays that a meteorite has been exposed to indicates how long it has 'been on its own', or at least near the surface of an asteroid.

Cosmic-ray exposure ages are determined using measurements of the abundances of certain rare nuclides that in meteorites are almost exclu-sively produced by cosmic rays. Some of these nuclides are noble gases, e.g., ^{21}Ne and ^{38}Ar; oth-ers are short-lived radionuclides such as ^{10}Be and ^{26}Al.

Typical cosmic-ray exposure ages are 10^5–10^7 years for carbonaceous chondrites, 10^6–10^8 years for other stones, 10^8 years for stony-irons and 10^8–10^9 years for irons. The differences in age are attributable to material strength, which gov-erns how quickly a body breaks up or a surface erodes. The observed clustering of cosmic ray ages implies that certain meteorite groups experienced major breakups that generated a large fraction of the members of each of these groups in a single event.

The **terrestrial age** of a meteorite is the time since it fell, i.e., how long the meteorite has been on Earth. Weathered appearance is correlated with terrestrial age of hot desert meteorites, but for Antarctic meteorites, no such correlation exists. Terrestrial ages of hot desert meteorites are typ-ically <50 kyr, although a few achondrites are up to $\sim$0.5 Myr, and some irons have been on Earth even longer. Terrestrial ages of Antarctic meteorites are generally <0.5 Myr, although a few Antarctic chondrites and irons are up to a few Myr old. The distribution in terrestrial ages can be used to constrain possible variations in the influx of meteorites.

11.7 Meteorite Clues to Planet Formation

Small bodies in the Solar System have not been subjected to as much heat or pressure as planet-sized bodies, and they remain in a more pristine state. Meteorites thus provide detailed information about environmental conditions and physical and chemical processes during the epoch of planet for-mation. This information pertains to timescales, thermal and chemical evolution, mixing, magnetic fields and grain growth within the protoplanetary disk. Processes identified include evaporation, condensation, localized melting and fractionation, both of solids from gas and among different solids.

The age of the Solar System, based on dating of CAIs in the Efremovka and NWA 2634 CV3 meteorites, is $4.568 \pm 0.001 \times 10^9$ years; other nuclide systems and other CAIs yield similar ages. Meteorites thus definitively date the origin of the Solar System with a fractional uncertainty of ~ 2 parts in 10^4. Chondritic solids formed within a period of $\lesssim 5$ million years at the beginning of Solar System history. Whereas some types of chondrules have ages indistinguishable from those of CAIs, others appear to have solidified a few million years later. Most meteorites from differentiated parent bodies are a bit younger, but usually not very much. Almost all meteorites are thus older than known Moon rocks (~ 3–4.45×10^9 yr) and terrestrial rocks ($\lesssim 4 \times 10^9$ yr, although some contain grains of the durable mineral zircon up to 4.4×10^9 yr old).

The vast majority of elements in most meteorite groups are identical in isotopic composition, aside from variations that may plausibly be attributed to mass fractionation, radioactive decay or cosmic ray irradiation. Thus, matter within the solar nebula must have been relatively well mixed. Differences in isotopic composition between individual meteorites and between meteorites and the Earth yield information on the place of formation of the individual molecules and grains out of which the meteorites have formed. Some grains have very high D/H ratios (compared with cosmic values); such fractionation implies formation in (very) cold interstellar molecular clouds. Other grains have noncosmic isotopic ratios in many elements that imply condensation in outflows from stars, such as occur in the ejecta from supernova explosions (§3.4.2). Hence, meteorites seem to contain stellar outflow and interstellar condensates in addition to material that formed or was significantly processed within our own Solar System.

11.7.1 Meteorites from Differentiated Bodies

Isotopic anomalies found in some achondrites imply rapid differentiation and recrystallization of planetesimals. Excess ^{60}Ni, which is the stable decay product of ^{60}Fe ($t_{1/2} \sim 2.4 \times 10^6$ yr), is correlated with iron abundances in **HED** (howardite–eucrite–diogenite) achondrite meteorites. HEDs originate from the asteroid 4 Vesta (or possibly another differentiated planetesimal) and were once molten. Live ^{60}Fe must thus have been present when the planetesimal resolidified. This, together with signatures left by ^{26}Al, implies that the HED parent body formed and differentiated within 3–5 Myr of the solidification of the oldest known Solar System materials (CAIs). Some iron meteorites have hafnium–tungsten (^{182}Hf–^{182}W) signatures that date differentiation of their parent bodies to $\lesssim 1.5$ Myr after CAIs formed. The oldest differentiated meteorites thus appear to be at least as old as most chondrules. This suggests that large planetesimals and small grains existed at the same time within the protoplanetary disk, albeit not necessarily in the same location.

The cooling rate of a rock, as well as the pressure and the gravity field that it was subjected to while cooling, can be deduced from the structure and composition of its minerals, e.g., the Widmanstätten pattern that is apparent in some iron meteorites (Fig. 11.5). Thus, we can estimate the size of a meteorite's original parent body. By knowing what size bodies melted in the early Solar System (and possibly where they accreted relative to other bodies by the presence or lack of volatiles in the meteorite), we get a better idea of the heat sources responsible for differentiation.

Spontaneous fission of heavy radioactive nuclei, such as ^{244}Pu, causes radiation damage within crystalline materials in the form of **fission tracks**. This damage tends to be annealed out at high temperatures. The annealing temperature varies among minerals. The difference in fission track density among a set of meteoritic minerals with different annealing temperatures can be interpreted in terms of a cooling history. The retention of radiogenic noble gases in meteoritic minerals is analogous to fission track retention and can likewise be used to estimate a cooling history.

The composition and texture of differentiated meteorites such as achondrites, irons and pallasites reflect igneous differentiation processes (i.e., large-scale melting) within asteroid-size parent bodies. It appears as if some small bodies $\lesssim 100$ km in radius differentiated. Neither accretion energy nor long-lived radioactive nuclides provides adequate sources of heat. Possible heat sources include the extinct radionuclides ^{26}Al and ^{60}Fe, as well as **electromagnetic induction heating**. Electromagnetic induction heating occurs when eddy currents are generated within an object and dissipated via Joule heating. Meteorite parent bodies passing through currents that may have been produced by the massive T-Tauri phase solar wind (§15.2) would have been subjected to electromagnetic induction heating. Although there are many uncertainties in this mechanism, maximum heating would occur in bodies nearest the Sun and probably for bodies between 50 and 100 km in radius. The largest ^{26}Al/^{27}Al concentrations observed in primitive meteorites would be sufficient to melt chondritic composition planetesimals as small as 5 km in radius. (The largest abundance of ^{60}Fe deduced in achondrites is a substantially weaker heat source.) Short-lived radionuclides would not, by themselves, have provided enough energy to melt chondritic planetesimals of any radius formed $\gtrsim 2$ Myr (three half-lives of ^{26}Al) after CAI formation, but they could have led to less drastic thermal processing. The formation ages of many iron meteorites suggest that their parent bodies accreted early enough for melting via ^{26}Al decay.

11.7.2 Primitive Meteorites

Chondrites have never been molten and thus preserve a better record of conditions within the protoplanetary disk than do the differentiated meteorites. Indeed, some of the grains in chondrites predate the Solar System and thus also preserve a record of processing in stellar atmospheres, winds, explosions and the interstellar medium. These grains may have been affected by passage through a hot shocked layer of gas during their entry into the protoplanetary disk. The precursors to chondrules and CAIs formed by agglomeration of presolar grains and solar nebula condensates. These agglomerates were subsequently heated to the point of melting. Many had their rims melted at a later time and/or were fragmented as the result of high-speed collisions. Ultimately, they were incorporated into planetesimals in which they were subjected to nonhydrous processing at 700–1700 K (especially petrographic types 4–6) and/or hydrothermal processing at lower temperatures (primarily types 1 and 2).

The differences in bulk composition among chondrites are closely related to the volatilities of the constituent elements. Figure 11.19 shows the gradual depletion pattern with increasing volatility that occurs in most chondrites. If each meteorite was formed in equilibrium at a unique temperature, then relative abundances of elements with condensation temperatures above this value would be solar, and more volatile elements would be almost absent. The gradual depletion patterns observed imply that the constituents of individual meteorites condensed in a broad range of environments. Grains could be brought together from a variety of locations to produce such admixtures, or most of the material that formed the terrestrial planets and asteroids cooled to around 1300 K while the gaseous and solid components remained well mixed and then gas was subsequently lost as material cooled further. Significant condensation in the asteroid region continued down to $\lesssim 500$ K before the gas was completely removed. Elemental depletions in bulk terrestrial planets and differentiated asteroids are consistent with this conclusion.

Remanent magnetism in carbonaceous chondrites suggests that a magnetic field of strength

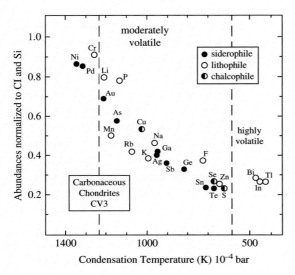

Figure 11.19 The abundances of moderately volatile elements in bulk CV chondrites compared to their abundances relative to silicon in CI chondrites are plotted against the condensation temperatures of the elements in a solar composition gas. The gradual decrease in abundance with decreasing condensation temperature implies that the components of individual meteorites condensed or were altered in a variety of environments. The lack of dependence of abundances on the geochemical character of the elements shows that the meteorites have not been fully melted subsequent to accretion. (Palme and Boynton 1993)

1–10 G existed at some locations within the protoplanetary disk. The magnetic field recorded in the high-temperature component is anisotropic from chondrule to chondrule, so magnetization presumably occurred before the incorporation of the chondrules into their meteorite parent body.

The general high degree of uniformity of isotope ratios implies the solar nebula was for the most part well mixed, but the small violations of this rule tell us that some things did not mix or never vaporized. Oxygen isotopic ratios show relatively large variations that cannot be explained by mass-dependent fractionation within primitive chondrites and between groups of meteorites (Fig. 11.16); these data are usually taken to imply distinct reservoirs that were incompletely mixed

during nebular processes, although non-mass-dependent fractionation, via e.g., photochemical processes, can occur in certain circumstances (§11.4.2).

11.7.3 Presolar Grains

Some isotopic anomalies found in chondritic meteorites indicate survival of grains that solidified before the formation of the protoplanetary disk. Almost pure concentrations of the rare heavy isotope of the noble gas neon, ^{22}Ne, probably from sodium decay (^{22}Na, $t_{1/2} = 2.6$ years!) have been detected in small, carbon-rich phases within some primitive meteorites. Other pockets with enhanced but not pure ^{22}Ne may have been created by implantation of neon from stellar winds. The grains containing heavy neon enhancements probably condensed within outflows from carbon-rich stars, supernova explosions and possibly nova outbursts. Decay products from the short-lived nuclides ^{41}Ca, ^{44}Ti and ^{99}Tc have also been identified within presolar grains.

There is no strong evidence that any macroscopic ($\gtrsim 10$ μm) grains are of presolar origin. Tiny carbon-rich interstellar grains such as the one pictured in Figure 11.20 have been found in some chondrites. Unambiguous proof of the presolar origin of these grains comes from their isotopic compositions, which are anomalous both in trace elements such as Ne and Xe and in the more common elements C, N and Si. Although carbon-rich silicon carbide and graphite grains stand out most clearly, the majority of interstellar grains found in meteorites are silicates. The survival of interstellar grains constrains the thermal and chemical environments that they experienced on their journey from interstellar cloud to meteorite parent bodies. Some of the grains were clearly never heated above 1000 K and must have been much cooler during any episode in which they were exposed to an oxygen-rich environment.

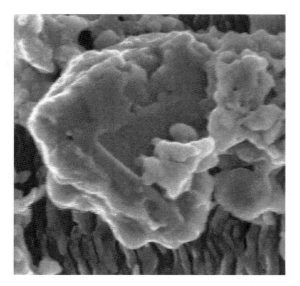

Figure 11.20 Image of a tiny presolar silicon carbide (SiC) grain (1 μm across) extracted from the Murchison meteorite. This very-high-resolution secondary electron image was obtained using a scanning electron microscope. The wormlike background shows a foil substrate that is not part of the grain. (Courtesy Scott Messenger)

Key Concepts

- A meteorite is a rock from another world.
- Primitive meteorites (chondrites) have never been completely molten since they grew from solid grains within the protoplanetary disk. They are agglomerations of chondrules, matrix and CAIs. Chondritic meteorites contain elemental and isotopic abundances of refractory elements similar to that of the Sun's atmosphere.

- Differentiated meteorites, which include achondrites, irons and stony-irons, come from parent bodies that had melted and segregated metal from rock. Most differentiated meteorites come from asteroids, but a few dozen meteorites from the Moon and a similar number from Mars have also been collected.
- Meteoroids arrive at the top of Earth's atmosphere moving at 10 km/s or faster. Collisions with air molecules heat these bodies, transforming them into glowing meteors. Most meteors vaporize completely; only a small fraction produce meteorites.
- In addition to being fascinating in their own right, meteorites provide us with a wealth of data on conditions during the planet-forming epoch.
- The ages of many meteorites have been estimated using radiometric dating. The oldest meteoritic components of Solar System origin were formed 4.568×10^9 years ago. Most meteorites formed within a few million years of these oldest inclusions.
- The near but not total homogeneity of isotopic composition among meteorites tells us that substantial mixing of presolar material occurred, but some interstellar grains survived.
- The local mineralogical and compositional heterogeneity of primitive meteorites implies an active dynamic environment within the protoplanetary disk.

Further Reading

Nice nontechnical summaries, including many good photographs, can be found in:

Wasson, J.T., 1985. *Meteorites: Their Record of Early Solar-System History*. W.H. Freeman, New York. 274pp.

McSween, H.Y., Jr., 1999. *Meteorites and Their Parent Planets*, 2nd Edition. Cambridge University Press, Cambridge. 322pp.

Zinner, E., 1998. Stellar nucleosynthesis and the isotopic composition of presolar grains from primitive meteorites. *Annu. Rev. Earth Planet. Sci.*, **26**, 147–188.

Taylor, S.R., 2001. *Solar System Evolution*, 2nd Edition. Cambridge University Press, Cambridge. 484pp.

Lipschutz, M.E., and L. Schultz, 2007. Meteorites. In *Encyclopedia of the Solar System*, 2nd Edition. Eds. L. McFadden, P.R. Weissman, and T.V. Johnson, Academic Press, San Diego, pp. 251–282.

A very useful collection of review chapters can be found in:

Lauretta, D.S., and H.Y. McSween, Eds., 2006. *Meteorites and the Early Solar System II*. University of Arizona Press, Tucson. 942pp.

A compendium of articles written from various different viewpoints can be found in:

Krot, A.N., E.R.D. Scott, and B. Reipurth, Eds., 2005. *Chondrites and the Protoplanetary Disk. ASP Conference Series* **341**, Astronomical Society of the Pacific, San Francisco. 1029pp.

Additional information on radiometric dating can be found in:

Tilton, G.R. 1988. Principles of radiometric dating. In *Meteorites and the Early Solar System*. Eds. J.F. Kerridge and M.S. Matthews. University of Arizona Press, Tucson, pp. 249–258.

Three good articles discussing meteorite clues to the formation of the Solar System are:

Kerridge, J.F., 1993. What can meteorites tell us about nebular conditions and processes during planetesimal accretion? *Icarus*, **106**, 135–150.

Palme, H., and W.V. Boynton, 1993. Meteoritic constraints on conditions in the solar nebula. In *Protostars and Planets III*. Eds. E.H. Levy and J.I. Lunine. University of Arizona Press, Tucson, pp. 979–1004.

Podosek, F.A., and P. Cassen, 1994. Theoretical, observational, and isotopic estimates of the lifetime of the solar nebula. *Meteoritics*, **29**, 6–25.

Two very different models of the formation of chondritic meteorites are given in:

Shu, F.H., H. Shang, and T. Lee, 1996. Toward an astrophysical theory of chondrites. *Science*, **271**, 1545–1552.

Scott, E.R.D., and A.N. Krot, 2005. Thermal processing of silicate dust in the solar nebula: Clues from primitive chondrite matricies. *Astrophys. J.*, **623**, 571–578.

Problems

11-1. Calculate the speed at which meteoroids with the following heliocentric orbits encounter the Earth's atmosphere:
  (a) An orbit very similar to that of Earth, so $v_{inf} \ll v_e$, and thus $v_{impact} \approx v_e$
  (b) A parabolic orbit with perihelion of 1 AU and $i = 180°$
  (c) A parabolic orbit with perihelion of 1 AU and $i = 0°$
  (d) An orbit with $a = 2.5$ AU, $e = 0.6$ and $i = 0°$
  (e) An orbit with $a = 2.5$ AU, $e = 0.6$ and $i = 30°$

11-2. (a) Calculate the equilibrium temperature of a meteoroid of mass M, density ρ and albedo A in the vicinity of the Earth.
  (b) Evaluate your result for a chondrite with $M = 10^6$ kg, $\rho = 2500$ kg m^{-3} and albedo $A = 0.05$ and for an achondrite with $M = 10^3$ kg, $\rho = 3000$ kg m^{-3} and albedo $A = 0.3$.

11-3. Differentiation brings the densest liquids down to the core of a body. Elements and compounds mix with chemically compatible compounds, so even though uranium is very dense, most of it ends up in a planet's crust. However, some 'superheavy' elements not yet found in nature are located near the same column as iron in the Periodic Table (Appendix D) and thus are likely to be siderophile.

(a) Which elements are these?

(b) All known isotopes of these elements have short half-lives and thus are exceedingly unlikely to be present naturally in the Solar System. However, nuclear physics models suggest that more neutron-rich isotopes of some superheavy elements may have half-lives of order 10^9 years. Such isotopes might be produced in r-process nucleosynthesis (§3.4.2). If they do exist in our Solar System, in which type of meteorites are these superheavy siderophiles likely to be concentrated?

11-4. **(a)** Calculate the pressure on a meteor moving at a speed of 10 km s^{-1} at an altitude of 100 km above the Earth's surface.

(b) Repeat for a meteor at the same speed 10 km above Earth's surface.

(c) Repeat your calculations in parts (a) and (b) for a meteor traveling at 30 km s^{-1}.

(d) The tensile strengths of comets are of order 10^3 pascals, the strengths of chondrites are roughly 10^7 pascals, stronger stony objects have strengths approximately 10^8 pascals, but iron impactors have effective strengths of about 10^9 pascals. Compare these tensile strengths to the pressures calculated in parts (a)–(c) and comment.

11-5. **(a)** Calculate the size of an iron meteor (density $\rho = 8000$ kg m^{-3}) that passes through an amount of atmospheric gas equal to its own mass en route to the surface of the Earth. You may assume a spherical meteorite, vertical entry into the atmosphere and neglect ablation.

(b) Repeat your calculation for a chondritic meteorite of density $\rho = 4000$ kg m^{-3}.

(c) Repeat your calculation in part (a) for an entry angle of 45°.

11-6. How deep must a (H_2O) lake or ocean be to substantially shield the underlying bedrock from the impact of a 100-m-radius iron meteoroid?

11-7. Calculate the terminal velocity near the Earth's surface for falling rocks of the following sizes and densities:

(a) $R = 0.1$ m, $\rho = 8000$ kg m^{-3}

(b) $R = 0.1$ m, $\rho = 2000$ kg m^{-3}

(c) $R = 1$ m, $\rho = 2000$ kg m^{-3}

(d) $R = 100$ μm, $\rho = 2000$ kg m^{-3}

11-8. Calculate the fractional abundance of ^{234}U in naturally occurring uranium ore. (Hint: Use the first decay chain shown in Figure 11.17.)

11-9. **(a)** Use the decay chains given in Figure 11.17 to estimate lower bounds on the abundance of elements 84–91 in terrestrial uranium ore.

(b) Why are your values only lower bounds?

11-10. In this problem, you will calculate the age of a rock using actual data on the abundances of rhenium and osmium, which are related via the decay
$$^{187}\text{Re} \underset{t_{1/2}=4.16\times10^{10}\text{y}}{\longrightarrow} {}^{187}\text{Os}.$$

(a) The following list summarizes some measurements of Re and Os isotope ratios for different minerals within a particular rock:

$^{187}Re/^{188}Os$	$^{187}Os/^{188}Os$
0.664	0.148
0.669	0.148
0.604	0.143
0.484	0.133
0.512	0.136
0.537	0.138
0.414	0.128
0.369	0.124

Plot the results on a piece of graph paper with $^{187}Re/^{188}Os$ along the horizontal axis and $^{187}Os/^{188}Os$ along the vertical axis.

(b) Draw a straight line that goes as closely as possible through all the points and extend your line to the vertical axis to determine the initial ratio of $^{187}Os/^{188}Os$.

(c) Draw and label several lines representing theoretical isochrones for a rock with the same initial ratio of $^{187}Os/^{188}Os$ as the rock being studied. Use these lines to estimate the age of the rock.

11-1. Calculate the abundance of ^{26}Al (in kilograms per kilogram of chondritic material and as a ratio to the abundance of ^{27}Al in chondrites) required to generate sufficient heat to melt a chondritic mixture of magnesium silicates and iron initially at 500 K. You may assume that the asteroid is sufficiently large that negligible heat is lost during the period in which most of the ^{26}Al decays.

11-2. State the size, age and composition of a typical chondrule.

CHAPTER 12

Minor Planets and Comets

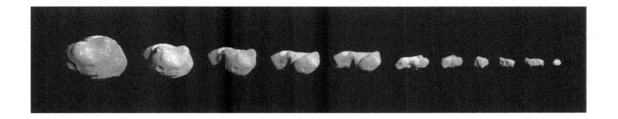

However, the small probability of a similar encounter [of the Earth with a comet] can become very great in adding up over a huge sequence of centuries. It is easy to picture to oneself the effects of this impact upon the Earth. The axis and the motion of rotation changed; the seas abandoning their old position to throw themselves toward the new equator; a large part of men and animals drowned in this universal deluge, or destroyed by the violent tremor imparted to the terrestrial globe.

Pierre-Simon Laplace, *Exposition du Système du Monde*, 2nd edition (1799)

In addition to the eight known planets, countless smaller bodies orbit the Sun. These objects range from dust grains and small coherent rocks with insignificant gravity to dwarf planets that have sufficient gravity to make them quite spherical in shape. Most are very faint, but some, the **comets**, release gas and dust when they approach the Sun and can be quite spectacular in appearance, as displayed in Figure 12.1. In this chapter, we describe the orbital and physical properties of the great variety of small bodies, ranging in radius from a few meters to more than 1000 km, that orbit the Sun. We refer to these bodies as comets if a coma and/or tail has been detected and as **minor planets** if not.

Asteroids is a term generally used for rocky minor planets that orbit the Sun at distances ranging from interior to Earth's orbit to a bit exterior to the orbit of Jupiter. More than 500 000 asteroids have been permanently cataloged as of 2018, and tens of thousands more are added each year. Asteroids exhibit a large range of sizes, with the largest asteroid, 1 Ceres, being ∼475 km in radius. The next largest asteroids are 2 Pallas, 4 Vesta and 10 Hygiea, ranging in radius from about 270 to 220 km (Table E.6). The total mass in the asteroid belt is $\sim 5 \times 10^{-4} \, M_\oplus$.

The Kuiper belt, beyond the orbit of Neptune, is analogous to the asteroid belt but on a grander scale. **Kuiper belt objects** (KBOs) are icy bodies, and the largest KBOs are an order of magnitude more massive than 1 Ceres. The total mass of the Kuiper belt exceeds that of the asteroid belt by about two orders of magnitude. Yet because the Kuiper belt is located much farther from both the Earth and the Sun than is the asteroid belt, more is known about asteroids than about KBOs.

The Kuiper belt is also the primary source of the **short-period** or **ecliptic comets (ECs)**, comets that are on eccentric orbits near the ecliptic plane and return with regularity (orbital periods <200 years) to the inner Solar System. The Oort cloud at distances ≳10 000 AU is the reservoir of **long-period** or **nearly isotropic comets (NICs)**, comets with orbital periods >200 years whose orbits are nearly isotropically distributed in inclination. NICs also contain **dynamically new comets**, comets whose orbits suggest that they have entered the inner Solar System for the first time.

We first present the nomenclature used to refer to individual minor planets and comets in §12.1. In §12.2, we give a summary of the orbital groupings of minor planets and comet reservoirs. The size distribution and collisional evolution of minor planets and comets, including families and multiple systems, are discussed in §12.3. Section 12.4 describes the classification scheme of minor planets based primarily on their bulk composition and spectroscopy. Several minor planets and comets have been imaged in detail by interplanetary spacecraft and/or the *Hubble Space Telescope* (*HST*); these bodies are described in §§12.5 and 12.6. The formation of a comet's coma and tail is discussed in §12.7; said section concludes with a summary of cometary composition. Orbital and physical changes in the population of minor planets and comets over the past 4 Gyr are addressed in §12.8.

12.1 Nomenclature

All minor planets with well-determined orbits are designated by a number, in chronological order, followed by a name, e.g., 1 Ceres, 324 Bamberga or 136199 Eris. After an object is discovered but before it has a well-determined orbit, it gets a provisional name, which is related to the date of the object's discovery.

Comets are named after their discoverer(s). Numbers follow names when one person or group discovers multiple comets. The names of long-period comets are preceded by C/, e.g., Comet C/Kohoutek. The names of short-period comets are preceded by a P/, e.g., P/Halley and P/Encke. Deceased comets, i.e., comets that have collided with the Sun or one of the planets or simply disintegrated, are preceded by D/. The most famous deceased comet is (was) D/Shoemaker–Levy 9

Figure 12.1 COLOR PLATE Comet Hale–Bopp (C/1995 O1) as viewed in April 1997 above Natural Bridges National Monument in Utah. (Courtesy Terry Acomb/John Chumack/PhotoResearchers)

(§§8.1.2, 12.3.5). When a comet splits (§12.3.5), each fragment is given the designation and name of the parent comet followed by an upper case letter, beginning with the letter A for the fragment that passes perihelion first, e.g., 73P/Schwassmann–Wachmann 3A. If a fragment splits further, the pieces receive numerical indices, as e.g., components Q_1 and Q_2 from Comet D/Shoemaker–Levy 9.

Comets are also given a designation that includes the year of their (re)discovery or perihelion passage. The form of these designations changed in 1995, so the literature contains both formats. In the old system, the name of the comet is followed by the year of its perihelion passage, together with a Roman numeral based sequentially on perihelion date to distinguish it from other comets passing perihelion that same year, e.g., Comet C/Kohoutek (C/1973 XII). In the current

system, the year of discovery (or recovery) is followed by a letter indicating the half-month in which the comet was first observed followed by a number to distinguish it from other comets seen during the same period. For example, 1P/1682 Q1 represents Halley's comet during its 1682 apparition, and indicates that it was initially spotted in the second half of August of 1682 (the letter I is not used in this system). Note that short-period comets receive a different designation for each apparition.

12.2 Orbits

Minor planets and comets occupy a wide variety of orbital niches (see Fig. 1.2). Most travel in the relatively stable regions between the orbits of Mars and Jupiter (known as the **asteroid belt**) exterior to Neptune's orbit (the **Kuiper belt** and the **Oort**

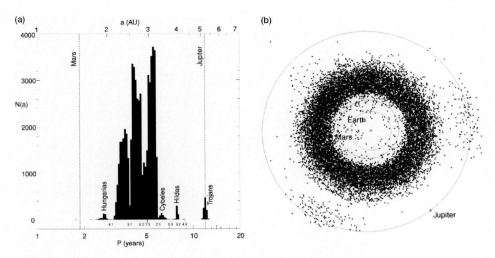

Figure 12.2 (a) Histogram of asteroids versus orbital period (with corresponding semimajor axes shown on the upper scale); the scale of the abscissa is logarithmic. All of the asteroids represented have $H_v < 15$; the 100 000 such asteroids with the smallest numbers are included. The planets Mars and Jupiter are shown by *dashed vertical lines*. Note the prominent gaps in the distribution for orbital periods 1/4, 1/2, 2/5, 3/7 and 1/3 that of Jupiter. One asteroid, Thule, is located at the 4:3 resonance. (Courtesy A. Dobrovolskis) (b) Locations projected onto the ecliptic plane of approximately 7000 asteroids on 7 March 1997. The orbits and locations of the Earth, Mars and Jupiter are indicated, and the Sun is represented by the dot in the center. (Courtesy Minor Planet Center)

cloud) or near the triangular Lagrangian points of Jupiter (the **Trojan asteroids**). The Kuiper belt and the Oort cloud are by far the most massive of these reservoirs and contain the largest objects.

Smaller numbers of minor planets are found in unstable regions. Most of these cross or closely approach the orbits of one or more of the eight planets, which control their dynamics. Those that come near our home planet are known as **near-Earth objects** (NEOs); those orbiting among the giant planets are called **centaurs**.

12.2.1 Asteroids

Main Belt Asteroids

The absolute magnitude of an asteroid or comet is equal to the apparent magnitude (§1.2.5) if it were illuminated by the amount of sunlight at 1 AU and observed at zero phase angle from a distance of 1 AU. Figure 12.2a shows the distribution of the semimajor axes for the orbits of the first 100 000 numbered asteroids with absolute magnitude $H_v <$

15. The absolute magnitude of a body in our Solar System is equal to the apparent magnitude (§1.2.5) if the body were at 1 AU from both the observer and the Sun, as seen at phase angle $\phi = 0$ (for definition ϕ see Fig. 4.4 and §4.1.2). Figure 12.2b displays the location of ∼7000 asteroids at one particular time. Most asteroids are in the **main asteroid belt** (the MBAs), at heliocentric distances between 2.1 and 3.3 AU. The mean inclination of asteroid orbits to the ecliptic plane is 15°, and the mean eccentricity is ∼0.14.

Several gaps and concentrations of asteroid semimajor axes can be distinguished in Figure 12.2a. The gaps were first noted in 1867 by Daniel Kirkwood and are known as the **Kirkwood gaps**. The Kirkwood gaps coincide with resonance locations with the planet Jupiter, such as the 4:1, 3:1, 5:2, 7:3 and 2:1 resonances. As discussed in §2.3.4, if an asteroid orbits the Sun with a period commensurate to that of Jupiter, the asteroid's orbit is strongly affected by the cumulative gravitational influence of Jupiter. Perturbations by the giant planet produce chaotic zones around the resonance locations, where asteroid eccentricities can be forced to values high enough to cross the orbits of Mars and Earth. These asteroids may then be removed by gravitational interactions and/or collisions with the terrestrial planets. In some cases, eccentricities can be excited to such high values that the asteroids may ultimately collide with the Sun.

The opposite situation occurs in the outer asteroid belt, where asteroid orbits are very strongly perturbed by nearby Jupiter. An asteroid in the outer asteroid belt may acquire such a high eccentricity that it suffers a close approach to Jupiter and is scattered to interstellar space. Asteroids that orbit the Sun at the 3:2 and 4:3 resonances with Jupiter are protected from such encounters, and we find enhancements in the asteroid population at these resonances: The **Hilda asteroids** complete three orbits during two jovian years, and 279 Thule orbits the Sun four times every three jovian years.

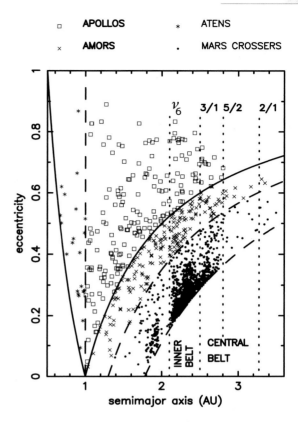

Figure 12.3 The orbital distribution of near-Earth objects (NEOs). The *solid* and *long-dashed lines* mark the boundaries of the different populations among the NEOs that are listed above the plot. The *short-dash lines* mark the location of three major mean motion resonances with Jupiter and the ν_6 secular resonance, as indicated. (Adapted from Morbidelli 2002)

Near-Earth Objects

Objects that venture close to Earth attract much attention because of the danger posed by the possibility of their colliding with our planet (§6.4.5). Such potential impactors belong to the population of NEOs, a name collectively given to all asteroids and (inert/dormant) comets that have perihelia inside of 1.3 AU. As of 2018, ~18 000 NEOs have been detected.

The NEO population is subdivided into four categories based on the perihelia (q) and aphelia of their orbits, as shown in Figure 12.3. About 40% of the NEO population, with 1.017 AU $< q <$ 1.3 AU, are called **Amor** asteroids, named after one of the prominent members of this group, 1221 Amor; their radii range up to ~15 km. About 50% of the NEOs have $q <$ 1.017 AU and semimajor axes $a >$ 1 AU; these are referred to as **Apollo** asteroids, after the archetype of this group, 1862 Apollo. The largest detected Apollo asteroids have radii of 4–5 km. The **Atens**, $\lesssim$10% of the NEOs,

have $a <$ 1 AU and aphelia greater than Earth's perihelion, 0.983 AU. Objects with orbits completely interior to that of our planet are referred to as **Apohele** asteroids. Apoheles are difficult to detect, and not many are known.

The dynamical lifetime of NEOs is relatively short, $\lesssim 10^7$ years, and hence the NEO population needs replenishment from more stable orbits. Numerical models show that the primary NEO source regions are the chaotic zones near the resonance locations in the main asteroid belt, discussed earlier (Kirkwood gaps). The ultimate fate of most NEOs is either ejection into interstellar space (the dominant loss mechanism for NEOs originating in the outer asteroid belt) or collisions with or tidal or thermal destructions by the Sun (the dominant loss mechanism for NEOs that originate in the inner belt). A small but significant fraction of NEOs collides with planets and moons (§6.4.5).

The resonance zones that are the source regions for NEOs require repopulation to balance losses. One possibility is resupply via collisions within the asteroid belt. Such collisions are often disruptive (§12.3.2), and the orbits of the smaller fragments are typically altered the most compared to that of the parent asteroid. The Yarkovsky effect (§2.8.3) can also significantly change the orbital parameters of small asteroids. Typically, kilometer-sized bodies drift in semimajor axis by $\sim 10^{-4}$ AU in one million years. Although this is significant, at the same time, it is slow enough that populations have time to evolve collisionally. Because much of the NEO population is resupplied by sporadic events, the total number of NEOs may fluctuate significantly over time.

Another potential source of NEOs is extinct comets that have developed nonvolatile crusts and ceased activity. A few bodies in Earth-crossing orbits that have been classified as asteroids are associated with meteor streams, suggesting a cometary origin. Also, comet 2P/Encke is in an orbit typical of Earth-crossing asteroids. A total of perhaps 10–15% of the NEOs may be inert comets.

Trojan Asteroids

As of 2018, a total of ~ 7000 asteroids have been discovered near Jupiter's L_4 and L_5 triangular Lagrangian points. These bodies are known as the **Trojan asteroids**. The total population of Trojan asteroids larger than 15 km in size is estimated to be roughly half that for MBAs. The largest Trojan, 624 Hektor, has a mean radius of ~ 100 km. Neptune may also have a considerable population of Lagrange point librators, but because of their

faintness, only 18 are known (as of late 2018). The bodies sharing Neptune's orbit, the Trojans, the 4/3 and 3/2 librators, along with the MBAs that are not near resonances, occupy the only known stable (older than the Solar System) orbits of known asteroids between the major planets.

Eight kilometer-sized asteroids have been found librating on tadpole orbits about the triangular Lagrangian points of Mars. Several near-Earth asteroids, e.g., 3753 Cruithne and 2002 AA$_{29}$, are coorbital horseshoe librators with Earth, but those orbital locks are most likely of geologically recent origin. The dynamics of the Trojans and other coorbitals are discussed in §2.2.2.

12.2.2 Trans-Neptunian Objects, Centaurs

The overwhelming majority of small bodies within the Solar System are trans-neptunian objects (TNOs), whose orbits lie (entirely or in part) beyond the distance of Neptune. The first known TNO, 134340 Pluto, was discovered in 1930; Pluto's large (and rarely used) minor planet number was not assigned until Pluto was reclassified in 2006. The existence of a disk of numerous small bodies exterior to the major planets was postulated by K.E. Edgeworth in 1949 and (more prominently) by G.P. Kuiper in 1951, based on a natural extension of the original solar nebula beyond the orbit of Neptune. This ensemble is therefore referred to as either the **Kuiper belt** or the **Edgeworth–Kuiper belt**. No Kuiper belt object other than Pluto and its large moon Charon was known until 1992. The discovery of (15760) 1992 QB$_1$ marked the onset of a flurry of search activities, and almost 3000 TNOs have been detected as of late 2018. The orbits of these bodies fall into several dynamical groupings, as shown in Figure 12.4. Orbits and sizes of the six largest known TNOs are given in Table E.7, and masses and densities of several TNOs are listed in Table E.8. The TNOs show a clear trend of density increasing with size; TNOs with radii $R < 200$ km have densities

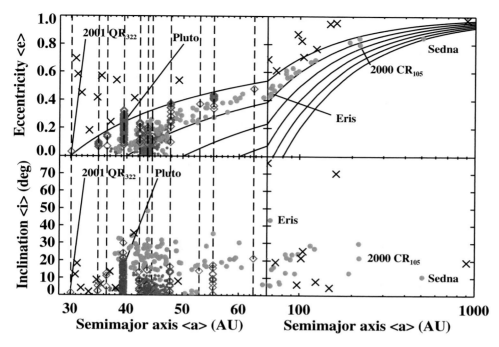

Figure 12.4 COLOR PLATE Orbital elements for TNOs, time averaged over 10 Myr. The *dashed vertical lines* indicate mean motion resonances with Neptune that are occupied: the 1:1, 5:4, 4:3, 3:2, 5:3, 7:4, 9:5, 2:1, 7:3, 5:2 and 3:1 in order of increasing heliocentric distance. *Solid curves* trace loci of constant perihelia $q = a(1-e)$. (Adapted from Chiang et al. 2007)

under $1000 \, \text{kg m}^{-3}$, whereas those with radii $R > 400$ km have densities exceeding $1000 \, \text{kg m}^{-3}$.

Classical Kuiper Belt Objects

About half of the known TNOs are **classical KBOs** (CKBOs), which travel on low eccentricity ($e \lesssim 0.2$) orbits exterior to Neptune. Most CKBOs have semimajor axes between 37 and 48 AU. The total mass of the classical Kuiper belt is a few times that of its largest member, Pluto.

Many of the CKBOs are locked in one or more mean motion resonances with Neptune. Pluto occupies a 2:3 mean motion resonance with Neptune. This resonance appears to be chaotic, but the chaos is so mild that Pluto's orbit is stable for billions of years and maybe much longer. In addition to Pluto, many small objects ($\sim$10%–30% of the CKBO population) are trapped in this same 2:3 resonance with Neptune; these objects are often

referred to as **plutinos**. Apart from the 2:3, the most populated resonances are 3:5, 4:7, 1:2 and 2:5 (Fig. 12.4).

Scattered Disk Objects

An increasing number of TNOs are being detected on high-eccentricity, nonresonant orbits with perihelia beyond the orbit of Neptune. These objects are referred to as **scattered disk objects** (SDOs). The largest known SDO is 136199 Eris, whose mass is slightly larger than that of Pluto (Table E.8). The number of known SDOs is several times smaller than that of known CKBOs. However, many of the SDOs travel on highly eccentric orbits and spend most of the time near aphelion, where they are quite faint. From the observed populations, the total mass of SDOs is estimated to be (very roughly) an order of magnitude larger than that of the classical Kuiper belt.

The vast majority of SDOs are on orbits with perihelia 33 AU $\lesssim q \lesssim$ 40 AU. These bodies come close enough to the giant planets that they could have been placed in their current orbits by planetary perturbations, but they are far enough from Neptune that their orbits are stable on billion-year timescales. Nonetheless, some occasionally get close enough to Neptune to be scattered inwards on planet-crossing trajectories (discussed later), and this reservoir is the primary source of EC comets.

The orbit of the TNO 90377 Sedna, with a perihelion at 76 AU and aphelion of ~900 AU, is exceptional. Sedna is often considered to be a member of the inner Oort cloud (§12.2.3). Its orbit could have resulted from perturbations by a passing star (probably when the Solar System was very young and had not left its crowded stellar nursery) or by an unknown planet that may still orbit in the outer Solar System or may have escaped to interstellar space eons ago.

Centaurs

Centaurs orbit between (and in some cases also cross) the orbits of Jupiter and Neptune. Dozens of such objects are known, and many are on highly eccentric and/or inclined orbits. Centaurs are in chaotic planet-crossing orbits, which have dynamical lifetimes of 10^6–10^8 yrs. Dynamical calculations suggest that they are transitioning from the trans-neptunian region (primarily the scattered disk but also the classical and resonant portions of the Kuiper belt) and that some are destined to become short-period comets (§12.7).

Centaur 2060 Chiron ($a = 13.7$ AU) has a dark neutral color, similar to many asteroids, and an orbit that crosses those of both Saturn and Uranus. Calculations show that Chiron must pass close to Saturn every 10^4–10^5 years. When this happens, the orbit is perturbed significantly, and hence Chiron orbits the Sun on a highly chaotic trajectory (Fig. 2.11). In 1987–1988, Chiron developed a coma that was spotted through a brightening of the object. Since then Chiron has also been classified as a comet, 95P/Chiron.

While evolving inwards, objects usually preserve their orbital inclination. Most of the short-period comets have low orbital inclinations and a prograde sense of revolution. This was the primary reason that most ecliptic comets were suggested to have originated in the Kuiper belt. The discovery of many centaurs that travel on unstable transitional paths crossing the orbits of the giant planets is convincing evidence for this hypothesis.

The TNOs, centaurs and other planet crossers blur the distinction between minor planets and comets. Volatile-rich minor planets can become comets if they are brought close enough to the Sun, and comets look like minor planets if they outgas all of their near-surface volatiles and become dormant or inert. When an object is not outgassing over at least part of its orbit, it is usually considered to be a minor planet. However, many dormant comets may be hidden among the minor planets. In this text, we adopt the traditional observational definition that an object is a comet if, and only if, a coma and/or tail has been observed.

12.2.3 Oort Cloud

To deduce the source region of dynamically new comets, Jan Oort plotted (in 1950) the distribution of the inverse semimajor axes, $1/a_0$, for 19 long-period comets. Based on this small sample, Oort postulated the existence of about 10^{11} 'observable' comets in what is now known as the **Oort cloud**.

Figure 12.5 shows the distribution of the original $1/a_0$ for a much larger distribution of long-period comets. The inverse semimajor axis is a measure of the orbital energy per unit mass, $GM_\odot/(2a_0)$. The original orbit is that of the comet before it entered the planetary region and became subject to planetary perturbations and nongravitational forces. Positive values in $1/a_0$ indicate bound orbits. Although negative values denote hyperbolic orbits, the few hyperbolic orbits shown in Figure 12.5 have very

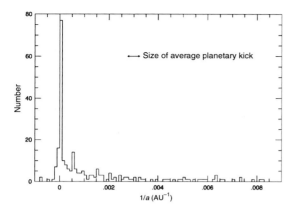

Figure 12.5 Distribution of the original (inbound) inverse semimajor axis, $1/a_0$, of all long-period comets in the 2003 version of Marsden and Williams's *Catalogue of Cometary Orbits*. The typical perturbation on $1/a$ due to a comet's passage through the inner Solar System is indicated on the graph. (Adapted from Levison and Dones 2007)

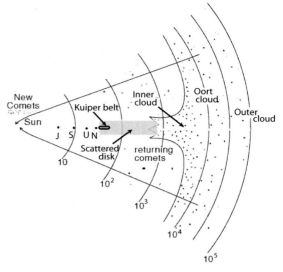

Figure 12.6 Schematic diagram of the structure of the inner and outer Oort cloud. The location of the giant plants and the Kuiper belt is indicated. Note that the distance scale is logarithmic. (Adapted from Levison and Dones 2007)

small values for $1/a_0$ and are almost certainly caused by errors in the calculation of the orbital elements.

The evidence for the existence of the Oort cloud at heliocentric distances $\gtrsim 10^4$ AU is based on the large spike in Figure 12.5 between 0 and 10^{-4} AU^{-1}. The spike, which is much narrower than the typical perturbation on $1/a$ due to a passage through the inner Solar System ($\sim 5 \times 10^{-4}$ AU^{-1}), represents comets from the Oort cloud that have entered the planetary region for the first time. These comets have semimajor axes of $(1-5) \times 10^4$ AU and are randomly oriented on the celestial sphere. The original orbits of these comets, when they were still confined to the Oort cloud, were probably perturbed by the tidal field of the galactic disk, by nearby stars or by close encounters with giant molecular clouds. The number of comets in the classical Oort cloud can be estimated from the observed flux of dynamically new comets and is expected to be $\sim 10^{11} - 10^{12}$.

Dynamical models of the formation of our Solar System show that ejection of planetesimals from the planetary region results in an inner Oort cloud, which is initially 5–10 times more populated than the outer (classical) Oort cloud. A schematic of

the structure of the Oort cloud is shown in Figure 12.6. Large perturbations caused by stellar encounters and/or giant molecular clouds penetrating into the inner Oort cloud could produce comet showers lasting a few million years about every 100 million years, and repopulate the outer Oort cloud.

When comets pass through the planetary region of the Solar System, gravitational perturbations by the planets scatter their orbits in $1/a$ space. It may take a comet ~ 1000 returns to the planetary region before its orbit is changed to that of a short-period comet. Dynamical calculations show that fewer than 0.1% of new (long-period) comets actually become short-period comets.

12.2.4 Nongravitational Forces

Although to lowest order comet trajectories are determined by the gravitational pull of the Sun and the planets, the observed orbits of most active comets deviate from the paths predicted from these gravitational tugs in small but significant ways. These deviations can advance or retard the time of perihelion passage of a comet by many days from one orbit to the next. These variations in a comet's orbit result from **nongravitational forces**, caused by the momentum imparted to a comet's nucleus by the gas and dust that escape as the comet's ices sublime. The process is analogous to rocket propulsion (§F.1), but the magnitude of the effects is much smaller because only a tiny fraction of the comet's mass is lost per orbit, mass escapes at a slower speed and forces exerted at differing phases of the comet's orbit produce opposing effects. Nongravitational forces have some similarities to the Yarkovsky effect on small bodies, which is described in §2.8.3.

The motion of a comet is thus intertwined with its evolution as a physical object. Cometary activity depends strongly on heliocentric distance. Outgassing, in turn, changes the orbit of a comet, albeit in a less profound manner.

12.3 Size Distribution and Collisions

Collisions have played a major role in shaping the asteroid and Kuiper belts, as well the population of observed comets. Such collisions were frequent events in the early history of our Solar System, and evidence for geologically recent collisions is growing. In this section, we discuss phenomena that have shaped our view of the disruptive environment in which minor planets and comets formed and evolved. These topics range from the overall size distribution of these bodies to their bulk densities and the presence of interplanetary dust.

12.3.1 Size Distribution

The size distribution of minor planets and comets can be approximated by a power law valid over a finite range in radius. Size distributions can be given in differential form:

$$N(R)dR = \frac{N_0}{R_0}\left(\frac{R}{R_0}\right)^{-\zeta} dR (R_{min} < R < R_{max}),$$

$$(12.1a)$$

where R is the object's radius and $N(R)dR$ the number of bodies with radii between R and $R + dR$. The size distribution can also be presented in cumulative form:

$$N_>(R) \equiv \int_R^{R_{max}} N(R')dR' = \frac{N_0}{\zeta-1}\left(\frac{R}{R_0}\right)^{1-\zeta},$$

$$(12.1b)$$

where $N_>(R)$ is the number of bodies with radii larger than R and R_{max} is the radius of the largest body. In equations (12.1a) and (12.1b), ζ is the power-law index of the distribution, R_0 is the fiducial radius and N_0 is a constant that depends on the choice of R_0.

Theoretical calculations imply that a population of collisionally interacting bodies evolves towards a power-law size distribution with $\zeta = 3.5$, provided the disruption process is self-similar (i.e., depends only on the speed and size ratio of the colliding bodies). In such a steady state, the number of objects that leaves a certain mass bin is equal to the number of objects that enters this mass bin. A slope of $\zeta = 3.5$ implies that most of the mass is in the largest bodies, and most of the surface area is in the smallest bodies (Problem 12-4). Size distributions for the MBAs and NEOs are displayed in Figure 12.7 and show good agreement with theoretical predictions. Similarly, the size distribution for TNOs with $R \lesssim 130$ km matches a power law with $\zeta \approx 3.5$, but larger objects display a somewhat steeper slope, indicating that the larger bodies have not (yet) reached a steady state.

The size distribution for ecliptic comets is best fit by a power law with $\zeta = 2.9 \pm 0.3$ at radii $R > 1.6$ km. This slope is slightly flatter than the size distribution for asteroids and TNOs and than that expected for a collisionally evolved population of bodies ($\zeta = 3.5$). This shallower slope suggests a larger loss of bodies than expected based on collisional processes. Indeed, one might expect a shallower slope for comets because (*i*) noncollisional fragmentation, such as splitting (§12.3.5), is common among comets and (*ii*) the nuclei are

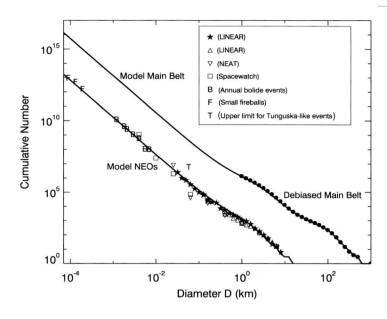

Figure 12.7 Comparison of a model of the dynamical evolution (*solid lines*) with the observationally debiased size distribution of MBAs and near-Earth objects (NEOs). Most bodies with $D \lesssim 100$ km are fragments (or fragments of fragments) derived from a limited number of breakups of bodies with $D \gtrsim 100$ km. The NEO model population is compared with estimates derived from telescopic surveys, spacecraft detections of bolide detonations in Earth's atmosphere and photographs of fireballs. The symbol T is an estimate on the upper limit of 50 m NEOs, derived from the uniqueness of the Tunguska airblast in 1908 (§6.4.5). (Adapted from Bottke et al. 2005b)

eroded through sublimation processes during each perihelion passage.

12.3.2 Collisions and Families

Typical random (relative) velocities in the Kuiper belt are $\gtrsim 1$ km s^{-1}, and for asteroids, they are several km s^{-1}. These numbers are much larger than the escape velocities of most minor planets. (The escape velocity from Ceres is $v_e \approx 0.5$ km s^{-1}.) Thus, most collisions between minor planets should be erosive or disruptive. The final outcome of a collision depends on the relative velocity and strength/size of the object (§15.4.2). Large minor planets and iron-nickel bodies have the greatest resistance to disruption. In **super-catastrophic collisions**, the colliding bodies are completely shattered, and the fragments are dispersed into independent yet similar orbits. Immediately after an impact, the resulting grouping of bodies is tightly clustered in space. However, asteroid fragments rapidly spread out in orbital longitude as a consequence of small differences in orbital period, as do the particles within a planetary ring (Problem 13-7).

An intriguing pattern becomes visible when the inclinations of MBAs are plotted against the semi-major axis, as shown in Figure 12.8. The numerous groupings, or collections of asteroids with nearly the same **proper orbital elements** (elements averaged to remove the effects of periodic perturbations, principally from Jupiter), are referred to as **families**. Individual families are named after their largest member. More than 50 families have been discovered in the asteroid belt; some have been identified among the Trojans and one family (Haumea) of KBOs is known. Members of individual families often share similar spectral properties as well, further supporting a common origin in a single body that has undergone a catastrophic collision.

Backward integrations of the orbits of individual family members show that at least seven asteroid families were formed by collisions in the last 10 Myr. Such catastrophic collisions ultimately provide a new influx to the NEO population. The catastrophic disruption of an ~170-km-diameter MBA ~160 Myr ago led to the creation of the Baptistina family. Over time, dynamical processes (i.e., Yarkovsky effect) changed the orbits of fragments

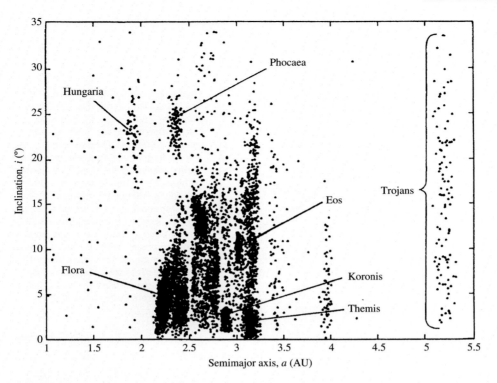

Figure 12.8 This plot of the inclination versus semimajor axis of MBAs reveals many groupings of asteroids with similar proper orbital elements. These groups, or asteroid families, represent remnants of collisionally disrupted large asteroids. (Kowal 1996)

produced in this collision such that they could strike the terrestrial planets. It has been suggested that one such fragment may have hit Earth and led to the extinction of the dinosaurs 66 Myr ago (K – T boundary, §6.4.5).

12.3.3 Collisions and Rubble Piles

The presence of numerous families in the asteroid belt is indicative of a violent collisional environment. Although the (super-)catastrophic collisions that lead to the formation of asteroid families are rare, less energetic impacts, where bodies simply get fractured or shattered but are not dispersed, happen much more frequently (§6.4). The latter case occurs when the velocity of the individual fragments is less than their mutual escape velocity,

and some or most of them coalesce back into a single body, forming a **rubble pile**. Rubble piles are composed of a gravitationally bound collection of smaller bodies and internal void spaces. Collisional fragments may also form binary or multiple systems, where bodies are of similar size and mass, or one larger object is orbited by one or more small satellites. Such systems, however, are usually not long lived. Tidal interactions between a bound pair of asteroids (§2.7.2) lead to orbital evolution on a timescale of $\sim 10^5$ years. Satellites interior to the synchronous orbit evolve inwards, so that the system ultimately becomes a rubble-pile compound object. Satellites outside the synchronous orbit evolve outwards.

Impacts by bodies much smaller than the target object, including (micro)meteoroids, pulverize

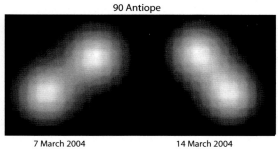

90 Antiope

7 March 2004   14 March 2004

Figure 12.9 *Galileo* image of S-type asteroid 243 Ida and its moon Dactyl. Ida is about 56 km long. Dactyl, the small object to Ida's right, is about 1.5 km across in this view and probably ~100 km away from Ida. (NASA/*Galileo* PIA000136)

Figure 12.10 The binary C-type asteroid 90 Antiope as imaged with the adaptive optics system on the very large telescope (VLT) on two different dates in 2004. The two components are almost equal in size, each with a radius of 43 km, separated by 171 km. (Adapted from Descamps et al. 2007)

the near-surface rock and thereby create a layer of regolith as on our Moon (§9.1). We hence expect the larger minor planets to be covered by a thick layer of regolith. Larger bodies retain a greater fraction of impact ejecta and they have longer lifetimes against super-catastrophic collisions, so there probably are systematic differences in the quantity of regolith produced on bodies as a function of size.

12.3.4 Binary and Multiple Systems

In 1993, the *Galileo* spacecraft, en route to Jupiter, took the photograph of the asteroid 243 Ida shown in Figure 12.9. In addition to an exquisite image of the asteroid, the big surprise was a tiny moon in orbit about Ida. This detection launched a flurry of observational campaigns to search for asteroid companions.

As of 2018, more than 300 minor planets have been identified as multiple systems. About 2% of all main belt and Trojan asteroids are binaries or multiplets. More than 10% of TNOs and ~15% of NEOs are in multiple systems. An example of a binary system with two similarly sized bodies is shown in Figure 12.10. A handful of asteroids and KBOs are triple systems. Pluto has five known satellites, displayed in Figure 12.11.

The distribution of mass ratios and orbital characteristics provides constraints on the origin, collisional history and tidal evolution of minor planets. Purely two-body gravitational interactions cannot convert unbound orbits to bound ones, nor vice versa. Temporary captures can be caused by three-body interactions with the Sun (Figs. 2.6 and 2.7) or gravitational forces exerted by a nonspherical body, but for long-term stability, energy must be

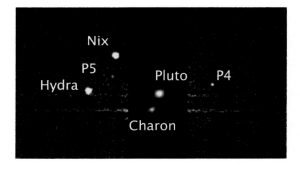

Figure 12.11 Sextuple system: *HST* image of Pluto, its large moon Charon, the small moons Hydra and Nix, as well as the most recently discovered tiny moons, P4 and P5, as indicated. This image has been processed to substantially enhance the brightness of the four little moons relative to that of the much larger bodies Pluto and Charon. Pluto-Charon can be considered a close binary encircled by four small satellites. The four linear features emanating from Pluto and from Charon are diffraction spikes from the telescope. (*HST*/NASA, ESA, Mark Showalter, SETI Institute)

dissipated as heat or removed permanently from the system by another body or bodies. Stable bound pairs can be formed when disruptive or large cratering collisions (or strong tidal encounters with a planet) produce debris that subsequently interacts gravitationally or collisionally. The circularizing second stage in such a process is known as the **second burn** in rocketry. Alternatively, bodies that approach one another slowly can become bound when a physical collision or gravitational interaction with a third body removes energy.

Binary systems may be more common among the NEOs than MBAs because close encounters with Earth and other terrestrial planets can tidally disrupt fragile bodies, similar to the tidal disruption by Jupiter witnessed for Comet D/Shoemaker–Levy 9 (§8.1.2). Debris from such a disruptive event could evolve into a binary system.

Considering the various scenarios to create binary and multiple systems and the differences between main belt and Trojan asteroids, NEOs and TNOs, different processes appear to dominate in the various minor body reservoirs. The impact scenario is the most likely process in the main belt and Trojan asteroid region, but the capture scenario is favored for TNOs. Because the dynamical lifetime of NEOs ($\sim$10 Myr) is much shorter than the collisional timescale against disruption ($\sim$100 Myr), binary formation in the NEO populations may be dominated by rotational disruption, such as caused by, e.g., tidal interactions with planets or cometary jetting.

12.3.5 Comet-Splitting Events

Some comets possess multiple nuclei that are spatially separated. The gravitational fields of comets are too weak for these nuclei to be bound binary or multiple systems. Rather, they were presumably formed by recent breakup (**splitting**) events. More than 40 split comets have been observed over the past 170 years. The first clear case was 3D/Biela in 1845. After comet Biela broke up, the brightest fragment was left with a large companion, which

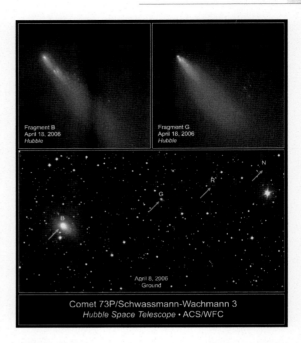

Figure 12.12 Breakup of comet 73P/Schwassmann–Wachmann 3. The top frames show the 'second generation' fragmentation of fragments B and G shortly after large outbursts in activity. The original fragments were created during a splitting event in 1995. The *bottom panel* displays a wider field of view, showing several of the original fragments. (Courtesy Hal Weaver and NASA/*HST*)

evolved as a separate comet. On its next return, in 1852, 3D/Biela appeared as a double comet. Neither piece has been seen since then, but in 1872, an intense meteor shower occurred when the Earth crossed the orbit of D/Biela, indicative of the 'death' of this comet. This shower gradually diminished in intensity over the next century.

In many splitting events, only tiny fragments separate from the principal nucleus; such small pieces last for at most a few weeks. Images of a splitting event are shown in Figure 12.12. Because the splitting of a comet releases a substantial amount of dust and exposes fresh ice, it is typically accompanied by a flare-up in the comet's brightness and a (temporary) increase in dust emission, such as shown in the visible lightcurve of C/West in Figure 12.13. The most extraordinary flare-up ever

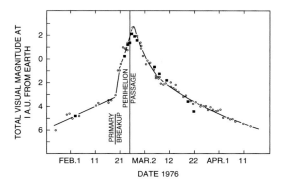

Figure 12.13 The lightcurve of C/West, which shows evidence of a splitting event (indicated in the figure). (Sekanina and Farrell 1978)

observed was an overnight brightening of comet 17P/Holmes in 2007, from a magnitude of ~17 to 2.8, i.e., by a factor of almost a million. This same comet also showed a major outburst in 1892, when Edwin Holmes discovered the comet (hence the comet's name).

The best understood cause for the breakup of a cometary nucleus is tidal disruption during a close encounter with the Sun or a planet. An extraordinary example of a split comet was D/Shoemaker–Levy 9, which was initially found to orbit and later to crash into Jupiter (July 1994; §8.1.2). Ground-based and *HST* images revealed more than 20 cometary nuclei, strung out like pearls on a string (see Fig. 8.9).

Sun-grazing comets have a perihelion distance <2.5 $R_\odot$. More than 3000 Sun-grazing comets have been discovered using *Solar and Heliospheric Observatory* (*SOHO*) images; most were first identified by amateur astronomers. The majority of these have similar orbital properties and belong to the **Kreutz family** of Sun-grazing comets. This comet family is named after Heinrich Kreutz, who made the first extensive observations of Sun-grazing comets in the nineteenth century. He suggested that these comets are fragments of a single object that broke up about 2000 years ago. Most Sun-grazing comets do not survive perihelion passage – they evaporate or disrupt completely or collide with the Sun. Figure 12.14 shows a comet

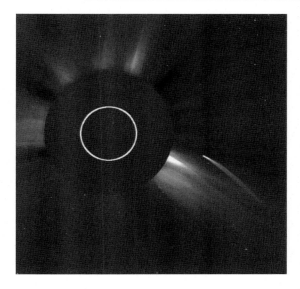

Figure 12.14 A Sun-grazing comet caught by the LASCO coronagraph on the *SOHO* spacecraft as it moved toward the Sun on 5–6 July 2011. The inner few solar radii are blocked by a coronagraph. The circle is drawn in to represent the size and location of the solar disk. (Courtesy SOHO/LASCO consortium; ESA and NASA) The associated movie taken by the *Solar Dynamics Observatory* (*SDO*) shows the comet to evaporate completely.

approaching the Sun; in the associated movie, one can see the comet completely vaporize because of the heat from the Sun.

12.3.6 Mass and Density

To measure a celestial body's mass, one generally needs to observe a gravitational interaction. After a body's mass is known, its density can be determined if its size and shape are known. A body's density yields invaluable information on its internal structure if its composition is known (e.g., via spectroscopy). In some cases, e.g., 20 000 Varuna, when the shape and rotational period are known, the object's density can be estimated by assuming the shape to agree with that of a body in hydrostatic equilibrium (§6.2.2).

Although with the ever-increasing number of confirmed binaries, our list of asteroid and KBO masses and densities has expanded significantly over the past years, only a tiny fraction of the minor

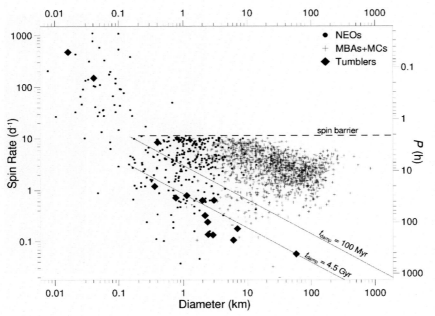

Figure 12.15 Plot of the rotational period (*right vertical axis*) and spin rate (*left vertical axis*) versus diameter for main belt (MBA) and Mars-crossing (MC) asteroids, near-Earth objects (NEO) and tumbling asteroids. The *dashed line* denotes the maximum rotation rate for objects of density 3000 kg m^{-3} that are bound gravitationally. (Adapted from Pravec et al. 2007)

planet population has good density measurements. These values show that the densities of minor planets vary substantially from object to object, with measured values ranging from $\sim$500 kg m^{-3} to almost 4000 kg m^{-3}. A comparison of asteroid densities with those of meteorites with similar spectra gives information on the asteroid's porosity. The three largest asteroids, 1 Ceres, 2 Pallas and 4 Vesta, appear to be coherent objects without substantial porosity, as expected for such large bodies. Many of the smaller bodies have porosities greater than 20% and in some cases exceeding 50%. The most porous objects probably have a loosely consolidated rubble-pile structure.

The three largest known KBOs, Eris, Pluto and Haumea (Table E.7), which are too massive to retain substantial porosities, have densities of $\sim$2000 kg m^{-3}, indicative of coherent bodies composed of comparable amounts of rock and ice. The densities of smaller objects are closer to 1000 kg m^{-3}. Because the composition of these bodies is presumably similar to that of the larger objects, with a rock/ice mass ratio of $\sim$0.5, they must be highly porous, like the primitive asteroids. As for

asteroids, the relatively high observed multiplicity together with the low internal densities of KBOs also hints at an early intense collisional epoch. In addition, numerical calculations show that the Kuiper belt must once have been much (perhaps several hundred times) more massive than at present to explain the accretion of the number of observed large ($R \gtrsim 100$ km) KBOs.

12.3.7 Rotation

More than 80% of minor planets with known spin rates rotate with a period between 4 and 16 hours. Asteroids spinning faster than about 2 hours would throw loosely attached material off their equator and hence are unlikely to survive (Problem 12-10). There is a clear correlation between rotation period and asteroid size: asteroids with radii less than about 100 m typically spin faster than larger bodies, as shown in Figure 12.15. The shortest rotation period observed for asteroids more than 1 km in size is 2.2 hours, essentially equal to the theoretical limit. Many of the smallest NEOs, $R \lesssim 100$ m, rotate much faster. The tiny ($\sim$15-m radius) asteroid 1998 KY$_{26}$ has a rotation period of only 10.7 minutes. Such rapidly rotating bodies must be coherent rocks without a regolith. A few asteroids with exceptionally long rotation periods may have been tidally despun (§2.7.2) by (as yet undetected) moons.

Much less is known about the rotation rates of TNOs. For similarly sized TNOs and MBAs, TNOs appear to have longer rotation periods, $\sim$8.2 hours versus 6.0 hours for MBAs. Smaller TNOs may spin somewhat faster than the larger ones. A notable exception is the 3.9-hour spin period of the large TNO Haumea. This high spin rate may have been imparted during the collision that created the Haumea family.

The physics of cometary rotation is analogous to that of asteroid rotation, with the added complication that outgassing of comets can produce torques that alter the rotational angular momentum vector. Comets that are (or were in the geologically recent past) active are thus more likely to exhibit complex (non-principal axis) rotation states than are asteroids. Moreover, outgassing-induced spin-up of cometary nuclei may cause some comets to split into two or more pieces, which can instantaneously alter the rotation rate.

12.3.8 Interplanetary Dust

The interplanetary medium contains countless microscopic dust grains, visible through faint reflections of sunlight, producing the **zodiacal light** and the **gegenschein**. The zodiacal light is visible in clear moonless skies, particularly in the spring and fall just after sunset and before sunrise in the direction of the Sun. Tiny dust particles, concentrated in the local **Laplacian plane** (the mean or reference plane about which satellite orbits precess), scatter sunlight in the forward direction, and the resulting zodiacal light is about as bright as the Milky Way. The gegenschein is visible in the antisolar direction as a faint glow caused by backscattered light from interplanetary dust. The total volume of the zodiacal dust corresponds to one $\sim$5 km-radius sphere.

The presence of interplanetary dust is further evidenced in the form of meteors, streaks of light in the night sky. Meteors are caused by centimeter-sized dust grains that, when falling down, are heated to incandescence by atmospheric friction (§§6.4.3, 11.3). Under excellent conditions, one may see five to seven meteors per hour. On rare nights, many more appear to come from a single point in space; such events are called **meteor showers** and are generally named after the stellar constellation that contains the **radiant point**, e.g., the Perseids on 11 August and the Leonids on 17 November. Many of the meteor showers are associated with cometary orbits: The debris left behind by the outgassing comet is intercepted by the Earth when it intersects the comet's path. Interplanetary dust particles (Fig. 11.11) collected by high-altitude aircraft in the Earth's stratosphere are similar in composition to chondritic meteorites.

(Sub)micrometer- to centimeter-sized material is removed from the Solar System by radiation pressure, or by Poynting–Robertson and/or solar wind (corpuscular) drag (§2.8). Thus small grains are lost from the interplanetary medium. The main source of interplanetary dust is collisions between asteroids and outgassing by comets. Many spacecraft have made *in situ* observations of the interplanetary dust particles, and several spacecraft have also detected **interstellar dust grains**.

12.4 Bulk Composition and Taxonomy

Comets, asteroids and meteorites provide key information on the origin of our Solar System, because they are relics of the environment in which the planetesimals that ultimately accreted into the eight known planets formed and evolved. Our main means to extract information on composition is via spectroscopy.

Table 12.1 **Asteroid Taxonomic Types**

C	Carbonaceous asteroids; similar in surface composition to CI and CM meteorites. Dominant in outer belt (>2.7 AU).
D	Extreme outer belt and Trojans. Red featureless spectrum, possibly due to organic material.
P	Outer and extreme outer belt. Spectrum is flat to slightly reddish, similar to M types, but lower albedo.
S	Stony asteroids. Major class in inner–central belt.
M	Stony-iron or iron asteroids; featureless flat to reddish spectrum.
W	Visible light spectra similar to those of M types but have an absorption band near 3 μm (indicative of hydration).
V	Similar to basaltic achondrites. Type example: 4 Vesta.

Asteroids, similar to meteorites, appear to be a compositionally diverse group of rocks. Some asteroids contain volatile material such as carbon compounds and hydrated minerals, but others appear to be almost exclusively composed of refractory silicates and/or metals. Whereas some asteroids resemble primitive meteorites, others have undergone various amounts of thermal processing.

Comets originate in the outer Solar System where solar heating is minimal, and they are relatively small ($\lesssim$ tens of km across). Comets, therefore, are among the most primitive objects in our Solar System, and as such they yield key information about the thermochemical and physical conditions of the regions in which they formed.

In §§12.4.1 and 12.4.2, we summarize the asteroid classification scheme and recent advances in our understanding of the relationships between these classes, asteroid composition and the effect of space weathering. In §12.4.3, we discuss the spectroscopic diversity of TNOs. We defer a detailed discussion of cometary composition until §12.7.6, after covering the physics of coma formation.

12.4.1 Asteroid Taxonomy

Histograms of albedos (at 0.55 μm) for asteroids with radii $R > 20$ km show a bimodal distribution, with pronounced peaks at $A_0 \approx 0.05$ and 0.18. These albedos together with reflectance spectra at visible and near-infrared wavelengths have historically been used to sort asteroids into **taxonomic classes** (Table 12.1). However, **space weathering** – the interaction of solar wind particles, solar radiation and cosmic rays with planetary bodies – induces chemical alterations in the surface material and therefore can obscure the real composition of an asteroid. This complicates the matching of asteroid types with meteorite classes.

The largest class of asteroids contains the **carbonaceous** or **C-type** asteroids. They are dark bodies with typical geometric albedos $A_0 \sim 0.04$–0.06 and flat spectra (neutral in color) longwards of 0.4 μm, similar to those of carbonaceous chondritic meteorites (CI, CM) (§11.1). They appear to be low-temperature condensates, primitive objects that have undergone little or no heating. Roughly 40% of the known asteroids are C types.

The next largest class of cataloged asteroids (30%–35%) are the **S-type** or **stony** asteroids. These objects are fairly bright, with geometric albedos ranging from 0.14 to 0.17, and reddish.

About 5%–10% of the asteroids have been classified as **D and P types**. The D- and P-type asteroids are quite dark, with $A_0 \approx 0.02$–0.07, and on average are somewhat redder than S types. Neither D nor P types exhibit spectral features. They may represent even more primitive bodies than the carbonaceous C-type asteroids. The red color of P and D types has been attributed to organic compounds, perhaps created via space weathering.

M-type asteroids have a spectrum similar to P-type asteroids but with a higher albedo, $A_0 \approx 0.1$–0.2. The M-type visible wavelength spectra lack silicate absorption features and are reminiscent of metallic nickel–iron. Their spectra are analogous to iron meteorites and enstatite chondrites, meteorites composed of grains of nickel–iron embedded in enstatite, a magnesium-rich silicate.

The meteoritic analog of the M type suggests these asteroids have undergone substantial thermal processing via a melt phase. They have often been interpreted as fragments of the cores of disrupted large asteroids. This picture became blurred with the discovery of absorption bands near 3 μm on many of the larger ($\sim$75% with $R \gtrsim 30$ km) M-type asteroids. These objects have been reclassified into a **W type**, for 'water' in the form of hydration, features indicative of primitive rather than igneous objects.

Asteroid 4 Vesta (**V type**) is unique among large objects in that it appears to be covered by basaltic material. Its spectrum resembles that of **howardite** meteorites (classes/types): HED (howardite–eucrite–diogenite), eucrite (§11.1) and **diogenite (HED)** meteorites. In addition to Vesta, several small V-type asteroids have been detected in orbits similar to that of Vesta. These asteroids and HED meteorites may have been blasted off Vesta in the impacts that produced the two giant (400 km and 500 km) craters observed in *Dawn* images of Vesta.

12.4.2 Taxometric Spatial Distribution

There appears to be a strong trend among the taxonomic classes with heliocentric distance (Fig. 12.16). S-type asteroids prevail in the inner parts of the main belt, M types are seen in the central regions of the main belt, and the dark C-type objects are primarily found near the outer regions of the belt. D and P asteroids are found

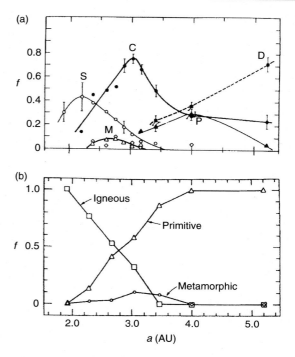

Figure 12.16 (a) Graph showing the relative distribution of the asteroid taxonomic classes as a function of heliocentric distance. The classes S, C, M, P and D are shown. Smooth curves are drawn through the data points for clarity. (Adapted from Gradie et al. 1989) (b) Distribution of igneous, primitive and metamorphic classes as a function of heliocentric distance. (This figure assumes that S-type asteroids are igneous bodies.) (Adapted from Bell et al. 1989)

only in the extreme outer parts of the asteroid belt and among the Trojan asteroids. Figure 12.16b shows the distribution of igneous (assuming S-type asteroids to be igneous bodies) and primitive asteroids as a function of heliocentric distance. This figure shows a correlation with heliocentric distance: Igneous asteroids dominate at heliocentric distances $r_\odot < 2.7$ AU and primitive asteroids at $r_\odot > 3.4$ AU. Metamorphic asteroids, which must have undergone some changes because of heating as characterized by their spectra, have been detected throughout the main belt. Spectral evidence for hydrated phases is particularly strong for asteroids in the main belt and is essentially absent beyond 3.4 AU.

The strong correlation of asteroid classes with heliocentric distance cannot be a chance occurrence. It must be a primordial effect, although it is likely modified by subsequent evolutionary or dynamical processes. If asteroids are indeed remnant planetesimals, their distribution in space might provide insight into the temperatures, pressures and chemistry of the solar nebula. The difference in spatial distribution between the C-type carbon-rich (primitive) and S-type carbon-poor asteroids is qualitatively consistent with the temperature structure expected in the primitive solar nebula.

It is not known to what degree post-formation evolution, including space weathering and dynamical processes, masks the original distribution of asteroid compositions. However, the orderly arrangement of taxonomic classes with heliocentric distance clearly contains important clues to the formation history of our Solar System. While there is a trend in taxonomic classes with heliocentric distance in the main and outer asteroid belt, NEOs have representatives from almost all the taxonomic classes, indicating that many locations in the asteroid belt feed the NEO population, as expected based upon dynamical arguments (§§12.2, 2.3.4).

12.4.3 Trans-Neptunian Object Spectra

The visible-light spectra of TNOs and centaurs are relatively featureless and vary from being neutral in color with respect to the Sun to being extremely red. It is not clear whether these diverse colors reflect different primordial compositions or different degrees of surface processing. Whereas red objects may be red as a consequence of space weathering, gray bodies may have had their 'clock' reset by collisions or cometary-like activity (as for 2060 Chiron) dredging up primordial materials that are neutral in color.

In contrast to the relatively featureless reflection spectra of most TNOs and centaurs at visible wavelengths, their spectra are very rich at near-infrared wavelengths. The largest known TNOs, Pluto and Eris, show absorption bands of the most volatile ices CH_4 and N_2.

Both Pluto and Neptune's largest moon Triton, likely a captured KBO, have tenuous atmospheres dominated by N_2 gas, with traces of CH_4 and CO. Other large TNOs may have atmospheres as well but so far have escaped detection. As discussed in §5.7.1, the continued presence of an atmosphere depends upon a body's gravity and on the temperature of its atmosphere. Similarly, the retention of volatile ices also depends on gravity and temperature. Calculations of Jeans escape (§5.7.1) from large KBOs show that Pluto, Triton, Eris and Sedna are large and cold enough for the continued presence of CH_4, N_2 and CO ices, but Haumea and Makemake are borderline candidates for volatile retention. Smaller KBOs would have long lost all CO, CH_4 and N_2 that are not trapped in H_2O ice via atmospheric escape.

12.5 Individual Minor Planets

Detailed images, spectra and other *in situ* data have been obtained for several asteroids. These spacecraft observations aid substantially in our overall understanding of these bodies, both as individual objects and as a group of bodies/planetesimals that contain information on the formation of our Solar System. In this section, we summarize the characteristics of some of the minor planets that have been investigated in detail, ordered by increasing orbital semimajor axis. We conclude this section with short discussions of the famous TNO Pluto and the most massive known TNO, Eros.

12.5.1 Near-Earth Asteroids

25143 Itokawa

In September 2005, the *Hayabusa* spacecraft went into orbit around the small ($268 \times 147 \times 105$ m in radius) near-Earth asteroid 25143 Itokawa. For a period of ∼3 months, *Hayabusa* hovered between

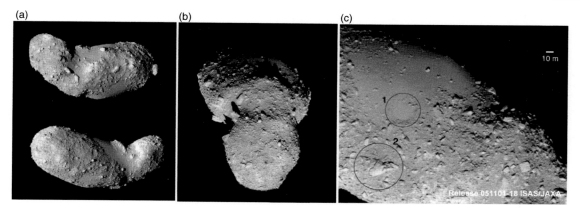

Figure 12.17 Images of 25143 Itokawa obtained by the *Hayabusa* spacecraft. Full views from different perspectives are shown in panels (a) and (b); panel (c) shows a detailed view from a distance of 4 km (see size bar of 10 m on the image). As shown, the surface is covered with huge boulders and seemingly 'naked' regions. At higher resolution, these naked regions are covered with finer grained regolith. In panel (c) a subdued crater is encircled (1), as is a huge boulder (2). (ISAS/JAXA)

7 and 20 km above the asteroid's surface and performed two touchdowns to collect materials, which were returned to Earth in 2010.

The images shown in Figure 12.17 reveal a body with a shape and surface morphology unlike any other known. The surface shows no evidence of craters but is instead littered with boulders, a few up to 50 m in size. The largest boulder is about one-tenth the size of the asteroid itself. Whereas 80% of the asteroid's surface is rough and littered with boulders, the remaining 20% is extremely smooth and featureless.

Spectra show that Itokawa's bulk composition is similar to that of LL ordinary chondrites (§11.1). This, combined with its low density (1900 ± 130 kg m^{-3}), suggests a porosity of $\sim$40%, i.e., the body is a rubble pile of material, held together by gravity. Because even a small impact would give ejecta velocities well over the escape speed of 0.1 m s^{-1}, perhaps most of the smallest grains have left the asteroid over time, but large boulders continue to accumulate.

Analysis of 10- to 100-μm-sized regolith particles from Itokawa that were brought back to Earth shows they resemble thermally metamorphosed LL chondritic material. Overall, the particles' size and

shape suggest that they were primarily formed on Itokawa's surface by meteoroid impact and that they suffered abrasion by seismic-induced shaking, hence rolling across the surface. The effect of space weathering is clearly seen on the grains. This weathering seemed to have produced the 20- to 50-nm thick layer of nanometer-sized iron particles seen on these grains. This is the layer that probably masks the chondritic composition of S-type asteroids. This metallic layer is overlain by a much thinner (5–15-nm thick) layer that contains tiny (1–2 nm in size) specks of iron sulfide. This layer likely formed when micrometeorites and/or solar wind particles vaporized nearby minerals upon impact, and the vapor condensed onto the rock particle.

The grains are indicative of intensive thermal metamorphism, which implies temperatures at $\sim$800°C, possibly resulting from decay of a short-lived radionuclide ^{26}Al during the early Solar System (§§11.6 and 15.8). The relatively slow rate of cooling implies an originally much larger ($R > 10$ km) body, which underwent catastrophic collisions, some pieces of which reaccreted into the present rubble-pile Itokawa. Based on the abundance of cosmic-ray-produced ^{21}Ne (§11.6), the

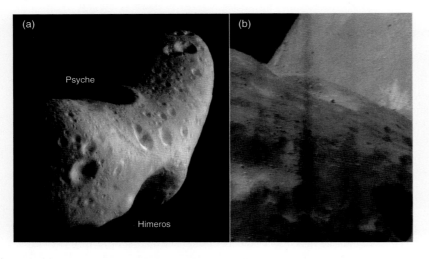

(a) Psyche

Himeros

(b)

Figure 12.18 Images of the near-Earth asteroid 433 Eros obtained by the *NEAR Shoemaker* spacecraft. (a) Mosaic taken from an altitude of ~200 km. The ~10-km saddle Himeros and 5.3-km-diameter crater Psyche are indicated. (NASA/NEAR, PI02923) (b) A close-up of Himeros, taken from a distance of 51 km, showing a region about 1.4 km across. (NASA/NEAR, PI02928)

grains have resided less than 8 Myr on Itokawa's surface, implying that the asteroid is losing its surface materials into space at a rate of tens of centimeters per million years, and hence the asteroid will disappear within a billion years (Problem 12-14).

433 Eros

The S-type asteroid 433 Eros (mean $R = 8.4$ km, $A_0 = 0.25 \pm 0.05$) is the largest NEO that has been studied extensively by spacecraft; images are shown in Figure 12.18. On a global scale, irregularly shaped Eros is dominated by convex and concave forms, including a depression extending more than ~10 km (Himeros) and a bowl-shaped crater 5.3 km in diameter (Psyche). Eros's surface is heavily cratered. Furthermore, about a million ejecta blocks with diameters between 8 and 100 m lie on the surface. The different orientations of sinuous and linear depressions, ridges and scarps indicate that these features were formed in multiple events.

Eros's elemental composition and spectral properties are similar to ordinary chondrites. As discussed earlier for Itokawa, space weathering appears indeed to 'mask' the surfaces of many asteroids, and the common S-type asteroids are likely the parent bodies of ordinary chondrites. Eros appears to be a primitive undifferentiated

body with a density of 2670 ± 30 kg m^{-3}, suggesting an ~20% internal porosity.

12.5.2 Main Belt Asteroids

21 Lutetia

Figure 12.19 shows an image of 21 Lutetia, obtained by ESA's *Rosetta* spacecraft. Although

Figure 12.19 Asteroid 21 Lutetia ($R \approx 62 \times 50 \times 47$ km) imaged at closest approach, a distance of ~3200 km, by the *Rosetta* spacecraft. (ESA and OSIRIS Team MPS/UPD/LAM/IAA/RSSD/INTA/UPM/DASP/IDA)

Figure 12.20 Image mosaic (four images) of C-type asteroid 253 Mathilde. The size of the asteroid as shown is 59 × 47 km. (NASA/NEAR, PIA02477).

classified as an M-type asteroid, Lutetia does not display much evidence of metals on its surface. Moreover, its spectrum resembles that of carbonaceous chondrites and C-type asteroids and is not at all like that of metallic meteorites.

Lutetia has a complex and morphologically diverse surface. There are several big (tens of kilometers) craters and many smaller craters; it clearly is a very old object. Lutetia is covered by an ~600-m-thick layer of regolith, which is revealed by unique landslide structures along the walls of some craters. Many other structures can be recognized, such as pits, craters, chains, ridges, scarps and grooves, some radially aligned and others concentric around relatively young craters.

253 Mathilde

253 Mathilde, shown in Figure 12.20, is a C-type asteroid with a geometric albedo of ~0.04 and R ~25 km. It displays few albedo variations, suggesting that the entire body is probably homogeneous and undifferentiated, as expected for primitive bodies. Mathilde's color is similar to that of carbonaceous chondrites, but Mathilde's density (1300 ± 200 kg m^{-3}) is much lower than that of the

meteorites, implying a 40%–60% internal porosity, i.e., a rubble-pile compound body.

About 50% of Mathilde's surface was imaged by the *Near-Earth Asteroid Rendezvous* (*NEAR*) spacecraft. Four craters with diameters that exceed the asteroid's mean radius of 26.5 ± 1.3 km were found. Crater morphologies range from relatively deep fresh craters to shallow, degraded ones. This abundance of large craters and a range of morphologies are consistent with a surface in equilibrium with the cratering process.

243 Ida and Dactyl

243 Ida, an irregularly shaped S-type asteroid (long axis: $R \approx 28$ km), shown in Figure 12.9, is a member of the Koronis family. This *Galileo* image also shows the small (mean radius of 0.7 km), almost round satellite Dactyl, which orbits Ida at a distance of 85 km. Ida and Dactyl have very similar photometric properties but distinct differences in spectral properties. Hence, their textures must be similar, but Dactyl has a slightly different composition; specifically, it may contain more pyroxene than Ida. The differences are within the range of variations reported for other members of the Koronis family and argue for compositional inhomogeneities of the Koronis parent body, possibly resulting from differentiation.

4 Vesta

The *Dawn* spacecraft, launched in September 2007, arrived at 4 Vesta in July 2011; a full-disk image of the asteroid is shown in Figure 12.21. *Dawn* orbited Vesta for more than one full year before continuing its journey to 1 Ceres. Vesta's mean radius is 265 km, which makes this asteroid the largest one after Ceres. Vesta, a V-type asteroid, is unique among large objects because it has a basaltic surface. Two prominent pyroxene absorption bands dominate its spectrum, just like that of HED achondrite meteorites (§11.7.1).

Vesta's surface topography is characterized by a 460-km-wide crater at Vesta's south pole. The

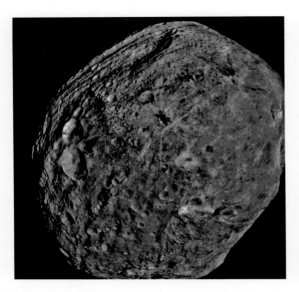

Figure 12.21 This full view of the giant asteroid Vesta was taken by NASA's *Dawn* spacecraft from a distance of ~5000 km. The northern hemisphere (*upper left*) is heavily cratered, in contrast to the south. The cause of this contrast, as well as the origin of grooves that circle the asteroid near its equator, shown best in the movie, is unknown. The movie also shows the ~460-km-diameter crater near Vesta's south pole. The resolution of this image is about 500 m per pixel. (NASA/JPL-Caltech/UCLA/MPS/DLR/IDA, PIA14894)

Dawn spacecraft identified a pronounced central peak, ~100 km across, that rises 20–25 km above the relatively flat crater floor. A set of circumferential troughs associated with the south pole crater are seen near the equator. An older basin offset from the south pole crater has been identified; this ancient basin has its own set of troughs. Numerous other depressions on Vesta's surface may also be remains of large impact basins. Figure 12.22a shows a more detailed view of part of the surface, revealing a scarp with landslides and craters in the scarp wall. As shown, these smaller craters have a simple bowl-shaped morphology; some craters have central peaks. The asteroid is uniformly bright, with many brighter and extremely dark spots across its surface, some of which are shown in Figures 12.22b and c. The bright areas occur mostly in and around craters, but the dark materials also seem to be related to impacts and include carbon-rich compounds.

Several small V-type, or **vestoid**, asteroids have been detected in orbits similar to that of Vesta. Most of these asteroids and the HED meteorites may have been created in the impact that produced the south pole crater. Assuming this to be the case, these HED meteorites were used to develop a model of the origin and interior structure of Vesta. This model suggests that Vesta formed within the first ~2 Myr of the formation of our Solar System, trapping short-lived radionuclides in its interior, the heat of which led to melting, fractionation and

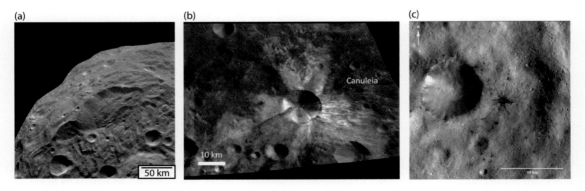

Figure 12.22 (a) A detail of asteroid 4 Vesta imaged by NASA's *Dawn* spacecraft. This image was taken through the camera's clear filter and shows a steep scarp with landslides and vertical craters in the scarp wall. (b) Image of bright material that extends out from the crater Canuleia on Vesta. The bright material appears to have been thrown out of the crater during the impact that created it, and extends 20–30 km beyond the crater's rim. (c) Image of a dark-rayed impact crater and several dark spots on Vesta. (NASA/JPL-Caltech/UCLA/MPS/DLR/IDA, a: PIA14716, b: PIA15235, c: PIA15239)

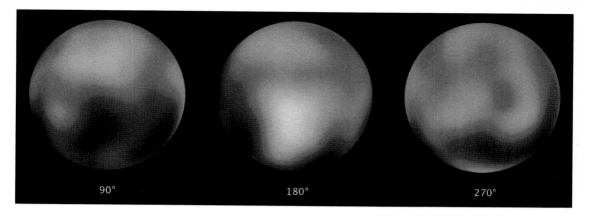

Figure 12.23 *HST* images of the surface of Pluto taken in 2002 and 2003. Most of the surface features, i.e., the variety of white and black colors, are likely produced by the complex distribution of frosts that migrate across Pluto's surface with its orbital and seasonal cycles. (Courtesy Marc Buie, NASA/*HST* and ESA)

the formation of an iron core. *Dawn*'s gravity data indeed suggest the presence of an $\sim$110-km-radius iron core. This, together with the mineral composition and crater chronology of Vesta's surface, is consistent with the aforementioned formation scenario.

12.5.3 Trans-Neptunian Objects

134340 Pluto

Because it is by far the brightest KBO and it has been known for far longer than any other object orbiting the Sun beyond Neptune, we have more information about Pluto than any other TNO. Pluto is about two-thirds the radius of our Moon, and its largest moon, **Charon**, is about half Pluto's radius. Pluto's mean density, 2030 ± 60 kg m^{-3}, is larger than expected for a rock/water-ice mixture in cosmic (50:50) proportion and suggests that Pluto is composed of roughly 70% rock by mass. This dwarf planet is probably differentiated, with the heaviest elements (rock, iron) comprising its core, overlain by a mantle of water-ice, and topped off with the most volatile ices. Infrared spectra suggest Pluto's surface to be covered by nitrogen-ice with traces of methane, ethane and carbon monoxide.

Pluto occulted a twelfth magnitude star in 1988, and the gradual rather than abrupt disappearance

and later reappearance of the star revealed the presence of an atmosphere with a surface pressure between 10 and 18 μbar. Given Pluto's surface temperature, such a pressure is consistent with an atmosphere in sublimation equilibrium with a surface covered with N_2-ice. The atmosphere's main constituent is likely nitrogen gas; traces of CH_4 ($<1\%$) and CO ($<0.5\%$) have been observed.

Spatially resolved images of Pluto taken with *HST* are shown in Figure 12.23. Images taken over the past two decades suggest that the spatial variation of colors and changes therein over time result from frosts that migrate across Pluto's surface with its orbital and seasonal cycles and chemical byproducts deposited out of Pluto's nitrogen–methane atmosphere. The orange-reddish color is probably the carbon-rich residue from methane being broken up by ultraviolet (UV) sunlight.

In contrast to the nitrogen-ice that covers Pluto's surface, its moon Charon is covered by water-ice, probably because all methane and nitrogen have escaped over time (§12.4.3). Not much is known about the surface composition of Pluto's four smaller moons, except that Nix and Hydra exhibit the same, essentially gray, colors as the larger moon, Charon. The orbits and colors of these small moons reinforce the theory that Charon was produced in a giant impact, similar to the formation of our Moon (§15.10).

136472 Makemake

Makemake, the third largest TNO, is an icy dwarf planet with a spectrum similar to that of Pluto and Eris, although its methane-ice absorption bands are even stronger than for either Pluto or Eris. Occultation measurements reveal a body with radii of 717×710 km. Makemake's rotation period is 7.77 hr; its density is poorly constrained, and is likely between 1500 and 2500 kg m^{-3}. Its visible albedo, $A_0 \approx 0.8$, is in between that of Pluto and Eris, and points at an icy surface. However, the dwarf planet also contains a small, much warmer region, characterized by an albedo $A_0 \sim 0.12$. Stellar occultation measurements, when Makemake was at a heliocentric distance of 52.2 AU, revealed an upper limit of $\sim$100 nbar to the surface pressure produced by an atmosphere. The absence of an atmosphere suggests a lack of nitrogen ice on the body's surface, as the vapor pressure of N_2 ice is at microbar levels, assuming the body's temperature is equal to its equilibrium temperature.

136199 Eris

The dwarf planet and TNO Eris was discovered in 2005. Eris's orbit is quite eccentric, with $e = 0.44$ and $i = 44°$. The orbit of its satellite, Dysnomia, indicated a mass of Eris that is $\sim$27% larger than that of Pluto (Table E.7). Eris's size has been determined via a multichord stellar occultation experiment, which showed the dwarf planet to be quite spherical with a radius of 1163 km, so that its density is $\sim$2520 kg m^{-3}. Eris has a very high visible geometric albedo, $A_0 = 0.96 \pm 0.07$. At present, Eris is near aphelion, near 97 AU. At this distance, one would not expect Eris to have an atmosphere. However, one might expect ices to sublime off its surface when Eris approaches its perihelion, which is located at 37.8 AU from the Sun.

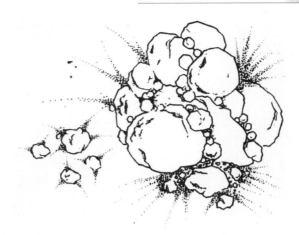

Figure **12.24** A schematic representation of a cometary nucleus according to the rubble-pile version of the dirty snowball model in which the individual fragments are lightly bonded by thermal processing or sintering. (Weissman 1986)

12.6 Shape and Structure of Comet Nuclei

Comets are best known for the spectacular tails that some of them produce. Nonetheless, we begin our discussion with comet nuclei, which are analogous to minor planets and contain the overwhelming majority of a typical comet's mass. We consider comas and tails in §12.7 and discuss the composition of cometary nuclei in §12.7.6.

Comet nuclei are very small and dark. They are quite faint when far from the Sun and are lost in the light of the coma when the comet is active. Thus they are difficult to observe from the ground. Before the space age, the composition and structure of cometary nuclei were determined from observations of the material that they released into their comas and tails, as well as from dynamical observations, particularly nongravitational forces and breakup of nuclei. Such observations led Fred Whipple to propose his **dirty snowball** theory in 1950. In his model, a version of which is depicted in Figure 12.24, a cometary nucleus is a loosely bound agglomeration of frozen volatile material interspersed with meteoritic dust.

Because comets are rather small and rich in volatiles, they cannot have undergone much thermal evolution and are therefore regarded as the most pristine objects observed in our Solar System. The outer layers of a comet, however, may have undergone significant processing while in the Kuiper belt or Oort cloud. In particular, exposure to energetic particles and cosmic rays for several billion years must have darkened and reddened a comet's surface, an extreme version of space weathering (§12.4). This outer layer of a comet is probably about a meter thick and is referred to as the **irradiation mantle**. Comets in the Oort cloud may, in addition, also be heated by passing stars and supernova explosions.

Several comets have been imaged at close range by spacecraft. The images obtained have led to a deeper understanding of both the surface and the interior structure of cometary nuclei.

1P Halley

The first comet that was imaged in detail was 1P/Halley, displayed in Figure 12.25. *Giotto* images revealed Halley's extremely low albedo, $A_v = 0.04$, which now appears to be a typical characteristic of comets. Halley is an elongated potato-shaped object covered with craters, valleys and hills.

9P/Tempel 1

Figure 12.26 shows images of comet 9P/Tempel 1 before (panel a) and while (panel b) being impacted by the *Deep Impact* mission. Several dozen circular features, 40–400 m in diameter, cover the comet's surface. Most intriguing are the smooth areas, some bounded by scarps tens of meters high. One such area appears to be eaten away at the edges, revealing another, older layer underneath. The nucleus of 9P/Tempel 1 seems to be layered with geologic strata of uncertain origin. Impacts, sublimation, mass wasting and ablation are all

Figure 12.25 The nucleus of 1P/Halley photographed by the *Giotto* spacecraft from a distance of ~600 km. The nucleus is ~16 × 8 km in extent. The (dust) jets visible in the image point in the sunward direction. (Courtesy Halley Multicolor Camera Team; ESA)

important in shaping the morphological features that are so prominent on this and other comets.

81P/Wild 2

Figure 12.27 shows several images of comet 81P/Wild 2 taken by the *Stardust* mission. Wild 2's roundish shape suggests that this comet may not be a collisional fragment in contrast to other comet nuclei, which are more potato shaped. In addition to craters, the nucleus of comet Wild 2 is dominated by depressions. Features referred to as **pinnacles**, tens of meters to more than 100 m high, are visible on the limb. Pinnacles might be erosional remnants, where the environment has been eroded away through, e.g., sublimation.

12.7 Comas and Tails of Comets

The generally unexpected and sometimes spectacular appearances of comets have triggered the

(a)   (b)

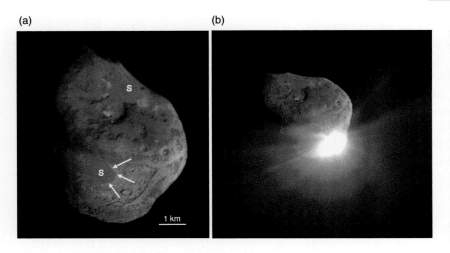

1 km

Figure 12.26 *Deep Impact* imaged comet 9P/Tempel 1 before (a) and during (b) impact. (a) This composite image has a resolution of 5 m/pixel; the comet's dimensions are 7.6 × 4.9 km. Smooth areas are indicated with the letter S, and the *arrows* highlight a bright (due to viewing geometry) scarp, which shows that the smooth area is elevated above the rough terrain. (NASA/JPL/UMD, PIA02142, PIA02137)

(a)   (b)

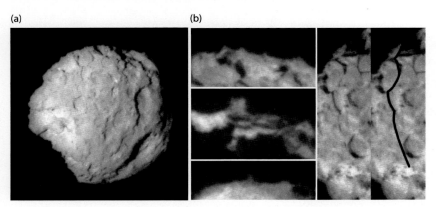

Figure 12.27 Images of comet 81P/Wild 2 taken by the *Stardust* spacecraft highlight the diverse features that make up its surface. The comet's dimensions are 5.5 × 4.0 × 3.3 km. (a) A full view of the comet shows the numerous depressions. (b) These higher resolution images show a variety of small pinnacles and mesas on the limb of the comet (*left side*), and a 2-km-long scarp on the right (outlined by the *black line* on the rightmost image). (NASA/JPL-Caltech, PIA06285, PIA06284)

interest of many people throughout history. A bright comet can easily be seen with the naked eye. Comets are usually not discovered until after a coma and tail have formed. Depending on the apparent size of the coma and tail, a comet can be very bright and easily seen with the naked eye. Some comets have a tail extending more than 45° on the sky. The earliest records of comet observations date to ∼6000 BCE in China.

The first detailed scientific observations of comets were made by Tycho Brahe in 1577.

Edmond Halley used Newton's gravitational theory to compute parabolic orbits of 24 comets observed up to 1698. He noted that the comet apparitions in 1531, 1607 and 1682 were separated by 75–76 years and that the orbits were described by roughly the same parameters. He hence predicted the next apparition to be in 1758. It was noticed much later that this Comet Halley, as it was named subsequently, has returned 30 times from 240 BCE to 1986; records of all of these apparitions have been found with the exception of 164 BCE.

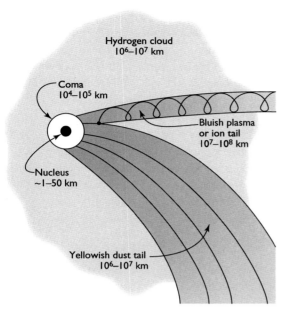

Figure 12.28 Schematic diagram of a comet, showing its nucleus, coma, tails and hydrogen cloud. (de Pater and Lissauer 2010)

Up through the early twentieth century, three or four comets were discovered each year. The discovery rate increased to 20–25 comets per year with the development of more powerful cameras (CCDs) in the 1980s and rose to 40–50 per year from the ground in the first decade of the twenty-first century.

Figure 12.28 shows a schematic illustration of a comet. The small **nucleus**, often only a few kilometers in diameter, is usually hidden from view by the large **coma**, a cloud of gas and dust roughly 10^4–10^5 km in diameter. Not seen with the naked eye, but shown schematically in Figure 12.28, is the large **hydrogen coma**, between 1 and 10 million km in extent, which surrounds the nucleus and visible gas/dust coma. Two tails are often visible, both in the antisolar direction: a curved yellowish **dust tail** and a straight **ion tail**, usually of a blue color.

Comets are usually inert at large heliocentric distances and only develop a coma and tails when they get closer to the Sun. When the sublimating gas **evolves off** the surface of a comet's nucleus, dust is dragged along. The gas and dust form a comet's coma and hide the nucleus from view. Most comets are discovered after the coma has formed when they are bright enough to be seen with relatively small telescopes.

In the following subsections, we discuss a comet's coma and tail in more detail. We describe the changes in a comet's brightness while gas and dust evolve off its surface (§12.7.1) and then address the ultimate fate of both the gas and dust in this process (§§12.7.2 and 12.7.3). We discuss the formation and make-up of a comet's tails in §§12.7.4 and 12.7.5 and conclude with a more in-depth description of the cometary nucleus and its composition in §12.7.6.

12.7.1 Brightness

The apparent brightness of a comet, B_ν, varies with heliocentric distance, $r_\odot$, and the distance to the observer, r_Δ, a behavior usually approximated by

$$B_\nu \propto \frac{1}{r_\odot^\zeta r_\Delta^2}. \tag{12.2}$$

An inert object, such as an asteroid, has an index $\zeta = 2$, but comets typically show $\zeta > 2$, attributed to the fact that a comet's **gas production rate** (i.e., the amount of gas being released per second) increases with decreasing heliocentric distance. Although some comets follow a power law in $r_\odot$ over a large range of heliocentric distances, others deviate significantly from a power law. Most comets brighten considerably when their heliocentric distance drops below 3 AU. From the observed relationship between a comet's brightness and its heliocentric distance, it has been deduced that water-ice is the dominant volatile in most comets. The sharp dropoff in the magnitude of nongravitational forces beyond 2.8 AU provides further evidence for the predominance of H_2O-ice.

The maximum brightness of a comet is usually reached a few days after perihelion, and the brightness variation shows asymmetries between the branches before and after perihelion. Pronounced differences have been observed in the brightening of old and new comets. Dynamically new comets often brighten gradually on their inbound journey, beginning at large heliocentric distances (at $r_\odot \gtrsim 5$ AU, with $\zeta \approx 2.5$). In contrast, most short-period comets do not brighten much while inbound at large distances but may 'flare up' when they get closer to perihelion ($\zeta \approx 5$ is typical).

Because the side of the comet facing the Sun is hotter than the anti-sunward side, gas evolves predominantly from the sunward side, as indeed suggested from Earth-based observations of many comets. Spacecraft images of several comets confirm activity primarily from the sunlit side, such as the picture of comet Halley shown in Figure 12.25. The images further show that the release of volatiles and entrained dust is usually confined to a few small areas on the surface. Such asymmetric outgassing and jets give rise to the nongravitational forces, which distort a comet's orbit, as discussed in §12.2.4.

The gas production rate, Q, of water molecules inside $r_\odot < 2.5$ AU can usually be approximated by:

$$Q \approx \frac{1.2 \times 10^{22} \pi R^2}{r_{\odot \text{AU}}^2} \qquad (12.3)$$

(molecules s^{-1}), with the comet radius, R, in meters. Equation (12.3) was derived assuming $A_b = 0.1$.

12.7.2 Ultimate Fate of Coma Gas

Except for the inner $\sim$100 km, a comet's coma stays optically thin at most wavelengths. Hence, essentially all molecules in the coma are irradiated by sunlight at visible and UV wavelengths. The mean lifetimes of the molecules and radicals against dissociation and ionization therefore vary as $r_\odot^2$. The primary volatile constituent of a cometary nucleus is water-ice. The typical lifetime of water molecules at $r_\odot \approx 1$ AU is $\sim$5–8$\times 10^4$ s. The main ($\sim$90%) initial photodissociation products of water are H and OH. The typical lifetime for OH radicals at $r_\odot = 1$ AU is $\sim$1.6–1.8 $\times 10^5$ s. The OH is dissociated into O and H.

The H atoms produced by dissociation of H_2O and OH form a large hydrogen coma, several $\times 10^7$ km in extent. Hydrogen is ultimately lost either by photoionization or by charge exchange reactions (§5.7.2) with solar wind protons.

12.7.3 Dust Entrainment

Dust grains of various sizes (typically microns across) and compositions are entrained in the outflowing cometary gases. The largest particles that may be dragged off the surface at $r_\odot = 1$ AU are of order 0.1 m in size. Submicrometer-sized grains almost reach the speed of the escaping gases ($\sim$1 km s^{-1}), but larger particles barely attain the gravitational escape velocity of the nucleus ($\sim$1 m s^{-1}). The dust decouples from the gas and becomes subject to solar radiation forces at a distance of a few tens of nuclear radii. Larger particles may escape during cometary outbursts, where part of the 'dust crust' is thrown into space.

Comets thus 'pollute' the interplanetary environment significantly, with both gas and dust. Whereas the gases have a short lifespan, dust particles may stay around longer. In particular, the larger particles, which are affected least by solar radiation pressure, may share the orbital properties of the comet for a long time. These particles are seen as **meteor streams** or cometary **dust trails** (§12.3.8).

Dynamically new comets, which have never entered the inner Solar System before, are likely to have highly volatile ices on their surfaces. This volatile material sublimates and evolves off the surface as soon as its sublimation temperature is reached. This, together with the explosive release of unstable species created by 4.5 Gyr of irradiation by galactic cosmic rays outside the heliosphere, explains the relatively high activity in new comets at large heliocentric distances. In contrast, an old, periodic comet has lost much or all of its volatile surface material. Such a comet is covered by a **dust crust**. This crust is built up by dust grains too heavy to be dragged off the surface by the sublimating gases. Such a dust crust usually forms when the comet recedes from the Sun, and sublimation gradually ceases. On the comet's return to the inner Solar System, gas pressure builds up in cavities as soon as subsurface ices reach their sublimation temperature. When the gas pressure is high enough, a portion of the dust crust is blown off, exposing fresh ice. This effect produces the sudden increase in activity as seen for many periodic comets when they approach perihelion. During each perihelion passage, a typical comet sublimates away a layer of ice only ~ 1 m thick (Problem 12-19), which is small compared with the size of a comet. However, periodic comets can develop thick dust crusts, and their activity can drop so low that no coma can be observed even near perihelion. Such extinct comets are classified as minor planets.

12.7.4 Morphology and Composition of Dust Tails

The dust/gas ratio in comets is usually between 0.1 and 10. Dust grains of many sizes have been detected. Thermal infrared measurements show that the bulk of the grains are less than ~ 5 μm in radius, and reflected sunlight in the visible implies many grains down to 0.1 μm in size. Radar observations reveal the presence of much larger (centimeter-sized) grains. Although the total number of macroscopic particles is much smaller than that of the micrometer-sized grains, most of the mass is contained in the larger particles.

As discussed in §2.8.1, solar radiation pressure on (sub)micrometer-sized dust grains 'blows'

Figure 12.29 Comet C/West in March 1976 showing both streamers and striae. About a week after this picture was taken, the nucleus broke into four pieces. (Courtesy Akira Fujii)

the particles outwards from the Sun relative to the trajectory of the nucleus. Particles of different sizes and with different release times become spatially separated in the dust tail. Thus the dust orbits depend on size, shape, composition and initial velocity. The ensemble of particles released at different times together form a curved tail, as depicted by the shaded region in Figure 12.28.

The simultaneous ejection of dust grains of different sizes gives rise to inhomogeneities in the dust tail. These inhomogeneities include **dust jets**: Highly collimated structures, which at larger distances become **streamers** – straight or slightly curved bands that converge at the nucleus. Observed infrequently are **striae**, which are parallel narrow bands at large distances from the nucleus that do not converge at the nucleus. Examples of striae and streamers are shown in Figures 12.29 and 12.30.

Figure 12.30 Comet C/McNaught (C/2006 P1) had an extremely long and unusual dust tail. Striae are visible over a large fraction of the sky. This image was taken from the Andes Mountains in Chile, looking down on the city lights of Santiago. In the *lower right*, the crescent moon is visible. (Courtesy Stéphane Guisard, ESO PR Photo 05h/07)

Dust Composition

The structure and mineralogy of cometary dust grains are determined by the chemical and physical processes prevalent at the time when they formed. The composition of dust grains has been measured *in situ* by the *Giotto* and *Vega* spacecraft. They revealed that, in addition to regular silicate grains, many grains were composed primarily of the light elements C, H, O and N, collectively referred to as **CHON** particles.

The *Stardust* mission captured thousands of particles, 5–300 μm across, from the coma of comet 81P/Wild 2 and brought them back to Earth. The *Stardust* particles were trapped in **aerogel**, a highly porous silica 'foam' with a density comparable to that of air; a picture is shown in Figure 12.31a. Grains that impact the aerogel are gradually slowed down without suffering substantial melting or vaporization. The track of a cometary grain as captured in the aerogel is shown in Figure 12.31b, and pictures of recovered grains are shown in Figure 12.31c. These grains are assemblages of different minerals, particularly the crystalline silicate minerals olivine and pyroxene. Isotopic analysis shows that most minerals are similar to those found in the inner Solar System, and only a few appear to be anomalous presolar grains.

The isotopic compositions of H, C, N, O and Ne in *Stardust* grains show that the cometary grains are unequilibrated aggregates composed of materials that originated in different reservoirs. In particular, the presence of high-temperature minerals as found in meteoritic calcium-aluminum inclusions (CAIs), which condense at temperatures greater than 1400 K, provide convincing evidence of strong radial mixing in the early solar nebula.

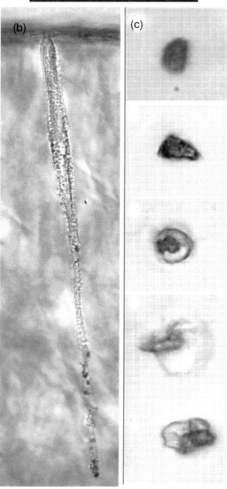

Grains that formed near the Sun must have been transported radially outwards to ice-rich regions of the protoplanetary disk, where they were assembled into cometesimals.

12.7.5 Ion Tails

Although sublimating gases are neutral when released by a comet, ultimately all atoms and molecules get ionized. The dominant ionization processes are photoionization and charge exchange with solar wind protons. Cometary ions and electrons interact with the interplanetary magnetic field and 'drape' the field lines around the comet, as discussed in Chapter 7 and depicted in Figure 7.13. This process induces a magnetic field similar to the field around Venus.

Cometary ions form an ion or plasma tail in the antisolar direction. The length of this tail often exceeds 10^7 km, and the main tail is roughly 10^5 km in diameter. The ion tail usually appears blue as a result of fluorescent transitions of the abundant, long-lived CO^+ ions, although occasionally a reddish ion tail has been seen caused by emissions of H_2O^+. The tail often consists of filaments, rays and bright knots, as shown in the series of images of Comet C/Hyakutake in Figure 12.32. The structure of ion tails changes on timescales of minutes to hours. The knots are caused by enhanced densities, and their motion can be followed down the tail. Typical speeds are $\lesssim 100$ km s^{-1}; this is much more than the cometary outflow speed but less than the solar wind velocity of ~ 400 km s^{-1}.

The specific structure of a cometary plasma tail depends on the interplanetary medium and its magnetic field. One can often see large disturbances in a plasma tail (Fig. 12.32). Sometimes the comet appears to lose its tail and starts forming a new one. These events have been attributed to **disconnection** events caused by a sudden reversal of the interplanetary magnetic field. Such reversals take place when the comet meets an interplanetary sector boundary or when it crosses the heliospheric current sheet.

Figure 12.31 (a) A piece of aerogel, similar to that used in the *Stardust* mission. (Courtesy NASA Photographer Maria Garcia 1997) (b) Track of a particle from comet 81P/Wild 2 as captured in the aerogel of the *Stardust* mission. The track is about 1 mm long. The particle entered at the top. The force of the impact broke up the tiny rock, and pieces can be seen as black dots all along the track. These particles are (sub)micrometer sized. (c) A close-up of several grains captured in and recovered from the aerogel. (Adapted from Brownlee et al. 2006)

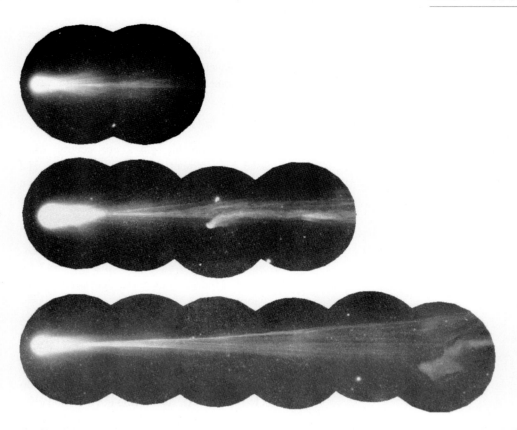

Figure 12.32 A spectacular disconnection event was photographed in C/Hyakutake from 24 to 26 March 1996. The tail is more than 10 million km long. (Courtesy Shigemi Numazawa)

12.7.6 Comet Composition

Because a cometary nucleus is often shrouded by a coma of gas and dust, it is difficult to determine its composition directly from remote observations. Instead, spectra of atoms/molecules and dust grains are used together with 'outflow' models to indirectly deduce composition. Complementary techniques are *in situ* measurements, which have been obtained for a few periodic comets. C/Hale–Bopp was an exceptionally bright comet, which enabled observations of the evolution in the production rates as a function of heliocentric distance (from 0.9 to 14 AU) for nine different molecules, as displayed in Figure 12.33. As expected, within a heliocentric distance of 3 AU, OH (from H_2O) is the most abundant species. Carbon monoxide is the most volatile species and dominated the production rate at large distances.

Numerous emission lines have been observed from carbon species, including C, C_2, C_3, CH, CH_2, CN and CO. All of these molecules are ultimately derived from ices in the comet's nucleus. The identification and characterization of these ices provide information on the conditions under which cometary ices formed.

Likely candidates for parent materials of the observed carbon species, in addition to dust grains, are carbon dioxide (CO_2), hydrogen cyanide (HCN), methane (CH_4) and more complex molecules such as formaldehyde (H_2CO) and methanol (CH_3OH), all of which have been detected in at least several comets. The CO_2

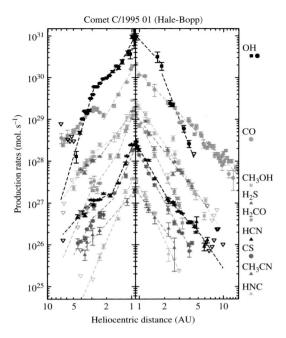

Comet C/1995 01 (Hale-Bopp)

Figure 12.33 COLOR PLATE Production rates of nine different molecules in C/Hale–Bopp as a function of distance from the Sun. Power-law fits (*dashed lines*) are superposed on the data. (Biver et al. 2002)

production rate is a few percent that of water, and the CO production rate varies from $\lesssim 1\%$ up to over 30% that of water in different comets. *In situ* measurements of comet 1P/Halley by the *Giotto* spacecraft revealed that about one-third of the CO was released directly from the nucleus (referred to as the 'native' source), and the remaining molecules came from an 'extended' source region, which could be dust grains or complex molecules in the coma. Formaldehyde also has a native and an extended source.

After adding up all the available nitrogen, the N/C and N/O abundance ratios appear to be depleted relative to the solar ratios by a factor of two to three.

Although there is much variability from comet to comet, taken together, the observations imply that on average, the overall cometary composition (gas + dust) is, within a factor of two, similar to solar values except for noble gases, hydrogen (deficient by a factor of $\sim$700) and nitrogen (deficient by a factor of $\sim$3). Because all known meteorites are strongly deficient in all volatile materials compared with solar values (Fig. 11.4), cometary dust can be considered as the most 'primitive' early Solar System material ever sampled.

12.8 Temporal Evolution of the Population of Asteroids and Comets

Minor planets presumably grew from planetesimals, as did the terrestrial planets. Asteroids can be viewed as remnant planetesimals that failed to accrete into a single body.

Collisions in the past and at present gradually grind the bodies to smaller and smaller fragments, the smallest (dust grains) being removed by Poynting–Robertson drag and radiation forces. The orbits of asteroid fragments $\gtrsim 10$ km across are modified by the Yarkovsky effect and can replenish the chaotic resonance zones in the main asteroid belt. After being placed into chaotic orbits, they are subsequently driven into planet-crossing trajectories, feeding the NEO population. Modifications to orbits of some TNOs similarly lead to planet-crossing objects, the centaurs, which ultimately feed the population of short-period comets.

There is broad agreement that comets have spent the past four billion years in the Kuiper belt (short-period comets) and the Oort cloud (long-period and dynamically new comets). The Oort cloud and Kuiper belt have survived as comet reservoirs for more than four billion years. Icy bodies in these reservoirs occasionally are perturbed into orbits that bring them into the planetary region. Whereas the galactic tide, passing stars and giant molecular clouds provide the primary perturbations for bodies in the Oort cloud, resonant perturbations of the outer planets (sometimes aided by orbit-altering collisions) dominate for Kuiper belt objects or more precisely for KBOs in the scattered disk, SDOs (§12.2).

When these icy bodies approach the Sun, their most volatile constituents sublimate and evolve off the nucleus, taking along with them more refractory dust and producing comets that may be spectacular in their visual appearance. Most active comets are quickly ejected from the Solar System as a result of gravitational perturbations by the planets. Some comets crash into the Sun, and others end their active lives releasing all of their volatiles or completely disintegrating. A small minority of comets collide with planets. The source regions of comets are gradually being depleted, but the average rate at which comets are being supplied to the planetary region will probably drop by at most a factor of a few between the present epoch and the end of the Sun's main sequence (hydrogen burning) lifetime six billion years hence.

Key Concepts

- Minor planets are solid objects smaller than planets in orbit about the Sun (or another star).
- The asteroid belt, located between the orbits of Mars and Jupiter, is the best-studied reservoir of minor planets and contains the largest collection of small rocky bodies in our Solar System.
- The Kuiper belt, located exterior to the orbit of Neptune, is far more massive than is the asteroid belt and contains larger objects. It is the primary reservoir of short-period comets.
- The Oort cloud, located ~10 000 AU from the Sun, is a giant reservoir of long-period and dynamically new comets. Oort-cloud comets (OC) formed in the giant planet zone and have been stored in the Oort cloud for billions of years.
- Additional reservoirs of minor planets include dynamically stable Trojan asteroids that share an orbit with Jupiter or Neptune, as well as some bodies (NEOs, centaurs) on unstable, planet-crossing orbits.
- Small minor planets are more common than are large ones. The size–frequency distributions of asteroids and KBOs are well approximated by power laws over a wide range in mass.
- Similar to meteorites, asteroids can be subdivided into several classes, the most common ones being stony (S), carbonaceous (C) and metal (M). Space weathering may mask the true composition of an asteroid, though.
- Collisions have played a major role in shaping the asteroid and Kuiper belts. Evidence for this is abundant (e.g., size–frequency distribution, moonlets, porosity of small bodies, families, zodiacal dust).
- Cometary activity is triggered by solar heating: When the temperature of the nucleus exceeds the sublimation temperature of a particular ice, the subliming gas evolves off the nucleus, dragging dust grains with it.
- Comets have two types of tails: a dust tail, 'shaped' by solar radiation pressure and an ion or plasma tail 'shaped' by the solar wind.
- Some cometary dust grains are composed of the most volatile materials (CHON), and others of the most refractory matter (CAIs) in our Solar System, indicative of efficient radial mixing in the early solar nebula.
- A comet's nucleus is essentially a dirty snowball. Close-up images of cometary nuclei reveal craters, pinnacles, scarps, as well as extremely smooth areas.

Further Reading

Good review papers on asteroids, TNOs, interplanetary dust and comets can be found in:

McFadden, L., P. R. Weissman, and T. V. Johnson, Eds., 2007. *Encyclopedia of the Solar System*, 2nd Edition. Academic Press, San Diego. 982pp.

More in-depth topical papers are contained in:

Bottke, W.F., Jr., A. Cellino, P. Paolicchi, and R.P. Binzel, Eds., 2002. *Asteroids III*. University of Arizona Press, Tucson. 785pp.

Barucci, M.A., H. Boenhardt, D.P. Cruikshank, and A. Morbidelli, Eds., 2008. *The Solar System Beyond Neptune*. University of Arizona Press, Tucson. 592pp.

The University of Arizona Press series on *Protostars and Planets* has several good reviews, including the papers by Cruikshank et al., Chiang et al., Jewitt et al. and Wooden et al. in B. Reipurth, D. Jewitt, and K. Keil, Eds., 2007. *Protostars and Planets V*. University of Arizona Press, Tucson. 951pp.

We recommend the following book on comets:

Huebner, W.F., Ed., 1990. *Physics and Chemistry of Comets*. Springer-Verlag, Berlin. 376pp.

Problems

12-1. Calculate the locations of the 3:1, 5:2, 2:1 and 3:2 resonances with Jupiter. (Hint: See Chapter 2.) Use Figure 12.2a to determine which of these resonances produce gaps and which have led to a concentration in the population of asteroids.

12-2. Draw a histogram of $1/a$ for short-period comets, using data on the web, such as those available at http://cfa-www.harvard.edu/iau/ or http://pdssbn.astro.umd.edu/.

12-3. Suppose a comet has a velocity of 40 km s^{-1} at perihelion. The perihelion distance is 1 AU. Calculate the aphelion distance, the velocity of the comet at aphelion and the orbital period of the comet.

12-4. **(a)** Show that if the exponent $\zeta = 4$ in equation (12.1a), then the mass is divided equally among equal logarithmic intervals in radius.
(b) Show that if $\zeta < 3$, then most of the mass in the asteroid belt is contained in a few large bodies in the sense that the largest factor of two in radius contains more mass than all smaller bodies combined.

(c) For which value of ζ does one find equal integrated cross-sectional areas in equal logarithmic size intervals?

12-5. If the differential size–frequency distribution of a group of objects that have equal densities can be adequately described by a power law in radius of the form $N(R)dR \propto R^{-\zeta}$, then it can also be described as a power law in mass of the form $N(m)dm \propto m^{-x}$. Derive the relationship between ζ and x.

12-6. Estimate the number of asteroids in the main belt with radii $R > 1$ km by using the observed number of large asteroids (Table E.6) and the observed slope of the size–frequency distribution, $\zeta = 3.5$.

12-7. **(a)** Calculate the mean transverse optical depth of the main asteroid belt. (Hint: Divide the projected surface area of the asteroids by the area of the annulus between 2.1 and 3.3 AU. Use the size–frequency distribution given by equation (12.1b) and Problem 12-6 and assume (and justify) reasonable values for R_{min} and R_{max} if necessary.)

(b) Calculate the fraction of space (i.e., volume) in the main asteroid belt near the ecliptic that is occupied by asteroids.

12-8. A 'typical' asteroid orbits the Sun at 2.8 AU and has an inclination of 15° and an eccentricity of 0.14. Calculate the typical collision velocity of two asteroids. (Hint: Compute the speed of a single asteroid relative to a circular orbit in the Laplacian plane of the Solar System and multiply by $\sqrt{2}$ to obtain the mean encounter velocity.)

12-9. **(a)** Calculate the gravitational binding energy (in Joules) of a spherical asteroid or KBO of radius R (km) and density ρ (kg m^{-3}).
(b) For what size asteroid is the gravitational binding energy equal to the physical cohesion (the fracture stress of rock is $\approx 10^8$ pascal)?
(c) Are nonspherical asteroids more or less tightly bound gravitationally than spherical ones of the same mass?
(d) What is the radius of the largest coherent spherical asteroid that can be disrupted by 1000 MT of TNT equivalent explosives (1 MT TNT = 4.18×10^{15} J)?
(e) What if the asteroid is a rubble pile (no strength)?
(f) Repeat parts (d) and (e) for 1 000 000 MT TNT.

12-10. Consider a spherical asteroid with a density $\rho = 3000$ kg m^{-3} and radius $R = 100$ km. The asteroid is covered by a layer of loosely bound regolith. What is the shortest rotation period this asteroid can have without losing the regolith from its equator?

12-11. **(a)** Estimate the size of the largest asteroid from which you could propel yourself into orbit under your own power.
(b) What variables other than asteroid size must be considered?
(c) How would you be able to launch yourself into a stable orbit?

12-12. The Earth intercepts about 4×10^7 kg of interplanetary dust per year. Estimate the equivalent radius of a spherical body with the above mass (assume a density of 3000 kg m^{-3}).

12-13. **(a)** What features distinguish the spectra of S-, C- and D-type asteroids from one another?
(b) Where do most members of each of these taxonomic classes orbit?

12-14. **(a)** If Itokawa continues to lose surface material at a rate of 0.5 m per Myr, calculate how much longer it may stay around in our Solar System.
(b) Discuss your result above together with typical dynamical timescales of NEOs.

12-15. **(a)** Calculate the surface temperature of Pluto and Charon at perihelion, assuming both bodies are rapid rotators and in equilibrium with the solar radiation field. (Hint: See §4.1.2.)
(b) Calculate the escape velocity from Pluto and Charon and compare these numbers with the velocity of N$_2$, CH$_4$ and H$_2$O molecules.

(c) Given your answers in (a) and (b), explain qualitatively the differences in surface ice coverage for Pluto and Charon.

12-16. Estimate the maximum rotation rate of a strengthless asteroid (held together by gravity alone) by setting the rotation speed at the equator equal to the escape velocity.
(a) Approximate the asteroid by a spherical body. (Hint: Your answer should depend only on the density of the body, not its size.)
(b) Is the maximum rotation rate for a nonspherical asteroid faster or slower than that for a spherical asteroid? Why?

12-17. An asteroid and a comet have the same apparent brightness while both are at $r_\Delta = 2$ AU and $r_\odot = 3$ AU. At a later time, they are both observed at $r_\Delta = 2$ AU and $r_\odot = 2$ AU. Which object is brighter, and by approximately how much?

12-18. Calculate the surface temperature of a comet with Bond albedo $A_b = 0.1$ at a heliocentric distance of 15 AU.

12-19. A comet's perihelion distance is 1 AU, and its aphelion distance is 15 AU. In the following, we make a very crude calculation of the average rate of shrinkage of the comet.
(a) Calculate the comet's orbital period.
(b) Estimate how many meters of ice the comet will lose each time it orbits the Sun. (Hint: To simplify the calculations, you may assume that ice sublimates off the comet's surface during one-tenth of its orbital period, that the average cometary distance over that period is 1.5

AU and that the density of the cometary ice is 600 kg m^{-3}.)

12-20. A comet consists primarily of water-ice; when the ice sublimates, the water molecules flow off the surface at the thermal expansion velocity, ~ 0.5 km s^{-1}. The typical lifetime of H_2O molecules is 6×10^4 s; for OH, it is 2×10^5 s; and for H, it is 10^6 s. Assume the outflow velocity of OH to be equal to that of H_2O, and for H it is on average 12 km s^{-1}. Calculate the typical sizes of the H_2O, OH and H comas.

12-21. Comet X is on a nearly parabolic orbit with a perihelion at 0.5 AU. Calculate the (generalized) eccentricity of grains released near perihelion as a function of β. Give a formula for the separation between the grains and the nucleus as a function of β and t that is valid for the first few days after release. (Hint: See §2.8.)

12-22. Show that if comets lose mass in proportion to their surface area, then their radius shrinks at a rate independent of mass and that the slope of the size distribution flattens out. You may use analytic and/or numerical techniques.

12-23. The striking photos of Pluto returned by the *New Horizons* spacecraft (e.g., Fig. G.27) appear to show mountains of water-ice floating in a sea of frozen nitrogen. Compute the equilibrium temperature of Pluto, assuming that Pluto's Bond albedo is 0.7. Then look up the melting point of N_2 and densities of water-ice, liquid nitrogen and nitrogen-ice at these temperatures. Comment on whether you think that this interpretation for the landforms of Pluto is plausible.

CHAPTER 13

Planetary Rings

The only system of rings which can exist is one composed of an indefinite number of unconnected particles revolving around the planet with different velocities according to their respective distances

James Clerk Maxwell, Adams Prize Essay, 1857

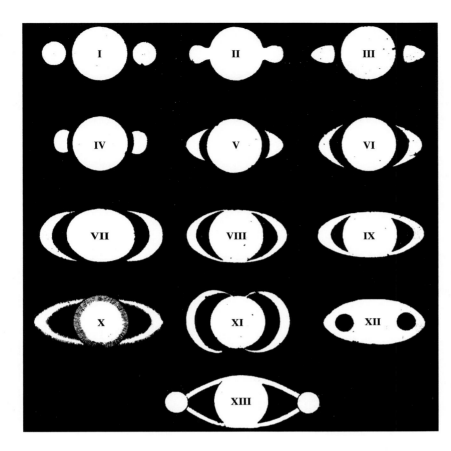

Figure 13.1 Seventeenth-century drawings of Saturn and its rings. I: Galileo, 1610; II: Scheiner, 1614; III: Riccioli, 1641 and 1643; IV–VII: Havel, theoretical forms; VIII, IX: Riccioli, 1648–50; X: Divini, 1646–48; XI: Fontana, 1636; XII: Biancani, 1616, Gassendi, 1638–39 and XIII: Fontana and others at Rome, 1644–45. As indicated in diagram I, Galileo believed Saturn's appendages to be two giant moons in orbit about the planet. However, these 'moons' appeared fixed in position, unlike the four satellites of Jupiter that he had previously observed. Moreover, Saturn's 'moons' had disappeared completely by the time Galileo resumed his observations of the planet in 1612. (Huygens 1659)

Each of the four giant planets in our Solar System is surrounded by flat, annular features known as **planetary rings**. Planetary rings are composed of vast numbers of small satellites, which are unable to accrete into large moons because of their proximity to the planet.

Saturn's rings were first observed by Galileo Galilei in 1610. Telescopes were rather primitive at that time, and photography had not been invented. Figure 13.1 shows various drawings of Saturn and its 'strange appendages' that were sketched by Galileo and other telescopic observers during the first half of the seventeenth century. In this pre-Newtonian era, the workings of the cosmos were poorly understood, and many explanations were put forth to explain Saturn's strange appendages, which grew, shrank and disappeared

every 15 years. In 1656, Christiaan Huygens finally deduced the correct explanation, illustrated in Figure 13.2, that Saturn's appendages are a flattened disk of material in Saturn's equatorial plane and they appear to vanish when the Earth passes through the plane of the disk.

For more than three centuries, Saturn was the only planet known to possess rings. Saturn's rings are quite broad, but photographs of Saturn's rings such as those displayed in Figure 13.3 do not show much detail, and little structure within the ring system was detected until the late twentieth century.

In 1977, observations of an occultation of the star SAO 158687 revealed the narrow opaque rings of Uranus. Figure 13.4 shows these data, which launched a golden age of planetary ring exploration. The *Voyager* spacecraft first imaged

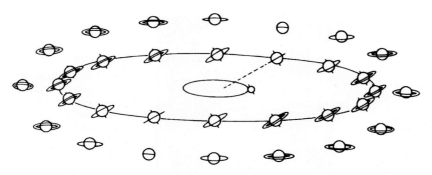

Figure 13.2 Schematic views of Saturn and its rings over one saturnian orbit according to Huygens's model.

and studied the broad but tenuous ring system of Jupiter in 1979 (§13.3.1). *Pioneer 11* and the two *Voyagers* obtained close-up images of Saturn's spectacular ring system in 1979, 1980 and 1981 (§13.3.2). Neptune's rings, whose most prominent features are azimuthally incomplete arcs, were discovered by stellar occultation in 1984. *Voyager 2* obtained high-resolution images of the rings of Uranus in 1986 (§13.3.3) and the rings of Neptune in 1989 (§13.3.4). The *Galileo* spacecraft and *New Horizons* obtained close-up images of Jupiter's

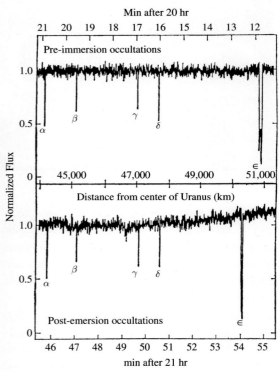

Figure 13.4 Lightcurve of the star SAO 158687 as it was observed to pass behind Uranus and its rings. Dips in the lightcurves corresponding to the occultation of the star by five rings are clearly seen both before immersion and after emersion of the star from behind the planet. Four of these pairs of features are symmetric about the planet, but the location, depth and duration of the outermost pair imply that the ϵ ring is both noncircular and nonuniform. (Adapted from Elliot et al. 1977)

Figure 13.3 Saturn and its rings over one-half of a saturnian orbit, as seen in ground-based photographs taken in the middle of the twentieth century.

Figure 13.5 COLOR PLATE This approximately natural-color image shows Saturn and its rings as seen by the *Cassini* spacecraft. The pronounced gap in the rings, the Cassini division, is a 3500-km-wide region that is much less populated with ring particles than are the brighter B and A rings to either side. The shadows of the rings darken parts of Saturn's cold, blue northern hemisphere, and Saturn shadows part of the rings. (NASA/JPL/CICLOPS, PIA06193)

rings in the late 1990s and 2007, respectively. The *Cassini* spacecraft began an intensive and multifaceted study of Saturn's rings from saturnian orbit in 2004. *Cassini* has returned vast amounts of data on Saturn's ring system, including spectacular images such as the one shown in Figure 13.5. *Hubble Space Telescope* (*HST*) observations and improved telescopes on the ground have provided new information on all four planetary ring systems.

In this chapter, we summarize our current observational and theoretical understanding of planetary rings. We focus on processes that are applicable to planet formation, which we cover in depth in Chapter 15. We begin in §13.1 with an explanation of how tides would rip apart large moons that were located too close to a planet and why the tiny satellites that make up rings can exist. The reasons that these tiny particles organize themselves into a flat, often broad, disk in the planet's equatorial plane are explained in §13.2. A more detailed observational summary is then presented in §13.3,

with each giant planet's ring system treated in turn and most discussion paid to the spectacular and diverse rings of Saturn. We then turn to models for ring–moon interactions (§13.4), which produce some of the most interesting features observed. We conclude in §13.5 with a discussion of the evolution of planetary ring systems and models of planetary ring formation.

13.1 Tidal Forces and Roche's Limit

Figure 13.6 shows that planetary rings are generally located closer to planets than are satellites that are large enough to be designated as moons. The strong tidal forces close to a planet lead orbital debris to form a planetary ring rather than a moon. The closer a moon is to a planet, the stronger the tidal forces that it is subjected to. If it is too close, then the difference between the gravitational force exerted by the planet on the point of the

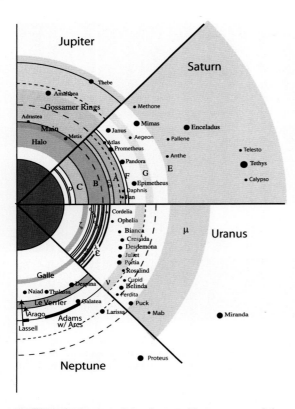

Figure 13.6 Diagram of the rings and inner moons of the four giant planets. The systems have been scaled according to their planetary equatorial radius. The *long-dashed curves* denote the radius at which orbital motion is synchronous with planetary rotation. The *short-dashed curves* show the location of Roche's limit for particles of density 1000 kg m⁻³. (Courtesy Judith K. Burns)

moon nearest to the planet from that exerted on the center of the moon is stronger than the moon's self-gravity. Under such circumstances, the moon is ripped apart unless it is held together by mechanical strength, and a planetary ring results.

To understand tidal disruption more quantitatively, we make the following assumptions:

(1) The system consists of one large primary body (the planet) and one small secondary body (the moon).
(2) The orbit is circular, the rotational period of the moon is equal to its orbital period and

the moon's obliquity is nil. (These assumptions make the analysis far simpler because the problem becomes stationary in a rotating frame.)
(3) The moon is spherical, and the planet can be treated as a point mass.
(4) The moon is held together by gravitational forces only.

The 'external' forces per unit mass on material in orbit about a planet of mass M_p are gravity:

$$\mathbf{g}_\rho = -\frac{GM_p}{r^2}\hat{\mathbf{r}}, \tag{13.1}$$

and centrifugal force:

$$\mathbf{g}_n = n^2 r\hat{\mathbf{r}}, \tag{13.2}$$

where the origin is at the center of the planet and n is the angular velocity of the system. Steady state in the frame rotating with the system gives

$$n^2 r\hat{\mathbf{r}} - \frac{GM_p\hat{\mathbf{r}}}{r^2} = 0, \tag{13.3a}$$

therefore:

$$n^2 = \frac{GM_p}{r^3}. \tag{13.3b}$$

Note that equation (13.3b) implies Kepler's third law for the case of circular orbits.

The sum of the gravitational force and the effect of the rotating frame of reference ('centrifugal force') is referred to as the **effective gravity**; the local effective gravity vector points normal to the **equipotential surface** in the rotating frame. The effective gravity, $\mathbf{g}_{\text{eff}}$, felt by an object that is at a distance r from the planet's center and traveling on a circular orbit at semimajor axis a is

$$\mathbf{g}_{\text{eff}} = GM_p\left(\frac{r}{a^3} - \frac{1}{r^2}\right)\hat{\mathbf{r}}. \tag{13.4}$$

The (effective) tidal force upon such a body is

$$\frac{d\mathbf{g}_{\text{eff}}}{dr} = GM_p\left(\frac{1}{a^3} + \frac{2}{r^3}\right)\hat{\mathbf{r}} \approx \frac{3GM_p}{a^3}\hat{\mathbf{r}}, \tag{13.5}$$

where the approximation $r^3 \approx a^3$ has been used in the last step. (The approximation $r \approx a$ is valid at this stage in the derivation as long as the size of the moon is much smaller than that of its orbit, $R_s \ll a$. It could not have been used before taking the derivative because this would have omitted the gradients that are the essence of the tidal force.) Note that equation (13.5) differs from equation (2.57) because it also includes a contribution from the centrifugal force. The moon's self-gravity just balances the tidal force at the surface of the moon when

$$\frac{GM_s}{R_s^2} = \frac{3GM_pR_s}{a^3}, \qquad (13.6)$$

where the subscript s refers to the satellite (moon). This occurs at a planetocentric distance of

$$\frac{a}{R_p} = 3^{\frac{1}{3}}\left(\frac{\rho_p}{\rho_s}\right)^{\frac{1}{3}} = 1.44\left(\frac{\rho_p}{\rho_s}\right)^{\frac{1}{3}}. \qquad (13.7)$$

In the above derivation, we made a number of simplifying assumptions. Let us now review the accuracy of these assumptions in order to assess the applicability of our calculations:

(1) We used the small moon/large planet approximation in order to neglect the influence of the moon on the planet and to neglect terms containing higher powers of the ratio of the radius of the moon to its orbital semimajor axis. This assumption is thus very accurate for bodies within our Solar System.

(2) All known inner moons have low eccentricities, so the approximation of circular orbits is very good. All moons near planets for which rotation rates have been measured are in synchronous rotation and have low obliquity. Young moons that have not had time to be tidally despun, and thus rotate rapidly, would be less stable.

(3) Although the giant planets are noticeably oblate, the departures of their gravitational potentials from those of point masses have only an order 1% effect on these tidal stability calculations. A much larger effect results from

moons being stretched out along the planet–moon line because of the planet's gravitational tug (Fig. 2.19). This stretching brings the tips of the moon farther from its center, which both decreases the magnitude of self-gravity and increases the tidal force. In 1847, Édouard Roche performed a self-consistent analysis for a liquid (fully deformable) moon and obtained

$$\frac{a_R}{R_p} = 2.456\left(\frac{\rho_p}{\rho_s}\right)^{\frac{1}{3}}. \qquad (13.8)$$

Such a marginally gravitationally bound fluid moon would fill its entire Roche lobe, extending to the inner Lagrangian point, L_1, a distance of one Hill radius, R_H, from the moon's center (§2.2.3). The shape of such a moon would be intermediate between that of an almond and a sphere, with a volume equal to about one-third that of a sphere of radius R_H. The location a_R is known as **Roche's limit** for tidal disruption. Note that the location of the Roche limit of a given planet depends on the density (and thus the composition) of a moon (ρ_s).

(4) Most small bodies have significant internal coherence, e.g., small moons are not always spherical. Internal friction and/or tensile strength of small bodies allows moons smaller than $\sim$100-km radius to be stable somewhat inside Roche's limit. Ring particles, which typically are so small that internal strength exceeds self-gravity by orders of magnitude provided the particles are not loose aggregates, can remain coherent well inside Roche's limit.

The concept of Roche's limit explains in a semiquantitative manner why we observe rings near giant planets, small moons a bit farther away and large moons only at greater distances (Fig. 13.6). However, the interspersing of some rings and moons implies that other factors are important in determining the precise configuration of a planet's satellite system. We return to theories on the origin and evolution of ring/moon systems in §13.5.

13.2 Flattening and Spreading of Rings

A ring particle orbiting a planet passes through the planet's equatorial plane twice each orbit unless its trajectory is diverted by a collision with another particle or by the ring's self-gravity. The average number of collisions that a particle experiences during each vertical oscillation is a few times as large as the optical depth of the rings, τ (Problem 13-6). The optical depth of a ring is defined analogously to the optical depth in other media (eq. 4.38), and when a path is not specified, it generally refers to the **normal optical depth**, i.e., the optical depth on a path perpendicular to the plane of the ring. Typical orbital periods for particles in planetary rings are 6–15 hours. Because τ is of order unity in the most prominent rings of Saturn (and Uranus), collisions are very frequent. Collisions dissipate energy but conserve angular momentum. Thus, the particles settle into a thin disk on a timescale

$$t_{\text{flat}} = \frac{\tau}{\mu}, \tag{13.9}$$

where μ is the frequency of the particles' vertical oscillations. An oblate planet exerts torques that alter the orbital angular momenta of orbiting particles. These torques cause inclined orbits about oblate planets to precess (see §2.6.2). When coupled with collisions among ring particles, a secular transfer of angular momentum between the planet and the ring can result. Only the component of the ring's angular momentum that lies along the planet's spin axis is conserved. Any net angular momentum of the disk parallel to the planet's equator is quickly dissipated (i.e., returned to the planet via the torque between the planet's equatorial bulge and the ring). Because collisions at high speeds

damp relative motions rapidly, the ring settles into the planet's equatorial plane on a timescale of a few orbits (or $\sim \tau^{-1}$ orbits, if $\tau \ll 1$).

Several mechanisms act to maintain a nonzero thickness of the disk. Finite particle size implies that even particles on circular orbits in a planet's equatorial plane collide with a finite velocity when an inner particle catches up with a particle farther out that moves less rapidly. Unless the collision is completely inelastic, i.e., unless the two particles stick, some of the energy involved goes into random particle motions. The ultimate consequence of these collisions is a spreading of the disk. Viewed in another way, spreading of the disk is the source of the energy required to maintain particle velocity dispersion in the presence of inelastic collisions. Gravitational scatterings between slowly moving particles is another process that converts energy from ordered circular motions to random velocities, in this case without the losses resulting from the inelasticity of physical collisions.

As a result of continuing collisions, rings spread in the radial direction. Because diffusion is a random walk process, the diffusion timescale is

$$t_{\text{d}} = \frac{\ell^2}{\nu_{\text{v}}}, \tag{13.10}$$

where ℓ is the radial length scale (the width of the ring or of a particular ringlet). The viscosity, ν_{v}, depends on the spread in particle velocities, c_{v}, and the local optical depth approximately as

$$\nu_{\text{v}} \approx \frac{c_{\text{v}}^2}{2\mu} \left(\frac{\tau}{1 + \tau^2} \right) \approx \frac{c_{\text{v}}^2}{2n} \left(\frac{\tau}{1 + \tau^2} \right). \tag{13.11}$$

Equation (13.11) was derived assuming that ring particles behave like a diffuse gas and so is valid only if particles move several times their diameters relative to one another between collisions. If the filling factor is large, i.e., if typical distances between particles are not much larger than particle sizes, then this approximation is not valid. The viscosity of dense, high τ, rings depends on other factors, including particle size and the **coefficient**

of restitution (ratio of the relative speed of two particles immediately after a collision to that just prior to impact) for inelastic collisions.

Equation (13.10) with $\ell = 6 \times 10^7$ m (the approximate radial extent of Saturn's main rings), and an observationally motivated estimate of the viscosity of $\nu_v = 0.01$ m^2 s^{-1}, yields a diffusion timescale comparable to the age of the Solar System. For smaller ℓ, timescales are much shorter. Even in regions of planetary rings where the viscosity is substantially lower than the value quoted above, viscous diffusion should be able to rapidly smooth out any fine-scale density variations unless other processes counteract viscosity. The structure in optically thick planetary ring systems must therefore be actively maintained except on the largest length scales, where it may have resulted from 'initial' conditions.

Collective gravitational effects are important when the velocity dispersion of the particles is so small that **Toomre's stability parameter**,

$$Q_T \equiv \frac{\kappa c_v}{\pi G \sigma_\rho}, \tag{13.12}$$

is less than unity. Here κ is the particles' epicyclic (radial) frequency (for orbits near the equatorial plane of an oblate planet, κ is slightly smaller than n, see eqs. 2.51 and 2.52), and σ_ρ is the surface mass density of the rings. The dispersion velocity of the ring particles is related to the Gaussian scale height of the rings, H_z, via the formula

$$c_v = H_z \mu. \tag{13.13}$$

If $Q_T < 1$, the disk would be unstable to axisymmetric clumping of wavelength:

$$\lambda = \frac{4\pi G \sigma_\rho}{\kappa^2}. \tag{13.14}$$

For parameters typical of Saturn's rings, λ is on the order of 10–100 m. However, numerical simulations show that clumping occurs for $Q_T \lesssim 2$, and that these clumps stir particle velocities and thereby keep Q_T above unity.

13.3 Observations

Planetary rings are a diverse lot. The differences in dynamical structure of the planetary ring systems are apparent in casual inspection of Figures 13.7–13.9, 13.12, 13.16–13.18. Whereas Saturn's ring particles have high albedo, particles in other ring systems are generally quite dark. Jupiter's rings are broad but extremely tenuous, and we see primarily micrometer-sized silicate dust. Saturn's main rings are broad and optically thick, with most of the optical depth provided by (area covered by) centimeter–meter bodies composed primarily of water-ice; Saturn's outer rings are tenuous and consist of micrometer-sized, ice-rich particles. Most of the material in the uranian rings is confined to narrow rings of particles similar in size to the bodies in Saturn's main ring system. Neptune's brightest ring is narrow and highly variable in longitude; more tenuous, broader rings have also been observed around Neptune. The fraction of micrometer-sized dust in Neptune's rings is larger than in Saturn's rings and Uranus's rings but smaller than in Jupiter's. The particles in Neptune's rings have very low albedos.

In this section, we summarize the properties of planetary ring particles and structure in planetary rings. Theoretical explanations for some of this structure are presented in §13.4; the mechanisms responsible for creating and maintaining the particle size distribution are not well understood theoretically, but a few general principles are discussed together with the observations.

13.3.1 Jupiter's Rings

The jovian ring system is extremely tenuous, so many of the best images are taken when the rings are edge-on, when all particles merge into a single line in the camera. Under such viewing geometries, optically thin rings ($\tau \ll 1$) are much brighter compared with conditions when the rings are partially open. Figure 13.7 shows

(a)

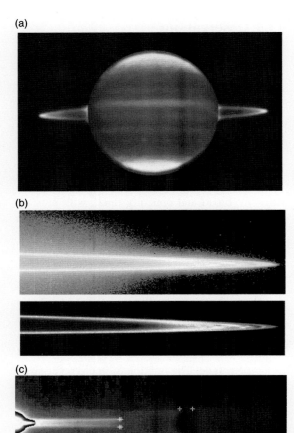

(b)

(c)

Figure 13.7 Images of Jupiter's ring system. (a) Mosaic of Jupiter and its ring system as imaged by I. de Pater at 2.27 μm with the 10-m Keck telescope in 1994. (b) *Galileo* images of Jupiter's ring system processed to emphasize the ring halo (*upper panel*) and the main ring (*lower panel*). (NASA/*Galileo*, PIA01622) (c) A mosaic of Jupiter's gossamer rings made from four visible light *Galileo* images. Images were obtained in near-forward scattered light (phase angle of 177–179° at an elevation of 0.15° – i.e., almost edge-on). The two gossamer rings have crosses showing the four extremes of the eccentric and inclined motions of Amalthea and Thebe. (Adapted from Burns et al. 1999)

Jupiter's ring system, which consists of four principal components: The main ring, the halo and two gossamer rings. Because Jupiter's rings appear much brighter in forward scattered light compared with backscattered light, the surface area and optical depth are dominated by dust even if most of the mass is in larger bodies. (See §4.1.2 for a discussion of forward vs. backscattered light.)

Jupiter's **main ring** is the most prominent component of this planet's ring system, especially in backscattered light, the latter being indicative of a greater fraction of macroscopic material (particle sizes much larger than the wavelength of visible light) in this component of the rings. The physically thin (30–100-km thick) main ring stops at 1.71 $R_{2\!\!+}$, interior of which is the **halo**, which extends inwards to 1.4 $R_{2\!\!+}$. Both the main ring and the halo have normal optical depths of $\tau \approx$ few $\times$ 10^{-6}. Although most halo particles are within a few thousand kilometers of the ring plane, the halo's full vertical extent is close to 40 000 km. The much fainter gossamer rings ($\tau \sim 10^{-7}$) lie exterior to the main ring.

At Jupiter's distance from the Sun, icy ring particles would evaporate rapidly, so Jupiter's ring particles must be composed of more refractory materials. But the lifetime of such grains is short as well. The primary formation mechanism for the dust grains is thought to be erosion (probably by micrometeorites) from the small moons bounding the rings.

The mass of Jupiter's rings is poorly constrained. The dust component of the ring system is very low in mass. The macroscopic particles observed in backscattered radiation clearly provide a substantially larger contribution but one that is uncertain by several orders of magnitude (Problem 13-11).

13.3.2 Saturn's Rings

Saturn's ring system is the most massive, the largest, the brightest and the most diverse in our Solar System (Figs. 13.5, 13.8 and 13.12). Most ring phenomena observed in other systems are present in Saturn's rings as well. The large-scale structure and bulk properties of Saturn's rings are listed in Table 13.1. A schematic illustration of

Table 13.1 Properties of Saturn's Rings[a]

	D Ring	C Ring	B Ring	Cassini Division	A Ring	F Ring	G Ring	E Ring
			Main Rings					
Radial location ($R_{\hbar}$)	1.08–1.23	1.23–1.53	1.53–1.95	1.95–2.03	2.03–2.27	2.32	2.73–2.90	3.7–11.6
Radial location (km)	65 000–74 500	74 500–91 975	91 975–117 507	117 507–122 340	122 340–136 780	140 219		180 000–700 000
Vertical thickness		<4 m	<100 m	<50 m	<100 m			10^3–2×10^4 km (increases with radial location)
Normal optical depth	$\sim 10^{-4}$–10^{-3}	0.05–0.2	1–10	0.1–0.15	0.4–1	1	10^{-6}	10^{-7}–10^{-5}
Particle size	μm–100 μm	mm–m	cm–10 m	1–10 cm	cm–10 m	μm–cm	μm–cm	~1 μm

[a] Data for main rings primarily from Cuzzi et al. (1984); data for ethereal rings primarily from Burns et al. (1984), de Pater et al. (2004b) and Horányi et al. (2009); data for the D ring from Showalter (1996). See text for the few known properties of the (recently discovered) Phoebe ring.

Saturn's rings and inner moons is shown in Figure 13.6.

Radial Structure

As seen through a small- to moderate-sized telescope on Earth, Saturn appears to be surrounded by two rings (Fig. 13.3). The inner and brighter of the two is called the **B ring** (or Ring B), and the outer one is known as the **A ring**. The dark region separating these two bright annuli is named the **Cassini division** after Giovanni Cassini, who discovered it in the 1670s. The Cassini division is not a true gap; rather, it is a region in which the optical depth of the rings is only about 10% of that of the surrounding A and B rings. A larger telescope with good seeing can detect the faint **C ring**, which lies interior to the B ring. The **Encke gap**, a nearly empty annulus in the outer part of the A ring, can also be detected from the ground under good observing conditions. Rings A, B and C and the Cassini division are known collectively as **Saturn's main rings** or **Saturn's classical ring system**. Interior to the C ring lies the extremely tenuous **D ring**, which was imaged by the *Voyager* and *Cassini* spacecraft but has not (yet) been detected from the ground. The narrow, multistranded, kinky **F ring** is 3000 km exterior to the outer edge of the A ring, with the region between the A and F rings known as the **Roche division**.

Several tenuous dust rings lie well beyond Saturn's Roche limit (assuming a particle density equal to that of nonporous water-ice); the most prominent by far being the fairly narrow **G ring** and the extremely broad **E ring** (Fig. 13.12). Although the E ring is quite ethereal, it is so broad that it can readily be observed from Earth when the ring system appears almost edge-on. The inner boundary of the E ring is fairly abrupt, just ~12 000 km inside the orbit of Enceladus. The peak intensity of the E ring is located ~10 000 km exterior to Enceladus's orbit. Geysers at Enceladus's south pole (§10.3.3) provide the bulk of the E ring's material.

The largest known ring of Saturn is associated with the planet's largest irregular moon, Phoebe. The **Phoebe ring** extends at least over the range 100–270 R_h. Impacts on Phoebe presumably eject the particles that make up this enormous but extremely tenuous ring.

The classically known components of Saturn's ring system are quite inhomogeneous upon close examination, displaying both radial and azimuthal variations. Figure 13.8 displays *Cassini* images of the lit and unlit faces of the rings. As can be seen in the close-up images shown in Figure 13.9, the character of this structure is correlated with the overall optical depth of the region in which it exists.

Observations of starlight and (spacecraft) radio signals that have passed through partially transparent rings, i.e., **stellar occultations** and **radio occultations**, provide direct measurements of the optical depth of the rings along the observed line of sight. Figure 13.10 shows an optical depth profile of Saturn's main rings obtained by *Cassini* observations of stellar occultation.

Azimuthal Variations

To a first approximation, Saturn's rings are uniform in longitude, i.e., the character of the rings varies much more substantially with distance from the planet than with longitude. This is, presumably, a consequence of the much shorter timescale for wiping out azimuthal structure via Kepler shear compared with radial diffusion times (Problem 13-7). However, various types of significant azimuthal structure have been observed in Saturn's rings.

The most spectacular longitudinal structures seen in Saturn's rings are the nearly radial features known as **spokes**, which are shown from two viewing geometries in Figure 13.11. Electric and/or magnetic effects are responsible for spokes, the only known planetary ring features that are predominantly radial in shape. They are centered within the B ring, where particles orbit

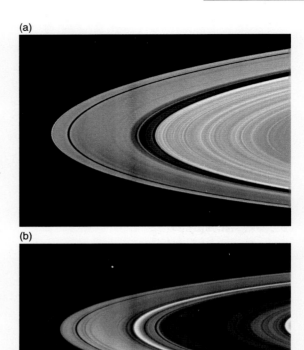

(a)

(b)

Figure 13.8 Clear-filter images of Saturn's rings taken by the *Cassini* spacecraft, at a distance of ~900 000 km from Saturn with 48 km/pixel resolution. (a) The optically thick B and A rings appear brightest in this view of the sun-lit face of the rings from 9° south of the ring plane. (b) Regions of moderate optical depth, such as the C ring and Cassini division, are most prominent in this perspective of the unlit face of the rings taken 8° north of the ring plane. In this geometry, the rings are illuminated primarily via **diffuse transmission** of sunlight, so both optically thick parts of the rings (which do not allow sunlight to pass through) and very optically thin regions (which do not scatter sunlight) appear dark. (Movie) The movie consists of 34 images, beginning with the one shown in panel (a) and ending with the one in panel (b), that were taken over the course of 12 hours as *Cassini* pierced the ring plane. Additional frames were inserted between the spacecraft images to smooth the motion in the sequence. Six moons move through the field of view during the sequence. The first large one is Enceladus, which moves from the upper left to the center right. The second large one, seen in the second half of the movie, is Mimas, going from right to left. (*Cassini* Imaging Team and NASA/JPL/CICLOPS; PIA08356)

(a)

(c)

(b)

(d)

Figure 13.9 Close-up images of portions of Saturn's A, B, C and D rings taken by the *Cassini* spacecraft are shown in panels (a)–(d), respectively. (a) This image of Saturn's A ring shows the gap-forming moon Pan within the Encke gap. The image was taken as Saturn's equinox approached and Pan cast a long shadow on the A ring. Three stars are visible through the rings. A high-resolution image of structure within the A ring is shown in Figure 13.19. (NASA/JPL/Space Science Institute) (b) The lit face of the mid to outer B ring 107 200–115 700 km from Saturn is viewed here at a resolution of 6 km/pixel. (NASA/JPL/CICLOPS, PIA07610). (c) The characteristic plateau and oscillating structure of Saturn's inner C ring are shown at a resolution of 4.7 km/pixel. This *Cassini* image views the lit face of the rings from an elevation of 9° above the ring plane. (NASA/JPL/CICLOPS, PIA06537) (d) Saturn's D ring, imaged from a distance of 272 000 km with resolution of 13 km/pixel. The inner edge of the C ring is seen as the bright area in the lower right corner of the image. (NASA/JPL/CICLOPS, Portion of PIA07714)

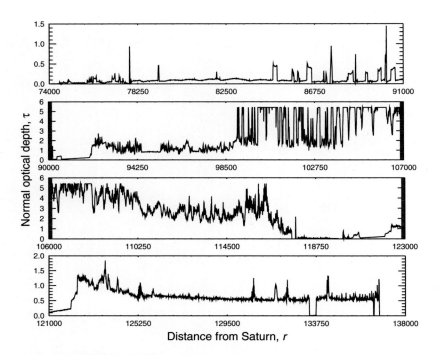

Figure 13.10 Optical-depth profile of Saturn's main rings obtained by observing the star α Arae through the rings with the *Cassini* Ultraviolet Imaging Spectrograph (UVIS) in November 2006. The angle between the direction of starlight and the plane of the rings is $B_{oc} = 54.43°$. The plot shows normal optical depth, τ, computed by multiplying the directly measured slant optical depth averaged at 10-km resolution by sin B_{oc}. The observed starlight is in the wavelength range 110–190 nm. Note that the optical depth scale differs among the panels, radial ranges of the panels overlap by a small amount and the regions in the B ring shown with $\tau = 5.5$ do not allow enough light to pass to allow for anything more than a lower bound on the optical depth. (Josh Colwell and the *Cassini* UVIS team)

synchronously with Saturn's magnetic field. Figure 13.11 shows that spokes appear darker than their surroundings in backscatter but brighter in forward scatter. The strongly forward scattered appearance of spokes implies they contain a significant component of micrometer- and submicrometer-sized dust grains (§4.1.2). Spokes exhibit the greatest contrast in backscatter when the tilt angle of the ring plane to the Sun is small; this enhanced visibility at low tilt angle implies that the vertical thickness of the dust is greater than that of the macroscopic particle layer.

Several narrow rings and ring edges are eccentric. Some of these features are well modeled by Keplerian ellipses that precess slowly as a result of the planet's quadrupole (and higher order) gravitational moments (eq. 2.54). However, a few features, such as the outer edges of rings B (Fig. 13.13) and A and the edges of Encke's gap (Fig. 13.21), are multilobed patterns that are controlled by satellite resonances; the dynamical mechanisms responsible for such features are described in §13.4.3.

Saturn's narrow F ring, shown in Figure 13.24, exhibits several types of unusual features that vary on timescales of hours to years. The ring consists of a relatively optically thick central core surrounded by a fairly diffuse multistranded structure, a variety of clumps, some local variations caused by small embedded moons and a regular series of longitudinal channels produced by the nearby moon Prometheus. The F ring lies near the Roche limit

(a)   (b)

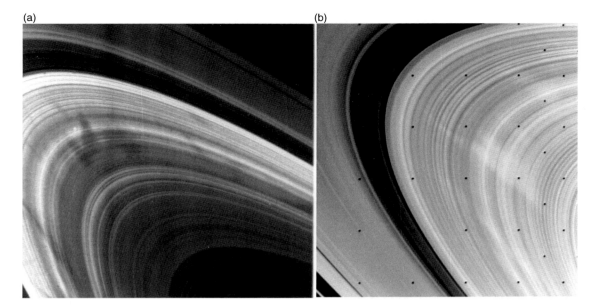

Figure 13.11 Two *Voyager* images of spokes in Saturn's B ring. Both images show the lit face of the rings, but the spokes look quite different because of the differing phase angles of the observations. The spokes appear dark in frame (a), which was taken in backscattered light, but they are brighter than the surrounding ring material in frame (b), which was imaged in forward scattered light with a resolution of ~80 km/pixel. The *black dots* in the images are reseau markings. (Within the optics of the *Voyager* cameras was a grid of black dots called 'reseau markings', which were used to correct for geometric distortions in the cameras of many early spacecraft.) (NASA; a: PIA02275; b: *Voyager 1*, FDS 34956.55)

for moderately porous ice, and it appears that a wide range of accretion, disruption and ring–moon interactions are occurring within it.

Thickness

Saturn's rings are extremely thin relative to their radial extent. Upper limits to the local thickness of the rings of ~150 m at several ring edges were obtained by the abruptness of some of the ring boundaries detected in stellar occultations by the rings observed from *Voyager 2* and from diffraction patterns in the radio signal transmitted through the rings by *Voyager 1*. Estimates of ring thickness from the viscous damping of spiral bending waves and density waves, models of self-gravitating clumps and from the characteristics of the outer edge of the B ring yield values ranging from one meter to tens of meters.

Particle Properties

The particles in Saturn's main ring system are better characterized than are particles in other planetary rings. Spectra of infrared light reflected off Saturn's main rings are similar to that of water-ice, implying that water-ice is a major constituent. The high albedo of Saturn's rings suggests that impurities are few and/or not well mixed at the microscopic level.

The frequent collisions among ring particles cause particle aggregation as well as erosion. Scale-independent processes of accretion and fragmentation lead to power-law distributions of particle number versus particle size. Such power-law distributions are observed over broad ranges of radii in the asteroid belt and in most of those planetary rings for which adequate particle size information is available. Data on particle sizes are

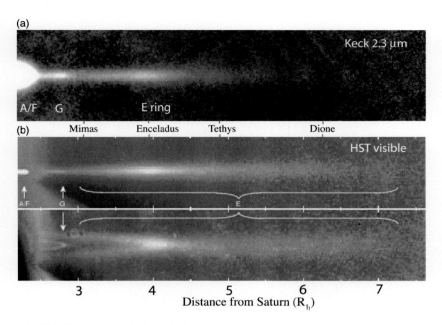

(a)

Keck 2.3 μm

A/F   G   E ring

(b)   Mimas   Enceladus   Tethys   Dione

HST visible

A F   G   E

3   4   5   6   7

Distance from Saturn (R_h)

Figure 13.12 Images of Saturn's tenuous E and G rings. Saturn is off to the left. (a) Infrared photograph ($\lambda = 2.3$ μm) taken on 8–10 August 1995 with the Keck telescope. Even in this view of the dark face of the rings, seen nearly edge-on, the main rings still appear substantially brighter than the E and G rings. (de Pater et al. 2004b) (b) *HST* images of Saturn's G and E rings at visible wavelengths, seen edge-on in August 1995 (*top panel*) and the unlit face open by 2.5° in November 1995 (*bottom panel*). The G ring is the relatively bright and narrow annulus whose ansa appears in the leftmost portion of the image. The E ring is much broader and more diffuse. (Courtesy J.A. Burns, D.P. Hamilton and M.R. Showalter)

thus often fit to a distribution of the form (see eqs. 12.1*a* and 12.1*b*)

$$N(R)dR = \frac{N_0}{R_0}\left(\frac{R}{R_0}\right)^{-\zeta} dR \; (R_{\min} < R < R_{\max})$$

(13.15)

and zero otherwise, where $N(R)dR$ is the number of particles with radii between R and $R + dR$ and N_0 and R_0 are normalization constants. The distribution is characterized by the values of its power-law index, ζ, and the minimum and maximum particle sizes, $R_{\min}$ and $R_{\max}$, respectively. A uniform power law over all radii implies infinite mass in either large or small radii particles (Problem 13-10); thus, such a power law must be truncated at large and/or small radius. Note that the value of the upper size limit is not very important for the total mass or surface area of the system if the distribution is sufficiently steep (ζ significantly larger than 4) and that the lower limit is not a major factor provided the distribution is shallow enough (ζ significantly smaller than 3). An imagined view from within Saturn's rings is shown in Figure 13.14.

Radar signals, with wavelengths of several centimeters, have been bounced off Saturn's rings. The high radar reflectivity of the rings implies that a significant fraction of their surface area consists of particles with diameters of at least several centimeters. Radio signals sent through rings by *Voyager 1* and *Cassini* give information on particle sizes from a comparison of optical depths at two wavelengths and from diffraction patterns of the signal. The combination of these data implies a broad range of sizes from ∼5 cm to 5–10 m in the optically thick B ring and inner A ring, with approximately equal areas in equal logarithmic size intervals (i.e.,

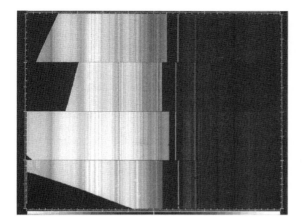

Figure 13.13 The eccentric outer edge of Saturn's B ring as imaged at four different longitudes by *Voyager 2*. The lit face of the rings is seen and Saturn is off to the left in all images. The left portion of the images shows the outermost region of the B ring, and the right part shows the inner part of the Cassini division. The middle two slices were taken from high-resolution images of the east ansa and the outer two slices are from the west ansa. The width of the gap separating the B ring from the Cassini division varies by up to 140 km. These variations are caused by perturbations exerted by Mimas near its 2:1 inner Lindblad resonance. Also visible are variations in fine structure in the B ring and the eccentric Huygens ringlet within the variable width gap. All slices were taken within about 7 hours. (Smith et al. 1982)

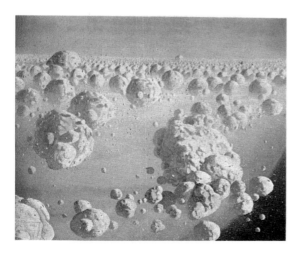

Figure 13.14 Hypothetical view from within Saturn's rings, showing macroscopic particles as loosely bound agglomerates. (Painted by W.K. Hartmann)

$\zeta \approx 3$) and most of the mass being in the largest particles. The power-law index in the size distribution is $2.8 < \zeta < 3.4$ for 5 cm $< R < 5$ m and $\zeta > 5$ for $R > 10$ m. The C ring and outer A ring also contain a substantial quantity of particles somewhat smaller than 5 cm.

Micrometer-sized dust particles are comparable in size to the wavelength of visible light, so they preferentially scatter in the forward direction. Micrometer-sized particles are most common in the dusty outer rings and the F ring but also dominate the spokes in the B ring (see Fig. 13.11) and are apparent in the very outer part of the A ring exterior to the Keeler gap. It is likely that the micrometer-sized dust seen in the spokes is rapidly reaccumulated by larger ring particles, causing the spoke to vanish.

Moonlets several kilometers in radius clear gaps in the rings, with gap width several times the moonlet's own radius (§13.4.3). The embedded moonlets Daphnis and Pan (Fig. 13.9a), with mean radii of 4 and 14 km, respectively, have very low densities, $\lesssim 500$ kg m^{-3}, implying high porosities. The gravitational effect by smaller moonlets is not sufficient to clear a gap. Instead, very small bodies are surrounded by a region of low density, which is flanked by density enhancements a few kilometers in extent.

Propeller-shaped features, such as the one shown in Figure 13.15, provide indirect evidence of many bodies of radii $\sim$20–250 m, assuming densities comparable to that of water-ice. These bodies are intermediate in size between the ring particles at the upper limits of the power-law size distribution and moonlets that are massive enough to clear complete gaps around their orbits. Although the propeller-producing objects are too small to be detected directly, their gravitational effect on the surrounding ring material betrays their presence by perturbing particle trajectories in a manner analogous to that shown in Figure 2.6.

Spectra of Saturn's rings are affected by both the composition and the size distribution of grains,

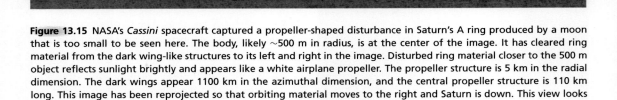

Figure 13.15 NASA's *Cassini* spacecraft captured a propeller-shaped disturbance in Saturn's A ring produced by a moon that is too small to be seen here. The body, likely ~500 m in radius, is at the center of the image. It has cleared ring material from the dark wing-like structures to its left and right in the image. Disturbed ring material closer to the 500 m object reflects sunlight brightly and appears like a white airplane propeller. The propeller structure is 5 km in the radial dimension. The dark wings appear 1100 km in the azimuthal dimension, and the central propeller structure is 110 km long. This image has been reprojected so that orbiting material moves to the right and Saturn is down. This view looks at the sunlit side of the rings, and resolution is ~1 km/pixel. (NASA/JPL-Caltech/SSL)

including that of ring particle regolith, i.e., the grains and ice crystals that cover the larger (centimeter- to meter-sized) boulders in the rings.

The E ring is distinctly blue in color, in contrast to Jupiter's dusty rings and Saturn's G ring, which are red. The blue color indicates that the particle size distribution is dominated by tiny grains. In Saturn's E ring, the particle radii cluster near 1 μm; hence, this population is not collisionally evolved. The E ring particles are ice crystals formed in the plumes of water-ice volcanoes or geysers on the moon Enceladus (§10.3.3).

Mass

The mass of Saturn's rings is much larger than that of any other ring system within the Solar System but is too low to have been measured by its gravitational effects on moons or spacecraft. Thus, it must be deduced from more circuitous theoretical arguments. Several different techniques have been used, and all give similar answers, lending confidence to the results.

The wavelengths of spiral density waves and spiral bending waves are proportional to the local surface mass density of the rings; therefore, we can deduce the mass where we see waves. The theory behind this analysis is presented in §13.4. The surface density, σ_ρ, at the two wave locations observed in the B ring is ~500–800 kg m^{-2}. Analysis of dozens of waves within the A ring reveals surface densities of ~500 kg m^{-2} in the inner to middle A ring, dropping to $\lesssim$200 kg m^{-2} near this ring's outer edge. Measured values in the optically thin C ring and Cassini division are ~10 kg m^{-2}. Because observable spiral waves cover only a small fraction of the area of Saturn's rings, we assume that the **opacity**, σ_ρ/τ, is constant within a given region of the rings in order to estimate the mass of the ring system. This approximation is fairly good wherever there are several waves near one another (so that it can be checked). It must be noted, though, that the energy input by the moons into the waves may make regions in which strong waves propagate somewhat anomalous. The particle flux produced from cosmic ray interactions with the ring system suggests an average surface density of at least 1000–2000 kg m^{-2}, a few times as large as the average from density wave estimates. These values can be reconciled if the surface density in the optically thick parts of the B ring (Fig. 13.10), where no density waves have been observed, is significantly higher than the surface density in the wave regions. The total mass of Saturn's ring system is estimated to be roughly $M_{\text{rings}} \sim 5 \times 10^{-8} \, M_\hbar \sim M_{\text{Mimas}}$, where Mimas, the innermost and smallest of Saturn's nearly spherical moons, has a mean radius of 196 km (Table E.5).

13.3.3 Uranus's Rings

The rings of Uranus are shown in Figure 13.16. Most of the material in the uranian ring system

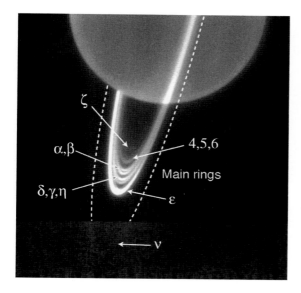

Figure 13.16 Composite image at 2.3 μm taken with the adaptive optics system on the Keck telescope in August 2005. Only the south side of the rings is shown here to emphasize the main ring system, the inner ζ and outer ν rings. (Adapted from de Pater et al. 2006b)

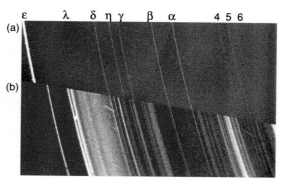

Figure 13.17 The main rings of Uranus, as imaged by *Voyager 2* in 1986. The region shown ranges from ~40 000 km from the planet's center (*right*) to ~50 000 km away (*left*). (a) Mosaic of two low phase angle (21°), high-resolution (10 km/pixel) images. The planet's nine narrow optically thick rings are clearly visible, and the very narrow moderate optical depth λ ring is marginally detectable. (NASA/*Voyager 2*, PIA00035) (b) High phase angle (172°) view. The forward scattering geometry dramatically enhances the visibility of the micrometer-sized dust particles. The streaks are trailed star images in this 96-second exposure. (PIA00142)

is confined to nine narrow annuli whose orbits lie between 1.64 and 2.01 R_{O} from the planet's center; close-up images of this region are shown in Figure 13.17. These nine optically thick rings were discovered from Earth-based observations of stars whose light was seen to diminish as they were occulted by Uranus's rings in the late 1970s. The data leading to the discovery of Uranus's five most prominent rings are shown in Figure 13.4. Eight of the nine optically thick rings are 1–10 km wide and have eccentricities of order 10^{-3} and inclinations of $\lesssim 0.06°$. The main uranian ring system also includes the narrow, moderate optical depth, dusty λ **ring**.

The outermost annulus, **Ring** ϵ, is the widest and most eccentric optically thick uranian ring, with $e = 8 \times 10^{-3}$ and width ranging from 20 km at periapse to 96 km at apoapse. Because the ϵ ring is optically thick, this difference in width leads to a pronounced asymmetry in ring brightness, with the ring at apoapse more than twice as bright as

at periapse. The majority of ring edges, including both the inner and outer boundaries of the ϵ ring, are quite sharp compared with ring width, but in a few cases, a more gradual dropoff in optical depth is observed.

The mass of the ϵ ring estimated from the particle size distribution and an assumed particle density of 1000 kg m^{-3} is 1–5 $\times$ 10^{16} kg. Dynamical models for the maintenance of the ϵ ring's eccentricity by ring self-gravity yield a mass estimate of ~5 $\times$ 10^{15} kg, but this estimate may not be very accurate because these models do not adequately reproduce certain aspects of ring structure. The combined mass of all of Uranus's other rings is probably a factor of a few smaller than that of Ring ϵ.

The particles in the nine optically thick uranian rings have a similar size distribution to those in Saturn's main ring system except that the lower limit is closer to ~10 cm, roughly an order of magnitude larger than in Saturn's rings. Particle sizes thus range from ~10 cm to ~10 m. Uranus's ring particles are extremely dark, with ring particle reflectivity at visible and near-infrared wavelengths being ~0.04. They appear as dark as the darkest asteroids and carbonaceous chondrite meteorites. However, they probably consist of **radiation-darkened ice**, which is a mixture of complex hydrocarbons embedded in ice that includes CH_4, CO and/or CO_2 produced as a result of the removal of H atoms via sputtering processes. Such radiation-darkened ice may also account for the low albedos of cometary nuclei (§12.6).

Wide, radially variable, low optical depth sheets of submicrometer- and micrometer-sized particles are interspersed with the optically thick uranian rings (Fig. 13.17). Interior to the main rings lies the broad, tenuous ζ **ring** (Fig. 13.16).

Two broad, low optical depth, rings of Uranus located well exterior to the planet's nine main rings were discovered by *HST*. The outermost, the μ **ring**, is more than 15 000 km wide. Similar to Saturn's E ring, the μ ring is distinctly blue, indicative of a particle size distribution dominated by submicrometer-sized material. Its peak intensity coincides with the orbit of the tiny moon Mab, which presumably is the source of its particles. The other ring, **ring** ν, is less than 4000 km wide. It has a 'normal' red color, suggestive of a much larger fraction of micrometer and larger particles. It does not coincide with any known moons.

13.3.4 Neptune's Rings

Neptune's ring system is quite diverse. Images such as the *Voyager* view displayed in Figure 13.18 show structure in both radius and longitude. The most prominent feature of the ring system is a set of **arcs** of optical depth $\tau \approx 0.1$ within the **Adams ring**. The arcs vary in extent from ~1° to ~10° and are grouped in a 40° range in longitude; they are about 15 km wide. At other longitudes within the Adams ring, $\tau \approx 0.003$; a comparable

Figure 13.18 Neptune's two most prominent rings, Adams (which includes higher optical depth arcs) and Le Verrier, as seen by *Voyager 2*. This forward scattered light (phase angle 134°) image, FDS 11412.51, was obtained using a 111-second exposure with a resolution of 80 km/pixel. (NASA/*Voyager 2*, PIA01493)

optical depth has been measured for the **Le Verrier ring**. Neptune's other rings are even more tenuous. Several moderately large moons orbit within Neptune's rings; these satellites are thought to be responsible for much of the radial and longitudinal ring structure that has been observed. Images obtained with *HST* and the Keck telescope have revealed that the arrangement of arcs is changing over time.

The particles in Neptune's rings are very dark and (at least in the arcs) red. They may be as dark as the particles in the uranian rings, but the properties of Neptune's ring particles are less well constrained by current data. The fraction of optical depth due to micrometer-sized dust is very high, ~50%, and appears to vary from ring to ring. The limited data available are not sufficient to make even an order of magnitude estimate of the mass of Neptune's rings, although they suggest that the rings are significantly less massive than the rings of Uranus unless they contain a substantial population of undetected large ($\gtrsim 10$ m) particles, which is unlikely given the paucity of smaller macroscopic ring particles.

13.4 Ring–Moon Interactions

Observations of planetary rings reveal a complex and diverse variety of structure, mostly in the radial direction, over a broad range of length scales, in contrast to naive theoretical expectations of smooth, structureless rings (§13.2). The processes responsible for some types of ring structure are well understood. Partial or speculative explanations are available for other features, but the causes of many structures remain elusive. The agreement between theory and observations is best for ring features thought to be produced by gravitational perturbations from known moons, which we discuss in this section.

13.4.1 Resonances

A common process in many areas of physics is resonance excitation: When an oscillator is excited by a varying force whose period is very nearly equal to the oscillator's natural frequency, the response can be quite large even if the amplitude of the force is small (eq. 2.30). In the planetary ring context, the perturbing force is the gravity of one of the planet's moons, which is generally much smaller than the gravitational force of the planet itself.

Resonances occur where the radial (or vertical) frequency of the ring particles is equal to the frequency of a component of a satellite's horizontal (or vertical) forcing, as sensed in the frame rotating at the frequency of the particle's orbit. In this case, the resonating particle is repeatedly near the same phase in its radial (vertical) oscillation when it experiences a particular phase of the satellite's forcing. This situation enables continued coherent 'kicks' from the satellite to build up the particle's radial (vertical) motion, and significant forced oscillations may thereby result. Particles nearest resonance have the largest eccentricities (inclinations) because they receive the most coherent kicks; the forced eccentricity (inclination) is inversely proportional to the distance from resonance for noninteracting particles in the linear regime. Collisions among ring particles and the self-gravity of the rings complicate the situation, and resonant forcing of planetary rings can produce a variety of features, including gaps and spiral waves.

The strongest **horizontal resonances** are found near the location where the ratio of the orbital period of the forcing moon to that of the ring particles is of the form $m_\theta:(m_\theta - 1)$, where the integer m_θ is the azimuthal symmetry number of the disturbance. The strongest **vertical resonances** are of the form $(m_\theta + 1):(m_\theta - 1)$. The precise locations of these resonances are affected by apsidal precession and nodal regression (§2.6.2), which cause horizontal resonances to lie somewhat exterior to analogous vertical resonances (Figs. 13.19 and 13.21). As can be seen in Figure 13.19, by far the lion's share of the strong resonances in Saturn's ring system lies within the outer A ring, near the orbits of the moons that excite them.

Resonant forcing leads to a secular transfer of orbital angular momentum from Saturn's rings to its moons. These torques produce two classes of structure in Saturn's rings: gaps/ring boundaries and spiral density and bending waves. The outer edges of Saturn's two major rings are maintained by the two strongest resonances in the ring system. The outer edge of the B ring is located at Mimas's 2:1 horizontal resonance and is shaped like a two-lobed oval *centered* on Saturn. The A ring's outer edge is coincident with the 7:6 resonance of the coorbital moons Janus and Epimetheus and has a seven-lobed pattern consistent with theoretical expectations. For a resonance to clear a gap or maintain a sharp ring edge, it must exert enough torque to counterbalance the ring's viscous spreading. In the low optical depth C ring, resonances with moderate torques create gaps, but in the higher optical depth A and B rings, resonances with similar strengths excite spiral density waves.

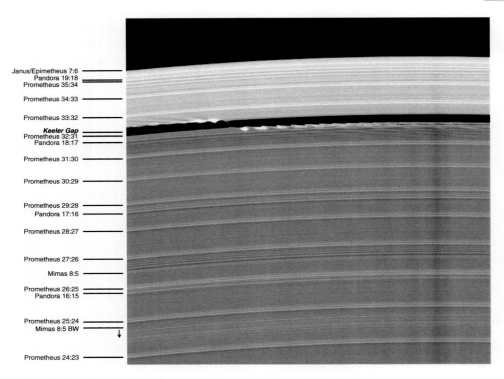

Janus/Epimetheus 7:6
Pandora 19:18
Prometheus 35:34

Prometheus 34:33

Prometheus 33:32

Keeler Gap
Prometheus 32:31
Pandora 18:17

Prometheus 31:30

Prometheus 30:29

Prometheus 29:28
Pandora 17:16

Prometheus 28:27

Prometheus 27:26

Mimas 8:5

Prometheus 26:25
Pandora 16:15

Prometheus 25:24
Mimas 8:5 BW

Prometheus 24:23

Figure 13.19 *Cassini* image of the lit face of the outer portion of Saturn's A ring, with the locations of strong satellite resonances and the Keeler gap marked on the left. The Janus/Epimetheus 7:6 ILR confines the outer edge of the A ring. The first few crests and troughs of the inwardly propagating Mimas 8:5 bending wave and the outwardly propagating Mimas 8:5 density wave can be seen on this image, whereas the wavelengths of the higher m_θ density waves at the resonances of the nearby small moons are shorter and not clearly discernible in this version. The tiny moon Daphnis is to the left of center, and its effects on the edges of the Keeler gap are quite prominent. (Image PIA07809 from NASA/JPL/CICLOPS, annotated by Matt Tiscareno)

13.4.2 Spiral Waves

Spiral density waves are horizontal density oscillations that result from the bunching of streamlines of particles on eccentric orbit. **Spiral bending waves**, in contrast, are vertical corrugations of the ring plane resulting from the inclinations of particle orbits. Schematic diagrams of spiral density and bending waves are shown in Figure 13.20. Both types of spiral waves are excited at resonances with moons and propagate as a result of the collective self-gravity of the particles within the ring disk (Fig. 13.21). Ring particles move along paths that are very nearly Keplerian ellipses with one focus at the center of Saturn. However,

small perturbations caused by the wave force a coherent relationship between particle eccentricities/periapses (in the case of density waves) or inclinations/nodes (in the case of bending waves), which produces the observed spiral pattern.

Spiral waves in planetary rings are extremely tightly wound, with typical **winding angles** (departures from circularity) being 10^{-5}–10^{-4} radian (compared with $\gtrsim 10^{-1}$ radian in most spiral galaxies). Such waves have very short wavelengths, of the order of 10 km. Figure 13.21 shows brightness contrasts between crests and troughs of density waves; a bending wave is made visible on this spacecraft image as a result of the dependence of

(a)

(b)

(c)

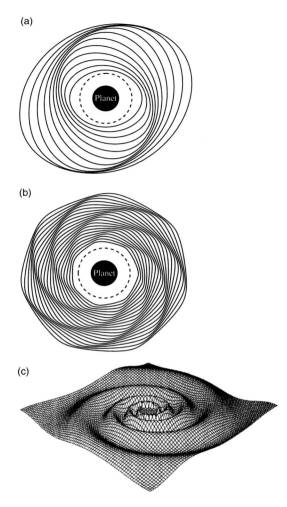

Figure 13.20 Schematic diagrams of the coplanar particle orbits that give rise to trailing spiral density waves near a resonance with an exterior satellite are shown in panels (a) and (b). (a) The two-armed spiral density wave associated with the 2:1 ($m_\theta = 2$) inner Lindblad resonance. (b) The seven-armed density wave associated with the 7:6 ($m_\theta = 7$) inner Lindblad resonances. The pattern rotates with the angular velocity of the satellite and propagates outwards from the exact resonance (denoted by a *dashed circle*). (Murray and Dermott 1999) (c) Schematic of an inward-propagating spiral bending wave showing variation of vertical displacement with angle and radius for a two-armed spiral. Spiral waves observed in Saturn's rings are much more tightly wound. (Shu et al. 1983)

brightness on local solar elevation angle. Figure 13.22 shows density waves observed by monitoring the light of a star passing through Saturn's partially transparent rings.

Ring self-gravity causes the perturbation excited at resonance to propagate as a wave. The strength of the gravitational interaction depends on the mass of the ring and the separation between the peaks and troughs of the wave. A higher surface density allows the disturbance to have a longer wavelength at a specified distance from resonance.

Six bending waves and $\sim$100 density waves in Saturn's rings have thus far been identified with the resonances responsible for exciting them and have been analyzed to determine the local surface mass density of the rings. The surface density at most wave locations in the optically thick A and B rings is $\sim$300–600 kg m^{-2}. Measured values in the optically thin C ring and Cassini division are $\sim$10 kg m^{-2}. Because the ratios of optical depths are $\sim$10, the larger magnitude of the difference in surface densities implies that the average particle size in the C ring and Cassini division is smaller than that in the B and A rings.

Inelastic collisions between ring particles act to damp bending waves. Larger velocities lead to more rapid damping. The damping rate of bending waves can be used to estimate the ring viscosity, which can be converted into an estimate of the ring thickness using equations (13.11) and (13.13). The A ring appears to have a local thickness of a few tens of meters or less; the thickness of the C ring is $\lesssim$ 5 m.

The back torque that the rings exert upon the inner moons (according to Newton's third law) causes these moons to recede on a timescale that is short compared with the age of the Solar System; current estimates suggest that Atlas, Pandora, Prometheus and Janus/Epimetheus should all have been at the outer edge of the A ring within the past $\sim$2 $\times$ 10^8 years, with the journey of Prometheus, a relatively large moon located quite close to the A ring, occurring on a timescale of $\lesssim$ 20 million

Figure 13.21 A portion of the lit face of Saturn's A ring is seen in this mosaic of *Cassini* images that were taken with resolution of 940 m/pixel just after the spacecraft entered orbit about Saturn. Saturn is off to the left. Various features produced by gravitational perturbations of moons are seen against an otherwise uniform background. The most prominent features on the left side are the Mimas 5:3 bending wave, which propagates inwards towards the planet, and the Mimas 5:3 density wave, which propagates away from Saturn. The separation between the locations of the two waves results from the nonclosure of orbits caused by Saturn's oblateness. The other density waves are excited by the moons Janus/Epimetheus, Pandora and Prometheus. The dark region on the right side is Encke's gap; the scalloped inner edge of the Encke gap and the associated satellite wake to the interior are produced by the moonlet Pan. (Lovett et al. 2006 and NASA/JPL)

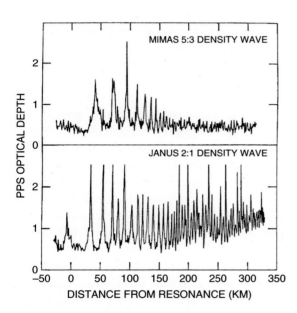

Figure 13.22 The Mimas 5:3 and Janus 2:1 density waves as viewed from the *Voyager 2* PPS stellar occultation of the star δ Sco, plotted so that $\tau(r)$ increases upwards. Note the sharp peaks and broad flat troughs caused by nonlinearities. (Esposito 1993)

years. Resonance locking to outer, more massive moons could slow the outward recession of the

small inner moons; however, angular momentum removed from the ring particles should force the entire A ring into the B ring in $<10^9$ yr. If the calculations of torques are correct and if no currently unknown force counterbalances them, then the small inner moons and/or the rings must be 'new', i.e., much younger than the age of the Solar System. However, a 'recent' origin of Saturn's rings appears to be a priori highly unlikely. We return to this problem when we discuss the origins of planetary ring systems in §13.5.

13.4.3 Shepherding

Moons and rings repel one another through resonant transport of angular momentum via density waves. As is the case for viscous spreading of a ring, the majority of the angular momentum is transferred outwards and most of the mass inwards (in this case, the ring and moon are considered together as parts of the same total reservoir of energy and angular momentum). This is a general result for dissipative astrophysical disk systems because a spread-out disk of material on circular (nearly) Keplerian orbits has a lower energy state for fixed total angular momentum than does more radially concentrated material. Analogously, whereas resonant transfer of energy from an inner moon to a moon on an orbit farther out frees up energy for tidal heating, transfer in the opposite direction is almost always unstable (i.e., the resonance lock is only temporary).

We now consider the process of **shepherding**, by which a moon repels ring material on nearby orbits. The essence of the interaction is that ring particles are gravitationally perturbed into eccentric orbits by a nearby moon and collisions among ring particles damp these eccentricities. The net result is a secular repulsion between the ring and the moon. Details of the interactions depend on whether a single resonance dominates the angular momentum transport or the moon and ring are so close that individual resonances do not matter, either because the resonances overlap or because the synodic period between the ring and the moon is so long that collisions damp out perturbations between successive close approaches. However, many aspects of the basic qualitative picture are the same in both cases.

A moon pushes material away from it (on both sides). The system is sketched in Figure 13.23. Ring material between two moons is thus forced into a narrow annulus between them. The annulus should be located nearer the smaller moon so that the torques balance. Viscous diffusion maintains a

Figure 13.24 The sunlit face of Saturn's multistranded F ring and its companion moons, as imaged by the *Cassini* spacecraft from 6°–7° above the ring plane. Prometheus, the inner moon, is more massive than the outer moon, Pandora, and orbits closer to the ring, so it exerts stronger perturbations. Note that the long axes of these decidedly nonspherical moons are aligned with the direction to the planet, which is expected because this is the lowest energy state for a synchronously rotating moon. (NASA/JPL/CICLOPS, PIA07712)

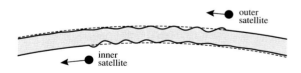

Figure 13.23 Schematic illustration of the shepherding of a planetary ring by two moons. Ring particles' eccentricities are excited as the particles pass by a moon. Interparticle collisions subsequently damp eccentricities and leave the particles orbiting farther from the moon than before the encounter. Particles orbiting closest to the moon are affected most strongly. (Murray and Dermott 1999)

finite width of this annulus. Figure 13.24 shows an image of Saturn's F ring together with its two shepherding moons Prometheus and Pandora. The entire 360° extent of the F ring is visible in the projected mosaic of *Cassini* images shown in Figure 13.25. The basics of confinement are explained earlier, but additional processes are also clearly operating to create the complicated features observed. The shepherding moons Cordelia and Ophelia confine the uranian ϵ ring, although in this case, the ring edges are maintained by individual 'isolated' resonances of these moons. Uranus's other narrow rings are also believed to be confined by shepherding torques; a few ring edges may be maintained by resonances with Cordelia and Ophelia, but the small moons that this model requires to hold most of Uranus's narrow rings in place have yet to be discovered.

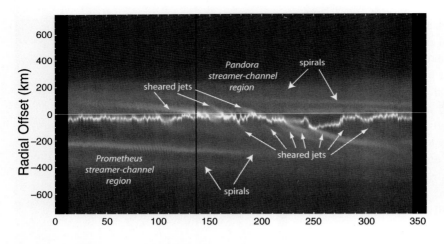

Figure 13.25 Mosaic of *Cassini* spacecraft images of Saturn's multistranded F ring, annotated to show the prominent jets, spirals and channels that are produced by the nearby moons Prometheus and Pandora. (Colwell et al. 2009)

What about a small moon embedded in the middle of a wide ring? Material is cleared on both sides, so a gap is formed around the moon. Diffusion acts to fill in the gap and blur the edges, but optical depth gradients can occur on length scales comparable to typical particle collision distances, allowing sharp edges to be produced. Figure 13.9a shows the moonlet Pan located within Encke's gap, which Pan keeps clear. If a moon is too small, then it cannot clear a gap larger than its own size, so no gap is formed. Propellers (Fig. 13.15) are an intermediate case.

13.5 Origins of Planetary Rings

Are ring systems primordial structures (dating from the epoch of planetary formation $\sim 4.56 \times 10^9$ years ago) that are remnants of protosatellite accretion disks, or did they form more recently as the result of the disruption of larger moons or interplanetary debris? Various evolutionary processes that occur on timescales far shorter than the age of the Solar System imply that tenuous, dust-dominated and/or narrow rings must be geologically very young. However, such a recent origin of Saturn's main rings is a priori quite unlikely.

Micrometer-sized dust is removed from rings quite rapidly. Some loss mechanisms lead to permanent removal of grains from ring systems. Processes leading to permanent loss of grains include sputtering, which is dominant for Jupiter's dust; gas drag, which is important for particles orbiting Uranus because of that planet's hot extended atmosphere; and Poynting–Robertson drag. Other dust removal mechanisms, such as re-accretion by large particles, which dominates for the dust comprising Saturn's spokes, allow for recycling. In all ring systems, the dust requires continual replenishment, which means macroscopic parent particles. The dust mass is so small that in most cases a quasi-steady state could exist over geologic time.

If rings do not last long, why do we observe them around all four giant planets? This question is most relevant to the case of the saturnian ring system because it is far more massive than the rings of other planets. In the cases of Jupiter, Uranus and Neptune, it is easier to envisage the current ring–moon systems as remnants of eons of disruption, re-accretion (to the extent possible so near a planet) and gradual net losses of material. In the previous section, we showed that Jupiter's ring system is formed from impact ejecta from small moons. By extension, it has been suggested that all small moons orbiting near planets, including the martian satellites Phobos and Deimos, should lead to the formation of rings.

Let us examine some specific origin scenarios for the macroscopic particles present within planetary rings. A stray body passing close to a planet may be tidally disrupted (e.g., Comet D/Shoemaker–Levy 9; see §8.1.2). However, under most circumstances, the vast majority of the pieces escape from or collide with the planet. Moreover, such an origin scenario does not explain why all four planetary ring systems orbit in the prograde direction. And this mechanism does not easily solve the 'short timescale' problems because the flux of interplanetary debris has decreased substantially over geologic time (at least in the inner Solar System, where we have 'ground truth' from radioisotope dating of lunar craters; see §6.4.4). Ring particles are thus most likely (possibly second or later generation) products of circumplanetary disks.

Planetary rings may be the debris from the disruption of moons that got too close to their planets and were broken apart by tidal stress or that were destroyed by impacts and did not re-accrete because of tidal forces. There are two difficulties with the hypothesis that the rings that we observe are the products of recent disruptions of moons: Why did the rings form recently, i.e., why is now special? Can moons form at or move inwards to the radii where planetary rings are seen? It is more difficult for a body to accrete than to remain held together (Problem 13-2). A ring parent moon must form beyond the planet's Roche limit and subsequently drift inwards. Tidal decay towards a planet only occurs for moons inside the orbit that is synchronous with the planet's spin period (unless the moon's orbit is retrograde), but Roche's limit is outside the synchronous orbit for Jupiter and Saturn. The presence and characteristics of small moons interspersed with ring particles in all four planetary ring systems argue strongly for a

model in which the rings vary rapidly compared with geological time, and there is a substantial amount of recycling of material between rings and nearby moons. A recent origin for the particles within the rings of Jupiter, Uranus and Neptune does not present theoretical difficulties because these rings are less massive than (or, in the case of Uranus possibly of comparable mass to) the nearby moons.

Thus, a major outstanding issue in planetary rings is the origin and age of Saturn's main ring system. Three proposed formation models involve unaccreted remnants of Saturn's protosatellite disk, debris from a destroyed moon and a tidally disrupted comet. The strongest evidence for a geologically recent origin of Saturn's rings is orbital evolution of rings and nearby moons resulting from resonant torques (responsible for density wave excitation) and ring pollution by accretion of interplanetary debris. Testing the validity of satellite torques is especially important because models also suggest that angular momentum transport via resonant torques and density waves is a significant factor in the evolution of other, less well-observed astrophysical disk systems, such as protoplanetary disks and accretion disks in binary star systems. For example, density wave torques may lead to significant orbital evolution of young planets within the protoplanetary disks on timescales of $\sim 10^5$–10^6 years (§15.7.1).

Key Concepts

- Planetary rings are collections of tiny satellites whose collective appearance is that of a flat annulus.
- All massive planetary rings orbit close enough to their planet that tidal forces from the planet inhibit accretion into one or a few moons, i.e., inside of Roche's limit. Interparticle collisions force these rings to be quite thin.

- Tenuous dust rings orbit in various locations. The dust in these rings is short lived and needs to be replenished on geologically short timescales.
- Jupiter's rings are broad and quite tenuous. Small particles provide most of the optical depth; the source of some of these particles is erosion from larger ring particles; others come from multi-kilometer moons.
- Saturn has by far the most massive rings in the Solar System. Saturn's rings are also the broadest and most diverse.

- Most of the mass in the uranian rings is in narrow, eccentric rings, but Uranus also has broad dust rings.
- The most prominent ring of Neptune contains a few relatively bright arcs that are confined to a small region in longitude.
- Interactions between moons and rings transport angular momentum, confine ring edges and excite waves within rings.
- Many of the processes that occur in rings are also likely to occur within protoplanetary disks.

Further Reading

The following review book provides a comprehensive overview of our knowledge of planetary rings at that date:

Greenberg, R., and A. Brahic, Eds., 1984. *Planetary Rings*. University of Arizona Press, Tucson. Particular attention should be given to the chapters by Cuzzi et al. (Saturn's rings), Burns et al. (ethereal rings) and Shu (spiral waves).

More recent reviews are available for the individual ring systems:

Burns, J.A., D.P. Simonelli, M.R. Showalter, D.P. Hamilton, C.C. Porco, H. Throop, and L.W. Esposito, 2004. Jupiter's ring–moon system. In *Jupiter: Planet, Satellites and Magnetosphere*. Eds.

F. Bagenal, T. E. Dowling, and W. McKinnon. Cambridge University Press, Cambridge, pp. 241–262.

Esposito, L.W., 2010. Composition, structure, dynamics, and evolution of Saturn's rings. *Ann. Rev. Earth Planet. Sci.*, **38**, 383–410.

French, R.G., P.D. Nicholson, C.C. Porco, and E.A. Marouf, 1991. Dynamics and structure of the uranian rings. In *Uranus*. Eds. J.T. Bergstrahl, E.D. Miner, and M.S. Matthews. University of Arizona Press, Tucson, pp. 327–409.

Porco, C.C., P.D. Nicholson, J.N. Cuzzi, J.J. Lissauer, and L.W. Esposito, 1995. Neptune's ring system. In *Neptune*. Ed. D.P. Cruikshank. University of Arizona Press, Tucson, pp. 703–804.

Problems

13-1. **(a)** Derive equation (13.7) from equation (13.6). (Hint: Recall that the analysis assumes that the planet is spherical.)
(b) Derive equation (13.7) from equation (2.28).

13-2. In §13.1, we estimated the limits for tidal stability of a spherical satellite orbiting near a planet by equating the self-gravity of the satellite to the tidal force of the planet at a point on the satellite's surface that lies along the line connecting the planet's center to the satellite's center. The results of this study are directly applicable to the ability of a spherical moon to accrete a much smaller particle that lands on the appropriate point on its surface.

(a) Perform a similar analysis for the mutual attraction of two spherical bodies of equal size and mass whose centers lie along a line that passes through the planet's center. The result is known as the **accretion radius**.

(b) Comment on the qualitative similarities and quantitative differences between the accretion radius, the tidal disruption radius estimated for a single spherical body (eq. 13.7) and Roche's tidal limit (eq. 13.8), which was calculated for a deformable body.

13-3. Calculate the Roche limit of each of the giant planets as a function of satellite density. Compare your results with the observed positions and sizes of some of these planets' inner satellites and rings and then comment.

13-4. Calculate the densities required for Neptune's six inner satellites, assuming each is located just exterior to the Roche limit for its density. Are these densities realistic? What holds these moons together?

13-5. Would a retrograde ring be stable? Justify your answer.

13-6. If a ring has a normal optical depth τ, then using the geometrical optics approximation, a photon traveling perpendicular to the ring plane has a probability of $e^{-\tau}$ of passing through the rings without colliding with a ring particle. For this problem, you may assume that the ring particles are well separated and that the positions of their centers are uncorrelated.

(a) What fraction of light approaching the rings at an angle θ to the ring plane will pass through the rings without colliding with a particle?

(b) The cross-section of a ring particle to collisions with other ring particles is larger than its cross-section for photons. Assuming all ring particles are of equal size, what is the probability that a ring particle moving normal to the ring plane would pass through the rings without a collision?

(c) On average, how many ring particles does a line perpendicular to the ring plane pass through?

(d) On average, how many ring particles does a particle passing normal to the ring plane collide with?

As ring particles on Keplerian orbits pass through the ring plane twice per orbit, two times your result is a good estimate for the number of collisions per orbit of a particle in a sparse ring. The in-plane component of the particles' random velocities increases the collision frequency by a small factor, the value of which depends on the ratio of horizontal to vertical velocity dispersion. For a ring that is thin and massive enough, local self-gravity can increase the vertical frequency of particle orbits, and combined with the gravitational pull of individual particles, greatly increase particle collision frequencies.

13-7. **(a)** Calculate the time necessary for Kepler shear to spread rings with the following parameters over 360° in longitude:

(i) Width, 1 km; orbit, 80 000 km from Saturn

(ii) Width, 100 km; orbit, 80 000 km from Saturn

(iii) Width, 1 km; orbit, 120 000 km from Saturn

(iv) Width, 2 km; orbit, 63 000 km from Neptune

(b) Calculate the radial diffusion time for the doubling of the widths of the rings in part (a) assuming a viscosity $\nu_v = 0.01$ m^2 s^{-1}. How do these times vary with viscosity (give the functional form)?

(c) Compare your results in parts (a) and (b) and comment on the observation that rings are generally observed to vary much more substantially with radius than with longitude.

13-8. Show that an ensemble of material on circular orbits within a Keplerian potential has a lower energy state for a given total angular momentum if it is spread out in radius rather than concentrated in a narrow annulus. Consider a body of mass $M = 3$ on a circular orbit about a much more massive primary at a distance $r = 1$. Take one-third of the body and move it to a circular orbit at $r = 2$ and move the remaining two-thirds inwards to conserve angular momentum.

(a) To what radius must the remaining two-thirds of the body be moved to conserve angular momentum?

(b) Calculate the total energy of the final state and compare it with that of the initial state.

(c) Comment on the relevance of your calculation to the spreading of a planetary ring or a protoplanetary disk.

13-9. Assume that a spoke in Saturn's rings forms radially and stretches out from $r = 1.6$ R$_\hbar$ to $r = 1.9$ R$_\hbar$. Calculate the orbital period for particles at the two ends of the spoke. Sketch the spoke after the particles at $r = 1.6$ R$_\hbar$ have completed one-quarter of an orbit.

13-10. Assume that the differential size distribution of particles in a planetary ring is given by a power law over a finite range (eq. 13.15).

(a) Compute the critical values of ζ for which equal amounts of (i) mass and (ii) surface area are presented by particles in each factor of two interval in radius. For larger ζ (steeper distributions), most of the mass or surface area is contained in small particles, but for smaller ζ, most is in the largest bodies in the distribution. (Hint: This problem requires you to integrate over the size distribution.)

(b) Prove that it is impossible to have both $R_{\min} = 0$ and $R_{\max} = \infty$ for any value of ζ.

13-11. Estimate (very crudely) the mass of Jupiter's ring system and associated inner moons. In parts (a) to (c) of this problem, you may assume that ring particles have a density of 1000 kg m^{-3}.

(a) Estimate the mass of the dust in each of the three parts of the ring system, assuming particle radii of 0.5 μm in the halo and 1 μm in the other regions.

(b) Estimate the mass of the macroscopic particles in the main ring, assuming particle radii are

(i) all 5 cm

(ii) all 5 m

(iii) distributed as a power law with $N(R) \propto R^{-3}$ from 5 cm to 5 m

(iv) distributed as a power law with $N(R) \propto R^{-2}$ from 5 cm to 5 m

(v) distributed as a power law with $N(R) \propto R^{-3}$ from 1 cm to 500 m

(vi) distributed as a power law with $N(R) \propto R^{-2}$ from 1 cm to 500 m

(c) Estimate the mass of the moon Metis using the size given in Table E.5.

(d) Compare the uncertainties introduced by the assumed density to those resulting from uncertainties in the particle size distribution.

13-12. The moonlet Pan orbits within the Encke gap in Saturn's rings.

(a) Calculate the location of Pan's 2:1 inner Lindblad resonance using the Keplerian approximation for orbits.

(b) Is this resonance within Saturn's rings? If yes, state which ring.

(c) What type of wave is such a resonance capable of exciting? How many spiral arms would it have?

13-13. Imagine you were to construct a scale model of Saturn's rings 5 km in radius (about the size of San Francisco). How thick would this model be locally? What would be the height of the corrugations corresponding to the highest bending waves? How wide would the A ring be? What would be the width of the F ring?

13-14. Briefly explain how shepherding of planetary rings works and why shepherding does not occur in the asteroid belt.

CHAPTER 14

Extrasolar Planets

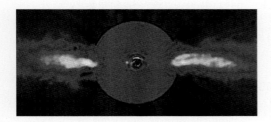

He, who through vast immensity can pierce
See worlds on worlds compose one universe
Observe how system into system runs
What other planets circle other suns
What varied Being peoples every star
May tell why Heaven has made us as we are

Alexander Pope, musing about 'worlds unnumbered', in *An Essay on Man*, 173?

The first 13 chapters of this book cover general aspects of planetary properties and processes and describe specific objects within our Solar System. We now turn our attention to far more distant planets. What are the characteristics of planetary systems around stars other than the Sun? How many planets are typical? What are their masses and compositions? What are the orbital parameters of individual planets, and how are the paths of planets orbiting the same star(s) related to one another? How are stellar properties such as mass, composition and multiplicity related to the properties of their planetary systems? These questions are hard to answer because extrasolar planets, often referred to as **exoplanets**, are far more difficult to observe than are planets within our Solar System.

Before the 1990s, our ability to understand how planets form was constrained because we had observed only one planetary system, our own Solar System. Thousands of extrasolar planets have been discovered within the past 25 years, and even larger numbers are likely to be found in the upcoming decades.

Radial velocity surveys, which measure changes in stellar motion that occur in response to the gravitational pull of a star's planets, first demonstrated that planets orbit many stars other than our Sun. Transit surveys, which detect periodic dips in stellar brightness resulting from partial obscuration of the stellar disk by those planets whose orbits bring them between the star and the observer, have revealed an even larger number of exoplanets within the past decade. These two leading exoplanet discovery techniques are most sensitive to large, short-period planets. Most of the known planets have combinations of size and orbit quite different from those within our own Solar System. More than 99% of the extrasolar planets known as of early 2019 are larger than any Solar System planet with comparable or smaller orbital period, but this may be entirely due to biases in the detection methods.

The sample of known exoplanets is both rapidly growing and strongly biased by detectability. We therefore present a detailed summary of detection techniques in §14.1. The heterogeneous and somewhat arcane nomenclature of stars and their planets is explained in §14.2. Highlights of exoplanet findings to date, which demonstrate the broad range of classes of planets and planetary systems, are given in §14.3. The relationship between the size, mass and composition of an exoplanet is discussed in §14.4. Population studies of exoplanets and exoplanetary systems are now possible, and results are presented in §14.5. We conclude this chapter with an assessment of the status of the field of exoplanets and future prospects.

14.1 Detecting Extrasolar Planets

Our Sun's nearest stellar neighbor is more than 270 000 AU away, almost 10 000 times as distant as Neptune and about 2000 times as far as the most distant spacecraft is from the Sun. Various methods for detecting planets around other stars are being used or studied for possible future use. Because exoplanets appear extremely faint, most methods are indirect in the sense that the planet is detected through its influence on the light from the star that it orbits. These diverse methods are sensitive to different classes of planets and provide us with complementary information about the planets they find, so many of them are already making valuable contributions to our understanding of the properties of planets and planetary systems. A brief review of detection techniques is presented in this section.

We begin with techniques that measure the reflex motion of a star or stellar remnant caused by the gravitational force exerted by orbiting planets (§§14.1.1–14.1.3). In §§14.1.4 and 14.1.5, we consider the patterns of dips in a stellar lightcurve produced by a planet whose orbit is viewed nearly edge-on and periodically blocks some of its star's light. This is followed in §14.1.6 by a description of how the gravity of one star's planets can affect the amount of light reaching us from a more distant star that appears to lie in almost exactly the same direction. Imaging, spectroscopy and other exoplanet detection and characterization techniques are covered in the subsequent subsections.

14.1.1 Timing Pulsars and Pulsating Stars

The first confirmed detection of extrasolar planets was provided by **pulsar timing**. Pulsars, which are magnetized rotating neutron stars (§3.3.2), emit radio waves that appear as periodic pulses to an observer on Earth. The pulse period can be determined especially precisely for old, rapidly rotating millisecond pulsars, whose frequent pulses provide an abundance of data; the most stable pulsars rank among the best clocks known.

Even though pulses are emitted periodically, because the speed of light is finite, the times at which they reach the receiver are not equally spaced if the distance between the pulsar and the telescope varies nonlinearly. Earth's motion around the Sun and Earth's rotation cause such variations, which can be calculated and subtracted from the data. If periodic variations are present in these reduced data, they may indicate the presence of companions orbiting the pulsar.

Pulsar timing measures variations in the distance to the pulsar, relative to a trajectory with constant velocity with respect to the barycenter of our Solar System. It thus reveals only one dimension of the pulsar's motion. The easiest planets to detect via pulsar timing are massive planets whose orbital planes lie close to the line of sight and with orbital periods comparable to or somewhat less than the length of the interval over which timing measurements are available.

Some variable stars pulsate with very regular periods. Such pulsations produce periodic variations in stellar brightness. The time intervals at which these oscillations are observed on Earth vary as a pulsating star moves in response to the gravitational tugs of orbiting planets. Pulsation times can be measured to deduce the presence of planets using the same principles as pulsar timing. However, the precision of **timing stellar pulsations** is not nearly as good as that of pulsar timing, so the minimum detectable planet masses are substantially larger.

14.1.2 Radial Velocity

Radial velocity (**RV**) surveys have been one of the most successful methods for detecting planets around main sequence stars. By fitting the Doppler shift of a large number of features within a star's spectrum, the velocity at which the star is moving towards or away from the observer can be precisely measured. After removing the motion of the observer relative to the barycenter of the Solar System and other known motions, radial motions of the target star resulting from planets that are orbiting the star remain.

The amplitude, K, of the radial velocity variations of a star of mass $M_\star$ that are induced by an orbiting planet of mass M_p is

$$K = \left(\frac{2\pi G}{P_{orb}}\right)^{1/3} \frac{M_p \sin i}{(M_\star + M_p)^{2/3}} \frac{1}{\sqrt{1 - e^2}}$$
$$= \left(\frac{G}{a}\right)^{1/2} \frac{M_p \sin i}{(M_\star + M_p)^{1/2}} \frac{1}{\sqrt{1 - e^2}}, \quad (14.1)$$

where P_{orb} is the orbital period, i is the angle between the normal to the orbital plane and the line of sight, e is the orbit's eccentricity and a is the orbital semimajor axis. In most cases, the star's mass can be estimated to an accuracy of $\sim$3% from its spectral characteristics and astrometrically measured distance.

As in the case of pulsar timing, radial velocity measurements yield the product of the planet's mass and the sine of the angle between the orbital plane and the plane of the sky, as well as the period and the eccentricity of the orbit. The RV technique

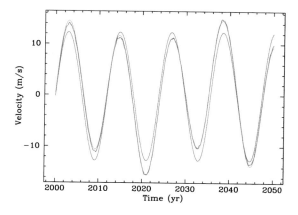

Figure 14.1 COLOR PLATE Velocity variations of the Sun in response to Jupiter (nearly sinusoidal *narrow blue curve*), Jupiter plus Saturn (*faint green curve*) and all eight planets plus Pluto (*thick red curve*). Jupiter's tug dominates the variations, with Saturn having much less influence than Jupiter but still far more than all of the remaining planets combined. The pull of Earth and Venus is evident in the short-period variations seen in the *thick red curve*. (Courtesy Elisa V. Quintana)

is most sensitive to massive planets and to planets in short-period orbits.

Figure 14.1 graphs the velocity variations of our Sun that would be seen by a distant observer fortuitously located near the plane of the planets' orbits. The best instruments are now achieving a precision of better than $\sim$0.5 m s^{-1} (representing a Doppler shift of less than two parts in 10^9) on spectrally stable stars. With this precision, Jupiter-like planets orbiting Sun-like stars are 'easily' detectable, although these detections require a long baseline of observations (comparable to the planet's orbital period). Planets as small as the Earth orbiting very close to stars can also be detected; however, Earth-like planets orbiting at 1 AU are beyond the current capabilities of this technique.

Astronomers classify stars, brown dwarfs and hot self-luminous giant planets according to their spectra. Photospheric temperature is the most important factor in determining both the general shape of spectra (because to first approximation hot bodies emit as blackbodies, see Fig. 4.2) and

the ionization states of elements that produce spectral lines, which also depend on temperature. The hottest stars are classified as **spectral type** O, with spectral types B, A, F, G, K, M, L and T representing successively cooler objects. Bodies of the coolest spectral types, L's and T's, have complex atmospheres including clouds and a variety of molecules, and represent a continuum of objects that stretches almost to Jupiter's temperature.

Precise radial velocity measurements require a large number of narrow spectral lines. Thousands of narrow absorption lines of depth $\gtrsim 1\%$ are present in most main sequence stars similar to the Sun and cooler (spectral types G, K and M). Stellar rotation, granulation from near-surface convection, and intrinsic variability such as starspots (including variations in starspot coverage on decadal timescales analogous to solar cycles; see Fig. 7.5) represent major sources of noise for radial velocity measurements of cool stars; note that these stars lose rotational angular momentum (it is carried away by stellar winds) and generally become less active as they age. The hottest stars (spectral types A, B and O) have few features in their optical spectra. Also, because hot main sequence stars usually rotate rapidly (equatorial velocities of $\sim$ 200 m/s compared to $\sim$ 2 m/s for Sun-like stars), the spectral lines that are present are Doppler broadened (§4.5.2) by a large amount.

14.1.3 Astrometry

Planets may be detected via the wobble that they induce in the motion of their stars projected onto the plane of the sky. This **astrometric** technique is most sensitive to massive planets orbiting about stars that are relatively close to Earth. The amplitude of the wobble, $\Delta\theta$, is given by the formula

$$\Delta\theta \leq \frac{M_\mathrm{p}}{M_\star} \frac{a}{r_\odot},   (14.2)$$

where $r_\odot$ is the distance of the star from our Solar System and a is the semimajor axis of the orbit. If

$r_\odot$ and a are measured in the same units, then the value of $\Delta\theta$ in equation (14.2) is in radians; if $r_\odot$ is measured in parsecs and a in AU, then the units of $\Delta\theta$ are arcseconds. For example, a 1 M♃ planet orbiting 5 AU from a 1 M$_\odot$ star located 10 parsecs away (1 parsec = 3.2616 light-years = 2.063 × 10^5 AU = 3.0857 × 10^{16} m) would produce an astrometric wobble with an amplitude of only 0.5 milliarcseconds (mas) $\approx 1.4 \times 10^{-7}$ degrees. The path of the star on the plane of the sky depends on all of the planet's orbital elements (§2.1.5), but the equality in equation (14.2) holds for orbits that are circular and/or lie in the plane of the sky. Because the star's motion is detectable in two dimensions, the plane of the planet's orbit can be measured, so there is no $\sin i$ ambiguity analogous to that in equation (14.1), and thus a better estimate of the planet's mass can, in principle, be obtained astrometrically than by using radial velocities.

Figure 14.2 graphs the positional variations of our Sun that would be seen by a distant observer fortuitously located near axis normal to the plane of the planets' orbits. Planets on more distant orbits are ultimately easier to detect using astrometry because the amplitude of the star's motion is larger, but finding these planets requires a longer baseline of observations because of their greater orbital periods. Astrometric systems require considerable stability over long times to reduce the noise that can lead to false detections. The best long-term precision demonstrated by single ground-based telescopes using adaptive optics is a bit under 1 mas. The *GAIA* space telescope launched by ESA in December 2013 has already achieved 1.5 orders of magnitude better astrometric precision, and is expected to detect exoplanets when more observations over a longer timespan become available. No astrometric claim of detecting an extrasolar planet has yet been confirmed, but data obtained by the *Hipparcos* satellite in the early 1990s have shown that several candidate brown dwarfs observed in radial velocity surveys are actually low-mass stellar companions whose orbits are viewed almost face-on.

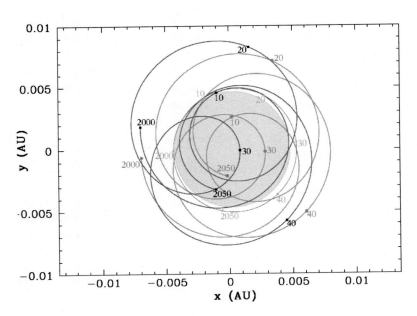

Figure 14.2 COLOR PLATE Motion of the Sun during the first half of the twenty-first century in response to Jupiter (*narrow blue ellipse, faint dates*), Jupiter plus Saturn (*light green curve*) and all eight planets plus Pluto (*thick dark red curve, dark dates*). The solar disk (*shaded yellow*) is shown for comparison. The Sun moves counterclockwise in this perspective, completing slightly less than one trip around the elliptical curve per decade. Jupiter's tug dominates the variations on short timescales, but since $\Delta\theta$ increases with a, Saturn, Uranus and Neptune have more influence on the Sun's position than they do on the Sun's velocity (see Fig. 14.1). The amplitude of the Sun's motion induced by the terrestrial planets is very small. (Courtesy E. V. Quintana)

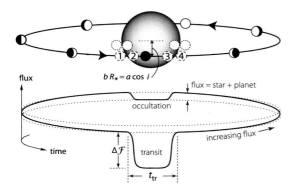

Figure 14.3 Schematic illustration of a transiting planet on a circular orbit. The transit begins at **first contact**, where the planet is at position (1) in the *upper diagram*. The entire disk of the planet blocks light from the star from the time of **second contact** (2) through that of **third contact** (3), and the transit concludes at **fourth contact** (4). (Adapted from Perryman 2018)

14.1.4 Transit Photometry

If Earth lies in or near the orbital plane of an extrasolar planet, that planet passes in front of the disk of (**transits**) its star once each orbit as viewed from Earth. Transiting planets are particularly important because they allow one to readily measure key planetary properties that cannot be found through other means. Figure 14.3 illustrates the geometry of a transiting planet. Precise stellar **photometric time series** (measurements of brightness as a function of time) can reveal such transits, which can be distinguished from rotationally modulated starspots and intrinsic stellar variability by their periodicity, approximately square-well shapes and relative spectral neutrality. Transit observations provide the size and orbital period of the detected planet. Although geometrical considerations allow only a small fraction of planets to be detectable by this technique, thousands of stars can be surveyed within the field of view of one telescope, so surveys using transit photometry can be quite efficient.

For a transit to be observed, the orbit normal must be nearly 90° from the line of sight,

$$\cos i < \frac{R_\star + R_p}{r},$$ (14.3)

where $R_\star$ and R_p are the stellar and planetary radii, respectively, and r is the distance between the two bodies when the planet is nearest the observer. The probability of observing a transit of a randomly oriented planet, $\mathcal{P}_{tr}$, is given by

$$\mathcal{P}_{tr} = \frac{R_\star + R_p}{a(1 - e^2)}.$$ (14.4)

The duration of a planetary transit is given by:

$$\mathcal{T}_{tr} = \frac{R_\star + R_p}{\pi a} \left(1 - b^2 \left(\frac{R_\star}{R_\star + R_p}\right)^2\right)^{1/2}$$
$$\times \frac{1 - e^2}{1 + e \cos \varpi} P_{orb},$$ (14.5)

where b, the **impact parameter**, is defined as the closest approach distance between the center of the planet and the center of the star on the sky plane in units of the stellar radius, and the longitude of periapse, ϖ, is measured relative to the line of sight. **Central transits**, wherein the center of the planet blocks light from the center of the stellar disk, last longer than transits with $b \neq 0$. Excluding transits with a duration less than half that of a central transit, which are more difficult to detect, the probability of a small planet orbiting 1 AU from a 1 $R_\odot$ star transiting across the stellar disk is 0.4%, but the transit probability of a planet at 0.05 AU from the same star is 8%.

Neglecting variations of brightness across the stellar disk, the depth of a transit, i.e., the fractional decrease in the observed flux of the star's light, is given by

$$\frac{\Delta \mathcal{F}}{\mathcal{F}} = \left(\frac{R_p}{R_\star}\right)^2.$$ (14.6)

If the star had a uniform brightness, the transit lightcurve would appear to be flat from the end of ingress (second contact in Fig. 14.3) to the beginning of egress (third contact).

However, the shape of the transit curve is affected by the limb darkening (§4.5.2) of the star, which results from the greater path length observed through the cooler upper portion of the star's photosphere near the stellar limb (edge). Limb darkening causes the edges of the star to appear fainter than the central region. Thus, the planet blocks a smaller fraction of the star's light than the ratio of the areas of the two bodies (eq. 14.6) when it obscures light from near the star's limb. As the planet moves towards the central region of the stellar disk, where the intensity is greatest, it occults an increasingly large fraction of the star's light. This means that the center of the transit is generally deeper than the edges (Figs. 14.13 and 14.16), even though the geometric area occulted by the planet remains the same for most of the event. Starspots can also affect the shape of a transit lightcurve, causing irregular variations on short timescales.

Scintillation in and variability of Earth's atmosphere typically limit photometric precision to roughly one-thousandth of a magnitude (1 **millimagnitude** or **mmag**). This precision allows for the detection of Jupiter-sized planets transiting Sun-like stars by using wide-field surveys conducted with small ground-based telescopes. Smaller planets can be detected around small stars. Better precision can sometimes be obtained by focused observations of a small number of stars from larger telescopes, enabling the observations of transits of Sun-like stars by planets smaller than Neptune. Far greater precision, in some cases as good as ~10 ppm (parts per million), has been achieved above the atmosphere, and planets even smaller than Earth have been detected around Sun-like (as well as smaller) stars (Fig. 14.29).

One major advantage of the transit technique is that many planets detected in this manner are observable via the radial velocity method as well, yielding a mass (as the inclination is known from the transits). Such combined and complementary measurements provide the density of the planet, an especially valuable datum for inferring the composition of the planet and providing a constraint on its formation.

14.1.5 Transit Timing Variations

For a planet traveling on a Keplerian orbit, the time interval between successive transits remains constant. However, planets that are perturbed by other planets (or by a companion star or brown dwarf) do not travel on purely Keplerian orbits. Transits of these perturbed planets are not strictly periodic. The amplitude of such **transit timing variations** (**TTVs**) depends on the mass of the perturber and relative orbits of the two bodies. Thus, if two or more planets within a system are observed to transit, then measuring TTVs can in some cases provide good estimates of planetary masses and orbital eccentricities.

Figure 14.4 shows an example of how the amplitude of TTVs can vary with the ratio of the period of the perturber to that of the planet whose transit times are being measured. Closely spaced orbits can lead to perturbations that are large enough to cause observable TTVs in single encounters. Near resonant orbits, for which the frequency of the perturbations is close to an integer multiple planet's orbital frequency, can produce especially large variations in transit time series ($\omega_f \approx \omega_o$ in equation 2.30). This makes many of the pairs of transiting planets whose orbital periods place them near low-order resonances with one another (§2.3.2) excellent candidates for mass estimation using TTVs.

Observing irregularities in transit times can also betray the existence of non-transiting planets. However, uniquely determining the masses and orbital properties of these unseen companions is much more difficult.

Similarly, planets whose orbits are inclined relative to one another can alter each other's orbital plane. For transiting planets, this effect can be observed via measurement of **transit duration variations** (**TDVs**). As with TTVs, analysis of TDVs can reveal nontransiting planets and constrain the masses of pairs of transiting planets; TDVs also provide information on the relative inclinations of transiting planets.

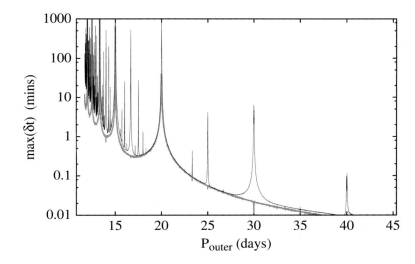

Figure 14.4 Maximum theoretical transit time advancement or delay for an exoplanet on a 10-day circular orbit around a solar-mass star, as a function of the perturber's orbital period. Both planets have mass $M_p = 1\,M_\oplus$. The *thick gray curve* corresponds to a perturber with zero initial eccentricity. The *thin black curve* shows a model with $e_0 \cos \omega = 0.01$. Both curves show the general trend towards larger TTVs when the planets' orbits are closer. Note also the spikes upwards in both curves near first-order resonances and the additional upward spikes at higher-order resonances in the curve for the eccentric perturber. Simulated data covered a baseline of 4000 days. (Courtesy Daniel Jontof-Hutter)

14.1.6 Microlensing

According to Einstein's general theory of relativity, the path of the light from a distant star that passes by a massive object (lens) between the source and the observer is bent. The bending angle is typically very small, and the effect is known as **microlensing**. The light from a source located directly behind the lens is bent such that it appears to come from a circular region known as the **Einstein ring**. The Einstein ring is generally so small that it is unresolvable. However, the lens magnifies the light from the source by a substantial factor when it passes closer to the line of sight than the radius of the Einstein ring, R_E, which is given by

$$R_E = \sqrt{\frac{4GM_L r_{\Delta L}}{c^2}} \left(1 - \frac{r_{\Delta L}}{r_{\Delta S}} \right)^{1/2}, \qquad (14.7)$$

where M_L is the mass of the lens, c is the speed of light and $r_{\Delta L}$ and $r_{\Delta S}$ are the distances from the Earth to the lens and the source, respectively.

Microlensing is used to investigate the distribution of faint stellar and substellar mass bodies within our galaxy. The brightness of the source can increase severalfold for a few months during a microlensing event, and the pattern of brightening can be used to determine (in a probabilistic manner) properties of the lens. If the lensing star has planetary companions, then these less massive bodies can produce characteristic blips on the observed lightcurve provided the line of sight passes within the planet's (much smaller; see equation 14.7) Einstein ring. A microlensing event is displayed schematically in Figure 14.5. Under favorable circumstances, planets as small as Earth can be detected. Microlensing events are only observed when the source and lens are very well aligned, so millions of potential source stars need to be monitored regularly to have a high probability of detecting planets using this technique.

Microlensing provides information on the mass ratio and projected separation of the planet and star. This technique is capable of detecting systems with multiple planets and/or more than one star. The properties of individual microlensing planets (especially orbital eccentricities and inclinations) can often only be estimated in a statistical sense because of the many parameters that influence a microlensing lightcurve. However,

(a)   (b)

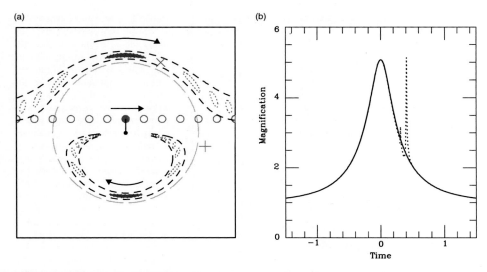

Figure 14.5 Schematic of a microlensing event illustrating the effect of the light from a distant source being bent by a lensing star that possesses a planetary companion. (a) The images (*dotted ovals*) are shown for several different positions of the source (*small circles*) along with the primary lens (black dot) and Einstein ring (*long-dashed circle*). The source is moving from left to right relative to the lens, and the images of its bent light move in a clockwise sense, as indicated by the *arrows*. The *filled ovals* correspond to the images of the source when it is at the position of the *filled circle*. If the primary lens has a planet near the path of one of the images, i.e., within the *short-dashed lines*, then the planet will perturb the light from the source, creating a deviation to the single lens lightcurve. (b) The observed amplification of the amount of light from the source received at the telescope as a function of time is shown for the case of a single stellar-mass lens (*solid line*) and a star with an accompanying planet located at the position of the × (*dotted line*). If the planet were located at the + instead, then there would be no detectable perturbation, and the resulting lightcurve would be essentially identical to the *solid curve*. The units of time are R_E/v, where v is the velocity of the source relative to the lens on the plane of the sky. (Courtesy Scott Gaudi)

additional information about the planets may be deduced under some special circumstances, such as very high magnification microlensing events and microlensing events that are viewed from two well-separated (of order 1 AU apart) locations. Follow-up observations of planets detected via microlensing (using other techniques because a given system's chance of producing a second observable microlensing event is exceedingly small) are very difficult because of the faintness of these distant systems. But when light from the lensing star can be distinguished from that of the source star, the mass of the planet's host star can be determined spectroscopically. Careful monitoring of many microlensing events is providing a very

useful data set on the distribution of planets within our galaxy.

14.1.7 Imaging

Extrasolar planets are very faint objects that are located near much brighter objects (the star or stars that they orbit), making them extremely difficult to image. As Figure 14.6 shows, the amount of starlight reflected from planets with orbits and sizes similar to those in our Solar System is $\lesssim 10^{-9}$ times as large as the stellar brightness, but the contrast is $\sim$3 orders of magnitude more favorable in the thermal infrared. Diffraction of light by telescope optics and atmospheric variability add to the difficulty of **direct detection** of extrasolar planets.

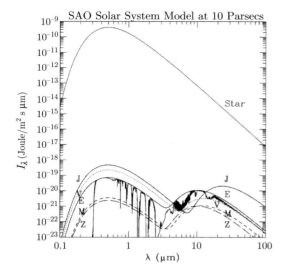

Figure 14.6 Spectral energy distributions of electromagnetic radiation emanating from the Sun, Jupiter, Venus, Earth, Mars and the zodiacal cloud. Each body is approximated by a blackbody of uniform albedo, with an additional curve showing Earth's atmospheric absorption features. This is a log-log plot, with the vertical axis spanning a factor of 10^{14} in intensity. (Adapted from Des Marais et al. 2002)

Nonetheless, several young giant planets orbiting stars in the solar neighborhood have been imaged in the thermal infrared (§14.3.9). Technological advances should eventually allow for imaging and spectroscopic studies of planets resembling those within our Solar System that are in orbit about nearby stars.

The brightness of a planet varies with the phase angle at which it is observed, and the magnitude of these variations depends on the characteristics of the material scattering the light (§§1.4.9 and 4.1.2). Whereas a bare solid surface such as that of Mercury is strongly backscattering, the optically thick clouds of Venus lead to smaller brightness variations with phase angle; Earth and Mars represent intermediate cases. Measurements of phase variations of exoplanets could thus provide constraints on the presence of atmospheres and clouds.

Young stars are often surrounded by dusty disks (§15.2.3, Fig. G.45). Light from such circumstellar disks can make direct imaging of exoplanets more difficult, although images of these disks can indirectly reveal the presence of planets that clear gaps within the disk analogous to those produced by moonlets within Saturn's rings (§13.4.3). Some (perhaps most) older stars are surrounded by low-mass second generation dust disks formed by erosion of small particles from asteroids and comets (Fig. 14.19a, §15.2.3). Such disks are referred to as **exozodiacal clouds** in analogy to the zodiacal cloud of particles near the ecliptic plane of our Solar System (§1.2.3). If exozodiacal clouds are typically significantly brighter than the Sun's zodiacal cloud, their presence will severely complicate efforts to image earthlike planets around other stars.

Many young substellar objects that do not orbit stars have been imaged in the infrared. This class of objects has members less massive than the deuterium burning limit (§3.4.2). These **free-floating planetary mass objects** appear to be more akin to low-mass stars and brown dwarfs than to the planets within our Solar System.

14.1.8 Other Techniques

Several other methods can, in principle, be used to discover extrasolar planets. Some of these have already been used to study exoplanets discovered using one of the methods discussed above.

- Spectra that include light from both the star and the planet with very high signal to noise ratio (SNR) and spectral resolution could be used to identify gases that would be stable in planetary atmospheres but not in stars, and Doppler variations of such signals could yield planetary orbital parameters.
- Planets transiting nearby stars could be detected as dark dots moving across high-resolution images of stellar disks obtained using interferometry.

- Radio emissions similar to those detected from Jupiter (§8.1.4) could reveal the presence of exoplanets.
- Highly circularly polarized radio emission has been observed to come from auroral regions on some young free-floating brown dwarfs with very strong magnetic fields. Some massive young exoplanets may emit similar radiation.
- **Artificial signals** (§16.13.2) from an alien civilization could betray the presence of the planets on which they live (and the aliens might be willing to provide us with substantially more information!).

14.1.9 Planets in Multiple Star Systems

Our Sun is a single star, lacking any bound stellar companions. But the majority of stars of mass $M_\star > 0.5\,M_\odot$ are bound to other stars in **multiple star systems**. Multiple star systems with just two components are known as **binary stars**. Planets can orbit around individual stars within a multiple system in what are known as **circumstellar** or **S-type** orbits, or they can orbit about two stars in a **circumbinary** (**P-type**) orbit.

Stellar multiplicity can affect planetary detectability. Radial velocity measurements are difficult unless the stars are sufficiently distant on the plane of the sky to allow their spectra to be measured separately or the primary star is much brighter than its neighbor(s). Transit detection is slightly more difficult if multiple stars fall into the same aperture because a similar planetary transit obscures a smaller fraction of the light received. Coronagraphic imaging is more complicated because the light from two stars must be blocked out rather than just that of a single stellar source.

Planets orbiting within eclipsing binary star systems can be revealed by **eclipse timing variations** (ETVs), i.e., departures from strict orbital periodicity that are produced by the planets' gravitational perturbations of the stellar orbit. This is analogous to planet detection via TTVs (§14.1.5).

Observing ETVs does not require the precise photometry needed for TTVs, but the lightcurve must be measured at a high temporal cadence.

14.1.10 Exoplanet Characterization

Short of making contact with an alien civilization, detailed studies of exoplanets, especially small ones resembling Earth, will require technological advances. Densities can be computed for planets detected both in transit and either by TTVs that they induce in other transiting planets (§14.1.5) or via radial velocity variations of their host star (§14.1.2). A planet's density is a powerful constraint on its composition. For instance, for a small planet such as Earth made mostly of rock and iron, the rock-to-iron ratio can be constrained because a mostly iron planet would be denser than a mostly rock planet. For a gas giant planet, a density larger than that expected for a cosmic mixture of elements (> 98% H and He) indicates that the planet has an excess of heavy elements, as is found within Jupiter (§§8.1.3, G.6) and Saturn (§8.2.2).

Transmission spectroscopy can reveal key properties of the atmospheres of transiting planets. During the transit, when the planet is passing in front of its parent star, some of the stellar light that one observes has passed through the planet's atmosphere. The tenuous upper reaches of planetary atmospheres transmit continuum radiation from the star yet absorb light in some spectral bands, producing deeper overall transits at these wavelengths. Clouds and hazes tend to absorb radiation over broad ranges in wavelength, producing muted spectra. Comparison of spectra taken during transit with those taken outside transit yields information about the composition and temperature of the planet's atmosphere, and can reveal the presence of high-altitude clouds and hazes.

Most planets that transit in front of their stars also pass behind their stars (Fig. 14.3). Because planets are usually much smaller than their stars, their disks are generally completely obscured in an event known as an **occultation** (but often referred to less precisely as a **secondary eclipse**). The surface brightness of a planet is far less than that of its star, so occultations are much less deep than are transits. The ratio of depths (transits/occultations) is smallest in the infrared, where the planet's brightness is a larger fraction of that of the star (Fig. 14.6).

A spectrum of the planet's thermal emission can be derived by subtracting a spectrum of the starlight seen during the occultation from that of the star + planet seen just prior to or following the occultation. Such a **difference spectrum** can be used to measure the planet's emitting temperature (§4.1) at a range of wavelengths. This information can help to estimate a planet's Bond albedo (§4.1.2).

Observations of the thermal emission from a planet over its entire orbit reveal variations in the planet's temperature with longitude. At the time of transit, the night hemisphere of the planet faces the observer, whereas near the time of occultation, one views the lit hemisphere. If the hottest and most luminous part of the atmosphere is seen a significant amount of time before or after the time of occultation, then the speed of winds in the atmosphere can be estimated.

Space-based photometric telescopes can also detect the sinusoidal phase modulation of light reflected by an inner giant planet as it orbits its star, provided the planet does not induce variations in the appearance of the star's photosphere that track its orbit. Photometric observations throughout the orbit of an exoplanet that both transits and is occulted yield the planet's albedo and phase function.

Starlight that is reflected by a planet is usually linearly polarized. The thermal radiation that a planet emits can become polarized upon scattering by clouds or a planetary ring. Detection and analysis of such polarized signals could help to characterize an exoplanet by, e.g., constraining cloud properties.

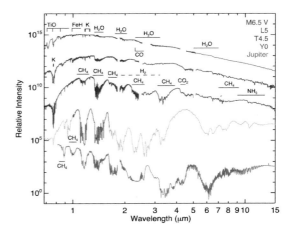

Figure 14.7 The observed spectra of five objects, from hot to cold (top to *bottom*): an M dwarf star, isolated brown dwarfs with three different types of spectra (L, T and Y) and Jupiter. The spectra are arranged with the brighter objects above the less luminous, but the curves have different normalizations, with vertical offsets between the curves chosen for clarity. The spectrum of the faint Y dwarf has only been observed over a short range in wavelength, so a faint gray curve representing the theoretical spectrum is shown at other wavelengths. Comparison of these spectra illustrates the changes in atmospheric chemistry and spectra seen as substellar hydrogen–helium dominated objects (e.g., giant planets) cool over time. Water and refractory diatomic species dominate the M dwarf spectrum. In L and T dwarfs, the water absorption bands get progressively deeper, methane appears and the refractory gases disappear as they condense into solid grains. At still cooler temperatures, ammonia appears, and all water is condensed into clouds at Jupiter. The effective temperatures are about 2750 K in the M dwarf, 1660 K in the L5 brown dwarf, 1300 K in the T4.5 brown dwarf, 420–450 K in the Y0 brown dwarf and 124 K for Jupiter. Jupiter's spectrum shortwards of 3.5 μm is entirely reflected sunlight. (Courtesy Mark Marley and Mike Cushing)

Within the next few decades, we should be able to take (unresolved) images of Earth-like exoplanets in both reflected starlight (using coronography) and the thermal infrared (using interferometry). Figure 14.7 shows spectral identifications of gases in the atmospheres of brown dwarfs. With improved telescopes and instrumentation, astronomers will be able to determine atmospheric and surface properties of 'nearby' exoplanets spectroscopically.

14.2 Exoplanet Nomenclature

Many different systems have been used to name individual stars. For very bright stars, classical names, such as Sirius, the brightest star in the night sky, or constellation name (often abbreviated using just the first three letters) with a Greek letter prefix, e.g., β Pictoris (§14.3.9), are often used. Numbers and constellation names such as 51 Pegasi (§14.3.2) are used for slightly less bright stars. Star catalog abbreviators followed by numbers, for example HD 209458 (§14.3.2), are used for hundreds of thousands of moderately bright or nearby stars. A few stars are named after their discoverer, e.g., Barnard's star (§14.3).

Exoplanets are generally referred to using a convention that is an extension of the system used for multiple star systems. It is standard to designate the primary star within a bound multiple star system with an 'A' following its name, the secondary with a 'B', etc. For instance, the primary star in the stellar system closest to our Sun is α Centauri A, whereas the second brightest star in this system is α Centauri B. Exoplanets are designated analogously, using lower case letters beginning with 'b' and assigned in the order in which the planets are detected; when multiple planets orbiting the same star are announced simultaneously, letters are ordered by increasing orbital period. Thus, 51 Pegasi's planet (§14.3.2) is known as 51 Pegasi b, or 51 Peg b for short.

Most of the stars observed by transit surveys are faint and only have long and obscure catalog designators. Therefore, when a transit survey finds that a particular star hosts one or more planets, this star is given the name of the observing instrument or collaboration followed by a number based on how many planet hosts have already been found or observed by that facility. For example, Kepler-11 is one of the first stars that the *Kepler* spacecraft (§14.3.5) identified to be a planet host, and the outermost of its six known planets is referred to as Kepler-11 g.

The planet-hosting stars found by microlensing surveys tend to be even more distant and faint. These surveys name all stellar lenses that they find. The names given to these stars, some of which are found to host planets, begin with the collaboration name and the year of the observations (see §14.3.8).

Pulsar names begin with 'PSR'. Three planets have been found to orbit the pulsar PSR B1257+12 (§14.3.1).

A few dozen exoplanets have been assigned official 'popular' names by the IAU. However, these popular names are rarely used in the scientific literature.

14.3 Observations of Extrasolar Planets

First one and then two jovian-mass planets in orbit about Barnard's star were announced with great fanfare in the 1960s. Only six light-years away but still impossible to see with the unaided eye because of its faintness, Barnard's star is the Sun's nearest isolated neighbor (only the α Centauri triple-star system is closer). However, the astrometric evidence for Barnard's star's purported planets was discredited in the 1970s. Subsequent claims for the discovery of the first extrasolar planet via astrometry continued to capture newspaper headlines but failed to stand up to further analysis or additional data.

14.3.1 Pulsar Planets

The first extrasolar planets were discovered in the early 1990s by Alexander Wolszczan and Dale Frail. Wolszczan and Frail found periodic variations in the arrival time of pulses from pulsar PSR B1257+12 (which has a 6 ms rotational/pulse period) that remained after the motion of the telescope about the barycenter of the Solar System had been accounted for, and they attributed these

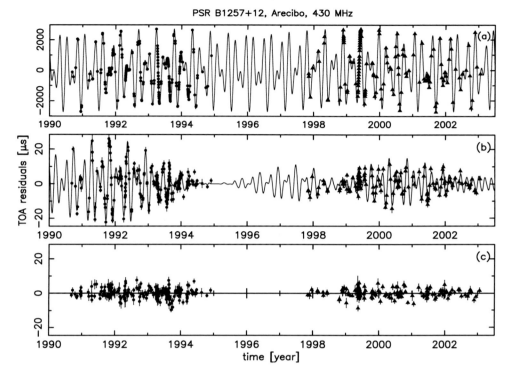

Figure 14.8 The best-fit residuals from modeling the times of arrival (TOAs) of pulses from PSR B1257+12 measured with the Arecibo radiotelescope at 430 MHz. (a) The *points* represent the residuals for a standard pulsar timing model without planets, and the *curve* shows the best-fit three-planet Keplerian model to these data. Note that the vertical scale on this panel is 100 times as large as are those on the two panels below it. (b) The *points* represent the residuals for the best-fit three-planet Keplerian model to these data, and the *curve* shows the changes to the residuals with the addition of gravitational perturbations between the two larger planets. (c) Residuals to the best-fit three-planet model with perturbations included. (Konacki and Wolszczan 2003).

variations to two companions of the pulsar. One companion has an orbital period of 66.54 days, and the product of its mass and orbital tilt to the plane of the sky is $M_p \sin i = 3.4$ M$_\oplus$; the other planet has a period of 98.21 days and $M_p \sin i = 2.8$ M$_\oplus$ (these masses assume that the pulsar is 1.4 times as massive as the Sun). Both planets have orbital eccentricities of $\sim$0.02. Subsequent observations, shown in Figure 14.8, revealed the effects of mutual perturbations of these two bodies on their orbits, thereby confirming the planet hypothesis and implying that both planets have $i \approx 50°$. Additionally, the data imply that there is a lunar mass

object with a period of 25 days orbiting interior to the two near-resonant planets.

The fourth pulsar planet to be detected is an $\sim$2.5 M$_{2\!\!\!+}$ object that travels on a circumbinary (P-type) orbit $\sim$23 AU from the close (191.4 day period) pulsar/white dwarf binary PSR B1620-26. This system lies within the low-metallicity globular cluster Messier 4 (M4).

Four other pulsar planets are known (as of March 2018). Each of them is the planetary-mass remnant of the core of a star that was the secondary of a binary system with the pulsar's massive stellar progenitor. This small star ended up too close to the pulsar, and the pulsar's extremely energetic wind and radiation stripped most of the mass from its companion. Pulsars responsible for destruction of their stellar companions are known as **black widow pulsars**.

Pulsar timing has been demonstrated to be a very sensitive detector of planetary objects, but it only works for planets in orbit about a rare and distinctly non-solar class of stellar remnants. The paucity of known pulsar planets is attributable to the combination of a small number of available search targets and a low occurrence rate of planets orbiting pulsars.

14.3.2 Radial Velocity Detections

In 1995, Michel Mayor and Didier Queloz discovered the first planet known to orbit a main sequence star other than the Sun, an $M_p \sin i = 0.47$ $M_{\jupiter}$, $P_{\text{orb}} = 4.23$ days companion to the star 51 Pegasi. Data from this historic discovery paper are shown in Figure 14.9.

Over the next 22 years, radial velocity surveys identified more than 600 objects with $M_p \sin i < 13$ $M_{\jupiter}$ (the maximum planetary mass, §1.3) in orbit about main sequence stars other than the Sun. This sample is highly biased by detectability factors. Most of these planets have masses exceeding that of Saturn, and the vast majority have orbital periods less than a decade. However, recent years have seen an increasing number of Neptune-mass planets, most with periods of less than a few months, and several planets whose masses are only a few times that of Earth having orbital periods of $\lesssim 1$ month.

More than 70 exoplanets found in radial velocity surveys, as well as > 1000 planets first detected via transit photometry (§§14.3.3 and 14.3.5), have periods less than one week. We refer to such short-period planets as **vulcans**, after the hypothetical planet once believed to travel about the Sun interior to the orbit of Mercury. Jupiter-sized vulcan planets such as 51 Peg b are usually referred to

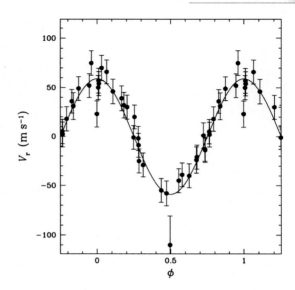

Figure 14.9 Radial velocity measurements of the star 51 Pegasi (*points with error bars*) as a function of phase of the orbital fit (*solid line*). These data include observations made over many orbits that have been folded at the observed orbital period. One and a half cycles are shown for clarity. (Adapted from Mayor and Queloz 1995; courtesy Didier Queloz)

as **hot jupiters,** a term that may well overemphasize the similarities of these planets with our Solar System's largest planet. The orbits of most of the vulcans are nearly circular, as expected because eccentric orbits this close to a star should be damped relatively rapidly by tidal forces (§2.7). A few vulcans have orbits with substantial eccentricity, probably caused by perturbations from a third body (another planet, a brown dwarf or a stellar companion, in some cases not yet observed) within the system.

Planets tend to be clustered into systems in the sense that there are more multiple planet systems (**multis**) than would be the case were planets randomly distributed among stars. For example, the star υ Andromedae has three jovian-mass planets, one of which is a hot jupiter. The other two planets are far more distant from υ And and travel on eccentric orbits. The radial velocities of the star after the effects of the perturbations by the inner

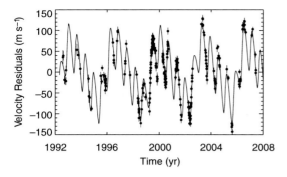

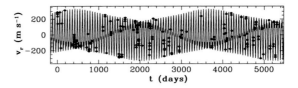

Figure 14.10 The variations in the radial velocity of the star υ Andromedae after subtracting off the star's motion caused by its hot jupiter (υ And b, which has an orbital period of only 4.6 days) are shown as a function of time. Uncertainties of individual measurements are indicated. The *solid curve* represents the model response of the star υ Andromedae to two additional giant planets with much longer orbital periods, υ And c and υ And d. (Courtesy Debra Fischer)

Figure 14.11 The variations in the radial velocity of the faint M dwarf star Gliese 876 observed at the Keck I telescope from 1997–2011 are plotted as a function of time. The *curve* passing through most of the points was calculated by varying the parameters of four mutually interacting planets in orbit about the star to best fit the data. The timescale is days since the first observation, which took place on 2 June 1997. (Courtesy Eugenio J. Rivera)

planet have been removed are shown in Figure 14.10. Dynamical calculations imply that at least the outermost two planets must have $\sin i > 1/5$ for the system to remain stable for υ And's 2.5×10^9 years age, thereby providing upper bounds to the planets' masses.

Some exoplanets are dynamically isolated from other known planets orbiting their star, as are most hot jupiters, including the inner planet of υ Andromedae. However, other giant planets have more similar periods, and some pairs of giant planets are in low-order mean motion resonances (§2.3.2) with one another.

The two giant planets orbiting the star Gliese 876 are locked in a 2:1 orbital mean motion resonance and have particularly strong interactions because of their large masses relative to that of their one-third $M_\odot$ star. Figure 14.11 shows a fit to the radial velocity data that accounts for these perturbations as well as the pull of the two known smaller planets in the system, one of which is locked in resonances with the giants. The amplitude of the mutual perturbations depends on the planets' actual masses, rather than the combination

$M_p \sin i$ that is derivable from radial velocity observations of the star's direct response to a planet's gravity (eq. 14.1). Therefore, measurement of the changes in planetary orbits produced by these perturbations (through the reflex motion of the star seen in RV data) breaks the mass/inclination ambiguity that is present for most planets detected exclusively by RV and allows for the determination of the inclinations of the planets' orbits as well as their masses.

Six stars, including Gliese 876 (often abbreviated as GJ 876), possess four or more planets that were discovered using the radial velocity method. Figure 14.12 shows the masses and orbits of the planets in these systems. Most of the planets in RV multis are more massive than those typical of high multiplicity systems found via transit photometry, and many of these planets orbit farther from their host star (Fig. 14.32). These differences are caused by the different biases of the two detection techniques.

14.3.3 Transiting Planets

Hundreds of vulcan planets, as well as dozens of more distant planets, have been observed both by radial velocity, giving $M_p \sin i$, and in transit, yielding i and R_p. The densities of these planets can be calculated from these data, and educated

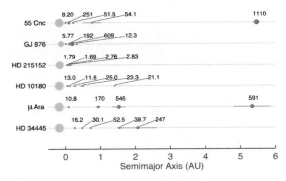

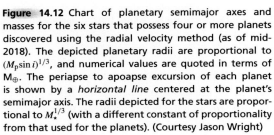

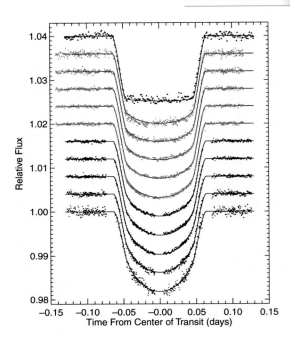

Figure **14.12** Chart of planetary semimajor axes and masses for the six stars that possess four or more planets discovered using the radial velocity method (as of mid-2018). The depicted planetary radii are proportional to $(M_p \sin i)^{1/3}$, and numerical values are quoted in terms of $M_\oplus$. The periapse to apoapse excursion of each planet is shown by a *horizontal line* centered at the planet's semimajor axis. The radii depicted for the stars are proportional to $M_\star^{1/3}$ (with a different constant of proportionality from that used for the planets). (Courtesy Jason Wright)

Figure **14.13** Data obtained from spaceborne observations of a total of five transits of the hot jupiter HD 209458 b. Lightcurves taken in 11 different bandpasses are shown, vertically offset from one another by just under 3% for clarity. The *uppermost lightcurve* (*black triangles*) was observed using the *Spitzer Space Telescope* on 23 December 2007 at an average wavelength of 8 μm. Moving downwards, the next five curves are superposed composites of two transits observed using *HST* on 31 May 2003 (*gray 'x' symbols*) and 5 July 2003 (*gray circles*) at average wavelengths of (from *top* to *bottom*) 971, 873, 775, 677 and 581 nm, respectively. The *lowest five curves* are composites of two transits observed using *HST* on 3 May 2003 (*black inverted triangles*) and 25 June 2003 (*black '+' symbols*) at average wavelengths of (from *top* to *bottom*) 540, 485, 430, 375 and 320 nm, respectively. The shape of each transit curve is determined by the limb darkening of the star, which is more pronounced at shorter wavelengths (see text). (Courtesy Heather Knutson)

guesses can be made about their compositions (see §14.4).

The hot jupiter HD 209458 b, which was discovered using the radial velocity method, was the first exoplanet to be observed in transit. This planet's orbital period is 3.525 days, its mass is 0.71 $M_{2\!\!\!+}$ and its radius is 1.38 ± 0.02 $R_{2\!\!\!+}$, which was larger than expected. We discuss the anomalously large sizes observed for many hot jupiters in §14.4.

After transits were detected by ground-based telescopes, transit lightcurves of HD 209458 b were obtained from space at many wavelengths, as illustrated in Figure 14.13. The shape of the transit lightcurves of HD 209458 b is affected by the limb darkening of the star (§14.1.4). At the center of the star, we see deeper into the stellar atmosphere, where the temperatures are high. At the edge of the star, where we are looking into the stellar atmosphere along a slant path, the elevation at which the optical depth equals one is high in the stellar atmosphere, where the temperatures are cooler. The difference in brightness between the limb of the star and its center can

be approximated as the difference between two Planck functions (eq. 4.3). At short wavelengths, a small change in temperature produces a large change in brightness, and the edges of the star appear to be significantly fainter than the center of the star. At these wavelengths, limb darkening causes the transit to have a smoothly curved shape. Moving towards longer wavelengths (upwards in

Fig. 14.13), observations shift onto the Rayleigh–Jeans tail (eq. 4.4), where the difference between the two blackbodies becomes smaller. As the amount of limb darkening diminishes, the transit has an increasingly angular, boxlike shape.

By integrating over the entire transit, the effective radius of a planet in a given bandpass can be measured. For the most favorable targets – generally planets with large atmospheric scale heights, radius ratio $R_p/R_\star$ not too small and orbiting bright stars – effective radii can be derived from light integrated over fairly narrow ranges in wavelength. In such cases, a transit spectrum can be produced by observing how the apparent radius of a planet varies with wavelength. The variation in apparent radius depends upon the abundances of the gaseous species and the sizes and distribution of atmospheric clouds and hazes.

Transit spectra have been obtained for about two dozen transiting exoplanets as of mid-2018. Figure 14.14 shows several examples. These spectra have revealed the presence of small particle hazes, opaque cloud decks and gases. Sodium, potassium and H_2O, all of which have strong absorption features at near-infrared and optical wavelengths, have been detected, as have a few other molecules. The spectra shown in Figure 14.14 hint at the diversity of exoplanet atmospheres owing to variations in atmospheric abundances, clouds and photochemical hazes.

Figure 14.15 shows the lightcurve of the combined luminosity of the star HD 189733 and its transiting hot jupiter for more than half an orbital period of the planet, including both the transit and the occultation. These measurements were taken in the thermal infrared, i.e., at a wavelength where the luminosity of the planet is primarily its own thermal radiation rather than reflected starlight. These data provide information on the temperature of the planet as a function of longitude relative to the substellar point. Figure 14.16 shows measurements near and during occultation at several infrared wavelengths. Such measurements can be used to

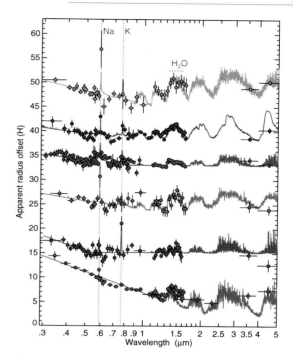

Figure 14.14 A selection of transit spectra of low-mass hot jupiters. The points show the observations, whereas the curves represent best fit theoretical spectra obtained by varying the assumed abundances of molecules and properties of the cloud/haze layers. The planets are arranged by cloud cover, with the cloudiest/haziest planets at the bottom; vertical offsets between the curves are chosen for clarity. The spectrum of HD 209458 b is shown in the red curve, third from the top. Many of the planets show larger radii at bluer wavelengths, likely a result of scattering by small, high altitude hazes. The amplitude of the variations arising from various gaseous species, such as Na, K and H_2O labeled here, differs between planets due to differences in atmospheric abundances and the distributions of clouds, which tend to flatten transit spectra, in planets with different atmospheric temperatures. Most of the data were obtained by the Hubble and Spitzer Space Telescopes. (Courtesy David Sing)

deduce (broadband) infrared emission spectra of HD 189733 b.

The infrared luminosities of several dozen hot jupiters have been deduced by differencing measurements taken near and during occultation. The observed luminosities of these planets are close to the values expected for dark (low-albedo) planets

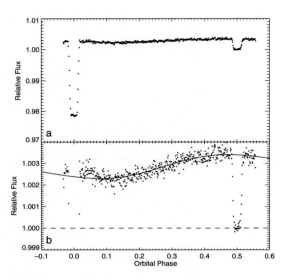

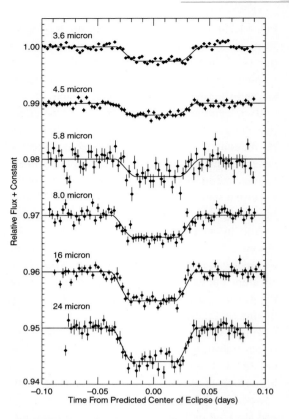

Figure 14.15 Photometric observation of the combined 8 μm radiation from the star HD 189733 and its transiting hot jupiter. The orbital phase is measured relative to the midpoint of the transit. The observed flux is normalized to that of the star alone, with the range in panel (a) being large enough to show the full depth of the transit and panel (b) showing a magnified view that emphasizes the smaller variations from the occultation of the planet (centered at orbital phase 0.5) and the increase in radiation emitted from the planet as the hemisphere facing the star comes into view. Note that the transit lightcurve is nearly flat bottomed because the star is barely limb darkened at 8 μm. (Knutson et al. 2007)

Figure 14.16 Photometric observation of the planet HD 189733 b before, during and after the epoch when it passes behind its star at various wavelengths in the infrared. The curves are vertically offset for clarity. The occultation depth is greater at longer wavelengths, where the planet's radiation is a greater fraction of the output of the star (see Fig. 14.6). Note that the occultation lightcurves are all nearly flat bottomed. (Charbonneau et al. 2008)

in equilibrium with the energy being absorbed from their stars and imply temperatures in the 1000–3000 K range.

The dynamics of atmospheres of planets that orbit very close to stars also differ from those of the planets within our Solar System. To illustrate the very different dynamical regimes present in the atmospheres of hot and cold giant planets, it is useful to introduce the concept of the **radiative time constant**. This is the timescale over which a given pressure level in the atmosphere can cool substantially. It depends on both the local temperature and the overlying optical depth. Radiative timescales in the Solar System's giant planets are significant fractions of their orbital timescales, and these planets complete $>10^4$ rotations per orbit, so

the atmospheres of these planets show no diurnal variations and muted seasonal changes.

In giant planets orbiting only ~0.05 AU from their stars, radiative timescales can be much shorter, on the order of hours, because thermal energy content is proportional to T whereas radiation is emitted in proportion to T^4 (see equations 4.1 and 4.11 and Problem 14-13). Moreover, these hot jupiters rotate synchronously, so they don't have day–night cycles. Energy is well redistributed across longitudes in some transiting hot jupiters, but other hot jupiters have a large day–night contrast. In these latter cases, the atmosphere cools so rapidly that winds cannot redistribute much energy to the night side of the synchronously rotating planet, so atmospheric temperatures vary substantially with longitude (and latitude). Hotter planets typically have higher day–night temperature contrasts than cooler ones. Winds and/or rotation move the hottest point on a planet away from the substellar point of many hot jupiters, and more for cooler planets than for the very hottest ones.

Direct measurements of visible light scattered off the atmosphere have been achieved for ~ 20 planets (as of 2018). Most of these planets are very dark, $\mathcal{A}_b \lesssim 0.05$, as expected for cloudless planets at red wavelengths. However, a few of them appear to be less dark, $\mathcal{A}_b \sim 0.2$, perhaps indicating scattering off cloud layers composed of silicates, which have the right condensation temperature to form clouds in the upper atmosphere of such hot planets.

Variations in the depth of the transit of HD 209458 b with wavelength reveal the presence of sodium in the upper atmosphere of this hot jupiter. The very large depth of the transit at the wavelength of the Lyman α line (Fig. 4.6) shows that hydrogen associated with the planet extends over an area larger than the size of the planet's Hill sphere (eq. 2.28); this implies that hydrogen is escaping the planet at a considerable rate, albeit not so rapidly as to have removed a substantial fraction of the planet's mass over its lifetime to date. Comparison of infrared spectra of this system during and surrounding the occultation (secondary eclipse) suggests the presence of silicates in the planet's atmosphere. Near-infrared transmission spectra of HD 189733 b reveal the presence of methane in this planet's atmosphere. In contrast, spectra taken of the much smaller and significantly cooler planet GJ 1214 b during transit are featureless, implying GJ 1214 b planet either has a high-altitude optically thick haze layer or a high mean molecular weight (thus low-scale height) upper atmosphere (eq. 3.16).

14.3.4 Rossiter–McLaughlin Effect

As a star rotates, gas in half of the stellar disk moves towards us, and gas in the other half moves away. Thus, planetary transits can affect the apparent radial velocity of the star by blocking light from either the rotationally blueshifted or redshifted half of the stellar disk. This is known as the **Rossiter–McLaughlin effect**. Planets orbiting in the prograde direction initially block a blueshifted portion of the stellar disk, leading to an apparent redshift, and then the opposite occurs; an example of this type of radial velocity curve is shown in the top panel of Figure 14.17. One component of the inclination of a transiting planet's orbit relative to the star's equator can be determined by measuring the planet's apparent radial velocity during transit.

The Rossiter–McLaughlin effect has been detected for more than one hundred planets, almost all of which are hot jupiters, as of 2018. The observed values provide clues to the origin of the population of hot jupiters. A clear majority of these planets have been found to orbit in the prograde direction near their star's equatorial plane, as in our Solar System. But some hot jupiters, such as those represented in the lower two panels of Figure 14.17, travel on substantially inclined (in some cases even retrograde) paths. Most of the high inclination planets orbit hot stars that have outer radiative zones (rather than the convective zones present in the corresponding regions of cooler stars such as the Sun). In contrast, most hot jupiters orbiting cooler stars have low orbital inclinations with respect to their star's equatorial plane. This difference between hotter and cooler stars has been interpreted as a consequence of the different rotational histories of the stars: cool stars have stronger magnetic braking and stronger tidal dissipation rates.

Most theories suggest that hot jupiters formed much farther from their stars than they are at present and subsequently migrated to their current

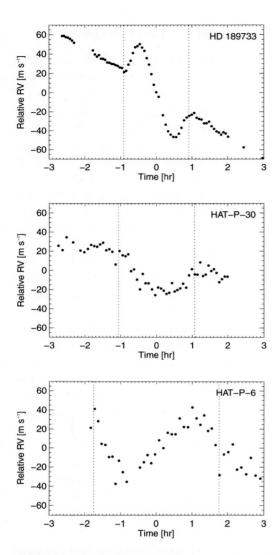

Figure 14.17 The apparent radial velocity variations of the stars HD 189733 (*top panel*, data from Winn et al. 2006), HAT-P-30 (*middle panel*, data from Johnson et al. 2011), HAT-P-6 (*bottom panel*, data from Albrecht et al. 2012), during and immediately surrounding a transit of each star's hot jupiter. In the *top panel*, the velocity anomaly, first high and then low, and symmetric around the time of mid-transit, shows that the stellar spin angular momentum and the orbital angular momentum (both projected onto the plane of the sky) are nearly aligned, with a displacement of only $1.4 \pm 1.1°$. The highly asymmetric anomaly in the *middle panel* implies a nearly perpendicular orbit with displacement of $73.5 \pm 0.9°$, and the reverse pattern in the *bottom panel* implies a nearly retrograde orbit ($165 \pm 6°$). Typical uncertainties in individual data points are $\sim 2\ \mathrm{km\,s^{-1}}$ in the *top panel* and $\sim 10\ \mathrm{km\,s^{-1}}$ in the others. (Courtesy Simon Albrecht and Josh Winn)

orbits. Orderly disk-induced migration (§15.7.1) would leave inclinations small, provided the star's equator was near the plane of the protoplanetary disk. In contrast, Kozai resonances (§2.3.3) and planet–planet scattering, can excite large inclinations. The overall inclination distribution suggests that at least one of the mechanisms that drive hot jupiters close to their stars randomizes inclinations. Thus, the Rossiter–McLaughlin data argue in favor of Kozai resonances and/or planet–planet scattering playing a major role in delivering at least some hot jupiters to their current orbits (§15.11).

14.3.5 NASA's Kepler Mission

The most successful instrument at finding exoplanets has been the ***Kepler*** spacecraft, which was launched by NASA in 2009. *Kepler*'s sole scientific instrument is a differential photometer with a wide field-of-view (105 square degrees) that has continuously and simultaneously monitored the brightness of approximately 130 000 main-sequence stars and 40 000 slightly more evolved stars, most with a duty cycle (fraction of the time that useful data were being collected) >85%, for four years. The *Kepler* mission was designed to detect Earth-sized planets in the habitable zones (§16.5) of Sun-like stars, necessitating a large sample size and long-duration, high duty cycle, very high precision observations.

Its sensitivity to small planets over a wide range of orbital periods gave *Kepler* the capability of discovering multiple planet systems. For closely packed planetary systems, nearly coplanar systems or systems with a fortuitous geometric alignment, *Kepler* was able to detect transits of more than one planet. For systems with widely spaced planets or large relative inclinations, not all planets transit, but some non-transiting planets were still detectable based on TTVs that their gravitational perturbations produced on one or more transiting planets (§14.1.5).

Kepler's primary goal was to conduct a statistical census of transiting exoplanets. We present initial results of this census in §14.5.2. Additionally, *Kepler* discovered many interesting planets and planetary systems. Some highlights are presented below.

Notable Planets & Planetary Systems

Kepler's first major exoplanet discovery was the Kepler-9 planetary system, which includes two transiting giant planets with orbital periods of 19.2 and 38.9 days. The nearby 2:1 mean motion resonance induces TTVs of several hours. Analysis of these TTVs enabled the planets to be confirmed and provided estimates of their masses. Each of these planets is slightly smaller than and less than half as massive as Saturn. The star also has a small transiting planet of unknown mass with an orbital period of 1.6 days. Kepler-18 is analogous to Kepler-9, with two Neptune-mass planets near the 2:1 orbital resonance and a smaller inner planet.

The first rocky planet found by *Kepler* was Kepler-10 b, which has $M_p \approx 4\,M_\oplus$, $R_p = 1.47\,R_\oplus$ and an orbital period of only 20 hours. This planet's high density results from compression caused by high internal pressure, and its bulk properties are consistent with an Earth-like composition.

Kepler-11 is a Sun-like star with six transiting planets that range in size from $\sim$1.8–4.2 $R_\oplus$. The known orbital and physical properties of these planets are listed in Table 14.1, which also gives the flux of stellar radiation intercepted by each of the planets in units of the solar constant (the mean solar flux intercepted by Earth; see equation 4.13), from which the planets' temperatures may be estimated (Problem 14-15). Orbital periods of the inner five of these planets are between 10 and 47 days, implying a very close-packed dynamical system; the outer planet, Kepler-11 g, has a period of 118.4 days. Transit timing variations have been used to estimate the masses of the inner five planets and provide an upper bound on Kepler-11 g's mass. None of these planets is rocky; most if not all have a substantial fraction of their volume occupied by the light gases H_2 and He.

Kepler-20 e was the first planet smaller than Earth to be confirmed around a main sequence star other than the Sun; its 6-day orbit means that it is far too hot to be habitable. Kepler-37 b, only slightly larger than Earth's Moon, was the first planet smaller than Mercury to be found orbiting a normal star; its orbital period is 13 days, and the stellar host is 80% as massive as the Sun.

Kepler-36 hosts two planets whose semimajor axes differ by little more than 10% but whose compositions are dramatically different. The inner, rocky, Kepler-36 b has a radius of 1.5 $R_\oplus$ and mass of $\approx 4.4\,M_\oplus$. In contrast, puffy Kepler-36 c has a radius of 3.7 $R_\oplus$ and mass of $\approx 8\,M_\oplus$, implying that most of its volume is filled with H/He.

Kepler-78 b has an extremely short period, orbiting its star in just 8.5 hours. This vulcan planet is slightly larger than Earth, and its mass, measured from the radial velocity variations it induces in its nearby host star, implies a rocky composition.

Kepler-88 b is a Neptune-sized planet with an orbital period of just under 11 days that exhibits very large TTVs (12 hour amplitude) and TDVs (5 minute amplitude). These TTVs and TDVs have been used to deduce the presence of its nontransiting jovian-mass companion, Kepler-88 c, which has an orbital period of just over 22 days. Kepler-88 b's TTVs and TDVs are so large because the ratio of the periods of Kepler-88's two known planets is 2.03, placing them just outside the 2:1 mean motion resonance (see §2.3 and Fig. 14.4), and Kepler-88 c is much more massive than other *Kepler* planets that orbit so close to a first-order mean motion resonance.

Kepler-90 is a star that is slightly larger and more massive than the Sun and hosts eight transiting planets. The three inner planets, whose periods range between 7 and 15 days, all have radii $\sim$1.5 $R_\oplus$. The next three planets, with periods from 60 to 125 days, have radii roughly twice as large as their inner siblings. The outer two planets are both larger than 7 $R_\oplus$ and have orbital periods

Table 14.1 **Kepler-11's Planetary System**

	Period (d)	a (AU)	e	i (°)	R_p ($R_\oplus$)	M_p ($M_\oplus$)	ρ (kg/m^3)	$\mathcal{F}$ ($\mathcal{F}_\odot$)
b	10.304	0.091	0.04	89.6	1.8	1.9	1720	125
c	13.024	0.107	0.03	89.6	2.9	2.9	660	92
d	22.685	0.155	0.004	89.7	3.1	7.3	1280	44
e	32.000	0.195	0.01	88.9	4.2	8.0	580	27
f	46.689	0.250	0.01	89.5	2.5	2.0	690	17
g	118.381	0.466	<0.15	89.9	3.3	<25	–	4.8

Source: Lissauer et al. 2013.

exceeding 210 days. None of the planets appear to be locked in resonance with one another.

Kepler-62 f is the first known exoplanet whose size (1.4 $R_\oplus$) and orbital position (in the middle of its star's habitable zone, §16.5) suggests that it could well be a rocky world with stable liquid water at its surface. Kepler-186 f is a habitable zone planet just a bit larger than Earth, so it is more likely to be rocky but because it orbits a smaller, cooler, more variable star, it might not be as good a candidate for habitability as Kepler-62 f. No mass measurements have been made for either of these potentially habitable exoplanets.

Kepler-1520 b produces an asymmetric transit lightcurve when it passes between its star and the Solar System every 15.7 hours. This lightcurve has been interpreted as a tail of dust being released from a **disintegrating planet**. This disintegrating planet, which can also be thought of as a giant exocomet, is substantially smaller than the Earth.

Kepler has also found one dozen circumbinary transiting planets. Some of these planets are discussed in §14.3.10.

Orbital Inclination to Stellar Equator

The tilt of the rotation axis of a star relative to the plane of the sky can be estimated by using a combination of stellar mass and radius (from spectroscopy), rotation period (from spot modulation of the star's brightness in the time series observed by *Kepler*) and rotational broadening of spectral lines (maximized for rotation axis in the plane of the sky, essentially zero for a star observed pole-on). This technique, as well as estimations using the patterns of starspot crossings during transits and the Rossiter–McLaughlin effect (§14.3.4), has been used to measure tilts of the orbits of some planets discovered by *Kepler* relative to their star's equator. Most of these planets are Neptune-size and smaller and have orbital periods of weeks or months, so as tidal torques are inversely proportional to the sixth power of orbital distance (eq. 2.58) minimal tidal damping is expected. Nonetheless, most single planet and multi-planet *Kepler* systems orbit near the plane of their star's equator, although high inclinations have been observed for a few singles and at least one multiple-planet system (Kepler-56).

The *K2* Mission

The *Kepler* spacecraft observed the same target stars for 4 years during its primary mission. The field of view targeted by *Kepler* was well above the plane of the spacecraft's orbit around the Sun (which is nearly identical to Earth's orbital plane, the ecliptic), so that observations could be made year-round. But *Kepler* required at least three of its four gyroscopic reaction wheels to point stably, and when two wheels failed, it could not continue its original mission. However, two reaction wheels were sufficient to point the telescope within the plane of *Kepler*'s orbit (because the net torque on the spacecraft from solar radiation pressure (§2.8.1) is small for this orientation), so the spacecraft was repurposed to make ~80 day long observations of fields near the ecliptic plane, and its new mission was named *K2*.

With much shorter observations of individual targets, fewer targets observed at the same time and somewhat poorer photometric precision because only two reaction wheels were available to stabilize pointing, *K2* could not match *Kepler* either in terms of finding small and/or long-period planets or assessing planetary abundances. However, over the course of more than four years, *K2* observed an order of magnitude more area on the sky, and therefore was able to search for short-period planets around more of the best targets for follow-up observations by other telescopes, bright Sun-like stars and moderately bright small stars.

Hundreds of planets have been found using *K2* data. These include small neighbors orbiting both interior (with $P = 0.79$ days) and exterior (with $P = 9.03$ days) to the previously known hot jupiter, WASP-47 b (which has an orbital period $P = 4.16$ days), making this planet an exception to the 'rule' that hot jupiters tend to lack close planetary neighbors.

The *K2* mission also detected several small asteroids transiting the white dwarf WD 1145+017 with periods close to 4.5 hours. Heavy elements should rapidly sink below the viewable photosphere of white dwarfs, but these elements have been observed in spectra of WD 1145+017 and of more than 25% of other white dwarfs. These **polluted white dwarfs** must have recently accreted refractory material such as tidally disrupted asteroidal fragments, presumably remnants of the progenitor star's planetary systems.

14.3.6 Small Nearby Exoplanets

Several temperate, roughly Earth-sized, planets have been discovered around very small stars that are located $\lesssim 40$ light years from Earth. The stellar characteristics of the two nearest temperate

planet hosts as well as those of the star hosting the largest number of temperate planets are listed in Table 14.2. Table 14.3 gives properties of the planets orbiting these three stars. Two of these systems, a single planet orbiting the Solar System's nearest stellar neighbor and the system containing seven transiting planets, are described in more detail below.

Proxima Centauri b

A planet has been discovered orbiting the closest star to our Solar System, Proxima Centauri. Proxima Cen, which is also known as α Centauri C, is a small, faint M dwarf star. Most of the radiation emitted from this cool star is in the near infrared and red regions of the spectrum, but Proxima also emits powerful high-energy flares several times per year. Proxima Cen travels on a wide orbit ($a \approx 9000$ AU, $e \approx 0.5$) about a pair of much more closely spaced Sun-like stars and is located 4.224 light years = 267 000 AU from the Sun.

The planet, Proxima Cen b, was discovered using the radial velocity method; it has a minimum mass $M_p \sin i = 1.3$ M$_\oplus$. Its orbital period is 11.2 days, implying a semimajor axis of ~ 0.05 AU. While little is known about this small world, the very presence of a planet about our Solar System's nearest stellar neighbor provides additional evidence for a high abundance of planets within our galaxy. A dust disk located ~1.3−4 AU from the star has recently been detected.

TRAPPIST-1's Planetary System

Seven planets have been observed to transit TRAPPIST-1, which is an ultra-cool dwarf star at the very lower end of the stellar main sequence (Fig. 3.5). TRAPPIST-1 is just under 40 light years from the Sun, making it a close stellar neighbor, albeit almost ten times as far away as Proxima Cen.

TRAPPIST-1's seven known planets are all similar in size to the Earth, with the planetary radii ranging from $\sim 0.7 - 1.3$ R$_\oplus$. Their orbital periods are between 1.5 and 19 days, and all neighboring

Table 14.2 Small Nearby Planet-Hosting Stars

Name	$R_\star$ ($R_\odot$)	$M_\star$ ($M_\odot$)	$L_\star$ ($L_\odot$)	T_{eff} (K)	Distance (parsecs)
Proxima Cen	0.14	0.15	0.0015	3050	1.295
Ross 128	0.20	0.17	0.0036	3190	3.38
TRAPPIST-1	0.12	0.09	0.00052	2520	12.1

Source: Anglada-Escudé et al. (2016), Bonfils et al. (2018), Van Grootel et al. (2018).

Table 14.3 Small Planets Orbiting Small Nearby Stars

	Period (d)	a (AU)	e	i (°)	R_p ($R_\oplus$)	M_p ($M_\oplus$)	ρ (kg/m^3)	$\mathcal{F}$ ($\mathcal{F}_\odot$)
Proxima Cen b	11.2	0.05	< 0.35	–	–	$\gtrsim 1.3$	–	~0.65
Ross 128 b	9.87	0.05	$\lesssim 0.2$	–	–	$\gtrsim 1.4$	–	~1.4
TRAPPIST-1 b	1.5109	0.0115	0.006	89.6	1.12	1.02	4000	4.3
TRAPPIST-1 c	2.4218	0.0158	0.007	89.7	1.10	1.16	4900	2.5
TRAPPIST-1 d	4.0498	0.0223	0.008	89.9	0.78	0.30	3400	1.1
TRAPPIST-1 e	6.0996	0.0293	0.005	88.7	0.91	0.77	5600	0.66
TRAPPIST-1 f	9.206	0.0385	0.010	89.7	1.05	0.93	4500	0.38
TRAPPIST-1 g	12.353	0.0469	0.002	89.7	1.15	1.15	4200	0.26
TRAPPIST-1 h	18.766	0.0619	0.006	89.8	0.77	0.33	4000	0.13

Source: Anglada-Escudé et al. (2016), Bonfils et al. (2018), Delrez et al. (2018), Grimm et al. (2018).

planet pairs appear to be locked in low-order mean-motion resonances (§2.3.2). These resonances stabilize the tightly-packed TRAPPIST-1 planetary system, allowing it to survive, and they also lead to substantial TTVs, which can be used to estimate the masses of the planets. Density constraints are weak at present, but more transits are being observed and good estimates of the planets' masses and densities should be available soon.

The middle planets in the TRAPPIST-1 system lie within the star's habitable zone. Because their star is so small and nearby, these planets offer by far the best known observational opportunity to study the atmospheres of Earth-size exoplanets orbiting within their star's HZ using transit spectroscopy. However, although the planets are likely to be temperate, they orbit very close to a small, active star, and thus they are subjected to far more extreme ultraviolet and X-rays than is the Earth, especially when TRAPPIST-1 flares.

14.3.7 Planets Orbiting Pulsating Stars

The first planet to have been observed around a pulsating star is V391 Pegasi b. This planet has $M_p \sin i \approx 3.2$ M$_{2\!+}$, $a \sim 1.7$ AU and small eccentricity. The stellar host is a post-red-giant helium-burning star with a current mass of ~0.5 M$_\odot$; models suggest that the star had a mass of ~0.85 M$_\odot$ when it was on the main sequence, implying that the planet used to have a significantly smaller orbit (§2.9; Problem 14-11).

14.3.8 Microlensing Detections

The first planet to be detected via microlensing is an ~2.6 M$_{2\!+}$ object seen ~4.3 AU away (on the plane of the sky) from an ~0.63 M$_\odot$ star. The lightcurve analyzed to achieve this detection is presented in Figure 14.18.

Almost 40 microlensing planets, some with masses only a few times that of the Earth, have

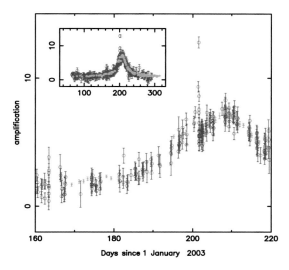

Figure 14.18 Lightcurve of the OGLE-2003-BLG-235/MOA 2003-BLG-53 microlensing event. The OGLE and MOA measurements are shown as *small filled* and *large open circles*, respectively. The *main panel* presents the complete data set during August 2003, and the inset at the upper left shows the data during all of 2003. The overall rise in brightness was primarily due to light being bent by the lensing star, but the high narrow peak one week before the midpoint of the event was caused by the combined gravitational tugs of the star and its 2–3 $M_{2\!\!\!+}$ planet. (Courtesy Ian Bond)

been detected as of early 2018. The OGLE-2006-109L system is especially interesting because it has two planets and resembles a somewhat smaller version of our Solar System: the star and its inner and outer planets are about half as massive as the Sun, Jupiter and Saturn, and their separations (on the plane of the sky) are just under half those of the three largest members of our Solar System.

14.3.9 Images and Spectra of Exoplanets

Planets on Distant Orbits

Massive young exoplanets and brown dwarfs located several astronomical units or farther from the star they orbit have been imaged in the thermal infrared. Two examples are shown in Figure 14.19.

The star β Pictoris is massive ($M_\star \sim 1.8$ M$_\odot$), nearby ($r_\odot \approx 19$ pc ≈ 62 ly) and young (age ≈ 20 Myr); β Pic has been studied extensively

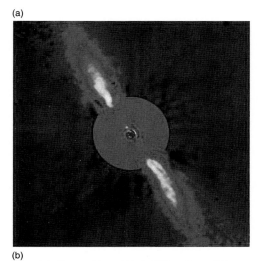

(a)

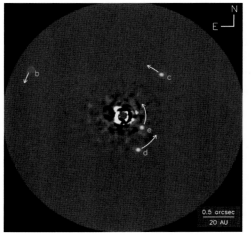

(b)

Figure 14.19 COLOR PLATE (a) This composite image represents the close environment of β Pictoris as seen in near infrared light. Very faint structure is revealed after subtraction of the much brighter stellar halo. The outer part of the image shows the reflected light from the dust disk; the inner part shows β Pic b as observed at 3.6 μm. The companion is less than 10^{-3} times as bright as β Pic, aligned with the disk, at a projected distance of 8 AU. (ESO/A.-M. Lagrange et al.) (b) A near-infrared (3.8 μm) image of the ~30 Myr-old HR 8799 system acquired using Keck 2 data taken in 2009. *Color* represents intensity. Observations at earlier epochs show counterclockwise Keplerian orbital motion for all four planets, as indicated by the *arrows*. (Courtesy NRC Canada/C. Marois) The orbit movie was produced by using motion interpolation on seven years of Keck data (2009–2016). Although the starlight was suppressed with a coronagraph, residual glare of the star creates a speckled halo of light. (Courtesy Jason Wang, UC Berkeley/C. Marois)

because it hosts a dust disk that we view nearly edge-on, numerous comets (whose tails produce varying absorption lines in the star's spectrum) and a substellar companion whose mass places it near the planet/brown dwarf boundary. Figure 14.19a combines images of β Pictoris's dust disk with high-resolution coronagraphic images showing the $M \sim 13\,M_{2\!\!\!+}$, $R \approx 1.46\,R_{2\!\!\!+}$, $T_{\mathrm{eff}} = 1724 \pm 15$ K substellar secondary, β Pic b, which orbits ≈ 9 AU from the star. The star β Pictoris and most stars around which planets have been imaged to date are more massive and younger than most of the stellar hosts of exoplanets observed using other techniques. The youth of β Pic b explains why its radius is inflated and its effective temperature is high despite its large distance from its star (see Fig. 3.7). Spectra of β Pic b show clear evidence for both CO and H_2O. The spectral lines of CO and H_2O are broadened (§4.5.2) by 25 km s^{-1}, presumably by the body's rotation, implying that β Pic b's rotational period is $\lesssim 8$ hr.

In most cases, including the β Pic system discussed above, only one substellar object has been imaged about a given star. But the HR 8799 system shown in Figure 14.19b has four planets, each of radius ~ 1.2–1.3 R$_{2\!\!\!+}$ and $T_{\mathrm{eff}} \sim 1000$ K. The planets' masses are ~ 4–8 M$_{2\!\!\!+}$, based on their thermal luminosities and the system's estimated age of tens of millions of years (Fig. 3.7). The innermost planet has $a \approx 15$ AU and the outermost planet is ~ 70 AU from the star. At least the middle pair of planets, HR 8799 c and d, must travel on nearly circular orbits for the system to have survived for the millions of years since the star has formed. The portions of the orbits shown in the movie associated with Figure 14.19b all appear to have small eccentricities and to be viewed nearly face-on. Dynamically, this is a **tightly packed** group of planets (i.e., if the planetary orbits were much closer to one another, the system would be quite unstable), with orbits that are likely to be nearly circular and coplanar. But the planets' masses and the large distances of the outer planets from their star present substantial challenges to planet

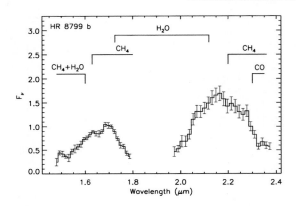

Figure 14.20 Near-infrared spectrum of the exoplanet HR 8799 b plotted with 1σ uncertainties. The locations of prominent water, methane and carbon monoxide absorption bands are indicated. (Courtesy Travis Barman)

formation theories. Additionally, HR 8799 hosts a dust disk located ~ 145–400 AU from the star, well beyond the orbits of all four known planets.

New instrumentation and observing techniques allow the thermal emission spectra of some extrasolar planets to be measured directly. All four of HR 8799's planets are cloudy. Figure 14.20 shows a moderate resolution spectrum of the outermost of HR 8799's four planets. The hot, young planets show absorption features from H_2O, CO and CH_4. Gaseous H_2O is seen in these hot young objects because, unlike in the case of Jupiter, the atmosphere is too warm for water clouds to form; thus, H_2O is found throughout the atmosphere. Although methane is seen in this object, its atmosphere is warm enough and atmospheric mixing is vigorous enough (§5.6) to also allow for a sizable quantity of CO to be present as well. The shape of the spectrum can be compared with models to provide an estimate of the planet's gravity and temperature.

The pre-main sequence star PDS 70 is accreting material from a surrounding disk. This disk has a 'gap' (region of substantially reduced surface density) extending from $\lesssim 17$ AU to 60 AU. The giant planet PDS 70 b orbiting within this gap is also still accreting material, albeit at the rate of only $\sim 10^{-8}$ M$_{2\!\!\!+}$/year.

Vulcan Planets

The atmospheres of hot jupiters are being studied via ground-based, high-resolution spectroscopy. At very high resolution, molecular bands are resolved into the individual lines, allowing robust identification of molecular species. Furthermore, while the absorption by molecules in Earth's atmosphere is static in wavelength, the exoplanet's molecular lines are Doppler-shifted by an amount that varies by ~ 100 km s^{-1}.

For transiting planets, the deepest planetary spectral lines are seen during transit (transmission spectroscopy) and around – but not during – occultation (day-side spectroscopy). The day-side spectroscopy views the planet's thermal emission, so non-transiting planets can be studied as well, and their masses and orbital inclinations can be measured by determining the radial component of the orbital velocity.

The Very Large Telescope (VLA) has detected CO absorption by several hot jupiters in transmission and in day-side spectroscopy, and H_2O absorption was found near 3.2 μm in the day-side spectrum of HD 189733 b. The refractory compound TiO has also been detected in the atmospheres of some hot jupiters. Iron and titanium vapor have both been observed in the atmosphere of the transiting ultra-hot jupiter KELT-9 b, which orbits a (very hot) A star with $P = 1.5$ days; this planet has an equilibrium temperature $T_{eq} > 4000$ K.

Clouds are common in the atmospheres of hot jupiters. Emission spectra of some hot jupiters show evidence for thermal inversions, implying that these planets have stratospheres (§5.1). Wind shear of several km s^{-1} is observed in the upper atmospheres of some hot jupiters, which means that there must be significant redistribution of heat at high altitudes on these planets.

14.3.10 Planets in Multiple Star Systems

Kepler-16(AB) b is an approximately Saturn-sized ($R_p = 0.7538 \pm 0.0025$ R$_{\text{2+}}$), Saturn-mass ($M_p =$ 0.333 $\pm$ 0.016 M$_{\text{2+}}$) planet that moves on a P-type orbit about two stars. This **circumbinary planet**, the first of its kind observed to transit, travels on a nearly circular orbit of period 229 days about a 41-day period eclipsing binary composed of one star that is about two-thirds the size and mass of the Sun and another less than one-quarter as large and massive as our Sun. The planet is observed to transit both of the stars, allowing for good size estimates.

The only pair of stars known to host more than one circumbinary planet is Kepler-47 AB, which hosts at least three transiting circumbinary planets. The primary star, Kepler-47 A, is Sun-like, but the secondary is only about one-third the size and mass of our Sun. The stars orbit one another with a period of 7.45 days. The planets have orbital periods from 49.5 to 303 days. The inner planet is the smallest, with $R_p \approx 3$ R$_\oplus$ and the outer planet the largest, with $R_p \approx 4.6$ R$_\oplus$.

Many planets have been detected on S-type orbits around one of the stars in a binary/multiple star system with binary orbital semimajor axes exceeding a few dozen AU. Radial velocity surveys suggest that giant planets within a few AU of individual stars are about as common around stars with stellar companions separated by more than 100 AU as they are around single stars; and binaries with semimajor axis 20 AU $< a_b <$ 100 AU are somewhat less likely to host such planets. Kepler-132(AB) has two stars, each a bit larger than our Sun, separated by $\sim$500 AU. Each of the stars has a transiting planet with a period of slightly over six days, and there are two longer period transiting planets in the system, although it isn't clear which star either of these planets orbit.

14.4 Mass–Radius Relationship

Measurements of both the mass and radius of a planet yield its density, and thereby provide constraints on its composition. The planet's temperature and whether or not it is differentiated (§§6.1.4, 15.5.2) can also affect its radius.

14.4.1 Theory

The factors determining the radii of stars and stellar remnants are discussed in §3.3.3. Section 3.3.4 presents a similar analysis for isolated planets, focusing on giant planets. Here we provide more information about hot jupiters, which are subjected to intense stellar irradiation, as well as planets of mass $M_p < 10\ M_\oplus$ made primarily of common compounds that are condensable within protoplanetary disks.

Inflated Hot Jupiters

Planets orbiting very close to stars are subjected to intense stellar heating. This heating retards convection in the upper envelope of close-in gas-rich planets. The high equilibrium temperatures of hot jupiters resulting from the large stellar fluxes impinging on them thus means that their entire envelopes can lie on a higher adiabat than that of Jupiter itself. The excess entropy retained throughout hot jupiters as a result of the intense stellar radiation that they receive can produce radii that are of order 10% larger than the radii of colder planets of the same mass, composition and age. The magnitude of this effect is greater for lower mass planets. Note that if a planet migrates close to a star after radiating away much of its initial accretion energy and shrinking to near $1\ R_{2\!\!\!\perp}$ (a process that requires tens of millions of years), its radius would likely grow only slightly larger than when the planet was in colder environs, because the star's heating would only affect and expand the outer part of the planet's atmosphere that had cooled below the planet's new T_{eff}.

The maximum possible radius for any 'cool' cosmic composition (i.e., H–He dominated; see Table 3.1) body is $\sim 1.1\ R_{2\!\!\!\perp}$ (Fig. 3.8). Hot jupiters thus were expected to have radii $R_p \lesssim 1.2\ R_{2\!\!\!\perp}$.

But the radii of many $\sim 1\ M_{2\!\!\!\perp}$ vulcans, including the first exoplanet to be observed in transit, HD 209458 b, are significantly larger than $1.2\ R_{2\!\!\!\perp}$. These hot jupiters must thus be even hotter than can be accounted for by the radiation they receive from their stars. The leading theory for this anomalous inflation is that slightly ionized atmospheric winds blowing across the planetary magnetic field produce currents penetrating the interior and giving rise to ohmic heating in the deeper layers, thereby limiting the loss of the planet's thermal energy more severely than the energy from the star at the surface alone.

Planets Made of Heavy Elements

Planets of order $1\ M_\oplus$ that are composed primarily of silicates, iron and H_2O are subject to pressure-induced compression that increases their average densities by tens of percent. For given proportions of silicates, iron and H_2O, planetary radius varies with mass roughly as:

$$R_p \propto M_p^{0.31} \qquad (1\ M_{\mathbb{C}} < M_p < 1\ M_\oplus), \qquad (14.8a)$$

and

$$R_p \propto M_p^{0.27} \qquad (1\ M_\oplus < M_p < 10\ M_\oplus). \qquad (14.8b)$$

Throughout these ranges, the radius of a planet composed of an Earth-like mixture is expected to be $\sim 20\%$ less than that of a planet of the same mass containing equal amounts (by mass) of Earth-like mixture and H_2O. Figure 14.21 shows models of the radial structure of rocky and rock/H_2O planets of mass 1–10 $M_\oplus$.

Figure 14.22 presents the relationship between composition and mass for $1\ M_\oplus$ and $5\ M_\oplus$ planets that lack H/He envelopes. Many different mixtures of iron+silicates+ice share the same mass and radius, and hence the composition of a planet cannot be uniquely determined from mass and radius measurements. The maximum radius for a rocky planet of a specified mass corresponds to the 'silicate mantle' vertex in Figure 14.22. A planet

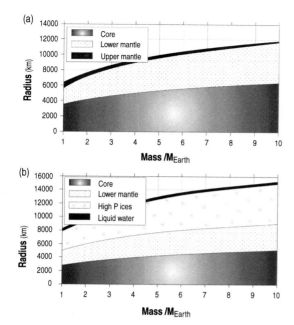

Figure 14.21 Radii of (a) Earth-composition planets and (b) ocean planets that are equal mixtures (by mass) of H_2O and Earth composition. The *inset keys* show the various layers of the planets. The crust is included in the upper mantle portion of the Earth-composition planets, and the amount of water in Earth is insignificant on the scale of these plots. For hot H_2O-rich planets, the upper part of the region shown as high-pressure ice is replaced by liquid water, and for cold H_2O-rich planets, the upper part of the region shown as liquid water is replaced by ice I (ordinary, low-pressure, water-ice – see Fig. 6.4. (Courtesy C. Sotin; see Sotin et al. 2007 for details)

with a larger radius necessarily implies that it has H_2O or other light components. The presence of H/He envelopes, which clearly must be voluminous for the largest planets known in this mass range (Fig. 14.24), further complicates efforts to deduce planetary composition.

14.4.2 Observations

Figure 14.23 shows the masses and radii of several dozen transiting planets for which both quantities are well determined. Most transiting extrasolar giant planets are predominantly hydrogen and helium, as are Jupiter and Saturn. Many are

substantially larger than cool planets made of H and He in cosmic proportions (Table 3.1). The factors responsible for these large sizes are discussed in §14.4.1.

The vulcan planet HD 149026 b has $M_p = 0.36\,M_{\mathbf{4}}$, and $R_p = 0.725 \pm 0.03\,R_{\mathbf{4}}$, which together imply that more than half of its mass consists of elements heavier than helium (assuming that H and He are present in cosmic ratio). The bulk composition of HD 149026 b is intermediate between those of Saturn and Uranus, and HD 149026 b is more richly endowed in terms of total amount of 'metals' than is any planet in our Solar System. More quantitative statements about the compositions of the hot jupiters will not be possible until the physical process that keeps HD 209458 b and some of its brethren from shrinking is known, and it can be determined whether, and if so to what extent, this process affects other transiting hot jupiters.

Figure 14.24 shows the mass–radius–temperature relationship for exoplanets with masses $M_p < 20\,M_\oplus$. The only exoplanets with mass and radius measurements that imply a purely rocky composition have $R_p \lesssim 1.7\,R_\oplus$, and many have orbital periods under 1 day. All planets with known masses and radii $R_p > 3\,R_\oplus$ have a significant fraction of their volumes occupied by H_2 and/or He. Most planets with $1.7\,R_\oplus < R_p < 3\,R_\oplus$ could be rich in water (and perhaps other astrophysical ices); alternatively, planets in this size range could be primarily rocky worlds overlain with low-mass H/He envelopes that significantly increase their volumes.

14.5 Exoplanet Demographics

Exoplanet discoveries have expanded our planet inventory by a factor of several hundred. The distribution of known extrasolar planets is highly biased towards planets that are most easily detectable

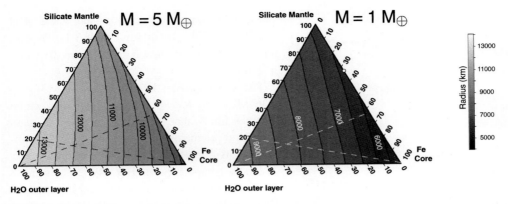

Figure 14.22 Ternary diagrams showing the relationship between composition and radius for 1 $M_\oplus$ and 5 $M_\oplus$ planets. *Solid curves* representing constant radius are shown in *black* at increments of 500 km. Different mixtures of the three most likely end-member components (iron cores, silicate mantles and H_2O outer regions) yield planets of different sizes. Each point in the ternary diagrams depicts a unique composition with a corresponding radius shown by the *shade of gray color*. The three vertices correspond to pure compositions of H_2O, iron or silicates, and the opposite sides of the triangle correspond to 0% of that end member. Thus, the side that connects Fe core and silicate mantle represents waterless planets. Earth's composition is shown by a *circle* essentially on this line in the 1 $M_\oplus$ diagram. Planets that formed in disks of solar nebula composition that are composed of all substances that condense above any specified temperature (Fig. 15.5), or mixtures of materials that condense at a range of temperatures, lie above both of the *dashed lines*. (Courtesy Diana Valencia)

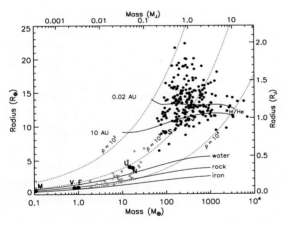

Figure 14.23 Mass–radius relationships for theoretical planets and observed planets as of 2017. The planets of the Solar System are shown as *large black circles with colored halos*. Transiting exoplanets are represented by *smaller colored circles*. *Green* is used for planets with mass $M_p \leq M_\oplus$, *blue* for planets with mass $M_\oplus < M_p < 50\ M_\oplus$ and *dark red* for planets with mass $50\ M_\oplus < M_p$. The *solid curves* represent models of 4.5-Gyr-old planets of the specified compositions. Two curves are given for H/He planets (73% H, 27% He by mass) to illustrate the effects of stellar radiation; they are labeled by distance from their star, which is assumed to be 1 $M_\odot$. The dotted curves are for constant values of density at 10^2, 10^3 and 10^4 kg m^{-3}. (Courtesy Jonathan Fortney)

using transit photometry and the Doppler radial velocity technique, which have been by far the most productive methods of discovering exoplanets. These extrasolar planetary systems are quite different from our Solar System; however, it is not yet known whether our planetary system is close to the norm, quite atypical or somewhere in between.

Nonetheless, some unbiased statistical information can be distilled from available exoplanet data. One of the most fundamental statistics is the planetary **occurrence rate**, which is the average number of planets with a specified range of characteristics around members of a specified class of stars (not to be confused with the fraction of stars with planets, which is smaller).

Consider, for example, a transit survey, which measures planetary sizes and orbital periods. The probability distribution

$$f(R_p, P)\, d\ln R_p\, d\ln P, \tag{14.9}$$

gives the likelihood that a star possesses a planet in the infinitesimal volume of radius-period phase space $d\ln R_p\ d\ln P$. Radial velocity surveys can

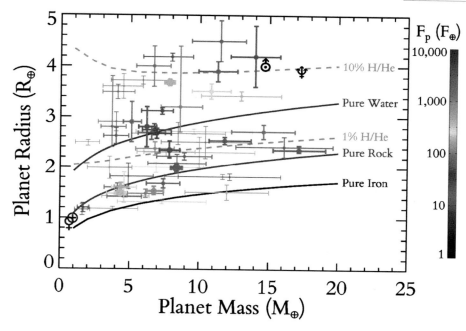

Figure 14.24 COLOR PLATE Mass–radius diagram for small planets with measured masses at least twice as large estimated uncertainties as of mid-2017. The transiting exoplanets are represented by points and error bars; the sizes of the points and thickness of the error bars vary inversely with the fractional uncertainty of the mass measurement. Standard planetary symbols (Table E.1) are used to represent the four Solar System planets that lie within the plot's mass and radius ranges. The colors represent the flux of stellar radiation intercepted by the planets in units of the solar constant, $\mathcal{F}_\odot$, as indicated by the scale bar on the right (Uranus and Neptune, which receive far less radiant energy than any of the other planets graphed, are shown in black). Model mass-radius curves for various compositions are plotted for comparison. The red curve represents silicate rock. The dashed orange curves show models that are 1% and 10% H/He by mass atop an Earth-like (2/3 silicate mantle and 1/3 iron core) interior. (Courtesy Eric Lopez)

be used to derive occurrence rates as a function of planetary mass and period (a probability distribution analogous to equation (14.9) with M_p replacing R_p), and eccentricity can also be included as a third variable.

The integral of the distribution given by equation (14.9) over a range in planetary radius and orbital period is the average number of planets per star. Typically, occurrence rates are given on a grid in period-radius or period-mass phase space that shows the average number of planets per star (sometimes multiplied by 100 to give percentages, as in Fig. 14.30) with properties in the specified range. The symbol $\eta_\oplus$ (pronounced **eta-earth**) is used to denote the number of Earth-like planets per star; however, the range of planetary properties appropriate to consider Earth-like is controversial.

We consider in turn radial velocity surveys, which provide the most information on giant planets within a few AU of their stars; *Kepler* data that give statistical information for planets as small as Earth with orbital periods of up to several months, as well as even smaller planets on very short-period orbits and planets larger than $\sim 2\,R_\oplus$ with periods of up to one year; microlensing surveys, which yield the best constraints on planets located several AU from their stars; and imaging surveys, which provide the best information about massive planets orbiting far from their stars.

14.5.1 Radial Velocity Surveys

Roughly 0.7% of single Sun-like stars (late F, G and early K spectral class main sequence stars

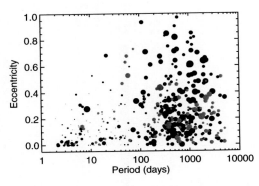

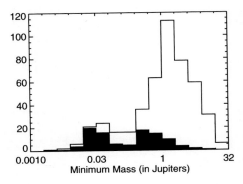

Figure 14.25 The eccentricities of the 472 exoplanets discovered by the radial velocity method that have well-determined orbits (as of 2016) are plotted against orbital period. The *dot size* is proportional to $(M_p \sin i)^{1/3}$, and the *gray points* represent planets in systems known to possess more than one planet. Whereas the eccentricities of almost all of those planets with periods of less than a week are quite small (consistent with 0), presumably as a result of tidal damping, the eccentricities of planets with longer orbital periods are generally much larger than those of the giant planets within our Solar System. (Courtesy Jason Wright, data from exoplanets.org)

Figure 14.26 Histogram of the number of planets observed as a function of minimum planet mass, $M_p \sin i$, obtained from the same data set used to produce Figure 14.25. Each bin encompasses a factor of 2 in minimum mass. The *black region* at the bottom represents planets with orbital periods of less than 30 days, and the *white region* shows the number of planets with longer periods. Because more massive planets are easier to detect, the tail-off in the distribution above 1–2 M_{24} is real, but the drop-off at smaller masses is a consequence of observational selection effects. Likewise, the shift in the distribution to smaller masses at shorter periods results from the larger radial velocity perturbations by planets of a given mass orbiting closer to the star (eq. 14.1) and the greater number of times that these planets have orbited since the highest precision radial velocity surveys have been underway. (Courtesy Jason Wright, data from exoplanets.org)

that are chromospherically quiet, i.e., have inactive photospheres) have planets more massive than Saturn within 0.1 AU. Approximately 7% of Sun-like stars have planets more massive than Jupiter within 3 AU. Only about 1% of low-mass stars (M dwarfs with masses 0.3–0.5 $M_\odot$) are orbited by giant planets within 2 AU.

Figure 14.25 plots the eccentricities of exoplanets versus their orbital periods. Most planets orbiting interior to ∼0.1 AU, a region where tidal circularization timescales are less than stellar ages, have small orbital eccentricities. The median e of giant planets with $0.1 < a < 3$ AU is ∼0.26, and some of these planets travel on very eccentric orbits.

Within 5 AU of Sun-like stars, Jupiter-mass planets are more common than planets of several Jupiter masses, and substellar companions of mass ≳10 M_{24} are rare. A histogram of the number of planets detected via radial velocity variations in various mass ranges is shown in Figure 14.26. There is a paucity of objects of mass ∼20–60 M_{24} near Sun-like stars, which is referred to as the **brown dwarf desert**.

Occurrence rates of planets with orbital periods less than 50 days are presented as a function of planetary mass in Figure 14.27. The numbers of planets detected supplemented by likely candidates and the estimated fraction of such objects that would have gone undetected represent the mean number of short-period planets in the quoted mass range for Sun-like stars.

Figure 14.28 shows that stars with higher metallicity (heavy element content) are much more likely to host giant planets within a few AU than are metal-poor stars, with the probability of hosting such a planet varying roughly as the square of stellar metallicity. The Sun itself has a higher metallicity than do most ∼1 $M_\odot$ stars in the solar neighborhood. At least over the range 0.3–1.5 $M_\odot$, more massive stars are likelier to host giant planets orbiting within a few AU. In contrast,

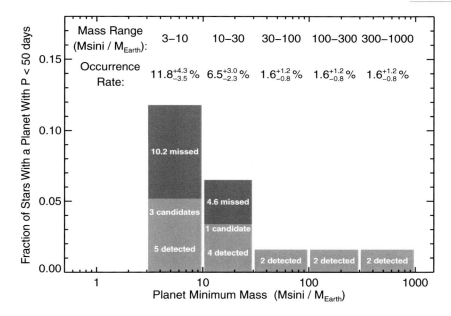

Figure 14.27 Occurrence rates of planets with orbital period $P < 50$ days as a function of minimum mass based on radial velocity observations. The *bottom* (*green*) portions of the histogram represent confirmed planets, and the *middle* (*yellow*) and *upper* (*blue*) rectangles in the two lowest mass bins represent unconfirmed but likely candidate planets and a correction factor to account for an estimate of the fraction of planets that were not detected, respectively. (Howard et al. 2010)

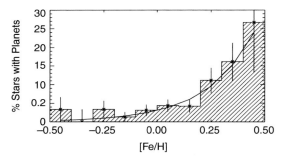

Figure 14.28 The fraction of Sun-like stars possessing giant planets with orbital periods of less than four years is shown as a function of stellar metallicity. Metallicity is measured on a logarithmic scale, with [Fe/H] = 0 corresponding to the solar value. (Adapted from Fischer and Valenti 2005)

the likelihood of a star to host a Neptune-mass planet with orbital period <2 months does not appear to depend strongly on stellar metallicity or mass.

Multiple planet systems are more common than if detectable giant planets were randomly distributed among stars (i.e., than if the presence of a giant planet around a given star was not correlated with the presence of other giant planets around that same star). However, few additional planets have been found about stars that host hot jupiters.

14.5.2 Kepler Planet Candidates

Kepler has detected more than 4500 exoplanet candidates, the vast majority of which are likely to be true exoplanets. Figure 14.29 shows the radii and periods of *Kepler* planet candidates, as well as the number of candidates in the system in which they reside. When interpreting the data displayed in Figure 14.29, it is important to keep in mind that planets orbiting closer to their star have a higher probability of transiting (eq. 14.4), and the signatures of larger and shorter-period planets are easier to detect in folded transit lightcurves. Thus, the lack of small planets and paucity of all planets at the longest periods represented in the plot does not imply that such planets are uncommon. Indeed, statistical studies need to correct for planetary detectability to calculate occurrence rates.

Although little is known about most *Kepler* planet candidates individually, the statistical properties of the ensemble provide key information about planets orbiting within ~ 1 AU of their star.

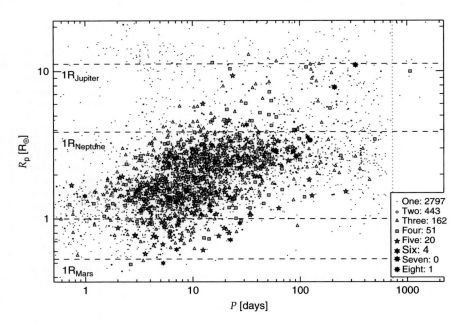

Figure 14.29 COLOR PLATE Planet period versus radius for *Kepler* planetary candidates. Planets that are the only candidate for their given star are represented by *black dots*, those in two-planet systems as *dark blue circles*, those in three-planet systems as *green triangles*, those in four, five and six planet candidates systems are displayed using *green squares*, *yellow five-pointed stars* and *orange six-pointed stars*, respectively, and the eight planets orbiting Kepler-90 as *red eight-pointed stars*. It is immediately apparent that there is a paucity of giant planets in multi-planet systems. The upward slope in the lower envelope of these points is caused by the difficulty in detecting small planets with long orbital periods, for which transits are shallow and few were observed. The dotted vertical line in the right portion of the plot marks half the time between the start and end of *Kepler*'s prime mission; only two transits were observed for planets to the right of this line. (Courtesy Daniel Fabrycky, Rebekah Dawson and Jason Rowe)

Note that statistical studies generally use the ensemble of *Kepler* planet candidates rather than that of verified *Kepler* planets. This is because the techniques used to verify that planet candidates are indeed planets, such as detection of the planet in RV data, TTV analysis and validation by demonstrating that the planetary hypothesis is by far the most likely cause of the observed pattern in the lightcurve, produce a very heterogeneous sample that is extremely difficult to de-bias.

Neptune-sized planets are far more common than are Jupiter-sized ones, and planets similar in size to Earth have even larger occurrence rates. Generally, the number of planets in a given logarithmic range in radius increases with decreasing planet size down to at least ~1 $R_\oplus$, below which

the survey is incomplete because of inadequate signal-to-noise. The number of planets per logarithmic period bin increases with increasing period; one exception to this trend is a small excess of Jupiter-sized planets with periods of ~4 days, the hot jupiter population referred to above; smaller planets do not have a similar concentration of periods. Figure 14.30 shows the occurrence rate of planets of various sizes and periods orbiting M dwarf stars (most with masses 0.3 $M_\odot$ < $M_\star$ < 0.6 $M_\odot$) derived from *Kepler* data.

Figure 14.31 shows the size distribution of planets orbiting Sun-like stars with P_{orb} < 100 days. The numbers shown were derived using *Kepler* data and have been adjusted to account for transit geometry and planet detection efficiencies. Note

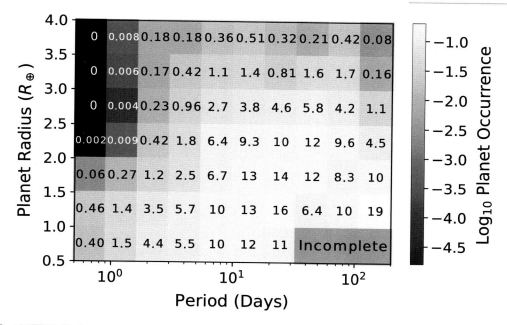

Figure 14.30 Occurrence rate of planets smaller than Neptune around small (M dwarf) stars as a function of planet radius and orbital period derived from data taken during *Kepler's* primary mission. The grid used is linear in planet radius and logarithmic in orbital period, with individual cells representing intervals of 0.5 $R_\oplus$ in R_p and 0.26 in $\log P_{orb}$. The shading of each cell indicates the planet occurrence (here expressed as the mean number of planets per 100 stars) within the range of radius and period covered by the cell. The occurrence rates shown are corrected for geometric factors and the difficulty of detecting small planets transiting faint and/or noisy stars. Typical fractional uncertainties range from ∼20% in the middle of the grid to order unity near the edges. No values are given for planets smaller than Earth with orbital periods longer than 33 days because such planets are very difficult to detect in the data. Most of the M dwarf stars observed by *Kepler* have masses in the range 0.3 $M_\odot \lesssim M_\star \lesssim$ 0.5 $M_\odot$ because stars less massive than this are very faint. (Courtesy Courtney Dressing; see Dressing and Charbonneau 2015 for details)

the paucity of planets with radii $R_p \approx$ 1.7 $R_\oplus$. Most planets with radii $R_p <$ 1.6 $R_\oplus$ whose masses have been measured have densities consistent with an Earth-like composition, whereas most larger planets with measured masses have densities lower than that of Earth-composition planets of their size, implying that they also possess low-density constituents. Low mass H/He envelopes of short-period planets are susceptible to photoevaporation induced by high-energy photons from the nearby star, so this dearth of planets has been referred to as the **evaporation valley**.

Almost 40% of the transiting planet candidates detected in *Kepler* data are members of multiple candidate systems. More than half of these planet candidates orbit stars with two candidate transiting planets, and most of the rest are in systems of three or four planet candidates, but four stars have six candidates and one (Kepler-90, §14.3.5) has eight. Figure 14.32 illustrates the characteristics of the 25 systems with five or more planet candidates.

Population statistics imply that many of the *Kepler* targets with one or more observed transiting planet candidate(s) must possess additional planets that are either not transiting or were too small to be detected. A few non-transiting planets have been discovered by analyzing the pattern of TTVs that they produced in their transiting neighbors (§14.1.5). In many more cases, TTVs reveal that

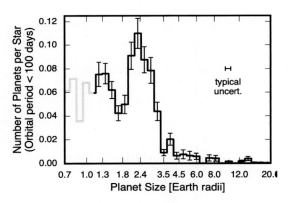

Figure 14.31 The size distribution of close-in planets derived using *Kepler* data. The histogram shows the number of planets per star with orbital periods less than 100 days as a function of planet size. The radius distribution exhibits a deep trough centered at $R_p \approx 1.7\ R_\oplus$. The solid line is shown in black for sizes where detection efficiency is high, and in gray for small sizes where such a large fraction of transiting planets escape detection that it is difficult to correct for the incompleteness of the sample. From Fulton and Petagura (2018).

an additional planet is present, but do not contain sufficient information to specify any of its properties. When geometrical considerations concerning transit probabilities as well as completeness (planet detectability using currently available data) considerations are factored in, approximately 10% of *Kepler* target stars with masses of $0.4\ M_\odot < M_\star < 1.2\ M_\odot$ have multiple planets of radius $1\ R_\oplus \lesssim R_p$ and $3 < P_{orb} < 200$ days.

The large number of candidate multiple transiting planet systems observed by *Kepler* shows that many of the multi-planet systems within 1 AU of their stars are nearly coplanar, with typical relative orbital inclinations of only $1-2°$. Most planets in multi-planet systems travel on nearly circular orbits ($e \lesssim 0.05$); a much wider distribution of eccentricities is observed for planets with no observed companions.

As shown in Figure 14.33, the distribution of observed period ratios implies that the vast majority of candidate pairs are neither in nor near low-order mean motion resonances. Nonetheless, there are statistically significant excesses of candidate pairs both in first-order resonance (§2.3.2) and spaced slightly too far apart to be in resonance, as well as deficits at period ratios slightly lower than those of resonance, particularly near the 2:1 and 3:2 resonances. Virtually all candidate systems are stable, as tested by numerical integrations that assume a nominal mass–radius relationship derived from planets within our Solar System.

14.5.3 Microlensing and Imaging

With fewer than 40 confirmed planetary detections and very limited characterization of most of the planets and host stars, microlensing discoveries do not rival the extensive results of statistical characterization of exoplanets as do radial velocity planets and *Kepler* transiting planet candidates. Nonetheless, because of the complementarity of the techniques, microlensing allows the best estimates of the frequencies of planets orbiting several AU from low-mass stars, which are the most common stars in our galaxy. The average number of planets (per star) with masses $5\ M_\oplus < M_p < 10\ M_{\jupiter}$ at distances of 0.5–10 AU is $1.6^{+0.7}_{-0.9}$. Within these ranges, smaller planets are more common than larger ones (measured in planets per logarithmic mass range), but large planets contain a greater fraction of the total planetary mass.

Imaging searches imply that few exoplanets have masses $M_p > 5\ M_{\jupiter}$ and apoapse distances more than $\sim 10-20$ AU from their host stars. Such distant massive planets are especially uncommon around stars of mass $M_\star \lesssim 1\ M_\odot$.

14.6 Conclusions

Before the discovery of extrasolar planets, models of planetary growth suggested that most single solar-type stars possess planetary systems

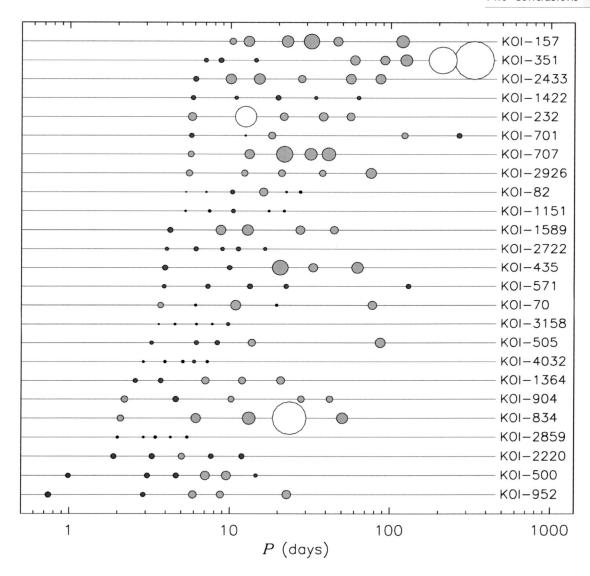

Figure 14.32 *Kepler* candidate planetary systems of five or more planets. Each *line* corresponds to one system, as labeled on the *right side*. Systems are ordered by the orbital period of the innermost planet. Planet radii are to scale relative to one another and planets are colored by size, with planets smaller than Earth represented as *filled black circles*, those with $R_\oplus \leq R_p < 1.7\ R_\oplus$ using *dark blue circles with black boundaries*, those with $1.7\ R_\oplus \leq R_p \leq 6\ R_\oplus$ by *light blue circles with black boundaries* and those larger than 6 $R_\oplus$ as *open white circles*. Each system is identified by its **KOI** (*Kepler Object of Interest*) number. Each target star found to possess a transit-like signature by the *Kepler* project was assigned an integer KOI number. Individual planet candidates were assigned two digit decimal designations; e.g., the sixth planet candidate found at KOI-351 is designated KOI-351.06. (Courtesy Daniel Fabrycky and Jason Rowe)

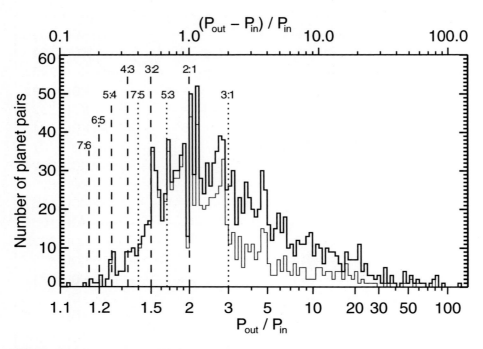

Figure 14.33 Histogram of period ratios for planet pairs within multiple planet systems observed by *Kepler*. The *thin lower curve* shows the number of neighboring pairs of planets, whereas the *thick upper curve* gives the total number (non-neighboring as well as neighboring) of planet pairs within the specified ranges of period ratio. First-order mean motion resonances are marked by *long-dashed lines*; *short-dashed lines* show the locations of second-order mean motion resonances. (Courtesy Daniel Fabrycky)

that are grossly similar to our Solar System. Observations have subsequently demonstrated that nature is far more creative than the human imagination.

Models developed to explain the formation of our Solar System showed that stochastic factors are important in planetary growth, so that the number of terrestrial planets (as well as the presence or absence of an asteroid belt) was expected to vary from star to star even if their protoplanetary disks were initially very similar. The difficulty in accreting giant planet atmospheres prior to dispersal of circumstellar gas suggested that many systems might lack gas giants. The low eccentricities of the giant planets in our Solar System (especially Neptune) are difficult to account for, so systems with planets on highly eccentric orbits were viewed as possibilities, although researchers did not hazard

to estimate the detailed characteristics of such systems. A maximum planetary mass similar to that of Jupiter was suggested as a possibility if Jupiter's mass was determined by a balance between a planet's gap-clearing ability and viscous inflows (§15.6), although it was noted that the value of the viscosity could well vary from disk to disk. Orbital migration of some giant planets towards their parent star (§15.7) was also envisioned, but because migration rates were expected to increase as the planet approaches the star, almost all of these planets were expected to be accreted by their star, and the existence of numerous giant planets with orbital periods of a few days was not predicted.

Carbon dioxide on Earth cycles between the atmosphere, the oceans, life, fossil fuels and carbonate rocks. This cycling has been important for moderating long-term climate variations and thereby allowing advanced life to develop and flourish over the past few billion years on Earth (§16.6.1). Carbonates are not readily recycled on a geologically inactive planet such as Mars; in contrast, they are not formed on planets that lack surface water, e.g., Venus (§5.8). Larger planets of a given composition remain geologically active for longer because they have smaller surface-area-to-mass ratios, enabling them to retain heat from accretion and radioactive decay longer.

It must thus be admitted that theoretical models based on observations within our Solar System failed to predict many of the types of planets that have been detected by radial velocity and transit surveys. However, fewer than half of the stars surveyed have planets of the types detected in significant numbers to date. Radial velocity surveys are biased in favor of detecting massive planets orbiting close to stars, and transit surveys are even more biased to planets near stars; planets similar to those in our own Solar System would in most cases not yet have been detected.

Radial velocity surveys now have enough data to detect Jupiter analogs around many stars and find that such planets are uncommon. But Saturn-mass planets in Jupiter-like orbits and Jupiter-mass planets in Saturn-like orbits would generally not have been detected. Thus, it is possible that tens of percent of single Sun-like stars possess planetary systems fairly similar to our own. Alternatively, although theoretical considerations suggest that terrestrial planets are likely to grow around most Sun-like stars, many of these planets may be lost if most systems also contain giant planets that migrate into the central star.

Discovering an exoplanet with size, mass, star and orbit similar to those of our own may be possible using photometric data of some of the brightest and least variable stars observed during the *Kepler* mission, but is beyond the present capabilities of any other planet-finding technique. Although an Earth analog at a distance of 10 parsecs (33 light-years) would be brighter than the faintest objects observed by the *HST*, the adjacent overwhelmingly brighter Sun-like star makes detecting such a planet exceedingly challenging. The Sun's radius is 100 times that of Earth, its mass is 300 000 times as large and its brightness is 10^6–10^{10} that of our home planet (Fig. 14.6).

The number of variables involved in determining a planet's habitability precludes a complete discussion, but some of the major issues are summarized in Figure 14.34. Remedial measures that could improve the habitability of the mini-Earth shown on the left in Figure 14.34 include (1) move it closer to the star to reduce the amount of greenhouse warming needed to keep surface temperatures comfortable, (2) add extra atmospheric volatiles and (3) include a larger fraction of long-lived radioactive nuclei than on Earth to maintain crustal recycling. Some remedial measures that could improve the habitability of the super-Earth shown on the right in Figure 14.34 are: (1) Move it farther from the star. (2) Include a smaller fraction of atmospheric volatiles. It is not clear that more active crustal recycling would be a problem, within limits, but crustal activity would be lessened if the planet had a smaller inventory of radioactive isotopes. (3) Give it a wide, optically thick ring. Provided the planet has a moderate to large obliquity, such a ring would shadow a significant portion of the planet for much of its 'year' (see Figs. 13.3 and 13.5).

We still do not know whether terrestrial planets on which liquid water flows are rare, are the norm for solar-type stars or have intermediate abundances. Nonetheless, even if planetary migration destroys some promising systems, planets qualifying as continuously habitable for long periods of time by the liquid-water criterion (§16.5) are expected to be sufficiently common that if we are the only advanced life form in our sector of the Galaxy, biological and/or local planetary factors (see Chapter 16) are much more likely to be the principal limiting factor than are astronomical causes.

Figure 14.34 COLOR PLATE Theoretical comparison of planets of different sizes with the same composition as Earth. (*Left*) A smaller planet would be less dense because the pressure in the interior would be lower. Such a planet would have a larger ratio of surface area to mass, so its interior would cool faster. Its lower surface gravity and more rigid crust would allow for higher mountains and deeper valleys than are seen on Earth. Most important to life is that the atmosphere of the mini-Earth would have a much smaller surface pressure as a result of four factors: larger surface area to mass, lower surface gravity, more volatiles sequestered in crust because there would be less crustal recycling and more atmospheric volatiles escaping to space. This would imply, among other things, lower surface temperature, resulting from less greenhouse gas in a column of the atmosphere. (*Center*) Earth – home sweet home. (*Right*) A larger planet made of the same material as Earth would be denser and have a hotter interior. Its higher surface gravity and more ductile crust would lead to muted topography. It would have a much greater atmospheric pressure, and unless its greenhouse was strong enough to boil away the planet's water, much thicker oceans, probably covering the planet's entire surface. (Lissauer 1999)

Key Concepts

- Thousands of planets have been discovered in orbit around stars other than our Sun.
- Several complementary techniques are used to detect and characterize extrasolar planets. The most successful of these methods to date have been transit photometry and Doppler radial velocity measurements.
- Most exoplanets detected to date are more than five times as massive as the Earth and/or orbit within 0.5 AU of their star. This is caused primarily by biases in search methods, but it is clear that many and probably most planetary systems are quite different from our Solar System.

- Giant planets are more common around stars with high abundances of heavy elements relative to hydrogen and around more massive stars.
- NASA's *Kepler* spacecraft has identified more than 4000 candidate transiting planets, the overwhelming majority of which are likely to be real planets. *Kepler* data are providing an excellent statistical census of Earth-sized and larger planets orbiting near their stars.
- Nearly planar systems of multiple planets orbiting within 1 AU of their stars are common.
- Exoplanet studies are advancing at a very rapid pace.

Further Reading

For a nice popular-level discussion of comparative planetology, including the properties of hypothetical planets of various masses and compositions, see:

Lewis, J.S., 1999. *Worlds Without End: The Exploration of Planets Known and Unknown*. Helix, Reading, MA. 264pp.

Good reviews of many aspects of extrasolar planets are available in:

Seager, S., 2010, Ed. *Exoplanets*. University of Arizona Press, Tucson. 526pp.

More comprehensive and up-to-date reviews, some of which are considerably more technical, are available in:

Deeg, H.J., and J.A. Belmonte, 2018, Eds. *Handbook of Exoplanets*. Springer International Publishing, in press.

A nice review of the abundances of planets of various sizes and periods and the properties of observed multi-planet systems is given in:

Winn, J.N., and D.C. Fabrycky, 2015. The occurrence and architecture of exoplanetary systems. *Ann. Rev. Astron. Astrophys.*, **53**, 409–447.

Exoplanet science results from the *Kepler* mission are summarized in:

Lissauer, J.J., R. Dawson, and S. Tremaine, 2014. Advances in exoplanet science from Kepler. *Nature*, **513**, 336–344.

A comprehensive review of exoplanet detection techniques and a catalog of information on all exoplanets known as of 2018 are provided in:

Perryman, M., 2018. *The Exoplanet Handbook*. Cambridge University Press, 2nd edition. 974pp.

Three websites that have general information on exoplanets as well as extensive exoplanet catalogs are:

the Extrasolar Planet Encyclopedia: http://exoplanet .eu,

the NASA Exoplanet Science Institute: http://nexsci .caltech.edu,and: http://exoplanets.org.

The website of the *Kepler* mission is: https://www .nasa.gov/mission_pages/kepler/main/ index.html.

Problems

14-1. A planet of mass $M_p = 2\,M_{2\!\!+}$ travels on a circular orbit of radius 4 AU about a 1-$M_{\odot}$ star. The Solar System lies in the plane of the orbit. Write the equation for the star's radial velocity variations caused by the planet and sketch the resulting curve.

14-2. What is the amplitude of the astrometric wobble induced by a planet of mass $M_p = 2\,M_{2\!\!+}$ that travels on a circular orbit of 4 AU radius about a 1-$M_{\odot}$ star located 4 parsecs from the Sun?

14-3. **(a)** Calculate the probability of transits of the planets Venus and Jupiter being observable from another (randomly positioned) planetary system.
(b) Discuss semiquantitatively the probability that transits of both planets could be detected by a single observer. Hint: The inclination of the orbit of Venus relative to that of Jupiter is about $2.3°$.

14-4. **(a)** Derive an expression for the duration, in hours, of a central transit of an exoplanet on a circular orbit, as a function of $M_{\star}/M_{\odot}$ and $R_{\star}/R_{\odot}$, and orbital period (in days).
(b) What is the duration of Earth's transit of the Sun as observed by a distant observer in the ecliptic plane?
(c) A transiting exoplanet with an orbital period of 1 year orbits a Sun-like star and has a transit duration 2 hours longer than your expectation from part (b). What can one infer about the planet's orbit?
(d) A transiting exoplanet with an orbital period of 1 year orbits a Sun-like star and has a transit duration 2 hours shorter than your expectation from part (b). Give two different explanations of how this could arise.

14-5. Calculate the depth, in ppm (parts per million), of the transit of a 1 $R_{\odot}$ star by a 1 $R_{\oplus}$ planet.

14-6. The star Kepler-4 has mass $M_{\star} = 1.1$ $M_{\odot}$ and radius $R_{\star} = 1.55\ R_{\odot}$. Its vulcan planet Kepler-4 b has a radius $R_p = 3.9\ R_{\oplus}$ and an orbital period of 3.2 days.
(a) Estimate the transit depth in ppm.
(b) Estimate the occultation depth (i) assuming a geometric albedo $A_0 = 0.3$, and (ii) assuming $A_0 = 0.05$.
(c)* Estimate the occultation depth at 2 μm assuming a Bond albedo $A_b = 0$.

14-7. Consider two planets, each much less massive than their star, that have small eccentricities and orbits that place them close to a first order mean motion resonance.
(a) Use conservation of energy and the generalized form of Kepler's third law (eq. 2.18) to show that the changes in the orbital periods of the planets, $\Delta P_{orb,i}$ (assumed to be $\ll P_{orb,i}$), are related to one another by:

$$\Delta P_{orb,2} = -\Delta P_{orb,1}\frac{m_1}{m_2}\left(\frac{P_{orb,2}}{P_{orb,1}}\right)^{5/3}.$$
$$(14.10)$$

(b) Use equation (14.10) to compute the ratio of the amplitudes of the TTVs of the two planets. Neglect the (usually small) TTVs resulting from orbital precession. Hint: The TTVs of the two planets are periodic with the same periodicity but opposite in phase. Write down a formula for the ratio of the TTVs in terms of

$P_{orb,1}$, $P_{orb,2}$, $\Delta P_{orb,1}$ and $\Delta P_{orb,2}$, and then use equation (14.10) to eliminate $\Delta P_{orb,1}$ and $\Delta P_{orb,2}$.

14-8. **(a)** Calculate the ratio of the light reflected by Earth at 0.5 μm to that emitted by the Sun at the same wavelength.
(b) Calculate the ratio of the thermal radiation emitted by Earth at 20 μm to that emitted by the Sun at the same wavelength.
(c) Repeat the above calculations for Jupiter.

14-9. What types of extrasolar planets are most easily detected by the following methods, and why?
(a) Radial velocity (Doppler) surveys
(b) Astrometry
(c) Transit photometry
(d) Coronagraphy

14-10. As demonstrated in Problems 2-4 and 2-5, Newton's theory of gravity is quite accurate for most Solar System situations, and the most easily observable effect of general relativity is the precession of Mercury's orbit. Those extrasolar planets that orbit much closer to their star than Mercury's distance from the Sun travel much faster, and they are also subjected to a greater gravitational field, so relativistic deviations from the trajectories predicted by Newton's laws should be larger. The first-order (weak field) general relativistic corrections to Newtonian gravity imply the periapse precession at the rate given by equation (2.67). Calculate the general relativistic precession of the periapse of:
(a) The transiting planet Gliese 436 b, for which $M_\star = 0.56\,M_\odot$, $M_p = 25\,M_\oplus$, $e = 0.16$, $P_{orb} = 2.644$ days.

(b) The highly eccentric planet HD 80606 b, for which $M_\star = 1.0\,M_\odot$, $M_p \sin i = 4\,M_{\text{♃}}$, $e = 0.93$, $P_{orb} = 111.44$ days.

14-11. Estimate the semimajor axis of the planet V391 Pegasi b (§14.3.7) when its star was on the main sequence.

14-12. Consider a Sun-like star with radius of 1 $R_\odot$ and effective temperature of 6000 K that has a close-in giant planet with Jupiter's radius and effective temperature of 1500 K.
(a) Calculate the ratio of the (bolometric) luminosity of the planet to that of its star using the wavelength-integrated blackbody radiation formula (eq. 4.14).
(b) Calculate the ratio of the infrared flux emitted by the two bodies at a wavelength of 24 μm. You may use the Rayleigh–Jeans approximation to the blackbody radiation formula, which is given in equation (4.4).
(c) Imagine that you measure, with a space telescope, the combined flux emitted by both the planet and the star. The planet then passes behind the star so that you no longer see its contribution to the total system flux. What is the percentage drop in the total flux that you would measure? Modern infrared detectors can measure changes on the order of 100 ppm. Do you think the passage of such a hot jupiter behind its star would be detectable?
(d) Redo part (b) but for wavelengths of 5 and 0.5 μm. Note that at these shorter wavelengths, the full blackbody formula (eq. 4.3) must be used. Are all wavelengths equally good for detecting the planet passing behind the star? Why or why not?

14-13. Estimate the ratio of the radiative timescale of the atmosphere (the time it would take for the thermal energy to fall by half) of a planet whose temperature is 1500 K to that of a planet whose atmosphere has $T = 100$ K at a comparable pressure level. You may assume that the heat capacities and radiative efficiencies are the same, so you only need to consider the thermal energy contents above the 1-bar pressure level and the blackbody luminosities of the planets.

14-14. Consider a terrestrial planet with Earth's radius orbiting 0.03 AU away from a cool M dwarf star with luminosity $\mathcal{L} = 10^{-3}\mathcal{L}_\odot$.

(a) Calculate the stellar flux intercepted by this planet (in J m^{-2}). What is this flux in units of the solar flux intercepted by the Earth?

(b) Planets orbiting this close to their star are likely to become 'tidally locked' and keep one side always facing the star (like the Moon keeps one side facing Earth). The flux computed in (a) is at the substellar point. Assuming that this planet does not possess an atmosphere, describe qualitatively how the surface temperature varies with location on the globe. Discuss which locations on the planet might be 'habitable' and 'uninhabitable'.

(c) Discuss what other attributes the planet has that would affect the surface temperature of the planet.

14-15. Estimate the equilibrium temperature of Kepler-11 g using the data given in Table 14.1:

(a) Assuming $A_b = 0.5$ and efficient global redistribution of heat.

(b) Assuming $A_b = 0.05$ and efficient global redistribution of heat.

(c) At the subsolar point assuming $A_b = 0.5$ and no redistribution of heat.

(d) Suppose that the *James Webb Space Telescope* measured the brightness temperature of this planet in a bandpass near 3 μm to be 400 K. Discuss what this information tells you about the planet.

14-16. Radii and masses measured for six hypothetical transiting planets are listed below. State the ranges in composition consistent with these measurements. In some cases, the ranges allowed by the error bars include some unlikely or unphysical compositions. Which cases are these, and why?

(a) $R_p = 3 \pm 0.1\ R_\oplus$, $M_p = 3 \pm 1\ M_\oplus$;
(b) $R_p = 3 \pm 1\ R_\oplus$, $M_p = 3 \pm 1\ M_\oplus$;
(c) $R_p = 1 \pm 0.5\ R_\oplus$, $M_p = 3 \pm 1\ M_\oplus$;
(d) $R_p = 12 \pm 1\ R_\oplus$, $M_p = 300 \pm 80\ M_\oplus$;
(e) $R_p = 3 \pm 1\ R_\oplus$, $M_p = 30 \pm 10\ M_\oplus$;
(f) $R_p = 2 \pm 0.1\ R_\oplus$, $M_p = 9 \pm 1\ M_\oplus$.

14-17. Under certain circumstances, a planet's composition can be well constrained without measurement of its mass (nor the composition of its atmosphere or surface). Describe such a planet.

14-18. Under certain circumstances, a planet's composition can be well constrained without measurement of its radius (nor the composition of its atmosphere or surface). Describe such a planet.

CHAPTER 15

Planet Formation

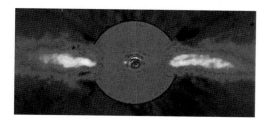

The origin of the Earth and planets is one of the most involved problems facing science today, and one that can only be solved by recourse to many disciplines.

Victor S. Safronov, *Evolution of the Protoplanetary Cloud and Formation of the Earth and Planets*, 1969

The origin of the Solar System is one of the most fundamental problems of science. Together with the origin of the Universe, galaxy formation and the origin and evolution of life, it is a crucial piece in understanding where we come from. Because planets are difficult to study at interstellar distances, we have detailed knowledge of only one planetary system, the Solar System. Data from other planetary systems are now beginning to provide further constraints (see Chapter 14). But even though 99% of known planets orbit stars other than the Sun, the bulk of the data available to guide modelers of planet formation is from objects within our Solar System. Models of planetary formation are developed using the detailed information of our own Solar System, supplemented by astrophysical observations of extrasolar planets, circumstellar disks and star-forming regions. These models are used together with observations to estimate the abundance and diversity of planetary systems in our galaxy, including planets that may harbor conditions conducive to the formation and evolution of life (see Chapter 16).

If this text were from a chronological perspective, Solar System formation would have been covered prior to our discussion of planetary processes and properties. But because modeling of planet formation relies heavily on observations and models of planetary bodies, we have deferred our in-depth treatment of this important topic until now, the last chapter of this book that is devoted to the physical sciences.

We begin in §15.1 by summarizing those observations within our Solar System that have been most useful in constraining models of the formation of planetary systems. Complementary observations of young stars and their circumstellar (likely in many cases to be protoplanetary) disks are then summarized in §15.2.

Sections 15.3–15.11 describe models of various aspects of the planet formation process. In §15.3, we discuss the formation and evolution of the protoplanetary disk. Growth of solid objects

from microscopic grains to lunar-sized and larger bodies is described in §15.4. The final stages of terrestrial planet formation are considered in §15.5 and models for the accretion of gaseous envelopes of giant planets in §15.6. Planets can move towards (or under some circumstances away from) their star during and after formation (§15.7). Giant planets play a major role in the formation of asteroids and comets (§15.8). Section 15.9 presents models for the sources of the rotational angular momentum of terrestrial and giant planets. Several different mechanisms described in §15.10 are thought to have produced the diverse ensemble of moons observed in our Solar System. Models that have been developed to explain those exoplanets that are quite different from any bodies orbiting the Sun are presented in §15.11.

We conclude the chapter by assessing the capability of planetary models to explain the Solar System properties described in §15.1 and exoplanet characteristics presented in Chapter 14.

15.1 Solar System Constraints

Any theory of the origin of our Solar System must explain the following observations:

Orbital Motions, Spacings and Planetary Rotation: The orbits of most planets and asteroids are nearly coplanar, and this plane is near that of the Sun's rotational equator. The planets orbit the Sun in a prograde direction (the same sense as the Sun rotates) and travel on nearly circular trajectories. Most planets rotate around their axis in the same direction in which they revolve around the Sun and have obliquities of $<30°$. Venus and Uranus are exceptions to this rule (see Tables E.2 and E.3). Major planets are confined to heliocentric distances $\lesssim 30$ AU, and the separation between orbits increases with distance from the Sun (Table E.1). Most of the smaller bodies orbiting the Sun (asteroids, Kuiper belt objects, etc.) move on somewhat more eccentric and inclined paths (Tables E.6 and

E.7), and their rotation axes are more randomly oriented. Aside from the asteroid belt between 2.1 and 3.3 AU and the regions centered on the stable triangular Lagrangian points of Jupiter and Neptune, interplanetary space contains very little stray matter. But this does not provide a significant constraint on the planet formation epoch because most orbits within these empty regions are unstable to perturbations by the planets on timescales short compared with the age of the Solar System (Fig. 2.15). I.e., bodies initially in orbits traversing these regions would likely have collided with a planet or the Sun or been ejected from the Solar System. Thus, in a sense, the planets are about as closely spaced as they could possibly be.

Angular Momentum Distribution: Although the planets contain $\lesssim 0.2\%$ of the Solar System's mass, more than 98% of the angular momentum in the Solar System resides in the orbital motions of the giant planets. In contrast, the orbital angular momenta of the satellite systems of the giant planets are far less than the spin angular momenta of the planets themselves.

Age: Radioisotope ^{207}Pb/^{206}Pb dating of refractory inclusions (calcium-aluminum inclusions) found within chondritic meteorites, the oldest Solar System solids known, yields an age of 4.568 Gyr. Dating with other isotope systems yields similar ages. Chondrules, as well as most differentiated meteorites that originated within small bodies, solidified only a few million years later (§11.7). Rocks formed on the Moon and Earth are younger: Lunar rocks are typically between 3 and 4.4 Gyr old, and terrestrial rocks are $\lesssim 4$ Gyr old.

Sizes and Densities of the Planets: The relatively small terrestrial planets and the asteroids, which are mainly composed of rocky material, lie closest to the Sun. The uncompressed (zero pressure) density (§6.2.2) decreases with heliocentric distance, which suggests a larger fraction of heavier elements, such as metals and other refractory (high condensation temperature) material, in planets closer to the Sun. At larger distances we find the giants Jupiter and Saturn and farther out the somewhat smaller Uranus and Neptune. The low densities of these planets imply lightweight material. Whereas Jupiter and Saturn are primarily composed of the two lightest elements, hydrogen and helium (Jupiter has $\sim 90\%$ H and He by mass and Saturn $\sim 80\%$), Uranus and Neptune contain relatively large amounts of ices and rock (they contain ~ 10–15% H and He by mass).

Shapes and Densities of Small Bodies: Smaller bodies tend to be more irregularly shaped. This is a consequence of their weaker gravity, but it also implies that either they were never molten or have sustained disruptive collisions subsequent to their resolidification. Bodies of radii $R \lesssim 100$ km tend to be of lower density (for a given surface composition), implying substantial porosity on microscopic and/or macroscopic scales.

Asteroid Belt: Countless minor planets orbit between Mars and Jupiter. The total mass of this material is $\sim 1/20$ the mass of the Moon. Except for the largest asteroids ($R \gtrsim 100$ km), the size distribution of these objects is similar to that expected from a collisionally evolved population of bodies (see Chapter 12).

Kuiper Belt: Most small bodies within the Solar System orbit beyond Neptune. The greatest concentration of such bodies is within a flattened disk at heliocentric distances between 35 and 50 AU.

Comets: There is a 'swarm' of ice-rich solid bodies orbiting the Sun at $\gtrsim 10^4$ AU, commonly referred to as the Oort cloud. There are roughly 10^{12}–10^{13} objects larger than 1 km in this 'cloud'. The bodies are isotropically distributed around the Sun, aside for a slight flattening produced by galactic tidal forces. The Kuiper belt and the scattered disk represent a second comet reservoir and provide most of the Jupiter family comets (JFC).

Moons: Most planets, including all four giant planets, have natural satellites. Almost all close-in satellites orbit in a prograde sense in a plane closely aligned with the planet's rotational equator.

They are locked in synchronous rotation, so their orbital periods are equal to their rotation periods. Many of the smaller, distant satellites (as well as Triton, Neptune's large and not so distant moon) orbit the planet in a retrograde sense and/or on orbits with high eccentricity and inclination (Table E.4). Planetary satellites are primarily composed of a mixture of rock and ice in varying proportions. Jupiter's Galilean moons imitate a miniature planetary system, with the density of the satellites decreasing with distance from their planet.

Planetary Rings: All four giant planets have ring systems orbiting in their equatorial planes. Ring particles travel on prograde trajectories, and most rings lie interior to most sizable moons.

Satellites of Minor Planets: Numerous small bodies orbiting the Sun have satellites. In some cases, the primary and secondary are similarly sized, but in others, one body dominates. The distribution of satellite sizes and orbits within the Kuiper belt is different from that in the asteroid belt (§12.3.4).

Meteorites: Meteorites display a great deal of spectral and mineralogical diversity. The crystalline structure of many inclusions within primitive meteorites indicates rapid heating and cooling events. Interstellar grains contained in chondrites imply that some parts of the protoplanetary disk remained cool, whereas high-temperature inclusions that clearly formed in the Solar System show that other parts were subjected to far hotter conditions. The intermingling of grains with different thermal histories within individual meteorites indicates substantial mixing of solid material within the disk. The small spread in ages among most meteorites indicates that the accretion epoch was brief, and the presence of decay products of various short-lived nuclei in chondritic meteorites demonstrates that solid material accreted rapidly. There is also evidence for (local) magnetic fields of the order of 10^{-4} T during the planet formation epoch.

Isotopic Composition: Although elemental abundances vary substantially among Solar System bodies, isotopic ratios are remarkably uniform. This is true even for bulk meteorite samples. Most isotopic variations that have been observed can be explained by mass fractionation (§11.4) or as products of radioactive decay. Some of these decay products imply that short-lived radionuclides were present when the material in which they now reside solidified. The similarity of the isotopic ratios suggests a well-mixed environment. However, small-scale variations in the isotopic ratios of oxygen and a few trace elements in some primitive meteorites imply that the protoplanetary nebula was not completely mixed on the molecular level, i.e., that some presolar grains did not vaporize.

Differentiation and Melting: The interiors of all of the major planets, many asteroids and most if not all large moons are differentiated, with most of the heavy material confined to their cores. This implies that each of these bodies was warm at some time in the past.

Composition of Planetary Atmospheres: The elements that make up the bulk of the atmospheres of both the terrestrial planets and planetary satellites can form compounds that are condensable at temperatures that prevail on Solar System bodies; hydrogen and noble gases are present in far less than solar abundances. Giant-planet atmospheres consist primarily of H_2 and He, but have enhanced abundances of most if not all ice-forming elements; this enhancement increases from Jupiter to Saturn to Uranus/Neptune.

Surface Structure: Most planets and satellites show many impact craters, as well as past evidence of tectonic and/or volcanic activity. A few bodies are volcanically active at the present epoch. Other surfaces appear to be saturated with impact craters. At current impact rates, such a high density of craters could not have been produced over the age of the Solar System.

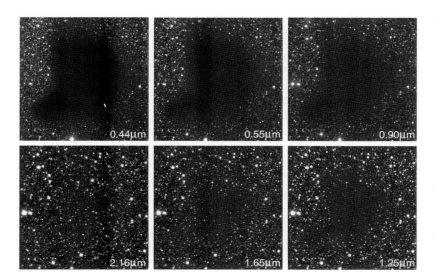

Figure 15.1 The sky area of the globule Barnard 68 in the Ophiuchus star-forming region, imaged in six different wavebands, clockwise from the blue to the near-infrared spectral region. The obscuration caused by the cloud diminishes dramatically with increasing wavelength, implying that most of the dust is in the form of sub-μm grains. Because the outer regions of the cloud are less dense than the inner ones, the apparent size of the cloud also decreases as wavelength increases, with more background stars shining through the outer parts. (European Southern Observatory PR Photo 29b/99)

15.2 Star Formation: A Brief Overview

In analogy with current theories on star formation, it is generally thought that our Solar System was 'born' in a dense (by the standards of interstellar space) molecular cloud as the result of gravitational collapse. In the remainder of this chapter, we review current ideas on star formation; the formation of a disk around a (proto)star; and, finally, the evolution of such a disk and the accretion (growth) of planets, moons, etc.

15.2.1 Molecular Cloud Cores

Our Milky Way galaxy contains a large number of cold, dense molecular clouds, one of which is shown at several wavelengths in Figure 15.1. Molecular clouds are the densest component of the **interstellar medium (ISM)** and range in size from giant systems with masses of $\sim 10^5$–10^6 M$_\odot$ to small ~ 0.1–10 M$_\odot$ cores. The small cores are usually embedded in the larger complexes. The cores from which stars form have densities of $\sim 10^{11}$

molecules per cubic meter and temperatures of only ~ 10 K. Molecular clouds consist mostly of H$_2$ and presumably He (helium is extremely difficult to detect remotely unless it is quite hot because this noble gas is chemically inert and holds its electrons quite tightly). Many other molecules are present, including numerous combinations of H, C, N, O and S. All of these more massive molecules combined, however, make up only a small fraction of the total mass of the cloud.

The typical interstellar cloud is stable against collapse. Its internal pressure (ordinary gas pressure augmented by magnetic fields, turbulent motions and rotation) is more than sufficient to balance the inward pull of self-gravity. This excess pressure would cause the cloud to expand, were it not for the counterbalancing pressure of surrounding gas of higher temperature ($\sim 10^4$ K) and lower density ($\sim 10^5$ atoms m^{-3}).

In equilibrium bound self-gravitating systems where magnetic pressure and external pressure can be ignored, the **virial theorem** (eq. 3.22) is applicable, so the (gravitational) potential energy of the core is twice the kinetic energy. The kinetic energy of a gas cloud is primarily thermal energy unless the cloud is highly turbulent or rapidly rotating. When $|E_G| > 2E_K$, the cloud may collapse under its own self-gravity. The potential energy of a uniform density mass scales as the square of the cloud's mass divided by the size of the cloud, M^2/R, which implies a dependence on the fifth power of size. In contrast, thermal energy is linearly proportional to the cloud's mass, i.e., the third power of size. Thus, the ratio of gravitational potential energy to thermal energy, $|E_G|/E_K$, increases with size in proportion to R^2, and a sufficiently massive uniform density cloud must collapse. Problem 15-1 explains how to solve for the critical mass at which instability occurs, which is known as the **Jeans mass**, M_J. The result is given by:

$$M_J \approx \left(\frac{kT}{G\mu_a m_{amu}} \right)^{3/2} \frac{1}{\sqrt{\rho}}. \tag{15.1}$$

A cloud with $M > M_J$ will collapse if its only means of support is thermal pressure. Note that the critical mass, M_J, decreases if the density in the cloud increases. Observed cores within molecular clouds appear to be dense enough to collapse gravitationally into objects of stellar masses. However, the density in a small, cold (10 K) cloud would need to exceed $\sim 10^{-8}$ kg m^{-3} ($\sim 10^{19}$ molecules m^{-3}) to form Jupiter-mass objects from gravitational collapse. This is much larger than the observed densities of interstellar clouds.

15.2.2 Collapse of Molecular Cloud Cores

For a molecular cloud that is in equilibrium, pressure gradients (and sometimes magnetic forces) balance gravitational forces. In contrast, if there were no forces to balance gravity, the cloud would collapse in **free fall**. The **free-fall timescale**, t_{ff}, i.e., the amount of time that it would take an unsupported core to collapse, can be calculated by considering the trajectory of a molecule that begins with zero velocity at a distance r from the center of the cloud. As shown in §2.5.2, the gravitational force exerted by a uniform spherical shell of material on matter exterior to the shell is equal to that exerted by the same mass of material located at the center of the shell. Isaac Newton demonstrated using triple integrals that no net force is exerted on matter inside the shell. So the molecule under consideration is pulled inwards by a mass of

$$M_c(r) = \frac{4}{3}\pi r^3 \rho_c, \tag{15.2}$$

where $M_c(r)$ is the mass of the core lying within a distance r from the center and ρ_c is the mean density within this region. The trajectory of a gas parcel initially at rest at a distance r from the center of the cloud can be approximated as a very eccentric ellipse with apoapse at r and periapse at the center. Thus, the time it takes the molecule to fall to the center is equal to half the period of an orbit that has semimajor axis $a = r/2$. From the generalized form of Kepler's third law (eq. 2.18), this is given by:

$$t_{ff}(r) = \left(\frac{\pi^2 r^3}{8G(\frac{4}{3}\pi r^3 \rho_c)} \right)^{1/2} = \left(\frac{3\pi}{32G\rho_c} \right)^{1/2}. \tag{15.3}$$

Note that the free-fall time for a uniform density core is independent of location. In other words, the matter in the various parts of the core would reach the center at the same time in free fall. Because cores are observed to be densest near their centers, the interiors cave in most quickly, producing an inside-out collapse.

In our galaxy, equation (15.3) predicts collapse timescales of order 10^5 yr, which would produce an average star formation rate far greater than is observed. The free-fall solution does not directly apply to star formation because pressure is important, at least initially, but it both illustrates the physics involved and yields a lower bound on the timescale of collapse.

Rotation can prevent material from continuing its collapse, and lead to the formation of a disk around the **protostar** growing at the center and/or to fragmentation. Rotation becomes a dominant effect when the centrifugal force balances the gravitational force. Unless angular momentum is redistributed during collapse, material with initial specific angular momentum L_s joins the disk at the **centrifugal radius**, r_{cen}, which is given by

$$r_{cen} \approx L_s^2/GM, \qquad (15.4)$$

where M is the mass interior to the radius under question (the relationship in eq. 15.4 would be exact if this mass were distributed in a spherically symmetric manner). If the core rotates rapidly, it may break up into two or more subclouds, where the angular momentum is taken up by the individual fragments orbiting one another. Each of the subclouds may collapse into a star, forming a binary or multiple star system. The majority of stars are observed to be in such binary or multiple systems. Cores with less angular momentum may form only a single star.

Because a molecular cloud core must contract by orders of magnitude to form a star, even an initially very slowly rotating clump contains much more angular momentum than the final star can take without breaking up. We expect, therefore, that virtually all single stars and probably many binary/multiple systems are surrounded by a flat disk of material at some stage during their formation. Although the star may contain most of the core's initial mass, most of the angular momentum is in the disk. Recall that in the Solar System today, 99.8% of the mass is in the Sun, and more than 98% of the angular momentum resides in planetary orbits.

Collapse converts gravitational potential energy into kinetic energy of the collapsing material, and intermolecular collisions transfer ordered motion into random thermal motions. If this energy is retained, the core's temperature rises substantially, and the collapse ceases. However, if this energy is lost, e.g., via radiation, then the cloud becomes even more unstable.

Molecular cloud cores tend to start out tenuous and sufficiently transparent at infrared wavelengths that most of the thermal energy is radiated away, so the core stays relatively cool. But the increasing density eventually makes the core opaque, which prevents thermal energy from readily escaping. Released gravitational energy then heats the protostar growing at the center of the core, building up the internal pressure until hydrostatic equilibrium (balance between gravity and the pressure gradient; §3.2) is reached. When the temperature inside the protostar gets hot enough ($\sim10^6$ K), nuclear reactions (the conversion of deuterium into helium; eq. 3.32b) start. The energy generated by this process is sufficient to temporarily forestall further contraction. When the supply of deuterium becomes exhausted, the star shrinks and heats up until the central temperature reaches the $\sim10^7$ K value required for ^{1}H fusion (eqs. 3.32) at a rate sufficient to prevent further collapse.

15.2.3 Young Stars and Circumstellar Disks

Numerous young stars have been observed within several molecular clouds. Young stars that are still contracting towards the main sequence are called **pre-main sequence stars**. Excess emission at infrared wavelengths, indicative of circumstellar material extending out to tens or hundreds of AU from the star, is observed in 25%–50% of pre-main sequence solar mass stars. *Hubble Space Telescope* (*HST*) images, such as the ones shown in Figure 15.2, have revealed disklike structures of order 100 AU in radius around young stars. Millimeter data indicate disklike structures with masses between ~0.001 and 0.1 $M_\odot$ around several protostars. Such massive **protoplanetary disks** are also observed around some young stars, but only much less massive disks have been seen around older stars.

Cool **debris disks** of solid particles have been observed around the nearby star β Pictoris (Fig. 14.19a) and many other main sequence stars. These circumstellar disks typically extend a few hundred

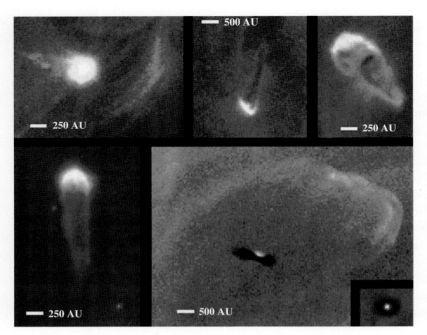

Figure 15.2 Young stars with disks in the Orion Nebula. The *top row* and the image at the *lower left* show disks of gas and dust that are being photo-evaporated by ultraviolet radiation from nearby massive stars. The other two images in the *bottom row* show silhouettes where disks associated with young stars obscure light from background hot gas. Note that the sizes of these disks are considerably larger than the planetary region of our Solar System. (NASA/*HST* images by J. Bally, D. Devine and R. Sutherland)

AU from the star, but their optical depths are small, and the observed particles may contain as little as ~1 $M_{\mathbb{C}}$ of material. Although small dust dominates the radiating area of these disks, such small particles could not have survived for the lifetimes of the stars, so larger (source) particles must also be present. These disks are typically more prominent around younger main sequence stars, but some older stars have fairly bright disks. *HST* images of the 1500-AU-wide β Pictoris disk, such as the one shown in Figure 15.3, reveal that the inner part of the disk is warped, a shape that may be produced by the gravitational pull of one or more nearby planets or a brown dwarf on an inclined orbit.

15.3 Evolution of the Protoplanetary Disk

Based on observations of star formation in our galaxy at the present epoch, we assume that our Sun and planetary system formed in a molecular cloud. The growing Sun together with its surrounding disk are referred to as the **primitive solar nebula**; the planetary system formed from the **protoplanetary disk** within this nebula. A **minimum mass** of ~0.02 $M_{\odot}$ for the protoplanetary disk can be derived from the present abundance of refractory elements in the planets and the assumption that the abundances of the elements throughout the nebula were solar. The actual mass was probably significantly larger because some (perhaps most) of the refractory component of this mixture was not ultimately incorporated into planets. The history of our solar nebula can be divided into three stages: infall, internal evolution and clearing.

15.3.1 Infall Stage

When a molecular cloud core becomes dense enough that its self-gravity exceeds thermal, turbulent and magnetic support, it starts to collapse. Because the core is densest near the center, collapse proceeds from the inside out. Infall continues until the reservoir of cloud material is

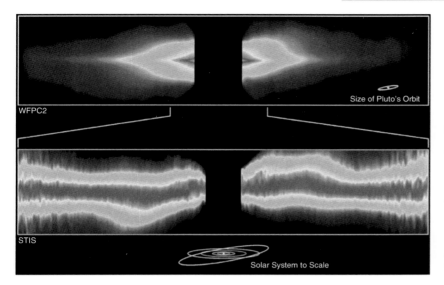

WFPC2

Size of Pluto's Orbit

STIS

Solar System to Scale

Figure 15.3 *HST* images of the inner portion of the dust disk around the star β Pictoris. The bright glare of the central star is blocked by a coronagraph. The warps in the disk might be caused by the gravitational pull of one or more unseen (planetary?) companions. (Courtesy Al Schultz, *HST*/NASA)

exhausted or until a strong stellar wind reverses the flow. The duration for the infall stage is comparable to the free-fall collapse time of the core, $\sim 10^5$–10^6 yr.

Initially, gas and dust with low specific angular momentum relative to the center of the core falls towards the center, forming a protostar. Eventually, matter with high specific angular momentum falls towards the protostar but cannot reach it because of centrifugal forces. Essentially, the material is on orbits that do not intersect the central, pressure-supported star. However, as the gas and dust mixture falls to the equatorial plane of the system, it is met by material falling from the other direction, and motions perpendicular to the plane cancel. The energy in this motion is dissipated as heat in the forming disk. Significant heating can occur, especially in the inner portion of the disk, where the material has fallen deep into the potential well. The equatorial plane of the resultant disk is roughly perpendicular to the rotation axis of the initial collapsing molecular cloud core. The direction of the core's angular momentum determines the plane of the disk, whereas the magnitude of the angular momentum governs how the material is divided between the protostar and its disk.

Consider a parcel of gas that falls from infinity to a circular orbit at $r_\odot$. Half of the gravitational energy per unit mass is converted to orbital kinetic energy:

$$\frac{GM_{\text{protostar}}}{2r_\odot} = \frac{v_c^2}{2};$$ (15.5)

the other half is available for heat. At 1 AU, the circular velocity $v_c = 30$ km s^{-1} if $M_{\text{protostar}} = 1$ M$_\odot$. If no energy escapes the system, it follows that the temperature in a hydrogen gas would be $\sim 7 \times 10^4$ K (Problem 15-3). However, this very high temperature is never actually attained because the timescale for radiative cooling is much shorter than the heating time.

Gas reaches supersonic velocities as it descends towards the midplane of the nebula. The gas slows abruptly when it passes through a **shock front** as it is accreted onto the disk. Models of protoplanetary disk formation suggest that typical postshock temperatures for the protoplanetary disk are ~ 1500 K at 1 AU and ~ 100 K at 10 AU. Equilibrium is reached when all forces balance, i.e., the gravitational force towards the center balances with the centrifugal force outward and the gravitational force toward the midplane balances with the pressure gradient outwards.

Near the midplane of the disk, $z = 0$, the vertical component of the star's gravity is well approximated by:

$$g_z = \frac{GM_\odot}{r_\odot^3} z. \tag{15.6}$$

The situation is somewhat analogous to a planetary or stellar atmosphere in that gas pressure supplies support against gravity in the direction perpendicular to the midplane. But unlike the case of an atmosphere wherein the gravitational field can often be approximated by a constant, the z-component of the star's gravity vanishes where $z = 0$, and varies linearly with height above the midplane. Provided stellar gravity dominates that of the disk and temperature variations in the z-direction can be ignored, the equation of hydrostatic equilibrium (eq. 3.13b) becomes:

$$\frac{dP}{dz} = -\rho \frac{GM_\odot}{r_\odot^3} z. \tag{15.7}$$

Using the ideal gas law (eq. 3.14) to eliminate P from equation (15.7) leaves:

$$\frac{d\rho}{dz} = -\frac{\mu_a m_H}{kT} \rho \frac{GM_\odot}{r_\odot^3} z, \tag{15.8}$$

which can be rearranged as:

$$\int \frac{d\rho}{\rho} = \int -\frac{\mu_a m_H}{kT} \frac{GM_\odot}{r_\odot^3} z \, dz \tag{15.9}$$

and then integrated to give:

$$\ln \rho = -\frac{\mu_a m_H}{2kT} \frac{GM_\odot}{r_\odot^3} z^2 + c_o, \tag{15.10}$$

where c_o is a constant of integration. Exponentiating equation (15.10) and evaluating the constant of integration by noting that $\rho = \rho_0$ at $z = 0$ yields:

$$\rho = \rho_0 e^{-z^2 / \left(\frac{2kT}{\mu_a m_H} \frac{r_\odot^3}{GM_\odot}\right)}. \tag{15.11}$$

The gas density and pressure variations perpendicular to the midplane of the disk are thus given by

$$\rho_{g_z} = \rho_{g_{z_0}} e^{-z^2 / H_z^2}, \tag{15.12a}$$

$$P_z = P_{z_0} e^{-z^2 / H_z^2}, \tag{15.12b}$$

where the **Gaussian scale height**, H_z, is given by:

$$H_z = \sqrt{\frac{2kT r_\odot^3}{\mu_a m_{amu} GM_\odot}}. \tag{15.13}$$

Note that the Gaussian form of the density and pressure variations with distance from the midplane of the disk has a fundamentally different shape from the exponential dropoff characteristic of a uniform temperature planetary atmosphere given in equation (3.18).

15.3.2 Disk Dynamical Evolution

Unless the collapsing cloud has negligible rotation, a significant amount of material lands within the disk. Redistribution of angular momentum within the disk can then provide additional mass to the star. Processes affecting the disk at this epoch are illustrated schematically in Figure 15.4. The structure and evolution of the disk are primarily determined by the efficiency of the transport of angular momentum and heat. Angular momentum and mass can be transported in the following ways.

Magnetic Torques: If magnetic field lines from the star thread through the disk, then there is a tendency towards corotation, i.e., material orbiting more rapidly than the star's spin period loses angular momentum and that orbiting less rapidly gains angular momentum. The field lines couple the star to the disk if the gas in the disk is sufficiently ionized, which it tends to be in the innermost parts of the disk, where temperatures are high. Angular momentum is transferred outwards, from the star to the disk, but transfer to larger radii is inhibited by the lack of ionized gas in the disk and the weakness of stellar magnetic field lines at greater distances.

Gravitational Torques: Local or global gravitational instabilities may lead to rapid transport of

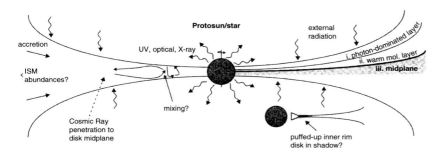

Figure 15.4 Schematic diagram of a protoplanetary disk that is subject to radiation from both its central star and more distant but brighter stars. (Pudritz et al. 2007)

material within the protoplanetary disk. As discussed in §13.2, a thin rotating disk is unstable to local axisymmetric perturbations if Toomre's parameter $Q_T < 1$ (eq. 13.12; note that for a gaseous disk, the sound speed replaces the velocity dispersion). Nonaxisymmetric local instabilities also occur when $Q_T \lesssim 1$. These instabilities can produce spiral density waves, which transport mass and angular momentum on a dynamical timescale until a stable configuration is again reached. This limits the disk mass to a value less than or comparable to the protostar's mass.

Large protoplanets may clear annular gaps surrounding their orbits and excite density waves at resonant locations within the protoplanetary disk (see Fig. 15.12). These density waves transfer angular momentum outward. Such processes have been observed in Saturn's rings (§13.4), albeit on a much smaller scale.

Viscous Torques: Because molecules revolve around the protosun in roughly Keplerian orbits, those closer to the center move faster than those farther away. Collisions among the gas molecules speed up the outer molecules, hence driving them outward, and slow down the inner molecules, which then fall towards the center. The net effect is that most of the matter diffuses inward, angular momentum is transferred outward and the disk as a whole spreads. The disk evolves on a diffusion timescale, given by equation (13.10), where the length scale, ℓ, is equal to the radius of the disk (or that portion of the disk under consideration).

15.3.3 Chemistry in the Disk

The chemical composition of protoplanetary disks determines what raw materials are available for planetesimal formation. The initial chemical state of the disk depends on the composition of the gas and dust in the interstellar medium, as well as subsequent chemical processing during the collapse phase. Because the Sun formed from the same reservoir of raw materials as its protoplanetary disk, the abundances of the elements in the Sun tell us what the original *elemental* composition of the disk was. But because the Sun is too hot for molecules to be stable, it does not provide information about the chemical compounds within which these elements resided while in the disk.

Equilibrium Condensation

The chemical evolution of interstellar matter as it is incorporated into planetesimals determines the compositions of planets. Gas cools after passing through the shock front that it encounters while entering the protoplanetary disk. The chemical composition can be calculated within those regions in which nebular material has experienced temperatures high enough to completely evaporate and dissociate all incoming interstellar gas and dust (>2000 K). At such high temperatures, the chemistry can be assumed to be in thermodynamic equilibrium because the chemical reaction rates are rapid compared with the cooling rate of

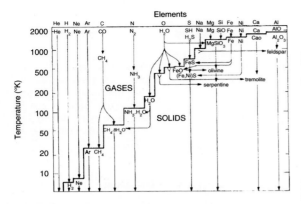

Figure 15.5 Flow chart of major reactions during fully equilibrated cooling of solar nebula material from 2000 to 5 K. The 15 most abundant elements are listed across the top, and directly beneath are the dominant gas species of each element at 2000 K. The staircase curve separates gases from condensed phases. (Barshay and Lewis 1976)

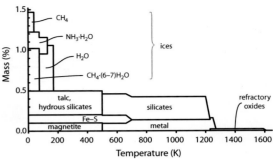

Figure 15.6 Amount and composition of major condensed components formed during fully equilibrated cooling of solar nebula material. (Lodders 2010)

the disk. This situation is likely to occur close to the protostar.

As a protoplanetary disk cools, elements condense out of the gas and undergo chemical reactions at different temperatures. Figure 15.5 illustrates the major compounds formed by this process. Refractory minerals such as rare-earth elements and oxides of aluminum, calcium and titanium (e.g., corundum, Al_2O_3, and perovskite, $CaTiO_3$) condense at a temperature of $\sim$1700 K. At $T \sim$ 1300 K, iron and nickel condense to form an alloy; at slightly lower temperatures, magnesium silicates appear. If chemical equilibrium is maintained, chemical reactions in the gas and with the dust take place as the temperature drops, such as the reactions of iron with H_2S to form troilite (FeS) at $\sim$700 K and with water to form iron oxide (Fe + H_2O $\rightarrow$ FeO + H_2) at $\sim$500 K. If grain growth is rapid compared with cooling, then these molecules cannot react with iron, and they remain in gaseous form until it is cold enough for them to condense as ices. As shown in Figure 15.6, the bulk of condensation occurs at temperatures between 1200 and 1300 K and below 200 K.

Water plays an extremely important role below 500 K and condenses as pure water-ice at temperatures below 200 K. Assuming equilibrium is maintained, ammonia and methane gas condense as hydrates and clathrates, respectively ($NH_3 \cdot H_2O$, $CH_4 \cdot 6H_2O$), at temperatures somewhat lower than that at which ice condenses. At temperatures of below $\sim$40 K, CH_4 and Ar ices form.

Disequilibrium Processes

The presence of complex disequilibrium compounds and both refractory and volatile grains of presolar origin (§§11.7.2, 12.7.4) implies that the basic equilibrium condensation models are too simplistic. Reactions such as CO + $3H_2$ $\rightarrow$ CH_4 + H_2O, N_2 + $3H_2$ $\rightarrow$ $2NH_3$ and the formation of hydrated silicates are thermodynamically favored at low temperatures, but they have high activation energies. Such reactions are **kinetically inhibited** because they take a very long time to reach equilibrium, longer than nebular evolution allows. Because equilibrium cannot be assumed, the condensation sequence at low temperatures is quite uncertain.

When the nebula cools below a temperature at which the chemical reaction times become comparable to the timescale of cooling, the chemistry becomes more complicated. This **freeze-out temperature** is different for different species. For example, the CO/CH$_4$ and N$_2$/NH$_3$ ratios are sensitive functions of the temperature and pressure in the nebula. At the low pressures given by models of the solar nebula, carbon is thermodynamically most stable in the form of CO at $T \gtrsim 700$ K and in the form of CH$_4$ at lower temperatures. Nitrogen is most stable as N$_2$ at $T \gtrsim 300$ K and as NH$_3$ at lower temperatures. Thus, if the protoplanetary disk were in thermodynamic equilibrium, CO and N$_2$ would dominate in the warm inner nebula, but in the cold outer nebula, CH$_4$ and NH$_3$ would be the favored forms of C and N, respectively.

Several lines of evidence imply that the solar nebula was not in equilibrium. The existence of N$_2$ and CO ices on Pluto and Triton, for example, suggests that the outer solar nebula did not have enough time to equilibrate chemically. The depletion of the N/C ratio in comets relative to the solar abundances of these elements (§12.7.6), even though NH$_3$ has a significantly higher condensation temperature than CH$_4$, also indicates that these substances did not achieve chemical equilibrium.

The *Galileo* probe discovered enhancements of $\sim$3–6 times solar in the moderately volatile elements C, N and S, as well as the noble gases Ar, Kr and Xe in Jupiter's atmosphere (at a pressure of $\sim$10 bar). Because these elements have a broad range of condensation temperatures, the small range in enhancements suggests that these elements were brought in by planetesimals that condensed at temperatures low enough for these elements to either be trapped within H$_2$O-ice or be stable as solids.

15.3.4 Clearing Stage

Because there is no gas left between the planets, the gas must have been cleared away at some stage during the evolution process. The gas may have been removed via ablation from the faces of the protoplanetary disk by ultraviolet radiation emanating from nearby hot stars (Fig. 15.2) and/or from the young, not yet settled and highly active Sun. The timing of the gas loss is a crucial issue concerning the growth of giant planets but is not directly constrained; it is therefore usually assumed that the lifetime of the gaseous protoplanetary disk was the same as that typical of massive dust disks around young stars, $\lesssim 10^7$ yr.

15.4 Growth of Solid Bodies

As a disk of gaseous matter cools, various compounds condense into microscopic grains. For a disk of solar composition, the first substantial condensates are silicates and iron compounds. At lower temperatures, characteristic of the outer region of our planetary system, large quantities of water-ice and other ices can condense (§15.3.3). These cooler regions may have also retained a significant amount of preexisting condensates from the interstellar medium and stellar atmospheres. Growth of solid particles then proceeds primarily by mutual collisions.

15.4.1 Planetesimal Formation

The motions of small grains in a protoplanetary disk are strongly coupled to the gas. Nonetheless, the vertical component of the star's gravity (eq. 15.6) causes the dust to slowly sediment out towards the midplane of the disk. Collisional growth of grains during their descent to the midplane of the disk shortens sedimentation times, and sedimentation in turn increases growth rates. This process accounts for particulate growth up to the millimeter to centimeter size range.

Growth from centimeter-sized particles to kilometer-sized planetesimals depends primarily on the relative motions between the various bodies. The motions of (sub)centimeter-sized material in the protoplanetary disk are strongly coupled to the gas. The gas in the protoplanetary disk is partially supported against stellar gravity by a pressure gradient in the radial direction, so gas circles the star slightly less rapidly than the Keplerian rate for circular orbits (eq. 2.21). The 'effective' gravity felt by the gas is given by equation (2.64). The second term on the right-hand side of equation (2.64) is the outward force (per unit mass) produced by the pressure gradient. For circular orbits, the effective gravity must be balanced by centrifugal acceleration, $g_{eff} = -r_\odot n^2$. Because the pressure gradient is much smaller than the gravity, we can approximate the angular velocity of the gas, n_{gas}, as

$$n_{gas} \approx \sqrt{\frac{GM_\odot}{r_\odot^3}}(1-\eta), \qquad (15.14)$$

where

$$\eta \equiv \frac{-r_\odot^2}{2GM_\odot \rho_g}\frac{dP}{dr_\odot} \approx 5 \times 10^{-3}. \qquad (15.15)$$

For estimated protoplanetary disk parameters, the gas rotates ~0.5% slower than the Keplerian speed.

Large particles moving at (nearly) the Keplerian speed thus encounter a headwind that removes part of their orbital angular momentum and causes them to spiral inwards towards the star. Small grains drift less because they are so strongly coupled to the gas that the headwind they encounter is very slow. Kilometer-sized planetesimals also drift inwards very slowly because their ratio of surface area to mass is small. As a consequence of the difference in (both radial and azimuthal) velocities, small (sub)centimeter grains can be swept up by the larger bodies.

Peak rates of inward drift occur for particles that collide with roughly their own mass of gas in one orbital period. Meter-sized bodies in the terrestrial planet region of the solar nebula drift inwards at the fastest rate (Problem 15-4), up to ~10^6 km yr^{-1}. Thus, a meter-sized body at 1 AU would spiral inwards, approaching the Sun in ~100 years! This radial migration can remove solids from the planetary region or bring particles of various sizes

together to enhance accretion rates. Thus, the material that survives to form planets must complete the transition from centimeter to kilometer size rather quickly.

The growth of solid bodies from millimeter size to kilometer size within a turbulent protoplanetary disk presents particular problems. The physics of interparticle collisions in this size range is not well understood. Furthermore, the high rate of orbital decay caused by gas drag for meter-sized particles implies that growth through this size range must occur very rapidly. Various hypotheses describe this phase of growth. Differing models of planetesimal formation yield a wide variety of size distributions for the initial population of planetesimals. In some scenarios, planetesimals much larger than 1 km are expected.

15.4.2 From Planetesimals to Planetary Embryos

The primary factors controlling the growth of planetesimals into planets differ from those responsible for the accumulation of dust into planetesimals. Solid bodies larger than ~1 km in size face a headwind only slightly faster than that experienced by 10-m objects (for parameters thought to be representative of the terrestrial region of the solar nebula). Since their mass-to-surface-area ratios are much larger than those of small particles, their orbits are not significantly affected by interactions with the gas in their path. The primary perturbations on the Keplerian orbits of kilometer-sized and larger bodies in protoplanetary disks are mutual gravitational interactions and physical collisions. These interactions lead to accretion (and in some cases erosion and fragmentation) of planetesimals.

The size distribution of planetesimals evolves principally via physical collisions among its members. Physical collisions between solid bodies can lead to accretion, fragmentation or inelastic rebound of relatively intact bodies; intermediate outcomes are possible as well. The outcome of a collision depends on the internal strength of the planetesimals, the coefficient of restitution of the bodies and most sensitively on the kinetic energy of the collision. Conversion of gravitational potential energy into kinetic energy as the bodies approach one another implies that the speed at which two bodies of radii R_1 and R_2 and masses m_1 and m_2 collide is given by

$$v_i = \sqrt{v_\infty^2 + v_e^2}, \qquad (15.16)$$

where v_∞ is the speed of m_2 relative to m_1 far from encounter and v_e is their mutual escape velocity from the point of contact:

$$v_e = \left(\frac{2G(m_1 + m_2)}{R_1 + R_2} \right)^{1/2}. \qquad (15.17)$$

Note that equation (15.17) is equivalent to the first equality in equation (2.24) with $M = m_1 + m_2$ and $r = R_1 + R_2$.

The impact speed is thus at least as large as the escape velocity, which for a rocky 10-km sized object is ~ 6 m s^{-1}. The rebound speed is equal to ϵv_i, where the **coefficient of restitution** $\epsilon \leq 1$. If $\epsilon v_i < v_e$, then the bodies remain bound gravitationally and soon recollide and accrete. Net disruption requires both fragmentation, which depends on the internal strength of the bodies, and post-rebound velocities greater than the escape speed. Because relative velocities of planetesimals are generally less than the escape velocity from the largest common bodies in the swarm, the largest members of the swarm are likely to accrete the overwhelming bulk of the material with which they collide unless ϵ is very close to unity. Very small planetesimals are most susceptible to fragmentation. The largest bodies in the swarm accrete at a rate essentially identical to the collision rate. Subcentimeter-sized grains corotating with the gas may impact kilometer-sized planetesimals at speeds well above the escape velocities of these planetesimals. This process could lead to erosion of planetesimals via 'sandblasting'.

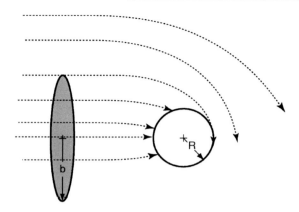

Figure 15.7 Schematic diagram of the gravitational focusing of planetesimal trajectories by an accreting planetary embryo or planet. The critical trajectory that collides tangentially with the planet has an **unperturbed impact parameter**, b, larger than the radius of the planet, $b > R$. (Adapted from Brownlee and Kress 2007)

The simplest model for computing the collision rate of planetesimals ignores their motion around the Sun completely. A collision occurs when the separation between the centers of two particles equals the sum of their radii. The largest bodies in a region of the disk are known as **planetary embryos**. The mean rate of growth of a planetary embryo's mass, M, is:

$$\frac{dM}{dt} = \rho_s v \pi R^2 \mathcal{F}_g, \qquad (15.18)$$

where v is the average relative velocity between the large and small bodies, ρ_s the volume mass density of the swarm of planetesimals and the planetary embryo's radius, R, is assumed to be much larger than the radii of the planetesimals. The last term in equation (15.18) is the gravitational enhancement factor, $\mathcal{F}_g$, which results from the focusing process illustrated in Figure 15.7.

As long as relative velocities are large enough that the **Kepler shear** of the disk (the gradient in circular velocities in the radial direction, i.e., at differing heliocentric distances) can be ignored, the expression for $\mathcal{F}_g$ can be calculated in a straightforward manner. To do so, we first determine the maximum unperturbed impact parameter of a planetesimal that leads to a collision with the planetary embryo. Let b_{max} be the unperturbed impact parameter corresponding to a planetesimal just barely caught by the planetary embryo. Such a planetesimal can be viewed as being on a hyperbolic orbit of the embryo with periapse distance equal to R_p. Conservation of the planetesimal's angular momentum relative to the planet as it moves from a large distance to the surface implies:

$$b_{max} v_\infty = v_i R_p. \tag{15.19}$$

Employing conservation of energy using equation (15.16) yields

$$b_{max} = \frac{v_i R_p}{v_\infty} = \frac{\sqrt{v_\infty^2 + v_e^2}}{v} R_p$$

$$= R_p \sqrt{1 + \frac{v_e^2}{v_\infty^2}}. \tag{15.20}$$

Thus, the gravitational enhancement is given by:

$$\mathcal{F}_g = \left(\frac{b_{max}}{R_p}\right)^2 = 1 + \left(\frac{v_e}{v_\infty}\right)^2. \tag{15.21}$$

When the relative velocity between planetesimals is comparable to or larger than the escape velocity, $v \gtrsim v_e$, the growth rate is approximately proportional to R^2, and the evolutionary path of the planetesimals exhibits an orderly growth of the entire size distribution. When the relative velocity is small, $v \ll v_e$, one can show, by rewriting the escape velocity in terms of the protoplanet's radius, that the growth rate is proportional to R^4 (eqs. 15.16–15.21). In this situation, the planetary embryo rapidly grows larger than any other planetesimal, which can lead to **runaway growth**. A numerical simulation of runaway growth is shown in Figure 15.8.

The growth rate of an embryo of mass M in a disk whose velocity dispersion is governed by planetesimals of mass m scales as $\dot{M} \propto M^{4/3} m^{-2/3}$. If embryos dominate the stirring, then $\dot{M} \propto M^{2/3}$; in this circumstance, if individual embryos control the velocity in their own zones,

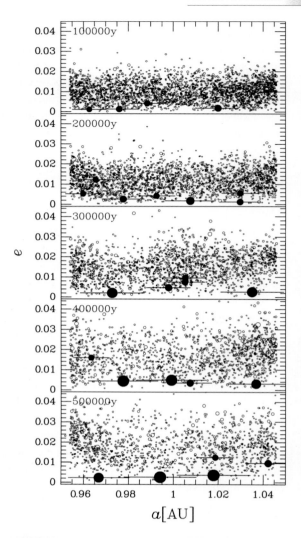

Figure 15.8 Snapshots of a simulation of a system of interacting and merging planetesimals, shown in the a–e plane. The *circles* represent planetesimals, and their radii are proportional to the radii of planetesimals. The system initially consists of 4000 planetesimals whose total mass is 1.3×10^{24} kg. The *filled circles* represent planetary embryos with mass larger than 2×10^{22} kg, and *lines* from the center of each planetary embryo extend 5 R_H outward and 5 R_H inward (see eq. 2.28 for evaluation of the Hill sphere radius, R_H). (Kokubo and Ida 2000)

larger embryos take longer to double in mass than do smaller ones, although embryos of all masses continue their runaway growth relative to surrounding planetesimals; this phase of rapid accretion of planetary embryos is known as **oligarchic**

growth. A runaway embryo can grow so much larger than the surrounding planetesimals that its $\mathcal{F}_g$ can exceed 1000; however, three-body stirring by the embryo (see Fig. 2.16) prevents $\mathcal{F}_g$ from growing much larger than this.

Runaway and oligarchic growth require low random velocities and thus small radial excursions of planetesimals. The planetary embryo's feeding zone is therefore limited to the annulus of planetesimals that it can gravitationally perturb into intersecting orbits. Thus, rapid growth ceases when a planetary embryo has consumed most of the planetesimals within its gravitational reach. Planetesimals whose orbits come within ∼4 times the planetary embryo's Hill sphere eventually come close enough to the planetary embryo during one of their orbits that they may be accreted (unless their semimajor axis is very similar to that of the embryo, in which case they may be locked in tadpole or horseshoe orbits that avoid close approaches; §2.2.2).

The vertical profile of the planetesimal swarm is similar to that of the gas given by equation (15.11), with the velocity dispersion of the solids, v, replacing the thermal velocity of the gas, $(\frac{2kT}{\mu_a m_H})^{1/2}$. The **surface mass density** of solids, σ_ρ, is given by the integral of the density of the swarm over the thickness of the disk:

$$\sigma_\rho \equiv \int_{-\infty}^{\infty} \rho_s dz = \int_{-\infty}^{\infty} \rho_{s_0} e^{-z^2 \frac{GM_\star}{v^2 r_\star^3}} dz$$

$$= \sqrt{\frac{\pi v^2 r_\star^3}{GM_\star}} \rho_{s_0}. \tag{15.22}$$

The mass of a planetary embryo that has accreted all of the planetesimals within an annulus of width $2\Delta r_\odot$ is:

$$M = \int_{r_\odot - \Delta r_\odot}^{r_\odot + \Delta r_\odot} 2\pi r' \sigma_\rho(r') dr' \approx 4\pi r_\odot \Delta r_\odot \sigma_\rho(r_\odot). \tag{15.23}$$

Setting $\Delta r_\odot = 4 R_H$, where R_H is given by eq. 2.28, and generalizing to a star of any mass, $M_\star$, we obtain the **isolation mass**, M_i (in kilograms), which is the largest mass to which a

planetary embryo orbiting can grow by runaway accretion:

$$M_i \approx 5 \times 10^{20} (r_{AU}^2 \sigma_\rho)^{3/2} \left(\frac{M_\odot}{M_\star}\right)^{1/2} \tag{15.24}$$

where σ_ρ is in kg m^{-2}. For a minimum-mass solar nebula, the mass at which runaway growth must have ceased in Earth's accretion zone would have been ∼6 $M_{\mathbb{C}}$ and in Jupiter's accretion zone ∼1 $M_\oplus$.

Runaway growth can persist beyond the isolation mass given by equation (15.24) only if solid bodies can diffuse into the planet's accretion zone. Three plausible mechanisms for such diffusion are scattering between planetesimals, perturbations by planetary embryos in neighboring accretion zones and gas drag. Alternatively, radial motion of the planetary embryo may bring it into zones not depleted of planetesimals. Gravitational torques resulting from the excitation of spiral density waves in the gaseous component of the protoplanetary disk have the potential of inducing rapid radial migration of planets (§15.7.1), as can gravitational interactions with a massive disk of planetesimals (§15.7.2). Gravitational focusing of gas could also vastly increase the rate of inward drift of planetary embryos.

The limits of runaway growth are less severe in the outer Solar System than in the terrestrial planet zone, so runaway growth of Jupiter's core may have continued until it attained the mass necessary to rapidly capture its massive gas envelope. The 'excess' solid material in the outer Solar System could have been subsequently ejected to the Oort cloud or to interstellar space via gravitational scattering by the giant planets.

In contrast, the small terrestrial planets, orbiting deep within the Sun's gravitational potential well, could not have ejected substantial amounts of material. This implies that a high-velocity growth phase subsequent to runaway accretion was required to yield the present configuration of terrestrial planets.

15.5 Formation of the Terrestrial Planets

We now consider the late stages of the growth of the inner planets in our Solar System. We begin with dynamical models and then turn to the implications of these models for the structures of the planets' interiors and the compositions of their atmospheres.

15.5.1 Dynamics of the Final Stages of Planetary Accumulation

The self-limiting nature of runaway and oligarchic growth implies that massive planetary embryos form at regular intervals in semimajor axis. The agglomeration of these embryos into a small number of widely spaced terrestrial planets necessarily requires a stage characterized by large orbital eccentricities, significant radial mixing and giant impacts. At the end of the rapid-growth phase, most of the original mass is contained in the large bodies, so their random velocities are no longer strongly damped by energy equipartition with the smaller planetesimals. Mutual gravitational scattering can pump up the relative velocities of the planetary embryos to values comparable to the surface escape velocity of the largest embryos, which is sufficient to ensure their mutual accumulation into planets. The large velocities imply small collision cross-sections and hence long accretion times.

After the planetary embryos have perturbed one another into crossing orbits, their subsequent orbital evolution is governed by close gravitational encounters and violent, highly inelastic collisions. This process has been studied using N-body integrations of planetary embryo orbits. Because the simulations endeavor to reproduce our Solar System, they generally begin with about 2 $M_\oplus$ of material in the terrestrial planet zone, typically divided (not necessarily equally) among hundreds of bodies. The end result is the formation of two to five terrestrial planets on a timescale of about 10^8 years. Sample results are shown in Figure 15.9.

Some of these systems look quite similar to our Solar System, but most have fewer terrestrial planets, and these planets travel on more eccentric orbits. It is possible that the Solar System is by chance near the quiescent end of the distribution of terrestrial planets. Alternatively, processes such as fragmentation and gravitational interactions with a remaining population of small debris, thus far omitted from the calculations because of computational limitations, may lower the characteristic eccentricities and inclinations of the ensemble of terrestrial planets.

An important result of these N-body simulations is that planetary embryo orbits execute a random walk in semimajor axis as a consequence of successive close encounters. The resulting widespread mixing of material throughout the terrestrial planet region diminishes any chemical gradients that may have existed when planetesimals formed, although some correlations between the final heliocentric distance of a planet and the region where most of its constituents originated are preserved in the simulations. Nonetheless, these dynamical studies imply that Mercury's high iron abundance is unlikely to have arisen from chemical fractionation in the solar nebula.

The mutual accumulation of numerous planetary embryos into a small number of planets must have entailed many collisions between protoplanets of comparable size. Mercury's silicate mantle was probably partially stripped off by one or more of such giant impacts, leaving behind an iron-rich core. Accretion simulations also lend support to the giant impact hypothesis for the origin of the Earth's Moon (§15.10.2); during the final stage of accumulation, an Earth-sized planet is typically found to collide with several objects as large as the Moon and frequently one body as massive as Mars.

15.5.2 Accretional Heating and Planetary Differentiation

Impacting planetesimals provide a planet with energy as well as mass. This energy heats a

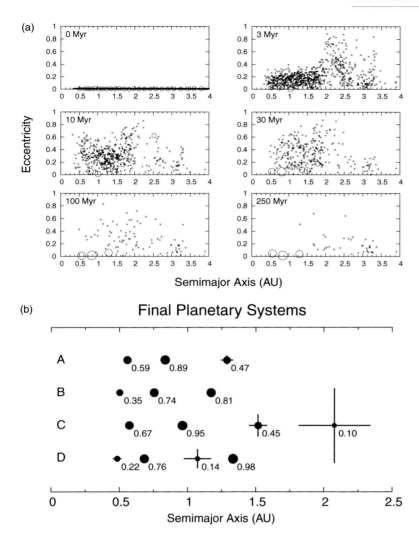

Figure 15.9 (a) Simulation of the final stages of terrestrial planet growth in our Solar System using an *N*-body code that assumes all physical collisions lead to mergers. The simulation begins with 25 planetary embryos as massive as Mars, ~1000 planetesimals each of mass 0.04 M$_{\circ}$ and Jupiter and Saturn on their current orbits. The planetary embryos and planetesimals are represented as *circles* whose radii are proportional to the body's radius and whose locations are displayed in $a–e$ phase space at the times indicated. (b) Synthetic terrestrial planet systems produced by four different *N*-body simulations of the final stages of planetary accretion. The final planets are indicated by *filled circles* centered at the planet's semimajor axis. The *horizontal line* through each circle extends from the planet's perihelion to its aphelion; the length of the *vertical line* extending upward and downward from a planet's center represents its excursions perpendicular to the invariant plane at the same scale. The numbers to the *lower right* of each circle represent the planet's final mass in M$_{\oplus}$. For example, the outermost planet in simulation A has $a = 1.29$ AU, $e = 0.035$, $i = 1.55°$ and $M_{\mathrm{p}} = 0.47$ M$_{\oplus}$. The results of the simulation shown in part (a) are presented in row A. The initial disks used for these four simulations are very similar, and the different outcomes arise from stochastic variations of accretion dynamics. See O'Brien et al. (2006) for particulars of the calculations. (Courtesy David O'Brien)

growing planet. Decay of radioactive elements (§11.6.1) also heats planetary bodies, with short-lived nuclides such as ^{26}Al being the most important for growing bodies during the first few million years of Solar System history, and potassium, thorium and uranium dominating over billion-year timescales (Problem 15-10). A heated planet or even a small planetesimal if it contains a sufficient quantity of short-lived radioisotopes may become warm enough that portions melt, allowing denser material to sink and the planet to differentiate.

The nonradiogenic energy available to a growing planet is supplied by accreted planetesimals (which contribute both their kinetic energy 'at infinity' and the potential energy released as the planetesimal falls onto the planet's surface), gravitational potential energy released as the planet contracts (in response to increased pressure) or differentiates, decay and exothermic chemical processes. The planet loses energy via radiation to space. Energy may be transported within the planet via conduction, or if the planet is (partially or fully) molten, via convection. Energy transport within a planet is important to the global as well as local heat budgets because radiative losses can only occur from the planet's surface or atmosphere.

Conduction is rather slow over distances of thousands of kilometers that are characteristic of planetary interiors, and convection operates only in regions that are sufficiently molten to allow fluid motions to occur (§4.4). Thus, to a first approximation, a growing solid planet's temperature is given by a balance between accretion energy deposited at the planet's surface, radioactive decay and radiative losses from the surface. The temperature of a given region changes slowly when it becomes buried deep below the surface (unless short-lived radionuclides are sufficiently abundant or large-scale melting and differentiation occur). For gradual accretion, the temperature at a given radius can thus be approximated by balancing the accretion energy source with radiative losses at the time when the material was accreted. For the 10^8-year accretion times estimated for the terrestrial planets,

not enough accretion energy would be retained to melt and differentiate an $M_p \lesssim 1\ M_\oplus$ planet.

However, modern theories of planetary growth imply that terrestrial planets accumulate most of their mass in planetesimals of radius 100 km and larger. Impactors deposit $\sim$70% of their kinetic energy as heat in the target rocks directly beneath the impact site, with the remaining $\sim$30% being carried off with the ejecta. If an impactor is large, it may raise deeply buried heat to near the surface, where energy may be radiated away. A more important effect is that heat may become buried by deep ejecta blankets. The ejecta blankets produced by such large impactors are thick enough that most of the heat from the impacts remains buried. Planets can thus become quite warm, with temperature increasing rapidly with radius. Accretion energy can lead to the differentiation of planetary (but not asteroidal) sized bodies (Problems 15-13 and 15-15).

A planetary embryo can form a protoatmosphere as it accretes solid bodies. When the mass of a growing planet reaches $\sim$0.01 $M_\oplus$, impacts are energetic enough for water to evaporate. Complete degassing of accreting planetesimals occurs when the radius of the planetary embryo reaches about 0.3 $R_\oplus$. A massive protoatmosphere that is optically thick to outgoing radiation can trap energy provided by the impacting planetesimals. This process, known as the **blanketing effect**, is capable of increasing the surface temperature of the protoplanet by even more than the greenhouse effect. Solar radiation determines the temperature at the top of the atmosphere and is scattered and absorbed at lower altitudes. The atmosphere provides a partially insulating blanket that retains much of the heat released from impacting planetesimals, so the surface becomes quite hot.

Calculations show that the protoatmosphere's blanketing effect becomes important when the growing planet's mass exceeds 0.1 $M_\oplus$. The surface temperature exceeds $\sim$1600 K, the melting temperature for most planetary materials, when the planet's mass is 0.2 $M_\oplus$. As a result, the

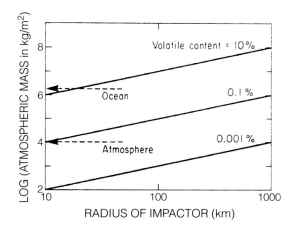

Figure 15.10 The mass per unit surface area of an atmosphere of a 1 $M_{\oplus}$ planet that is in equilibrium between the rate of addition of volatiles to the planet by accretion of material with the indicated volatile content and impact erosion of the atmosphere by impactors of the indicated radii. The *arrows* show the mass per unit surface area of the present terrestrial ocean and atmosphere. (Hunten et al. 1989)

surface melts, and newly accreting planetesimals on the molten surface will also melt. Heavy material migrates downwards, and lighter elements float on top. This process of differentiation liberates a large amount of gravitational energy in the planet's interior. Enough energy can be released to cause melting of a large fraction of the planet's interior, allowing the planet to differentiate throughout.

15.5.3 Accumulation (and Loss) of Atmospheric Volatiles

Atmospheric gases form a tenuous veneer surrounding many of the smaller planets and moons in the Solar System, amounting to far less than 1% of the mass of each body. These atmospheres consist primarily of high-Z ($Z \geq 3$) elements. The atmospheres of the terrestrial planets and other small bodies were probably outgassed from material accreted as solid planetesimals. The problem of the origin of terrestrial planet atmospheres is not simply bringing the required volatiles to the planets because losses were also important. Impacting planetesimals on a growing planet surrounded by

a proto-atmosphere may lead to the following phenomena (§6.4.3):

(1) If the planetesimals are small enough to be stopped by atmospheric drag or disrupted by ram pressure, all of their kinetic energy is deposited in the atmosphere. Most rocky objects smaller than a few dozen meters in radius are stopped in an atmosphere similar to that of Earth at the present time and deposit all of their energy in the atmosphere.

(2) Ejecta excavated by larger impacting planetesimals are slowed down by the atmosphere and transfer kinetic energy to it. The interaction is quite complicated. But note that an atmosphere has a large compressibility, in contrast to a solid surface, and that the gas can be briefly raised to very high temperatures and pressures. Additionally, the energy from atmospheric impacts is released over an extended area and over an interval of tens of seconds.

(3) If the impactor is large, the energy transferred to the atmosphere may be sufficient to blow off part of the atmosphere via hydrodynamic escape (§5.7.3). If the size of the impactor is comparable to or larger than the atmospheric scale height, impact erosion blows off a large portion of the atmosphere, i.e., an atmospheric mass equal to the mass intercepted by the impactor. The same impactor may also add volatiles to the accreting planet. Whether this mass is more than that blown off from the atmosphere depends on the size of the impactor, its volatile content and the density of the atmosphere. Impactors with radii of ~100 km and volatile content of 1% would yield a balance between impact erosion and accretion of volatiles for an atmospheric mass per unit area similar to that of the terrestrial ocean. A similarly sized impactor population with a volatile content of 0.01% would keep a present-day Earth's atmosphere in equilibrium. Figure 15.10 graphs the mass per unit area of an atmosphere in equilibrium between

impact erosion and addition of volatiles as a function of impactor radius and volatile content. Atmospheric blowoff is more likely to occur on smaller planets, such as Mars. A growing planet may lose its atmosphere several times during the accretion period because impacts with large planetesimals are quite common.

In addition to impact erosion, atmospheric gases may be lost via Jeans escape (§5.7.1). In particular, whereas light elements such as H and He easily escape from the top of a terrestrial atmosphere, heavier gases may have escaped this way in the early hot protoatmospheres. The present-day terrestrial planet atmospheres were probably formed towards the end of the accretion epoch by outgassing of the hot planet and impacts by small planetesimals.

15.6 Formation of the Giant Planets

The large amounts of H_2 and He contained in Jupiter and Saturn imply that these planets formed within $\sim 10^7$ years of the formation of the Sun, while a significant amount of gas remained within the protoplanetary disk. Any formation theory of the giant planets must account for these timescales. In addition, formation theories should explain the elemental and isotopic composition of these planets and variations therein from planet to planet, their presence and/or absence of internal heat fluxes, their axial tilts and the orbital and compositional characteristics of their ring and satellite systems. In this section, we discuss the formation of the planets themselves; the formation of their moons and ring systems is addressed in §15.10.1.

Elements heavier than helium constitute <2% of the mass of a solar composition mixture. The giant planets are enriched in heavy elements relative to the solar value by roughly 5, 15 and 300 times for Jupiter, Saturn and Uranus/Neptune, respectively.

Thus, all four giant planets accreted solid material much more effectively than gas from the surrounding nebula. Moreover, whereas the total mass in heavy elements varies by only a factor of a few between the four planets, the mass of H and He varies by about 2 orders of magnitude between Jupiter and Uranus/Neptune.

Table E.13 shows the composition of the giant planet atmospheres. The enhancement in heavy elements increases from Jupiter to Neptune. This gradual, nearly monotonic relationship between mass and composition argues for a unified formation scenario for all of the planets and smaller bodies. Moreover, the continuum of observed extrasolar planetary properties, which stretches to systems not very dissimilar to our own, suggests that extrasolar planets formed in a similar way to the planets within our Solar System.

The D/H ratio in the giant planet atmospheres may provide important clues to the formation history of these planets. The D/H ratios in Jupiter and Saturn are equal to the interstellar D/H ratio of 2×10^{-5}. Because 90% of Jupiter's mass and 75% of Saturn's mass consist of H and He, one would indeed expect the D/H ratios to agree with the interstellar value. Uranus and Neptune are only about 10% H and He by mass. The observed D/H values on Uranus and Neptune are higher than the interstellar value, which can be attributed to exchange of deuterium with an icy reservoir.

Various classes of models have been proposed to explain the formation of giant planets and brown dwarfs. The mass function (abundance of objects as a function of mass) of young compact objects in star-forming regions extends down through the brown dwarf mass range to below the deuterium-burning limit. This observation, together with the lack of any convincing theoretical reason to think that the collapse process that leads to stars cannot also produce substellar objects, strongly implies that most isolated (or distant companion) brown dwarfs and isolated high planetary mass objects form via the same collapse process as do stars.

By similar reasoning, the brown dwarf desert, a profound dip in the mass function over the range $\sim$10–50 M$_{2+}$ for companions orbiting within several AU of Sun-like stars (§14.5.1), strongly suggests that the vast majority of extrasolar giant planets formed via a mechanism different from that of stars. Within our Solar System, bodies up to the mass of Earth consist almost entirely of condensable material, and even bodies of mass $\sim$15 M$_{\oplus}$ consist mostly (by mass) of condensable material. Observations of low-density sub-Neptune exoplanets such as those in the Kepler-11 system (Table 14.1) imply that H/He can dominate the volume of a planet that is only a few times as massive as the Earth.

The theory of giant planet formation favored by most researchers is the **core nucleated accretion model**, in which the planet's initial phase of growth resembles that of a terrestrial planet, but when the planet becomes sufficiently massive (several M$_{\oplus}$), it is able to accumulate substantial amounts of gas from the surrounding protoplanetary disk. Aside from core nucleated accretion, which we describe in detail later, the only giant planet formation scenario receiving significant attention is the **disk instability hypothesis**, in which a giant gaseous protoplanet forms directly from the contraction of a clump that was produced via a gravitational instability in the protoplanetary disk. The disk instability model may account for some massive exoplanets imaged far from their stars, but it does a poor job of explaining the planets in our Solar System and most observed exoplanets.

The core nucleated accretion model relies on a combination of planetesimal agglomeration and gravitational accumulation of gas. According to this scenario, the initial stages of growth of a gas giant planet are identical to those of a terrestrial planet. Dust settles towards the midplane of the protoplanetary disk and agglomerates into kilometer-sized or larger planetesimals, which continue to grow into bigger solid bodies via pairwise inelastic collisions. As the (proto)planet grows,

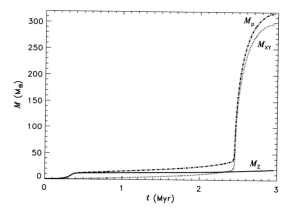

Figure 15.11 The mass of a giant planet that grows to 1 M$_{2+}$ is shown as a function of time according to one particular simulation based on the core nucleated accretion model. The planet's total mass is represented by the *dot-dashed curve*, the mass of the solid component is given by the *solid curve* and the *dotted curve* represents the gas mass. The solid core grows rapidly by runaway accretion in the first 4×10^5 years. The rate of solid body accumulation decreases when the planet has accreted nearly all of the condensed material within its gravitational reach. The envelope accumulates gradually, with its settling rate determined by its ability to radiate away the energy of accretion. Eventually, the planet becomes sufficiently cool and massive that gas can be accreted rapidly. This simulation is for growth at 5.2 AU from a 1 M$_{\odot}$ star, with a local surface mass density of solids equal to 100 kg/m^2. (Lissauer et al. 2009)

its gravitational potential well deepens, and when its escape speed exceeds the thermal velocity of gas in the surrounding disk, it begins to accumulate a gaseous envelope. While the planet's gravity pulls gas from the surrounding disk towards it, thermal pressure from the existing envelope limits accretion. Eventually, increases in the planet's mass and radiation of energy allow the envelope to shrink rapidly. At this point, the factor limiting the planet's growth rate becomes the flow of gas from the surrounding protoplanetary disk.

Figure 15.11 illustrates the growth of a giant planet in one numerical realization of the core accretion model. During the runaway planetesimal accretion epoch (§15.4.2), the (proto)planet's mass increases rapidly. The internal temperature and thermal pressure increase as well, preventing nebular gas from falling onto the protoplanet. When the feeding zone is depleted, the planetesimal accretion rate and therefore the temperature and thermal pressure decrease. This allows gas to fall onto the planet much more rapidly. Gas accumulates at a gradually increasing rate until the mass of gas contained in the planet is comparable to the mass of solid material. The rate of gas accretion then accelerates rapidly, and runaway gas accretion occurs.

When a planet has a mass large enough for its self-gravity to compress the envelope substantially, its ability to accrete additional gas is limited only by the amount of gas available. Hydrodynamic limits allow quite rapid gas flow on a planet of mass $10\ M_{\oplus} \lesssim M_{\mathrm{p}} \lesssim 1\ M_{2\!\!\!+}$. As the planet grows, it alters the disk by accreting material from it and by exerting gravitational torques on it. Numerical calculations such as those illustrated in Figure 15.12 show that these processes can lead to gap formation and, eventually, to isolation of the planet from the surrounding gas. Gaps that small moons clear via a similar process are observed within Saturn's rings (Figs. 13.9a and 13.19).

The planet starts to contract when the factor limiting the rate at which the planet accumulates gas transitions from internal thermal pressure to the disk's ability to provide gas. Initially, contraction takes place rapidly on a Kelvin–Helmholtz timescale, t_{KH}, which is the ratio of the planet's gravitational potential energy, E_{G}, to its luminosity, $\mathcal{L}$:

$$t_{\mathrm{KH}} \equiv \frac{E_{\mathrm{G}}}{\mathcal{L}} \sim \frac{GM^2}{R\mathcal{L}}. \qquad (15.25)$$

Contraction slows down as the fluid envelope becomes denser and less compressible, and the temperature and luminosity decrease with time. The slow cooling of the envelope is a major source of the excess thermal energy emitted into space by the giant planets.

The fact that Uranus and Neptune contain less H_2 and He than do Jupiter and Saturn suggests that

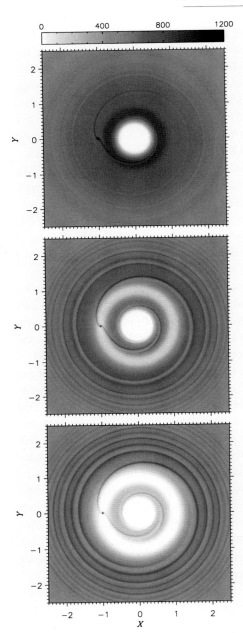

Figure 15.12 The surface density of a gaseous circumstellar disk containing an embedded planet on a circular orbit located 5.2 AU from a 1 $M_{\odot}$ star. The ratio of the scale height of the disk to the distance from the star is $H_z/r = 1/20$, and the viscosity is $\nu_{\mathrm{v}} = 1 \times 10^{11}$ m^2/s. The distance scale is in units of the planet's orbital distance, and the scale bar gives the surface density in units of kg/m^2. The planet is located at $(-1, 0)$ and the star at $(0, 0)$. (a) $M_{\mathrm{p}} = 10$ $M_{\oplus}$. (b) $M_{\mathrm{p}} = 0.3\ M_{2\!\!\!+}$. (c) $M_{\mathrm{p}} = 1\ M_{2\!\!\!+}$. Details of the calculations can be found in D'Angelo et al. (2003). (Courtesy Gennaro D'Angelo)

our Solar System's two outermost planets never quite reached runaway gas accretion conditions, possibly because of a slower accretion of planetesimals. The rate of accretion of solids depends on the surface density of condensates and the orbital frequency, both of which decrease with heliocentric distance.

15.7 Planetary Migration

Growing planets do not remain on fixed orbits. Exoplanets that orbit very close to their stars (§14.3.2) imply that in some cases, these changes are substantial. We first discuss interactions between planets and the gaseous components of protoplanetary disks and subsequently scattering of planetesimals by planets.

15.7.1 Torques from Protoplanetary Disks

Planetary orbits can **migrate** towards (or in some circumstances away from) their star as a consequence of angular momentum exchange between the protoplanetary disk and the planet. As is the case for moons near planetary rings (§13.4), protoplanets drift away from the disk material with which they interact. Planets beyond a disk's edge are pushed in only one direction, but others are subjected to partially offsetting torques. For conditions thought to exist in most protoplanetary disks, the net torque is negative, and therefore planets lose angular momentum to the disk and drift towards the central star.

If a planet's mass is small enough that it does not clear a gap around its orbit, then the planet's migration regime is referred to as **Type I migration**. For small bodies, Type I migration torques occur as a result of small perturbations to the disk structure, and the magnitude of motion is linearly proportional to the perturbing body's mass. But for planet-sized bodies, Type I migration involves complicated nonlinear gravitational and fluid mechanical interactions; rates may be very

large but are very difficult to calculate, and this is currently an active research area.

When the mass of a (proto)planet has reached ~ 1 $M_{2\!+}$, i.e., when $R_H/H_z \gtrsim 1$, the planet perturbs the disk so strongly that it clears a gap in the disk surrounding its orbit. The planet is then dragged along by the disk as the disk viscously evolves. Orbital migration of a planet becomes unavoidable after it has opened up a gap in the disk. The speed of this **Type II migration** does not vary with planetary mass unless the planet's mass is comparable to or larger than the mass of the local disk. When the planet's mass becomes similar to that of the protoplanetary disk, inertial effects become important, and the rate of migration slows.

15.7.2 Scattering of Planetesimals

Planets can also migrate as a back-reaction to clearing large amounts of planetesimals from the regions within their gravitational reach. The distribution of orbits within the Kuiper belt and the existence of the Oort cloud provide strong evidence for **planetesimal-induced migration** of the four giant planets within our Solar System. This process is distinct from the interactions with the gaseous disk responsible for the types of migration discussed in §15.7.1. The mechanism operates as follows: Gravitational stirring by Uranus and Neptune excites high eccentricities in the surrounding planetesimals. Those that acquire sufficiently small perihelia can be 'handed off' to the neighboring planet with a smaller semimajor axis, with a resultant gain in angular momentum for the first planet. In this way, planetesimals get passed inward from Neptune to Uranus to Saturn and finally to Jupiter, which is massive enough to readily eject them from the Solar System or onto nearly parabolic paths about the Sun. When bodies on nearly parabolic paths reach distances of $\sim 10^4$ AU from the Sun, their heliocentric velocities are so slow that the tidal forces of our galaxy and the tugs of nearby stars can raise their perihelia out of the planetary region, placing them into the Oort cloud. The other giant planets are also massive enough to eject planetesimals from the Solar System or to the Oort cloud; however, the characteristic timescales for direct ejection are longer than for passing the planetesimals inwards to the control of Jupiter.

For a planet on a circular orbit, change in angular momentum L is related to change in semimajor axis a according to

$$\Delta L = \frac{1}{2} M_{\mathrm{P}} \sqrt{\frac{GM_\odot}{a}} \Delta a. \tag{15.26}$$

Thus Jupiter, being the innermost and most massive giant planet, migrates the shortest distance, and Neptune migrates farthest.

As Neptune migrated outward, it pumped the eccentricities of objects carried along in its outer mean motion resonances. The orbital distribution of Kuiper belt objects (KBOs), especially the many plutinos (including Pluto) with $e \sim 0.3$, provides strong evidence for the outwards migration of Neptune by several AU. This amount of migration requires a planetesimal disk of a few dozen $M_\oplus$.

Jupiter migrated inward, probably by a few tenths of an AU, as a back-reaction from expelling small bodies (as well as, perhaps, not-so-small bodies) from the outer Solar System to the Oort cloud and to interstellar space. In contrast, the other giant planets, which perturbed more planetesimals inward to Jupiter-crossing orbits than directly outward to the Oort cloud and beyond, migrated away from the Sun, with Uranus and Neptune each moving outward by several AU.

15.8 Small Bodies Orbiting the Sun

The planetesimal model of planet formation implies that much of the material that wound up in the terrestrial planets once resided in planetesimals of sizes ranging from kilometers to hundreds of kilometers. Questions of the origins of the major populations of small bodies in our Solar System therefore focus on the locations of these reservoirs and why the bodies within them did not grow larger.

15.8.1 Asteroid Belt

Thousands of minor planets of radii >10 km orbit between Mars and Jupiter (see Chapter 12), yet the total mass of these bodies is $<10^{-3}$ $M_\oplus$. This is 3 to 4 orders of magnitude less than would be expected for a planet accreting at ~ 3 AU within a smoothly varying protoplanetary disk. Why is there so little mass remaining in the asteroid region? Why is this mass spread among so many bodies? Why are the orbits of most asteroids more eccentric and inclined to the invariable plane of the Solar System than are those of the major planets? Why are the asteroids so diverse in composition, as indicated by their spectra and the wide variety of meteorites found on Earth?

Many small asteroids are differentiated. However, accretional heating and long-lived radionuclides could not have supplied sufficient energy to cause the melting required for differentiation. Proposed energy sources are electromagnetic induction heating (§11.7.1) and the decay of short-lived radionuclides, especially ^{26}Al (Problem 15-12). The observed segregation of asteroidal spectral types by semimajor axis (Fig. 12.16) places an upper limit on the amount of planetesimal mixing that could have occurred within the asteroid belt.

The heliocentric distribution of the various asteroid classes follows the general condensation sequence: Whereas high-temperature condensates are found in the inner regions of the asteroid belt, lower temperature condensates typically orbit the Sun at larger distances. The dependence on heliocentric distance of igneous, metamorphic and primitive asteroids (Fig. 12.16) suggests a heating mechanism that declined rapidly in efficiency with heliocentric distance. Bodies in the Kuiper belt presumably formed in a manner analogous to asteroids, and larger KBOs are likely differentiated. These bodies have a higher ice:rock mass ratio than asteroids because they formed farther from the Sun.

Proximity to Jupiter is almost certainly responsible for the mass depletion in the asteroid belt, the material remaining in this region not having accreted into a single (small) planet, as well as for the distribution of orbital properties of asteroids. Large planetary embryos scattered into the asteroid zone by Jupiter and/or direct resonant perturbations of Jupiter are capable of exciting eccentricities and inclinations of asteroid zone planetesimals and planetary embryos to values at which most collisions do not lead to accretion (Problem 15-18). Much of the material once contained in small bodies orbiting between Mars and Jupiter could thereby have been scattered into Jupiter-crossing orbits, from which it would have been ejected from the Solar System or accreted by Jupiter. Other planetesimals could have been ground to dust or even partially vaporized by high-velocity collisions. Planetary embryos that formed within the present asteroid belt near resonances with Jupiter may have been resonantly pumped to high eccentricities and perturbed their nonresonant neighbors; orbital migration, as well as the dispersal of the gaseous component of the protoplanetary disk, could have enhanced these perturbations by sweeping resonance locations over a large portion of the asteroid region.

Changes in Jupiter's orbit caused by gravitational interactions with the remnant planetesimal disk and/or other planets played a major role in clearing some parts of the asteroid belt. Such variations in Jupiter's orbit may have been instrumental in moving some asteroids (perhaps bodies originating exterior to Jupiter's orbit) into the Trojan regions, 60° ahead of and behind Jupiter in its orbit.

15.8.2 Comet Reservoirs

The highly volatile composition of comets places their origin in the outer regions of the planet-forming disk. Kuiper belt objects on nearly circular orbits likely formed close to their present locations. Because models of star formation imply that the densities of gas and dust in the Oort cloud were much too small for planetesimals to form, Oort

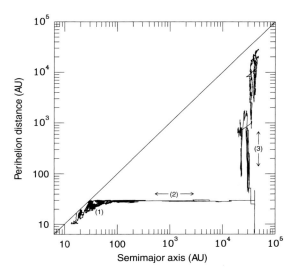

Figure 15.13 The dynamical evolution of an object as it evolves into the Oort cloud. The object began on a nearly circular orbit between the giant planets. In the initial phase of the evolution (1), the object remains in a moderate eccentricity orbit in the giant planet region. Neptune eventually scatters it outward, after which it undergoes a random walk in inverse semimajor axis (2). When the orbit becomes almost parabolic, the galactic tidal force can raise its perihelion above the planetary region (3). (Levison and Dones 2007)

cloud comets likely formed in or near the region now inhabited by the giant planets.

Dynamical simulations, such as the one presented in Figure 15.13, show that as long as the giant planets provide the dominant perturbations, the small body's perihelion remains within the planetary region and the inclination of its orbit does not change much, i.e., the body stays near the ecliptic plane. When the body reaches distances of greater than 10 000 AU, perturbations from the tidal pull of the galaxy can lift its perihelion out of the planetary region, and the body can thus be 'stored' in the Oort cloud.

Current theories of Oort cloud formation imply that substantial quantities of small planetesimals that formed between ~3 and 30 AU from the Sun were ejected from the planetary region by gravitational perturbations from the giant planets.

Some Oort cloud comets may have formed around other stars in the Sun's birth cluster and subsequently been captured by the Solar System before the dispersal of this star cluster. Tidal perturbations from the gravitational field of the galaxy, passing stars and giant molecular clouds have randomized the orbits of Oort cloud comets over the past 4.5×10^9 years. Aside from a small flattening caused by the galactic tide, the Oort cloud is nearly spherical, with prograde as well as retrograde objects. Accounting for the inefficiency in transporting bodies from the planetary region into bound Oort cloud orbits and for losses over the age of the Solar System, the mass of solid material ejected from the planetary region could have been 10–$1000\ M_\oplus$.

The Kuiper belt requires that planetesimals existed beyond the orbit of Neptune. Thus, the abrupt cutoff of observed massive planets beyond the orbit of Neptune (the masses of Pluto and Eris are each $<2 \times 10^{-4}$ times that of Neptune; see Tables E.3 and E.8) cannot be explained solely by the lack of material in this region of the Solar System.

As comets contain both highly volatile CHON particles and extremely refractory mineral grains, radial mixing throughout the solar nebula at the time of formation must have been efficient. Moreover, some interstellar grains must have entered the solar nebula without being vaporized.

15.9 Planetary Rotation

Planets accumulate rotational angular momentum from the relative motions of accreted material. Jupiter and Saturn are predominantly composed of hydrogen and helium, which they must have accreted hydrodynamically in flows quite different from those that govern the dynamics of planetesimals. Such flows lead to prograde rotation. In contrast, very little net spin angular momentum is accumulated by a planet that accretes while on a circular orbit within a uniform surface density disk of small planetesimals.

The stochastic nature of planetary accretion from planetesimals allows for a random component to the net spin angular momentum of a planet in any direction. Because planets might accumulate a significant fraction of their mass and spin angular momentum from only a very few impacts, stochastic effects may be very important in determining planetary rotation (Problem 15-9). Stochastic impacts of large bodies may be the primary source of the rotational angular momentum of the terrestrial planets, with the observed preference of low obliquities being a chance occurrence (and, in the case of Mercury, tidal torques exerted by the Sun, §2.7.2). The nonzero obliquities of the giant planets might have been produced by giant impacts. Spin-orbit resonances also might have tilted the rotation axes of some or all of the giant planets. From the observed rotational properties of the planets, the size of the largest bodies to impact each planet during the accretionary epoch has been estimated to be 1%–10% of the planet's final mass.

15.10 Satellites of Planets and of Minor Planets

The population of planetary satellites within our Solar System is very diverse, and differing origins scenarios explain various subsets of these bodies.

15.10.1 Giant Planet Satellites

Satellites orbiting closest to giant planets (near the Roche limit) are generally small. Planetary rings dominate where tidal forces from the planet are sufficient to tear apart a moon held together solely by its own gravity. Larger moons orbit at distances ranging from a few planetary radii to several dozen planetary radii. The outer regions of the satellite systems of all four giant planets contain small bodies on highly eccentric and inclined orbits. The diversity of planetary satellites suggests that they are formed by more than a single mechanism.

The satellite systems of the giant planets consist of **regular** and **irregular** satellites. Regular satellites move on low-eccentricity prograde orbits near the equatorial planes of their planets. They orbit close to the planets, well within the bounds of the planet's Hill sphere. These properties imply that regular satellites formed within a disk orbiting in the planet's equatorial plane. Irregular satellites generally travel on high-eccentricity, high-inclination orbits lying well exterior to a planet's regular satellite system; most irregular satellites are quite small. Most, if not all, irregular satellites were captured from heliocentric orbits.

The regular satellites of the giant planets likely formed by a solid body accretion process in a gas/dust disk surrounding the planet. Such disks may consist of material from the outer portions of the protoplanet's envelope or (more likely) matter that was directly captured from the protoplanetary disk. Solid-body accretion rates within a minimum mass 'subnebula' disk surrounding a giant planet are very rapid (Problem 15-19). The density of gas within the giant planets' circumplanetary disks exceeded that of the nearby protoplanetary disk, and temperatures were also higher. Thus, chemical reactions proceeded further towards equilibrium. When youthful Jupiter's high luminosity (from radiation of accretional energy) is included, the model naturally accounts for the decrease in density of the Galilean satellites with increasing distance from Jupiter. The densities of the moons around Saturn and Uranus do not vary in such a systematic manner with distance from the planet; however, these lower mass planets were never as luminous as was young Jupiter. Because tidal forces prevent material from accreting within a planet's Roche limit (§13.1), rings formed around the giant planets. Note, however, that most if not all of the ring systems that we see at present are not primordial (§13.5).

The wide variety of properties exhibited by the satellite systems of the four giant planets in our Solar System suggests that stochastic processes may be even more important for satellite formation than current models suggest them to be in planetary growth. A possible explanation for this difference is that satellite systems are subjected to a very heavy bombardment of planetesimals on heliocentric orbits, which may fragment moons and also produce them. Deterministic models of satellite formation must thus be interpreted with caution.

Terrestrial planets and smaller objects presumably never possessed gas-rich circumplanetary disks; thus, other explanations are required for the origins of the moons of Mars, Earth, asteroids and KBOs.

15.10.2 Formation of the Moon

The Earth's Moon is a very peculiar object. The Moon/Earth mass ratio greatly exceeds that of any other satellite/planet (although Charon/Pluto and various other satellite/minor planet ratios are larger; see Table E.8), raising the question of how this much material was placed into orbit about Earth. The density of the Moon is ~25% smaller than the uncompressed density of the Earth (Table E.14). Yet the Moon is severely depleted in volatiles (which tend to have low densities), having less than half the potassium abundance of Earth and very little water. The combination of low mean density and lack of volatiles implies that the Moon is not simply an amalgam of solar composition material that is able to condense above a certain temperature. Rather, the Moon's bulk composition resembles Earth's mantle, albeit depleted in volatiles. The bulk composition of the lunar crust and mantle could be understood if the Moon equilibrated with a large iron core, but the Moon's core is quite small (§9.1.3). Capture, coaccretion and fission models of lunar origin have all been studied

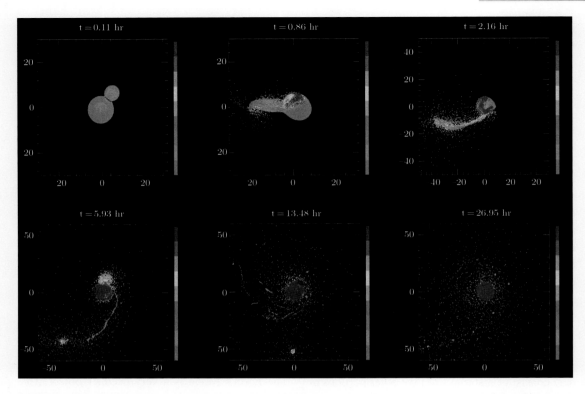

Figure 15.14 COLOR PLATE These computer-generated images illustrate the first day after a Mars-sized protoplanet and the proto-Earth collide with a velocity upon contact of 9 km s^{-1}. This contact velocity corresponds to a nil relative approach velocity, $v_\infty = 0$, and is less than the escape velocity from Earth because the collision occurred when the centers of the two bodies were separated by almost 1.5 $R_\oplus$. This collision produced a circumterrestrial disk 1.62 times as massive as the Moon; only 5% of the mass of the disk was metallic iron. The particles are color-coded by temperature, and time is indicated within each panel. See Canup (2004) for details on the calculation. (Courtesy Robin Canup)

in great detail, but none satisfies both the dynamical and chemical constraints in a straightforward manner.

The favored theory of lunar formation is the **giant impact model**, in which a collision between the Earth and a Mars-sized or larger planetary embryo ejects more than 1 $M_\mathbb{C}$ of material into Earth's orbit. Results of a numerical simulation of such a collision are shown in Figure 15.14. Some of the orbiting material ultimately falls back to Earth as the disk spreads radially because of viscosity and gravitational torques, but material in the outer part of the disk moves outwards and remains in orbit.

When the material in the circumterrestrial disk is cool enough to form condensed bodies, it can quickly accumulate into a single large moon. A numerical integration of this accumulation is shown in Figure 15.15. Provided both bodies were differentiated before the impact, this model explains the apparent similarities between the lunar composition and that of the Earth's mantle and at the same time the lack of volatile material on the Moon. Volatile material was completely vaporized by the impact, and it remained in a gaseous state within the circumterrestrial disk, allowing most of the volatiles to escape into interplanetary space.

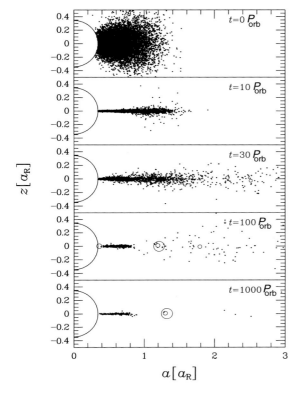

Figure 15.15 Snapshots of the protolunar disk in the r–z plane at times $t = 0, 10, 30, 100, 1000\ P_{orb}$, where P_{orb} is the Keplerian orbital period at the Roche limit. The initial number of disk particles is 10 000, and the disk mass is four times the present lunar mass. The *semicircle* centered at the coordinate origin stands for the Earth. *Circles* represent disk particles, and their sizes are proportional to the physical sizes of the disk particles. The *horizontal scale* shows the semimajor axis of disk particles in units of the Roche limit radius, a_R (see eq. 13.8). Note the very massive transient ring around the Earth. (Kokubo et al. 2000)

15.10.3 Satellites of Small Bodies

Mars's moons Phobos and Deimos are similar in composition to primitive, C- and D-type asteroids (§§10.1, 12.4.1). Because these satellites orbit in the plane of the martian equator, they most likely accreted from a small disk formed by the **disruptive capture** of one or more planetesimals that fragmented as a result of tidal stress upon close approach of Mars or that collided with a small object while passing close to Mars.

The relative sizes and orbits of asteroidal and Kuiper belt binaries imply that some of these pairs formed as the result of collisions. Others probably originated when three objects came into close proximity and the current binary pair was able to gravitationally transfer mechanical energy to the object that escaped.

A giant impact origin analogous to that described in §15.10.2 also appears likely for Pluto's satellite system, which is dominated by the moon Charon that has more than 10% as much mass as Pluto itself.

15.11 Exoplanet Formation Models

The orbits of most of the extrasolar giant planets thus far observed are quite different from those of Jupiter, Saturn, Uranus and Neptune (§14.5), and new models have been proposed to explain them. It has been suggested that many of the planets orbiting close to their star formed substantially farther from the star and subsequently migrated inwards to their current short-period orbits. Most mechanisms for altering orbits fall into one of three classes (which sometimes act in consort): disk–planet interactions, planet–planet (or, in the case of binary stars, star–planet) perturbations and scatterings and tidal forces from the central star.

Disk-induced planetary orbital decay (§15.7.1) had been studied before the discovery of extrasolar planets. But no one predicted giant planets near stars because migration speeds were expected to increase as the planet approached the star, so the chance that a planet moved substantially inwards and was not subsequently lost was thought to be small.

Two mechanisms have been proposed for stopping a planet less than one-tenth of an AU from the star: Tidal torques from the star counteracting disk torques or a substantial reduction in disk torque when the planet was well within a nearly empty zone close to the star. However, the substantially larger abundance of giant planets with orbital periods ranging from 15 days to 3 years, which feel negligible tidal torque from their star and are exterior to the region of the disk expected to be cleared by magnetic accretion onto the star (§15.3.2), is more difficult to explain by this model. Perhaps (at least the inner few AU of) protoplanetary disks are cleared from the inside outward, leaving migrating planets stranded.

The wide range in orbital inclinations relative to their star's equator exhibited by the hot jupiters, particularly those that orbit stars hotter than our Sun (§14.3.3), point toward a dynamical mechanism that randomizes planetary inclinations in addition to transporting planets to orbits close to their star. Whereas disk-induced migration tends to keep orbits planar, planet–planet interactions, especially the Kozai mechanism (§2.3.3), produce a wide range of inclinations. Such planet–planet perturbations could place the planets in high-eccentricity orbits with periapses lying close to their stars. In the absence of dissipative forces, orbits would hardly ever circularize very near the stars. But tidal forces become significant when periapses are in the range of typical hot jupiter semimajor axes. Planets like HD 80606, which has $e = 0.93$ and periapse $q = 0.03$ AU, might be transitioning from an extremely eccentric orbit to a circular one typical of hot jupiters as a result of tidal torques from the star near periapsis. Most hot jupiters orbiting cooler stars have low orbital inclinations relative to the star's equatorial plane, but tidal damping timescales for i are much shorter in these stars because they have convective envelopes, so planetary inclinations do not constrain origins.

Many giant exoplanets move on orbits that are more eccentric than those of any of the major planets within our Solar System, although typical eccentricities are much smaller than that of HD 80606 (Fig. 14.25). These eccentric orbits may be the result of stochastic gravitational scatterings among massive planets (which have subsequently merged or been ejected to interstellar space), perturbations from a stellar binary companion (which might no longer be present if the now-single stars were once members of unstable multiple star systems) or the complex and currently ill-constrained interactions between the planets and the protoplanetary disk (§15.7.1).

The sample of known extrasolar planets contains strong biases. Most solar-type stars could well have planetary systems that closely resemble our own. Nonetheless, if giant planets (even of relatively modest Uranus masses) orbiting near or migrating or being scattered through 1 AU are the norm, then terrestrial planets in habitable zones (§16.5) may be scarcer than they were previously thought to be.

15.12 Confronting Theory with Observations

The current theory of planetary growth via planetesimal accretion within a circumstellar disk provides excellent explanations of the causes of many of the observed Solar System and exoplanet properties, but less complete or less satisfactory explanations for several others.

15.12.1 Solar System's Dynamical State

Dynamical models of planetary accretion within a flattened disk of planetesimals produce moderately low-eccentricity, almost coplanar orbits of planets, except at the outer fringes of the Solar System. The ultimate sizes and spacings of solid planets are determined by the ability of protoplanetary embryos to gravitationally perturb one another into crossing orbits. Such perturbations are often caused by weak resonant forcing and occur on timescales much longer than the bulk of planetesimal interactions discussed in §15.4.2. A more massive protoplanetary disk probably produces larger but fewer planets. Stochastic processes are important in planetary accretion, so nearly identical initial conditions could lead to quite different outcomes, e.g., the fact that there are four terrestrial planets in our Solar System as opposed to three or five is probably just the luck of the draw.

The angular momentum distribution of the Solar System resulted from outward transport of mass and angular momentum (via poorly characterized viscous, gravitational and/or magnetic torques) within the protosun/protoplanetary disk, plus a subsequent removal of most of the Sun's spin angular momentum by the solar wind.

Jupiter played a major role in preventing the formation of a planet in the asteroid zone. Jovian resonances could have directly stirred planetesimals in the asteroid zone; Jupiter could have scattered large failed planetary embryos inwards from 5 AU; or Jupiter may even have migrated temporarily inwards itself, yielding the same effect. The resulting stirring could have prevented further planetary growth and/or ejected an already-formed planet from the Solar System.

The giant planets ejected a substantial mass of solid bodies from the planetary region. The majority of these planetesimals escaped from the Solar System, but $\gtrsim 10\%$ ended up in the Oort cloud.

The Kuiper belt likely formed *in situ* from planetesimals orbiting exterior to Neptune's orbit. The dynamical structure of the Kuiper belt suggests that Neptune slowly migrated outwards by several AU during the final stages of planetary formation. During this gradual migration, Neptune could have trapped objects in resonance and excited their eccentricities, thereby producing the observed populations of plutinos in the 2:3 resonance as well as resonant KBOs (Fig. 12.4).

The prograde rotation of Jupiter and Saturn can be explained as a deterministic result of gas accretion, and the excess of prograde rotation among the other planets may have been produced in a systematic way via expansion of their accretion zones or may just be a chance result. Planetary obliquities result from stochastic impacts of large bodies and/or spin-orbit resonances between planets.

The gross features of the regular satellite systems of the giant planets can be understood if these planets were circumscribed by disks during their youth; various models for the formation of such disks exist.

The high cratering rate in the early Solar System and the lower rate at the current epoch are a consequence of the sweep-up of debris from planetary formation. The early high bombardment rate caused ancient surfaces on planets, asteroids and satellites to be covered with craters. Huge impacts led to the formation of the Moon, stripped off the outer layers of Mercury and may have changed the spin orientation of Uranus. Some large planetesimals probably were captured into planetocentric orbits, e.g., Triton about Neptune.

15.12.2 Composition of Planetary Bodies

The masses and bulk compositions of the planets can be understood in a gross sense as resulting from planetary growth within a disk whose temperature and surface density decreased with distance from the growing Sun. The terrestrial planets are rocky because the more volatile elements could not condense (nor survive in solid form) so close to the Sun, but comets and the moons of the giant planets retain ices because they grew in a colder environment. The condensation of water-ice beyond ~ 4 AU provided the outer planets with enough mass to gravitationally trap substantial amounts of H_2 and He from the solar nebula. Longer accretion times at greater heliocentric distances together with the timely disappearance of gas may account for the decrease in the gas fractions of the giant planets with increasing semimajor axis.

Some solids were transported over significant radial distances within the protoplanetary disk, leading to mixing of material that condensed in different regions of the solar nebula and/or had survived from the presolar era. This mixing helps explain the bulk compositions of those chondritic meteorites that contain both refractory inclusions and volatile-rich grains, as well as the very refractory Solar System condensates included among the samples of comet P/Wild 2 returned by *Stardust* (§12.7.4). Radial and vertical mixing could have brought together material for meteorites that are progressively more depleted in volatiles over too large a range in condensation temperature to be explained by equilibrium solidification at any one time and place. Explanations of many detailed characteristics of meteorites (especially the formation of chondrules and remanent magnetism) remain controversial. The (well-established) planetesimal hypothesis explains the similarity in ages among primitive meteorites and the fact that all other Solar System rocks, whose components presumably at one point passed through a stage similar to primitive meteorites, are the same age as primitive meteorites or younger.

All planets were hot during the accretionary epoch. The present terrestrial planets show evidence of this early hot era in the form of extensive tectonic and/or volcanic activity. Jupiter, Saturn and Neptune have excess thermal emissions resulting from accretional and differentiation heating. Outgassing of the hot newly formed planets, combined with late accretionary veneers from the asteroid belt and comet reservoirs, led to the formation of atmospheres on the terrestrial planets (§5.8.1).

15.12.3 Extrasolar Planets

Although many more planets are now known outside of our Solar System than within it, we have far fewer data on these planets, and detection statistics are highly biased. The orbits of exoplanets provide strong evidence that radial migration in protoplanetary disks is an important process. Gravitational interactions among planets, as well as between planets and the disks in which they formed, can provide the torques required for these orbital changes. Disk–planet interactions can indeed be so powerful that it is difficult to explain why so many giant planets have not migrated all of the way inwards and been consumed by their stars. The high eccentricities of many exoplanets came as a surprise and, in general, the discoveries of exoplanets to date have shown that nature

is more creative than theorists, making predictions quite difficult!

15.12.4 Successes, Shortcomings and Predictions

The planetesimal hypothesis provides a viable theory of the growth of the terrestrial planets, the cores of the giant planets and the smaller bodies present in the Solar System. The formation of solid bodies of planetary size should be a common event, at least around young stars that do not have binary companions orbiting at planetary distances. Planets could form by similar mechanisms within circumpulsar disks if such disks have adequate dimensions and masses.

The formation of giant planets, which contain large quantities of H_2 and He, requires rapid growth of planetary cores, so that gravitational trapping of gas can occur before the dispersal of the gas from the protoplanetary region. According to the scenario outlined in this chapter, the largest body in any given zone is the most efficient accreter, and its mass 'runs away' from the mass distribution of nearby bodies in the sense that it doubles in mass faster than typical bodies. Such rapid accretion of a few large solid protoplanets can lead to giant planet core formation in $\sim 10^6$ years, provided disk masses are a few times as large as those given by 'minimum mass' models of the solar nebula. Thus, we appear to have a basic understanding of giant planet formation, although our models of the origin of giant planets must be regarded as somewhat more uncertain than those of terrestrial planet accretion because a wider variety of physical processes needs to be considered to account for both the massive gaseous components and the solids' enrichments of giant planets.

Key Concepts

- A wide variety of observations within and beyond our Solar System are combined with theoretical models to draw a picture of planetary formation.

- Stars form from the collapse of molecular cloud cores. Almost all stars form together with circumstellar disks. Much of the material in these disks is accreted by the growing star, and in most cases, some accumulates into planets.
- Most of the material in our Sun's protoplanetary disk was well mixed on the molecular level, but this mixing was not complete.
- Terrestrial planets as well as other solid rocky and icy bodies form by accretion of solid bodies, primarily via pairwise physical collisions.
- Silicates and metal-rich condensates existed throughout almost all of the Sun's protoplanetary disk, but ices existed only in the outer parts. Well

inside Mercury's orbit, the temperature was too high for solids to exist.
- Gas giant planets probably formed by the growth of a solid core followed by gravitational accumulation of hydrogen and helium.
- Earth's Moon formed by the collision of a roughly Mars-sized body with Earth 4.5 billion years ago; both bodies had differentiated prior to the impact.
- The high eccentricities of most known extrasolar giant planets, as well as the close-in orbits of many gas giant exoplanets, imply that considerable planetary migration has occurred. Planet–disk interactions as well as planet–planet scattering can lead to migration.

Further Reading

The series of *Protostars and Planets* books contains review papers on molecular clouds and star and planet formation. The latest volume in the series is:

Reipurth, B., D. Jewitt, and K. Keil, Eds., 2007. *Protostars and Planets V*. University of Arizona Press, Tucson. 951pp.

A comprehensive textbook on stellar formation is:

Stahler, S.W., and F. Palla, 2005. *The Formation of Stars*. Wiley-VCH, Weinheim, Germany. 865pp.

Several papers related to different aspects of the formation of our Solar System are given by:

Lin, D.N.C., 1986. The nebular origin of the Solar System. In *The Solar System: Observations and Interpretations*. Ed. M.G. Kivelson. Rubey Vol. IV. Prentice Hall, Englewood Cliffs, NJ, pp. 28–87.

Lissauer, J.J., 1993. Planet formation. *Annu. Rev. Astron. Astrophys.*, **31**, 129–174.

Lissauer, J.J., 1995. Urey Prize lecture: On the diversity of plausible planetary systems. *Icarus*, **114**, 217–236.

Lissauer, J.J., O. Hubickyj, G. D'Angelo, and P. Bodenheimer, 2009. Models of Jupiter's growth incorporating thermal and hydrodynamics constraints. *Icarus*, **199**, 338–350.

A good chapter on the formation of terrestrial planet atmospheres is:

Ahrens, T.J., J.D. O'Keefe, and M.A. Lange, 1989. Formation of atmospheres during accretion of the terrestrial planets. In *Origin and Evolution of Planetary and Satellite Atmospheres*. Eds. S.K. Atreya, J.B. Pollack, and M.S. Matthews. University of Arizona Press, Tucson, pp. 328–385.

Equilibrium chemistry in the solar nebula is described in:

Prinn, R.G., and B. Fegley, Jr., 1989. Solar nebula chemistry: Origin of planetary, satellite and cometary volatiles. In *Origin and Evolution of Planetary and Satellite Atmospheres*. Eds. S.K. Atreya, J.B. Pollack, and M.S. Matthews. University of Arizona Press, Tucson, pp.78–136.

Problems

15-1. **(a)** Calculate the gravitational potential energy of a uniform spherical cloud of density ρ and radius R.
(b) Determine the Jeans mass, M_J, of an interstellar cloud of solar composition with density ρ and temperature T. (Hint: Set the gravitational potential energy equal to negative twice the cloud's kinetic energy and solve for the radius of the cloud.)
(c) Show that if the cloud collapses isothermally, it becomes more *unstable* as it shrinks.
(d) Show that if the cloud retains the gravitational energy of its collapse as heat, it becomes more *stable* as it shrinks.

15-2. The molecular cloud Sgr B2 has a diameter of 30 light-years and a mass of $5 \times 10^5 \, M_\odot$. Its temperature is 20 K. Will it collapse, and if so, how quickly?

15-3. Consider an H_2 molecule that falls from ∞ to a circular orbit at 1 AU from a 1 $M_\odot$ star.
(a) Calculate the circular velocity at 1 AU and determine the total mechanical (kinetic + potential) energy of a molecule on a circular orbit at 1 AU. Note that the total energy of the molecule at rest at infinity is zero.
(b) Calculate the temperature increase of the hydrogen gas assuming it has not suffered radiative losses.

15-4. **(a)** Calculate the amount of gas that a particle R meters in radius orbiting at 1 AU from a 1 $M_\odot$ star passes through (collides with) during one year. You may assume that the density of the protoplan-

etary disk is 10^{-6} kg m^{-3} and $\eta = 5 \times 10^{-3}$ and use equation (5.14).
(b) Assuming a particle density of 3000 kg m^{-3}, calculate the radius of a particle that passes through its own mass of gas during one orbit.

15-5. **(a)** Calculate the rate of growth, dR/dt, of a planetary embryo of radius $R = 4000$ km and mass $M = 10^{24}$ kg, in a planetesimal disk of surface density $\sigma_\rho = 100$ kg m^{-2}, temperature $T = 300$ K and velocity dispersion $v = 1$ km s^{-1} at a distance of 2 AU from a star of mass 3 $M_\odot$. You may use the two-body approximation for planetesimal/planetary embryo encounters.
(b)* What will halt (or at least severely slow down) the accretion of such a planetary embryo? What will its mass be at this point?

15-6. Calculate the growth time for Neptune assuming *in situ* ordered growth (i.e., not runaway accretion; use $\mathcal{F}_g = 10$) in a minimum-mass nebula. (Hint: Determine the surface density by spreading Neptune's mass over an annulus from 25 to 35 AU.) Is this model realistic? Why, or why not?

15-7.* Compare a hypothetical planetary system that formed in a disk with the same size as the solar nebula but only half the surface mass density with our own Solar System. Assume that the star's mass is 1 $M_\odot$ and that it does not have any stellar companions. Concentrate on the final number, sizes and spacings of the planets. Explain your reasoning. Quote formulas and be quantitative when possible.

15-8.* According to the core-nucleated model of giant planet growth, the initial phases of accumulation of hydrogen–helium atmospheres are regulated by the radiation of accretion energy (§15.6). However, when a giant planet is sufficiently massive, its accretion rate becomes limited by access to gas from the surrounding protoplanetary disk. Consider a planet moving in a disk of cold gas with a relative velocity v_∞. Gas, similar to particles, is focused by the gravity of the planet. But unlike solid particles, gas is highly collisional, even on short timescales.

(a) Trajectories of gas parcels are bent in a convergent manner by the gravity of the planet (see Fig. 15.7). In a cylindrically symmetric approximation of the problem, these trajectories eventually collide on the symmetry axis behind the planet. Assume that the collision converts the transverse velocity of the gas into heat that is radiated away and that material bound to the planet following this damping falls onto the planet and is accreted. Calculate the rate at which the planet accumulates gas as a function of M_p, ρ_g and v_∞.

(b) Incorporate the effects of the thermal motions of the gas particles with sound speed (thermal velocity) c_s by replacing v_∞ in your formula by $(v_\infty^2 + c_s^2)^{1/2}$.

(c) Note that your result does not depend on the radius of the planet, in contrast to the formula for solid body accretion rates (eqs. 15.16–15.21). What physics has been omitted, and how can your formula be improved to account for the planet's size?

(d) Would you expect v_∞ to be constant? If not, how does it vary within the disk?

(e) What other factors should be accounted for in a more sophisticated analysis of the problem?

15-9.* A planet of mass M_p and radius R_p initially spins in the prograde direction with zero obliquity and rotation period P_{rot}. It is impacted nearly tangentially at its north pole by a body of mass m, whose velocity before encounter was small compared with the escape speed from the planet's surface.

(a) Derive an expression for the planet's spin period and obliquity after the impact. You may assume that the projectile was entirely absorbed.

(b) Numerically evaluate your result for $M_p = 1\,M_\oplus$, $R_p = 1\,R_\oplus$, $P_{rot} = 10^5$ seconds and $m = 0.02\,M_\oplus$.

(Of course, a truly tangential impactor is likely to 'skip off' rather than being absorbed, but even for a trajectory only $\sim 10°$ from the horizontal, most ejecta can be captured at the velocities considered here. The case of a polar impactor is also a singular extremum, but both of these effects together only add a factor of a few to the angular momentum provided by a given mass impacting with random geometry, and they make the algebra much easier.)

15-10.* Most of the heating by radioactive decay in our planetary system at the present epoch is due to decay of four isotopes, one of potassium, one of thorium and two of uranium. Chondritic elemental abundances are listed in Table 3.1. Isotopic fractions and decay properties are given in the *CRC Handbook*.

(a) What are these four major energy-producing isotopes? What energy is released per atom decayed? What energy is released per kilogram decayed? What is the *rate* of energy produced by 1 kg of the pure isotope? What is the rate released per kilogram of the element in its naturally occurring isotopic ratio? Note: Some of these isotopes decay into other isotopes with short half-lives (e.g., radon). The decay chain must be followed until a stable (or very long-lived) isotope is reached, adding the energy contribution of each decay along the path.

(b) What is the rate of heat production per kilogram of chondritic meteorite (or, equivalently, per kilogram of the Earth as a whole, neglecting the fact that the volatile element potassium is less abundant in the Earth than in CI chondrules) from each of these sources?

(c) What was the heat production rate from each of these sources 4.56×10^9 years ago?

(d) There are very many radioactive isotopes known. What characteristics do these isotopes share that make them by far the most important? (Hint: There are two very important characteristics shared by all four and one other by three of the four.)

15-11. **(a)** How long would it take radioactive decay at the early Solar System rate calculated in part (c) of the previous problem to generate enough heat to melt a rock of chondritic composition, assuming an initial temperature of 300 K and no loss of energy from the system?

(b) How long would it take radioactive decay to generate as much energy as the gravitational potential energy obtained from accretion for an asteroid of radius 500 km? How long for an asteroid 50 km in radius?

15-12. Find the *smallest* body for which ^{26}Al heating could have produced internal melting. Assume an initial radiogenic heating rate of 10^{-3} J m^{-3} s^{-1}, decay constant of 8×10^{-7} yr^{-1}, specific heat $= 700$ J kg^{-1} K^{-1}, diffusivity 10^{-6} m^2 s^{-1} and a melting temperature of 1800 K. (Hint: Compare the timescales for thermal diffusion with that for heating up to the melting point.)

15-13. Find the initial temperature profile of the Earth, assuming that it was homogeneous and it accreted so fast (or that large impacts buried the heat so deep) that radiation losses were negligible. Do this for both the zero and infinite conductivity cases.

15-14. Calculate the rise in temperature if the Earth differentiated from an initially homogeneous density distribution to a configuration in which one-third of the planet's mass was contained in a core whose density was twice that of the surrounding mantle. You may assume infinite conductivity.

15-15.* Repeat the two previous problems for an asteroid of radius 100 km.

15-16. Describe and sketch the temperature profiles you would expect after the accretion of Mars caused by:
(a) Accretion heating only
(b) Radioactive heating only

Suppose Mars had accreted very slowly ($>10^8$ years) from tiny planetesimals devoid of radioactive material.

(c) Would you expect Mars to be differentiated? Why or why not?

15-17. The surface mass density of protoplanetary disk A is twice as large as that of protoplanetary disk B. The disks orbit identical stars. Neglecting migration, which disk produces more planets? Which forms larger planets? Be quantitative when possible.

15-18. (a) Calculate the escape speed from the asteroids 1 Ceres and 243 Ida using the data provided in Table E.8.

(b) Compare these escape speeds with the typical asteroid encounter velocities that you calculated in Problem 12-8 and comment on the outcome of collisions among asteroids.

15-19. (a) Compute the surface density of solids in a minimum-mass circumjovian protosatellite disk by spreading the masses of Jupiter's four large moons over a region comparable to their current orbits.

(b)* Determine the growth time of Io and Callisto using equation (15.18) with $\mathcal{F}_g = 2$.

CHAPTER 16

Planets and Life

Where the telescope ends, the microscope begins. Who
can say which has the grander view

Victor Hugo, *Les Misérables*, 1862 [IV.3.ii]

One of the most basic questions that has been pondered by natural philosophers concerns humankind's place in the Universe: Are we alone? This question has been approached from a wide variety of viewpoints, and similar reasoning has led to widely divergent answers. Some theologians have considered Earth and humanity to be God's special place and beings, unique in the entire universe, but others saw no reason why God would have bothered creating stars other than the Sun and not also have surrounded them with planets teeming with life. Aristotle believed that earth, the densest of the four elements[1] that comprised our environment, fell towards the center of the Universe, so no other worlds could possibly exist; in contrast, Democritus and other early atomists surmised that the ubiquity of physical laws implies that innumerable Earth-like planets must exist in the heavens.

Some aspects of the question of human uniqueness remain ill constrained, but others have yielded to scientific investigation. Copernicus, Kepler, Galileo and Newton convincingly demonstrated that the Earth is not the center of the Universe and that other worlds qualitatively similar to Earth orbit the Sun. Telescopic observations – and more recently interplanetary spacecraft – have told us a great deal about these neighboring worlds (see Chapters 8–10). In the past two decades, hundreds of planets have been discovered in orbit about stars other than our Sun (see Chapter 14).

A major scientific debate concerning the possibility of life (advanced or otherwise) on Mars was ongoing at the beginning of the twentieth century. The martian climate is more Earth-like than that of any of our other neighbors. Early Space Age findings about the current martian surface conditions implied that the environment was far less hospitable than previously conjectured. More recently obtained data suggest that early Mars may have been as hospitable to life as was early Earth (§9.4). Descendants of such life may survive deep under Mars's surface and/or may have traveled to Earth within meteorites and be our very distant ancestors!

In this chapter, we present a brief summary of the relationship between life and the planet(s) that it forms and develops upon. Even a very long book on planets and life would need to be selective, a single chapter much more so. We begin by introducing the factors that must be considered in order to address the question. 'Are we alone?' After that, we assess what life is and the composition of life on Earth. Our next topics are astrophysical restrictions on the setting of planets hosting life analogous to that on Earth and planetary factors that are important to life. We then address life's origins followed by techniques that have been proposed to detect and study extraterrestrial life. We conclude with a discussion of the likely abundance of life in the galaxy.

16.1 Drake Equation

Although microbial life might exist on Mars and/or some moons of the outer planets (§16.13), we are confident that no body in the Solar System other than Earth has 'intelligent' life such as ourselves. But what are the chances of finding an advanced civilization elsewhere in our galaxy? The answer depends on how many, if any, other civilizations exist. To provide a format in which to make quantative estimates, Search for Extra-Terrestrial Intelligence (SETI) pioneer Frank Drake developed a conceptual technique to estimate the number of communicating civilizations in our Galaxy, N_{cc}. The **Drake equation** reads:

$$N_{cc} = \mathcal{R}_\star f_{pl} n_{hab} f_\ell f_i f_{cc} L_{cc}. \tag{16.1}$$

---

[1] The standard ancient Greek world view was that the Universe consisted of four elements: air, water, fire and earth. Aristotle reasoned that because heavenly material traveled in circles about the Earth, it must be different from more familiar matter, so he incorporated a fifth element, **quintessence**.

The symbols on the right hand side of equation (16.1) represent a wide variety of physical and biological processes:

- $\mathcal{R}_\star$ is the rate of star formation.
- f_{pl} is the fraction of stars with planetary systems.
- n_{hab} is the average number of habitable planets per planetary system.
- f_ℓ is the fraction of habitable planets on which life actually forms.
- f_i is the fraction of life-bearing planets on which intelligent life develops.
- f_{cc} is the fraction of intelligence-bearing planets that develop the technology to communicate over interstellar space.
- L_{cc} is the average lifetime of a technological civilization.

The focus of this chapter is on the n_{hab} term in the Drake equation, but factors contributing to the f_ℓ and f_i terms are also considered.

16.2 What Is Life?

As living organisms, we are intrigued by life. We consider life forms to be fundamentally different from non-living matter, more than simply the sum of its parts. However, defining **life** turns out to be extremely difficult, and many current definitions of life attempt to do so by enumerating life's key properties.

The basic unit of all life on Earth is the **cell**, which can be defined as a membrane-bounded molecular system of replicating catalytic polymers containing organic material and chemically coded instructions. Complex organisms, including animals and plants, are composed of **eukaryotic** cells, which are characterized by having nuclei containing most of their genetic information. Simple cells, such as those of bacteria, lack nuclei and are referred to as **prokaryotic**. Eukaryotic cells are generally much larger and more complex than are prokaryotic cells. The primary components of these two types of cells are shown in Figure 16.1.

The cell acts in two critical life processes: as a factory to produce proteins for structural and other function and as a design model in cell division, after which there are two factories with the same capabilities as the original cell. All living organisms on Earth are composed of one or more cells. Sub-life forms that are not composed of cells, such as **viruses**, cannot function and reproduce on their own. Nonetheless, although life on Earth uses cells, the essence of what we think of as living does not involve being organized as cells, so our definition should not include cellular structure as a requirement. Analogously, even though all life on Earth is composed of complex, carbon-based **organic molecules**, there may be other chemical reactions leading to enough similarity in function that it would be parochial to exclude them from our definition of life (§16.4).

Living organisms process energy and have the capability to grow and reproduce. But fire shares these same attributes. Cells and ecosystems are able to harvest energy, metabolize, replicate and evolve; this ensemble of properties can be considered as fundamental to life. The ability of organisms to undergo **Darwinian evolution** (§16.10) is an important aspect of life on Earth. However, a mule cannot reproduce, yet it is clearly alive. The same can be said of worker bees and individuals of whatever species that/who are (for their entire lives or the remainder of their lives) sterile. Most macroscopic organisms cannot survive independently; for instance, we as animals require foods produced by photosynthetic plants and other organisms and, even more fundamentally, bacteria living in our guts perform chemical synthesis essential for our digestive processes. Nonetheless, especially looking at the level of ecosystems rather than individual organisms, defining life to be a self-sustaining system capable of undergoing Darwinian evolution has considerable merit.

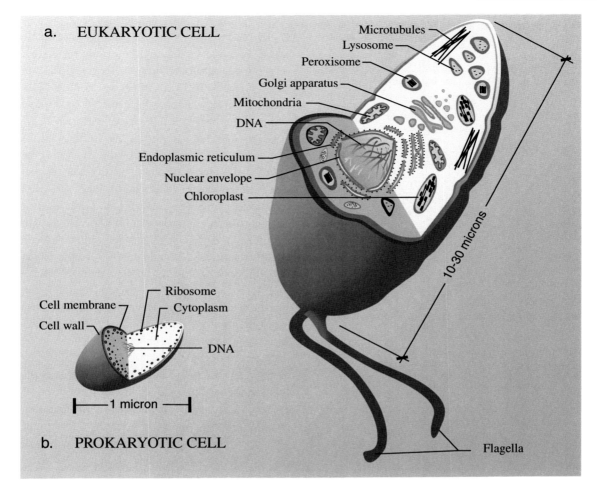

a. **EUKARYOTIC CELL**

Microtubules
Lysosome
Peroxisome
Golgi apparatus
Mitochondria
DNA
Endoplasmic reticulum
Nuclear envelope
Chloroplast

10-30 microns

Ribosome
Cell membrane
Cytoplasm
Cell wall
DNA

├— 1 micron —┤

b. **PROKARYOTIC CELL**

Flagella

Figure 16.1 COLOR PLATE Schematic views of the two types of cells that comprise life on Earth. Note the difference in scales between the two parts of the diagram. Not all eukaryotes have every feature illustrated here. For instance, animals do not have chloroplasts. Flagella are used for locomotion by many single-celled organisms. (Drawing by Cynthia Lunine. From Lunine 1999)

Some computer codes can replicate and adapt, but few people would consider them to be alive. As robotics and artificial intelligence advance, machines are able to mimic more and more of life's functions. Would a robot that was able to reproduce itself from materials and energy found in nature be considered alive? What if that robot also had the ability to adapt and evolve, producing more and more fit descendants indefinitely rather than simply identical replicas of itself? Are humans a junction between unintelligent carbon-based life and intelligent silicon chip-based life?

Thus, life is difficult to define. Although we have a sense that 'we know it when we see it', the semantic difficulties in defining life are related to the more practical problem of finding extraterrestrial life (§16.13).

16.3 Biological Thermodynamics

Life requires energy to survive, and that energy must be in a 'usable' form. An environment that is in complete thermodynamic equilibrium is not suitable for life – disequilibrium is essential. For example, a system that is hot possesses thermal energy, but if the system is isolated and in equilibrium at a uniform temperature, there is no way to extract this energy to perform useful work. By using some of the free energy provided by a disequilibrium environment, life hastens the progression toward thermodynamic equilibrium.

Living organisms maintain almost the same temperature throughout their bodies and with their environment, i.e., living organisms, including cells, are nearly isothermal systems. This means that work cannot be done through transfer of heat. Instead, energy that is used for work in biological systems is transferred biochemically. Living systems are always open systems, i.e., both matter and energy are freely exchanged between the organism and its environment.

Total energy is conserved, as required by the first law of thermodynamics, and the second law of thermodynamics requires that the net entropy of the Universe increases. The Gibbs free energy (§3.1.4) measures the useful work obtainable from any system. If the change in the Gibbs free energy is negative, $\Delta G < 0$, reactions go spontaneously (unless kinetically inhibited), but those with $\Delta G > 0$ require feeding energy into the system to proceed. Living organisms construct microenvironments such as cells that permit chemical reactions that have $\Delta G < 0$ but are otherwise kinetically inhibited to proceed.

Life on Earth uses complex sequences of chemical reactions, some individually having $\Delta G > 0$, but whose sum results in an overall decrease in G. By assembling carbon and other elements into complex molecules, life creates local decreases in entropy that represent stored energy, a state

of disequilibrium. The simultaneous production of O_2 by photosynthesis and CH_4 by **methanogenic** organisms creates a disequilibrium atmosphere that may provide a telltale sign of extraterrestrial life (§§16.6.1 and 16.13). Ultimately, however, life produces heat and other wastes that more than compensate for the local decrease in entropy.

Given that life requires some form of available energy, the search for habitable environments includes the search for abiological processes that can maintain chemical disequilibrium. The most dynamic environment for life on Earth is near our planet's surface, where short-wavelength light from the 5700 K solar photosphere interacts with matter at $\sim$300 K. More than 99% of the Gibbs free energy in our biosphere comes from the Sun. The remaining <1% is produced through oxidation of inorganic matter by microorganisms (e.g., **chemolithotrophs**). Below ground, subduction and volcanism (much of which is driven by plate tectonics, see §6.3.1) bring the hot reducing environment of the mantle into contact with the oxidized sediments of the surface, thereby creating environments in chemical disequilibrium that help to support our planet's massive subsurface biosphere. The most important source of chemical energy is **redox** reactions occurring at the surfaces of minerals. Redox reactions refer to reduction-oxidation reactions in which atoms have their oxidation state changed.

Photosynthesis uses sunlight to convert carbon dioxide and water into glucose ($C_6H_{12}O_6$, sugar) and oxygen. The overall chemical reaction can be written:

$$6CO_2 + 6H_2O + \text{light}$$
$$\rightarrow C_6H_{12}O_6 + 6O_2 + \text{energy}. \quad (16.2)$$

Most commonly, sunlight is absorbed by the pigment **chlorophyll**, which sets off a chain of chemical reactions that results in the synthesis of **adenosine triphosphate (ATP)**, a small organic compound known as the energy currency of a cell.

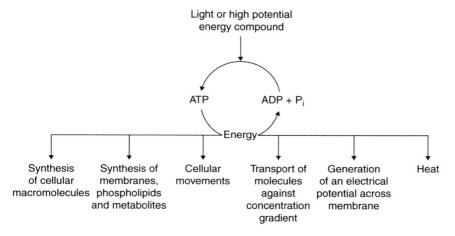

Figure 16.2 Schematic diagram of the adenosine triphosphate (ATP) cycle. ATP is formed from adenosine diphosphate (ADP) and phosphate through photosynthesis and metabolism of energy-rich compounds in most cells. Hydrolysis of ATP to ADP and P$_i$ releases energy that is trapped as usable energy. (Fig. 1.6 from Haynie, 2001, which was redrawn from Lodish et al. 1995)

Complete metabolism of 1 mole of glucose produces up to 38 moles of ATP. The change in free energy for these complete redox reactions is $\Delta G = -2823$ kJ mole^{-1}.

ATP provides the chemical energy for many biochemical processes, including synthesis of DNA (discussed later), muscle contraction and chemical communications within and between cells. Figure 16.2 shows the ATP cycle, where ATP is formed from **adenosine diphosphate (ADP)** through photosynthesis in plants and metabolism of energy-rich compounds, starting with glucose, in most cells. Hydrolysis of ATP to ADP and P$_i$ (phosphate) releases energy that is trapped as usable energy:

$$ATP + H_2O \longleftrightarrow ADP + P_i + H^+. \qquad (16.3)$$

All terrestrial life is based on carbon-chain chemistry and uses water as a solvent. Two types of molecules are the basis for all known life: **proteins**, which form life's structural basis, and **nucleic acids**, which encode the instructions required to synthesize proteins. Proteins consist of one or more **polypeptides**. A polypeptide is a chain of amino acids linked together by a particular kind of covalent bond that is referred to as a **peptide bond**.

A protein's minimum energy state is being folded. Such a state is sometimes compared to an organic crystal: Rigid (though still flexible) and held together by **van der Waals** (noncovalent, electrostatic and dipolar) forces. This stability is important because more than 50% of the dry mass (mass without water) of a human body is in proteins. A protein in its **denatured** or unfolded state is more fluidlike and flexible. The transition from a folded to an unfolded state can be induced by heat, similar to the way water-ice would melt when exposed to heat, i.e., unfolding of a protein can be compared to a phase change in a material. The stability of a protein can be evaluated by calculating the change in Gibbs free energy necessary to unfold the protein at a given temperature.

Proteins catalyze most of the metabolism within cells. **Nucleic acids** are polymers of **nucleotides**, each of which contains a sugar, a phosphate and a nitrogenous base. The genetic code that contains

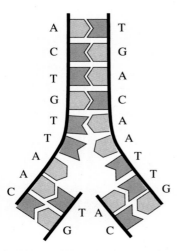

Figure 16.3 Sketch representing the duplication of DNA, which is the repository of genetic information in modern cells. Each strand maintains exactly the same information in the form of a base four code consisting chemically of the nucleotides adenine (A), which links only with thymine (T), and guanine (G), which links exclusively with cytosine (C). Duplication occurs by separating the two strands and copying each one. (From Sneppen and Zocchi 2005)

the information required for terrestrial life is maintained one dimensionally in a double-stranded polymer called deoxyribonucleic acid (**DNA**). The replication of DNA is represented schematically in Figure 16.3. Ribonucleic acid (**RNA**) mediates the transfer of genetic information from DNA to the formation of proteins. Virtually all proteins are composed of 20 types of **amino acids**. These amino acids, which form the building blocks for terrestrial life, all contain the element nitrogen in addition to carbon, hydrogen and oxygen, the four most cosmically abundant chemically active elements. Some amino acids contain sulfur. Phosphorus is also required to prevent DNA from folding up: The phosphates are negatively charged and attractive to water; the repulsion between the two chains of phosphates produces the DNA ladder. Humans require many other elements, including substantial amounts of calcium for our bones and iron for our blood.

16.4 Why Carbon and Water?

Apart from free energy, the most basic requirements for life as we know it are reactive carbon compounds and water. Life uses a chemical system to store, read and write very complex genetic information. Such a system must be able to synthesize stable, complex molecules with structures that permit them to be read and transcribed. The need for free energy to create and maintain life's highly non-equilibrium structures seems impossible to avoid, but are carbon and water as inevitable?

Carbon atoms have four valence electrons, which can be shared with as many as four other atoms (including other carbon atoms) to produce an extremely diverse set of chemicals. Carbon atoms are also able to form **multiple bonds** (sharing two or more electrons with another atom) as well as **single bonds** (one shared electron). A molecule having multiple bonds can free electrons to add additional connections to new atoms without having the original attachments completely severed. Thus, long chain molecules can be built up gradually. Among these molecules are repetitive carbon-chain configurations of unlimited sizes. Many of these molecules react with water, the most common volatile on Earth.

Silicon lies just below carbon on the periodic table (Appendix D), and these two elements share various chemical properties, including the availability of four valence electrons. However, several chemical differences appear to make silicon far less suitable for life than is carbon. Bonds that silicon makes with oxygen or hydrogen are stronger than silicon–silicon bonds, so long chains of silicon atoms analogous to many carbon compounds are unlikely. Silicon does not usually make multiple bonds, so it lacks an important path that carbon uses to build up large organic molecules. When silicon reacts with oxygen, it forms SiO_2. Silicon dioxide is a crystal lattice solid; SiO_2 combines with metallic elements (especially Mg) to form rock. In contrast, carbon and oxygen form CO_2, a gas that readily reacts with many other compounds.

Liquid water is a polar molecule that is an excellent solvent because of its ability to form hydrogen bonds. We know that water works well for the carbon-based life found on Earth. It is composed of elements that have many other functions for terrestrial life. Water is common on planets and moons within our Solar System. Ammonia (NH_3) is also a very good solvent, but it is less abundant than water, and ammonia tends to mix with water to form a solution. At 1 bar pressure, ammonia and most other polar solvents are liquid only at temperatures well below 273 K; thus, as chemical reaction rates increase steeply with temperature, biochemistry within these sovents would proceed very slowly. Additionally, we do not know of any naturally occurring reservoir of liquid ammonia in the Solar System. Polar liquids such as H_2S, PH_3 and HCl are conceivable solvents for extraterrestrial life. However, none of these molecules is good at forming hydrogen bonds, and all are rarer than ammonia, in part because sulfur, phosphorus and chlorine are easily bound up in stable minerals on rocky planets. The compounds HCN, CH_3OH and HF are polar molecules that can form hydrogen bonds, but HCN polymerizes, and HF is both rare and attacks rocks. Life that used a nonpolar solvent would require even more fundamentally different chemistry from that on Earth, making it difficult to envision.

In sum, although life based on exotic chemistry cannot be excluded, carbon and water offer major advantages of stability, flexibility and abundance over all alternatives that have been studied. Moreover, our example of one type of life demonstrates that this combination can work.

16.5 Circumstellar Habitable Zones

Life on Earth has been able to evolve and thrive thanks to billions of years of benign climate. The defining ecological requirement for life on our planet is the presence of liquid water. Many terrestrial organisms can survive extreme desiccation, but they all share the absolute requirement for liquid H_2O to grow and reproduce. Mars appears to have had a climate sufficiently mild for liquid water to have rained, flowed and pooled on its surface for extended periods of time when the Solar System was roughly one-tenth its current age (§9.4.5). But at the present epoch, Mars's low atmospheric pressure and (at most times and places) low temperature imply that liquid water quickly freezes and/or boils on the martian surface. Venus is too hot, with a massive carbon dioxide–dominated atmosphere (§5.2). Stellar evolution models predict that the young Sun was about 25% less luminous than at present (Fig. 3.6); nonetheless, Earth, and probably Mars, was warm enough to be covered by liquid oceans 4 billion years ago.

Liquid water is essential for life on Earth. Based on the hypothesis that extraterrestrial life shares this requirement, the **habitable zone** (HZ) of a star has been conventionally defined to be the orbital region in which an Earth-like planet can possess stable liquid water on its surface. This is a very conservative (but observationally useful) definition because a planet's surface temperature also depends on its inventory of volatiles such as CO_2 through the greenhouse effect, the albedo, atmospheric/oceanic circulation, etc. Energy sources such as radioactive decay (§11.6.1, Problem 15-10), remnant accretion energy (§15.5.2, Problem 15-13) and tidal heating (§2.7.3) can warm a planet's surface to the melting point of water. These energy sources can also maintain subsurface reservoirs of liquid water. Earth has a thriving subsurface biosphere, albeit one that is composed almost exclusively of simple prokaryotic organisms that can survive in oxygen-poor environments. Additionally, although terrestrial life requires liquid H_2O, other life forms may not (§16.4).

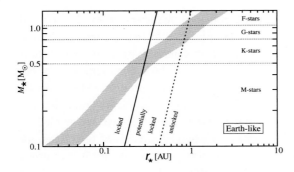

Figure 16.4 The *shaded region* represents the location of the circumstellar habitable zone (for Earth-like planets) as a function of stellar mass for zero-age main sequence stars. A rocky planet to the left of the *solid line* would have its rotation tidally locked in a (geologically) short period of time (§2.7.2) and henceforth always present the same hemisphere to its star, a planet to the right of the *dashed line* would likely not be in such a rotation state, but planets located between these two lines might or might not be in synchronous rotation. (Grießmeier et al. 2009)

The orbital distance of the HZ depends on the star's luminosity, which, for main sequence stars, depends primarily on stellar mass (§3.3). The shaded region in Figure 16.4 shows the extent of the HZ of a young main sequence star as a function of stellar mass. The inner boundary of the HZ is taken to be the location at which water is lost as a result of a moist greenhouse or runaway greenhouse effect (§5.8.2). The outer boundary is where greenhouse warming is inadequate to maintain surface temperatures above freezing anywhere on the planet. Although young brown dwarfs and planets can emit substantial amounts of energy, these objects are not encircled by long-lived HZs because (unlike stars) they do not have long-lived epochs of slowly varying luminosity (§3.3 and Fig. 3.7). Note that a planet orbiting within the HZs of a low-mass star would have its rotation tidally locked (Problem 16-6), and its nearby, but dim, star would perpetually shine on the same hemisphere of the planet.

The boundaries of a star's habitable zone can be roughly approximated by the region where a planet's surface (ground) temperature allows for the presence of liquid water at 1 bar pressure,

$$273° < T_g < 373°. \tag{16.4}$$

The ground temperature is given by the sum of the equilibrium temperature and the warming caused by the greenhouse effect (255 K and 33 K, respectively, for Earth). The equilibrium temperature, T_{eq}, in turn depends primarily upon the flux of stellar radiation received by the planet and the planet's Bond albedo (eq. 4.17).

The albedo and the amount of greenhouse warming are extremely difficult to measure for potentially habitable exoplanets, but the flux of stellar radiation intercepted by the planet can be calculated from the properties of its star and its orbital period. If this flux is close to the solar constant, which is the mean flux of solar radiation intercepted by Earth (eq. 4.13), then the planet is within its star's habitable zone. As smaller stars emit a larger fraction of their energies in the IR and planetary albedo tends to be smaller in the IR than in the visible, the HZ boundaries for smaller stars are located at a little lower stellar flux, but this effect is far less than luminosity changes, so the boundaries move inwards in actual distance.

Equation (16.4) represents a good approximation of the HZ boundaries for a mostly dry planet, i.e., a planet whose surface is not dominated by oceans. (Such a dry planet can maintain small reservoirs of surface water.) In contrast, evaporation of water on an ocean-covered planet would lead to increased atmospheric opacity and a runaway greenhouse effect for surface temperatures $T_g \gtrsim 310$ K, so the inner boundary of the HZ for a planet with large oceans is substantially outwards of that given by equation (16.4).

Because a star's luminosity increases with time (§3.3), both the inner and outer boundaries of its HZ move outwards. Thus, a planet that is in the HZ when a star is young may subsequently become too hot. Venus may have been an example, although its current surface is too young to show any evidence of a more clement climate that may have existed billions of years ago (§9.3). Other planets could be too cold for liquid water when young but warm up enough to have liquid water on their surface later as their star's luminosity increases; this may happen to Mars a few billion years hence. However, the formation of life might require liquid water early in the planet's existence, prior to the escape of some light gases. Thus, the **continuously habitable zone** (CHZ), where liquid water could have been present for a specified period of time, usually from early in the star's life to the current epoch, is the most promising region for life as we know it.

Earth has had liquid water on its surface for at least the vast majority of the past 4 billion years. Four billion years ago, the Sun's luminosity was only ~75% as large as at present. Climate models suggest that the Earth should have been frozen over with such a low solar luminosity; this disagreement between theory and observations is known as the **faint young Sun problem**. Another aspect of the faint young Sun problem is that the oldest regions of the martian surface show signs of large-scale running water, but younger regions do not (§9.4.5). This suggests that at least episodically, early Mars had a warmer and thicker atmosphere than the planet presently has (§5.8.2). Because the luminosity of a star is a rapidly increasing function of stellar mass (eq. 3.23, Figure 3.4) and planets drift away from a star as it loses mass (eq. 2.65), one conceivable solution to the faint young Sun problem is that the Sun was more massive in its youth (Problem 16-8). However, Sun-like stars lose mass too slowly for this process to solve the problem. Rather, the warmth of early Earth and Mars is attributed to abundant greenhouse gases in the atmospheres of both planets, with carbon dioxide, water and possibly ammonia and/or methane suggested to have played major roles. In the case of Mars, episodic warm periods after moderate to large impacts may have temporarily provided climatic conditions that allowed enough rain to carve the observed channels (§16.7).

Low-mass stars survive far longer than do high-mass stars (§3.3). The main sequence phase of low- to moderate-mass stars provides regions where planets may maintain liquid water on their surfaces for long periods of time. The lifetimes of these continuously HZs are shorter for more massive stars. Advanced life took billions of years to develop on Earth. Thus, even if Earth-like planets form around high-mass stars at distances where liquid water is stable, it is unlikely that benign conditions exist long enough on these planets for life to form and evolve into advanced organisms. Short-lived ($\lesssim 10^8$ years) HZs also occur at substantial distances from post-main sequence stars when these stars burn helium in their cores and shine brightly as red giants.

Although the quantity of radiation received by planets in the nominal HZs is the same regardless of stellar type, the 'quality' differs. More massive stars are hotter and therefore radiate a larger fraction of their energy at shorter wavelengths (§4.1.1). This greater flux of ultraviolet radiation could conceivably speed up biological evolution enough to compensate for the massive star's shorter lifetime. But ionizing radiation can harm living organisms and there is an optimum rate of genetic evolution beyond which more rapid rates of mutation are (in a statistical sense) harmful to the advancement of the genome (§16.10). Note that differences in planetary atmospheres substantially affect the amount of ultraviolet (UV) radiation reaching a planet's surface (§5.5.1).

At the other end of the spectrum, the smallest, faintest stars can live for trillions of years. But these cool dwarf stars emit almost all of their luminosity at infrared wavelengths that may be difficult for life to harness, and they typically display larger temporal variations in luminosity than do solar-type stars. Also, an HZ planet would orbit so close to its faint star that the same hemisphere would always face the star (as the Moon's nearside always faces the Earth; see §2.7.2 for a discussion of tidal synchronization and Problem 16-6 for a quantitative estimate of the time required for rotation of a planet orbiting near its star to become tidal synchronized with the planet's orbit). Thus, there would be no day–night cycle, and unless it was sufficiently thick, the atmosphere would freeze out on the planet's cold, perpetually dark, hemisphere. Moreover, the higher orbital velocities in the HZ of an M star (Problem 16-7) lead to higher impact velocities during the accretion epoch and thereby greater impact erosion of atmospheres (§5.7.3), so such planets are likely to be deficient in the volatiles required by life.

The minimum separation of Earth-mass planets on low-eccentricity orbits required for the system to be stable for long periods of time (§2.4) is comparable to the width of a star's continuously HZ zone. Thus, orbital stability arguments support the possibility that most stars could have one or even two planets with liquid water on their surfaces, but unless greenhouse effects conspire to substantially compensate for increasing distance from the star, larger numbers of 'habitable' planets are unlikely (Problem 16-3).

The nominal definition of a habitable planet is one with liquid water on its surface. By this criterion, early Mars was habitable, at least on occasion, but current Mars is not. But temperature increases with depth within Mars, and liquid water can exist at a moderate depth below the martian surface. Thus, life may have formed when it was warm enough for liquid water to exist in the energetically rich surface, and a subsurface biosphere could still exist even though present Mars does not pass the nominal surface liquid criterion for habitability.

Other heat sources can create reservoirs of liquid water even well outside of nominal HZs. Dissipation of tidal energy probably provides sufficient heating to maintain a liquid water ocean tens of kilometers below the surface of Jupiter's moon Europa (§10.2.2). Given a sufficient thermal blanket from, e.g., a moderately thick H_2 atmosphere, internal sources of radioactive energy could maintain liquid water reservoirs near the surface of a mostly rocky planet that orbits very far from its star or even a **rogue planet** with these characteristics that has escaped from its star and wanders by itself through interstellar space. The above examples all involve extra heat input to nominal cold locations. Planets closer to a star than its HZ could have niches cool enough for liquid water on or near their dark sides if they were tidally locked or near the poles if their obliquity was close to 0° or 180°. Thus, circumstellar HZs should be a reference point where life is most likely, rather than the only regions where life as we know it could possibly reside.

Near the center of the Milky Way galaxy, the distances between stars are much smaller than typical stellar separation at the location of the Sun's orbit. Thus, supernovae, gamma ray bursts, etc. present a greater potential hazard to any life that may exist close to the galactic center. The bulk of a terrestrial planet is composed of heavy elements that were produced within stars (§3.4.2). In the outer regions of our galaxy, stars are very far apart, and the material out of which new stars are being formed may not have enough heavy elements for Earth-like planets to grow. Considerations of this type have led to the concept of a **galactic habitable zone**, analogous to circumstellar HZs. The concept of a galaxy's HZ may well be viable, but the extent of such a region is far more difficult to quantify than is a star's HZ, and its boundaries are likely to be much fuzzier.

16.6 Planetary Requirements for Life

Just because a planet is located in an HZ does not mean that it is suitable for life. Earth offers a benign environment because it was endowed with enough volatiles to produce an ocean and a significant atmosphere but not so much water as to be blanketed by oceans tens of kilometers deep. Our planet is large enough to retain an atmosphere

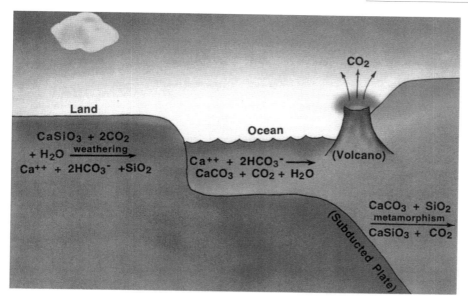

Figure 16.5 COLOR PLATE The Earth's carbonate-silicate cycle. Note the importance of plate tectonics (§6.3.1) in returning CO_2 to the atmosphere. (From Catling and Kasting 2007)

(§5.7), can recycle crust via plate tectonics (§6.3.1) and is far enough away from the asteroid belt that it is not frequently hit by projectiles energetic enough to cause mass extinctions.

Does the origin of life require land masses? And if land masses are required for life to develop, were the small, probably short-lived island volcanoes, which may well have been the only land masses of early Earth (§6.3.1), sufficient for the origin of life? Or did life's origin await the growth of larger more stable 'continents' on Earth? What about large oceans? Tidal zones to have cycling and pumping of environments to allow molecular self-assembly in lakes or ponds (§16.6.3)? Perhaps our ancestors began on Mars and traveled to Earth within rocks ejected by impacts (§16.12)? Must life ascend to land in order to develop technology (§16.10.3)?

Clearly, there is a wide variety of criteria that a planet must meet in order to be a suitable location for life, but many of these criteria are difficult to understand in detail and to quantify. We examine a few of these planetary properties in this section.

16.6.1 Biogeochemical Cycles

Carbon dioxide on our planet cycles between the atmosphere, the oceans, life, fossil fuels and carbonate and other rocks on a wide range of timescales. Carbonate rock, the majority of which is located in Earth's mantle, forms the largest reservoir. Silicate rocks and atmospheric CO_2 react with the aid of liquid water to form, ultimately, silica and carbonate rocks (see eq. 5.17); in some cases, living organisms are involved; in other cases, they are not. Plants and many other photosynthetic organisms remove CO_2 from the atmosphere. This carbon is first stored in living organisms; some is returned directly to the atmosphere via respiration, and a portion remains isolated for long periods of time as **fossil fuels**. However, the part of the carbon cycle most important to long-term climate is the **carbonate-silicate cycle** that is shown schematically in Figure 16.5 (and discussed in §5.8.2). The products of silicate weathering, including calcium (Ca^{++}) and bicarbonate (HCO_3^-) ions and dissolved silica (SiO_2), are transported by rivers to the ocean. Organisms such as foraminifera use these products to make shells of calcium carbonate ($CaCO_3$); other organisms make shells out of silica. Most of these shells eventually dissolve, but some are buried in sediments on the seafloor. On a lifeless planet, calcium carbonate would be incorporated in ocean sediments through nonbiological chemistry when this mineral reached a high enough concentration to become saturated in seawater. The combination of silicate weathering plus carbonate precipitation yields the net transformation $CO_2 + CaSiO_3 \rightarrow CaCO_3 + SiO_2$. Thermal processing (§6.1.2) of buried carbonate rocks returns CO_2 to the atmosphere.

Feedback mechanisms play an important role in determining the ranges of many parameters (§5.8.2). **Positive feedback** amplifies the effect of a change in a primary input. For example, during the past few million years, the climate cycle on Earth has alternated between ice ages and interglacials. When the climate cools, ice sheets form; because ice is highly reflective (has a high albedo), this change leads to further cooling, as diagrammed in Figure 5.11. A quantitative example of this effect is worked out in Problem 16-19. Evidence suggests that most or all of Earth's surface was covered by ice about 700 Myr ago. This state is referred to as **snowball Earth**, and it may have had a substantial influence on the evolution of life leading to the Cambrian explosion (§16.10.2). The runaway greenhouse effect (§5.8.2) is another example of positive feedback, which involves the atmospheric abundance of water vapor (Fig. 5.12). Small changes in input parameters in which positive feedback occurs can have large effects on habitability. In contrast, variations of parameters such as atmospheric CO_2 abundance can be damped by negative feedback (§5.8.2).

The CO_2 cycle shown schematically in Figure 16.5 acts as a **negative feedback loop** to buffer terrestrial climate on timescales of 10^8 years (§5.8.2). Carbon dioxide is an important greenhouse gas (§5.1.2). The reaction rate of CO_2 with silicate rocks increases exponentially with temperature,

but the other phases of this CO_2 cycle are virtually temperature independent (Fig. 5.14). Thus, if it gets hot, weathering rates increase and the CO_2 abundance drops; during colder times, CO_2 increases. Variations in the atmospheric abundance of CO_2 are thought to have played a major role in maintaining moderate climates on Earth for most of our planet's history.

Another aspect of the carbon cycle limits the abundance of atmospheric CO_2 on billion-year timescales. Ocean basalts remove CO_2 from seawater at a rate that increases with the CO_2 abundance in seawater, which varies in sync with its atmospheric abundance. These carbonates descend into the Earth's mantle as plates are subducted (§6.3.1). Heating within the Earth's mantle frees the CO_2, which is subsequently returned to the atmosphere via volcanic activity (§6.3.2; Problem 16-11). Although this is also a negative feedback loop, it does not depend on climate and thus acts to buffer the atmospheric CO_2 abundance rather than climate variations.

Carbonates are not readily recycled on a geologically inactive planet such as modern Mars; in contrast, they are not formed on planets such as Venus, which lack surface water (§5.8.2). For a given composition, a larger planet ought to remain geologically active for a greater amount of time, because it has a smaller ratio of surface area to mass and thus retains heat from accretion and radioactive decay longer. The number of variables involved in determining a planet's habitability precludes a complete discussion, but some of the major issues are summarized in Figure 14.34.

Despite abundant atmospheric N_2, nitrogen is often a limiting nutrient for life on Earth. Nitrogen **fixation** is the process by which nitrogen is taken from its relatively inert molecular form in the atmosphere and converted into soluble nitrogen compounds such as ammonia, nitrate and nitrogen dioxide. Few organisms can metabolize nitrogen unless it has been **fixed** from the atmosphere. In the anoxic prebiotic atmosphere of the young Earth, lightning oxidized N_2 via the reaction

$N_2 + 2CO_2 \rightarrow 2NO + 2CO$. NO was then converted to nitrosyl hydride (HNO), which is soluble in water. Lightning thereby provided a modest flux of fixed nitrogen, which may have been important to early life. In an atmosphere containing some CH_4, nitrogen fixation can progress via HCN, which is hydrolyzed in solution to form NH_4^+ (ammonium).

Biological fixation of nitrogen arose when fixed N became scarce, which may have been prior to the last common ancestor of all modern life on Earth, and it is a strictly anaerobic process. Biological fixation of nitrogen also occurs via HCN, and when anaerobes developed the capability of nitrogen fixation, NH_4^+ probably became the dominant form of combined nitrogen in Earth's oceans. Once our planet's atmosphere contained significant quantities of O_2 (Fig. 5.16), NH_4^+ ions were oxidized to nitrite (NO_2^-) and then converted to nitrate (NO_3^-) by bacteria; this process is referred to as **nitrification**. **Denitrification** is the microbial reduction of NO_3^- to N_2O and N_2.

16.6.2 Gravitational and Magnetic Fields

Because biomechanics depends on material strength, gravity and other physical properties related to viscosity, organisms do not scale simply with size. For example, larger animals require more support, so elephants need to have more bulky legs than a scaled-up mouse; birds use a fundamentally different mechanism to fly than do insects; single-celled organisms have little inertia (compared with viscous drag) moving in liquid water, so they use different locomotion strategies than do larger creatures. These factors must be taken into account for planets with different surface gravity from Earth. However, planetary gravity is probably most important through its effect on escape of atmospheric gases (§5.7).

Earth has a much larger magnetic field than do any of the other terrestrial planets within our Solar System (Table E.18). Are magnetic fields necessary for life? Direct effects are minor – a few species are known to use magnetic fields as aids for navigation, but most organisms do not. Earth's magnetic field also plays a protective role by stopping charged particles from eroding the atmosphere or reaching the surface. Such particles could have profound direct and indirect effects on life, but they are unlikely to sterilize a planet that has a sufficiently massive atmospheric blanket. Indeed, during magnetic field reversals, the dipole component of Earth's magnetic field drops to near zero (§7.4.1). A subsurface biosphere would be little affected by charged particles hitting the planet's surface and thus does not require the protection of either a planetary magnetic field or an atmosphere.

16.6.3 Can Moonless Planets Host Life?

The romantic mood set by a full Moon has led to countless human births. It may be no coincidence that women's menstrual cycles are similar in length to the lunar month. Some species of sea turtles lay their eggs on beaches only at full Moon. The nocturnal behavior of various species requires the light of the Moon, and clearly different ecological niches would be available on a moonless planet. Although these factors would affect some aspects of life significantly, they cannot be regarded as implying a large moon is required for life to exist. However, other effects of the Moon have been suggested to be more essential to life on Earth.

The Moon stabilizes Earth's obliquity (Fig. 2.18) and thereby its climate. In contrast, perturbations from the other planets produce substantial variations in the martian obliquity (Fig. 2.17), which probably are responsible for the patterns observed in Mars's layered polar terrains (§9.4). However, planets with retrograde spins, more rapid spin rates or those in certain planetary systems configured differently from our own can have stable obliquities even without a moon. And the climate of a planet with more ocean and less continental mass than Earth would depend little on planetary obliquity.

The Moon is the primary body responsible for ocean tides on Earth in the present era, and it produced substantially larger tides when it was closer to Earth billions of years ago (§2.7.2). Tidal zones at the boundaries between sea and land on Earth are subject to cyclic variations in conditions, with alternating wet and damp/dry periods. Very productive ecosystems currently exist within tidal zones. Organic molecules can be concentrated by evaporation in tide pools and be supplied with repeated addition of nutrients; thus, life on Earth may have originated in tidal regions. If so, then a large moon is advantageous to the formation of life. Nonetheless, because solar tides are non-negligible, a moon cannot be viewed as essential to life on this account.

The leading model for the Moon's formation is the giant impact theory (§15.10.2). This large impact removed volatile elements and compounds from the Earth. Such a devolatilization may have been required for Earth itself to have been suitable for life. However, models of planetary growth suggest that most terrestrial volatiles came from a small fraction of the material composing our planet, so an otherwise Earth-like planet could simply accrete fewer volatile compounds than Earth. Alternatively, a planet could be devolatilized by a nearly head-on mega-impact that would not loft enough material into orbit to produce a large moon.

16.6.4 Giant Planets and Life

Giant planets are unlikely abodes for life because any solid surface that they might have would be at extremely high pressure (see Chapter 8). Living organisms occupy essentially all known environments on or below the surface of our planet in which liquid water is present. However, despite billions of years of evolution, no known terrestrial organisms have adapted to a purely aerial life cycle. Moreover, although moderate temperature zones occur within giant planet atmospheres (Fig. 8.14), parcels of gas do not remain in place but rather are repeatedly mixed downwards by convection into regions that are too hot for organic molecules to survive.

Giant planets may, however, harbor habitable moons. To first approximation, moons of giant planets seem to have similar potential for habitability as do terrestrial planets of the same mass and distance from their star. Formation circumstances may result in different average compositions, but the distributions of composition of such similarly sized moons and planets probably overlap.

Tidal interactions between large moons and giant planets are likely to be much more substantial than between stars and planets within HZs because the distances involved are much less. This implies that such moons are probably in synchronous orbits. Assuming the primary source of energy is radiation from the star rather than from the planet, the consequences of such a spin-orbit resonance would be far less than for planets in synchronous rotation within HZs of faint, low-mass, M dwarf stars. But the length of the day on such a moon would be slowed to approximately (see Problem 1-4) the moon's orbital period about its planet. Additionally, if it was in orbital resonance with another moon, energy from their coupled tidal recession from the planet would be deposited in one or both moons (§2.7.3). This energy source could produce subsurface oceans, as is likely the case on Jupiter's moon Europa (§10.2.2).

Giant planets can also affect the habitability of terrestrial planets orbiting the same star. They can destabilize the orbit of a terrestrial planet, leading to large variations in eccentricity or even to collisions with another planet or the star or to ejection into interstellar space. They can cause obliquity variations affecting the climates of terrestrial planets. Giant planets affect the flux of impactors impinging on terrestrial planets. Such impacts can have a devastating effect on life – a fact that no dinosaur is likely to argue with!

Giant planets may also be important to the formation of habitable planets. It is likely that giant vulcan planets, which orbit very close to their star, migrated through the HZ (§15.7), and they may have cleared enough material from that region to prevent the formation of terrestrial planets large enough to retain an atmosphere. Giant planets that orbit farther from their star may have positive effects on terrestrial planet habitability. The small fraction of volatiles possessed by Earth probably were diverted from colder regions of the protoplanetary disk by perturbations of Jupiter and Saturn.

16.7 Impacts and Other Natural Disasters

Impacts, as with earthquakes, volcanic eruptions and storm systems, come in various sizes. Impacts of increasingly larger size have greater potential for killing individual organisms and wiping out entire species. Large violent events occur far less frequently than do small ones. But unlike internal processes such as earthquakes, volcanoes and storms, there is virtually no upper limit on the energy and the destructive potential of an impact. The smallest space debris to hit Earth's atmosphere is slowed to benign speeds by gas drag or vaporized before it hits the ground. The largest impactors can melt a planet's entire crust and annihilate life. How a planet's environment responds to an impact depends on the properties of that planet, especially its atmosphere. For example, a thin atmosphere cannot hold as much dust and aerosol as a thick one, so debris will fall out sooner and not obstruct sunlight for as long; a shallow ocean is much easier to evaporate than a deep one is. The remainder of this section examines the effects of impacts and other major natural disasters on Earth.

The effects that an impact has on life depend in a qualitative way on the impact energy. Strong rocky and iron impactors ranging in size from a toaster to a house may hit the ground at high velocity, killing living beings in their path. The rocky bolide that exploded over the Tunguska river basin in Siberia on 1908 June 30 was about the size of a football field (50–100 m across); it produced a blast wave

Figure 16.6 The devastation wrought by the Tunguska explosion is evident in this photograph of felled trees ~20 km from the epicenter. The picture was taken two decades after the bolide explosively disintegrated in the atmosphere above this remote part of Siberia in 1908.

that knocked over trees tens of kilometers away. Figure 16.6 is a photograph of a small portion of the flattened forest as it appeared two decades after the impact. The damage caused by impacts of comets and asteroids up to hundreds of meters in size is limited to a small region of our planet, although coastal flooding produced by **tsunamis** (seismic sea waves) can be fatal to organisms living near the shoreline of an entire ocean basin. Note that a significant fraction of the Earth's human population live on lowlands near the ocean, and some species are even more concentrated in these zones. Impactors of radius larger than ~500 m throw enough dust into the upper atmosphere to substantially darken the sky for much of a growing season, and they are thereby capable of destroying most of the human food supplies. If such an impact occurred today, it could lead to hundreds of millions, if not billions, of deaths.

Fossil records in rock strata imply that, on a few occasions during the past half billion years, a large number of species of animal and plant life have disappeared nearly simultaneously. These major prunings of our portion of the tree of life (Fig. 16.13) are referred to as **mass extinctions** (§16.11). Mass extinctions can be produced by large impacts, which load the atmosphere with dust and chemicals (from vapor and pulverized matter originating in both the impactor and the crater ejecta). Such impacts produce high-velocity ejecta that glow hot while reentering the atmosphere and cause global forest fires (Problem 16-13). Impacts that are even larger fill the atmosphere with so much hot ejecta and greenhouse gases that they can vaporize much or all of the planet's oceans.

Organisms that can live, grow and reproduce at temperatures above about 320 K are known as **thermophiles**. Phylogenetic evidence implies that the last common ancestor of all life on Earth was a thermophilic prokaryote (§16.9.2), which would have been most capable of surviving such a scalding impact. Still larger impacts would destroy all life on the planet, although it is possible that some organisms could survive, possibly in a dormant state, within meteoroids ejected by the impact and subsequently reestablish themselves on the planet by a fortuitously gentle return to the planet's surface when conditions there had improved.

At the other end of the impact hazards scale, meteor impacts have been observed to damage property. There are also (unconfirmed) eyewitness accounts that one of the fragments of the 10-kg meteorite Nakhla, which landed in Egypt on 1911 June 28, struck and killed a dog. Three-quarters of a century later, scientists realized that Nakhla had come from Mars. Thus, an Earthling may, indeed, have been killed by a martian invader.

Larger and more structurally complex life forms tend to place tighter requirements on their environment and thus have greater susceptibility to extinction via impact. Although some tiny organisms can thrive at temperatures exceeding 380 K, no eukaryotes can endure extended exposure to temperatures above 340 K, so all cells with nuclei could be wiped out by an impact that is not sufficiently energetic to kill all prokaryotes. Moreover, large organisms such as animals and plants require food sources, in many cases linked directly or indirectly to photosynthesis, and thus cannot survive extended periods of darkness.

The impact rate on the terrestrial planets of our Solar System was orders of magnitude larger four billion years ago than it is at present (§6.4.4). During that epoch, large impacts heated the upper portion of our planet sufficiently to kill any non-thermophilic organisms living on Earth at the time. In another planetary system, large impact fluxes could continue, making planets with Earth-like compositions, masses and radiation fluxes hostile abodes for living organisms, especially complex forms of life.

16.7.1 K–T Event

The largest mass extinction of the past 200 million years occurred 65 million years ago when half of the genera of multicellular organisms, including the entire dinosaur superorder, suddenly died off. This event marks the end of the Cretaceous (K) time period and the beginning of the Tertiary (T) period and is known as the K–T boundary. Sediments deposited at the K–T boundary strongly indicate that this mass extinction had an extraterrestrial origin. The geological record from this epoch, such as the cross-section photographed in Figure 16.7, shows a layer rich in impact-produced minerals, as well as much higher (10–100 times) than normal levels of iridium, an element rare in Earth's crust but more abundant in primitive meteorites.

The Chicxulub crater provides convincing evidence that the K–T boundary and the extinction of the dinosaurs were indeed triggered by an impact of a 5–10-km radius object, most likely a carbonaceous asteroid. The Chicxulub crater is located on the Yucatán Peninsula in Mexico and is visible in seismic maps of the area, such as the one shown in Figure 16.8. Chicxulub was produced 65 million years ago and is the largest crater on Earth known to have formed within the past 1.8 Gyr. The Chicxulub crater is no longer visible at the surface but can be studied by using gravity anomaly measurements (§6.2.2). The crater appears to have a multiring basin morphology: a peak ring with a

Figure 16.7 Cross-section of the geological record of the K–T boundary at Stevns Klint, Denmark. The lower chalky deposit from the Cretaceous period was produced by planktonic organisms that were wiped out as a consequence of the impact. Above this lies a 1-mm-thick oxidized layer that contains various signs of an impact: shocked quartz, tektite spherules (§8.2) and iridium. The shale on top was deposited in the Tertiary. The hammer head is shown for scale. (From Ward 2007)

diameter $D \approx 80$ km, an inner ring with $D \approx 130$ km and an outer ring with $D \approx 195$ km. Evidence exists for large impacts close to the times of other mass extinctions, although the connections are not anywhere near as clear as with the dinosaur-killer, and other causes, mostly geological or biological, are suspected.

The energy involved in an impact of an $R \approx$ 10-km sized body with a velocity of ~ 15 km s^{-1} is about two orders of magnitude larger than the energy involved in the impact of individual fragments of Comet Shoemaker–Levy 9. At the instant the bolide hit the Earth's surface, two shock waves must have propagated away from the impact site, as discussed in §6.4.2. One shock wave propagated into the bedrock, and the other went backwards, into the impactor. Immediately after this, a colossal plume of vaporized rock, the fireball, must have risen upwards into space, launching dust and rocks on ballistic trajectories that carried them far around the Earth. In the case of Chicxulub, this fireball is thought to have been followed by a second plume, driven by the sudden release of CO_2 gas from a layer of shocked limestone located about 3 km below the surface. The cavity itself may have

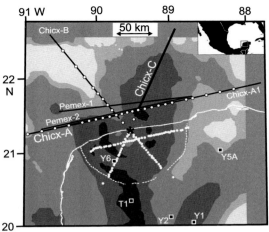

Figure 16.8 The Chicxulub seismic experiment. *Solid lines* show off-shore reflection lines, and *white dots* show wide-angle receivers. *Shading* shows measured gravity anomalies; the crater is marked by an ~ 30-mGal circular gravity low. The *dashed white lines* mark the positions of the sink-holes in the carbonate 'platform'. *Squares* show well locations; Y6 is ~ 1.6 km deep, and T1, Y1, Y2 and Y5a are 3–4 km deep. All radii are calculated using the *asterisk* as the nominal center. (Morgan et al. 1997)

reached a depth of about 40 km before the center rebounded to form a central peak. The peak grew so large and high that it collapsed, thereby triggering several outward expanding rings and ridges. Meanwhile, the crater walls continued to expand outwards. The transient cavity is thought to have had a diameter of ~ 100 km.

The heat from Chicxulub ejecta reentering the atmosphere probably ignited global forest fires. Also, large amounts of nitric acid (HNO_3) and sulfuric acid (H_2SO_4) were likely formed, raining down (acid rain), killing plants and animals and dissolving rocks over a large area around the impact site. Because the impact happened on a peninsula, a large tsunami wave (recorded in rocks found in Mexico and Cuba) spread outward and, upon hitting Florida and the Gulf coast, must have destroyed vast areas in what is now Mexico and the United States. Tsunami deposits from this event have been found in Texas and the Caribbean Sea. Fine dust, which had been brought up by the fireballs, stayed suspended in the atmosphere for

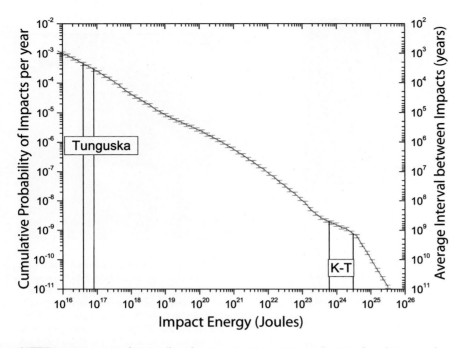

Figure 16.9 Frequency of projectiles of various sizes impacting on the top of Earth's atmosphere at the present era, based on the observed population of near-Earth objects (NEOs, see §12.2.1). The observed NEO population is represented by the curve, and the small circles show estimates of the entire population. Note that observations are only complete for the largest sizes. The scale on the left gives the total number of NEOs larger than a given diameter (shown in the lowest scale) or brighter than the absolute magnitude (§12.2.1) specified at the bottom of the box. The corresponding kinetic energy is given at the top and the average interval between impacts larger than a given energy is shown to the right. Arrows represent the energies of three notable impacts discussed in the text. (Courtesy Alan Harris, MoreData!)

many months before reaching the surface. This could have turned the sky dark over the entire Earth and prevented sunlight from reaching the surface, so the surface temperature dropped to well below freezing for many months. When the sky cleared, temperatures may have risen to uncomfortably high levels because of enhanced levels of greenhouse gases such as H_2O and CO_2. This global cycle of extreme hot–cold–hot temperatures would have killed off many animal and plant species in areas quite remote from the impact site. An alternative model on the consequences of the K–T impact suggests that the greenhouse warming effect is small, but that the slow (over many years) production of sulfuric acid kept the temperature low (by tens of degrees K) for many decades, which could have had a similarly destructive effect on plant and animal life.

16.7.2 Frequency of Impacts

Because of the destruction that impacts may produce, impact frequency is an important factor in planetary habitability. Figure 16.9 gives estimates of the frequency of impacts on Earth as a function of impact energy. It is clear that large impacts are much rarer than small ones. This is fortunate because big impactors are vastly more hazardous than are the little ones (Table 16.1).

If the amount of dust injected into the stratosphere is sufficient to produce an optical depth >2 worldwide for several months, the surface temperature can be suppressed by ~10 K globally. To produce an optical depth of 2 requires about 10^{13} kg of dust, about a hundred times more than has been lofted by any of the large volcanic eruptions from the past century (e.g., Mt. Pinatubo in 1991).

Table 16.1 **Impacts and Life**[a]

Impactor Size[b]	Example(s)[c]	Most Recent Impacts[d]	Planetary Effects	Effects on Life
Super colossal R > 2000 km	Moon-forming event	4.5×10^9 yr ago	Melts planet	Drives off volatiles Wipes out life on planet
Colossal R > 700 km	Pluto 1 Ceres (borderline)	$\gtrsim 4.3 \times 10^9$ yr ago	Melts crust	Wipes out life on planet
Mammoth R > 200 km	4 Vesta (large asteroid)	$\sim 3.9 \times 10^9$ yr ago	Vaporizes oceans	Life may survive below surface
Jumbo R > 70 km	95P/Chiron (largest active comet)	3.8×10^9 yr ago	Vaporizes upper 100 m of oceans	Pressure-cooks photic zone May wipe out photosynthesis
Extra large R > 30 km	Comet Hale–Bopp	$\sim 2 \times 10^9$ yr ago	Heats atmosphere and surface to ~ 1000 K	Continents cauterized
Large R $\gtrsim$ 10 km	K–T impactor 433 Eros (largest NEO)	65×10^6 yr ago	Fires, dust, darkness Atmosphere/ocean chemical changes Large temperature swings	Half of species extinct
Medium R > 2 km	1620 Geographos	$\sim 5 \times 10^6$ yr ago	Optically thick dust Ozone layer threatened Substantial cooling	Photosynthesis interrupted Significant extinction
Small R > 500 m	$\sim$1000 NEOs Lake Bosumtwi	$\sim$500 000 yr ago	High altitude dust for months Some cooling	Massive crop failures Many individuals die but few species extinct Civilization threatened
Midget R > 200 m	99942 Apophis (borderline)	$\sim$50 000 yr ago	Tsunamis	Coastal damage
Peewee R > 30 m	Tunguska event Meteor crater	30 June 1908	Major local effects Minor hemispheric dusty atmosphere	Newspaper headlines Romantic sunsets increase birth rate

[a] Adapted from Lissauer (1999b) and Zahnle and Sleep (1997).
[b] Based on U.S. Department of Agriculture classifications for olives, pecan halves and eggs.
[c] Actual events or bodies capable of producing events of this size.
[d] Some numbers are based on data concerning particular events and others (indicated by $\sim$ preceding time) on statistical arguments based on the mean impact frequency.
NEO = near-Earth object.

This much dust is injected into the stratosphere by an impact with an energy of $\gtrsim 10^5$ MT, i.e., an impact by a stony asteroid 500 m in radius at a velocity of 15–20 km s^{-1}. Such impacts occur roughly once per million years (Fig. 16.9). The effects of impacts of various sizes on planet Earth and on life are indicated in the last two columns of Table 16.1.

16.7.3 Volcanos and Earthquakes

Energetic volcanic eruptions, such as the explosion of Mt. Pinatubo in the Philippines on 1991 June 15, can throw up so much fine volcanic ash and sulfur dioxide (which reacts with water to produce sulfuric acid aerosols) that they significantly reduce the amount of sunlight reaching the ground. In April 1815, a violent volcanic explosion decimated

Mt. Tambora on Sumbawa Island in what is now Indonesia. Mt. Tambora induced so much cooling that 1816 was known as the 'year without a summer' in much of the northern hemisphere, resulting in substantial crop losses and starvation. The maximum magnitude of such cooling by volcanic debris, and thus the potential for volcanoes to cause large extinctions, is not known.

The greatest damage that very large earthquakes inflict on most ecosystems is probably the result of tsunamis that they generate. In contrast, human casualties are generally dominated by the shaking on land because we live and work in buildings that are too flimsy to withstand the ground motions but heavy enough to crush us when they collapse. Also, as occurred in the 1906 San Francisco earthquake, damage to cities from the ground motion can both ignite fires and destroy fire suppression capabilities.

16.8 How Life Affects Planets

Life and the environments where it flourishes on Earth have been closely intertwined for billions of years. The first microbes consumed the most readily available foods, and subsequent life had to develop new energy sources in order to survive. The most productive energy source developed has been to harness solar energy via photosynthesis, and the most efficient form of photosynthesis produces oxygen (O_2) as a byproduct. Oxygenic photosynthesis also provides a significant sink for atmospheric carbon dioxide. Microbes living in the digestive systems of cattle excrete methane into the atmosphere. Large forest fires can fill the air with soot, warming the atmosphere but shielding the surface from solar radiation. Humanity's use of fossil fuels for energy releases gases and particulates into the environment.

The presence of life has profound effects on the surface morphology of Earth. Forests cover large fractions of continental crust, changing the albedo (including seasonal variations), soil composition and the local climate. Micro-organisms also change the soil composition through metabolism. The roots of land plants stabilize soil and reduce erosion, beavers dam rivers, etc. Humankind has large effects on the surface morphology, e.g., through building and mining projects, pavement that affects drainage, and altering the composition of the atmosphere. The list goes on and on.

Some microbial metabolic processes fractionate the stable isotopes of sulfur, ^{32}S and ^{34}S. In these prokaryotes, sulfate reduction can impart a large (1%–4.5%) fractionation in favor of the lighter isotope. Sulfate reduction only occurs under anaerobic conditions, and fractionation requires moderate to high sulfate (SO_4^{2-}) abundances. This process is preserved in the geological record when the metabolic product hydrogen sulfide (H_2S) reacts with ferrous iron (Fe^{2+}) to form pyrite.

The second most abundant gas in Earth's present-day atmosphere is oxygen, O_2. This has not always been the case. Figure 5.16 shows striking variations in the abundance of oxygen in our planet's atmosphere with time. Many gaps exist within the geologic record, but it is clear that the atmosphere of the early Earth was reducing. The oxygenation of Earth's atmosphere occurred over billions of years, with an especially significant increase ~2.2 Gyr ago (§5.8.2).

Oxygenic photosynthetic bacteria produce O_2 as a waste product. Methanogenic organisms release CH_4. Some of this methane rises to the stratosphere and is photodissociated. A portion of the hydrogen formed by dissociation continues moving upwards to the exosphere and subsequently escapes (§5.7). For the first few billion years, Earth's atmosphere and crust were sufficiently reducing that they were capable of absorbing the released oxygen. Some processes, such as hydrogen escape and long-term sequestration of carbon within the crust and the mantle, diminished the environment's ability to absorb oxygen, and eventually the O_2 abundance in our planet's atmosphere increased substantially.

Ozone (O_3) is produced photochemically in an atmosphere containing O_2 (§5.5.1). The presence of a significant amount of ozone in an atmosphere greatly reduces the flux of ultraviolet light reaching the planet's surface.

Biological processes play an important role in the carbon cycle (§16.6.1). Fossil fuels, such as coal, crude oil and most of our planet's 'natural gas' (methane), were produced from partially decayed plants and other formerly living matter. Shells of some marine organisms sequester carbon and other elements for long timescales. The simultaneous long-term presence of abundant mutually reactive gases such as oxygen and methane in a planet's atmosphere requires a highly non-equilibrium process such as oxygenic photosynthesis.

Humanity's effects on the surface of our planet are readily apparent. The atmosphere has also been significantly altered by human behavior. We have added artificial gases to the atmosphere, such as the chlorofluorocarbons that significantly depleted the ozone layer during the latter part of the twentieth century. We have also substantially increased the abundances of naturally occurring gases such as carbon dioxide and methane, thereby amplifying the greenhouse effect and causing global warming (§5.8.1). Although recent human modifications to our environment are more substantial and widespread, agriculture and nomadic herding have affected climate for $\sim 10^4$ years and were major (possibly dominant) factors in the desertification of the Sahara. Note that humans are not the only species to adopt agriculture. Leaf-cutter ants have a very specialized social network that operates sophisticated fungus farms. The ants harvest the leaves on which the fungus grows, and they use pesticides to protect the fungus that is their food source.

Life has persisted on Earth for well over three billion years despite a significant increase in the Sun's luminosity (§3.3), a large reduction of the amount of escaping geothermal heat and many other changing conditions on our planet. This endurance suggests that life itself may play a role in regulating the environment to allow for its continued existence. The concept of a self-regulatory symbiotic relationship between life and the planet that it occupies is known as **Gaia**. For example, biogenic gases in Earth's atmosphere alter the magnitude of the greenhouse effect and thus the temperature of our planet. Gaia is an intriguing but unproven and controversial hypothesis.

The underpinnings of the laws of physics, including the values of physical constants, are not known. Life as we know it would not be possible were such laws and/or constants considerably different. Thus, our very existence constrains the structure and evolution of our Universe. This consistency between fundamental physics and biology and observation by intelligent creatures is referred to as the **anthropic principle**. The anthropic principle does not necessarily imply special luck or a supernatural creator. It is possible that other universes, not interacting with (i.e., not causally connected to) our own, also exist. Such other universes could be governed by very different physical laws than is our own, and the vast majority of them may thus be lifeless.

16.9 Origin of Life

Questions of origins are generally quite difficult to answer: It is almost always easier to figure out how something works than to understand how something got to be the way it is. This difficulty is what makes Darwin's theory of the origins of species by means of natural selection one of the crowning achievements of the human intellect, but no one has yet been able to extend this model back to understand the formation of living organisms from nonliving components.

Many complex molecules, including most amino acids and sugars, come in two forms that are mirror images of one another. Living organisms on Earth use exclusively left-handed amino acids and right-handed sugars. These molecules are chemically compatible with one another but not with their mirror image **isomers**. A 50/50 blend of left- and right-handed forms is called a **racemic mixture**, and a molecular preference for only one handedness is referred to as **homochirality**. Racemic polypeptides could not form the specific shapes required for enzymes because they would have side chains sticking out randomly. Also, a wrong-handed amino acid disrupts the stabilizing helix in proteins. DNA could not organize itself into a helix if even a single wrong-handed monomer were present, so it could not form long chains. This means it could not store much information, so it could not support life. Thus, there is a selective pressure to have a system that is basically homochiral, and we understand why terrestrial life uses a single set of isomers. But it is not clear how life chose left-handed amino acids and right-handed sugars and whether independently evolved life would always use the same set or if aliens might be made of right-handed amino acids and left-handed sugars.

The origin of life is studied both by examining the evolution of organisms that are or once were alive and by analyzing paths to the chemical synthesis of life. Techniques for working backwards include following the fossil record, examining commonalities in the genome of known life forms and trying to distill the essence of those processes shared by all living organisms on Earth. One concept is an 'RNA world' in which RNA performed the present-day functions of both proteins and DNA. But RNA is quite complex and may itself have been preceded by some other replicating system. Unfortunately, no traces of pre-cellular life have been identified, and that it ever existed is purely conjectural. Working forward uses techniques such as trying to reproduce the steps of chemical synthesis from inorganic molecules to primitive living organisms, computer simulations of Darwinian evolution and theoretical studies of the origin of homochirality. In this section, we look

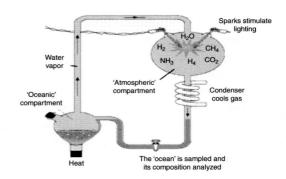

Figure 16.10 COLOR PLATE Schematic diagram of the Miller–Urey experiment. The Miller–Urey experiment demonstrated abiotic production of moderately complex organic molecules by sending electrical pulses through a mixture of simple molecules. (From Purves et al. 2001, 2004)

at the chemical synthesis questions first and then examine the properties of living organisms to look backwards.

16.9.1 Synthesis of Organic Molecules

How were the extremely complicated RNA and DNA molecules that code the chemical instructions for all living organisms produced from simple inorganic molecules? The first stages of this process are the best understood. In 1953, Stanley Miller and Harold Urey subjected a mixture of the reducing gases H_2, NH_3 and CH_4 (proposed by Urey at the time to represent the atmosphere of early Earth) to sparks in the presence of liquid H_2O. Using the laboratory apparatus diagrammed in Figure 16.10, they produced simple organic molecules, including four amino acids. The **Miller–Urey experiment** showed that under the appropriate conditions, non-equilibrium chemical precursors of life could be formed within a nonliving environment. Although the atmosphere of early Earth was probably not as reducing as that in the Miller–Urey experiment, it may have been sufficiently hydrogen rich for organic molecules to have been produced by lightning. Organic molecules can also be formed in molecular clouds and protoplanetary disks.

The next step in the development of complex organic molecules may have relied on catalysis by metal sulfides in a hot environment. Much of the chemistry in these **iron-sulfur schemes** occurs today near **hydrothermal vents** in the deep ocean, such as the **black smoker** pictured in Figure 16.11. This high-pressure environment allows water to remain liquid at temperatures well above 373 K, and chemical reactions occur rapidly at these high temperatures. **Extremophile** prokaryotes thrive in this 'extreme' environment and have access to a richer food source than is available to the less heat-tolerant organisms that live in more temperate nearby waters. But the fauna and associated prokaryotes within this ecosystem also require oxygen for survival, and this oxygen is produced by photosynthetic organisms that depend on sunlight for energy.

An alternative chemical path to the increasing complexity relies on clay. Concentrated solutions of simple organic monomers, when thermally cycled to temperatures close to the boiling point of water (as may have happened in warm tidal pools), produce long-chain polymers of organic molecules. **Clays** consist of layers of silicates with water between the layers. As water is added to a clay, the surfaces spread apart. As water is removed, molecules attached to the clay are brought closer together, and it has been shown that this process can produce long polymers of organic molecules.

The transition from polymers to a living organism that contains the molecular instructions to reproduce itself is not well characterized. Suffice it to say that this process is likely to involve more than one stage because the probability that a collection of polymers will *randomly* self-assemble into a living organism is vanishingly small. This step is key because once complex self-replicating structures are formed, a process akin to Darwinian evolution (§16.10) can proceed. Figure 16.12 shows the formation of life as the narrow point in an hourglass diagram representing

Figure 16.11 This black smoker is located at a hydrothermal vent in the mid-Atlantic ridge. Vents such as this release a chemically rich broth of hot fluids into the deep ocean and can thereby provide the primary food source for the prokaryotes at the base of a diverse ecosystem. (Courtesy the United States National Oceanic and Atmospheric Administration)

the steps in the transition from simple chemicals to advanced organisms.

16.9.2 The Phylogenetic Tree and Last Universal Common Ancestor

All life on Earth shares a common chemical basis and fundamental genetic attributes, implying a common ancestry. Even the simplest forms of life known are chemically quite complex, and we descend from a very sophisticated **last universal common ancestor** (**LUCA**), which may have been

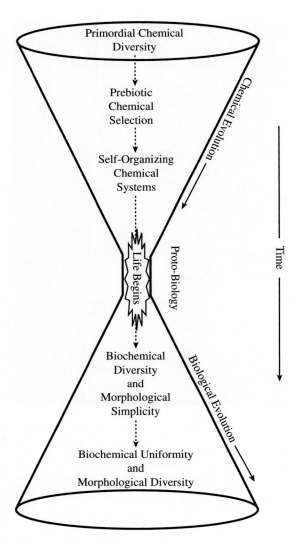

Figure 16.12 Schematic view of the transition from chemical evolution to biological evolution. Of many possible chemical processes, only a few are suitable for life. But once life exists, evolution by natural selection allows for great diversity. (Adapted from Lunine 1999)

an individual organism or an ecosystem of organisms that shared genes with one another. LUCA is likely to have been quite different from any organism alive today. An analogy is provided by animals, many varieties of which have descended from the **Cambrian explosion**, a geologically brief interval of time in which a diverse array of animals first

appeared. The ancestral animals that evolved into many **phyla** (the grossest division within the animal kingdom, categorizing basic body structure) now observed are extinct. A more recent example is that chimpanzees and humans share a common ancestor several million years ago, but this ancestor was neither like a chimp nor like a person. Evolution is not a forced march to increasing complexity, and most paths wind up being evolutionary dead ends.

The relationships among the various life forms found on Earth today are represented by the **phylogenetic tree**, which is displayed graphically in Figure 16.13. Terrestrial life is divided into three **domains**: **Bacteria**, **Archaea** and **Eukarya**. The first two of these domains are smaller cells that lack true nuclei, i.e., are prokaryotes. Eukarya appear to be chimeras that evolved from organisms such as Bacteria plus Archaea (§16.10.2).

From a phylogenetic perspective, Eukarya appear to be more closely related to Archaea than to Bacteria. In a morphological sense, eukaryotes are quite distinct from the two prokaryotic domains. Whereas the DNA of archaea and bacteria are contained in a single circular strand, eukaryotic cells have their DNA on several open-ended chromosomes that are located within a small area of the cell that is known as the **nucleus**. The vast majority of archaea and bacteria are similar in size (typically ~1 μm), but typical eukaryotic cells are larger by a factor of about 10 in linear dimension. Most importantly, because it is key to the existence of all advanced life on Earth, eukaryotes have the complexity to form truly multicellular organisms. (Some prokaryotes live in association with one another for mutual benefit, but the degrees of differentiation, specialization and interdependence are much less than for advanced eukaryotes.) Note that animals, plants and fungi represent a very small portion of phylogenetic diversity within the eukaryotic domain (Fig. 16.13).

The known organisms located near the root of the phylogenetic tree are the Archaea and

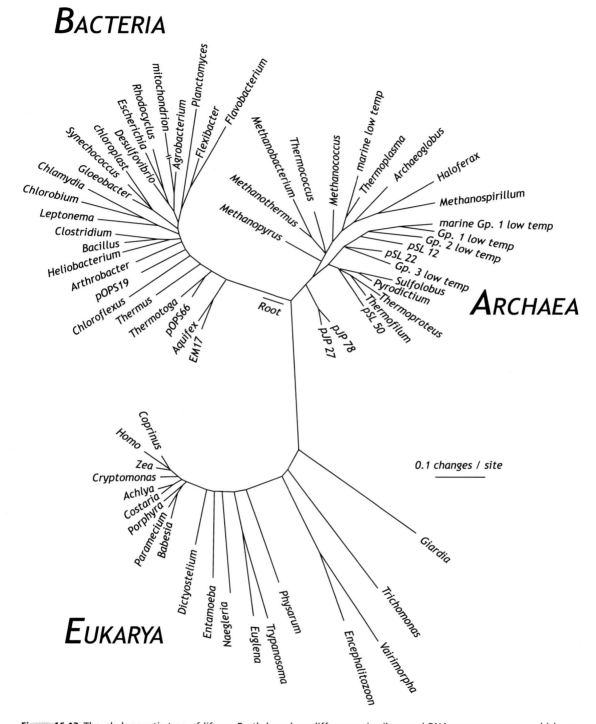

Figure 16.13 The phylogenetic tree of life on Earth based on differences in ribosomal RNA gene sequences, which are better conserved among species than is most other genetic information. Distances along the curves are proportional to genetic differences. The tree shows that evolutionarily there are three types of organisms: Eukaryotes, Bacteria and Archaea. (Pace 1997)

Bacteria that exhibit fewer genetic characteristics exclusive to their domains. These prokaryotes are thus probably the most similar modern life forms to LUCA. They are all thermophiles, i.e., they thrive at high temperatures (~330 K). This observation suggests that life may have formed at such high temperatures. Alternatively, environmental stresses on early Earth may have destroyed nonthermophilic organisms.

16.9.3 Young Earth and Early Life

Planet formation is a violent process, and large environmental variations on a young planet can be very hostile to life. Impact rates were extremely high during the Earth's early history (Fig. 6.33), and the substantial heating of the atmosphere, hydrosphere and upper crust by massive objects may have caused an **impact frustration of life** (Table 16.1). A late large impact may well have heated the upper portion of our planet sufficiently to have killed off all nonthermophilic organisms (§16.7). Thus, the thermophilic root of the phylogenetic tree might be the result of an impact-induced heating event that killed off all life not tolerant of high temperatures. In this case, life could have originated in either a cool or a hot environment. Life and the planet(s) that it occupies can affect each other in many ways, and it is not possible to have a complete understanding of one without knowledge of the other.

During its youth, Earth was hot and geologically active; thermal processing of rocks makes it difficult to find traces of life's origins. Fossils and uniquely biogenic organic compounds provide the clearest evidence for past life, but certain isotopic signatures are more durable. Fractionation leads to a concentration of the lightest isotope of carbon in organic molecules (vs. carbonates). The **enzyme** (biological catalyst) used by some bacteria produces organic carbon compounds with a ratio of $^{13}C/^{12}C$ that is 2.5%–3% smaller than that of the carbon in the air. Carbon enriched in its lighter isotope by this amount has been found in some

Figure 16.14 COLOR PLATE (a) Modern stromatolites formed by bacteria. (From Harnmeijer 2007) (b) Stromatolite from the Transvaal Dolomite, Republic of South Africa; this rock is ~0.2 m wide and 2.5 billion years old. (From Des Marais et al. 2003)

of the oldest rocks on Earth, which solidified up to 3.8 Gyr ago when Earth was less than 20% of its current age. Rock formations that are 3.5 Gyr old show more convincing C and S isotopic evidence. Stromatolites, such as those shown in Figure 16.14, are rocks consisting of sedimentary growth structures that are morphologically similar to those produced by modern prokaryotes. Analogous, even more ancient, rock formations provide significant evidence for life 3.5 Gyr ago and less convincing signs of life as far back as 3.7 billion years ago. Rocks 3.5 Gyr old also contain what appear to be microfossils. Clear evidence of microfossils is present in 3.2-billion-year-old rocks.

Some biogenic molecules, such as **terpanes**, **hopanes** and **steranes**, which are characteristic of cell membranes of Archaea, Bacteria and Eukarya, respectively, are stable over geologic time. (Bacteria produce some steranes, but others are unique to Eukarya.) Biogenic molecules provide chemical evidence for both **cyanobacteria** (also known, misleadingly, as blue-green algae), and eukaryotes in 2.7-Gyr-old rocks, although the steranes (evidence for Eukarya) may be the result of later contamination. Eukaryotic microfossils first appear 1.5 Gyr ago, and the oldest large fossils are from red algae that lived 1.2 billion years ago. Thus, many 'living' planets may well be occupied only by very simple organisms.

16.10 Darwinian Evolution

Terrestrial life has developed and flourished in a wide variety of environments. Major developments in the history of life on Earth are shown in Figure 16.15. A key to life's adaptability is the evolutionary advancement that results from occasional random mutations and recombinations that occur within the genetic code of organisms and are passed on to subsequent generations. **Mutations** are alterations of the genetic code; they can be caused by a wide variety of natural processes, including cosmic rays, ultraviolet photons, chemicals such as the free radical oxidizing agent peroxyl (-OOH) and mechanical damage. Mutations occur frequently on Earth. Most mutations are disadvantageous (often lethal) or neutral to their host, but occasionally a mutation improves the chances of an organism to reproduce. Darwin's great insight was to realize that evolution occurs via **natural selection**: The fittest organisms have the greatest likelihood to survive and pass on their genes to subsequent generations.

In evolution, as in architecture, form follows function. Dolphins have evolved to swim in a manner analogous to some very distantly related but similarly sized fish. The camera-like eyes that humans and other vertebrates possess are very similar to those present in advanced cephalopods such as octopuses and squids. Our eyes use the same principles as the eyes of the evolutionarily very distant (by animal standards) cubozoan jellyfish. But insects (which are much more closely related to us than are jellyfish) and various other animals use a very different compound eye, and many animals are sightless. Thus, camera-eye vision has developed independently several times. Such separate paths to the same result are referred to as **evolutionary convergence**. But the animal kingdom represents a very small part of the phylogenetic diversity, and this convergence may be the result of limited capabilities of animals to adapt, with function being limited by the forms that are accessible.

Whereas the size, shape and general complexity of both the genomes and the cells of Archaea and Bacteria are very similar, Eukarya are quite different. Yet according to phylogenetic relationships, the three domains are roughly equidistant from one another (Fig. 16.13). Is the superficial similarity between the two prokaryotic domains simply a coincidence? A relic of the (unknown to us) genome size of LUCA? An example of evolutionary convergence on a grander scale, where two very different genome sizes work well but intermediate sizes are less fit? Or are many Bacteria and Archaea minimal organisms, containing the smallest genomes possible for the carbon/water/DNA mode of life that is found on Earth?

The rate of evolution varies greatly among species. Whereas animals developed into a vast array of complex forms during the first few million years of the Cambrian period (§16.10.2), some 2-billion-year-old fossil cyanobacteria appear morphologically virtually identical to species that are alive today. In part, cyanobacteria are simpler organisms, so there are fewer genetic combinations for nature to try. But also, bacteria had been around for more than a billion years when these fossils were formed and have duplication times as

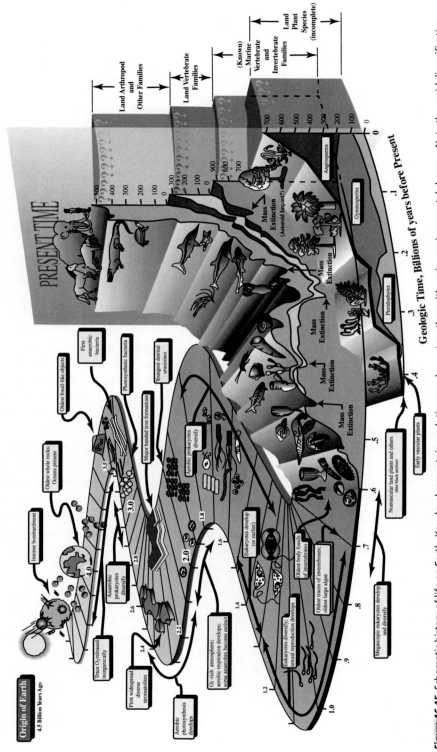

Figure 16.15 Schematic history of life on Earth. Key developments in evolution are shown together with major environmental changes. Note the rapid diversification of advanced life forms over the past 600 Myr. (From Lunine 1999)

short as 20 minutes, so they had already undergone of order 10^{13} generations. They have since been through another $10^{13} \approx 2^{43}$ or so generations, but if genetic experimentation is similar to particles trying to escape from marginally stable chaotic orbits (§2.4.2), then this is just one more factor of 2 compared with 40 that preceded it, and thus unlikely to lead to a great deal more change. On the other hand, when a major evolutionary breakthrough (or environmental change, such as the oxygenation of Earth's atmosphere) occurs, the rate of evolution may be fast and furious even for well-established species. For example, small mammals had been around for more than 100 million years when the K–T impact wiped out the dinosaurs and opened up ecological niches for mammals to adapt to fill.

A major cause of our uncertainty in the commonness of life in the Universe is that we have only a single example of the origin of life. Thus, we do not know whether all of the steps in the formation of life are likely given the proper planetary conditions or whether one or more of the steps are extremely unlikely, making life rare even if 'habitable planets' are common. As far as evolution is concerned, some of the many steps between the simple organisms on Earth over three billion years ago and our species have occurred many times (e.g., photosynthesis in prokaryotes, multicellularity in eukaryotes, vision in animals), but we only have evidence of others occurring once (e.g., the formation of eukaryotic cells, the capability of a species to develop written language, and probably the origin of oxygenic photosynthesis). Understanding the evolutionary events that occurred only once on Earth is important for us to estimate the fraction of inhabited planets that are likely to develop technological civilizations.

The remainder of this section describes the evolution of life on Earth in greater detail. We begin by explaining the mechanisms for evolution and the tree of life. Then we discuss the origin of structurally complex life and lastly the development of intelligence and technology.

16.10.1 Sex, Gene Pools, and Inheritance

A **gene** is a unit of heredity. In modern cells, genes are segments of DNA that specify the sequence of amino acids in particular proteins or polypeptide chains or the sequence of nucleotides in particular RNA molecules. Genes provide the information required for production and replication.

The fittest organisms are the ones with the best chance to survive and pass their genes on to subsequent generations. One of the key components in fitness is adaptability and, more specifically, genetic adaptability. Life on Earth uses several techniques to modify genomes, including random mutations, sexual reproduction, gene duplication, lateral gene transfer and endosymbiosis.

Random mutations allow the gene pool to 'try something new'. Mutations are crucial to the advancement of life, but the overwhelming majority of random mutations are bad or neutral. Thus, mutations are not a very efficient way of improving the **gene pool** of a species. Furthermore, because it is quite unlikely for the combination of two or more random mutations to confer a net benefit, environments and genomes in which multiple mutations occur within a generation are not conducive to survival, especially among species in which individuals have few offspring.

Advanced multicellular life forms, such as plants and animals, experiment with new combinations of genetic information that worked well in the past as part of the reproductive process. In **sexual reproduction** approximately half of the genetic information in the offspring comes from each of the two parents. Bacteria and archaea are single-celled life forms that reproduce by cell division. As discussed later, they are also able to share genetic information with one another via a process known as **lateral gene transfer**. Lateral gene transfer allows prokaryotes to maintain small genomes but still have the **genetic plasticity** to adapt to many environmental stresses.

In general, eukaryotic organisms have much larger genomes than do prokaryotes. One of the mechanisms of genetic growth is **gene duplication**, wherein the genome expands by repeating a sequence of amino acids that was present only once in the ancestor(s) of the organism. Two copies of the same gene are not necessarily of much use, but the 'extra' copy is a gene that has withstood the test of time and subsequently can be modified gradually for future use in a different function.

The correlations of genomes among living organisms enable the construction of the **phylogenetic tree**. Phylogeny refers to the evolutionary relationships among a group of organisms. The function of certain molecules, such as ribosomes (which are the sites of protein synthesis), is so critical in modern organisms that regions within the corresponding gene evolve very slowly. Such regions can be compared, and the number of differences between DNA sequences can be used to gauge the evolutionary distance between organisms. Major branches of the phylogenetic tree are shown in Figure 16.13. Genetic comparisons confirm the observation that species that look and act more similarly usually have more recent common ancestors. Genetic material is passed down from parents to offspring. However, some genes appear in organisms that are widely separated but not in 'intermediate' species. These genes are too similar, and this process occurs in too many cases, for it to just be that the genes emerged independently by chance. Rather, microbes use two mechanisms of producing new variants that disobey the rules of tree-like evolution: lateral gene transfer and endosymbiosis.

Lateral gene transfer involving the passage of genes among organisms of distantly related groups causes various branches in the tree of life to exchange bits of their genetic fabric. Three mechanisms lead to lateral transfer of genetic information: **Conjugation** is the transfer of genes from one prokaryotic organism to another by a mechanism involving cell-to-cell contact. **Transduction** refers to the transfer of host genes from one cell to another by a virus. **Transformation** is the transfer

of genetic information via free DNA. Lateral gene transfer primarily occurs among prokaryotes and generally involves genes coding specific functions rather than basic biochemical processing. Genes allowing bacteria to tolerate heavy metals and resist antibiotics appear to be relatively easy to acquire via lateral transfer.

Endosymbiosis is the process of one cell living within another. Eukaryotes appear to have originated when the genomes of endosymbiotic bacteria that lived within archaea fused to produce a more complicated form of life. Various features within eukaryotic cells, including double-membrane-bounded **organelles** (specialized subunits within the cells), mitochondria and chloroplasts (Fig. 16.1), are of endosymbiotic origins.

16.10.2 Development of Complex Life

Human beings are exceedingly complex structurally, with specialized cells and organelles that communicate with each other in a rapid and precise manner. The bacteria that live in our guts are quite simple by comparison, although they are nonetheless vastly more ordered and complicated than nonliving chemical mixtures. Our understanding of the origin and development of life on Earth is far from complete, yet we can identify several stages in the development that led to our species.

The genetic relationships among terrestrial life forms are represented by the phylogenetic tree (Fig. 16.13). All eukaryotes are closely related in terms of fundamental cellular structure and biochemistry and share many properties lacking in the bacteria and archaea. So, as with the origin of life, we have only a single example of the development of complex cells, and it could well involve an extremely unlikely process that rarely occurs within the Universe.

The next major advance in complexity was the development of **multicellularity**. Multicellular organisms are able to have division of labor that allows for far greater complexity than has been achieved by single-celled organisms. Although many bacteria species live in clumps or strands, only a few form structures in which the cells are **differentiated**, that is, in which different cells perform different functions for the mutual benefit of the group. Most Eukarya are multicellular, and this trait has emerged multiple times among Eukarya; some multicellular species are more closely related to specific single-celled eukaryotes than they are to certain other multicellular species. Thus, the transition to multicellularity has occurred many times on Earth and should not be regarded as a major bottleneck in evolution. In contrast, only the most advanced life forms on our planet – animals, plants and fungi – have developed the most sophisticated form of multicellularity that involves communications among cells to produce organisms consisting of very specialized parts. These three **kingdoms** of advanced life (animals, plants and fungi) are clustered together on the phylogenetic tree (Fig. 16.13).

The fossil record demonstrates that **metazoa** (multicellular animals) have been on Earth for at least 600 Myr. The genetic diversity of the animal kingdom suggests that the first animals appeared more than one billion years ago. Genetic **molecular clocks** assume that genetic divergence occurs at a linear rate, and they estimate this rate by fitting the genetic difference of organisms whose last common ancestor can be determined from the fossil records. Early animals were soft and did not make good fossils, so the lack of direct evidence does not rule out their presence 1 Gyr ago. However, the molecular clocks that suggest animals are this ancient could be erroneous because evolution can proceed at different rates as the result of changes in environmental stresses. The origin of animals represents a key single-example transition in the advancement of life because all thinking creatures on Earth are members of the animal kingdom.

Geological time parses the history of our planet into geologic **periods** and the other divisions that are based primarily on the types of fossils found in various layers of sediments. The major epochs of geological time are shown in Figure 16.16. Radiometric dating (§11.6) now provides an absolute chronology, but division of Earth's past into geologic periods remains extremely useful for paleontology.

Almost all animal phyla present on Earth today, plus some that have gone extinct, were formed just before or during the Cambrian period, which began 541 Myr ago (Fig. 16.16). Why did animal life evolve so rapidly during this relatively brief epoch of geological time?

There are several factors that may have contributed to the **Cambrian explosion** of animal forms. Were environmental conditions special in the Cambrian? Possibly. Animals need energy to survive, and breathing oxygen is essential to the path by which animals produce that energy. Early in Earth's history, the atmosphere had too little oxygen for any large organisms to have lived (see Fig. 5.16). Unfortunately, our knowledge of the temporal variations of atmospheric oxygen is not adequate to determine whether or not a large increase in oxygen was coincident with the Cambrian explosion. Another possible factor is that some of the species of animals living during the Cambrian may have been more genetically plastic (susceptible to genetic changes that are neutral or favorable to survival) than any of their descendants that survived the competition created by the Cambrian explosion. Evolution usually does not 'get it right' the first time – there are often valleys of adaptive inferiority en route to the evolutionary high ground. Subsequent to the Cambrian, the competition in all of the major ecological niches has been much greater, so the poorly competitive intermediate body plans would have had a smaller chance to survive and adapt. This **permissive ecology** may have occurred because advanced animals had never existed before; there had been a mass extinction; or new environments had been created by climate change, local geological factors

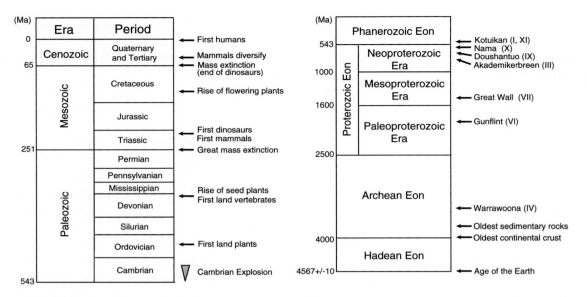

Figure 16.16 Geological time on Earth. The diagram on the *right* encompasses the entire history of our planet. The one on the *left* presents an expanded view of the Phanerozoic eon, during which complex life has diversified substantially and left abundant fossil records. Note that the beginnings of the Proterozoic eon and the Phanerozoic eon are contemporaneous with large increases in the abundance of atmospheric O_2 shown in Figure 5.16. (Courtesy Andrew Knoll)

or the rise of oxygen. Additionally, a single major evolutionary advance, possibly the development of hard shells, could have led to a host of evolutionarily much easier changes.

16.10.3 Intelligence and Technology

The human brain is a remarkable organ that enables humanity to do many things that no other species on Earth has ever accomplished. How difficult is it for life to develop such intelligence? Quantifying intelligence is difficult within our own species and becomes far more challenging when comparisons are being made with other species, especially species that are extinct. The best available indicator is **encephalization**, which is based on brain size. Larger animals require bigger brains to manage the basic tasks of bodily functions, so this needs to be factored in. Figure 16.17 plots the brain mass vs. body mass for various mammals and birds. Thus, the standard measurement of interspecies

intelligence is made using an **encephalization quotient**, EQ, which compares brain size with that of 'typical' animals of the same weight. A fit to numerous animals yields the formula

$$EQ_{0.67} = \frac{m_{\text{brain}}}{0.012 m_{\text{body}}^{0.67}}, \qquad (16.5)$$

where the masses of the brain, m_{brain}, and the body, m_{body}, are measured in kilograms and the subscript 0.67 on EQ refers to the power to which m_{body} is raised in equation (16.5). Animals with $EQ > 1$ are more intelligent than average. Among primates, EQ correlates well with innovative behavior, social learning and tool use; among birds, it correlates with behavioral flexibility.

Humans have the highest EQ of any extant (or known extinct) species, with $EQ_{0.67} = 7.1$, meaning that we are more than seven times as brainy as typical animals of our body weight. Next come dolphins, with some species having $EQ_{0.67}$ as high as 5, comparable to the $EQ_{0.67}$ of

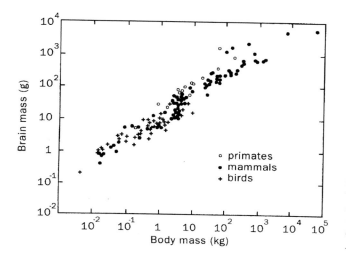

Figure 16.17 Comparison of the masses of the brains and bodies for a variety of birds, as well as primates and other mammals. Humans are represented by the open circle farthest above the general trend. (From Jakosky 1998)

tool-using human ancestors **Homo erectus** and **Homo habilis**. By comparison, great apes average only about $EQ_{0.67} = 1.9$, about the same as that of the more distant human ancestor **Australopithecus**, which lived 3 Myr ago. Although human EQ is substantially larger than that of our ancestors only a few million years ago, dolphins reached $EQ_{0.67} = 4$ at least 15 Myr ago and have advanced little since then. Many species have lower EQ than their ancestors in the fossil record; intelligence confers many evolutionary advantages, but large brains are very resource-intensive organs to grow and maintain, and thus intelligence comes at a high cost.

All early animals lived in water. Humans are not the only animals to make and use tools; some other apes make and use simple tools, as do a few species of birds. In contrast, no water-dwelling species, even the very intelligent marine mammals, is known to make tools. The use of tools may have been the primary reason that it was an advantage for our recent ancestors to develop larger brains but not evolutionarily worthwhile for dolphins to become more encephalized. The transition of vertebrate animals to land only occurred well after land plants had evolved and provided a food source. Thus, land plants may be a required link in the development of technological life.

16.11 Mass Extinctions

Many evolutionary changes are caused by the cumulative effects of gradual small mutations and by competition among species. Large changes in ecosystems can also occur very suddenly. The fossil record of the past half billion years shows abrupt boundaries in the types of species that inhabited the Earth. These sharp transitions occur roughly once per hundred million years and, as illustrated in Figure 16.18, they are clearly visible in the diversity of marine fauna in the fossil record. The most recent of these mass extinctions occurred at the boundary of the Cretaceous and Tertiary periods 65 million years ago, when a large impact killed off half of the genera then extant, including all of the dinosaurs (§§6.4.5 and 16.7.1). By opening up ecological niches, mass extinction allows new life forms to develop. The extinction of the dinosaurs opened the way for the present **age of mammals**.

The causes of the mass extinctions before the K–T boundary are less certain, but in addition to impacts, possible culprits include volcanic activity, changes in atmospheric composition, climate changes resulting from the previously mentioned processes, movement of the continents and/or changes in the Sun's luminosity.

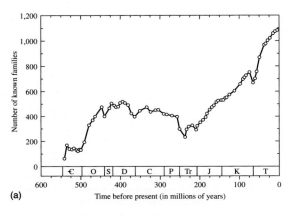

(a)

Time before present (in millions of years)

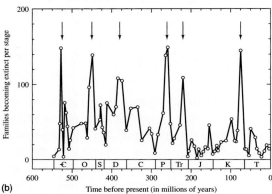

(b)

Time before present (in millions of years)

Figure 16.18 These diagrams show variations in marine animals with time during the Phanerozoic eon. Standard symbols of geologic periods as listed in Figure 16.16 are given near the bottom of the plots (the Carboniferous, C, period encompasses both the Mississippian and Pennsylvanian periods). Panel (a) shows the growth in the overall trend of increasing diversity, with some drops. Panel (b), which plots extinctions but not evolution of new families, clearly illustrates the effects of six major mass extinctions, the largest of which occurred at the Permian/Triassic boundary. (From Lunine 2005)

Changes in the Earth's obliquity and orbital eccentricity are implicated in the geologically recent cycle of ice ages and interglacials during the past few million years (§2.6.3). The disappearance or emergence of a group of organisms could also alter the environment enough to produce a mass extinction.

At present, Earth is undergoing a major (possibly developing into a mass) extinction caused by a single dominant species, **Homo sapiens**.

The average rate at which species become extinct outside of mass extinctions is one in one million species lost per year. The current rate is 1000 times as large as the average extinction rate. The (anthropogenic) causes of the current mass extinction are summarized in the acronym **HIPPO** that was coined by biologist E.O. Wilson. The letters stand for the factors that are contributing to this massive die-off, listed in decreasing order of importance: habitat destruction, introduced species, pollution, population growth and overconsumption.

Two factors that have the potential of severe and rapid habitat destruction are war and global warming. A major nuclear war could launch so much debris into the stratosphere via the blasts and smoke from ignited fires to create a substantial anti-greenhouse effect (analogous to that produced by stratospheric haze on Titan, §4.6, as well as to one of the effects of the K–T impact §16.7.1), resulting in a global cooling referred to as **nuclear winter**. Nuclear winter could destroy an entire growing season and thereby lead to mass starvation comparable to that induced by the dust lofted into the upper atmosphere by the impact of a 1-km body. Climate change, caused by the current anthropogenic increase in greenhouse gases (see Fig. 5.10), is a senseless experiment that humanity is wreaking on the world's environment. **Anthropogenic climate change** already contributes to the habitat destruction listed above. Although ecosystems can adjust to gradual climatic changes, trees only move during the seed stage, and they grow at finite rates; some animals cannot move very fast; and so on.

16.12 Panspermia

Just because life thrives on Earth at the present epoch does *not* mean that it originated here. Living (or dormant) organisms could have formed elsewhere and been transported to Earth via a process known as **panspermia**. Dozens of meteorites from Mars have been found on Earth (§11.2), demonstrating that such transport is possible. The characteristics of some of the martian meteorites imply that the interplanetary transport process does not have to shock rocks so severely that hardy microbes could not survive. These rocks spent a long time in space, but dynamical simulations show that ejecta from cratering events on Mars can reach Earth in less than one year. Models also suggest that rocks can be ejected from Earth and land on Mars, although the amount of material transported from Earth to Mars is likely to be far less than the amount of Mars rocks arriving at Earth, primarily because Earth's greater escape velocity reduces the amount of ejecta that escapes into space (Problem 4-16).

If convincing evidence for martian life, extant or extinct, is ever found, the biggest question will be whether or not these organisms are or were genetically related to life on Earth. Life composed of the opposite chirality molecules from terrestrial organisms would clearly be of different origins, whereas very similar life forms would almost certainly be related. However, it might be difficult to distinguish independently evolved organisms with the same chirality as used by terrestrial life from related life that diverged prior to Earth life's LUCA. If martian life turns out to be our relatives, this would confirm the viability of interplanetary transport of life. Because the martian surface retains a vastly superior record of the first billion years of Solar System history, studying martian fossils could help us understand the origin of terrestrial life, whether that origin took place on Mars or Earth. (The lunar regolith is also likely to contain well-preserved debris knocked off Earth and Mars during the early history of our Solar System.) However, if martian life were found to be unrelated to us, then it would represent a **second genesis** of life within our neighborhood and thereby be strong evidence that life is common in the Universe.

Invasive species inadvertently transported by airplanes and in the ballast waters of ships are a threat to many environments on Earth (§16.11).

Spacecraft transport of organisms from one planet to another is potentially an even greater danger. To prevent such contamination from occurring, precautions for **planetary protection** are undertaken. Interplanetary spacecraft that may land on or crash into the surface of a potentially habitable world are cleaned to reduce the likelihood of contamination with life from Earth. Substantially more precautions are planned when spacecraft bring martian rocks to Earth. Although such precautions are clearly prudent for scientific, ecological and safety reasons, impact ejecta presumably carry viable microbes from Earth to other planets, and if other planets are inhabited, vice versa, so measures need not be as extreme as if the planets were completely biologically isolated from one another.

Science fiction writers and some scientists have proposed that (at least if it is currently lifeless) we should transform Mars into a world suitable for terrestrial life, a process referred to as **terraforming**. The purposeful seeding of one planet with life from another, whether or not the environment of the colonized planet needs to be modified, is called **directed panspermia**. The technical challenges to terraforming Mars or Venus are vast. Analysis of these difficulties is beyond the scope of this book, but suffice it to say that we should not use the possibility that these planets could one day be made habitable as an excuse to allow our planet's environment to deteriorate!

Russian astrophysicist Nikolai Kardashev classified hypothetical advanced civilizations according to the resources that they use. A **type I civilization** uses virtually all energy incident upon its planet, a **type II civilization** uses virtually all energy emitted by its star and a **type III civilization** uses virtually all energy liberated by its galaxy (Problem 16-16).

Although interplanetary panspermia appears to be not only plausible but also almost inevitable, undirected interstellar panspermia is far, far more difficult and therefore enormously less likely (Problem 16-17). But if technological civilizations have arisen in the galaxy, there are no known fundamental physical or biological laws that would prevent interstellar colonization. At present, interstellar travel is well beyond the capabilities of humanity, but a mere century ago, the same could have been said about travel to the Moon. Even if only one civilization had decided to colonize the galaxy, it could have done so in a geologically short period of time (Problem 16-18). The fact that we have not seen clear evidence for extraterrestrials despite the potential for such colonization is known as **Fermi's paradox**, in honor of physicist Enrico Fermi, who first framed the question 'Where are they [aliens]?' in this manner.

16.13 Detecting Extraterrestrial Life

The bulk of this chapter has been devoted to the properties of living organisms and ecosystems and to interactions of life with planetary environments. We now turn to the search for life beyond Earth.

Six basic ways to search for extraterrestrial life are:

(1) Look for signs of extraterrestrial life, intelligent or otherwise, here on Earth. This ranges from studying martian meteorites for signs of microbial fossils to the (far less reputable) scientific analysis of the possibility that intelligent aliens pilot some unidentified flying objects (UFOs).

(2) Visit other worlds (using robotic probes or manned spacecraft) to perform *in situ* analysis or to return samples for laboratory analysis on Earth. For the foreseeable future, this is limited to planets and moons within our Solar System.

(3) Detect spectral **biosignatures** of molecules or combinations thereof that are unlikely to have formed by any process other than life. Substantial quantities of O_2 and CH_4 cannot coexist in equilibrium, but both of these gases can be produced by living organisms.

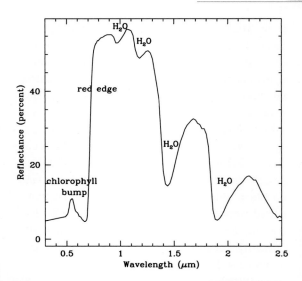

Figure 16.19 The reflection spectrum (albedo) of a deciduous leaf in the visible and near-infrared. The *small bump* just longword of 500 nm is a result of absorption by chlorophyll at 450 nm and 580 nm; this bump gives plants their green color. The sharp rise from 700 nm to 750 nm is known as the **red edge**. Vegetation's red edge is caused by the contrast between the strong absorption of chlorophyll and the otherwise reflective leaf. The red edge is considered to be a potential biomarker. (From Seager et al. 2005)

As shown in Figure 16.19, chlorophyll produces a very distinctive spectral signature that makes it a possible spectroscopic biosignature of extraterrestrial planets. Chlorofluorocarbons (CFCs) would indicate life that was 'intelligent' but not very wise.

(4) Search for evidence of **celestial engineering**, immense structures produced by advanced civilizations.

(5) Look for artifacts produced by distant intelligent organisms that were transported to the Solar System. Such objects could be anything from deliberate attempts to communicate with us (mail) to interstellar probes to space junk.

(6) Intercept radio or other (presumably electromagnetic) signals sent by a technological civilization.

Looking for signs of extraterrestrial life here on Earth seems at first to be the simplest option because it is easier to study evidence available locally. Stories of terrestrial visits by advanced extraterrestrial life forms have been used to sell newspapers, books and movies, but there is no scientifically convincing evidence of visits to Earth by extraterrestrial intelligence (ETI).

Observations of Jupiter's moon Europa by the *Galileo* spacecraft strongly suggest a liquid ocean below the visible ice crust (§10.2.2). Liquid water may also be present well below the surfaces of Ganymede, Callisto (§10.2.3) and Titan (§10.3.1). The geysers observed on Enceladus may be driven by liquid H_2O (§10.3.3). Some meteorites and asteroids contain minerals that were formed by liquid water, but it is extremely unlikely that small bodies would retain water in liquid form for geologically long periods of time.

Although at 95 K the surface of Titan is far too cold for liquid water, this large moon has a methane-rich atmosphere, in which photochemical reactions produce organic molecules (§5.5.1). Some analogous processes may have occurred within the Earth's early atmosphere, which is thought to have been highly reducing.

Spectra of distant worlds can be obtained from Earth and near-Earth space. But Figure 16.20 shows that the disk-integrated spectrum of an inhabited planet can be quite complicated, making unique interpretation difficult.

Hypothesized mega-engineering projects might produce structures that could be detected over the vast distances of interstellar space. One possibility is a **Dyson sphere** enclosing a star and trapping all of its radiation for the use of the civilization that constructed it. To prevent overheating, the Dyson sphere would radiate longwavelength infrared radiation from an immense surface around the star. Detecting such radiation would probably be far easier than proving that it could not be produced by a natural, nonliving process.

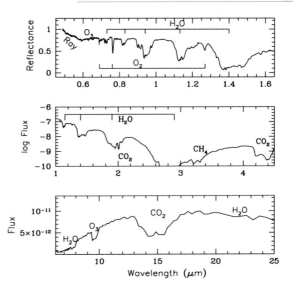

Figure 16.20 Observed disk-integrated reflection spectra of Earth over a wide range of wavelengths. The *top panel* shows the spectrum at visible wavelengths measured from earthshine reflected off the dark side of the Moon. The brightening at short wavelengths is caused by Rayleigh scattering (§5.5). The *middle panel* shows Earth in the near-infrared (IR) as measured by NASA's *Deep Impact* spacecraft, with flux in units of W m^{-2} μm^{-1}. The *bottom panel* displays a mid-IR spectrum from NASA's *Mars Global Surveyor* en route to Mars, with flux in units of W m^{-2} Hz^{-1}. Major molecular features are noted on the plots. (From Meadows and Seager 2010).

16.13.1 Signs of (Past) Life on Mars?

The circumstellar HZ is defined as the region where liquid water would be stable on an Earth-like planet (§16.5). Earth is the only planet in our Solar System with standing liquid water on its surface, but Mars displays clear evidence of physical and chemical attributes that were created by liquid water (§9.4), suggesting that young Mars may have been as suitable for life as was young Earth.

As discussed in §§5.8.2 and 9.4, Mars must have had a very different climate early in its history. When Mars had running water on its surface, the climate may have been suitable for life to develop. The *Viking* landers, therefore, searched for life via a number of different experiments.

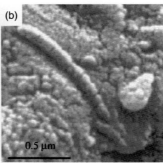

Figure 16.21 Small structures seen in these pieces of the martian meteorite ALH 84001 look similar to microfossils found on Earth, but they are smaller than all known terrestrial Bacteria and Archaea. (From Jakosky et al. 2007)

In addition to simple cameras, some instruments looked for organic chemicals and metabolic activity in the atmosphere and soil, for example, through the addition of nutrients to the soil and looking for chemical byproducts resulting from living organisms (life as we know it). No signs of life were detected. The soil was completely devoid of organic molecules. In retrospect, this should have been expected because the martian soil is directly exposed to solar UV radiation, which breaks up organic molecules. The experiments designed to search for metabolic activity gave some positive results, which are now attributed to unfamiliar reactive chemical states in martian minerals that were produced by solar UV radiation.

One meteorite from Mars, ALH 84001, has several chemical and morphological features that initially were attributed to possible traces of microbial life on ancient Mars. This 1.93-kg rock is so named because it was the first meteorite found in the Allen Hills area of Antarctica in 1984. ALH 84001 is clearly from Mars because its mineralogy is characteristic of volcanic material on planetary-sized bodies and its oxygen isotope ratios are the same as those of the SNC (shergottites, nakhlites and chassignites) meteorites. (Recall that some SNC meteorites have trapped gas abundances that closely resemble the composition of the martian atmosphere measured by the *Viking* spacecraft; §11.2.) The rock was probably ejected into space by an impact about 16 Myr ago and fell in Antarctica 13 000 years ago, as judged from the time that cosmic ray exposure stopped (§11.6.4).

Four putative biomarkers were found in ALH 84001: The rock contains carbonates that resemble terrestrial deposits formed where bacteria were active. The carbonates contain magnetite grains that are very similar to magnetite formed by terrestrial bacteria. Complex organic molecules that could have been produced from biological decay products are found in the carbonates. Lastly, and most spectacularly, the carbonates harbor very tiny round and rodlike structures, such as those shown in Figure 16.21, that look like microfossils.

More than a decade of careful studies have produced plausible nonbiological explanations for ALH 84001 and thereby cast severe doubt on the arguments for biomarkers. It is not clear that the carbonates even formed at temperatures cool enough to allow terrestrial-type life. Magnetite grains similar to those found in the Mars rock have been produced by nonbiological processes in the laboratory, although the analogy is not perfect, and this piece of evidence is currently thought to be the strongest evidence for life. Organic molecules similar to those found in ALH 84001 have been detected in meteorites from small bodies on which biological activity is highly unlikely, and some have even been seen in the interstellar medium. Finally, the putative microfossils are at least an order of magnitude smaller in volume than any self-sufficient organisms on Earth, and they are simple enough to have plausibly formed through nonbiological activity. Thus, although the evidence for life on Mars in ALH 84001 has

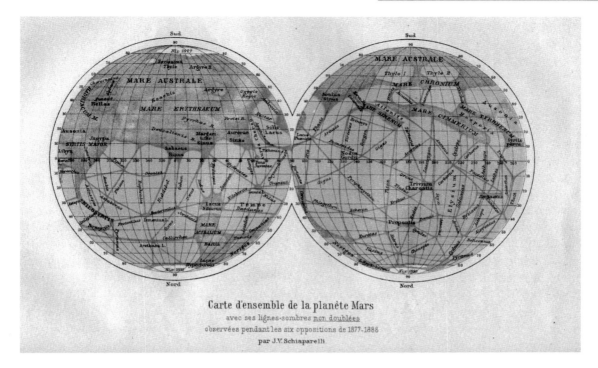

Carte d'ensemble de la planète Mars
avec ses lignes-sombres non doublées
observées pendant les six oppositions de 1877-1888
par J.V. Schiaparelli

Figure 16.22 Map of the planet Mars by Giovanni Schiaparelli based on his visual observations through a telescope at six epoch when Earth was nearest Mars during the period 1877–1888. This is the first map in which martian 'canals' appeared in strength. In Italian, 'canali' indicates either (natural) channels or (constructed) canals, but translations into English almost uniformly used the word 'canals'. (From Flammaron 1892)

not been completely refuted, it is now regarded as weak, and the episode serves as a caution to researchers searching for signs of extraterrestrial life.

Methane has been reported in Mars's atmosphere, though its reality has been heavily debated (§9.4). Because CH_4 in the martian atmosphere is photochemically destroyed on a timescale of a few hundred years, if it is present, there must be an active source. Subsurface methanogenic organisms could provide this source, although volcanic or hydrothermal processes may be capable of delivering an abiological source.

Many linear features appeared on some late nineteenth-century drawings of the surface of Mars, such as the one shown in Figure 16.22. These lines were interpreted by Percival Lowell and some other astronomers as canals built by

a civilization trying to survive a global drought. Much higher resolution images of Mars obtained by spacecraft (§9.4) show no evidence of such engineered structures.

16.13.2 Search for Extra terrestrial Intelligence

The Search for Extra-Terrestrial Intelligence (**SETI**) is an endeavor to detect signals from alien life forms. A clear detection of such a signal would likely change humanity's world view as much as any other scientific discovery in history. Because our society is in its technological infancy, another civilization capable of communicating over interstellar distances is likely to be enormously advanced compared with our own – compare our technology to that of a mere millennium ago and then extrapolate millions or billions of years into the future! Thus, a dialog with extraterrestrials could alter our society in unimaginable ways.

The primary instrument used by SETI is the radiotelescope. Most radio waves propagate with little loss through the interstellar medium, and many wavelengths also easily pass through Earth's atmosphere. Radio waves are easy to generate and to detect. Radio thus appears to be an excellent means of interstellar communication, whether data are being exchanged between a community of civilizations around different stars or are being broadcast to the galaxy to reach unknown societies in their technological infancy. Signals used for local purposes, such as radar and TV on Earth, also escape and can be detected at great distances.

The first deliberate SETI radiotelescope observations were performed by Frank Drake in 1960. Since that time, improvements in receivers, data processing capabilities and radiotelescopes have doubled the capacity of SETI searches roughly once per year. Although a betterment by a factor of $\sim 10^{11}$ is quite impressive, only a minuscule fraction of directions and frequencies have been searched, so SETI proponents are not discouraged at the lack of success to date.

Some SETI observations have been made for visible photons. No such signals have yet been identified, but our technical capabilities to conduct such searches continue to increase rapidly and may yet bear fruit. Looking for physical artifacts of remote civilizations within our Solar System is less promising unless and until someone determines how to conduct such a search in an effective manner.

16.14 Are We Alone?

How common is life in the Universe? This fundamental question can be divided into various parts, which are related to the terms in the Drake equation (eq. 16.1): How common are habitable planets? What are the chances that life can begin on a habitable planet? How long does life, once formed, typically survive? If life exists, what are the chances that advanced life forms exist? In this sense, advanced life can be defined at increasing levels of complexity and thus decreasing likelihood of formation. By analogy with life on Earth, such a sequence could be prokaryotes, eukaryotes, differentiated multicellular organisms, thinking beings, technological civilizations. Other paths, such as noncellular life, as well as extensions of this list to more sophisticated beings, may also be possible.

The first question listed above concerns planetary science and has been the primary subject of this chapter. Because our knowledge of both planetary processes and biological requirements is incomplete, it is also a philosophical question. The phase space of habitable planets has many dimensions, and its width is narrow in some of these parameters. Thus, even with hundreds of billions of stars in the galaxy, if only a small range of stellar mass, planetary mass, planetary volatile inventory, orbital semimajor axis, distribution of giant planets, etc. are suitable for life, then we may indeed be alone (Problem 16-21). However, even if the range of each of these parameters is small with the other parameters being held fixed, if changes in one parameter can be compensated for by changes in another (say increased orbital semimajor axis balanced by more volatiles to enhance greenhouse warming), then our galaxy could well be teeming with life (Problem 16-22). The range of each parameter is also uncertain and more difficult to estimate in some cases than in others.

The question of the likelihood of a habitable planet actually being inhabited is much more difficult to address quantitatively given the major gaps in our understanding of the origin of life. The fraction of inhabited planets on which advanced life develops is also highly uncertain because some of the very many steps involved are also poorly understood (§16.9.2). In the case of terrestrial life, simple prokaryotic cells can survive over a much broader range of conditions (temperature, acidity, salinity, oxygen abundance) than can the more sophisticated eukaryotes. Indeed, the Earth itself was not habitable for multicellular organisms until the oxygen abundance reached an adequate level, $\sim 10^9$ years after life first inhabited our planet. Planetary conditions that remain within the range hospitable to prokaryotes may extend outside the limits acceptable for advanced life, with extremes exceeding the conditions that have produced the mass extinctions of the past half billion years on Earth (§16.7). Returning to the concept of Gaia (§16.8), life forms that exert negative feedback on planetary climate are likely to have more time to evolve into advanced civilizations than those that accentuate climate variations. The final term in the Drake equation (16.1), the mean lifetime of civilizations, is especially difficult to estimate scientifically. The technological society that humanity has created in the past several millennia is a fundamentally different way of life than our planet had ever known before. 'Intelligent' beings such as humans are able to exert conscious control of their environment, including in principle the ability to prevent large impacts by deflecting asteroids and to leave the planet entirely if necessary, e.g., because its star has left the main sequence. Such abilities may provide an important, if not crucial, advantage to the survival of life in the very long term. But also among our marvelous inventions is the increasing capacity to bring about our own extinction via war, novel germs, climate change, etc.

Given the characteristic timescales of our Galaxy, the ETI that we would be statistically most likely to hear from would be of order one billion years more advanced than us. The brains and bodies of modern humans are similar to those of our ancestors who walked the Earth 10^5 years ago, yet our society has advanced substantially as a result of **cultural evolution**. Our knowledge base is now shared by technologies from written language to the internet. The pace of such cultural evolution is growing at an exponential rate. We soon will be able to control our genetic evolution and not need to rely on random mutations in order to 'improve' our genes. Even with these advances, artificial intelligence may soon surpass biological intelligence. It is difficult to say how cultural, technological and biological evolution will develop on Earth over the next thousand years, much more so for gigayears into the future. Compare our civilization with the simple prokaryotes that represented the most advanced lifeforms on our planet three billion years ago and then project a billion or so years into the future to lifeforms that might be as difficult for us to fathom as we are to primitive prokaryotes! Are the deities worshiped by the ancient Greeks and Romans better guesses than the extraterrestrials of modern science fiction? Is this what most ETI is like?

We end with two quotes. The first is from noted British science fiction writer Arthur C. Clark (1917–2008): 'Sometimes I think we're quite alone in the Universe, and sometimes I think we're not. In either case the idea is quite staggering.' The second, from Russian rocketry pioneer Konstantin Eduardovich Tsiolkovskii (1857–1935), reads. 'The Earth is the cradle of mankind. But one does not live in the cradle forever.'

Key Concepts

- The Drake equation encompasses the factors that determine the number of technologically advanced civilizations in our Galaxy.
- Life is difficult to define by physical characteristics. Evolution appears to be a fundamental property of a biosphere.
- Living organisms are never in equilibrium. All organisms must capture, store and use energy in order to live; if they do not, they decay and die.
- Complex carbon chemistry and liquid water are essential for all life on Earth. Other chemical combinations are conceivable, but none appears to be as good.

- The importance of liquid water to life on Earth has led to the concept of a circumstellar HZ, the region in which a planet must orbit for radiation from its star to maintain H_2O on its surface.
- The cycling of carbon and other elements between the biosphere and abiotic portions of the Earth is key to maintaining a flourishing environment for life.
- Life on a planet may be affected by the planet's magnetic field, large moons and other planets or stellar companions orbiting the same star.
- Large impacts can disrupt or even destroy life on a planet.
- Life can chemically alter a planet on short and long timescales. These changes influence the destiny of such life.
- All life on Earth is related and evolved from a common ancestor or group of ancestors.
- The steps leading to the formation of life on Earth are not well known. Therefore, we are uncertain as to whether or not life is common elsewhere.

- After life on Earth began, the key factor in its development has been evolution by means of natural selection. The fittest organisms are the ones most likely to survive and to produce off-spring that are similar, but often not identical, to themselves.
- Mass extinctions, the sudden die-off of large numbers of species, can have a profound effect on the course of evolution.
- Microbial organisms can be transported between neighboring planets such as Earth and Mars. However, such panspermia is quite unlikely over interstellar distances.
- Various techniques are being used to search for life on planets other than our Earth. Some searches look for chemical signatures that could come from simple life forms, and others seek out signals from technologically advanced civilizations. Success of such searches could have profound implications for the future of humanity.

Further Reading

A basic textbook that focuses on the quest to find extraterrestrial life and also covers the origin of life and planetary habitability at a level accessible to nonscience majors is:

Goldsmith, D., and T. Owen, 2001. *The Search for Life in the Universe*, 3rd Edition. University Science Books, Sausalito, CA. 580pp.

A nontechnical discussion of habitable environments focusing on terrestrial planets and satellites of giant planets in our Solar System is given by:

Jakosky, B. 1998. *The Search for Life on Other Planets*. Cambridge University Press, Cambridge, UK. 326pp.

A detailed but mostly nontechnical account of life and its interactions with our planet is presented by:

Lunine, J.I., 2013. *Earth: Evolution of a Habitable World*, 2nd Edition. Cambridge University Press, Cambridge, UK. 304pp.

A broad perspective of the sciences of life in the Universe at the advanced undergraduate/graduate student level is provided in:

Lunine, J.I., 2005. *Astrobiology: A Multidisciplinary Approach*, Pearson Education, San Francisco. 586pp.

An extensive collection of review articles covering many aspects of astrobiology is presented in:

Sullivan, W.T. III and J.A. Baross, Eds., 2007. *Planets and Life: The Emerging Science of Astrobiology*. Cambridge University Press, Cambridge, UK. 604pp.

Details on atmospheric factors that affect the boundaries of the habitable zone are provided by:

Kasting, J.F., D.P. Whitmire, and R.T. Reynolds, 1993. Habitable zones around main sequence stars. *Icarus* **101**, 108–128.

Life is reviewed in an astrophysical context by:

Chyba, C.F., and K.P. Hand, 2005. Astrobiology: The study of the living universe. *Ann. Rev. Astron. Astrophys.* **43**, 31–74.

A speculative, pessimistic assessment of the likely abundance of advanced life in the Universe is presented in the popular book:

Ward, P.D., and D. Brownlee, 2000. *Rare Earth: Why Complex Life is Uncommon in the Universe.* Copernicus Books, New York. 368pp.

The contrasting viewpoint that evolution is likely to lead not only to advanced organisms but also to intelligent ones is provided by:

Morris, S.C., 2003. *Life's Solution: Inevitable Humans in a Lonely Universe.* Cambridge University Press, Cambridge, UK. 464pp.

In this book, Simon Conway Morris advocates the viewpoint that humanlike intelligence is an evolutionary advantage that is likely to develop as a result of evolutionary convergence and that the most viable solutions to Fermi's paradox are the lack of earthlike planets and/or the difficulty of the formation of life.

A third perspective on life beyond Earth is given by:

Grinspoon, D., 2004. *Lonely Planets: The Natural Philosophy of Alien Life.* HarperCollins, New York. 441pp.

For details on thermodynamical processes important to life, see:

Haynie, D.T., 2008. *Biological Thermodynamics,* 2nd Edition. Cambridge Univ. Press. 422pp.

The effects of impacts on life are discussed by:

Toon, O.B, K. Zahnle, D. Morrison, R.P. Turco, and C. Convey, 1997. Environmental perturbations caused by the impacts of asteroids and comets. *Rev. Geophys.* **35**, 41–78.

Updated information on the search for near-Earth asteroids and impact hazards can be found at http://impact.arc.nasa.gov/

A discussion of ancient life on Earth is presented by:

Knoll, A.H., 2003. *Life on a Young Planet – The First Three Billion Years of Evolution on Earth.* Princeton University Press, Princeton, NJ. 277pp.

An easily accessible research-level account of planetary requirements for the emergence of life is given by:

Chyba, C.F., D.P. Whitmire, and R. Reynolds, 2000. Planetary habitability and the origins of life. In *Protostars and Planets IV*, 1365–1393, University of Arizona Press, Tucson, AZ. 1422pp.

A good book on global ecosystems, humanity's modifications of Earth and the hypothetical process of terraforming other planets is:

Fogg, M.J., 1995. *Terraforming: Engineering Planetary Environments.* Society of Automotive Engineers, Warrendale, PA. 544pp.

A nice review of proposed spectral signatures of extraterrestrial life is presented by:

Des Marais, D.J., M.D. Harwit, K.W. Jucks et al., 2002. Remote sensing of planetary properties and biosignatures on extrasolar terrestrial planets. *Astrobiology* **2**, 153–181.

An amusing and informative account of reproductive strategies by various animal species is presented by:

Judson, O., 2002. *Dr. Tatiana's Sex Advice to All Creation*. Henry Holt & Co., New York. 309pp.

A well-written novel about a truly bizarre living interstellar gas cloud is:

Hoyle, F., 1957. *The Black Cloud*. William Heinemann Ltd, London.

Although such a living organism might be able to exist, survive and even reproduce, it is much more difficult to conceive of how it could form out of very low-density matter by natural processes.

Problems

16-1. Estimate the distance to the nearest communicating civilization as a function of N_{cc}, the number of communicating civilizations in our Galaxy. You may assume that the Galaxy is a disk of full thickness 2000 light-years and radius 40 000 light-years and the Sun is located near the galactic midplane at a radius of 25 000 light-years. (Hint: Your answer should contain two formulas, one applicable if the nearest communicating civilization is within 1000 light-years and the other if the distance is between 1000 and 15 000 light-years. You need not consider the situation in which N_{cc} is only a few, in which case the distance is comparable to the radius of the galaxy.)

16-2. Why does carbon appear to be the best element for producing complex molecules suitable for life?

16-3. A simplistic definition of the boundaries of a star's HZ is where the planet's equilibrium temperature, T_{eq} (eq. 4.17), allows for the presence of liquid water at 1 bar pressure, $273° < T_{eq} < 373°$.
(a) Determine the blackbody limits of habitability, r_{in} and r_{out}, according to this definition as functions of stellar luminosity, L (in units of the Sun's luminosity, $L_{\odot}$), and planetary albedo, A_b. Express your answer in astronomical units.
(b) By this definition, which Solar System planets would be in the Sun's HZ assuming $A_b = 0.3$?
(c) Discuss, qualitatively, how including the greenhouse effect affects the answers to parts (a) and (b). What other neglected effects may be important?

16-4. Consider the planet-hosting star Gliese 876. The mass of Gliese 876 is $M_\star = 0.32$ $M_{\odot}$, and its luminosity only $0.012 \, \mathcal{L}_{\odot}$.
(a) At what distance does a planet orbiting Gliese 876 intercept the same flux of radiation from its star as the Earth receives from the Sun?
(b) What is the orbital period of a planet at this distance?

16-5. **(a)** For what value of orbital eccentricity, e, does a planet receive twice as much stellar radiation at periastron as at apastron?
(b) Calculate the latitude on Earth at which the amount of solar energy received at the summer solstice is twice that received at winter solstice. You may neglect the eccentricity of Earth's orbit.

(c) Comment on the similarities and differences that the above two types of annual variations in energy input would have on climate.

16-6. The rate at which a star's tidal torque removes rotational angular momentum from a planet's spin is proportional to the square of the star's mass and inversely proportional to the sixth power of its orbital distance (eq. 2.58). Approximate the star's luminosity to vary as the fourth power of the star's mass (eq. 3.23). Consider planets around stars of different mass, with each of the planets orbiting at the appropriate distance for the stellar radiation flux to be equal to that received by Earth. Show that the rate at which planetary rotation is slowed is proportional to a large negative power of the star's mass and determine the value of that power.

16-7. Calculate the orbital velocity of planets in the middle of the HZ as a function of their star's mass.

16-8. In this problem, you will quantify the requirements of the effects of solar mass loss on the faint young Sun problem. Because the solar wind and the energy emitted by the Sun's luminosity remove mass from the Sun (§2.9), billions of years ago, the Sun was more massive than it is today. The stellar mass–luminosity relationship (eq. 3.23) thus implies that the early Sun was more luminous than it would have been if its mass were only 1 $M_\odot$. Moreover, the orbit of the Earth must have expanded as the Sun shed mass (eq. 2.65), so the early Earth orbited closer to the Sun than our planet does at present.

(a) Combine the effects of the larger luminosity of a more massive early Sun with the closer proximity in the past and the $r_\odot^{-2}$ dependence of solar radiation flux (eq. 4.13) to show that the flux of radiation reaching early Earth varied with solar mass approximately as:

$$\mathcal{F} \propto M_\odot^6. \tag{16.6}$$

(b) Assume that the average rate of solar mass loss over the past 4×10^9 years is equal to the rate at present (§2.9, Problem 2-20) and compute the mass of the Sun 4 billion years ago.

(c) Use equation (16.6) to estimate the increase in estimated solar radiation reaching Earth 4 billion years ago as a result of the Sun's larger mass at that epoch.

(d) Does accounting for solar mass loss amplify or reduce the faint young Sun problem? Is this change large or small?

16-9. What would happen to the Earth if the Sun suddenly became twice as bright as it is at present?

16-10. If planets orbit too near one another, the planetary system is dynamically unstable (§2.3.5). This instability means that it is difficult for more than one or two Earth-like planets to orbit within a star's HZ. Consider a star with a giant planet in its HZ. Although such a planet would destabilize most orbits within its star's HZ, it could have Earth-sized moons or Lagrangian point (§2.2.1) companions. Discuss how many Earth-sized bodies could survive for eons within the HZ in such a system. (Hint: Consider tides and collisions, as well as gravitational perturbations.)

16-11. If all volcanic activity on Earth ceased, how would life on our planet be affected? Over what timescales would these changes occur?

16-12. The K–T ejecta layer in Europe consists of a 3-mm-thick layer of impact-generated **spherules** (spherical particles). Typical K–T spherules are about 100 μm in radius. Assume that each spherule is opaque.

(a) What is the volume of an individual 100-μm radius spherule?

(b) How many spherules are packed into a volume of 3 mm³?

(c) What is the cross-sectional area of an individual spherule?

(d) What is the total cross-sectional area of all of the spherules in a square millimeter of the ejecta layer?

The optical depth of the spherule layer is given by the cross-sectional area presented by the spherules per unit area. The optical depth of the spherule layer provides an extreme lower limit on the total optical depth of solids ejected into the atmosphere by the K–T impact event.

16-13. The K–T boundary layer contains abundant soot derived from fires. The global average is inferred to be 0.02 kg soot per square meter.

(a) How much soot was generated by the K–T event?

(b) Modern wildfires generate CO_2: CO:soot roughly in the ratio of 100:10:1. Assuming these ratios hold, how much CO_2 was generated? Note that the current atmosphere contains 2×10^{15} kg of CO_2.

(c) The mass of the biosphere is now 1×10^{16} kg. Roughly what fraction of the biosphere must burn to produce the K–T soot layer? Is this plausible? (Note: The K–T boundary layer soot was deposited during an interval that was effectively instantaneous compared with the geologic timescale, not necessarily rapid in human terms. Thus, both forests that were ignited by reentering impact ejecta and those that died as a result of impact effects and burned within the subsequent few years could have contributed to this soot layer.)

(d) For humans, CO has noticeable adverse health effects at about 200 ppm and becomes acutely toxic at 2000 ppm. Did CO pyrotoxins pose a significant threat to dinosaurs (assuming that their susceptibility was comparable to that of humans)?

16-14. Assume that mass extinctions occur once per 100 million years and eliminate half of the species and that, excluding mass extinctions, the average species survives for 10 million years.

(a) Estimate the probability of a given species to vanish as the result of a mass extinction.

(b) Estimate the average lifetime of a species.

(c) What do you conclude about the importance of mass extinctions for the survival of species?

(d) Why are mass extinctions more relevant to the long-term course of evolution than they are to the survival of most individual species?

16-15. Discuss the pros and cons of artificially introducing terrestrial life to other Solar System bodies with potentially habitable environments.

16-16. Calculate the amount of energy incident upon Earth and compare it with the energy used by humanity, the luminosity of the Sun and the total luminosity of the Milky Way Galaxy. Comment on the scales of hypothetical Kardashev type I, II and III civilizations.

16-17. For microbes to be transported from one planetary system to another without the aid of 'intelligent' life forms such as humans, several improbable things must occur. First, material containing living organisms (active or dormant) must be ejected from one planetary system. Then it must arrive in another while still in a viable form and subsequently be captured in a nondestructive manner by a hospitable planet.

(a) Consider the transport of dormant life (spores) in rocks. The first step to quantify this process is estimating how many rocks with viable spores are ejected from Earth and leave the Solar System. This is not an easy calculation because most ejection occurs during rare large impacts, and the integrated flux of such impactors is dominated by collisions early in Earth's history, so the result depends on the time at which living organisms on Earth first developed the ability to form hardy spores. Moreover, the next step requires an analysis of the scattering of ejected bodies by the planets; this calculation is better defined but mathematically complex. For the sake of definiteness, assume that the integrated total number of rocks with viable spores that leave the Solar System, N, is of order 10^{10}, i.e., a few per year. Estimate the probability that a given one of these rocks would impact an Earth-like planet orbiting another star within 250 Myr (the viability time of the stoutest spores known). You may assume that the rocks leave the Solar System at a velocity small compared with the typical velocity of the Sun relative to nearby stars of 30 km/s and that stars are 200 000 AU apart (about the distance to the nearest star today). Multiply your estimate by N and include a factor for the survivability of the spores in their impact onto their new home. Comment on the possibility that after forming, life would spread throughout the galaxy in this manner.

(b) Calculate the probability that the spore-containing rock is captured by a protoplanetary disk. Discuss qualitatively the subsequent steps needed for life to be spread in this manner.

(c) Another possible mode of interstellar panspermia is transport of active (possibly subsurface) life within a planet. The donor planet could have been ejected from circumstellar orbit or remain bound to its star. Discuss the various requirements for this process and the role that it could play in spreading life throughout the galaxy.

16-18. Consider the spread of a colonizing civilization throughout the Milky Way Galaxy. Make the (very optimistic) assumption that the number of habitable worlds in our galaxy is equal to the number of stars, 3×10^{11}.

(a) Assume that each colonized planet initiates a colony on a new planet every 10 000 years. How long would it take for every habitable planet in the galaxy to be colonized?

(b) If the only new planets to be colonized were the ones around stars neighboring already colonized planets, how long would it take for the civilization to cross the Galaxy?

(c) Comment on the differences between your results in (a) and (b). Which formalism do you believe is more realistic, and why?

(d) Compare your results with the age of the Galaxy and comment on Fermi's paradox.

16-19. **(a)** Calculate the mean equilibrium temperature of Earth with its current albedo, $A_b = 0.29$, but lacking any greenhouse warming. The difference between this temperature and Earth's mean temperature of 288 K is caused by greenhouse warming (§4.6).
(b) Repeat your calculation for an ice-covered Earth with $A_b = 0.55$, assuming the same greenhouse warming as deduced in part (a).
(c) Comment on how your answers apply to the theory of snowball Earth. How might a snowball Earth develop, and how might it end? (Hint: Consider the stability of each configuration both to small changes in the amount of ice cover and to the other factors that affect climate.)

16-20. Prognosticators of galactic habitability typically assume that the necessary ingredients and locales for life would not have been available when the galaxy was only a few billion years old. Explain this line of reasoning and discuss why habitable worlds with the necessary ingredients for life might be more common during more recent times. Mention any observational data on planet formation that bear on this discussion.

16-21. Estimate the number of habitable planets in our galaxy under the conservative (pessimistic) assumption that for it to be habitable, all of the planet's properties need to be similar to the one planet that we know to be habitable, Earth. Make a list of planetary requirements for life and assume that the allowable range in each quantitative parameter is ±10%. Note that this does not mean that 10% of the systems satisfy each requirement; for instance, far fewer than 10% of the stars in the galaxy are within 10% of the mass of our Sun. Then multiply the probabilities with the number of stars in the galaxy to arrive at your result. Repeat your calculations for ranges that are half as large and twice as large.

16-22. Estimate the number of habitable planets in our galaxy under the optimistic assumption that many of the planet's properties can be different from Earth provided other properties differ in a compensating manner. For example, a more massive star could be compensated for by the planet being on a more distant orbit, and a greater fraction of volatiles could compensate for a smaller planetary mass to yield a similar atmospheric density (and possibly also similar rates of plate tectonics/crustal recycling). As in the previous problem, assume that the allowable range in each quantitative parameter is ±10%.

16-23. Consider the Drake equation (eq. 16.1). Plug in your own numbers for the various variables, justifying your choices, to estimate the total number of communicating civilizations in the galaxy. Make three calculations, one optimistic, one realistic and one pessimistic (by your own standards but state your justification).

APPENDIX A: SYMBOLS USED

a	semimajor axis of an orbit	C_V	thermal heat capacity (or molecular heat) at constant volume
a_{AU}	semimajor axis of orbit in AU	D	diameter
A	surface area	D_i	molecular diffusion coefficient of species i
$\mathcal{A}$	area enclosed by orbit; cross-sectional area		
A_0	geometric albedo (head-on reflectance)	e	(generalized) eccentricity
A_b	Bond albedo	e, e^-	electron
A_v	albedo at visual wavelengths	e^+	positron
A_ν	albedo at frequency ν	E	total energy
b	impact parameter	$E, \mathbf{E}$	electric field strength and vector
b_m	semiminor axis of an orbit	E_G	gravitational potential energy
B	ring opening angle	E_K	kinetic energy
$B, \mathbf{B}$	magnetic field strength and vector	E_{rot}	kinetic energy of rotation
B, B_ν	brightness, at frequency ν	$\varepsilon_e, \varepsilon_v$	enhancement factor; evaporative; loading parameter
B_{oc}	angle between occulted signal and ring plane		
		EW	equivalent width
c	speed of light in vacuum	f	true anomaly (angle between planet's periapse and instantaneous position)
c_s	speed of sound		
c_v	velocity dispersion	$\mathbf{F}$	force
c_P	specific heat at constant pressure	$\mathbf{F}_c$	centripetal force
c_V	specific heat at constant volume	F_f	amplitude of the driving force
C	a constant	$\mathbf{F}_g$	force of gravity
C_D	drag coefficient	$\mathbf{F}_{g,eff}$	effective gravitational force
C_H	heat transfer coefficient	$\mathbf{F}_D$	drag force
C_J	Jacobi's constant	$\mathbf{F}_T$	tidal force
C_P	thermal heat capacity (or molecular heat) at constant pressure	$\mathbf{F}_{rad}$	radiation force
		$\mathcal{F}$	flux

$\mathcal{F}_g$	gravitational enhancement factor	m_v	(visual) apparent magnitude
$\mathcal{F}_\nu$	flux density at frequency ν	m_{amu}	mass of an atomic mass unit
$\mathcal{F}_\odot$	solar constant (incident flux at Earth)	m_{gm}	mass of one gram-mole
g_{eff}	effective gravitational acceleration	m_H	mass of hydrogen atom
g_p	gravitational acceleration	M	mass
g_r	radar backscatter gain	M_p	mass of planet
G	gravitational constant	M_v	absolute visual magnitude of a star
h	Planck's constant	M_J	Jeans mass
$\hbar$	normalized Planck's constant, $h/(2\pi)$	$M_\star$	mass of star
h	vertical scale length	$M_\odot$	solar mass
h_{cp}	height of crater peak	$M_\oplus$	Earth mass
H	scale height	$\mathcal{M}_B$	magnetic dipole moment
H	enthalpy	$\mathcal{M}_R$	Richter magnitude
H_v	absolute visual magnitude of an asteroid	n	neutron
H_z	Gaussian scale height	n	energy level of an electron
i	inclination angle	n	index of refraction
I	moment of inertia	n	mean angular velocity of body in orbit
I_ν	specific intensity	n_{po}	polytropix index
j	differential energy flux of particles	n_o	Loschmidt's number (Table C.3)
j_ν	mass emission coefficient	n_{qm}	principal quantum number
J	photodissociation rate	N	number density of particles (m^{-3})
J	electric current	N	Brunt–Väisälä (buoyancy) frequency
J, J_ν	mean intensity, at frequency ν	N_A	Avogadro's number (Table C.3)
J_i	action variable	**p**	momentum
J_n	gravitational moments	p	as subscript: polarization, planet, particle
k	Boltzmann constant	p, p$^+$	proton
k	wave vector	P	pressure
k_T	tidal Love number	P_n	Legendre polynomials
k_{ri}	chemical reaction rate for reaction i	P_L	linear polarization
K	radial velocity variation amplitude	P_{orb}	orbital period
K_{po}	polytropic constant	P_{rot}	rotation period
K_T	thermal conductivity	P_{yr}	orbital period in years
$\mathcal{K}$	eddy diffusion coefficient	q	electric charge
ℓ	characteristic length or depth scale	q	pericentric separation
ℓ_{fp}	mean free path	Q	amount of heat
$L, \mathbf{L}$	angular momentum magnitude and vector	Q_{pr}	radiation pressure coefficient
L_1–L_5	Lagrangian (equilibrium) points	Q_T	Toomre's stability parameter
L_s	latent heat of sublimation or condensation	$\mathcal{Q}$	gas production rate
L_s	Mars solar longitude	$r, \mathbf{r}$	distance, separation
$\mathcal{L}_\odot$	solar luminosity	r_c	corotational radius
$\mathcal{L}_\star$	stellar luminosity	r_{cen}	centrifugal radius
$\mathcal{L}$	luminosity	r_{AU}	distance in AU
m	mass	r_{Bohr}	Bohr radius

r_Δ	distance from observer (or Earth)	v_o	thermal velocity
$r_\odot$	heliocentric distance	v_r	radial component of the velocity
$r_{\odot AU}$	heliocentric distance (AU)	v_∞	terminal velocity
$r_{\Delta AU}$	distance from observer (or Earth) (AU)	V	volume
r_{CM}	position of center of mass	z_{ex}	altitude exobase
R	radius of object	Z	atomic number
R_e	equatorial radius		
R_p	polar radius	x, y, z	Cartesian coordinate axes
R_s	distance of closest approach	r, ϕ, θ	spherical coordinate system
R_E	radius of Einstein ring	α	pitch angle of a particle
R_H	radius of Hill sphere	α_R	Rosseland mean absorption coefficient
$R_\oplus$	radius of Earth	α_ν	mass extinction coefficient
$R_\star$	radius of star	β	ratio of radiation to gravitational force, F_{rad}/F_g
R_{gas}	universal gas constant		
R_{Sch}	Schwarzschild radius	β_{cp}	ratio of corpuscular to radiation drag
$\mathcal{R}$	Rydberg's constant	$\eta_\oplus$	number of Earth-like planets per star
S	entropy	δ_{jk}	Kronecker delta (along axes j, k)
S_ν	source function	γ	photon
t	time	γ	ratio of specific heats, C_P/C_V
t_d	diffusion timescale	γ_c	Lyapunov exponent
t_d	Ohmic dissipation timescale (§7.2)	γ_c^{-1}	Lyapunov timescale
t_m	mean lifetime	Δ	bow shock thickness
$t_{1/2}$	half-life of nuclide	ϵ	flattening (geometric oblateness) (($R_e -$
t_{ff}	free-fall timescale		$R_p)/R_e$)
t_{KH}	Kelvin–Helmholtz timescale	ϵ, ϵ_ν	emissivity at frequency ν
t_{pr}	decay time due to Poynting–Robertson drag	ζ	exponent in power-law distribution
		θ	angle between line of sight and normal to surface
t_ϖ	time of periapse passage		
T_a	atmospheric temperature	θ	colatitude
T_b	brightness temperature	Θ	potential temperature
T_e	effective temperature	κ	epicyclic (radial) frequency
T_s	surface temperature	κ_ν	mass absorption coefficient
T_g	ground temperature	λ	wavelength
T	temperature	λ_m	mean longitude
T_g	magnitude torque	λ_{esc}	escape parameter
T_{eq}	equilibrium temperature	λ_{III}	longitude based on the rotation period of Jupiter's magnetic field
T_{tr}	triple point		
$\mathcal{T}_{tr}$	duration of transit	μ	frequency of vertical oscillation
U	total energy	μ_a	molecular mass in amu
$v, \mathbf{v}$	velocity magnitude and vector	μ_θ	$\equiv \cos\theta$
v_c	circular orbit velocity	ν	frequency
v_e	escape velocity	ν_e	collisional frequency for electrons
v_i	impact velocity	ν_e	electron neutrino

$\bar{\nu}_e$	electron antineutrino	ω	argument of periapse
ν_v	kinematic viscosity	ω_f	forcing frequency
ρ	density	ω_o	frequency of oscillator, wave
ρ_g	gas density		frequency
ρ_p	density of (proto)planet	ω_{rot}	spin angular velocity
ρ_s	volume mass density of swarm planetesimals	Ω	longitude of ascending node
		Ω_s	solid angle
$\rho_\star$	density of star	ϖ	longitude of periapse
σ	Stefan–Boltzmann constant	ϖ_ν	single scattering albedo
σ_o	electrical conductivity		at frequency ν
σ_x	(molecular) cross-section		
σ_ν	mass scattering coefficient	$\odot$	Sun
σ_ρ	surface mass density	$\varmercury$	Mercury
τ	optical depth	$\venus$	Venus
ϕ	solar phase (Sun–target–observer) angle	$\oplus$	Earth
ϕ	longitude, azimuth	$\leftmoon$	Moon
Φ_c	centrifugal potential	$\mars$	Mars
Φ_g	gravitational potential	$\jupiter$	Jupiter
Φ_i	upward particle flux in an atmosphere	$\saturn$	Saturn
Φ_ℓ	limiting flux	$\uranus$	Uranus
Φ_B	magnetic flux	ψ	Neptune
Φ_J	Jeans escape rate	$\pluto$	Pluto
Φ_ν	line shape		
ψ	obliquity of a body (angle between rotation axis and orbit pole)		movie
			WebColor

APPENDIX B: ACRONYMS USED

ADP	adenosine diphosphate
ALH	Allen hills (meteorite recovery area in Antarctica)
ALMA	Atacama large millimeter array
ATP	adenosine triphosphate
AU	astronomical unit
BIF	banded iron formation
BLG	bulge (thick central portion of the Milky Way galaxy)
CA	closest approach
CAI	calcium–aluminum inclusion (found in chondritic meteorites)
CCD	charge coupling device
CI, CM, CO, CV, CR, CH, CB, CK	types of carbonaceous chondrite meteorites
CICLOPS	Cassini Imaging Central Laboratory for Operation
CIRS	Composite InfraRed Spectrometer
CME	coronal mass ejection
CNO	carbon nitrogen oxygen (hydrogen fusion catalyst) cycle
DNA	deoxyribonucleic acid
DS2	Dark Spot 2 (on Neptune)
EC	ecliptic comet

EH, EL	enstatite chondrite meteorite (with high and low iron abundances)
EL	equilibrium level
EQ	encephalization quotient
ESA	European Space Agency
ESO	European Southern Observatory
ETI	extraterrestrial intelligence
ETVs	eclipse timing variations
EUV	extreme ultraviolet wavelengths
EUVE	Extreme UltraViolet Explorer
FDS	flight data system
GCMS	Gas Chromatograph Mass Spectrometer (on the *Huygens* probe)
GDS	Great Dark Spot (on Neptune)
GRS	Great Red Spot (on Jupiter)
HED	howardite–eucrite–diogenite (achondrite meteorite types from asteroid 4 Vesta)
HF	higher frequency emissions
HIPPO	habitat loss, invasives, pollution, population and overexploitation
HiRISE	High Resolution Imaging Science Experiment (on *Mars Reconnaissance Orbiter*)

HRSC	High Resolution Stereo Camera (on *Mars Express*)	OGLE	Optical Gravitational Lensing Experiment
HST	*Hubble Space Telescope*	PIA	photo identification number (JPL)
HZ	habitable zone		
IAU	International Astronomical Union	ppm	parts per million
IDP	interplanetary dust particle	PPS	photopolarimeter subsystem (instrument on the *Voyager* spacecraft)
ILR	inner Lindblad resonance		
IMF	interplanetary magnetic field	REE	rare-Earth elements (elements with atomic numbers 57–70)
IR	infrared		
IRAS	*InfraRed Astronomical Satellite*	RKBO	resonant Kuiper belt object
IRTF	InfraRed Telescope Facility	RNA	ribonucleic acid
ISM	interstellar medium	ROSAT	Roentgen satellite
ISO	*InfraRed Space Observatory*	RV	radial velocity (method for observing exoplanets)
ISO	interstellar object		
JAXA	Japan Aerospace Exploration Agency	SAO	Smithsonian Astrophysical Observatory
JFC	Jupiter family comet	SDO	scattered disk object
JPL	Jet Propulsion Laboratory	SED	Saturn electrostatic discharges
KBO	Kuiper belt object	SETI	search for extraterrestrial intelligence
KOI	*Kepler* Object of Interest		
LT	local time (on planet)	SKR	Saturn's kilometric radiation
LTE	local thermodynamic equilibrium	SL9	Comet Shoemaker–Levy 9
		SMOW	standard mean ocean water
LUCA	last universal common ancestor	SNR	signal to noise ratio
MBA	main belt asteroid	*SOHO*	*Solar and Heliospheric Observatory*
MC	Mars crossing		
MER	*Mars Exploration Rover*	TDV	transit duration variation
MESSENGER	*MErcury Surface, Space ENvironment, GEochemistry and Ranging* spacecraft	TNO	trans-Neptunian object
		TRACE	Transition Region and Coronal Explorer
MGS	*Mars Global Surveyor*	TTV	transit timing variation
mmag	millimagnitude	UFO	unidentified flying object
MOC	Mars Orbiter Camera	UV	ultraviolet wavelengths
MOLA	Mars Orbiter Laser Altimetry	UVIS	UltraViolet Imaging Spectrograph (on the *Cassini* spacecraft)
MRO	*Mars Reconnaissance Orbiter*		
NASA	National Aeronautics and Space Administration (of the United States)	VIMS	Visual and Infrared Mapping Spectrometer (on the *Cassini* spacecraft)
NEAR	*Near-Earth Asteroid Rendezvous* spacecraft	VLA	Very Large Array radio telescope
NEO	near-Earth object		
OC	Oort cloud comets	VLT	very large telescope

APPENDIX C: UNITS AND CONSTANTS

Table C.1 **Prefixes**

Prefix	Value
y (yocto-)	10^{-24}
z (zepto-)	10^{-21}
a (atto-)	10^{-18}
f (femto-)	10^{-15}
p (pico-)	10^{-12}
n (nano-)	10^{-9}
μ (micro-)	10^{-6}
m (milli-)	10^{-3}
c (centi-)	10^{-2}
d (deci-)	10^{-1}
da (deca-)	10
h (hecto-)	10^{2}
k (kilo-)	10^{3}
M (mega- or million)	10^{6}
G (giga- or billion)	10^{9}
T (tera-)	10^{12}
P (peta-)	10^{15}
E (exa-)	10^{18}
Z (zetta-)	10^{21}
Y (yotta-)	10^{24}

Table C.2 **Units**

Symbol (Name in SI Units)	Value in cgs Units
μm (micrometer)	10^{-4} cm
m (meter)	100 cm
km (kilometer)	10^{5} cm
kg (kilogram)	10^{3} g
t (tonne)	10^{6} g
J (joule)	10^{7} erg
eV (electron volt)	1.602×10^{-12} erg
W (watt)	10^{7} erg s^{-1}
N (newton)	10^{5} dyne
atm (atmosphere)	1.013 25 bar
Pa (pascal)	10 dyne cm^{-2}
bar	10^{6} dyne cm^{-2}
Hz (hertz)	1 cycle s^{-1}
Ω (ohm)	1.1126×10^{-12} esu
mho (ohm^{-1})	8.988×10^{11} esu
A (ampere)	2.998×10^{9} esu
γ (gamma)	10^{-5} gauss
T (tesla)	10^{4} gauss
Jy (jansky)	10^{-23} erg cm^{-2} Hz^{-1} s^{-1}

Table C.3 Physical Constants

Symbol	Value in cgs Units	Value in SI Units	Quantity
c	$2.997\,925 \times 10^{10}$ cm s^{-1}	$2.997\,925 \times 10^{8}$ m s^{-1}	Velocity of light
G	6.674×10^{-8} dyn cm^2 g^{-2}	6.674×10^{-11} m^3 kg^{-1} s^{-2}	Gravitational constant
h	$6.626\,069 \times 10^{-27}$ erg s	$6.626\,069 \times 10^{-34}$ J s	Planck's constant
k	$1.380\,650 \times 10^{-16}$ erg deg^{-1}	$1.380\,650 \times 10^{-23}$ J deg^{-1}	Boltzmann constant
m_e	$9.109\,382 \times 10^{-28}$ g	$9.109\,382 \times 10^{-31}$ kg	Electron mass
m_p	$1.672\,622 \times 10^{-24}$ g	$1.672\,622 \times 10^{-27}$ kg	Proton mass
m_{amu}	$1.660\,539 \times 10^{-24}$ g	$1.660\,539 \times 10^{-27}$ kg	Atomic mass unit
n_o	2.686×10^{19} cm^{-3}	2.686×10^{25} m^{-3}	Loschmidt's number
N_A	$6.022\,142 \times 10^{23}$ mole^{-1}	$6.022\,142 \times 10^{23}$ mole^{-1}	Avogadro's number
r_{Bohr}	$5.291\,77 \times 10^{-9}$ cm	$5.291\,77 \times 10^{-11}$ m	Bohr radius or atomic unit
R_{gas}	8.3145×10^{7} erg deg^{-1} mole^{-1}	8.3145 J deg^{-1} mole^{-1}	Universal gas constant
$\mathcal{R}$	$1.097\,373 \times 10^{5}$ cm^{-1}	$1.097\,373 \times 10^{7}$ m^{-1}	Rydberg constant
q	4.803×10^{-10} esu	$1.602\,176\,6 \times 10^{-19}$ C	Electron charge
σ	5.6704×10^{-5} erg cm^{-2} deg^{-4} s^{-1}	5.6704×10^{-8} W m^{-2} deg^{-4}	Stefan–Boltzmann constant

Table C.4 Material Properties

Symbol	Value in cgs Units	Value in SI Units	Quantity
ρ	1.293×10^{-3} g cm^{-3}	1.293 kg m^{-3}	Density of air at STP[a]
ν_v	0.134 cm^2 s^{-1}	1.34×10^{-5} m^2 s^{-1}	Kinematic viscosity of air at STP[a]
c_P	1.0×10^{7} erg g^{-1} deg^{-1}	1.0×10^{3} J kg^{-1} deg^{-1}	Isobaric specific heat capacity of air at STP[a]
c_V	7.19×10^{6} erg g^{-1} deg^{-1}	7.19×10^{2} J kg^{-1} deg^{-1}	Isochoric specific heat capacity of air at STP[a]
c_P	1.2×10^{7} erg g^{-1} deg^{-1}	1.2×10^{3} J kg^{-1} deg^{-1}	Typical value for the specific heat of rock
L_v	2.50×10^{10} erg g^{-1}	2.50×10^{6} J kg^{-1}	Specific latent heat of vaporization for water
L_s	2.83×10^{10} erg g^{-1}	2.83×10^{6} J kg^{-1}	Specific latent heat of sublimation for water-ice

[a] STP: standard temperature (273 K) and pressure (1 bar).

Table C.5 **Astronomical Constants**

Symbol	Value in cgs Units	Value in SI Units	Quantity
AU	1.496×10^{13} cm	1.496×10^{11} m	Astronomical unit of distance
ly	9.4605×10^{17} cm	9.4605×10^{15} m	Light year
pc	3.086×10^{18} cm	3.086×10^{16} m	Parsec
$M_\odot$	1.989×10^{33} g	1.989×10^{30} kg	Solar mass
$R_\odot$	6.96×10^{10} cm	6.96×10^{8} m	Solar radius
$\mathcal{L}_\odot$	3.827×10^{33} erg s^{-1}	3.827×10^{26} J s^{-1}	Solar luminosity
$\mathcal{F}_\odot$	1.37×10^{6} erg cm^{-2} s^{-1}	1.37×10^{3} J m^{-2} s^{-1}	Solar constant
$M_\oplus$	5.976×10^{27} g	5.976×10^{24} kg	Earth's mass
$R_\oplus$	6.378×10^{8} cm	6.378×10^{6} m	Earth's equatorial radius
g_p(eq)	978 cm s^{-2}	9.78 m s^{-2}	Gravity at sea level on Earth's equator
g_p(pole)	983 cm s^{-2}	9.83 m s^{-2}	Gravity at sea level at Earth's poles

APPENDIX D: Periodic Table of Elements

Key to chart

Atomic number →	**97** (with oxidation states +3 +4)
Symbol →	**Bk**
Atomic mass →	(247)
Name →	Berkelium

Legend:
- Metals
- Non-metals
- Semi-metals
- Artificially prepared elements

Main Table

Atomic No.	Symbol	Oxidation states	Atomic mass	Name
1	H	+1 −1	1.008	Hydrogen
2	He	0	4.003	Helium
3	Li	+1	6.94	Lithium
4	Be	+2	9.01	Beryllium
5	B	+3	10.81	Boron
6	C	+2 +4 −4	12.01	Carbon
7	N	+1 +2 +3 +4 +5 −3	14.00	Nitrogen
8	O	−2	16.00	Oxygen
9	F	−1	19.00	Fluorine
10	Ne	0	20.18	Neon
11	Na	+1	22.99	Sodium
12	Mg	+2	24.31	Magnesium
13	Al	+3	26.98	Aluminum
14	Si	+2 +4 −4	28.09	Silicon
15	P	+3 +5 −3	30.97	Phosphorus
16	S	+4 +6 −2	32.06	Sulfur
17	Cl	+1 +5 +7 −1	35.45	Chlorine
18	Ar	0	39.95	Argon
19	K	+1	39.10	Potassium
20	Ca	+2	40.08	Calcium
21	Sc	+3	44.96	Scandium
22	Ti	+2 +3 +4	47.88	Titanium
23	V	+2 +3 +4 +5	50.94	Vanadium
24	Cr	+2 +3 +6	52.00	Chromium
25	Mn	+2 +3 +4 +6 +7	54.94	Manganese
26	Fe	+2 +3	55.85	Iron
27	Co	+2 +3	58.93	Cobalt
28	Ni	+2 +3	58.69	Nickel
29	Cu	+1 +2	63.55	Copper
30	Zn	+2	63.39	Zinc
31	Ga	+3	69.72	Gallium
32	Ge	+2 +4	72.59	Germanium
33	As	+3 +5 −3	74.92	Arsenic
34	Se	+4 +6 −2	78.96	Selenium
35	Br	+1 +5 −1	79.90	Bromine
36	Kr	0	83.80	Krypton
37	Rb	+1	85.47	Rubidium
38	Sr	+2	87.62	Strontium
39	Y	+3	88.91	Yttrium
40	Zr	+4	91.22	Zirconium
41	Nb	+4 +5	92.91	Niobium
42	Mo	+6	95.94	Molybdenum
43	Tc	+7	(98)	Technetium
44	Ru	+3 +4 +8	101.07	Ruthenium
45	Rh	+3	102.91	Rhodium
46	Pd	+2 +4	106.42	Palladium
47	Ag	+1	107.87	Silver
48	Cd	+2	112.41	Cadmium
49	In	+3	114.82	Indium
50	Sn	+2 +4	118.71	Tin
51	Sb	+3 +5	121.75	Antimony
52	Te	+4 +6 −2	127.60	Tellurium
53	I	+1 +5 +7 −1	126.91	Iodine
54	Xe	0	131.29	Xenon
55	Cs	+1	132.91	Cesium
56	Ba	+2	137.33	Barium
71	Lu	+3	174.97	Lutetium
72	Hf	+4	178.49	Hafnium
73	Ta	+5	180.95	Tantalum
74	W	+6	183.85	Tungsten
75	Re	+4 +6 +7	186.21	Rhenium
76	Os	+3 +4	190.2	Osmium
77	Ir	+3 +4	192.22	Iridium
78	Pt	+2 +4	195.97	Platinum
79	Au	+1 +3	196.97	Gold
80	Hg	+1 +2	200.59	Mercury
81	Tl	+1 +3	204.38	Thallium
82	Pb	+2 +4	207.2	Lead
83	Bi	+3 +5	208.98	Bismuth
84	Po	+2 +4	(209)	Polonium
85	At	−1	(210)	Astatine
86	Rn	0	(222)	Radon
87	Fr	+1	(223)	Francium
88	Ra	+2	226.03	Radium
103	Lr	+3	(262)	Lawrencium
104	Rf	+4	(267)	Rutherfordium
105	Db		(268)	Dubnium
106	Sg		(271)	Seaborgium
107	Bh		(270)	Bohrium
108	Hs		(277)	Hassium
109	Mt		(276)	Meitnerium
110	Ds		(281)	Darmstadtium
111	Rg		(280)	Roentgenium
112	Cn		(285)	Copernicium
113	Nh		(284)	Nihonium
114	Fl		(289)	Flerovium
115	Mc		(288)	Moscovium
116	Lv		(293)	Livermorium
117	Ts		(294)	Tennessine
118	Og		(294)	Oganesson

* Lanthanide

Atomic No.	Symbol	Oxidation states	Atomic mass	Name
57	La	+3	138	Lanthanum
58	Ce	+3 +4	140.12	Cerium
59	Pr	+3 +4	140.9	Praseodymium
60	Nd	+3	144.24	Neodymium
61	Pm	+3	(145)	Promethium
62	Sm	+2 +3	150.36	Samarium
63	Eu	+2 +3	151.96	Europium
64	Gd	+3	157.25	Gadolinium
65	Tb	+3	158.92	Terbium
66	Dy	+3	162.50	Dysprosium
67	Ho	+3	164.93	Holmium
68	Er	+3	167.26	Erbium
69	Tm	+3	168.93	Thulium
70	Yb	+2 +3	173.04	Ytterbium

◆ Actinide

Atomic No.	Symbol	Oxidation states	Atomic mass	Name
89	Ac	+3	227.02	Actinium
90	Th	+4	232.04	Thorium
91	Pa	+4 +5	231.04	Protactinium
92	U	+3 +4 +5 +6	238.03	Uranium
93	Np	+3 +4 +5 +6	237.05	Neptunium
94	Pu	+3 +4 +5 +6	(244)	Plutonium
95	Am	+3 +4 +5 +6	(243)	Americium
96	Cm	+3	(247)	Curium
97	Bk	+3 +4	(247)	Berkelium
98	Cf	+3	(251)	Californium
99	Es		(252)	Einsteinium
100	Fm		(257)	Fermium
101	Md		(258)	Mendelevium
102	No	+2 +3	(259)	Nobelium

Numbers in parentheses are mass numbers of most stable known isotope of radioactive elements that are rare or not found in nature.

APPENDIX E: SOLAR SYSTEM TABLES

Table E.1 **Planetary Mean Orbits and Symbols**

Planet	Symbol	a (AU)	e	i (deg)	Ω (deg)	ϖ (deg)	λ_m
Mercury	☿	0.3871	0.206	7.005	48.3309	77.4561	252.2509
Venus	♀	0.7233	0.007	3.394	76.6799	131.5637	181.9798
Earth	⊕	1.0000	0.017	0.0	0.0	102.9374	100.4665
Mars	♂	1.5237	0.093	1.850	49.5581	336.6023	355.4333
Jupiter	♃	5.203	0.048	1.303	100.464	14.331	34.351
Saturn	♄	9.543	0.056	2.489	113.666	93.057	50.077
Uranus	♅	19.192	0.046	0.773	74.01	173.01	314.06
Neptune	♆	30.069	0.009	1.770	131.78	48.12	304.35

λ_m is mean longitude. All data are for the J2000 epoch and were taken from Yoder (1995).

Pluto, which was classified as a planet from its discovery in 1930 until 2006, also has an official symbol, ♇. The symbol for Earth's Moon is ☾, and that used for the Sun is ☉.

Table E.2 **Terrestrial Planets: Geophysical Data**

	Mercury	Venus	Earth	Mars
Mean radius R (km)	2440	6051.8	6371.0	3389.9
Mass ($\times 10^{24}$ kg)	0.3302	4.8685	5.9736	0.64185
Density (kg m^{-3})	5427	5204	5515	3933
Flattening ϵ			1/298.257	1/154.409
Semimajor axis			6378.136	3397
Sidereal rotation period	58.6462 d	−243.0185 d	23.934 19 h	24.622 962 h
Mean solar day (in days)	175.9421	116.7490	1	1.027 490 7
Polar gravity (m s^{-2})			9.832 186	3.758
Equatorial gravity (m s^{-2})	3.701	8.870	9.780 327	3.690
Core radius (km)	~1600	~3200	3485	~1700
Obliquity to orbit (deg)	~0.1	177.3	23.45	25.19
Sidereal orbit period (yr)	0.240 844 5	0.615 182 6	0.999 978 6	1.880 711 05
Escape velocity v_e (km s^{-1})	4.435	10.361	11.186	5.027
Geometric albedo	0.106	0.69	0.367	0.150

Venus albedo from https://nssdc.gsfc.nasa.gov/planetary/factsheet/venusfact.html. All other data are from Yoder (1995).

Table E.3 **Giant Planets: Physical Data**

	Jupiter	Saturn	Uranus	Neptune
Mass (10^{24} kg)	1898.6	568.46	86.832	102.43
Density (kg m^{-3})	1326	687.3	1318	1638
Equatorial radius (1 bar) (km)	$71\,492 \pm 4$	$60\,268 \pm 4$	$25\,559 \pm 4$	$24\,766 \pm 15$
Polar radius (km)	$66\,854 \pm 10$	$54\,364 \pm 10$	$24\,973 \pm 20$	$24\,342 \pm 30$
Volumetric mean radius (km)	$69\,911 \pm 6$	$58\,232 \pm 6$	$25\,362 \pm 12$	$24\,624 \pm 21$
Flattening ϵ	0.064 87	0.097 96	0.022 93	0.0171
Sidereal rotation period	$9^h 55^m 29\overset{s}{.}71$	$10^h 32^m 35^s \pm 13$	$-17^h 14^m$	16^h
Hydrostatic flattening[a]	0.065 09	0.098 29	0.019 87	0.018 04
Equatorial gravity (m s^{-2})	23.12 ± 0.01	8.96 ± 0.01	8.69 ± 0.01	11.00 ± 0.05
Polar gravity (m s^{-2})	27.01 ± 0.01	12.14 ± 0.01	9.19 ± 0.02	11.41 ± 0.03
Obliquity (deg)	3.12	26.73	97.86	29.56
Sidereal orbit period (yr)	11.856 523	29.423 519	83.747 407	163.723 21
Escape velocity v_e (km s^{-1})	59.5	35.5	21.3	23.5
Geometric albedo	0.52	0.47	0.51	0.41

Most data are from Yoder (1995); the actual uncertainty in Saturn's rotation period is far larger than the formal error quoted above because different techniques give different answers. Values are from Anderson and Schubert (2007).

[a] Hydrostatic flattening as derived from the gravitational field and magnetic field rotation rate.

Table E.4 **Principal Planetary Satellites: Orbital Data and Visual Magnitude at Opposition**

Planet		Satellite	a (10^3 km)	Orbital Period (days)	e	i (deg)	m_v
Earth		Moon	384.40	27.321 661	0.054 900	5.15[a]	−12.7
Mars	I	Phobos	9.375	0.318 910	0.015 1	1.082	11.4
	II	Deimos	23.458	1.262 441	0.000 24	1.791	12.5
Jupiter	XVI	Metis	127.98	0.294 78	0.001 2	0.02	17.5
	XV	Adrastea	128.98	0.298 26	0.001 8	0.054	18.7
	V	Amalthea	181.37	0.498 18	0.003 1	0.388	14.1
	XIV	Thebe	221.90	0.674 5	0.017 7	1.070	16.0
	I	Io	421.77	1.769 138	0.004 1f	0.040	5.0
	II	Europa	671.08	3.551 810	0.010 1f	0.470	5.3
	III	Ganymede	1 070.4	7.154 553	0.001 5f	0.195	4.6
	IV	Callisto	1 882.8	16.689 018	0.007	0.28	5.6
	XIII	Leda	11 160	241	0.148	27[a]	19.5
	VI	Himalia	11 460	251	0.163	175.3[a]	14.6
	X	Lysithea	11 720	259	0.107	29[a]	18.3
	VII	Elara	11 737	260	0.207	28[a]	16.3
	XII	Ananka	21 280	610	0.169	147[a]	18.8
	XI	Carme	23 400	702	0.207	163[a]	17.6
	VIII	Pasiphae	23 620	708	0.378	148[a]	17.0
	IX	Sinope	23 940	725	0.275	153[a]	18.1
Saturn	XVIII	Pan	133.584	0.575 05	0.000 01	0.000 1	19.4
	XXXV	Daphnis	136.51	0.594 08	0.000 03	0.004	21
	XV	Atlas	137.670	0.601 69	0.001 2	0.01	19.0
	XVI	Prometheus	139.380	0.612 986	0.002 2	0.007	15.8
	XVII	Pandora	141.710	0.628 804	0.004 2	0.051	16.4
	XI	Epimetheus	151.47[b]	0.694 590[b]	0.010	0.35	15.6
	X	Janus	151.47[b]	0.694 590[b]	0.007	0.16	16.4
	I	Mimas	185.52	0.942 421 8	0.020 2	1.53f	12.8
	XXXII	Methone	194.23	1.009 58	0.000	0.02	23
	XLIX	Anthe	197.7	1.037	0.02	0.02	24
	XXXIII	Pallene	212.28	1.153 7	0.004	0.18	22
	II	Enceladus	238.02	1.370 218	0.004 5f	0.02	11.8
	III	Tethys	294.66	1.887 802	0.000 0	1.09f	10.3
	XIV	Calypso (T−)	294.66[b]	1.887 802[b]	0.000 5	1.50	18.7
	XIII	Telesto (T+)	294.66[b]	1.887 802[b]	0.000 2	1.18	18.5
	IV	Dione	377.71	2.736 915	0.002 2f	0.02	10.4
	XII	Helene (T+)	377.71[b]	2.736 915[b]	0.005	0.2	18.4
	XXXIV	Polydeuces (T−)	377.71[b]	2.736 915[b]	0.019	0.18	23
	V	Rhea	527.04	4.517 500	0.001	0.35	9.7
	VI	Titan	1 221.85	15.945 421	0.029 2	0.33	8.4
	VII	Hyperion	1 481.1	21.276 609	0.104 2f	0.43	14.4
	VIII	Iapetus	3 561.3	79.330 183	0.028 3	7.52	11.0[c]
	IX	Phoebe	12 952	550.48	0.164	175.3[a]	16.5
	XX	Paaliaq	15 198	687	0.36	45[a]	21.2
	XXVI	Albiorix	16 394	783	0.48	34[a]	20.4
	XXIX	Siarnaq	18 195	896	0.3	46[a]	20.0

(cont.)

Table E.4 (cont.)

Planet		Satellite	a (10³ km)	Orbital Period (days)	e	i (deg)	m_v
Uranus	VI	Cordelia	49.752	0.335 033	0.000	0.1	24.2
	VII	Ophelia	53.764	0.376 409	0.010	0.1	23.9
	VIII	Bianca	59.166	0.434 577	0.000 3	0.18	23.1
	IX	Cressida	61.767	0.463 570	0.000 2	0.04	22.3
	X	Desdemona	62.658	0.473 651	0.000 3	0.10	22.5
	XI	Juliet	64.358	0.493 066	0.000 1	0.05	21.7
	XII	Portia	66.097	0.513 196	0.000 5	0.03	21.1
	XIII	Rosalind	69.927	0.558 459	0.000 6	0.09	22.5
	XXVII	Cupid	74.393	0.612 825	~0	~0	25.9
	XIV	Belinda	75.256	0.623 525	0.000	0.0	22.1
	XXV	Perdita	76.417	0.638 019	0.003	~0	23.6
	XV	Puck	86.004	0.761 832	0.000 4	0.3	20.6
	XXVI	Mab	97.736	0.922 958	0.002 5	0.13	25.4
	V	Miranda	129.8	1.413	0.002 7	4.22	15.8
	I	Ariel	191.2	2.520	0.003 4	0.31	13.7
	II	Umbriel	266.0	4.144	0.005 0	0.36	14.5
	III	Titania	435.8	8.706	0.002 2	0.10	13.5
	IV	Oberon	582.6	13.463	0.000 8	0.10	13.7
	XVI	Caliban	7 231	580	0.16	141[a]	22.4
	XX	Stephano	8 004	677	0.23	144[a]	24.1
	XVII	Sycorax	12 179	1288	0.52	159[a]	20.8
	XVIII	Prospero	16 256	1978	0.44	152[a]	23.2
	XIX	Setebos	17 418	2225	0.59	158[a]	23.3
Neptune	III	Naiad	48.227	0.294 396	0.00	4.74	24.6
	IV	Thalassa	50.075	0.311 485	0.00	0.21	23.9
	V	Despina	52.526	0.334 655	0.00	0.07	22.5
	VI	Galatea	61.953	0.428 745	0.00	0.05	22.4
	VII	Larissa	73.548	0.554 654	0.00	0.20	22.0
	VIII	Proteus	117.647	1.122 315	0.00	0.55	20.3
	I	Triton	354.76	5.876 854	0.00	156.834	13.5
	II	Nereid	5 513.4	360.136 19	0.751	7.23[a]	19.7
	IX	Halimede	15 686	1875	0.57	134[a]	24.4
	XI	Sao	22 452	2919	0.30	48[a]	25.7
	XII	Laomedeia	22 580	2982	0.48	35[a]	25.3
	XIII	Neso	46 570	8863	0.53	132[a]	24.7
	X	Psamathe	46 738	9136	0.45	137[a]	25.1

Data are from Yoder (1995), with updates from Showalter and Lissauer (2006), Jacobson et al. (2009), Nicholson (2009), Jacobson (2010), http://ssd.jpl.nasa.gov, and other sources.

i, orbit plane inclination with respect to the parent planet's equator, except where noted.

T, Trojan-like satellite, which leads (+) or trails (−) by ~60° in longitude the primary satellite with same semimajor axis. f, forced eccentricity or inclination.

[a] Measured relative to the planet's heliocentric orbit because the Sun (rather than the planetary oblateness) controls the local Laplacian plane of these distant satellites.

[b] Varies because of coorbital libration; value shown is long-term average.

[c] Varies substantially with orbital longitude; average value is shown.

Table E.5 **Planetary Satellites: Physical Properties and Rotation Rates**

Satellite	Radius (km)	Mass (10^{20} kg)	Density (kg m^{-3})	Geom. Albedo	Rot. Period (days)
Earth	$6378^2 \times 6357$	59 742	5515	0.367	0.997
Moon	1737.53 ± 0.03	734.9	3304	0.12	S
Mars	$3396^2 \times 3376$	6419	3933	0.150	1.026
MI Phobos	$13.1 \times 11.1 \times 9.3$	1.063×10^{-4}	1900	0.06	S
MII Deimos	$7.8 \times 6.0 \times 5.1$	1.51×10^{-5}	1500	0.07	S
Jupiter	$71\,492^2 \times 66\,854$	1.8988×10^7	1326	0.52	0.414
JXVI Metis	$30 \times 20 \times 17$			0.06	S
JXV Adrastea	$10 \times 8 \times 7$			0.1	S
JV Amalthea	$125 \times 73 \times 64$			0.09	S
JXIV Thebe	$58 \times 49 \times 42$			0.05	
JI Io	1821.3	893.3	3530	0.61	S
JII Europa	1565	479.7	3020	0.64	S^a
JIII Ganymede	2634	1482	1940	0.42	S
JIV Callisto	2403	1076	1850	0.20	S
JVI Himalia	85	0.042			0.324
JVII Elara	40				0.5
Saturn	$60\,268^2 \times 54\,364$	5.6850×10^6	687	0.47	0.44
SXVIII Pan	$17 \times 16 \times 10$	5×10^{-5}	420 ± 150	0.5	S
SXXXV Daphnis	$4 \times 4 \times 3$	8×10^{-7}	340 ± 260		
SXV Atlas	$20 \times 18 \times 9$	7×10^{-5}	460 ± 110	0.9	S
SXVI Prometheus	$68 \times 40 \times 30$	0.0016	480 ± 90	0.6	S
SXVII Pandora	$52 \times 41 \times 32$	0.00137	490	0.9	S
SXI Epimetheus	$65 \times 57 \times 53$	0.00527	640	0.8	S
SX Janus	$102 \times 93 \times 76$	0.019	630	0.8	S
SI Mimas	$208 \times 197 \times 191$	0.38	1150	0.5	S
SXXXII Methone	1.6 ± 0.6				
SXXXIII Pallene	$3 \times 3 \times 2$				
SII Enceladus	$257 \times 251 \times 248$	0.65	1610	1.0	S
SIII Tethys	$538 \times 528 \times 526$	6.27	985	0.9	S
SXIV Calypso	$15 \times 11.5 \times 7$			0.6	
SXIII Telesto	$16 \times 12 \times 10$			0.5	
SIV Dione	$563 \times 561 \times 560$	11.0	1480	0.7	S
SXII Helene	$22 \times 19 \times 13$			0.7	
SXXXIV Polydeuces	$1.5 \times 1.2 \times 1.0$				
SV Rhea	$765 \times 763 \times 762$	23.1	1240	0.7	S
SVI Titan	2575	1345.7	1880	0.21	~S
SVII Hyperion	$180 \times 133 \times 103$	0.056	540	0.2–0.3	C
SVIII Iapetus	$746 \times 746 \times 712$	18.1	1090	0.05–0.5	S
SIX Phoebe	$109 \times 109 \times 102$	0.083	1640	0.08	0.387
Uranus	$25\,559^2 \times 24\,973$	8.6625×10^5	1318	0.51	0.718
UVI Cordelia	13 ± 2			0.07	
UVII Ophelia	16 ± 2			0.07	
UVIII Bianca	22 ± 3			0.07	
UIX Cressida	33 ± 4			0.07	
UX Desdemona	29 ± 3			0.07	

(cont.)

Table E.5 (*cont.*)

Satellite	Radius (km)	Mass (10^{20} kg)	Density (kg m^{-3})	Geom. Albedo	Rot. Period (days)
UXI Juliet	42 ± 5			0.07	
UXII Portia	55 ± 6			0.07	
UXIII Rosalind	29 ± 4			0.07	
UXIV Belinda	34 ± 4			0.07	
UXV Puck	77 ± 3			0.07	
UV Miranda	$240 \times 234.2 \times 232.9$	0.659	1200	0.27	S
UI Ariel	$581.1 \times 577.9 \times 577.7$	13.53	1670	0.34	S
UII Umbriel	584.7	11.72	1400	0.18	S
UIII Titania	788.9	35.27	1710	0.27	S
UIV Oberon	761.4	30.14	1630	0.24	S
Neptune	$24\,764^2 \times 24\,342$	1.0278×10^6	1638	0.41	0.671
NV Despina	74 ± 10			0.06	
NVI Galatea	79 ± 12			0.06	
NVII Larissa	104×89			0.06	
NVIII Proteus	$218 \times 208 \times 201$			0.06	
NI Triton	1352.6	214.7	2054	0.7	S
NII Nereid	170			0.2	0.48

Most data are from Yoder (1995), with updates from http://ssd.jpl.nasa.gov, Porco et al. (2007), Jacobson et al. (2008), Thomas et al. (1998, 2007) and Thomas (2010), Pilcher et al. (2012).

C, chaotic rotation; S, synchronous rotation.

[a] Europa's ice crust may rotate slightly faster than synchronous.

Table E.6 **Eight Largest Asteroids ($a < 6$ AU)**

#	Name	Tax. Class	M_v	Radius[a] (km)	A_0	a (AU)	e	i (deg)	P_{orb} (yr)	P_{rot} (hr)	Axial Tilt (deg)
1	Ceres	C/G	3.34	467.6	0.09	2.766	0.080	10.59	4.607	9.075	9
4	Vesta	V	3.20	264.5	0.42	2.362	0.090	7.13	3.629	5.342	32
2	Pallas	B	4.13	256	0.16	2.772	0.231	34.88	4.611	7.811	110
10	Hygiea	C	5.43	203.6	0.07	3.137	0.118	3.84	5.56	27.623	126
511	Davida	C	6.22	163	0.05	3.166	0.186	15.94	5.63	5.130	65
704	Interamnia	F	5.94	158.3	0.07	3.062	0.150	17.29	5.36	8.727	60
52	Europa	C	6.31	158	0.06	3.099	0.104	7.48	5.460	5.631	52
87	Sylvia	P/X	6.94	143.0	0.04	3.489	0.080	10.86	6.52	5.184	35

All orbital data are from http://ssd.jpl.nasa.gov/.

[a] Mean radius; most asteroids are substantially nonspherical.

Table E.7 Seven Largest Distant Minor Planets (Known as of 2012; $a > 6$ AU)

#	Name	Provisional Name	Dynamical Class	M_v	Radius[a] (km)	A_0	a (AU)	e	i (deg)	P_{orb} (yr)	P_{rot} (hr)
134340	Pluto		RKBO	-0.7^b	1188.3 ± 0.8	0.5^b	39.482	0.249	17.14	247.7	153.3
136199	Eris	2003 UB$_{313}$	SDO	-1.17	1163 ± 6	0.96	67.728	0.44	43.97	557.5	
136472	Makemake	2005 FY$_9$	RKBO	-0.48	710 ± 30	0.81	45.678	0.16	29.00	308.0	7.77
136108	Haumea	2003 EL$_{61}$	SDO	0.18	675 ± 125	0.84	43.329	0.19	28.21	284.8	3.92
225088		2007 OR$_{10}$	SDO	2.0	640 ± 110	0.19	67.21	0.50	30.7	551.0	
	Charon		moon	1.3	606 ± 0.5	0.375	39.482	0.249	17.14	247.7	153.3
50000	Quaoar	2002 LM$_{60}$	CKBO		555 ± 2.5	0.11	43.616	0.038	7.99	288.1	17.7
90377	Sedna	2003 VB$_{12}$	IOC	1.56	500 ± 40	0.32	489.6	0.84	11.93	10718	10.27

Data from Stansberry et al. (2008), Brown et al. (2010), Sicardy et al. (2011), Pál et al. (2012), Santos-Sanz et al. (2012), Braga-Ribas et al. (2013), Stern et al. (2018), and http://ssd.jpl.nasa.gov/.

CKBO: classical Kuiper belt object; IOC, inner Oort cloud; RKBO, resonant Kuiper belt object; SDO, scattered disk object.

[a] For all bodies, except Pluto/Charon and Eris, the radius and albedo were derived from Spitzer and Herschel (thermal IR) data.

[b] Changes with season between 0.44 and 0.61 as a consequence of variations in ice cover.

Table E.8 Masses, Radii and Densities of Selected Minor Planets

Body	Class[a]	Mass (10^{19} kg)	R (km)	ρ (kg m^{-3})	Method
Near-Earth Asteroids					
433 Eros	S	$6.7 \pm 0.3 \times 10^{-4}$	18.7	2670 ± 30	Orbiting spacecraft
25143 Itokawa	S	$3.5 \pm 0.1 \times 10^{-9}$	0.18 ± 0.01	1900 ± 130	Orbiting spacecraft
Main Belt Asteroids					
1 Ceres	C	94.3 ± 0.7	467.6 ± 2.2	2210 ± 40	Orbit perturbation
2 Pallas	C	23.9 ± 0.6	256 ± 3	3400 ± 900	Orbit perturbation
4 Vesta	V	26.7 ± 0.3	264.5 ± 5	3440 ± 120	Orbit perturbation
10 Hygiea	C	10 ± 4	203.6 ± 3.4	2760 ± 1200	Orbit perturbation
87 Sylvia	P	1.48 ± 0.01	143	1200 ± 100	Multiple system
90 Antiope	C	0.083 ± 0.002	42.9 ± 0.5	1250 ± 50	Binary system
216 Kleopatra	M	0.464 ± 0.002	67.5 ± 1	3600 ± 200	Multiple system
243 Ida	S	0.0042 ± 0.0006	15.7	2600 ± 500	Spacecraft encounter
253 Mathilde	C	0.0103 ± 0.0004	26.5	1300 ± 200	Spacecraft encounter
Trojan Asteroids					
617 Patroclus	P	0.136 ± 0.011	61×56	800 ± 200	Binary system
624 Hektor	D	1.0 ± 0.1	$190 \times 100 \times 100$	1600 ± 300	Binary system
Trans-Neptunian Objects					
20000 Varuna	CKBO		355^{+80}_{-65}	990^{+90}_{-20}	Shape equilibrium
134340 Pluto	RKBO	1303 ± 3	1188.3 ± 0.8	1854 ± 6	Multiple system
Charon	Moon	158.6 ± 1.5	606.0 ± 0.5	1702 ± 17	Multiple system
136108 Haumea	SDO	421 ± 10	$960 \times 770 \times 495$	2600	Shape equilibrium
136199 Eris	SDO	1670 ± 20	1163 ± 6	2520 ± 50	Binary system
50000 Quaoar	CKBO	160 ± 30	445 ± 35	4200 ± 1300	Binary system

Most data from Table 9.5 of de Pater and Lissauer (2010). Haumea characteristics from Lockwood et al. (2014); Pluto & Charon characteristics from Stern et al. (2018).

CKBO, classical Kuiper belt object; RKBO, resonant Kuiper belt object; SDO, scattered disk object.

Table E.9 **Atmospheric Parameters for the Giant Planets**[a]

Parameter	Jupiter	Saturn	Uranus	Neptune
Mean heliocentric distance (AU)	5.203	9.543	19.19	30.07
Geometric albedo (A_0)	0.52	0.47	0.51	0.41
Bond albedo	0.343 ± 0.032	0.342 ± 0.030	0.290 ± 0.051	0.31 ± 0.04
Effective temperature (K)	124.4 ± 0.3	95.0 ± 0.4	59.1 ± 0.3	59.3 ± 0.8
Equilibrium temperature (K)	110	81	58	46
Temperature ($P = 1$ bar) (K)	165.0	134.8	76.4	71.5
Tropopause temperature (K)	111	82	53	52
Mesosphere temperature (K)	160–170	150	140–150	140–150
Exobase temperature (K)	900–1300	800	750	750
Tropopause pressure (mbar)	140	65	110	140
Scale height (at 1 bar) (km)	24	47	25	23
Adiabatic lapse rate (K/km)	2.1	0.9	1.0	1.3
Energy balance[b]	1.63 ± 0.08	1.87 ± 0.09	1.05 ± 0.07	2.68 ± 0.21

[a] All values from Table 4.1 in de Pater and Lissauer (2010). References are provided therein.
[b] Ratio (energy radiated into space)/(solar energy absorbed).

Table E.10 **Atmospheric Parameters for Venus, Earth, Mars and Titan**[a]

Parameter	Venus	Earth	Mars	Titan
Mean heliocentric distance (AU)	0.723	1.000	1.524	9.543
Geometric albedo A_0	0.69	0.367	0.15	0.21
Bond albedo	0.77	0.306	0.25	0.20
Surface temperature (K)	737	288	215	93.7
Equilibrium temperature (K)	232	255	210	85
Exobase temperature (K)	270–320	800–1250	200–300	149
Surface pressure (bar)	92	1.013	0.00636	1.47
Scale height at surface (km)	16	8.5	11	20
Adiabatic lapse rate (K/km)	10.4	9.8	4.4	1.4

[a] Venus albedos from https://nssdc.gsfc.nasa.gov/planetary/factsheet/venusfact.html. All other values from Table 4.2 in de Pater and Lissauer (2010). References are provided therein.

Table E.11 **Atmospheric Parameters for Mercury, the Moon, Triton and Pluto**[a]

Parameter	Mercury	Moon	Triton	Pluto
Mean heliocentric distance (AU)	0.387	1.000	30.069	39.48
Geometric albedo A_0	0.138	0.113	0.76	0.08–1.0
Bond albedo	0.119	0.123	0.85	0.72
Surface temperature (K)	100–725	277	38	~40–60
Equilibrium temperature (K)	434	270	32	39
Exobase temperature (K)	600	270–320	100	58
Surface pressure (bar)	few $\times 10^{-15}$	3×10^{-15}	1.4×10^{-5}	1.5×10^{-5}
Scale height at surface (km)	13–95	65	14	33

[a] Most values from Table 4.3 in de Pater and Lissauer (2010). References are provided therein. Some parameters for Pluto from Buratti et al. (2017).

Table E.12 **Atmospheric Composition of Earth, Venus, Mars and Titan**[a]

Constituent	Earth	Venus	Mars	Titan
N_2	0.7808	0.035	0.027	~0.95
O_2	0.2095	0–20 ppm	0.0013	
CO_2	400 ppm	0.965	0.953	10 ppb
CH_4	2 ppm		10–250 ppb	0.049
H_2O	<0.03	30 ppm	<100 ppm	0.4 ppb
Ar	0.009	70 ppm	0.016	28 ppm
CO	0.2 ppm	20 ppm	700 ppm	45 ppm
O_3	~10 ppm		0.01 ppm	
SO_2	<2 ppb	100 ppm		

[a] From Table 4.4 of de Pater and Lissauer (2010), apart from Earth's CO_2, for which the 2013 value is given. References and details are given therein. All numbers are volume mixing ratios. ppb, parts per billion; ppm, parts per million.

Table E.13 **Atmospheric Composition of the Sun and the Giant Planets**[a]

Gas	Element[b]	Protosolar	Jupiter	Saturn	Uranus	Neptune
H_2	H	0.835	0.864	0.88	~0.83	~0.82
He	He	0.162	0.136	0.119	~0.15	~0.15
H_2O	O	8.56×10^{-4}	$>4.2 \times 10^{-4}$			
CH_4	C	4.60×10^{-4}	2.0×10^{-3}	4.5×10^{-3}	0.023	0.03
NH_3	N	1.13×10^{-4}	7×10^{-4}	5×10^{-4}		
H_2S	S	2.59×10^{-5}	7.7×10^{-5}			

[a] From Table 4.5 of de Pater and Lissauer (2010). References and details are given therein. All numbers are volume mixing ratios (i.e., mole fractions).
[b] The elements O, C and N are in the form of H_2O, CH_4 and NH_3 on the giant planets, respectively.

Table E.14 **Densities and Central Properties of the Planets and the Moon**[a]

Planet	Radius (Equatorial) (km)	Density (kg m^{-3})	Uncompressed Density (kg m^{-3})	Central Pressure (Mbar)	Central Temperature (K)
Mercury	2 440	5 427	5 300	~0.4	~2 000
Venus	6 052	5 204	4 300	~3	~5 000
Earth	6 378	5 515	4 400	3.6	6 000
Moon	1 738	3 340	3 300	0.045	~1 800
Mars	3 396	3 933	3 740	~0.4	~2 000
Jupiter	71 492	1 326		~80	~20 000
Saturn	60 268	687		~50	~10 000
Uranus	25 559	1 318		~20	~7 000
Neptune	24 766	1 638		~20	~7 000

[a] Data from Hubbard (1984), Lewis (1995), Hood and Jones (2000), Guillot (1999) and Yoder (1995).

Table E.15 **Gravitational Moments and Moment of Inertia Ratios**

Body	J_2 ($\times 10^{-6}$)	J_3 ($\times 10^{-6}$)	J_4 ($\times 10^{-6}$)	J_6 ($\times 10^{-6}$)	I/MR^2	References
Sun					0.059	
Mercury	22.5 ± 0.1		6.5 ± 0.8		0.353	5
Venus	4.46 ± 0.03	-1.93 ± 0.02	-2.38 ± 0.02		0.33	1
Earth	$1\,082.627$	-2.532 ± 0.002	-1.620 ± 0.003	-0.21	0.331	1
Moon	203.43 ± 0.09				0.393	1, 2
Mars	$1\,960.5 \pm 0.2$	31.5 ± 0.5	-15.5 ± 0.7		0.365	1
Jupiter	14696.57	-0.042 ± 0.010	-586.01	34.12	0.254	1, 6
Saturn	$16\,290.7 \pm 0.3$		-936 ± 3	86 ± 9	0.210	4
Uranus	$3\,343.5 \pm 0.1$		-28.9 ± 0.2		0.23	1
Neptune	$3\,410 \pm 9$		-35 ± 10		0.23	1
Io	$1\,860 \pm 3$				0.378	3
Europa	436 ± 8				0.346	3
Ganymede	128 ± 3				0.312	3
Callisto	33 ± 1				0.355	3

1: Yoder (1995) and http://ssd.jpl.nasa.gov/. 2: Konopliv et al. (1998). 3: Schubert et al. (2004). 4: Anderson and Schubert (2007). 5: Smith et al. (2012). 6: Iess et al. (2018).

Table E.16 **Heat-Flow Parameters**[a]

Body	T_e (K)	T_{eq} (K)	H_i (J m^{-2} s^{-1})	L/M (J kg^{-1} s^{-1})
Sun	5770		6.2×10^7	1.9×10^{-4}
Carbonaceous chondrites				4×10^{-12}
Earth		263	0.075	6.4×10^{-12}
Mars		222	0.04	9×10^{-12}
Jupiter	124.4	113	5.44	1.8×10^{-10}
Saturn	95.0	83	2.01	1.5×10^{-10}
Neptune	59.3	48	0.433	3.2×10^{-11}

[a] All values from Table 6.3 in de Pater and Lissauer (2010). References are provided therein.

Table E.17 **Solar Wind Properties at 1 AU**[a]

	Most Probable Value	5–95% Range
Density (protons cm^{-3})	5	3–20
Velocity (km s^{-1})	375	320–710
Magnetic field (γ)	5.1	2.2–9.9
Electron temp. (10^5 K)	1.2	0.9–2
Proton temp. (10^5 K)	0.5	0.1–3
Sound speed (km s^{-1})	59	41–91
Alfvén speed (km s^{-1})	50	30–100

[a] After Gosling (2007).

Table E.18 **Characteristics of Planetary Magnetic Fields**[a]

	Mercury	Earth	Jupiter	Saturn	Uranus	Neptune
Magnetic moment ($\mathcal{M}_\oplus$)	2.4×10^{-4}	1[b]	20 000	600	50	25
Surface B at dipole equator (nT)	190	3.1×10^4	4.28×10^5	2.2×10^4	2.3×10^4	1.4×10^4
Maximum/minimum[c]	2	2.8	4.5	4.6	12	9
Dipole tilt[d]	<0.8°	10.8°	9.6°	0.0°	59°	47°
Dipole offset (R_p)	0.16	0.08	0.12	~0.04	0.3	0.55
Magnetopause distance[e] (R_p)	1.45	10	42	19	25	24

[a] After Kivelson and Bagenal (2007), with updates for Mercury based on *MESSENGER* data.

[b] $\mathcal{M}_\oplus = 7.906 \times 10^{22}$ N m T^{-1}.

[c] Ratio of maximum to minimum surface magnetic field strength (equal to 2 for a centered dipole field).

[d] Angle between the magnetic and rotation axis.

[e] Typical standoff distance of the magnetopause at the nose of the magnetosphere, in planetary radii.

APPENDIX F: INTERPLANETARY SPACECRAFT

A substantial fraction of our data on many Solar System objects has been obtained by close-up studies conducted by spacecraft. Figure F.1 shows humankind's first view of the far side of the Moon taken by the Soviet spacecraft *Luna 3*. This Appendix starts with a short section on **rocketry** (how a rocket works). Section F.2 contains tables listing many of the most significant lunar and interplanetary spacecraft and astronomical observations in space.

F.1 Rocketry

The principles of 'rocket science' are actually quite simple, although many practical aspects of 'rocket engineering' are far more complicated. A rocket accelerates by expelling gas (or plasma) at high velocity. Conservation of momentum implies that the velocity, v, of a rocket of mass M (which includes propellant), expelling gas at velocity v_{exp} and rate dM/dt, satisfies:

$$M\frac{d\mathbf{v}}{dt} = -\mathbf{v}_{exp}\frac{dM}{dt} + \mathbf{F}_{ext},\qquad(\text{F.1})$$

where $\mathbf{F}_{ext}$ accounts for all external forces on the rocket. Equation (F.1) is known as the fundamental **rocket equation**.

In a uniform gravitational field that induces an acceleration $\mathbf{g}_p$ with no other external forces, the rocket equation reduces to

$$\frac{d\mathbf{v}}{dt} = -\frac{\mathbf{v}_{exp}}{M}\frac{dM}{dt} + \mathbf{g}_p.\qquad(\text{F.2})$$

Integrating equation (F.2) and setting $v = 0$ at $t = 0$ gives

$$\mathbf{v} = -\mathbf{v}_{exp}\ln\frac{M_0}{M} - \mathbf{g}_p t,\qquad(\text{F.3})$$

where M_0 is the mass at $t = 0$, and there is a minus sign in front of the last term in equation (F.3) because the gravitational force is directed downwards. Note that there is a premium to burning fuel rapidly – the shorter the burn time, the greater the velocity for given ejection speed and mass. This is why high-thrust rocket engines are used to attain escape velocity from Earth; at present, such large thrusts can only be obtained using **chemical propulsion**. Figure F.2 illustrates the principal components of a chemical propulsion rocket engine.

For acceleration in free space, the final velocity of the rocket does not depend on the ejection

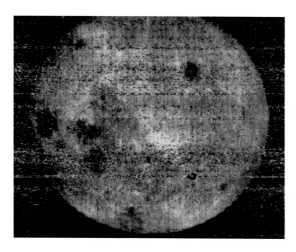

Figure F.1 This fuzzy image taken on 7 October 1959 provided humanity with its first view of the lunar far side.

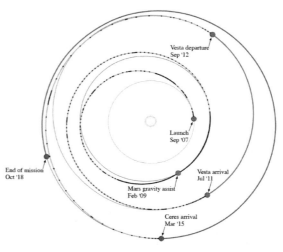

Figure F.3 *Dawn*'s interplanetary trajectory. The trajectory is light blue where the spacecraft was thrusting and black where it coasted. Thrusting in orbit around Vesta and Ceres is not shown. The regular interruptions in thrust were for conducting activities incompatible with optimal thrusting, usually pointing the high gain antenna to Earth. (Courtesy Marc D. Rayman)

rate. Thus, **electric propulsion**, which can achieve higher expulsion speeds (specific impulse) than chemical rockets, can be very efficient at modifying trajectories of orbiting bodies. Figure F.3 illustrates the trajectory of the *Dawn* spacecraft, which used electric propulsion to orbit the two largest asteroids, 4 Vesta and 1 Ceres. Results of the *Dawn* mission are presented in §§12.5.2 and G.11.3.

F.2 Tabulations

We list significant lunar missions in Table F.1 and interplanetary spacecraft in Table F.2. Table F.3 provides a list of space observatories. A more complete listing of spacecraft observation of Solar System objects is provided in the Appendix of the *Encyclopedia of the Solar System* (2007).

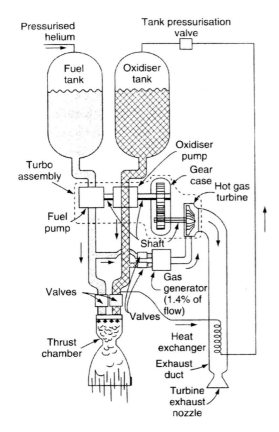

Figure F.2 Schematic of a chemical propulsion rocket engine.

Table F.1 **Selected Lunar Spacecraft**

Spacecraft	Sender	Launch Date	Type	Remarks
Luna 3	USSR	1959 Oct. 4	Flyby	Photographed far side of the Moon
Ranger 7	USA	1964 July 28	Impact	Returned 4308 photos
Ranger 8	USA	1965 Feb. 17	Impact	Returned 7137 photos
Ranger 9	USA	1965 Mar. 21	Impact	Returned 5814 photos
Luna 9	USSR	1966 Jan. 31	Lander	Soft landing; returned photos
Luna 10	USSR	1966 Mar. 31	Orbiter	First lunar orbiter
Surveyor 1	USA	1966 May 30	Lander	Soft landing; returned 11 150 photos
Lunar Orbiter 1	USA	1966 Aug. 10	Orbiter	Photographic mapping
Luna 11	USSR	1966 Aug. 24	Orbiter	Science return
Luna 12	USSR	1966 Oct. 22	Orbiter	Photographic mapping
Lunar Orbiter 2	USA	1966 Nov. 6	Orbiter	Photographic mapping
Luna 13	USSR	1966 Dec. 21	Lander	Surface science
Lunar Orbiter 3	USA	1967 Feb. 4	Orbiter	Photographic mapping
Surveyor 3	USA	1967 Apr. 17	Lander	Surface science
Lunar Orbiter 4	USA	1967 May 4	Orbiter	Photographic mapping
Lunar Orbiter 5	USA	1967 Aug. 1	Orbiter	Photographic mapping
Surveyor 5	USA	1967 Sep. 8	Lander	Surface science
Surveyor 6	USA	1967 Nov. 7	Lander	Surface science
Surveyor 7	USA	1968 Jan. 8	Lander	Surface science
Luna 14	USSR	1968 Apr. 7	Orbiter	Mapped gravity field
Apollo 8	USA	1968 Dec. 21	Manned orbiter	First humans in deep space
Apollo 10	USA	1969 May 18	Manned orbiter	Two-spacecraft undocking, docking
Apollo 11	USA	1969 July 16	Manned lander	First humans on Moon; 22-kg sample
Apollo 12	USA	1969 Nov. 14	Manned lander	34-kg sample return
Luna 16	USSR	1970 Sep. 12	Sample return	First robotic sample return (101 g)
Luna 17	USSR	1970 Nov. 10	Rover	First lunar rover
Apollo 14	USA	1971 Jan. 31	Manned lander	42-kg sample return
Apollo 15	USA	1971 July 26	Manned rover	77-kg sample return
Luna 19	USSR	1971 Sep. 28	Orbiter	Photographic mapping
Luna 20	USSR	1972 Feb. 12	Sample return	Returned 55-g sample
Apollo 16	USA	1972 Apr. 16	Manned rover	95-kg sample return
Apollo 17	USA	1972 Dec. 10	Manned rover	111-kg sample return
Luna 21	USSR	1973 Jan. 8	Rover	Lunokhod 2; traversed 39 km
Luna 22	USSR	1974 May 29	Orbiter	Photographic mapping
Luna 24	USSR	1976 Aug. 9	Sample return	Returned 170-g sample
Clementine	USA	1994 Jan. 25	Orbiter	Photographic mapping
Lunar Prospector	USA	1998 Jan. 6	Orbiter	Photographic mapping (plus impact)
SMART 1	ESA	2003 Sep. 27	Orbiter	Photographic mapping (plus impact)
SELENE/Kaguya	Japan	2007 Sep. 14	Orbiter	Photographic mapping
Chang'e 1	China	2007 Oct. 24	Orbiter	Photographic mapping
Chandrayaan 1	India	2008 Oct. 22	Orbiter	Photographic mapping and radar
LRO	USA	2009 June 18	Orbiter	Photographic mapping
LCROSS	USA	2009 June 18	Impactor	Impacted 9 October 2009
Chang'e 2	China	2010 Oct. 1	Orbiter	Photographic mapping
GRAIL	USA	2011 Sep. 10	Two orbiters	Gravity mapping
LADEE	USA	2013 Sep. 7	Orbiter	Atmosphere and dust
Chang'e 3/Yutu	China	2013 Dec. 1	Lander+rover	Explored Mare Imbrium
Queqiao	China	2018 May 20	Comms. relay	Near Earth-Moon L_2
Chang'e 4/Yutu 2	China	2018 Dec. 7	Lander+rover	First soft landing on farside

Table F.2 **Selected Interplanetary Spacecraft**

Spacecraft	Sender	Launch Date	Target	Type	Remarks
Mariner 2	USA	1962 Aug. 27	Venus	Flyby	First close-up data from Venus
Mariner 4	USA	1964 Nov. 28	Mars	Flyby	First 21 close-up photos of Mars
Venera 4	USSR	1967 June 12	Venus	Probe	Atmospheric measurements
Mariner 5	USA	1967 June 14	Venus	Flyby	Closest approach 3990 km
Venera 5	USSR	1969 Jan. 5	Venus	Probe	Operated for 53 min
Venera 6	USSR	1969 Jan. 10	Venus	Probe	Operated for 51 min
Mariner 6	USA	1969 Feb. 24	Mars	Flyby	Returned 75 photos
Mariner 7	USA	1969 Mar. 27	Mars	Flyby	Returned 125 photos
Venera 7	USSR	1970 Aug. 17	Venus	Lander	First soft landing
Mars 2	USSR	1971 May 19	Mars	Orbiter+lander	Orbiter succeeded; lander failed
Mars 3	USSR	1971 May 28	Mars	Orbiter+lander	Lander failed after 20 seconds
Mariner 9	USA	1971 May 30	Mars	Orbiter	Returned many photos
Pioneer 10	USA	1972 Mar. 3	Jupiter	Flyby	First Jupiter flyby, 1973 Dec. 3
Venera 8	USSR	1972 Mar. 27	Venus	Lander	Landed 1972 July 22
Pioneer 11	USA	1973 Apr. 6	Jupiter	Flyby	Closest approach 1974 Dec. 4
			Saturn	Flyby	First Saturn flyby, 1979 Sep. 1
Mars 5	USSR	1973 July 25	Mars	Orbiter	Orbited 1974 Feb. 12
Mariner 10	USA	1973 Nov. 3	Venus	Flyby	Closest approach 1974 Feb. 5
			Mercury	Flyby	Three flybys in 1974–1975
Venera 9	USSR	1975 June 8	Venus	Orbiter+lander	First images of surface
Venera 10	USSR	1975 June 14	Venus	Orbiter+lander	Landed 1975 Oct. 25
Viking 1	USA	1975 Aug. 20	Mars	Orbiter+lander	First long-term surface science
Viking 2	USA	1975 Sep. 9	Mars	Orbiter+lander	Landed 1976 Sep. 3
Voyager 2	USA	1977 Aug. 20	Jupiter	Flyby	Closest approach 1979 July 9
			Saturn	Flyby	Closest approach 1981 Aug. 26
			Uranus	Flyby	First Uranus flyby, 1986 Jan. 24
			Neptune	Flyby	First Neptune flyby, 1989 Aug. 24
Voyager 1	USA	1977 Sep. 5	Jupiter	Flyby	Closest approach 1979 Mar. 5
			Saturn	Flyby	Closest approach 1980 Nov. 12
Pioneer 12	USA	1978 May 20	Venus	Orbiter	Entered orbit 1978 Dec. 8
Pioneer 13	USA	1978 Aug. 8	Venus	Probe	Four probes
Venera 11	USSR	1978 Sep. 9	Venus	Lander	Landed 1978 Dec. 25
Venera 12	USSR	1978 Sep. 14	Venus	Lander	Landed 1978 Dec. 21
Venera 13	USSR	1981 Oct. 30	Venus	Lander	Landed 1982 Feb. 27
Venera 14	USSR	1981 Nov. 4	Venus	Lander	Landed 1982 Mar. 5
Venera 15	USSR	1983 June 2	Venus	Orbiter	Entered orbit 10 October 1983
Venera 16	USSR	1983 June 7	Venus	Orbiter	Entered orbit 11 October 1983
Vega 1	USSR	1984 Dec. 15	Venus	Balloon+lander	First Venus balloon
			1P/Halley	Flyby	First comet nucleus images
Vega 2	USSR	1984 Dec. 21	Venus	Balloon+lander	Landed 1985 June 15
			1P/Halley	Flyby	Closest approach 1986 Mar. 9
Sakigake	Japan	1985 Jan. 8	1P/Halley	Flyby	Closest approach 1986 Mar. 11
Suisei	Japan	1985 Aug. 18	1P/Halley	Flyby	UV imaging of H corona
Giotto	ESA	1985 July 2	1P/Halley	Flyby	Imaging and composition
Phobos 2	USSR	1988 July 12	Phobos	Lander	Mars + Phobos images; failed before landing

Table F.2 (*cont.*)

Spacecraft	Sender	Launch Date	Target	Type	Remarks
Magellan	USA	1989 May 5	Venus	Orbiter	Global radar mapper
Ulysses	ESA/USA	1990 Oct. 6	Jupiter	Flybys	Primary mission: Sun and solar wind
Galileo	USA	1989 Oct. 18	951 Gaspra	Flyby	First asteroid flyby, 1991 Oct. 29
			243 Ida	Flyby	Discovered Dactyl, first asteroid moon
			Jupiter	Orbiter+probe	First Jupiter probe, arrived 1995 Dec. 7
NEAR	USA	1996 Feb. 17	Mathilde	Flyby	Closest approach, 1997 June 2
			433 Eros	Orbiter	First asteroid orbiter; orbited 2000 Feb. 14
MGS	USA	1996 Nov. 7	Mars	Orbiter	Entered orbit 1997 Sep. 12
Pathfinder	USA	1996 Dec. 2	Mars	Lander+rover	First Mars rover
Cassini	USA	1997 Oct. 15	Jupiter	Flyby	Closest approach, 2000 Dec. 30
			Saturn	Orbiter	Entered orbit 2004 July 1
Huygens	ESA	1997 Oct. 15	Titan	Probe/lander	Travelled with Cassini
Deep Space 1	USA	1998 Oct. 24	9969 Braille	Flyby	Closest approach 1999 July 29
			19P/Borrelly	Flyby	Closest approach 2001 Sep. 22
Stardust	USA	1999 Feb. 6	5535 Annefrank	Flyby	Closest approach 2002 Nov. 2
			81P/Wild 2	Sample return	Flyby science 2004 Jan. 2
			9P/Temple 1	Flyby	Closest approach 2011 Feb. 14
Mars Odyssey	USA	2001 Apr. 7	Mars	Orbiter	Entered orbit 2001 Oct. 23
Hayabusa	Japan	2003 May 9	25143 Itokawa	Orbiter	Also returned a small sample
Mars Express	ESA	2003 June 2	Mars	Orbiter	Entered orbit 2003 Dec. 25
Spirit	USA	2003 June 10	Mars	Orbiter	Mars Exploration Rover I
Opportunity	USA	2003 July 7	Mars	Orbiter	Mars Exploration Rover II
Rosetta	ESA	2004 Mar. 2	2867 Šteins	Flyby	Closet approach 2008 Sep. 5
			21 Lutetia	Flyby	Closest approach 2010 July 10
			67P/C–G	Orbiter+Lander	67P/Churyumov–Gerasimenko
MESSENGER	USA	2004 Aug. 3	Mercury	Orbiter	Entered orbit 2011 March 17
Deep Impact	USA	2005 Jan. 12	9P/Temple 1	Flyby+impactor	Impacted 2005 July 4
			103P/Hartley 2	Flyby	Mission renamed EPOXI
MRO	USA	2005 Aug. 12	Mars	Orbiter	Entered orbit 2006 March 10
Venus Express	ESA	2005 Nov. 9	Venus	Orbiter	Entered orbit 2006 April 11
New Horizons	USA	2006 Jan. 19	Jupiter	Flyby	Perijove 2007 Feb. 28
			Pluto	Flyby	Closest approach 2015 July 14
			485968 Ultima Thule	Flyby	Closest approach 2019 Jan. 1
Phoenix Mars	USA	2007 Aug. 4	Mars	Lander	Explored polar region
Dawn	USA	2007 Sep. 27	4 Vesta	Orbiter	Orbited 2011–2012
			1 Ceres	Orbiter	Orbited 2015–2018
Akatsuki	Japan	2010 May 20	Venus	Orbiter	Atmospheric studies
Chang'e 2	China	2010 Oct. 1	4179 Toutatis	Flyby	After lunar mission
Juno	USA	2011 Aug. 5	Jupiter	Orbiter	Focus on gravity field and H_2O
Curiosity	USA	2011 Nov. 26	Mars	Rover	Mars Science Laboratory
Mangalyaan	India	2013 Nov. 5	Mars	Orbiter	Very low cost
Maven	USA	2013 Nov. 18	Mars	Orbiter	Atmospheric studies
Hayabusa 2	Japan	2014 Dec. 3	162173 Ryuku	Lander	Goal is to return sample
OSIRIS-REx	USA	2016 Sep. 8	101955 Bennu	Lander	Goal is to return sample
ExoMars TGO	ESA/Russia	2016 Oct. 19	Mars	Orbiter	Atmospheric trace gas analysis
InSight	USA	2018 May 5	Mars	Lander	Study planet's interior

Table F.3 **Selected Space Observatories**

Spacecraft	Sender	Launch Date	Orbit	Remarks
Explorer 1	USA	1958 Jan. 31	Low Earth	Discovered Van Allen Belts
IUE	USA/ESA	1978 Jan. 26	Low Earth	International Ultraviolet Explorer
ISEE 3/ICE	USA	1978 Aug. 12	Heliocentric	Monitored solar wind. Then flew through tail of 21P/Giacobini–Zinner
SMM	USA	1980 Feb. 14	Low Earth	Solar observatory
IRAS	USA/UK/NL	1983 Jan. 25	Low Earth	Infrared Astronomical Satellite
HST	USA/ESA	1990 Apr. 24	Low Earth	Hubble Space Telescope
ROSAT	Germany	1990 June 1	Low Earth	X-ray observatory
Ulysses	ESA/USA	1990 Oct. 6	Solar polar orbit	Jupiter flyby – observed B-field
EUVE	USA	1992 June 7	Low Earth	Extreme Ultraviolet Explorer
ISO	ESA	1995 Nov. 17	Low Earth	Infrared Space Observatory
SOHO	ESA/USA	1995 Dec. 2	Sun–Earth L_1	Solar observatory; discovered many Sun-grazing comets
Spitzer	USA	2003 Aug. 25	Heliocentric	Infrared telescope
CoRoT	France/ESA	2006 Dec. 27	Earth polar	Search for transiting exoplanets
Kepler	USA	2009 Mar. 6	Heliocentric	Search for transiting exoplanets
Herschel	ESA	2009 May 14	Sun–Earth L_2	Far-IR and sub-mm
WISE	USA	2009 Dec. 14	Earth polar	IR survey; discovered many NEOs
Van Allen Probes	USA	2012 Aug. 30	Earth eccentric	Two spacecraft studying radiation belts
GAIA	ESA	2013 Dec. 19	Sun–Earth L_2	Stellar astrometry
TESS	USA	2018 Apr. 18	Earth eccentric	Exoplanet transits

APPENDIX G: RECENT ADVANCES IN SOLAR SYSTEM STUDIES

G.1 Introduction

Planetary science is an active research field, and our knowledge of the planets and smaller bodies in the Solar System is increasing very rapidly. Thus, no compendium on this subject can be completely up to date. To give the student a flavor of this progress, we present in this Appendix new material to update the book. We focus on discoveries from the past few years, and provide both summaries and images obtained with ground-based telescopes and by spacecraft. Sections G.2–G.9 cover the planets and their satellites, and are arranged by object in order of increasing heliocentric distance. Sections G.10–G.13 include information on meteorites, minor planets, comets and an ~100 m radius interstellar visitor, respectively. Since exoplanets is a young and rapidly expanding field, Chapter 14 has been rewritten to incorporate knowledge acquired through mid-2018. The last two sections in this Appendix discuss a few recent results pertaining to planet formation and the history of life on Earth.

Good web sources to view recent planetary images include:

http://www.nineplanets.com/

http://www.nasa.gov/topics/solarsystem/index.html
http://hubblesite.org/gallery/album/solar_system
http://photojournal.jpl.nasa.gov/

G.2 Mercury

The *MESSENGER* spacecraft orbited Mercury from 2011 until 30 April 2015, at which time it made a controlled impact on the planet. Several of its findings were reported in §9.2. More recently, the spacecraft imaged small thrust fault scarps on Mercury's surface, with only tens of meters relief, extending several km in length. The paucity of impact craters on these faults, which crosscut older impact craters, suggests that they formed less than 50 Myr ago. These scarps are likely the smallest members of a continuum in scale of thrust fault scarps on Mercury that formed as the planet cooled and contracted. Their young age, together with evidence for recent activity on larger-scale scarps that formed over geological time, suggests a prolonged slow cooling of the planet's interior that is continuing today.

Figure G.1a shows that heavily cratered terrain (HCT) covers most of Mercury's surface. The northern smooth plains (NSP) are characterized by

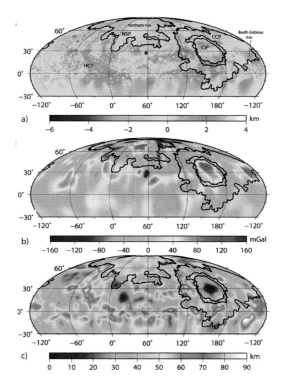

a)

b)

c)

Figure G.1 COLOR PLATE a) Topography, b) gravity field and c) crustal thickness of Mercury as determined by *MESSENGER*. See text for abbreviations. (Adapted from James et al. 2015)

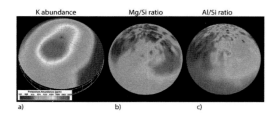

a)   b)   c)

Figure G.2 COLOR PLATE a) A map of the potassium abundance on Mercury, as measured with the Gamma-Ray Spectrometer (GRS) on *MESSENGER*. This region is also characterized by low Mg/Si, Ca/Si, S/Si and high Na/Si and Cl/Si abundances. b) A map of Mg/Si and c) a map of Al/Si on Mercury, obtained with *MESSENGER*'s X-Ray Spectrometer (XRS) and Mercury Dual Imaging System (MDIS). In both panels red indicates high values and blue low values. In both maps, the Caloris basin is located in the upper left, showing a low Mg/Si and high Al/Si ratio. An extensive region with high Mg/Si is also clearly visible in the maps but is not correlated with any visible impact basin. (NASA/Johns Hopkins University Applied Physics Laboratory/Carnegie Institution of Washington)

smooth volcanic terrain, with a 1.5 km high rise in the north (northern rise). The gravity map shown in Figure G.1b reveals several large-scale gravitational anomalies, attributed to mass anomalies in the subsurface structure. A combined inversion of the gravity and topography data results in the crustal thickness map in Figure G.1c. Mercury's crust (§6.2.1) is on average at least 38 km thick. Surface topography is supported through a combination of variations in crustal thickness and the deep (300–400 km depth) mass anomalies referred to above.

The Caloris basin is by far the largest impact basin on Mercury. Its interior consists of smooth, volcanically resurfaced, plains (CIP), and the basin is surrounded by extensive smooth plains (CCP). The center of the Caloris basin has an extremely thin crust, while the broad basin is likely underlain by a large negative mass anomaly.

Mercury is far less depleted in moderately volatile elements than is the Moon, since large abundances of potassium, sodium and chlorine are present, especially at high northern latitudes, as illustrated in Figure G.2. This high-potassium region is also distinct in its low Mg/Si, Ca/Si, S/Si and high Na/Si and Cl/Si abundances. The Caloris basin is characterized as a region with low Mg/Si and high Ca/Si and Al/Si. These deposits are younger than the heavily cratered terrains, and appear to be different in composition. The older terrains generally have higher Mg/Si, S/Si and Ca/Si ratios, and a lower Al/Si ratio. This compositional difference suggests crystallization of the (relatively young) smooth plains from a more chemically evolved magma source (§6.1.4).

G.3 Venus

As discussed in §9.3, Venus's surface is covered by numerous volcanic constructs. Based upon the random and rather sparse distribution of impact craters, it has been postulated that Venus may have undergone a global resurfacing event, perhaps only a few hundred million years ago. This hypothesis, together with the observation by the *Pioneer Venus* orbiter of a high SO_2 gas content in the venusian atmosphere in 1978, followed by a gradual decrease, suggests that Venus may still be volcanically active. Just after arrival at Venus in 2006, the *Venus Express* spacecraft observed another factor of $\sim$3 increase in the SO_2 abundance, followed by a gradual decrease, perhaps again caused by gases released by a large volcanic event.

Even more intriguing are observations of thermal emissions from potential hot spots. At a wavelength of 1 μm, the venusian atmosphere is partially transparent, and the planet's surface can be imaged. At three hot spots, anomalously high values in thermal emissivity were measured, interpreted as being caused by relatively recent volcanism (perhaps a few thousands to a few tens of thousands of years ago). In addition, a few bright, transient features were detected, including the apparent hot spot in the Ganiki Chasma region, shown in Figure G.3.

G.4 The Moon

Recent Craters

In addition to obtaining detailed topographic maps of the Moon, the multi-year, high-resolution, imaging by the *Lunar Reconnaissance Orbiter (LRO)* has also obtained before and after pictures of recent cratering events. In some cases, such as the pair of images shown in Figure G.4, the new crater can be associated with an impact flash observed from Earth. *LRO* has detected several hundred new impact craters with sizes > 10 m, which is roughly 1/3 more than expected based upon the standard production and chronology functions for the Moon (§6.4.4).

The Chinese spacecraft *Chang'e 4* touched down in the South Pole Aiken basin on the far side of the Moon (Fig. 9.2b) on 2 January 2019. To relay data to Earth, China launched a relay satellite, *Queqiao*, which orbits near the Earth–Moon L_2 Lagrangian point (§2.2.1). The stationary mother craft and the small rover *Yutu 2* are using several instruments to characterize their surroundings in great detail.

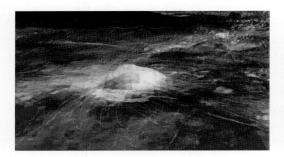

Figure G.3 Volcanic peak Idunn Mons in the Imdr Regio area of Venus. The topography is derived from *MAGELLAN* radar data, with a 30 times vertical exaggeration. The colored overlay shows the heat patterns derived from surface brightness data obtained by the *Venus Express* spacecraft. Temperature variations due to topography were removed. Red-orange is the warmest area, centered at the 2.5 km high summit, and purple is the coolest. Idunn Mons has a diameter of about 200 km. (ESA/NASA/JPL/S. Smrekar)

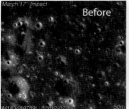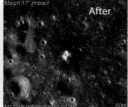

Figure G.4 Before and after images of the region centered on an 18 m diameter lunar crater that formed on 17 March 2013. Both the crater and its rays are much brighter than the surrounding terrain in Mare Imbrium. The scale bars at the lower right represent 50 m. The associated movie shows the flash from the impact of the ~40 kg meteoroid that produced the crater. (NASA/ASU/LROC) (Movie can be viewed at http://www.youtube.com/watch?v=IYIoGuUZCFM)

Gravity Field

The Moon's gravitational field has been mapped in detail using tracking data from the *Gravity Recovery And Interior Laboratory (GRAIL)* spacecraft. The *GRAIL* mission consisted of two spacecraft, each equipped with a gravity ranging system, which measured the change in distance by intersatellite ranging as the spacecraft flew above the lunar surface. *GRAIL* revealed an extremely high correlation between gravity and *LRO*-derived topography. Gravity and topography correlate better with increasing degree, because the lithosphere is increasingly able to support topographic loads at shorter wavelengths without compensating masses at depth. Most of the correlation at the 30–130 km scales is related to impact craters.

The *GRAIL* data suggest an average density of the highlands crust of 2550 ± 20 kg m^{-3}, which is much lower than the hitherto assumed density of 2800–2900 kg m^{-3} that is typical for the anorthositic crustal materials on the Moon. This reduced density has been attributed to impacts fracturing the crust and producing crustal porosities of up to ~20%. Lateral variations in crustal density of up to ± 250 kg m^{-3} have been identified. For example, the South Pole Aitken basin shows a density of 2800 kg m^{-3}, while regions with lower-than-average densities are seen around the impact basins Orientale and Moscoviense, two of the largest young impact basins on the Moon.

With *GRAIL*'s high spatial resolution, distinctive gravitational signatures can be recognized, such as impact basin rings, central peaks of complex craters, volcanic landforms and smaller simple bowl-shaped craters. The gravity field over lunar mascons reveals a bull's-eye pattern, with a central positive anomaly, i.e., the mascon (§9.1.3), surrounded by a negative collar, and a positive outer annulus. Numerical models show that this pattern is a natural consequence of excavation from the impact crater, followed by post-impact

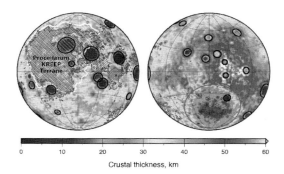

Figure G.5 COLOR PLATE Global map of crustal thickness of the Moon derived from gravity data obtained by NASA's *GRAIL* spacecraft. The lunar near side is shown on the left; the far side on the right. On the left, outlined in white, is the Procellarum KREEP Terrane, which contains high abundances of potassium, rare earth elements and phosphorus. In addition to the South Pole Aitken basin (the gray circle near the bottom of the far side map), the 24 black circles highlight impact basins with crustal thinning that have diameters over 200 km. (PIA17674; NASA/JPL-Caltech/S. Miljkovic)

isostatic adjustment and cooling and contraction of a voluminous melt pool.

Assuming everywhere a crustal porosity of 12% and a mantle density of 3220 kg m^{-3}, the data can be used to derive the map of the Moon's crustal thickness displayed in Figure G.5. The minimum crustal thickness of <1 km is in the interior of the far-side basin Moscoviense; the thickness at the *Apollo 12* and *14* landing sites is 30 km.

Atmosphere

The *Lunar Atmosphere and Dust Environment Explorer (LADEE)* spacecraft discovered Ne in the lunar exosphere, and mapped He and Ar over the equatorial regions. Helium was known to be part of the lunar atmosphere, but through observations with the neutral mass spectrometer on *LADEE* in conjunction with solar wind data obtained with the electrostatic analyzer instruments onboard the two *ARTEMIS (Acceleration, Reconnection, Turbulence and Electrodynamics of the Moon's Interaction with the Sun)* probes, a clear correlation between the lunar ^{4}He abundance and the solar wind flux of alpha particles has been established. *LADEE* made the first detection of Ne in the lunar atmosphere; as for ^{4}He, the bulk Ne abundance originates in the solar wind, as derived from isotopic measurements (^{20}Ne/^{22}Ne). The surface density of all three species is comparable, and is highest at night, just before sunrise, when it is coldest, as expected for noncondensable gases since the scale height is proportional to temperature. For ^{40}Ar there is a clear enhancement in abundance over the region that includes the Procellarum KREEP Terrane (§9.1.1; in contrast to early measurements, KREEP-containing rocks are concentrated underneath the Oceanus Procellarum and Mare Imbrium).

Putative detections of dust grains in the Moon's exosphere were reported during the *Apollo* era based on observations in forward scattered sunlight. *LADEE* confirmed the presence of such elevated dust grains in the form of a dust 'cloud' enveloping the Moon. This elevated dust is sustained by the continual bombardment of the lunar surface by interplanetary dust particles.

G.5 Mars

Atmospheric Escape

As discussed in §9.4, there is abundant evidence that liquid water was present on the ancient Mars surface, indicative of a thicker atmosphere at that time. This has led to speculations of potential habitability of ancient Mars. In the mid-1970s the *Viking* landers were sent to Mars to search for signs of metabolism, as described in §16.13.1. More recently, the *Curiosity* rover arrived at Gale crater to assess whether this crater has ever had or still has environmental conditions favorable to microbial life (see §9.4.6). In addition to the question of habitability, another key question is how H_2O and

CO_2 in the martian atmosphere have been removed (e.g., buried in the crust or lost to space).

Measurements of the isotope ratios of different elements contain information on isotopic fractionation, such as caused by atmospheric escape, where lighter isotopes may escape into space, leaving the heavier ones behind (§§5.7.1, 5.8). Using the mass spectrometer of the Sample Analysis at Mars (SAM) instrument on *Curiosity*, many isotopic ratios in the martian atmosphere have been measured, such as ^{15}N/^{14}N, ^{13}C/^{12}C, ^{38}Ar/^{36}Ar and the D/H ratio. The D/H ratio on Mars is about 6 times that on Earth (i.e., compared to SMOW, the Standard Mean Ocean Water on Earth). All these isotopic measurements are indicative of a substantial loss of Mars's atmosphere. Since isotopic ratios of C and O in carbonates within the 3.9 Gyr old martian meteorite ALH84001 (see §11.2 for details on this meteorite) show the same ratios as in the present martian atmosphere, most atmospheric loss must have happened prior to 4 Gyr ago. However, because the D/H ratio in ALH84001 is less enhanced than that in much younger martian meteorites, and because the *Curiosity* rover found that the D/H ratio in a 3–3.7 Gyr old clay sample in Gale crater was only 3 times that of SMOW, atmospheric loss, though at a reduced rate, likely continued over the past 3 Gyr. The measured D/H ratio in Gale crater requires escape rates about ten times as large as the present escape rates observed by the *Mars Atmosphere and Volatile EvolutioN* (*MAVEN*) mission.

Observations with *MAVEN* taken during a series of coronal mass ejections (CME; §7.2) in the direction of Mars showed that CMEs changed the overall morphology and dynamics of the martian magnetosphere and ionosphere, and induced widespread diffuse auroral emissions. *MAVEN* measured a typical escape flux during quiescent times of $\sim$4–5$\times 10^{24}$ s^{-1}, a flux dominated by O_2^+ ions; during a CME, the spacecraft measured that this rate can go up by more than an order of magnitude. Since such CME ejections were much more prevalent, and likely stronger, when the Sun was young, ion escape rates during the ancient Mars era may have been dominated by escape during CME events.

Observations by both *MAVEN* and the *Curiosity* rover suggest that atmospheric loss to space dominates over sequestering in surface reservoirs.

Methane Gas on Mars

Ground-based telescopic observations reported the discovery of a plume of methane gas in January–March 2003 on Mars with local abundances exceeding several tens of parts per billion volume (ppbv). In 2015, the Sample Analysis at Mars instrument suite (SAM) on the *Curiosity* rover reported that methane in Gale crater varied from between 0 and 10 ppbv. Using a different measurement technique, SAM reported in 2018 that methane cycles between 0.3 and 0.7 ppbv, peaking in late northern summer. In 2018, the ESA-Roscosmos *Exomars Trace Gas Orbiter* reported upper limits on methane of less than 0.1 ppbv in a survey of several dozen discrete locations over a period of months using the more sensitive technique of solar occultation spectroscopy. While debates regarding the reality of these spikes in methane abundance observed from Earth and the *Curiosity* rover are ongoing, no satisfactory explanations for the spikes nor the seasonal variations have been found.

Ice Deposits and Gullies

In §9.4.4 we discussed the presence of ice deposits on Mars, both in polar regions and at mid-latitudes. Images obtained with *MRO* have revealed scarps, produced by erosion, that expose deposits of water-ice that can extend down from 1–2 m below the surface to over 100 m. Since exposed water-ice sublimates away, these scarps are slowly disappearing, or retreating like glaciers on Earth. These deposits may have formed through snowfall during periods when Mars's obliquity was much higher than it is

at present (§2.6.3). In analogy to Earth, these ice-rich deposits must have preserved a record of ice deposition and past climate.

The presence of gullies on Mars had been attributed to a possible presence of liquid water (§9.4.5). However, *MRO* observations have shown an ongoing formation of gullies at places that are too cold for liquid water. Since the gullies appear to form at times when a thin layer of seasonal CO_2 frost covers the surface, the condensation and subsequent sublimation of CO_2 may play an important role in gully formation.

Current Cratering Rate

With spacecraft in orbit about Mars for decades now, we can determine the current impact rate on the planet by comparing images of the same region taken years apart and searching for new impact craters. Such studies measured an impact rate of 1.65×10^{-6} craters km^{-2} yr^{-1}, with effective diameters $D \geq 3.9$ m. This is only one-fourth as large as previous estimates, and implies that young martian surfaces whose ages have been estimated based upon crater counts may be ~4 times as old as previously thought. We say 'may be', because of various complicating factors, e.g., large variations in crater retention age depending on target strength, some craters may not have been identified, the effect of atmospheric ablation and the possibility that secondary craters produced by infrequent large impacts may dominate for the small size craters used for this study.

Crustal Density

The relationship between local topography, gravity, and crustal thickness was shown and discussed in §9.4.2. New techniques, using both topography and gravity maps, as have also been used for the Moon (§G.4), show that the bulk crustal density of the martian crust is 2582 ± 209 kg m^{-3}, much lower than the 2900 kg m^{-3} usually assumed. Strong lateral variations are found, e.g., the density is higher

18 km distance
2700 m above rover
triangle is 50 x 50 m

3.7 km distance
500 m above rover
triangle is 10 x 10 m

3.0 km distance
340 m above rover
triangle is 8 x 8 m

Figure G.6 High-resolution image of rocks in a region near Mount Sharp taken by NASA's *Curiosity* rover on Mars in November 2016. (PIA21256; NASA/JPL-Caltech/MSSS)

under volcanoes, consistent with earlier estimates. As on the Moon, the lower crustal density is indicative of a high porosity, such as can be caused by impact cratering.

Surface Features

Figure G.6 shows purple-hued rocks on the lower part of Mount Sharp. The purple color is suggestive of hematite. Winds and windblown sand in this season tend to keep rocks in this region relatively free of dust, revealing the rocks' true color. The colors in the image have been adjusted to resemble how rocks and sand would appear under daytime lighting conditions on Earth.

Orbital Decay of Phobos

The orbital motion of Mars's inner moon Phobos is observed to be accelerating at a rate of 1.27×10^{-3} degree yr^{-2}. This acceleration is caused by the lag in the tidal bulge that Phobos raises on Mars (§2.6.2), and it implies that $Q_{\male} = 83$.

G.6 Jupiter

The *Juno* spacecraft arrived at Jupiter in July 2016. The spacecraft is in a 53-day polar orbit. At perijove, *Juno* passes between Jupiter's radiation belts and the atmosphere, roughly 4000–6000 km above the cloud deck. With its passes over Jupiter's south pole, *Juno* was able to image the planet's polar regions in detail, as shown in Figure G.7. A most surprising finding is a set of persistent polygonal patterns of large circumpolar cyclones (8 in the north, 5 in the south) surrounding a single cyclone near the pole.

During these polar passes, *Juno* has been able to directly measure particle fluxes precipitating into the auroral zones, as well as fluxes in the upward direction.

Juno's polar orbits are also ideal to measure Jupiter's magnetic field near the planet, and its gravity field, in particular the high-order moments for both including higher-order moments than previous measurements for both of these fields. In addition, it was surprising to detect the odd gravitational moments J_3, J_5, J_7, and perhaps J_9. The improved gravity field, the lower order terms of which are given in Table E.15, was used to update models of Jupiter's interior structure. It was found that the planet's core is much more extended and dilute than expected, spanning ~0.3–0.5 $R_{\jupiter}$. The high-order gravitational moments are sensitive to the outer layers of the planet's envelope/atmosphere, and provide information on

Figure G.7 Jupiter's south pole as seen by NASA's *Juno* spacecraft from an altitude of 52 000 km. This is a composite image, taken with the JunoCam instrument on three separate orbits in stereographic projection. The oval features are cyclones, up to 1000 km in diameter. (PIA21641, NASA/JPL-Caltech/SwRI/MSSS/Betsy Asher Hall/Gervasio Robles). The associated movie uses images from *Juno*'s 11th perijove passage (PJ11). (NASA JPL SwRI MSSS Gerald Eichstädt)

differential rotation and to what depth the zonal winds extend. The non-zero values of the odd harmonic moments revealed a slight asymmetry between Jupiter's northern and southern hemispheres, caused by the asymmetry between jets in the two hemispheres. The high-order and odd harmonic moments have shown that Jupiter's winds must extend down to about 3000 km below the visible clouddecks.

Jupiter's magnetic field is surprisingly non-dipolar in character in the northern hemisphere. Most of the magnetic flux emerges in a narrow band in the northern hemisphere, some of which returns through an isolated patch near the equator, resembling a second magnetic south pole, i.e.,

in addition to the south magnetic pole near the jovigraphic south pole.

At radio wavelengths one probes levels in Jupiter's atmosphere below the visible cloud layers (§8.1). Using the upgraded Very Large Array, multi-frequency maps of the planet revealed that at radio-wavelengths between 1 and 4 cm Jupiter is very similar in appearance to *Hubble Space Telescope (HST)* images, despite the fact that at these wavelengths the planet's thermal (blackbody) radiation is measured, and not clouds as in images in reflected sunlight. At radio wavelengths the main source of opacity is ammonia gas, and hence the images shown in Figure G.8 essentially show the 3-dimensional distribution of ammonia gas: deeper warmer layers of the atmosphere are probed where the ammonia concentration is low (bright in the maps), whereas colder (dark in the images) altitudes are probed where the ammonia concentration is high. The images show a radio-hot belt that coincides with the southern edge of the North Equatorial Belt (NEB), and contains hot spots where Jupiter's deep atmosphere (>8 bar) is probed. These radio-hot regions are suggestive of a low ammonia abundance down to relatively deep levels ($\gtrsim$20 bar). The hot spots are likely the same spots as those seen at a wavelength of 5 μm. Interspersed, in the Equatorial Zone (EZ), are dark oval-shaped areas, interpreted as plumes where ammonia gas from the deep atmosphere is brought upwards. The plumes and hot spots form a wave pattern in the atmosphere. In contrast to these deep-seated features, most of the dynamics is confined to the upper 2–3 bar, i.e., above the NH_4SH cloud layer.

Juno's microwave radiometer (MWR) measures Jupiter's brightness temperature at six discrete wavelengths between 1.4 and 50 cm. The observations are performed along a narrow north–south track, shown schematically in Figure G.8a. Along these tracks Jupiter is probed down to over 100 bar, which corresponds to ~1000 km below the ammonia cloud deck, whereas the VLA probes down to

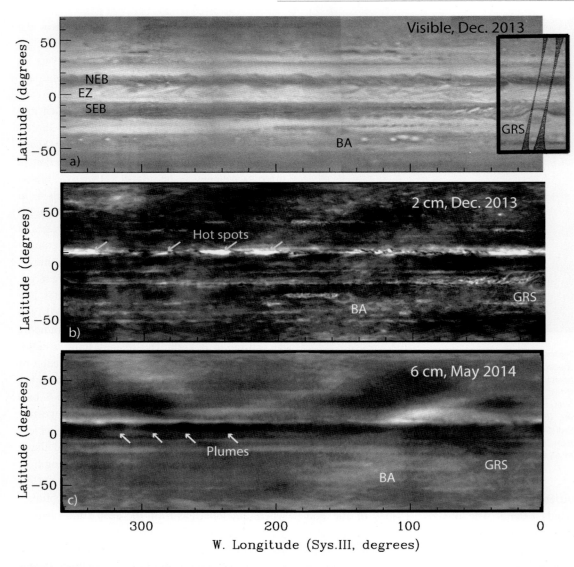

Figure G.8 Comparison of longitude-resolved VLA maps of Jupiter at 2 cm (13–18 GHz) and 6 cm (4–8 GHz) with a visible-wavelength map obtained by amateur astronomers. Arrows indicate the radio-hot belt with several hot spots (panel b) and a few of the dark plumes (panel c). The GRS and Oval BA are indicated on all three panels; the NEB, EZ and SEB are indicated on panel a. The inset with the GRS in panel a is from a different GRS to show the *Juno* MWR tracks overlain on the image. The narrow track corresponds to the field-of-view at 1.4 cm, and the wide one to that at 50 cm. The axes show planetographic latitude, and West longitude. The spatial resolution in panel b is of order 1000–1200 km, and 2450 km in panel c. One degree in longitude/latitude at disk-center corresponds to 1200 km. (Adapted from de Pater et al. 2019; the inset is from Janssen et al. 2017)

~10 bar, which corresponds to ~100 km below the ammonia cloud deck. The *Juno* data revealed that the low ammonia abundance in the NEB (Fig. G.8b–c) may extend down to deeper than the 100 bar pressure level. The Great Red Spot (GRS) exhibits a low brightness temperature at short wavelengths (1–6 cm; Fig. G.8b–c), but its brightness temperature may be higher than the environs at the longest MWR wavelengths (20–50 cm), probing depths of order 100 bar.

Europa

Hubble Space Telescope (*HST*) images revealed evidence of water plume activity over Europa's limb on several different occasions, where twice plumes were detected just north of Pwyll crater. The data are consistent with ~200 km high plumes of water vapor with line-of-sight column densities of about 10^{-5} m^{-2}. The plumes were detected above an area on Europa's surface that showed an excess in brightness temperature in *Galileo* PPR night-side data. ALMA observations show that this excess in brightness temperature is caused by a higher thermal inertia, likely caused by the fallout and condensation of plume material.

G.7 Saturn

Saturn's north pole is surrounded by a unique hexagonal pattern (see Fig. 8.20b) that has been observed since 1980. The *Cassini* spacecraft imaged the hexagon in detail on several occasions, during different seasons. The feature appears to be nearly stationary in a frame rotating with the planet (the hexagon rotates in 10 h 39 m 23.0 s, while the Voyager radio or System III rotation rate is 10 h 39 m 22.4 s). As illustrated in Figure G.9, *Cassini* images show that the entire interior of the hexagon in June 2013 was blue, but turned into a golden color by April 2017, a month before northern summer solstice. The center of the hexagon, a cyclone (similar to a tropical cyclone on Earth), remained blue, however. The change in color over the hexagon was probably caused by haze formation, due to photochemical reactions triggered by the Sun's UV light. The downward circulation in the eye of the hexagon may keep it clear from smog particles, maintaining its blue color.

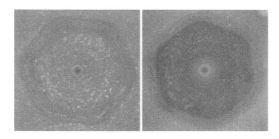

Figure G.9 Two *Cassini* views of Saturn's north polar hexagon: The left view was taken on 25 June 2013, and the right view on 25 April 2017, a month before northern summer solstice. Note the striking difference in colors: the 2013 image shows a blue color over the entire hexagon, whereas the 2017 image shows a golden color over the hexagon, except for the polar spot at the center, which remained blue. (PIA21611; NASA/JPL-Caltech/Space Science Institute/Hampton University)

As on Jupiter, at radio wavelengths we receive thermal (blackbody) radiation from the planet's warm atmosphere. The main source of opacity at these wavelengths is ammonia gas. Figure G.10 shows a radio map of the planet at 2 cm constructed from *Cassini* radiometer observations during the time that Saturn's large Northern Storm (Figure 8.19) was visible. In Figure G.10 the deviation in brightness temperature is plotted relative to that expected for a model in which the atmosphere is fully saturated (i.e., saturated in NH$_3$ gas). In many areas the atmosphere appears to be subsaturated in NH$_3$ gas, indicative of subsiding dry air. The large Northern Storm is visible near 40° N latitude. The radio-bright areas are suggestive of dry subsiding air, down to the 2–3 bar pressure level. *Cassini* observations at 2 cm and 5 μm show similar correlations as observed for Jupiter (5-μm and radio hot spots), although not all radio-bright areas correspond exactly to 5 μm bright regions.

In late 2016 the *Cassini* spacecraft's orbit was changed into a set of 20 ring-grazing orbits, before transitioning into its 'Grand Finale' tour. The ring-grazing orbits passed through the ring plane just outside the F ring, whereas the Grand Finale tour passed in between the rings and the planet's atmosphere. The latter orbits in particular enabled

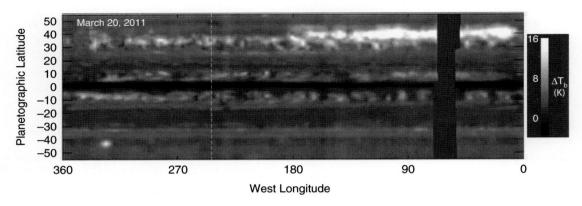

Figure G.10 Cylindrical map of Saturn's 2 cm brightness temperature constructed from *Cassini* radiometer observations taken in March 2011. The value plotted is the residual brightness relative to a model for a fully saturated atmosphere. The black stripe across the equator is where Saturn's rings obscure the planet's thermal radiation. The planet was mapped by continuous pole-to-pole scans during 14 hrs when the spacecraft was near periapse. Periapse (indicated by the white dashed line) was at 3.72 $R_\hbar$, where the resolution was best (1.6° in latitude). The resolution degrades linearly with spacecraft distance, out to 5.64 $R_\hbar$. The spacecraft motion along its trajectory combined with the planet's rotation-combined sweep through the longitude range is depicted. (Janssen et al. 2013)

a very accurate determination of Saturn's gravity field. The data revealed that Saturn's winds extend down to a depth of ~9000 km, i.e., such as would be expected if there is differential rotation on cylinders. The observations have not helped, yet, to determine the internal rotation rate of the planet. The internal rotation rate of a planet is usually derived from low-frequency radio observations, in particular the Saturn kilometric radio emissions for Saturn. However, this period varies between ~10.6 and 10.8 hrs, and is different between the northern and southern hemispheres, and each varies over time.

Measurements of the magnetic field this close to the planet confirmed its extreme axisymmetry with a small northward displacement of the center of the dipole (by 0.0466 ± 0.0002 $R_\hbar$).

In situ measurements in the gap between the D ring and Saturn's atmosphere revealed how rings interact with the planet. Nano-sized dust grains are transported along magnetic field lines from the rings to higher latitudes on Saturn, a phenomenon known as **ring rain**. A large fraction of material, dust and gases (e.g., H_2O, CH_4, N_2, CO_2, silicates and organics), falls directly in the atmosphere near the equator. Mass loss from the rings has been measured at a rate of up to a few tons per second, which may have implications for the longevity of the (inner) rings. A never-before-seen radiation belt of energetic particles was detected interior to the D ring, and an electric current system was discovered to connect Saturn and the D ring.

Rings, Propellers and Small Moonlets

As discussed in §13.5, the age of Saturn's rings is still being debated: Did the rings form with the planet, or did they form later, for example via breakup of a moon that ventured inside the Roche limit? Observations of the rings with *Cassini*'s passive radiometer revealed a small (≲0.5% in the A and B rings; 1–2% in the C ring) amount of non-icy (e.g., silicate) material in the overwhelmingly icy rings. When this small fraction of rocky material is attributed to pollution by micrometeorites, the rings cannot be much older than ~150 Myr. The middle part of the C ring shows an enhanced concentration of non-icy material (up to ~10%), which has been attributed to the breakup of a Centaur perhaps 10–20 Myr ago.

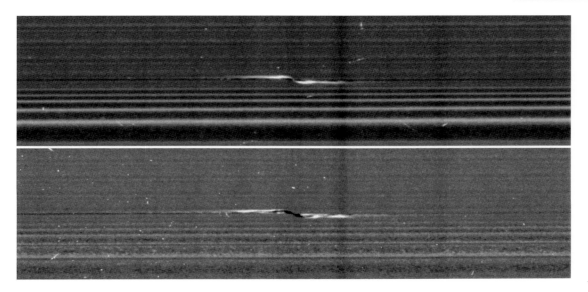

Figure G.11 Two views of the propeller 'Santos-Dumont': The top image shows the propeller on the rings illuminated by the Sun, while the bottom image shows the feature on the unilluminated side. The pixel scale is 207 m pixel^{-1}. (PIA21433, NASA/JPL-Caltech/Space Science Institute)

The mass of the rings can in principle be determined from precise gravity measurements obtained during the Grand Finale. A preliminary analysis suggests the ring mass to be roughly half that of the mass of the moon Mimas, which is slightly less than that derived from the *Voyager* data at the time. This ring mass is also indicative of relatively young rings, in agreement with the estimate above.

The ring-grazing and Grand Finale orbits of the *Cassini* spacecraft provided unique vantage points to image the rings close-up from different viewing angles. Figure G.11 shows views of propeller 'Santos-Dumont' from both the sunlit side and the unilluminated side, where sunlight filters through the backlit ring. On the lit side, dark regions indicate a lack of material, while on the unlit side, dark regions may indicate optically thick areas opaque to sunlight or regions devoid of material. The broad, dark band through the middle of the propeller on the unlit side is composed of a combination of both empty and opaque regions. On the lit side, the bright, narrow band of material connects the central (hidden) moonlet to the rings

in the ring plane. A 2-km wide gap (dark in both lit and unlit views) parallel to the rings is visible at the center of the propeller, flanked by wavy edges. The distance between the crests has been used to deduce the mass of the moonlet, which corresponds roughly to a snowball ~1 km in diameter. These large propellers appear to be long-lived, but also to have varying orbital periods, probably due to interactions with the ring.

Several small moons embedded within the rings have been imaged in detail with the *Cassini* spacecraft. Figure G.12 shows high-resolution images of Daphnis, Pan and Atlas. These moons are strikingly different in appearance from Saturn's larger moons. All three moons create gaps in the rings, while sculpting a wavy pattern along the edges of the gaps. The moonlets likely formed within the rings, from ring material accreting onto them (perhaps on a fragment from a previously broken moon). The centers of the moonlets look pretty round, which could be explained by a formation at a time when the ring system was quite young and vertically thicker. The thin ridges around the

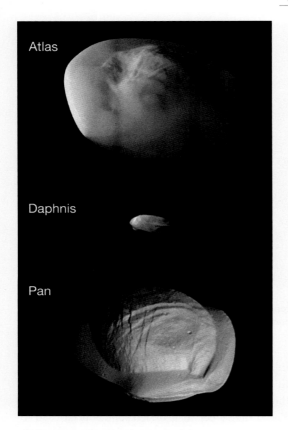

Atlas

Daphnis

Pan

Figure G.12 Three of Saturn's small ring moons: Atlas, Daphnis and Pan, at the same scale. Pan is 34 km across. (PIA21449, NASA/JPL-Caltech/Space Science Institute)

moons' equators can be explained by formation afterwards, after the moon had cleared the gap in the rings in which it resides today, and late enough in the formation process that the ring had already flattened into the very thin disk we see today, so that material was raining down onto the moon's equator, forming its 'skirt'.

Enceladus

As described in §10.3.3, Enceladus is geologically extremely active, with numerous geysers near its south pole. The origins of the geysers have been much debated. Two leading theories include the presence of a liquid ocean under the south pole driving the jets, or the presence of diapirs (§10.2.2) driving the geysers.

Using precise measurements of the orbit of the *Cassini* spacecraft through Doppler tracking, the satellite's quadrupole gravity field and harmonic coefficient J_3 have been determined. The moment of inertia is $0.335MR^2$, and the value of the J_3 coefficient implies a negative mass anomaly in the south-polar region, which is largely compensated by a positive subsurface anomaly compatible with a subsurface ocean 30–40 km below the surface and extending at least over southern latitudes from $\sim 50°$ to the south pole, if not globally.

Measurements of the rotation of Enceladus over a period of seven years revealed a forced physical libration of $0.120 \pm 0.014°$. This large value implies that Enceladus's core cannot be rigidly connected to its surface, but instead must be separated by a global liquid ocean, rather than a localized one. Through detailed modeling of the satellite's gravity, shape and libration, it has been shown that the liquid ocean is likely ~ 40 km thick, under an isostatic ice shell roughly 30 km thick over the equator, thinning to ~ 15 km over the north pole and 7 km over the south pole. The thin ice crust over the south pole facilitates the transport of water from the ocean to the surface.

The cosmic dust analyzer (CDA) on *Cassini*, sensitive to dust grains with radii >2 μm, measured an escaping flux of $\sim 10^{12}$ particles s^{-1}; many of these contribute to the E ring (§10.3.3). Amongst the grains are numerous pure water-ice particles, but many grains contain organics and Na, and a small percentage of the grains are very much enriched in sodium salts. The plume's gas composition is dominated by H_2O (volume mixing ratio of 95–99%), with 0.3–0.8% CO_2, 0.1–0.3% CH_4, 0.4–1.4% NH_3 and trace gases of, e.g., formaldehyde, methanol and various hydrocarbons; this composition is very similar to that seen in comets. Relatively high abundances of radiogenic ^{40}Ar have also been detected, but no primordial Ar (^{36}Ar, ^{38}Ar) has been seen. A relatively high abundance of native H_2 (0.4–1.4%) is present, which is likely produced through ongoing hydrothermal reactions of rock containing reduced minerals and organic materials. This relatively high H_2 abundance is suggestive of methanogenesis in Enceladus's ocean, where H_2 and CO_2 combine to form CH_4 and H_2O, a reaction that provides a source of chemical energy to support the synthesis of organic materials.

Dione

Enceladus's resonant companion, Dione, likely has a global ocean, of order 30–100 km deep, below ~100 km thick isostatic ice shell.

Titan

Titan's atmosphere and surface have been discussed in §10.3.1. A few more recent findings are discussed below. Figure G.13 shows a composite image of Titan's northern lakes and seas. The liquid in Titan's lakes and seas is mostly methane and ethane. Most of the bodies of liquid on Titan lie in the northern hemisphere. In fact nearly all the lakes and seas on Titan fall into a rectangle covering about 900 by 1800 km. Only three percent of the liquid 'rain' at Titan falls outside this area.

Images taken in 2006, when it was winter in the northern hemisphere, and 2012 during spring time, show no apparent change in the lakes, consistent with climate models that predict stability of liquid lakes over several years. This shows that the northern lakes are not transient weather events, in contrast to the temporary darkening of parts of the equator after a rainstorm in 2010 (Fig. G.14).

While clouds and rainfall were seen in 2005 in the south, near southern summer solstice, during subsequent (Earth) years clouds gradually appeared at mid-latitudes, and finally, after a hiatus of many years where no clouds were detected, clouds did appear in the north, as shown in Figure G.15.

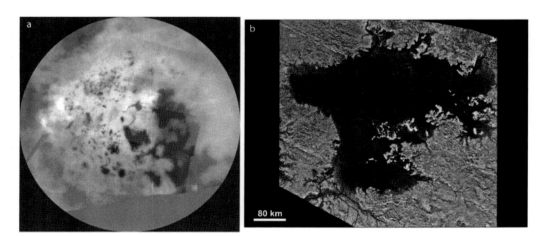

Figure G.13 a) Mosaic of Titan's northern lakes and seas. The data were obtained by *Cassini*'s radar instrument from 2004 to 2015. In this projection, the north pole is at the center. The view extends down to 60° N latitude. The area above and to the left of the north pole is dotted with smaller lakes. Lakes in this area are about 50 km across or less. (PIA19657; NASA/JPL-Caltech/ASI/USGS) b) Close-up of Titan's second largest known lake, Ligeia Mare. It is filled with liquid hydrocarbons, such as ethane and methane. The image is a false-color mosaic of synthetic aperture radar images obtained by the *Cassini spacecraft*. Dark areas signify a low radar return, as expected from lakes filled with liquid hydrocarbons. (PIA17031; NASA/JPL-Caltech/ASI/Cornell)

(a)   (b)

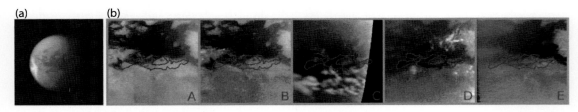

Figure G.14 (a) This huge arrow-shaped storm that blew across the equatorial region of Titan on 27 September 2010 produced large dark – likely wet – areas on Titan's surface. After this storm dissipated, *Cassini* observed significant changes on Titan's surface at the southern boundary of the dune field, shown in panel b. (b) This series of images shows the changes on Titan's surface, attributed to the rainstorm in panel a. These changes covered an area of 500 000 km^2. Image A in this montage was taken on 22 October 2007, and shows how this region had appeared before the storms. In image B, taken on 27 September 2010, the huge arrow-shaped cloud is on the left, just out-of-frame. The arrow-shaped cloud was quickly followed by extensive changes on the surface that can be seen in image C (14 October 2010) and image D (29 October 2010). By 15 January 2011 (image E), the area mostly appears dry and bright, with a much smaller area still dark, i.e., wet. The brightest spots in these images are methane clouds in the troposphere, the lowest part of the atmosphere. They are most visible on the left of image B, the lower half of image C and the right of image D. (NASA/JPL/SSI)

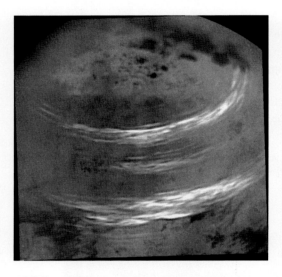

Figure G.15 Bands of bright, feathery methane clouds drifting across Titan on 7 May 2017. The dark regions at top are Titan's hydrocarbon lakes and seas. (PIA21450, NASA/JPL-Caltech/SSI)

Figure G.16 These two images of Uranus are composites of 117 images from 25 July 2012 (left) and 118 images from 26 July 2012 (right), all obtained with the near-infrared NIRC2 camera coupled to the adaptive optics system on the Keck II telescope. In each image, the north pole is on the right. The white features are high altitude clouds like Earth's cumulus clouds, while the bright blue-green features are thinner high-altitude clouds akin to cirrus clouds. Reddish tints indicate deeper cloud layers. (Lawrence Sromovsky, Pat Fry, Heidi Hammel, Imke de Pater and the Keck Observatory)

Gravity measurements of Titan support the notion of an internal, liquid water and ammonia ocean beneath its surface.

G.8 Uranus

Since Uranus's 2007 equinox, it has become much easier to view the planet's north polar region. Thanks to the development of new image processing techniques, stunning images of discrete cloud features in this region have been obtained, as shown in Figure G.16. Images were taken in two different near-infrared filters to determine the altitude of clouds and hazes. The broad H filter centered near 1.6 μm (blue and green colors) samples both weak and strong absorptions by methane gas, while a narrow band filter centered at 1.58 μm (red color) samples only regions of weak methane absorption. Most of the features in these images are quite subtle and require long exposures to be detectable above the background noise. But during long exposures, the features are smeared out by planetary rotation and zonal winds. To deal with that, many short-exposure images were taken and the effects of rotation and winds were removed before averaging. These polar features are visible while it is spring at the north pole; no such features were visible at the south pole during its summer and fall. This technique also revealed a never before seen scalloped wave pattern just south of the equator, similar to instabilities that develop in regions of horizontal wind shear.

G.9 Neptune

Images of Neptune, taken with the Keck telescope in June and July 2017, revealed an extremely large bright storm system near Neptune's equator, a region usually devoid of prominent clouds (§8.3.1). The images are shown in Figure G.17. The center of the storm complex is ∼9000 km across, roughly equal to one-third of Neptune's radius. The storm brightened considerably over a period of about one week.

Historically, very bright clouds have occasionally been seen on Neptune, but usually at latitudes polewards of ∼15°. *Voyager* and *HST* observations have shown that such clouds were usually companions of dark vortices (anti-cyclonic regions, like Jupiter's Great Red Spot), interpreted as **orographic clouds**, i.e., clouds that form when air near the vortex is forced to rise, and thereby cools. Due to Neptune's low temperature, these clouds are probably composed of methane ice. In the past such clouds, coupled to dark vortices, often persisted from one up to several years. The bright storm seen in Figure G.17 was seen until the end of 2017. No dark vortex was seen in association with this storm. The storm might be caused by a

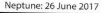
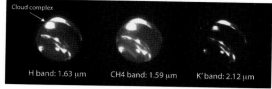
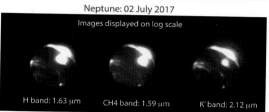

Figure G.17 Images of Neptune taken with the Keck II telescope at a wavelength of 1.6 μm in 2017 on 26 June and 2 July, revealed an extremely large, bright storm system near Neptune's equator (labeled 'cloud complex' in the upper figure). The center of the storm complex is 9000 km across, about 3/4 the size of Earth, or 1/3 of Neptune's radius. The storm brightened considerably between June 26 and July 2, as noted in the logarithmic scale of the images taken on July 2. (E. Molter, I. de Pater, C. Alvarez, W. M. Keck Observatory).

strong convection event, perhaps akin to the large 2011 storm on Saturn (§8.2.1).

G.10 Meteorites

NWA 7034

The basaltic clastic rock named NWA 7034 shown in Figure G.18 and its paired rocks, e.g. NWA 7533, are some of the most interesting martian meteorites known. These meteorites are polymict basaltic breccia that share geochemical linkages with the shergotites (§8.2) and have many geochemical similarities to the orbital, lander and rover mission data sets.

Four major age groups have been identified in NWA 7034, including zircons (§8.7) as old as 4.44 Gyr and phosphates formed 'only' 1.35 Gyr ago. Thus, this meteorite can be used to provide information about the conditions on Mars over a wide range of time.

Figure G.18 Views of the martian meteorite NWA 7034, informally referred to as 'Black Beauty'. The upper two images show two views of the largest piece of the meteorite, which has been cut to expose interior regions for a detailed study. The lower panel is a backscatter electron image showing the texture in NWA 7034. Large dark crystals are feldspar; large light-colored crystals are pyroxene. (Agee et al. 2013)

G.11 Minor Planets

G.11.1 Potentially Hazardous Asteroids

Near-Earth objects (§9.1.1) whose orbits come within 0.05 AU ($\sim$ 7.5 million km) of Earth's orbit and that are large enough to cause significant damage if they were to hit Earth are referred to as **potentially hazardous asteroids**, or **PHA**s. The number of PHAs with $R > 50$ m has been estimated at 4700 ± 1500.

G.11.2 Regolith on Small Asteroids

We discussed the production of regolith via impact cratering in §6.4.2 and §12.3.3. Impact ejecta, how-ever, often attain a speed exceeding the escape velocity of km-sized asteroids, and hence the efficiency of regolith production has been questioned. Laboratory experiments and modeling suggest that **thermal fatigue** may be the dominant mechanism of regolith production on asteroids, independent of size. Diurnal temperature variations cause rocks to break up and fragment on timescales that are much shorter than that caused by micrometeoroid impacts.

G.11.3 Individual Asteroids

1 Ceres

After a successful year orbiting asteroid 4 Vesta, the *Dawn* spacecraft used its ion drive engine (Appendix F) to escape from Vesta and travel farther from the Sun, arriving at the largest asteroid, 1 Ceres, in March 2015. Figure G.19 shows

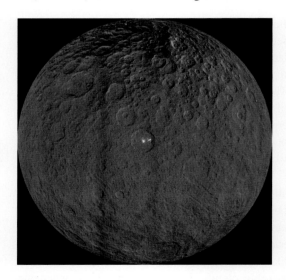

Figure G.19 Hemispheric view of 1 Ceres, constructed from high resolution images (35 m pixel^{-1}) obtained by NASA's *Dawn* mission. The associated movie was obtained during *Dawn*'s approach to Ceres on 19 February 2015. At that time *Dawn* observed Ceres for a full rotation, which lasts about nine hours. The first images in this sequence have a resolution of 4 km pixel^{-1}, increasing to 2 km pixel^{-1} as *Dawn* closed in on Ceres. The animation shows six hours and 45 minutes of the nine-hour rotation. (PIA21906; movie from PIA19546; NASA/JPL-Caltech/UCLA/MPS/DLR/IDA)

Ceres at high resolution, and the associated movie provides a global view of the asteroid through images of the lit hemisphere taken over one full rotation period. Ceres's mean radius is 469.7 km, and its polar axis is substantially smaller than the two equatorial ones, with the dimensions of the best-fitting triaxial ellipsoid being (483.1 × 481.0 × 445.9) ± 0.2 km). Ceres's mass is 9.384×10^{20} kg, with a mean density of 2162 kg m^{-3}, indicative of a mixture of rock and ice. Gravity measurements by the *Dawn* spacecraft indicate that Ceres is close to being in hydrostatic equilibrium, and imply Ceres is partially differentiated, with a denser core and low-density mantle. The crust of Ceres is ~40 km thick, and has a mean density of ~1280 kg m^{-3}. A low Bouguer gravity at high topographic areas, and vice versa, suggests isostatic equilibrium, i.e., the topography of Ceres appears to be compensated, which most likely can be attributed to a layer of low viscosity at depth.

Topographic maps, shown in Figure G.20, reveal a vertical relief of ~15 km, much less than the 41 km relief seen on Vesta, but large enough to provide evidence that the crust is quite strong. Ceres's mantle has a mean density of 2400 kg m^{-3}, implying a predominantly rocky composition. It is

possible that below the crust global or localized muddy briny water reservoirs could exist today. As shown, parts of Ceres's surface are heavily cratered, though there is a clear lack of the large impact basins that one might expect to find on bodies several billion years old. This suggests efficient relaxation of the topography at depth, whereas the upper mantle/lithosphere must be strong to explain the heavily cratered surface. The morphology of the craters on Ceres is in between that of Vesta, a mostly rocky object, and Rhea, one of Saturn's icy satellites. The craters on Ceres have, at least in part, smooth floors, indicative of the production of impact melt.

Many km-scale linear structures, including grooves, chains of pit craters, fractures and troughs, are seen on Ceres's surface. While features oriented radial to the craters are clearly related to craters, and thus were produced by impacts, other features may be tectonic in origin.

Figure G.21 shows an image of Ahuna Mons, a tall mountain that *Dawn* imaged on Ceres. This mountain is a large dome, and may have formed through viscous extrusion. It is probably geologically young, as implied by the scarcity of craters, absence of debris at its base and the bright streaks on its flanks.

Ceres's surface is quite dark, with a geometric albedo of 0.09, and most of it is homogenous in composition. But distributed over the surface are

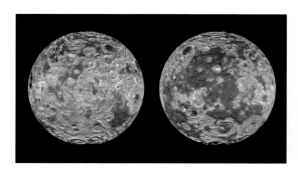

Figure G.20 COLOR PLATE Topographic maps of 1 Ceres's East and West Hemispheres. This pair of images shows color-coded maps from NASA's *Dawn* mission, revealing the highs and lows of topography on Ceres' surface. The map at left is centered on terrain at 60 degrees east longitude; the map at right is centered on 240 degrees West longitude. The color scale extends about 7.5 km below the mean radius in indigo to 7.5 km above the mean radius in white. (PIA19607; NASA/JPL-Caltech/UCLA/MPS/DLR/IDA)

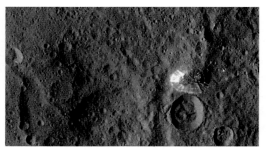

Figure G.21 The lonely 4-km high mountain on Ceres, Ahuna Mons, as imaged by the *Dawn* spacecraft. The spatial resolution is 140 m pixel^{-1}. (PIA19631; NASA/JPL-Caltech/UCLA/MPS/DLR/IDA)

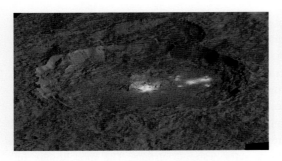

Figure G.22 This simulated perspective view shows Occator Crater, 92 km across and 4 km deep, which contains the brightest area on Ceres. A close-up view reveals a dome in a smooth-walled pit in the bright center of the crater. On and around the dome are numerous linear features and fractures. (PIA21913; NASA/JPL-Caltech/UCLA/MPS/DLR/IDA)

Figure G.23 COLOR PLATE Distribution of the concentration of hydrogen determined from data acquired by the gamma ray and neutron detector (GRaND) instrument aboard NASA's *Dawn* spacecraft, on both 4 Vesta and 1 Ceres. The instrument is sensitive to hydrogen in the upper meter of the asteroids' regoliths. The colors from red-to-blue on Vesta indicate a concentration of hydrogen from 0 to 0.04% (by weight), and on Ceres from 1.8 to 3.2%. If all hydrogen is in the form of water-ice, the water-equivalent hydrogen by weight varies from 16 (red) to 29% (blue) on Ceres. (PIA21081; NASA/JPL-Caltech/UCLA/MPS/DLR/IDA/PSI)

small bright sodium carbonate deposits, with one large complex of bright deposits on the (almost 100 km wide) floor of Occator crater (Figs. G.19 and G.22). This bright material most likely was brought up from below, through cryovolcanism in the geologically recent past.

Until the *Dawn* mission, no water-ice had been detected spectroscopically. However, ground-based spectroscopic data did show evidence of a surface covered in part by clays, which require water to form. Other inferences for wetness came from the detection of OH by the *International Ultraviolet Explorer* (*IUE*) in 1990. Twenty years later, the *Herschel Space Observatory* detected H_2O molecules on three out of four occasions, which, when detected, were typically ejected at a rate of $\sim 10^{26}$ mol s^{-1}. When *Dawn* arrived at Ceres, it not only confirmed the presence of hydroxylated silicates and ammoniated clays, but it also detected exposed water-ice in the Oxo crater and elsewhere, revealed by mass wasting on the side of craters. In addition, the gamma ray and neutron spectrometer provided evidence of widespread ice in the top meter of Ceres's surface, as shown in Figure G.23. Although this figure shows the spatial distribution of the concentration of hydrogen on Ceres and Vesta, this hydrogen can be present in the form of water-ice and hydrated minerals, and

as such provides information on the water content of a body. As shown, Ceres's poles are much wetter than the equatorial regions. For comparison, we also show the asteroid Vesta, which is much dryer than Ceres. The increase of subsurface ice on Ceres from the equator to the poles also agrees with the geomorphology of lobate flows on the asteroid's surface.

Water-ice on Ceres's surface is most likely the source of water products in Ceres's exosphere. A detailed comparison between water detections and non-detections in Ceres's exosphere with variations in solar wind parameters suggests that the production of H_2O and OH molecules might result from sputtering by high-energy solar protons, the flux of which is highly variable (e.g., solar flares and CMEs; §7.2).

951 Gaspra

951 Gaspra is a small ($R = 6.1 \pm 0.4$ km) irregularly shaped S-type asteroid, which is unusually red and olivine-rich. Its geometric albedo is typical for S-type asteroids, $A_0 = 0.22$. The object was most likely produced during the catastrophic disruption of the parent body ($R \gtrsim 100$ km) that

a)

b)

Figure G.24 a) Image of asteroid 162173 Ryugu taken on 26 June 2018 by the *Hayabusa 2* probe when it was 22 km away from the asteroid. The left side of the asteroid, at a higher spatial resolution, was taken when the spacecraft was 2 km above the surface. Ryugu's mean radius is slightly more than 400 m. (JAXA) b) Image of asteroid 101955 Bennu composed of 12 PolyCam images collected on 2 December 2018 by the *OSIRIS-REx* spacecraft when it was 24 km from the asteroid. The asteroid's mean radius is ~250 m. The accompanying movie, produced using a series of 36 images taken by *OSIRIS-REx* over an interval of four hours and 18 minutes, shows the asteroid over one full rotation from a distance of around 80 km. (NASA GSFC/University of Arizona)

produced the Flora family ~500 Myr ago. More than 600 craters have been identified on *Galileo* images of Gaspra's surface. Grooves on Gaspra's surface, together with subtle color/albedo variations on the surface, suggest that Gaspra is covered by a layer of regolith that could be many tens of meters thick.

162173 Ryugu and 101955 Bennu

The Japanese spacecraft *Hayabusa 2* is conducting an intensive study of the small (~430 m effective radius) NEA 162173 Ryugu, a PHA from the Apollo group. After *in situ* studies planned to include detailed mapping and touch-and-go landings, a sample will be returned to Earth for intensive study in the laboratory. Figure G.24a shows a closeup image of the asteroid. A large crater is visible on the side near the equator, and the entire surface is covered with boulders, somewhat akin to the NEA Itokawa (§12.5.1).

NASA's *OSIRIS-REx* spacecraft arrived at the even smaller (~250 m effective radius) NEA Bennu on 3 December 2018. It will bring back samples of the asteroid in 2023. Like Ryugu, Bennu is a carbonaceous PHA from the Apollo group. Figure G.24b shows an image of Bennu, which shows a striking resemblance with Ryugu. Both asteroids rotate in the retrograde direction, and are shaped like a spinning top. Their derived densities (1260 kg m^{-3} for Bennu) and the morphology of large craters on their surfaces suggest that both bodies are porous, composed of rubble piles. Such bodies, if rotating fast enough (faster than their present rotation periods of 7.6 hr for Ryugu and 4.3 hr for Bennu), may get shaped like a spinning top.

(514107) 2015 BZ$_{509}$

The small asteroid (514107) 2015 BZ$_{509}$ travels on a retrograde orbit that has the same average period as Jupiter. It is the first body observed to be locked in a **counter-orbital resonance**, which protects it from close approaches to the Solar System's largest planet for at least millions of years and perhaps much longer. The protection mechanism of this 1:1 orbital resonance is more complicated than that of the much more common Trojan asteroids (§§2.2.2, 12.2.1); it involves the asteroid having a sizable eccentricity, $e \approx 0.38$, and alternating between being near perihelion and aphelion on successive passages by Jupiter.

It is likely that (514107) 2015 BZ_{509} once resided in the Oort cloud, where very low orbital velocities make it possible for small perturbations from passing stars or the galactic tide to reverse orbital direction. It could then have been sent inwards to the planetary zone like most long-period comets and subsequently had its orbital semimajor axis and eccentricity reduced by a fortuitous combination of planetary perturbations, perhaps augmented by nongravitational forces produced by cometary activity or a collision with a small body.

G.11.4 Pluto and its Moons

NASA's *New Horizons* spacecraft sped past Pluto and its moons on 14 July 2015, and the data it sent back to Earth have revolutionized our understanding of this dwarf planet and its satellite system. Stunning views of Pluto and its one large moon, Charon, were obtained, as shown in Figures G.25 and G.26. Pluto exhibits a greater variety of landforms than ever previously seen on a single body that is smaller than Mars.

Pluto's radius is 1188.3 ± 1.6 km and its mass is $1.303 \pm 0.003 \times 10^{22}$ kg, implying a mean density of 1854 ± 11 kg m^{-3}, and a likely composition of $\sim 65\%$ rock and $\sim 35\%$ H_2O ice by mass. More volatile ices like N_2 and CO are also present, but probably only in much smaller amounts. Charon's radius is 606.0 ± 1.0 km and its mass is $1.59 \pm 0.02 \times 10^{21}$ kg, implying a density of 1701 ± 33 kg m^{-3} and a likely composition of $\sim 60\%$ rock and $\sim 40\%$ H_2O ice by mass, unless its upper layers have significant porosity. Neither Pluto nor Charon show any evidence for tidal or rotational distortions; upper bounds on their oblateness are $< 0.6\%$ and $< 0.5\%$, respectively.

Recognizable impact craters on Pluto range in diameter from ~ 0.5 km up to ~ 250 km.

Figure G.25 High-resolution image of Pluto taken by NASA's *New Horizons* spacecraft on 14 July 2015. This image is a color-enhanced view of the dwarf planet, where blue, red and infrared images taken by the Ralph/Multispectral Visual Imaging Camera (MVIC) were combined. As shown, many landforms have distinct colors, suggestive of a complex geological and climatological history. The smooth area near the center is Sputnik Planitia; this, together with the light-colored more rugged expanse to the east is the heart-shaped region, referred to as Tombaugh Regio. Image resolution is 1.3 km. (PIA19952; NASA/JHU-APL/SWRI)

The cumulative crater size-frequency distribution implies a wide range of surface ages, from less than 10 Myr to billions of years. (Accurate ages are difficult to obtain from crater counts in the outer Solar System because the impactors come from a population quite different from that impacting the Moon, where we can determine age by radiometric dating of lunar rocks modified by various impacts.) Crater morphologies vary from large, well-preserved craters with central peaks to highly degraded/eroded craters that are hardly recognizable.

Pluto's most striking geologic feature is a large heart-shaped area (Fig. G.25), named Tombaugh Regio. In the western portion of Tombaugh Regio lies a vast ($870\,000$ km^2 in area) plain known as Sputnik Planitia, parts of which are shown in

Figure G.26 High-resolution enhanced color view of Charon, taken by NASA's *New Horizons* spacecraft just before closest approach on 14 July 2015. The image combines blue, red and infrared images; the colors are processed to best highlight the variation of surface properties across Charon. Most striking is the reddish north polar region (top), informally named Mordor Macula. Image resolution is 2.9 km pixel^{-1}. (PIA19968; NASA/JHU-APL/SWRI)

Figures G.27 and G.28. Sputnik Planitia has no craters down to sizes of $\sim$1 km, indicative of a crater retention age in that area of $\sim$10–100 Myr; it is likely still being shaped by geological processes. Sputnik Planitia is 2–4 km below its environs, and located in the vicinity of the anti-Charon point. The

area, perhaps an ancient impact basin, appears to be covered by massive amounts of N_2-ice, mixed with lesser amounts of CO and CH_4 ice. The location of Sputnik Planitia near the Pluto–Charon tidal axis may be a natural consequence of the sequestration of volatile ices within its basin. Trapping of volatile ices is easiest in a deep basin, and these condensed ices would lead to an accumulation of mass and subsequent reorientation of Pluto's axis (true polar wander).

The cellular pattern decorating Sputnik Planitia, 20–30 km sized polygons surrounded by troughs of order 100 m deep, is indicative of active solid-state convection, at velocities of a few cm yr^{-1}, presumably of weak volatile ices like N_2 and CO. This pattern is prevalent on the west side, bordered on the outside by numerous isolated mountains, discrete angular blocks up to 40 km across and up to 5 km high. These mountains are composed of water-ice. As H_2O ice is buoyant with respect to N_2 and CO ice, the water-ice blocks rise isostatically. The smaller ones are probably icebergs 'floating' on an $\sim$5–10 km thick layer of N_2 ice.

The landscape in the south of Sputnik Planitia is covered by pits, as shown in Figure G.29. These pits are $\sim$100–1000 m across and up to a few hundred meters deep. A few glacial-like flows have been recognized, while farther upland to the east the landscape is characterized by roughly

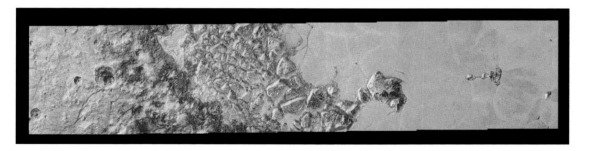

Figure G.27 A high-resolution view of Sputnik Planitia taken by the *New Horizons* spacecraft just before closest approach on 14 July 2015. Features as small as 250 m across can be discerned, from craters in the west to faulted mountain blocks and the textured surface of Sputnik Planitia further east. Enhanced color has been added from the global color image. This image is about 530 km across. (PIA19955; NASA/JHU-APL/SWRI)

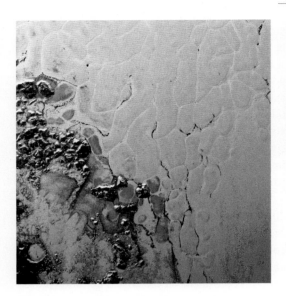

Figure G.29 The close-up image on the right shows an 80 × 80 km area on the east side of Pluto's Sputnik Planitia that is characterized by numerous pits. The large ring-like structure near the bottom right of the magnified view, as well as the smaller one near the bottom left, may be remnant impact craters. The upper-left quadrant of the close-up image shows the border between the relatively smooth Sputnik Planitia ice sheet and the pitted area, with a series of hills forming slightly inside this unusual shoreline. (PIA20212; NASA/JHU-APL/SWRI)

Figure G.28 A region of Sputnik Planitia showing churning ice cells that are geologically young and turning over due to convection. The numerous small pits on the surface are likely caused by sublimation. The image is ∼400 km across. (PIA20726; NASA/JHU-APL/SWRI)

N–S aligned blade-like ridges, several hundred meters high, and separated by 5–10 km. Both the pits and blade-like terrain may form through a combination of sublimation and undermining or collapse of ice due to melting at depth, while ridges could also grow by deposition of volatiles on their crests.

About 70° to the west of Sputnik Planitia the terrain is characterized by a group of dark-floored craters with bright rims and halos, creating a halo effect (likely caused by volatile deposition and sublimation), shown in Figure G.30. The crater floors and terrain between these bright-halo craters show signs of water-ice, while bright methane-ice has been identified on their rims and walls.

Numerous troughs and scarps, up to a few hundred km in length and several km high, have been interpreted as extensional fractures (graben and normal faults). Examples are shown in Figure G.31. The various states of degradation, together with the variety of faults, are suggestive

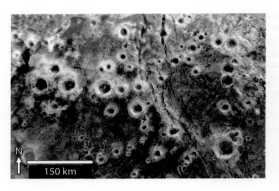

Figure G.30 This *New Horizons* image shows several dozen 'haloed' craters, the largest of which is ∼50 km across. The craters' bright walls and rims stand out from their dark floors and surrounding terrain. Bottom of online color image: Composite data from two of *New Horizons*'s instruments reveal a connection between the bright halos and distribution of methane ice, shown in purple (false-color). The floors and terrain between craters show signs of water-ice, colored in blue. (PIA20656; NASA/JHU-APL/SWRI)

of multiple deformation episodes and prolonged tectonic activity. Some extensional features are very young, which would be consistent with recent stresses induced by a late, possibly partial, freezing

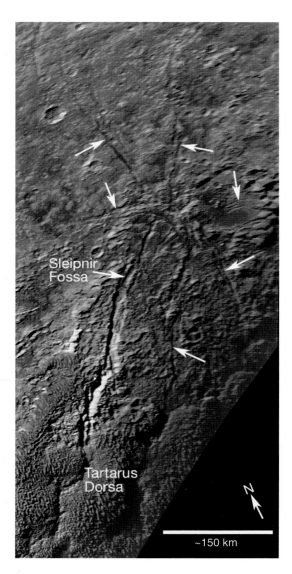

Figure G.31 The icy 'spider' on Pluto consists of at least six extensional fractures (indicated by white arrows) converging to a point near the center. They all expose red deposits below Pluto's surface. Most fractures on Pluto tend to be aligned parallel to each other and are probably caused by global-scale extension of Pluto's water-ice crust. The radiating pattern of the fractures seen on this image may result from a focused source of stress in the crust under the point where the fractures converge, perhaps caused by material welling up from underneath. The image resolution is approximately 680 m pixel^{-1}. (PIA20641; NASA/JHU-APL/SWRI)

of a subsurface ocean. Pluto most likely is underlain by a thick lithosphere of water-ice.

New Horizons measured an atmospheric pressure at Pluto's surface of 11.5 microbar, consistent with values derived from stellar occultations in 2015. This is remarkable, since it means that the pressure tripled since 1988, in contrast with the expectation that the atmosphere would start to collapse with Pluto moving away from the Sun after passing through perihelion in 1989. It should slowly collapse over the coming decades, however. Pluto's atmospheric composition is dominated by N_2, with traces of CH_4, CO, HCN and more complex hydrocarbons (C_xH_y), including acetylene, ethylene and ethane. Clearly, there is a complex interplay between the atmosphere and surface, with sublimation and condensation of volatile gases depending on the dwarf planet's detailed topography and ice coverage. Sputnik Planitia, with its vast expanses of volatile ices, must play a key role in the annual variations of Pluto's atmosphere.

The temperature increases with altitude from the surface upwards, but the detailed profile varies with location. A near-surface temperature of 37 ± 3 K was measured at the occultation entry side, near Sputnik Planitia, essentially at the condensation temperature of N_2 (37 K). It was ~45 K on exit, which was taken closer to the subsolar latitude of $-52°$. A temperature of ~110 K was measured at altitudes above ~25 km, although at the entry side this high a temperature was already reached at 10 km altitude. In the upper atmosphere (~850–1400 km), the temperature is only ~70 K, indicative of cooling through, e.g., C_2H_2 and HCN, although the abundances of the latter species may be too small to account for the amount of cooling.

Jeans escape rates (§5.7) of N_2 and CH_4 from Pluto's upper atmosphere are estimated from measured atmospheric properties to be $\sim 5 \times 10^{22}$ and 6×10^{25} mol s^{-1}, respectively. For N_2, this is about four orders of magnitude less than was expected prior to the *New Horizons* encounter. Since Charon is expected to intercept some of the escaping gases, the fact that no N_2 was detected on Charon is consistent with this measurement. In contrast, the much higher escape rate of CH_4 from Pluto can readily explain the reddish (hydrocarbons) polar cap over Charon.

As shown in Figure G.32, numerous distinct layers of haze are present up to altitudes exceeding 200 km. Individual layers appear to be contiguous over distances of at least 1000 km. These hazes are probably produced photochemically, analogous to the haze in Titan's atmosphere (§10.3.1). There are some low-lying isolated features in Pluto's atmosphere that might be (rare) condensation clouds.

Charon's surface is heavily cratered, implying that it is very old. Charon's northern and southern hemispheres are quite different, and are separated by a set of tectonic ridges and canyons, shown in Figure G.33. These canyons, seen along the entire equator on the hemisphere viewed by *New Horizons*, are 5–7 km deep. The ridges were likely formed by a tectonic extension of Charon's icy crust caused by freezing of a subsurface H_2O ice

ocean billions of years ago. Topographic highs and lows span almost 25 km in elevation, more than twice the range seen on Pluto. Charon's north polar region, referred to as Mordor Macula, is dark reddish. Its color is most easily explained by the freezing of small amounts of volatiles transported from Pluto's atmosphere, and transformed into tholins (§10.3.1) over timescales of millions of years. The southern hemisphere is somewhat younger than the north, but still quite old. It is relatively smooth, yet tectonic features are clearly seen, in some cases predating cratering. Features in the southern hemisphere are consistent with cryovolcanic resurfacing.

Pluto's small satellites are displayed in Figure G.34. All of these satellites are much smaller than Charon, which has a diameter of 1212 km. Nix and Hydra each are ~40 km across in their longest dimension. Kerberos and Styx are smaller still, each roughly 10–12 km across in their longest dimension. The four small moons have highly elongated shapes, a characteristic thought to be typical of small bodies in the Kuiper belt (and

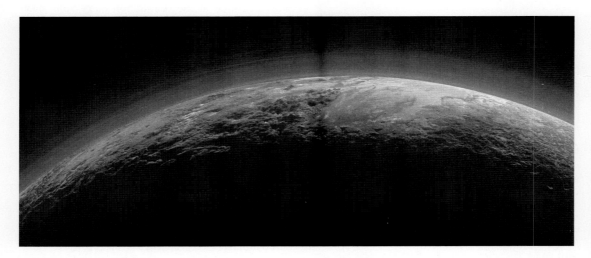

Figure G.32 A near-sunset view of the rugged, icy mountains and flat ice plains on Pluto, taken 15 minutes after *New Horizons*'s closest approach to Pluto (at a distance of 18 000 km), when the spacecraft looked back toward the Sun. The smooth expanse of Sputnik Planitia is flanked to the west (left) by rugged mountains up to 3.5 km high. To the right, east of Sputnik, rougher terrain is cut by apparent glaciers. The backlighting highlights more than a dozen layers of haze in Pluto's tenuous atmosphere. The image is 1250 km wide. (PIA19948; NASA/JHU-APL/SWRI)

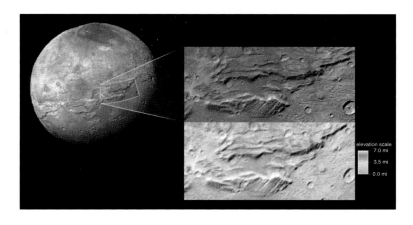

Figure G.33 High-resolution enhanced color view of Charon, taken by *New Horizons* on 14 July 2015. The colors, blue, red and infrared, are processed to best highlight the variation of surface properties across Charon. Most striking is the reddish north polar region, informally named Mordor Macula, and the apparent north-south dichotomy. An enlarged view of part of the canyon system dividing the hemispheres is shown on the right. The lower portion of this image shows color-coded topography of the same scene. The resolution of the image on the left is 2.9 km; that of the right is 400 m pixel^{-1}. (PIA20467; NASA/JHU-APL/SWRI)

elsewhere). All four moons rotate much faster than synchronous, with periods varying from ~1/2 day up to 5 days. The high crater density on these tiny moons indicates that their surfaces are several billion years old. They all have high albedos (0.5–0.8) and spectra indicative of water-ice. These small satellites are most likely the remnants of a giant collision that formed the Pluto system eons ago.

G.11.5 486958 Ultima Thule

After its successful flyby of Pluto and its moons, the *New Horizons* spacecraft encountered the TNO 486958 Ultima Thule on 1 January 2019. One of the best images is shown in Figure G.35. With an orbital period of 298 years and a low inclination and eccentricity, the body is classified as a classical Kuiper belt object. It is a contact binary, 31 km along the long axis, 14 km across the small body (Thule), and 19 km across the larger body (Ultima). The two objects probably were separate bodies initially, perhaps accreted from debris after a pre-existing body got shattered in a collision; while orbiting each other, over time they approached each other and became attached.

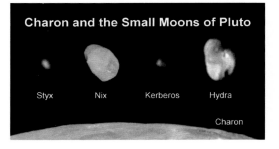

Figure G.34 *New Horizons* images of a sliver of Pluto's large moon, Charon, and all four of Pluto's small moons are displayed with a common intensity stretch and spatial scale (See Table G.1 for sizes). (PIA20033; NASA/JHU-APL/SWRI)

G.11.6 Rings of Minor Planets

The Centaur 10199 Chariklo has an equivalent radius of 124 ± 9 km and orbits between Uranus and Saturn, with aphelion near Uranus's orbit. When Chariklo occulted a star, the occultation profile shown in Figure G.36 revealed two rings. The rings, with respective widths of about 7 and 3 km, have (normal) optical depths of 0.4 and 0.06, and mean orbital radii of 391 and 405 km. The present orientation of the rings is consistent with an edge-on geometry in 2008, which provides a

Table G.1 Orbital Parameters and Physical Properties of Pluto's Moons

	Period (days)	a (km)	e	i (°)	R (km×km×km)	M (kg)	ρ (kg/m³)
Charon (P1)	6.3872	19 596	0.00005	0.0	606±1	1.587×10^{21}	1702
Styx (P5)	20.162	42 413	0.00001	0.0	16×9×8	$< 7.5\times10^{15}$	$\lesssim$1600
Nix (P2)	24.85	48 690	0.00000	0.08	50×35×33	4.5×10^{16}	~1700
Kerberos (P4)	32.17	57 750	0.00000	0.4	19×10×9	1.6×10^{16}	~2300
Hydra (P3)	38.20	64 721	0.00554	0.3	65×45×25	4.8×10^{16}	~900

Updated from Brozovic et al. (2015).

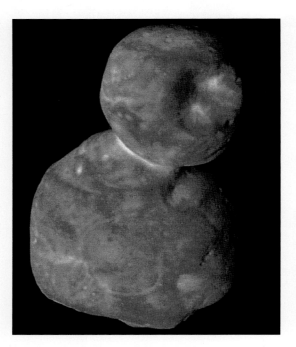

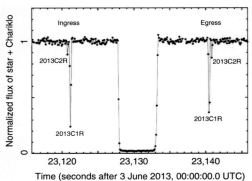

Figure G.36 Lightcurve of the occultation by the Chariklo system. The data were taken with the Danish 1.54 m telescope (La Silla) on 3 June 2013, at a rate of almost 10 Hz. The sum of the stellar and Chariklo fluxes has been normalized to unity outside the occultation. The central drop is caused by Chariklo, and two secondary events, 2013C1R and 2013C2R, are observed, first at **ingress** (before the main Chariklo occultation) and then at **egress** (after the main occultation). (Braga-Ribas et al. 2014)

Figure G.35 *New Horizons* Image of Ultima Thule, taken during its flyby with a resolution of 33 m/pixel. The KBO is 31 km across along its long axis. (NASA/JHU-APL/SWRI)

simple explanation for the dimming of the Chariklo system between 1997 and 2008, and for the gradual disappearance of ice and other absorption features in its spectrum over the same period. This implies that the rings are partly composed of water-ice. Note that for the rings to be located interior to Roche's limit (eq. 11.8), the ring particles must be significantly less dense than Chariklo itself.

Stellar occultation events revealed that the KBO 136108 Haumea is also surrounded by a ring. As described in §§12.3.2, 12.3.5, 12.3.7, Haumea is the largest member of the only family of KBOs discovered to date. Two moons also orbit Haumea. The largest of the two, with a radius of ~175 km (estimated from its brightness, assuming a similar albedo as the primary), orbits Haumea in 49 days at a distance of 49 503 km; the orbit has an eccentricity of 0.050. The smaller, inner moon (~85 km) orbits at a distance of ~ 25 150 km from the primary, in 18.3 days; the eccentricity of the orbit is

0.16, its orbit is inclined by $\sim17°$ relative to that of the larger moon. This newly discovered ring is 70 km wide, and almost circular at an orbital radius of 2287 km, close to the 3:1 mean motion resonance with Haumea's spin period. The ring is coplanar with Haumea's equator and the orbit of its larger satellite (Hi'iaka). The ring's apparent optical depth in the plane of the sky is ~0.5.

G.11.7 Planet 9 and TNOs on Sedna-like orbits

After the discovery of TNO 90377 Sedna on a highly eccentric orbit, with a perihelion at 75 AU and semimajor axis of ~500 AU (§12.2.2), several other such **Extreme TNO**s, collectively referred to as **ETNOs**, were discovered. These objects have perihelia between 35 and 80 AU, and semimajor axes between 250 and 500 AU. The orbits of most of the observed ETNOs cluster in their argument of perihelion as well as in physical space.

Such clustering is unlikely to occur by chance, suggesting a physical mechanism maintaining such orbits. An unseen planet with a mass $\gtrsim10$ $M_{\oplus}$, semimajor axis $\gtrsim700$ AU, in a very eccentric orbit in the same plane as the ETNOs but with a perihelion 180° away, could explain such clustering. Observers are attempting to photograph this hypothetical perturber, which has been nicknamed 'Planet 9'.

G.11.8 Manx Objects

Many bodies whose orbits imply they once resided in the Oort cloud have been observed over the past few centuries. All visitors from the distant reaches of our Solar System were comets that displayed comas and tails indicative of volatile material. Yet starting in the early 2010s, a few dozen asteroidal-like objects have been detected on orbits characteristic of long-period (or Oort cloud) comets. Most of these objects have been spotted by the Pan-STARRS1 telescope in Hawaii.

In contrast to bona-fide comets, these objects display zero or minimal activity, and are referred to as **Manxes** (after a breed of cats without tails). Manxes observed to date display a wide variety of surface materials, analogous to that of primitive (C, P and D) and igneous (S) asteroids.

Detection and characterization of numerous manxes might provide useful tests on formation and evolutionary models of our Solar System. Different dynamical models, depending in particular on variations in migration scenarios of the giant planets, predict different amounts and types of asteroids scattered into the Oort cloud and beyond, as well as into the inner Solar System.

G.12 Comets

G.12.1 67P/Churyumov–Gerasimenko

Comet 67P/Churyumov–Gerasimenko (67P/C–G) is a Jupiter-family comet, with an orbital period of 6.45 years, aphelion of 5.68 AU, and perihelion at 1.24 AU. The mean radius of 67P/C–G is ~2 km. The comet's rotation period is 12.4 hours, and torques from asymmetric outgassing caused the rotation period to drop by 20 minutes during its 2014–2016 passage through the inner Solar System.

The European Space Agency's *Rosetta* spacecraft rendezvoused with 67P/C–G in August 2014 and observed the comet at close proximity for over two years. During this time, the comet traveled inwards from 3.6 AU through perihelion and back out to 3.6 AU. *Rosetta* also deployed a small lander, *Philae*, in November 2014, and the spacecraft itself made a hard landing on the comet at the end of its mission in September 2016. *Rosetta* studied 67P/C–G with a combination of remote sensing and *in situ* measurements to characterize the comet's nucleus and its environs. These observations have led to substantial improvements in our

Figure G.37 Comet 67P/Churyumov–Gerasimenko imaged by the *Rosetta* spacecraft on 3 August 2014, as the spacecraft approached the comet. As shown, the comet has a double-lobed structure, probably caused by the merger of two bodies. (ESA/*Rosetta*)

Figure G.38 Comet 67P/C–G imaged by the *Rosetta* spacecraft on 11 September 2015 while the spacecraft was on the night side of the comet. This image shows the uneven distribution of activity, mostly on the sunward side, including a large collimated jet. (ESA/*Rosetta*)

knowledge of comets, and have made 67P/C–G by far the best-studied comet.

Figure G.37 shows that 67P/C–G's nucleus has two lobes, as do the majority of comets previously imaged at close range by interplanetary spacecraft, including 1P/Halley (Fig. 12.25).

The comet's decidedly non-spherical shape leads to variations in the amount of sunlit area by more than a factor of two. Moreover, the nucleus has substantial concavities. This oddly shaped nucleus leads to a coma with strong diurnal and seasonal variations in gas density, even though the distribution of ices on the surface of the nucleus of 67P/C–G is mostly uniform (variations of less than a factor of three at a resolution of 50 m), with only a single more active area that is several times (less than a factor of ten) as ice-rich as other surface regions. Figure G.38 shows the uneven distribution of jets.

Water vapor and other gases can be released at the surface of a comet and from the dust particles in the coma; in the latter case, they are referred to as **distributed sources**. At most 5% of the water leaving 67P/C–G is from distributed sources. This contrasts with some other comets,

such as 103P/Hartley 2, which release a substantial fraction of their water via distributed sources.

Comet 67P/C–G's water-loss rate varies with heliocentric distance as $r_\odot^{-4.2}$ from 3.8 AU to perihelion (see §12.7.1). Near perihelion, meter-sized chunks were observed to escape from 67P/C–G. The average integrated loss of material from the nucleus, during a single orbit, is roughly equivalent to a global layer ∼1 m deep (e.g., Problem 12.19), but may exceed 10 m in some locations.

As shown in Figure G.37, the comet's surface exhibits a wide variety of terrains, with different structures and texture. Figure G.39 shows a close-up color-enhanced view of the comet's neck region, connecting the two lobes. This region has shown much activity, including spectacular jets of dust and gas (Fig. G.38). The bluish color of the region suggests the presence of water-ice at or just below the surface.

Closeup images of both dusty smooth areas and jagged terrain on the comet are shown in Figures G.40 and G.41. The southern portion of the nucleus, which straddles both lobes, receives the

Figure G.39 False color image of Comet 67P/C–G, showing the smooth region connecting the comet's head and body. Broadband filter images centered at 989, 700 and 480 nm were combined using red, green, and blue colors, respectively, and the slight color differences have been enhanced. The spatial resolution in this image is 1.3 m per pixel. (ESA/*Rosetta*/MPS/UPD/LAM/IAA/SSO/INTA/UPM/DASP/IDA)

Figure G.40 Several boulders are strewn across a smooth terrain covered by dust, located on the lower side of the larger comet lobe. The large bright boulder in the upper left is ~45 m across and ~25 m high. The large boulder near the bottom of the frame is surrounded by many smaller boulders that seem to be appearing from beneath the smooth, dusty material. The terrain is layered near the margins, and smooth terrain is also seen at higher altitudes. The image was taken from a distance of 7.8 km from Comet 67P/C–G on 23 October 2014, and has a resolution of 83.4 cm pixel^{-1}. The size of the image is 854 x 854 m. (ESA/*Rosetta*/NAVCAM)

lion's share of the illumination near perihelion, when most of the dust is released. Some of this dust settles on the northern regions of the nucleus. This transfer may have produced the difference between the irregular surface common in the south and the smooth plains in the north.

A comparison of images taken over the course of the mission shows numerous small-scale changes, caused by processes including sublimation, ablation, jetting, dust transport, seismic shaking and mass wasting (§6.3). Examples are shown in Figures G.42 and G.43. Figure G.42 illustrates the effect of erosion on the comet, where a smooth layer of material was removed, exposing circular features (perhaps impact craters or vents). The exhumed boulder in this figure shows that a layer ~3 m deep had been removed in this process. Figure G.43 shows a jet of dust and water-ice grains emanating from a circular depression surrounded by a 10 m high wall. Although water-ice had been

detected at the location of the jet, the jet must have been powered by more energetic processes deeper in the comet than simple sublimation of surface ice.

The average bulk density of 67P/C–G's nucleus is only 533 kg m^{-3}, whereas the typical density of both pure water-ice grains and dust particles is much larger; hence the nucleus must be quite porous. Most dust mass is in the form of compact particles >1 mm in size, implying that much of the porosity is on larger scales than this, although other data imply that voids are not very large, at least down to a depth of 100 m below the surface. The thermal inertia near the surface of the comet is very low, probably resulting from the large porosity. Fluffy, very low density grains make up ~ 15% of the volume of non-volatiles in the coma, but <1% of the released mass.

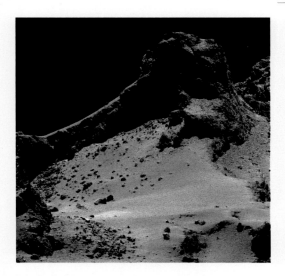

Figure G.41 View of a cliff and gravel field on the small lobe of Comet 67P/C–G, taken by *Rosetta*/OSIRIS from a distance of 8 km, on 14 October 2014. The resolution is 15 cm pixel^{-1}. (ESA/*Rosetta*/MPS/UPD/LAM/IAA/SSO/INTA/UPM/DASP/IDA)

Rosetta found that the dust mass in the coma of 67P/C–G is a few times as large as the mass of the gas (mostly water vapor), which is on the high side compared to that observed in many other comets. Note, though, that the measured dust/gas mass ratio is a lower limit to the overall non-volatile/volatile mass ratio of the comet, since larger dust grains may remain on the ground or fall back to the surface. In contrast to the relatively low water abundance, very volatile species such as N_2 and O_2 are more abundant in 67P/C–G than in most observed comets, and the percentages of CO and CO_2 are also significant ($\sim$5–10% each). The presence of this supervolatile material raises questions about the driving volatile for outbursts and nightside activity.

The D/H ratio in 67P/C–G has been measured by determining its isotopic ratios in water (HDO/H_2O, in both its ^{16}O and ^{18}O isotopes) and directly (D/H) using the mass spectrometer on *Rosetta*. The derived value of $(5.3 \pm 0.7) \times 10^{-4}$ is close to that observed in molecular clouds ($\sim 10^{-3}$) and higher than the value for many other Jupiter Family Comets (JFCs) (1.6×10^{-4} for 103P/Hartley 2 and 2.0×10^{-4} for 45P/Honda–Mrkos–Pajdusáková), and even higher than that measured in many Oort cloud comets. This suggests that the D/H ratio in comets, in particular in the JFCs, is highly heterogeneous. The combination of relatively low water

Figure G.42 Evidence of erosion in the Imhotep region: Smooth material seen in Nov. 2014 was removed by Feb. 2016, revealing now several circular features that were previously buried, as indicated by the lower arrow. The top arrow marks a boulder that was mostly covered in November 2014, and had been exhumed by Feb. 2016. A closeup on the right-most image shows the boulder in more detail. It is 4 m high, whereas in Nov. 2014 it stuck out only 1 m above the surface. The images were taken by *Rosetta's* OSIRIS camera on 22 November 2014 (left), 10 February 2016 (middle) and 25 May 2016 (right), with resolutions of 0.5 m pixel^{-1}, 0.9 m pixel^{-1} and 0.1 m pixel^{-1}, respectively. (ESA/*Rosetta*/MPS/UPD/LAM/IAA/SSO/INTA/UPM/DASP/IDA)

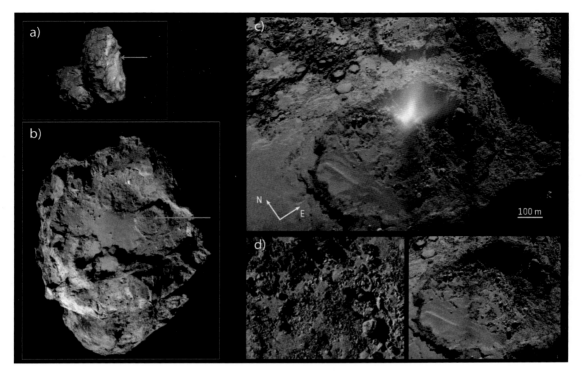

Figure G.43 a) Model of 67P/C–G highlighting the Imhotep region (indicated by the arrow). b) Image of the Imhotep region, taken by *Rosetta*'s navigation camera on 5 February 2016. c) A plume of dust imaged by ESA's *Rosetta* spacecraft on 3 July 2016. The plume appears to be powered from deep inside the comet, perhaps released from ancient gas vents or pockets of hidden ice. Later images revealed water-ice on the surface in this area. d) Comparison images of the same region: On the left, an image from 2 July, about 10 hours before the outburst, and on the right, the same region seen on 3 May. (a: comet model: ESA; b: ESA/*Rosetta*/NavCam, CC BY-SA 3.0 IGO; c,d: ESA/*Rosetta*/MPS/UPD/LAM/IAA/SSO/INTA/UPM/DASP/IDA)

with significant amounts of highly volatile species and a high D/H ratio is best explained if comet 67P/C–G is a mixture of volatile ices formed in very cold regions and minerals from the hot inner protosolar nebula.

Comet 67P/C–G has a layered structure similar to that seen on comet 9P/Tempel 1. Figure G.44 shows some of the more than 100 terraces or strata that have been identified on the comet's surface. The presence of these layers solves a long-standing conundrum as to the origin of this (and potentially other) double-lobed comet. The two leading theories in explaining the existence of such structures were the merging of two comets, or localized erosion of a single object, forming 67P/C–G's neck. Through a study of the layers of material all over the nucleus, it appears that 67P/C–G was likely formed by the merger of two comets. These two comets were probably pieces of a single comet that was split apart by an impact but separated at less than the escape velocity, so the two pieces collided at low-velocity and fused to produce 67P/C–G's double-lobe shape.

Rosetta's mass spectrometer detected numerous organic molecules, phosphorus and glycine, the simplest amino acid ($C_2H_5NO_2$), as well as its precursor molecules methylamine (CH_5N) and ethylamine (C_2H_7N). These molecules are key

Figure G.44 COLOR PLATE *Rosetta*/OSIRIS images show a multitude of terraces (green) or strata, parallel layers of material, which are outlined by red dot-dashed lines. These layers are seen in exposed cliff walls and pits all over the comet's surface. (ESA/*Rosetta*/MPS/UPD/LAM/IAA/SSO/INTA/UPM/DASP/IDA)

components of terrestrial life (§16.3). Glycine has also been detected in samples of Comet 81/P Wild 2 returned by NASA's *Stardust* spacecraft.

G.12.2 C/2017 K2 (PANSTARRS)

Comet C/2017 K2 (PANSTARRS) is inbound from the Oort cloud, heading towards a perihelion just exterior to Mars's orbit in 2022. Pre-discovery observations from 2013 show that this comet was active at a heliocentric distance $r_\odot = 23.7$ AU, the most distant location from which cometary activity has ever been observed. The measured properties are consistent with activity driven by sublimating supervolatile ices such as CO_2, CO, O_2 and N_2.

G.13 Interstellar Visitor

On 19 October 2017 the Pan-STARRS1 telescope system detected an object on a hyperbolic orbit (eccentricity $e \approx 1.20$), clearly coming from interstellar space. It was assigned a new type of designation, 1I/2017 U1, with I for Interstellar. The object was named 'Oumuamua, a Hawaiian word that conveys a 'scout' or 'messenger'.

Despite approaching within 0.25 AU from the Sun, this **InterStellar Object (ISO)** showed no sign of a dusty coma nor of released gases. However, an apparent acceleration of the object by $\sim 5 \times 10^{-6}$ m s^{-1} away from the Sun when its heliocentric distance was $r_\odot \sim 1.4$ AU is best explained by nongravitational forces, which suggests that 1I 'Oumuamua was outgassing a few kg s^{-1}.

This first ISO has a slightly reddish spectrum, similar to that of C- and D-type asteroids and cometary nuclei (Chapter 12). This spectral match suggests that 'Oumuamua is probably dark. The large amplitude of variations in its brightness implies that is very elongated, with an axis ratio of >3:1. Assuming an albedo of 0.04, the object's effective radius is of order 100 m.

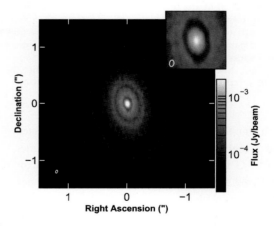

Figure G.45 ALMA (the Atacama Large Millimeter/submillimeter Array) image of the young star CI Tauri and its protoplanetary disk taken at 1.3 mm wavelength. The small white ellipse in the lower left represents the size of the beam (resolution), which corresponds to 7×4 AU. The inset at the upper right shows a $0.35''$ wide zoom on the innermost gap imaged with a slightly finer resolution. The disk is at least ~ 150 AU in radius. The color scale on the right gives the brightness in units of Jy/resolution element. (Clarke et al. 2018)

G.14 Planetary Formation

The Atacama Large Millimeter/submillimeter Array (ALMA) of telescopes in Chile is able to produce high-resolution images of disks around young stars. Figure G.45 shows a high-resolution ALMA image of the protoplanetary disk around the young ($\sim$2 Myr old) star CI Tau. The disk, which appears elongated because of our viewing angle, has three prominent gaps, the most distant of which is located $\sim$100 AU from the star. The gaps are likely produced by unseen giant planets. Radial velocity observations have detected an $\sim$10 M$_{24}$ hot jupiter orbiting CI Tau that orbits too close to the star to have an observable signature in the disk. ALMA observations provide a wealth of information on the detailed distribution of both the gas and dust in protoplanetary disks that will inform theories on planet formation.

G.15 Volcanic Activity, CO$_2$ and Mass Extinctions

The extinction rate of animal species is highly nonuniform (Fig. 16.18), and large, abrupt extinction events are referred to as mass extinctions (§16.11). The geological record of the Phanerozoic eon (the past 541 My) contains evidence for anywhere from four to six mass extinctions, depending on the threshold being used for their classification. Mass extinctions probably represent the high-magnitude tail of a broad size distribution of extinctions, rather than a separate class of events. Since the identification of the K–T extinction with a large impact on Earth (§16.7.1), geologists have searched in vain for evidence of large impacts that occurred at the same time as earlier mass extinctions, which suggests that other processes are capable of causing mass extinctions.

The largest extinction in the Phanerozoic marked the boundary between the Permian and Triassic Periods (and the Paleozoic and Mesozoic eras as well; Fig. 16.16), 252 Myr ago. This extinction occurred over a time interval of $\sim$10^4 to 10^5 years that was coincident with or immediately followed a period of extremely high volcanic activity that substantially increased the CO$_2$ content of the atmosphere. The boundary between the Triassic and Jurassic Periods, 202 Myr ago, is qualitatively similar in terms of enhanced volcanic activity, increased CO$_2$, and the types of species that went extinct. Although not as severe as the Permian–Triassic losses, the Triassic–Jurassic boundary still ranks among the largest four mass extinctions of the Phanerozoic.

The Paleocene–Eocene thermal maximum, which occurred $\sim$56 Myr ago, began with a large increase in atmospheric CO$_2$ over an interval of $\sim$5000 years. This heat spike caused many species to become extinct, but not enough were lost to classify this event as a mass extinction.

REFERENCES

Agee, C. + 15 co-authors, 2013. Unique meteorite from early Amazonian Mars: Water-rich basaltic breccia Northwest Africa 7034. *Science*, **339**, 780–785.

Albrecht, S. + 12 co-authors, 2012. Obliquities of Hot Jupiter Host Stars: Evidence for Tidal Interactions and Primordial Misalignments. *Astrophys. J.*, **757**, 18.

Anderson, J.D., and G. Schubert, 2007. Saturn's gravitational field, internal rotation, and interior structure. *Science*, **317**, 1384–1387.

Anglada-Escudé, G. + 21 co-authors, 2016. A terrestrial planet candidate in a temperate orbit around Proxima Centauri. *Nature*, **536**, 437–440.

Armstrong, J.C., C.B. Leovy, and T. Quinn, 2004. A 1 Gyr climate model for Mars: New orbital statistics and the importance of seasonally resolved polar processes. *Icarus*, **171**, 255–271.

Bagenal, F., 1992. Giant planet magnetospheres. *Annu. Rev. Earth Planet. Sci.*, **22**, 289–328.

Barshay, S.S., and J.S. Lewis, 1976. Chemistry of primitive solar material. *Annu. Rev. Astron. Astrophys.*, **14**, 81–94.

Bell, J.F., D.R. Davis, W.K. Hartmann, and M.J. Gaffey, 1989. Asteroids: The big picture. In *Asteroids II*. Eds. R.P. Binzel, T. Gehrels, and M.S. Matthews. University of Arizona Press, Tucson, pp. 921–945.

Biver, N. + 22 co-authors, 2002. The 1995–2002 long-term monitoring of Comet C/1995 O1 (HALE – BOPP) at radio wavelength. *Earth, Moon and Planets*, **90**, 5–14.

Blewett, D.T. + 17 co-authors, 2011. Hollows on Mercury: MESSENGER evidence for geologically recent volatile-related activity. *Science*, **333**, 1856–1859.

Bonfils, X. + 14 co-authors, 2018. A temperate exo-Earth around a quiet M dwarf at 3.4 parsecs. *Astron. Astrophysics*, **613**, A25, 9pp.

Bottke, W.F. + 6 co-authors, 2005b. Linking the collisional history of the main asteroid belt to its dynamical excitation and depletion. *Icarus*, **179**, 63–94.

Braga-Ribas, F. + 55 co-authors, 2013. The size, shape, albedo, density, and atmospheric limit of transneptunian object (50000) Quaoar from multi-chord stellar occultations. *Astrophys. J.* **773**, 26, 13pp.

Braga-Ribas, F. + 63 co-authors, 2014. A ring system detected around the Centaur (10199) Chariklo. *Nature*, **5-8**, 72–75.

Brown, M.E., D. Ragozzine, J. Stansberry, and W.C. Fraser, 2010. The size, density, and formation of the Orcus-Vanth system in the Kuiper belt. *Astron. J.*, **139**, 2700–2705.

Brownlee, D.E., and M.E. Kress, 2007. Formation of Earth-like habitable planets. In *Planets and Life: The Emerging Science of Astrobiology*. Eds. W.T. Sullivan III and J.A. Baross. Cambridge University Press, Cambridge, pp. 69–90.

Brownlee, D. + many co-authors, 2006. Comet 81 P/Wild 2 under a microscope. *Science*, **314**, 1711–1716.

Buratti, B.J. + 16 co-authors, 2017. Global albedos of Pluto and Charon from LORRI *New Horizons* observations. *Icarus*, **287**, 207–217.

Burns, J.A., P.L. Lamy, and S. Soter, 1979. Radiation forces on small particles in the Solar System. *Icarus*, **40**, 1–48.

Burns, J.A., M.R. Showalter, and G.E. Morfill, 1984. The ethereal rings of Jupiter and Saturn. In *Planetary Rings*. Eds. R. Greenberg and A. Brahic. University of Arizona Press, Tucson, pp. 200–272.

Burns, J.A. + 5 co-authors, 1999. The formation of Jupiter's faint rings. *Science*, **284**, 1146–1150.

Burrows, A. + 8 co-authors, 1997. A non-gray theory of extrasolar giant planets and brown dwarfs. *Astrophys. J.*, **491**, 856–875.

Canup, R.M., 2004. Simulations of a late lunar-forming impact. *Icarus*, **168**, 433–456.

Carr, M.H., 1999. Mars: Surface and interior. In *Encyclopedia of the Solar System*. Eds. L. McFadden, P.R. Weissman, and T.V. Johnson. Academic Press, San Diego, pp. 291–308.

Catling, D., and J.F. Kasting, 2007. Planetary Atmospheres and Life. In *Planets and Life: The Emerging Science of Astrobiology*. Eds. W.T. Sullivan III and J.A. Baross. Cambridge University Press, Cambridge, UK, pp. 91–116.

Chamberlain, J.W., and D.M. Hunten, 1987. *Theory of Planetary Atmospheres*. Academic Press, New York. 481pp.

Charbonneau, D. + 7 co-authors, 2008. The broadband spectrum of the exoplanet HD 189733b. *Astropyhs. J.*, **686**, 1341–1348.

Chiang, E.I. + 5 co-authors, 2007. A brief history of transneptunian space. In *Protostars and Planets V*. Eds. B. Reipurth, D. Jewitt, and K. Keil. University of Arizona Press, Tucson, pp. 895–911.

Clark, R.N., F.P. Fanale, and M.J. Gaffey, 1986. Surface composition of natural satellites. In *Satellites*. Eds. J.A. Burns and M.S. Matthews. University of Arizona Press, Tucson, pp. 437–491.

Colina, L., R.C. Bohlin, and F. Castelli, 1996. The 0.12–2.5 micron absolute flux distribution of the Sun for comparison with solar analog star. *Astron. J.*, **112**, 307–315.

Colwell, J.E., P.D. Nicholson, M.S. Tiscareno, C.D. Murray, R.G. French, and E.A. Marouf, 2009. The Structure of Saturn's Rings. In *Saturn from Cassini-Huygens*. Eds. M. Dougherth, L. Esposito, and T. Krimigis. Springer: Heidelberg, pp. 375–412.

Coustenis, A. + 24 co-authors, 2007. The composition of Titan's stratosphere from Cassini/CIRS mid-infrared spectra. *Icarus*, **189**, 35–62.

Cowley, S.W.H., 1995. The Earth's magnetosphere: A brief beginner's guide. *EOS*, **51**, 525–529.

Cuzzi, J.N. + 6 co-authors, 1984. Saturn's rings: Properties and processes. In *Planetary Rings*. Eds. R. Greenberg and A. Brahic. University of Arizona Press, Tucson, pp. 73–199.

D'Angelo, G., W. Kley, and T. Henning, 2003. Orbital migration and mass accretion of protoplanets in three-dimensional global computations with nested grids. *Astrophys. J.*, **586**, 540–561.

Del Genio, A. + 6 co-authors, 2009. Saturn atmospheric structure and dynamics. In *Saturn from Cassini-Huygens*. Eds. M. Dougherty, L. Esposito, and T. Krimigis. Springer-Verlag, Berlin. 805pp.

de Pater, I., and J.J. Lissauer, 2010. *Planetary Sciences*, 2nd Edition. Cambridge University Press. 647pp.

de Pater, I., M.H. Wong, P.S. Marcus, S. Luszcz-Cook, M. Ádámkovics, A. Conrad, X. Asay-Davis, and C. Go, 2010. Persistent rings in and around Jupiter's anticyclones – Observations and theory. *Icarus*, **210**, 742–762.

de Pater, I., F. van der Tak, R.G. Strom, and S.H. Brecht, 1997. The evolution of Jupiter's radiation belts after the impact of comet D/Shoemaker–Levy 9. *Icarus*, **129**, 21–47. Erratum (Fig. reproduction): 1998, **131**, 231.

de Pater, I., D. Dunn, K. Zahnle, and P.N. Romani, 2001. Comparison of Galileo probe data with ground-based radio measurements. *Icarus*, **149**, 66–78.

de Pater, I. + 6 co-authors, 2004a. Keck AO observations of Io in and out of eclipse. *Icarus*, **169**, 250–263.

de Pater, I., S. Martin, and M.R. Showalter, 2004b. Keck near-infrared observations of Saturn's E and G rings during Earth's ring plane crossing in August 1995. *Icarus*, **172**, 446–454.

de Pater, I. + 8 co-authors, 2005. The dynamic neptunian ring arcs: Evidence for a gradual disappearance of Liberté and a resonant jump of Courage. *Icarus*, **174**, 263–272.

de Pater, I., H.B. Hammel, S.G. Gibbard, and M.R. Showalter, 2006a. New dust belts of Uranus: One ring, two ring, red ring, blue ring. *Science*, **312**, 92–94.

de Pater, I. + 8 co-authors, 2006b. Titan imagery with Keck AO during and after probe entry. *J. Geophys. Res.*, **111**, E07S05.

de Pater, I. + 7 co-authors, 2010. Persistent rings in and around Jupiter's anticyclones – Observations and theory. *Icarus*, **210**, 742–762.

de Pater, I., R.J. Sault, M.H. Wong, L.N. Fletcher, D. DeBoer, and B. Butler, 2019. Jupiter's ammonia distribution derived from VLA maps at 3–37 GHz. *Icarus*, in press.

Dermott, S.F., and C.D. Murray, 1981. The dynamics of tadpole and horseshoe orbits. I: Theory. *Icarus*, **48**, 1–11.

Descamps, P. + 19 co-authors, 2007. Figure of the double asteroid 90 Antiope from adaptive optics and lightcurve observations. *Icarus*, **187**, 482–499.

Descamps, P. + 18 co-authors, 2008. New determination of the size and bulk density of the binary asteroid 22 Kalliope from observations of mutual eclipses. *Icarus*, **196**, 578–600.

Descamps, P. + 18 co-authors, 2010. Triplicity and physical characteristics of Asteroid (216) Kleopatra. *Icarus*, **211**, 1022–1033.

Desch, M.D. + 6 co-authors, 1991. Uranus as a radio source. In *Uranus*. Eds. J.T. Bergstrahl, A.D. Miner, and M.S. Matthews. University of Arizona Press, Tucson, pp. 894–925.

des Marais, D.-J. + 9 co-authors, 2002. Remote sensing of planetary properties and biosignatures on extrasolar terrestrial planets. *Astrobiology*, **2**, 153–181.

des Marais, D.J. et al. 2003. The NASA Astrobiology Roadmap. *Astrobiology*, **3**, 219–235.

Dressing, C.D., and D. Charbonneau, 2015. The occurrence of potentially habitable planets orbiting M dwarfs estimated from the full *Kepler* dataset and an empirical measurement of the detection sensitivity. *Astrophys. J.*, **807**, 45, 23pp.

Duncan, M.J., and T. Quinn, 1993. The long-term dynamical evolution of the Solar System. *Annu. Rev. Astron. Astrophys.*, **31**, 265–295.

Elliot, J.L., E. Dunham, and D. Mink, 1977. The rings of Uranus. *Nature*, **267**, 328–330.

Esposito, L.W., 1993. Understanding planetary rings. *Annu. Rev. Earth Planet. Sci.*, **21**, 487–521.

Etheridge, D.M. + 5 co-authors, 1996. Natural and anthropogenic changes in atmospheric CO_2 over the last 1000 years from air in Antarctic ice and firn. *J. Geophys. Res.*, **101**, 4115–4128.

Farrington, O., 1915. *Meteorites, Their Structure, Composition and Terrestrial Relations*. Chicago, published by the author.

Fedorov, A.V. + 7 co-authors, 2006. The Pliocene paradox (mechanisms for a permanent El Nino). *Science*, **312**, 1485–1491.

Fischer, D.A., and J. Valenti, 2005. The planet–metallicity correlation. *Astrophys. J.*, **622**, 1102–1117.

Flammaron, N.C., 1892. *La Planète Mars et ses Conditions d'Habitabilité* (2 vols). Gauthier-Villars, Paris.

Flasar, F.M. + 45 co-authors, 2005. Temperatures, winds, and composition in the saturnian system. *Science*, **307**, 1247–1251.

Fletcher, L.N. + 9 co-authors, 2007. Characterising Saturn's vertical temperature structure from Cassini/CIRS. *Icarus*, **189**, 457–478.

Fletcher, L.N. + 6 co-authors, 2008. Deuterium in the outer planets: New constraints and new questions from infrared spectroscopy. *AGU Fall Meeting Abstracts*, #P21B-04.

Forbes, J.M., F.G. Lemoine, S.L. Bruinsma, M.D. Smith, and X. Zhang, 2008. Solar flux variability of Mars' exosphere densities and temperatures. *Geophys. Res. Lett.*, **35**, L01201.

Formisano, V., S. Atreya, T. Encrenaz, N. Ignatiev, and M. Giuranna, 2004. Detection of methane in the atmosphere of Mars. *Science*, **306**, 1758–1761.

Fowler, C.M.R., 2005. *The Solid Earth: An Introduction to Global Geophysics*. 2nd Edition. Cambridge University Press, New York. 685pp.

Fulton, B.J., and E.A. Petagura 2018. The California *Kepler* Survey VII. Precise planet radii leveraging Gaia DR2 reveal the stellar mass dependence of the planet radius gap. eprint arXiv:1805.01453.

Ghil, M., and S. Childress, 1987. *Topics in Geophysical Fluid Dynamics: Atmospheric Dynamics, Dynamo Theory, and Climate Dynamics*. Springer-Verlag, New York. 485pp.

Goody, R.M., and J.C.G. Walker, 1972. *Atmospheres*. Prentice Hall, Englewood Cliffs, NJ. 160pp.

Gosling, J.T., 2007. The solar wind. In *Encyclopedia of the Solar System*, 2nd Edition. Eds. L. McFadden, P.R. Weissman, and T.V. Johnson. Academic Press, San Diego, pp. 99–116.

Gradie, J.C., C.R. Chapman, and E.F. Tedesco, 1989. Distribution of taxonomic classes and the compositional structure of the asteroid belt. In *Asteroids II*. Eds. R.P. Binzel, T. Gehrels, and M.S. Matthews. University of Arizona Press, Tucson, pp. 316–335.

Graham, J.R., I. de Pater, J.G. Jernigan, M.C. Liu, and M.E. Brown, 1995. W.M. Keck telescope observations of the Comet P/Shoemaker–Levy 9 fragment R Jupiter collision. *Science*, **267**, 1320–1323.

Greeley, R., 1994. *Planetary Landscapes*, 2nd Edition. Chapman and Hall, New York, London. 286pp.

Grießmeier, J.-M., A. Stadelmann, J.L. Grenfell, H. Lammer, and U. Motschmann, 2009. On the protection of extrasolar Earth-like planets around K/M stars against galactic cosmic rays. *Icarus*, **199**, 526–535.

Grimm, S.L. + 25 co-authors, 2018. The nature of the TRAPPIST-1 exoplanets. *Astron. Astrophys.*, **613**, A68, 21pp.

Guillot, T., 1999. Interiors of giant planets inside and outside the Solar System. *Science*, **286**, 72–77.

Guillot, T., G. Chabrier, D. Gautier, and P. Morel, 1995. Effect of radiative transport on the evolution of Jupiter and Saturn. *Astrophys. J.*, **450**, 463–472.

Hamblin, W.K., and E.H. Christiansen, 1990. *Exploring the Planets*. Macmillan Publishing Company, New York. 451pp.

Hamilton, D.P., 1993. Motion of dust in a planetary magnetosphere: Orbit-averaged equations for oblateness, electromagnetic, and radiation forces with application to Saturn's E ring. *Icarus*, **101**, 244–264.

Hammel, H.B. + 9 co-authors, 1995. HST imaging of atmospheric phenomena created by the impact of comet Shoemaker–Levy 9. *Science*, **267**, 1288–1295.

Hanel, R.A., B.J. Conrath, D.E. Jennings, and R.E. Samuelson, 1992. *Exploration of the Solar System by Infrared Remote Sensing*. Cambridge University Press, Cambridge. 458pp.

Harmon, J.K., M.A. Slade, and M.S. Rice, 2011. Radar imagery of Mercury's putative polar ice: 1999–2005. Arecibo results. *Icarus*, **211**, 37–50.

Hartmann, W.K., 1989. *Astronomy: The Cosmic Journey*. Wadsworth Publishing Company, Belmont, CA. 698pp.

Hartmann, W.K., 2005. *Moons and Planets*, 5th Edition. Brooks/Cole, Thomson Learning, Belmont, CA. 428pp.

Haynie, D.T., 2008. *Biological Thermodynamics*, 2nd Edition. Cambridge University Press. 422pp.

Holman, M.J., 1997. A possible long-lived belt of objects between Uranus and Neptune. *Nature*, **387**, 785–788.

Hood, L., and M.T. Zuber, 2000. Recent refinements in geophysical constraints on lunar origin and evolution. In *Origin of the Earth and Moon*. Eds. R. Canup and K. Righter. University of Arizona Press, Tucson, pp. 397–409.

Howard, A.W. + 9 co-authors, 2010. The occurrence and mass distribution of close-in super-Earths, Neptunes, and Jupiters. *Science*, **330**, 653–655.

Hubbard, W.B., 1984. *Planetary Interiors*. Van Nostrand Reinhold Company Inc., New York. 334pp.

Hubbard, W.B., M. Podolak, and D.J. Stevenson, 1995. The interior of Neptune. In *Neptune and Triton*. Ed. D.P. Cruikshank. University of Arizona Press, Tucson, pp. 109–138.

Hueso, R. + 16 co-authors, 2010. First Earth-based detection of a superbolide on Jupiter. *App. J. Lett.*, **721**, L129–L133.

Hundhausen, A.J., 1995. The solar wind. In *Introduction to Space Physics*. Eds. M.G. Kivelson and C.T. Russell. Cambridge University Press, Cambridge, UK, pp. 91–128.

Hunten, D.M., T.M. Donahue, J.C.G. Walker, and J.F. Kasting, 1989. Escape of atmospheres and loss of water. In *Origin and Evolution of Planetary and Satellite Atmospheres*. Eds. S.K. Atreya, J.B. Pollack, and M.S. Matthews. University of Arizona Press, Tucson, pp. 386–422.

Huygens, C., 1659. *Systema Saturnia*.

Iess, I. + 26 co-authors, 2018. Measurement of Jupiter's asymmetric gravity field. *Nature*, **555**, 220–222.

Jacobson, R.A. + 6 co-authors, 2008. Revised orbits of Saturn's small inner satellites. *Astron. J.*, **135**, 261–263.

Jacobson, R., 2010. Orbits and masses of the Martian satellites and the libration of Phobos. *Astron. J.*, **139**, 668–679.

Jacobson, R.M. + 5 co-authors, 2012. Irregular satellites of the outer planets: Orbital uncertainties and astrometric recoveries in 2009–2011. *Astron. J.*, **144**, 132–139.

Jakosky, B. 1998. *The Search for Life on Other Planets*. Cambridge University Press, New York. 326pp.

Jakosky, B.M., F. Westall, and A. Brack, 2007. Mars. In *Planets and Life: The Emerging Science of Astrobiology*. Eds. W.T. Sullivan III and J.A. Baross. Cambridge University Press, Cambridge, UK, pp. 357–387.

James, P.B., M.T. Zuber, R.J. Phillips, and S.C. Solomon, 2015. Support of long-wavelength topography on Mercury inferred from *MESSENGER* measurements of gravity and topography. *J. Geophys. Res. Planets*, **120**, 287–310.

Janssen, M.J., et al., 2017. MWR: Microwave radiometer for the Juno mission to Jupiter. *Space Sci. Rev.* **213**, 139–185. DOI 10.1007/s11214-017-0349-5.

Janssen, M.A., A.P. Ingersoll, M.D. Allison, et al., 2013. Saturn's thermal emission at 2.2 cm wavelength as imaged by the Cassini RADAR radiometer. *Icarus*, **226**, 522–535.

Johnson, J.A. + 22 co-authors, 2011. HAT-P-30b: A Transiting Hot Jupiter on a Highly Oblique Orbit. *Astrophys. J.*, **735**, 24.

Kary, D.M., and L. Dones, 1996. Capture statistics of short-period comets: Implications for Comet D/Shoemaker–Levy 9. *Icarus*, **121**, 207–224.

Kerridge, J.F., 1993. What can meteorites tell us about nebular conditions and processes during planetesimal accretion? *Icarus*, **106**, 135–150.

Kivelson, M.G., and F. Bagenal, 1999. Planetary magnetospheres. In *Encyclopedia of the Solar System*. Eds. P.R. Weissman, L. McFadden, and T.V. Johnson. Academic Press, Inc., New York, pp. 477–498.

Kivelson, M.G., and F. Bagenal, 2007. Planetary magnetospheres. In *Encyclopedia of the Solar System*,

2nd Edition. Eds. L. McFadden, P.R. Weissman, and T.V. Johnson. Academic Press, San Diego, pp. 519–540.

Kivelson, M.G., and G. Schubert, 1986. Atmospheres of the terrestrial planets. In *The Solar System: Observations and Interpretations*. Rubey Vol. IV. Ed. M.G. Kivelson. Prentice Hall, Englewood Cliffs, NJ, pp. 116–134.

Knutson, H.A. + 8 co-authors, 2007. A map of the day–night contrast of the extrasolar planet HD 189733b. *Nature*, **447**, 183–186.

Kokubo, E., and S. Ida, 2000. Formation of protoplanets from planetesimals in the solar nebula. *Icarus*, **143**, 15–27.

Kokubo, E., R.M. Canup, and S. Ida, 2000. Lunar accretion from an impact-generated disk. In *Origin of the Earth and Moon*. Eds. R.M. Canup and K. Righter. University of Arizona Press, Tucson, pp. 145–163.

Konacki, M., and A. Wolszczan, 2003. Masses and orbital inclinations of planets in the PSR B1257+12 system. *Astrophys. J.*, **597**, 1076–1091.

Konopliv, A.S., A.B. Binder, L.L. Hood, A.B. Kucinskas, W.L. Sjogren, and J.G. Williams, 1998. Improved gravity field of the Moon from Lunar Prospector. *Science*, **281**, 1476–1480.

Kowal, C.T., 1996. *Asteroids. Their nature and utilization*. Wiley, Chichester (UK), XVII + 153pp.

Kring, D., 2003. Environmental consequences of impact cratering events as a function of ambient conditions on Earth. *Astrobiology*, **3**, 133–152.

Laskar, J., T. Quinn, and S. Tremaine, 1992. Confirmation of resonant structure in the Solar System. *Icarus*, **95**, 148–152.

Levison, H.F., and L. Dones, 2007. Comet populations and cometary dynamics. In *Encyclopedia of the Solar System*, 2nd Edition. Eds. L. McFadden, P.R. Weissman, and T.V. Johnson. Academic Press, San Diego, pp. 575–588.

Lewis, J.S., 1995. *Physics and Chemistry of the Solar System*, Revised Edition. Academic Press, San Diego. 556pp.

Lillis, R.J. + 5 co-authors, 2008. An improved crustal magnetic field map of Mars from electron reflectometry: Highland volcano magmatic history and the end of the martian dynamo. *Icarus*, **194**, 575–596.

Lipschutz, M.E., and L. Schultz, 2007. Meteorites. In *Encyclopedia of the Solar System*, 2nd Edition. Eds. L. McFadden, P.R. Weissman, and T.V. Johnson. Academic Press, San Diego, pp. 251–282.

Lissauer, J.J., 1999. How common are habitable planets? *Nature*, **402**, C11–C14.

Lissauer, J.J., O. Hubickyj, G. D'Angelo, and P. Bodenheimer, 2009. Models of Jupiter's growth incorporating thermal and hydrodynamic constraints. *Icarus*, **199**, 338–350.

Lissauer, J.J. + 16 co-authors, 2013. All Six Planets Known to Orbit Kepler-11 Have Low Densities. *Astrophys. J.*, **770**, 131.

Lithgow-Bertelloni, C., and M.A. Richards, 1998. The dynamics of cenozoic and mesozoic plate motions. *Rev. Geophys.*, **36**, 27–78.

Lockwood, A.C., M.E. Brown, and J. Stansberry, 2014. The size and shape of the oblong dwarf planet Haumea. *Earth, Moon & Planets*, **111**, 127–137.

Lodders, K., 2003. Solar System abundances and condensation temperatures of the elements. *Astrophys. J.*, **591**, 1220–1247.

Lodders, K., 2010. Solar System abundances of the elements. In *Principles and Perspectives in Cosmochemistry*, Astrophysics and Space Science Proceedings. Springer-Verlag, Berlin, pp. 379–417.

Lovett, L., J. Horvath, and J. Cuzzi, 2006. *Saturn: A New View*. H.N. Abrams, New York. 192pp.

Luhmann, J.G., 1995. Plasma interactions with unmagnetized bodies. In *Introduction to Space Physics*. Eds. M.G. Kivelson and C.T. Russell. Cambridge University Press, Cambridge, UK, pp. 203–226.

Luhmann, J.G., C.T. Russell, L.H. Brace, and O.L. Vaisberg, 1992. The intrinsic magnetic field and solar-wind interaction of Mars. In *Mars*. Eds. H.H. Kieffer, B.M. Jakosky, C.W. Snyder, and M.S. Matthews. University of Arizona Press, Tucson, pp. 1090–1134.

Luhmann, J.G., and S.C. Solomon, 2007. The Sun–Earth connection. In *Encyclopedia of the Solar System*, 2nd Edition. Eds. L. McFadden, P.R. Weissman, and T.V. Johnson. Academic Press, San Diego, pp. 213–226.

Lunine, J.I., 2013. *Earth: Evolution of a Habitable World*, 2nd Edition. Cambridge University Press, Cambridge, UK. 304pp.

Lunine, J.I., 2005. *Astrobiology: A Multi-Disciplinary Approach*. Pearson Education, San Francisco. 586pp.

Lunine, J.I., and W.C. Tittemore, 1993. Origins of outer-planet satellites. In *Protostars and Planets III*. Eds. E.H. Levy and J.I. Lunine. University of Arizona Press, Tucson, pp. 1149–1176.

Lyons, L.R., and D.J. Williams, 1984. *Quantitative Aspects of Magnetospheric Physics*. Reidel Publishing Company, Dordrecht. 231pp.

Malin, M.C., and K.S. Edgett, 2000. Evidence for recent groundwater seepage and surface runoff on Mars. *Science*, **288**, 2330–2335.

Marsden, B.G., and G.V. Williams, 2003. *Catalogue of Cometary Orbits*, 15th Edition. The International Astronomical Union, Minor Planet Center and Smithsonian Astrophysical Observatory, Cambridge, MA.

Mayor, M., and D. Queloz, 1995. A Jupiter-mass companion to a solar-type star. *Nature*, **378**, 355–359.

Meadows, V., and S. Seager, 2010. Terrestrial planet atmospheres and biosignatures. In *Exoplanets*. Ed. S. Seager. University of Arizona Press, Tucson, pp. 441–470.

Melosh, H.J., 1989. *Impact Cratering: A Geologic Process*. Oxford Monographs on Geology and Geophysics, No. 11. Oxford University Press, New York. 245pp.

Mohanty, S., R. Jayawardhana, N. Hulamo, and E. Mamajek, 2007. The planetary mass companion 2MASS 1207-3932B: Temperature, mass, and evidence for an edge-on disk. *Astrophys. J.*, **657**, 1064–1091.

Morbidelli, A., 2002. *Modern Celestial Mechanics: Aspects of Solar System Dynamics*. Taylor and Francis/Cambridge Scientific Publishers, London. 368pp. (Out of print; see http://www.oca.eu/morby/.)

Morgan, J. + 18 co-authors + the Chicxulub Working Group, 1997. Size and morphology of the Chicxulub impact crater. *Nature*, **390**, 472–476.

Morrison, D., and T. Owen, 1996. *The Planetary System*. 2nd Edition. Addison-Wesley Publishing Company, New York.

Morrison, D., and T. Owen, 2003. *The Planetary System*, 3rd Edition. Addison-Wesley Publishing Company, New York. 531pp.

Murchie, S.L. + 10 co-authors, 2008. Geology of Caloris basin, Mercury: A view from MESSENGER. *Science*, **321**, 73–76.

Murray, C., and S. Dermott, 1999. *Solar System Dynamics*. Cambridge University Press, Cambridge, UK. 592pp.

Ness, N.F., J.E.P. Connerney, R.P. Lepping, M. Schulz, and G.-H. Voigt, 1991. The magnetic field and magnetospheric configuration of Uranus. In *Uranus*. Eds. J.T. Bergstrahl, E.D. Miner, and M.S. Matthews. University of Arizona Press, Tucson, pp. 739–779.

Nicholson, P.D., 2009. Natural satellites of the planets. In *Observer's Handbook*. Ed. P. Kelly, Royal Academic Society of Canada, pp. 24–30.

Pace, N.R., 1997. A molecular view of microbial diversity and the biosphere. *Science*, **276**, 734–740.

Pál, A. + 14 co-authors, 2012. "TNOs are Cool": A survey of the trans-Neptunian region. VII. Size and surface characteristics of (90377) Sedna and 2010 EK_{139}. *Astronomy & Astrophysics*, **541**, L6.

Palme, H., and W.N. Boynton, 1993. Meteoritic constraints on conditions in the solar nebula. In *Protostars and Planets III*. Eds. E.H. Levy and J.I. Lunine. University of Arizona Press, Tucson, pp. 979–1004.

Pasachoff, J.M., and M.L. Kutner, 1978. *University Astronomy*. W.B. Saunders Company, Philadelphia. 851pp.

Perryman, M., 2011. *The Exoplanet Handbook*. Cambridge University Press, Cambridge, UK. 410pp.

Perryman, M.A.C. + 22 co-authors, 1995. Parallaxes and the Hertzsprung–Russell diagram for the preliminary HIPPARCOS solution H30. *Astron. Astrophys.*, **304**, 69–81.

Pilcher, F., S. Mottola, and T. Denk, 2012. Photometric lightcurve and rotation period of Himalia (Jupiter VI). *Icarus*, **219**, 741–742.

Porco, C.C., P.C. Thomas, J.W Weiss, and D.C. Richardson, 2007. Saturn's small inner satellites: Clues to their origins. *Science*, **318**, 1602–1607.

Pravec, P., A.W. Harris, and B.D. Warner, 2007. NEA rotations and binaries. *Near-Earth Objects: Our Celestial Neighbors – Opportunity and Risk*, Proceedings IAU Symposium No. 236. Eds. A. Milani, G.B. Valsecchi, and D. Vokrouhlický, pp. 167–176.

Press, F., and R. Siever, 1986. *Earth*. W.H. Freeman and Company, New York. 626pp.

Pudritz, R., P. Higgs, and J. Stone, 2007. *Planetary Systems and the Origins of Life*. Cambridge University Press, Cambridge, UK. 315pp.

Purvis, W., G. Orians, C. Heller, and D. Sadava, 2004. *The Science of Biology*, 7th Edition. W.H. Freeman, New York.

Putnis, A., 1992. *Mineral Science*. Cambridge University Press, Cambridge, UK. 457pp.

Russell, C.T., 1995. A brief history of solar-terrestrial physics. In *Introduction to Space Physics*. Eds. M.G. Kivelson and C.T. Russell. Cambridge University Press, Cambridge, pp. 1–26.

Santos-Sanz, P. + 22 co-authors, 2012. "TNOs are cool": A survey of the trans-Neptunian region. IV. Size/albedo characterization of 15 scattered disk and detached objects observed with Herschel-PACS. *Astron. Astrophys.*, **541**, 92.

Schubert, G., J.D. Anderson, T. Spohn, and W.B. McKinnon, 2004. Interior composition, structure and dynamics of the Galilean satellites. In *Jupiter: Planet, Satellites and Magnetosphere*. Eds. F. Bagenal, T.E. Dowling, and W. McKinnon. Cambridge University Press, Cambridge, UK, pp. 281–306.

Seager, S., E.L. Turner, J. Schafer, and E.B. Ford, 2005. Vegetation's red edge: A possible spectroscopic biosignature of extraterrestrial planets. *Astrobiology*, **5**, 372–390.

Sekanina, Z., and J.A. Farrell, 1978. Comet West 1976. VI: Discrete bursts of dust, split nucleus, flare-ups, and particle evaporation. *Astron. J.*, **83**, 1675–1680.

Showalter, M.R., 1996. Saturn's D ring in the Voyager images. *Icarus*, **124**, 677–689.

Showalter, M.R., and J.J. Lissauer, 2006. The second ring–moon system of Uranus: Discovery and dynamics. *Science*, **311**, 973–977.

Shu, F.H., 1982. *The Physical Universe: An Introduction to Astronomy*. University Science Books, Berkeley, CA. 584pp.

Shu, F.H., J.N. Cuzzi, and J.J. Lissauer, 1983. Bending waves in Saturn's rings. *Icarus*, **53**, 185–206.

Sicardy, B. + 62 co-authors, 2011. A Pluto-like radius and a high albedo for the dwarf planet Eris from an occultation. *Nature*, **478**, 493–496.

Smith, B.A., et al., 1982. A new look at the Saturn system: The Voyager 2 images. *Science*, **215**, 504–537.

Smith, D.E., M.Y. Zuber, G.A. Neumann, and F.G. Lemoine, 1997. Topography of the Moon from the Clementine Lida. *J. Geophys. Res.*, **102**, 1591.

Smith, D.E. + 16 co-authors, 2012. Gravity field and internal structure of Mercury from MESSENGER. *Science*, **336**, 214–217.

Smith, M.D., 2004. Interannual variability in TES atmospheric observations of Mars during 1999–2003. *Icarus*, **167**, 148–165.

Smrekar, S.E., and E.R. Stofan, 2007. Venus: Surface and interior. *Encyclopedia of the Solar System*, 2nd Edition. Eds. L. McFadden, P.R. Weissman, and T.V. Johnson. Academic Press, San Diego, pp. 149–168.

Sneppen, K., and G. Zocchi, 2005. *Physics in Molecular Biology*. Cambridge University Press, Cambridge, UK. 311pp.

Solomon, S.C. + 10 co-authors, 2008. Return to Mercury: A global perspective on MESSENGER's first Mercury flyby. *Science*, **321**, 59–62.

Sotin, C., O. Grasset, and A. Mocquet, 2007. Mass-radius curve for extrasolar Earth-like planets and ocean planets. *Icarus*, **191**, 337–351.

Sromovsky, L.A., P.M. Frye, T. Dowling, K.H. Baines, and S.S. Limaye, 2001. Neptune's atmospheric circulation and cloud morphology: Changes revealed by 1998 HST imaging. *Icarus*, **150**, 244–260.

Sromovsky, L.A. + 7 co-authors, 2009. Uranus at Equinox: Cloud morphology and dynamics. *Icarus*, **203**, 265–286.

Stansberry, J. + 6 co-authors, 2008. Physical properties of Kuiper belt and Centaur objects: Constraints from Spitzer Space Telescope. In *The Solar System beyond Neptune*. Eds. M.A. Barucci et al. University of Arizona Press, Tucson, pp. 161–179.

Stern, S.A., W.M. Grundy, W.B. McKinnon, H.A. Weaver, and L.A. Young, 2018. The Pluto system after *New Horizons*. *Ann. Rev. Astron. Astrophys.*, **56**, 357–392.

Stevenson, D.J., 1982. Interiors of the giant planets. *Annu. Rev. Earth Planet. Sci.*, **10**, 257–295.

Stevenson, D.J., and E.E. Salpeter, 1976. Interior models of Jupiter. In *Jupiter*. Eds. T. Gehrels and M.S. Matthews. University of Arizona Press, Tucson, pp. 85–112.

Stix, M., 1987. In *Solar and Stellar Physics*, Lecture Notes Phys., **292**. Eds. E.H. Schröter and M. Schüssler. Springer, Berlin, p. 15.

Stuart, J.S., and R.P. Binzel, 2004. Bias-corrected population, size distribution, and impact hazard for the near-Earth objects. *Icarus*, **170**, 295–311.

Taylor, S.R., 1992. *Solar System Evolution: A New Perspective*. Cambridge University Press, Cambridge, UK. 307pp.

Taylor, S.R., 2007. The Moon. In *Encyclopedia of the Solar System*, 2nd Edition. Eds. L. McFadden, P. Weissman, and T.V. Johnson. Academic Press, San Diego, pp. 227–250.

Thomas, P.C., 2010. Sizes, shapes, and derived properties of the saturnian satellites after the Cassini nominal mission. *Icarus*, **208**, 395–401.

Thomas, P.C. + 7 co-authors, 1998. Small inner satellites of Jupiter. *Icarus*, **135**, 360–371.

Thomas, P. C. + 12 co-authors, 2007. Shapes of the saturnian icy satellites and their significance. *Icarus*, **190**, 573–584.

Tiscareno, M.S., P.C. Thomas, and J.A. Burns, 2009. The rotation of Janus and Epimetheus. *Icarus*, **204**, 254–261.

Turcotte, D.L., and G. Schubert, 2002. *Geodynamics*, 2nd Edition. Cambridge University Press, New York. 456pp.

Van Grootel, V. + 14 co-authors, 2018. Stellar parameters for TRAPPIST-1. *Astrophys. J.*, **853**, 30, 7pp.

Vasavada, A.R., and A.P. Showman, 2005. Jovian atmospheric dynamics: An update after Galileo and Cassini. *Rep. Prog. Physics*, **68**, 1935–1996.

Ward, P.D., 2007. Mass extinctions. In *Planets and Life: The Emerging Science of Astrobiology*. Eds. W.T. Sullivan III and J.A. Baross, Cambridge University Press, Cambridge, UK, pp. 335–354.

Weisberg, M.K., T.J. McCoy, and A.N. Krot, 2006. Systematics and evaluation of meteorite classification. In *Meteorites and the Early Solar System II*. Eds. D.S. Lauretta and H.Y. McSween Jr. University of Arizona Press, Tucson, pp. 19–52.

Weissman, P.R., 1986. Are cometary nuclei primordial rubble piles? *Nature*, **320**, 242–244.

Wilcox, J.M., and N.F. Ness, 1965. Quasi-stationary corotating structure in the interplanetary medium. *J. Geophys. Res.*, **70**, 5793–5805.

Williams, J., 1992. *The Weather Book*. Vintage Books, New York. 212pp.

Winn, J.N. + 11 co-authors, 2006. Measurement of the Spin-Orbit Alignment in the Exoplanetary System HD 189733. *Astrophys. J. Lett.*, **652**, L69–L72.

Wisdom, J., 1983. Chaotic behavior and the origin of the 3/1 Kirkwood Gap. *Icarus*, **56**, 51–74.

Yoder, C.F., 1995. Astrometric and geodetic properties of Earth and the Solar System. In *Global Earth Physics: A Handbook of Physical Constants*. AGU Reference Shelf 1, American Geophysical Union, pp. 1–31.

Zahnle, K., 1996. Dynamics and chemistry of SL9 plumes. In *The Collision of Comet Shoemaker–Levy 9 and Jupiter*. Eds. K.S. Noll, H.A. Weaver, and P.D. Feldman. Space Telescope Science Institute Symposium Series 9, IAU Colloquium 156. Cambridge University Press, Cambridge, UK, pp. 183–212.

Zahnle, K.J., and N.H. Sleep, 1997. Impacts and the early evolution of life. In *Comets and the Origin and Evolution of Life*. Eds. P.J. Thomas, C.F. Chyba, and C.P. McKay. Springer, New York, pp. 175–208.

Zuber, M.T. + 14 co-authors, 2000. Internal structure and early thermal evolution of Mars from Mars Global Surveyor topography and gravity. *Science*, **287**, 1788–1793.

Zurbuchen, T.H. + 13 co-authors, 2011. MESSENGER observations of the spatial distribution of planetary ions near Mercury. *Science*, **333**, 1862–1865.

INDEX

Printed in the United States
by Baker & Taylor Publisher Services